Feature Review Papers in Geological Oceanography

Feature Review Papers in Geological Oceanography

Guest Editor

Markes E. Johnson

MDPI

Basel • Beijing • Wuhan • Barcelona • Belgrade • Novi Sad • Cluj • Manchester

Guest Editor
Markes E. Johnson
Geosciences Department
Williams College
Williamstown, MA
USA

Editorial Office
MDPI AG
Grosspeteranlage 5
4052 Basel, Switzerland

This is a reprint of the Special Issue, published open access by the journal *Journal of Marine Science and Engineering* (ISSN 2077-1312), freely accessible at: https://www.mdpi.com/journal/jmse/special_issues/364815A386.

For citation purposes, cite each article independently as indicated on the article page online and as indicated below:

Lastname, A.A.; Lastname, B.B. Article Title. *Journal Name* **Year**, *Volume Number*, Page Range.

ISBN 978-3-7258-7925-0 (Hbk)
ISBN 978-3-7258-7926-7 (PDF)
https://doi.org/10.3390/books978-3-7258-7926-7

Cover image courtesy of Markes E. Johnson

Contents

About the Editor

Markes E. Johnson

Markes E. Johnson is the Charles L. MacMillan Professor of Natural Science Emeritus in the Geosciences Department at Williams College in Williamstown, Massachusetts. His interests broadly include the comparison of modern depositional environments with their counterparts in the rock record through the Phanerozoic Eon, spanning more than 500 million years of Earth history. He is the author of several books on modern and ancient environments in Mexico's Gulf of California, as well as a study on the geological imprint of islands through deep time.

Preface

The scope of the project was designed to cover a wide range of environments that host complex ecosystems alongside their ancient counterparts preserved in the Earth's rock record. Geological Oceanography examines a range of environments that physically extend outward from coastlines across continental shelves to the deepest parts of the ocean. Specific topics represented in the Reprint include rocky shores, reefs, coastal sand dunes, and tsunami deposits, as well as conservation efforts in the development of island geoparks. The exchange of ideas and observations between oceanographers and geologists is mutually beneficial because understanding contemporary environments helps to illuminate muted patterns from the past, whereas the geologic record offers to the present insights on the physical and ecological dynamics that shaped planet Earth over the long arc of time from its origins 4.5 billion years ago. The motivation for the project was to attract readers who would be inspired by the linkage of past and present ecosystems to extend such work. Geographic gaps in this field are particularly apparent in coastal regions surrounding the Indian Ocean and the islands it contains. It is hoped that this Reprint will have a particular impact on those beginning research in this part of the world.

Markes E. Johnson
Guest Editor

Journal of
*Marine Science
and Engineering*

MDPI

Editorial

Feature Review Papers in Geological Oceanography

Markes E. Johnson

Geosciences Department, Williams College, Williamstown, MA 01267, USA; mjohnson@williams.edu

Linkages between modern environments that host a panoply of complex ecosystems and their ancient counterparts preserved in the rock record are indispensable to our understanding and appreciation of the world we occupy. The bond is well encompassed by the concept of Geological Oceanography. Extending outward from all coastlines across continental shelves to the deepest parts of the ocean, there exist a multitude of inter-connected settings, all of which trace back in deep time through a web of development. Geological Oceanography is one of several sections sponsored by the *Journal of Marine Science and Engineering*. It actively promotes the exchange of ideas between oceanographers and geologists who study the same kind of phenomena but in parallel time frames: past and present. The intellectual tradeoff is mutually beneficial, because a better understanding of contemporary environments helps to reanimate some of the muted patterns from the past, whereas the geologic record lends a perspective to the present by offering insights on the physical and ecological dynamics that shaped planet Earth through the long arc of time from its origin 4.5 billion years ago.

The banner created at the launch in 2024 of the journal's Special Issue "Feature Review Papers in Geological Oceanography" acknowledges the rock cycle, which is well known by all beginner geology students. The interconnection of igneous, metamorphic, and sedimentary rocks is related to the ever-renewable processes of magma generation in production of solid products; the effects of heat and pressure in the alteration of those products under deep burial; and the consequences of surface weathering toward erosion of those same products as sediments typically deposited as strata under marine conditions. All are liable to variations in the endless manifestations of the rock cycle. For example, sedimentary rocks like sandstone derived from the weathering of granite may be altered through the heat and pressure of deep burial into quartzite. After exposure at the earth's surface, quartzite persists as one of the strongest components of rocky shores resistant to erosion [1].

Among the earliest historical awakenings to the rock cycle was the interpretation of sea cliffs at Siccar Point near Edinburgh, Scotland, by James Hutton (1726–1797) and a group of fellow intellectuals interested in the relationship of different rocks in the context of time [2]. The angular unconformity at Siccar Point represents a gap in geologic time, during which layers originally deposited in a horizonal plane were subsequently tilted by regional tectonics, eroded, beveled, and eventually buried by other sedimentary rocks. Different kinds of gaps in the rock record were recognized, thereafter, such as the exposure and reburial of igneous rocks, like basalt, by marine sediments as adjoining strata preserving topographic relief in the underlying rocks (technically a nonconformity). Observations by Hutton and others in the late eighteenth-century were eventually reframed in the context of plate tectonics, which, for example, explains the regional tectonics of the Caledonian Orogeny [3] that led to the tilting of the basal strata at Siccar Point. During the nineteenth century, geologists created a time scale giving convenient labels to the different intervals of geologic time. Hence, we conceptualize that the graywacke layers at the base of the cliffs at Siccar Point were deposited during the Silurian Period some 440 million years ago, and the overlying sandstone layers accumulated during the Devonian Period some 375 million years ago. The gap in time between the two amounts to roughly 65 million years, although that gap is fully occupied by the rock record in many localities, elsewhere.

Time gaps or disconformities in the rock record are highlighted in most of the dozen contributions assembled for the Special Issue on "Feature Review Papers in Geological Oceanography" (see the list of contributions). The papers were solicited from experts working all around the world from institutions in the United States; Russia; China; Australia; Europe (Poland, Spain, and Portugal); and North Africa (Algeria). The participants contributed reviews offering direct comparisons between modern-day environments where coastal and marine sediments are now being deposited and analogous settings from the past. For example, it is argued that rocky shores signify the oldest ecosystem on planet Earth, developed as nonconformities soon after the first islands emerged as igneous blocks. Former rocky shores are known by exceptional occurrences that date from a broad range of geologic time periods [4]. The organizer's lead paper, on the ecology of intertidal rocky shores, shows how modern granite shores provide data on marine invertebrates and algae that may be compared to former rocky shores with biotas that colonized unconformity surfaces across geologic time (Contribution 1). Dmitry Ruban (Southern Federal University, Rostov-on-Don, Russia) adds to this topic, describing how unconformities between Carboniferous granites and Triassic–Jurassic strata represent ancient rocky shores in the Western Caucasus (Contribution 2).

A group of researchers led by Chang Ma (Northwest University in Xi'an, China) offer a review of deposits in the Qiongdongnan Basin off Hainan Island in the South China Sea that resulted from deep-water gravity flows (Contribution 3). Such deposits are also known as turbidites or underwater avalanches. Their contribution enlists state-of-the-art technology in seismic stratigraphy that teases out subtle unconformities between each event related to successive underwater avalanches. This sort of study represents one of the fastest-growing areas of research being published through the *JMSE*'s Geological Oceanography Section.

A contribution authored by Ed Landing (New York State Museum, Albany, New York) and the organizer compares organo-sedimentary structures commonly called stromatolites that first were described by paleontologists from fossil deposits well before their living counterparts were recognized by visitors to Shark Bay in Western Australia (Contribution 4). These structures are built layer-by-layer by microbes that trap fine sediment in flat sheets or domes in shallow water. They represent the oldest known life-forms from the fossil record, dating back as far as 3.45 billion years [5], although their biology is almost entirely dependent on the study of living representatives. Ed Landing contributes a separate review on competing theories about organic-rich mudstones or black shales that appear throughout the rock record related to anoxia in marine settings (Contribution 5). Devoid of life due to diminished oxygen levels, the extreme depositional conditions of black shales represent a sedimentary environment defined by sharp stratigraphic disconformities in the rock record.

Coastal sand dunes, especially those composed of calcium-carbonate sediments tied to the disaggregation of organic materials from the sea, represent a fascinating topic, which is explored by David Backus (Haverford College, Haverford, Pennsylvania), in which the geographic distribution of today's active eolianites (wind-blown sand) help inform the distribution and source materials in "fossil" dunes traced to earlier times, as much as 5 million years ago (Contribution 6). More than a review, the paper introduces fresh outlooks on extensive deposits on Boa Vista Island in the Cabo Verde archipelago, both contemporary and ancient, in a subtropical setting, as well as unusual dunes in

https://doi.org/10.3390/jmse14080741

cool, temperate settings such as Scotland and Ireland. Under any scenario where marine sediments are transported onshore to beaches and farther inland as dunes, such deposits are disconformable with the underlying terrane.

Greggory Webb (University of Queensland, Brisbane) contributes a masterful review comparing the two most iconic contemporary and fossil reef structures in Australia (Contribution 7). The living Great Barrier Reef off the shores of Queensland extends over a distance of 2000 km and includes various styles of reef building dominated by corals. The Great Devonian Barrier Reef in Western Australia may be traced for more than 350 km with structures beautifully exposed on land composed both by corals and a diverse range of skeletal sponges called stromatoporoids. A key finding is that the Devonian reefs (from 370 million years ago) were just as reliant on framework binding by coralline algae and microbialites as contemporary reefs. The organizer followed up with a review on the topic of rhodoliths that take the form of spheroidal shapes grown by crustose coralline red algae. Johnson's piece (Contribution 8) compliments the reef story as a major, but relatively unappreciated, source of limestone in the rock record. As suppliers to beaches and coastal sand dunes due to storm over-wash, rhodoliths occur with a much greater latitudinal range than coral reefs today. Likewise, the juxtaposition of rhodolith deposits with underlying surfaces marks distinct disconformities both in modern settings and the rock record.

Alfred Uchman (Jagiellonian University, Kraków, Poland) is one of the leading academics studying trace fossils in sedimentary rocks of all ages from around the world, which is a specialization with a dedicated following [6]. As indicated by the term, no remains or traces of the original maker of tracks, burrows, or borings is left behind in fossil form. Uchman's review (Contribution 9) is focused on a single island in the Azores (Santa Maria Island) with an extraordinary record of marine fossils dating from the Pliocene and Pleistocene, beginning 5 million years ago. Typical of his carefully researched work, the interpretation of such traces depends on comparisons with living organisms found as active trace-makers worldwide. Sérgio Ávila (Universidade dos Açores, Punta Delgada, Portugal) is an expert not only on the marine invertebrates and fossils of Santa Maria Island, but also on those from other island groups in the northwest Atlantic Ocean, including the Canary and Cabo Verde islands. His decades of work on patterns of migration to and among those Atlantic islands takes into account changes in climate during the latest glacial and interglacial epochs of the Pleistocene (Contribution 10). This exacting line of study involves detailed records on the timing of arrivals and departures among thousands of marine invertebrate species and provides a model for further anticipated research in other island domains, particularly in the Pacific and Indian Oceans.

The geologic record of tsunami deposits, compared with recent events around the world, was compiled by a team led by Mohamed Amine Doukani (University of Oran, Algeria) and provides a much-anticipated review (Contribution 11). Caused by submarine earthquakes or by enormous landslides falling into the sea, tsunami deposits are recognized in rocks that date far back in geologic time. They represent the most violent disturbances witnessed in coastal areas around the world as well as classic examples of disconformable deposits with essentially instantaneous origins. The contribution is exceptionally thorough, including data on Pleistocene deposits from the author's home in Algeria, as well as those elsewhere dating far back in geologic time to the Archaean Eon more than 3 billion years ago.

The Special Issue's concluding entry (Contribution 12) breaks ground in Geological Oceanography by looking at the concept of geoheritage through the lens of the Global Geoparks Network under the auspices of the United Nations Educational, Scientific and Cultural Organization (UNESCO), specifically in regard to volcanic islands. Building on experience in the Canary Islands, Esther Martín-González (Museo Ciencias Naturales de

https://doi.org/10.3390/jmse14080741

Tenerife, Santa Cruz de Tenerife, Spain) led a team of fellow scientists reviewing not only their own island geoparks, but also two dozen others from around the world that share the goals of conservation and public education. The topic is relevant in the context of the *JMSE*, because volcanic islands are born from the sea and entail a range of eruptive styles that result in different landscapes clearly embedded in the rock cycle. Their account will help to promote similar contributions to the journal in regard to coastal geoparks of different kinds, as the UNESCO program continues to expand around the world.

The immensity of the rock cycle with elements expressed both in geologic time as well as contemporary physical settings allows yet more opportunities for comparison of present and ancient environments. Two such scenarios remain open for consideration from among the several addressed in our Special Issue devoted to Geological Oceanography. Large-scale delta systems represent one such setting with enormous ramifications for human populations dwelling in mega-cities on and around such coastal constructs. A review of 40 major deltas scattered around the world focusses on subsidence as a consequence of rising sea level [7]. Changes in sea levels contextualize one of the major thematic patterns preserved in the rock record. Additionally, it is worth thinking about the role of plate tectonics in the past assembly of mega-continents like Pangaea during the Triassic Period between 250 and 200 million years ago, when river systems would have discharged gargantuan sediment loads to coastal regions from distant source areas, far greater compared to the discharge of the present-day Amazon basin and related Amazon delta in South America. No such review with the same level of treatment of the examination of tsunamis (Contribution 11) is available, but this would be highly informative.

Another research area of much interest relates to the physical chemistry and associated biology of deep-sea thermal vents with their concentration of scarce minerals released via black smokers and unusual modes of life, such as tube worms capable of direct nutrient absorption through chemosynthesis. The surprising discovery of such ecosystems occurred in 1977, when the first such find was made exploring the Galápagos rift system on the East Pacific Rise [8]. Very quickly, it was realized that analogous ecosystems existed in the past, reaching back to the Silurian Period [9]. Summaries comparing modern and ancient deep-sea thermal systems were publicized as late as 2000 [10], but the topic is overdue a substantial review to include subsequent discoveries.

Conflicts of Interest: The author declares no conflicts of interest.

List of Contributions:

1. Johnson, M.E. Ecology of Intertidal Rocky Shores Related to Examples of Coastal Geology across Phanerozoic Time. *J. Mar. Sci. Eng.* **2024**, *12*, 1399. https://doi.org/10.3390/jmse12081399.
2. Ruban, D.A. Exposure of Carboniferous Granitoids on Triassic–Jurassic Seashores in the Western Caucasus: A Stratigraphical Review. *J. Mar. Sci. Eng.* **2024**, *12*, 1623. https://doi.org/10.3390/jmse12091623.
3. Ma, C.; Qu, H.; Liu, X. Indication of Deep-Water Gravity Flow Types by Shelf-Edge Trajectory Migration Patterns: A Case Study of the Quaternary Qiongdongnan Basin, South China Sea. *J. Mar. Sci. Eng.* **2024**, *12*, 2051. https://doi.org/10.3390/jmse12112051.
4. Landing, E.; Johnson, M.E. Stromatolites and Their "Kin" as Living Microbialites in Contemporary Settings Linked to a Long Fossil Record. *J. Mar. Sci. Eng.* **2024**, *12*, 2127. https://doi.org/10.3390/jmse12122127.
5. Landing, E. The Bakken Model: Deposition of Organic-Rich Mudstones and Petroleum Source Rocks as Shallow-Marine Facies Through the Phanerozoic. *J. Mar. Sci. Eng.* **2025**, *13*, 895. https://doi.org/10.3390/jmse13050895.
6. Backus, D.H. Contemporary Issues and Advancements in Coastal Eolianite Research. *J. Mar. Sci. Eng.* **2025**, *13*, 321. https://doi.org/10.3390/jmse13020321.

7. Webb, G.E. Australia's Two Great Barrier Reefs: What Can ~360 Million Years of Change Teach Us? *J. Mar. Sci. Eng.* **2025**, *13*, 1582. https://doi.org/10.3390/jmse13081582.
8. Johnson, M.E. Rhodoliths as Global Contributors to a Carbonate Ecosystem Dominated by Coralline Red Algae with an Established Fossil Record. *J. Mar. Sci. Eng.* **2026**, *14*, 169. https://doi.org/10.3390/jmse14020169.
9. Uchman, A. Trace Fossils on Oceanic Volcanic Islands of Macaronesia: Current State of Knowledge. *J. Mar. Sci. Eng.* **2025**, *13*, 1967. https://doi.org/10.3390/jmse13101967.
10. Ávila, S.P. Distributional Range Shifts Caused by Glacial–Interglacial Cycles: A Review on Timing, Main Processes, and Patterns of Late Pleistocene Marine Dispersal by Invertebrates in the NE Atlantic. *J. Mar. Sci. Eng.* **2025**, *13*, 2024. https://doi.org/10.3390/jmse13112024.
11. Doukani, M.A.; Madeira, J.; Satour, L.; Ávila, S.P. Tsunamites Versus Tempestites: A Comprehensive Review from the Precambrian to Recent Times. *J. Mar. Sci. Eng.* **2026**, *14*, 49. https://doi.org/10.3390/jmse14010049.
12. Martín-González, E.; Vegas, J.; Galindo, I.; Romero, C.; Sánchez, N. Volcanic Islands as Reservoirs of Geoheritage: Current and Potential Initiatives of Geoconservation. *J. Mar. Sci. Eng.* **2025**, *13*, 1420. https://doi.org/10.3390/jmse13081420.

References

1. Johnson, M.E. *Islands in Deep Time—Ancient Landscaptes Lost and Found*; Columbia University Press: New York, NY, USA, 2023.
2. Shipley, T.F.; Tikoff, B. Siccar Point Scotland and the role of mental models. *GSA Today* **2024**, *34*, 36–40. [CrossRef]
3. Searle, M.P. Tectonic evolution of the Caledonian orogeny in Scotland: A review based on the timing of magmatism, metamorphism and deformation. *Geol. Mag.* **2022**, *159*, 124–152. [CrossRef]
4. Johnson, M.E. Why are ancient rocky shores so uncommon? *J. Geol.* **1988**, *96*, 469–480. [CrossRef]
5. Schopf, J.W. *Cradle of Life—The Discovery of Earth's Earliest Fossils*; Princeton University Press: Princeton, NJ, USA, 1999.
6. Donovan, S.K. *The Palaeobiology of Trace Fossils*; John Wiley & Sons: Chichester, UK, 1994.
7. Ohenhen, L.O.; Shirzaei, M.; Davis, J.L.; Tiwari, A.; Nicholls, R.; Dasho, O.; Sadhasivam, N.; Seeger, K.; Werth, S.; Chadwick, A.J.; et al. Global subsidence of river deltas. *Nature* **2026**, *649*, 894–901. [CrossRef]
8. Corliss, J.B.; Dymond, J.; Gordon, L.I.; Edmond, J.M.; von Herzen, R.P.; Ballard, R.D.; Green, K.; Williams, D.; Bainbridge, A.; Crane, K.; et al. Submarine thermal springs on the Galapagos Rift. *Science* **1979**, *203*, 1073–1083. [CrossRef] [PubMed]
9. Little, C.T.; Herrington, R.J.; Maslennikov, V.V.; Morris, N.J.; Zaykov, V.V. Silurian hydrothermal-vent community from the southern Urals, Russia. *Nature* **1997**, *385*, 146–148. [CrossRef]
10. Van Dover, C.L. *The Ecology of Deep-Sea Hydrothermal Vents*; Princeton University Press: Princeton, NJ, USA, 2000.

https://doi.org/10.3390/jmse14080741

Journal of
*Marine Science
and Engineering*

MDPI

Review

Ecology of Intertidal Rocky Shores Related to Examples of Coastal Geology across Phanerozoic Time

Markes E. Johnson

Department of Geosciences, Williams College, Williamstown, MA 01267, USA; mjohnson@williams.edu;
Tel.: +1-413-2329

Abstract: The boundary between land and sea is among the most extensive and extreme ecological barriers on planet Earth. Intertidal organisms living at this junction are subject to potentially lethal conditions related to desiccation, temperature fluctuations, and wave shock. Although difficult to quantify at a global level, rocky coasts account for between a third and three-quarters of all shorelines today. Resistant to coastal erosion, rocky coasts also exhibit adaptations by marine invertebrates and marine algae within intertidal zones that are geographically widespread. Biotic composition is variable depending on the tidal range at any given place and on climatic differences between tropical and more temperate latitudes. Contemporary granite shores having low recession rates with well-established biotic zones are reviewed in three regions: Mount Desert Island in Maine, USA (44° N latitude), Mahé and Praslin islands in Seychelles (4°30′ S Latitude), and Lizard Island off the coast of Queensland, Australia (14°40′ S Latitude). The zonal composition and overall biotic diversity at these locations are compared, as well as the fossilization potential of key biota. Over the Phanerozoic time period during the last 538 million years, the ratio between land and ocean varied as a function of plate tectonics depending on the maximum dispersal or coalescence of continents and on relative changes in global sea level. Fossil biotas from a dozen paleoislands with dominant rocky shores that range from 485 million to 125,000 years in age are reviewed to show diversification through time, taking into account episodic mass extinctions. Relationships between storm tracks and volcanic eruptions that factor into physical disruptions are also considered as means of rapid burial and preservation. Themes related to rocky-shore ecology and paleoecology benefit from the cross-fertilization of ideas by marine biologists and marine paleontologists. It is hoped this review may attract a cohort of new students to these allied fields of study, especially in Asia, Africa, and South America.

Keywords: granite shores; inter-tidal zonation; fossilization potential; latitudinal diversity gradients; paleoislands; storm tracks

1. Introduction

A first-order overview of planet Earth's surface reveals that its geography is divided 70% by oceans against 30% by continents, as verified by any model globe or consultation of an atlas with full global coverage [1]. The world's shorelines are estimated to encompass about a million kilometers in total length [2]. From a biological perspective, the meeting point between land and sea constitutes the single most pervasive and equally restrictive barrier to all manner of life facing landward or seaward. From a geological standpoint, the tide line signifies an ecosystem of primeval heritage dating back in time nearly as old as the Earth itself [3]. It is surprising that rocky-shore paleoecology is absent as a topic from the earliest textbook on paleoecology from 1965 [4], going forward to a more current text covering a range of different ecosystems [5]. This review is aimed especially at students interested in marine biology and paleontology, who might overlook the attraction of studies related to the dynamics of rocky shores and their connection to coastal geology. Established scholars in both fields also may benefit from a wider outlook beyond specialization in a particular taxonomic group of marine invertebrates to include broader themes of plate tectonics, sea-level change, mass extinctions, and storm patterns that have shaped the rocky-shore ecosystem through geologic time. For those with a background in geology, the terminology related to these topics will be familiar but may be new for others more closely attuned to the present world. In either case, it is assumed that the concepts of ecology and paleoecology are mutually understandable in both camps and need not be defined. Here, coastal geology is understood to embrace a range of topics related to shore erosion and colonization by marine biotas as a distinct subset of geological oceanography.

Exactly when Earth's oceans became a permanent fixture remains under contention, but a working hydrosphere may have been in operation as early as 4.3 billion years ago but no later than 3.5 billion years ago based on samples from oceanic crust yielding oxygen and hydrogen isotopes from localities in Russia and Arctic Canada [6,7]. The earliest life to colonize a coastline was achieved between 3.3 and 3.5 billion years ago, as represented by fossil stromatolites with dome-shaped growths on the brecciated surface of a lava flow preserved in the Swaziland Supergroup of South Africa [8]. Contemporary stromatolites with the same shape that accrued by bacterial growth in matted layers of filamentous colonies were recognized and studied during the early 1960s at Hamlin Pool along the shores of Shark Bay in Western Australia [9]. This archaic biota has persisted in secluded settings up to the present day. Metazoan life with a more complex biology, including exoskeletons readily preserved as fossils, appeared at the start of the Cambrian period 538.8 million years ago and evolved in shallow, subtidal waters along continental shores in distinct waves that likewise suffered extinctions to episodically reshape the nature of the intertidal zone. The physical restraints against intertidal colonization by marine invertebrates and marine algae entailed desiccation and exposure to the atmosphere with fluctuations in temperature and high levels of ultraviolet radiation, as well as the stress of wave shock. Various accommodations in design and capabilities were acquired before life was able to expand and thrive in coastal settings under intertidal conditions, including rocky shores, sandy beaches, and mud flats [10].

Rocky coasts dominate today's ocean shores, with estimates sometimes quoted as high as 75% of the total, including shore platforms and sea cliffs [11,12]. However, such evaluations include cliff lines that sit back from the actual shore, having no contact with seawater. A detailed review of the massive compilation of data collected by Bird and Schwartz [2] involved calculations from 387,262 km of composite coastline, yielding a positive identification of 187,403 km of rocky shores for 48% overall [13]. Unsampled shores in Arctic Canada and Russia could only be compared to known shores in Alaska, where the prevalence of rocky shores was found to be 31%. Large island groups, such as those found in the Philippines and Indonesia, were left unsampled in the original database [2] but assumed to be comparable to studied shores in Sri Lanka, Thailand, and peninsular Malaysia, with 15% measurable rocky shores [13]. Given the potential inaccuracies for continental shores at high latitudes and major archipelagos elsewhere, it was predicted that future coastal surveys with greater precision would find no less than 33% of total rocky shores on a global basis [13].

Today, rocky shores are demonstrably extensive features that provide a stable substrate for a diverse biota consisting of marine invertebrates and marine algae that deploy a range of capabilities for attachment and subsequent growth in a rough-water setting [14]. Barnacles and oysters use organic cement to fix their exoskeletons to solid rock. Lithophagid bivalves occupy rock borings excavated by means of mechanical and chemical means. Other common bivalves, such as the blue mussel and horse mussel, deploy a byssus consisting of strong organic threads to anchor their shells to rocks. Chitons and limpets rely on the suction of pad-like muscular feet to cling to rocks. Coralline red algae may coat huge rock surfaces, and many species of green and brown algae are attached directly to rocks by a holdfast. The overall diversity of intertidal rocky shores is amplified by a host of species that take refuge within the protective tangle of mussel beds and algae. During the mid-twentieth century, marine biologists established a tripart division of zonation on rocky shores, consisting of an upper supralittoral fringe splashed by seawater during storms and the extremely high level of spring tides, a midlittoral zone exposed to the air all or in part on a daily basis, and an infralittoral fringe exposed only seldom by the extreme low-water level of spring tides [15]. The scheme takes into account that the maximum tidal range differs from place to place and is also affected by variable wave activity. The universality of the zonation was originally based on widespread observations around the rocky shores of the British Isles, North America, South Africa, India, and Australia.

The object of this review is two-part in organization, taking into account observations on rocky shores that conform to the patterns of universal zonation [15] at distinctly different latitudes north to south of the equator in the modern world compared to the fossil record of intertidal life across Phanerozoic time during the last 485 million years. In almost all cases, examples of sea cliffs with vertical relief provide the database utilized in this review to draw on contemporary and prehistoric rocky shores with evidence of biological colonization. A few examples of shore platforms that extend seaward are included in this review, but they are not counted as wave-cut platforms. Granite was chosen on the basis of its resistance to erosion as but one among many prospective rock types typically found on rocky shores. In a survey of coastal cliff erosion based on a full range of rock types at some 280 locations around the world, it was determined that granite suffered minimal erosion compared to other sedimentary and metamorphic rocks [16]. The rate of biologic colonization and renewal on granite rocky shores is considered to surpass the minimal rate of recession due to physical erosion. It means that the patchy clearance of granite rock due to excessive predation by sea stars or other kinds of predators (or by physical scraping of the substrate by mechanical means during storms) is readily recolonized by the intertidal biota at a rapid rate.

Examples of contemporary granite shores are reviewed and compared with respect to their biological zonation in three distinct regions: Mount Desert Island in Maine, USA (44° N latitude); Mahé and Praslin islands in the Seychelles (4°30′ S Latitude), and Lizard Island off the coast of Queensland, Australia on the Great Barrier Reef (14°40′ S Latitude). The potential for fossilization of key biological elements is based on the contents of associated sand bodies and cemented beach rock at Mount Desert Island and Lizard Island. Comparison with the modern record is based on analyses of fossilized rocky-shore biotas reported from a dozen paleoislands that demonstrate diversification from the late Cambrian 485 million years ago to the last interglacial maximum 125,000 years ago [17]. Key issues under consideration include evidence for the role of mass extinctions affecting the overall diversification of rocky-shore biotas through time; potential signatures of latitudinal diversity gradients at any given point in geologic time; the role of plate tectonics and sea-level fluctuations in the expansion or reduction of global rocky shores; and the significance of storm tracks for the rapid burial and preservation of rocky-shore biotas in settings usually thought of as vulnerable to net erosion based on the prevalence of normal wave activity.

2. Background on Coastal Geography and Geology

Rocky shores today and throughout the planet's geologic past are formed by all kinds of lithology, including sedimentary, igneous, and metamorphic rocks. Long stretches of shoreline may be composed of a single sedimentary rock type, such as the White Cliffs of Dover along the English Channel that rise to a height of 110 m and stretch laterally over a distance of 26 km. With a hardness of three on the Mohs Scale (from 1 to 10), chalk cliffs are soft and vulnerable to high rates of erosion measured between 0.22 and 0.46 m per year [16]. Exposed farther west along the English Channel and generally equal in exposed height and lateral extent to the White Cliffs of Dover, the interlayered limestone and shale cliffs around Lyme Regis have an average recession rate of only 0.07 m per year [16]. The partings in layered sedimentary rocks are susceptible to hydraulic pressure during wave impact, which acts to pry apart layers and hasten cliff retreat. Among metamorphic rocks, quartzite ranks seventh in hardness on the Mohs Scale and exhibits a density of 2.7 g/cm^3, which, in connection with outcrops that are massive in form, proves to be highly resistant to wave erosion. Among igneous rocks, andesite and basalt occur as rocky shores that extend co-laterally for great distances, especially along the Pacific coast of South America [1,13]. Solidified from repetitious lava flows, these igneous rocks also form distinct layers that are liable to marine erosion by prying and splitting action during wave impact. Distinct from andesite and basalt, granite is an igneous rock that forms at great depth in magma chambers, where it cools slowly into hard rock. It is subsequently exposed at the surface only after all overlying rocks have been unroofed by erosion. Granite dominates as basement rocks on continental shields widespread in Canada, Scandinavia, Siberia, Australia, and Brazil, accounting for approximately 15% of the Earth's surface [18,19]. Granite rocky shores are not as widely developed as other kinds of rock but are well represented along coastal Maine in the northeast USA, western Greenland, the Gulf of Bothnia between Sweden and Finland, Spain's Bay of Biscay, Egypt's Sinai Peninsula, and some islands widely distributed in the Seychelles archipelago in the Indian Ocean.

The granite shores selected for this review focus on Mount Desert Island along the coast of Maine, Mahé and Praslin islands in the Seychelles archipelago, and Lizard Island off the coast of Queensland in eastern Australia (Figure 1). In latitudinal spread, they represent a distinctly north-temperate setting with 48 degrees of separation to an equatorial setting and another 10 degrees of distance to a mid-tropical setting south of the equator.

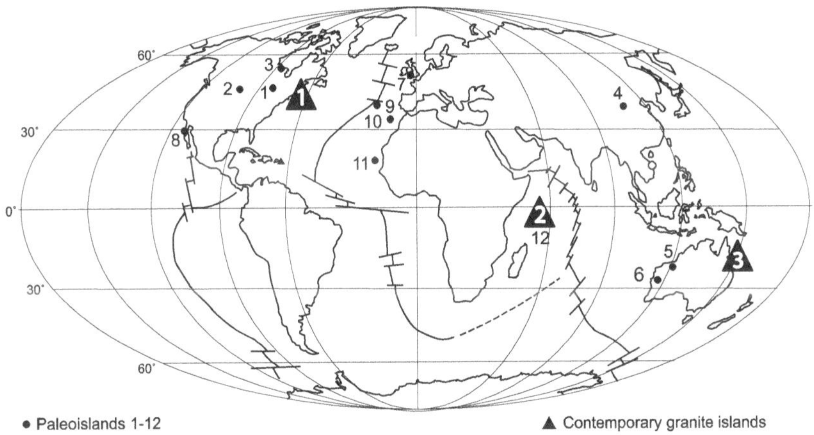

Figure 1. Global representation of today's continents and oceans on a Mollweide projection showing major ocean spreading zones. Numerals on black triangles denote the locations of distinct areas with prominent granite rocky shores on Mount Desert Island in coastal Maine, USA (1); the Seychelles Islands (2), and Lizard Island on the Great Barrier Reef off the coast of Queensland, Australia (3). The numbered black dots mark the location of a dozen paleoislands dominated by rocky shores with fossil biotas ranging in age through the Phanerozoic from the late Cambrian (1 and 2) to the last inter-glacial stage of the Pleistocene in the Seychelles Islands (12).

3. Operational Definitions and Study Methods

The identification of fossil biotas associated with rocky shores is based on a set of diagnostic criteria related to geological unconformities [10]. An unconformity is marked by the junction of two rock deposits separated by a gap in geologic time. In most cases, the two rock bodies are different from one another, for example, sedimentary limestone over igneous andesite. The difference in age between the two rock bodies may be on the order of a few million years to tens of millions or even hundreds of millions of years. Places where a conglomerate sits directly on an unconformity surface consisting of igneous, metamorphic, or even sedimentary rocks are most likely to represent former rocky shores. This is especially so where the abutting parent body of rock serves as the source of eroded clasts. Marine fossils may be found directly adhering to the unconformity surface or to individual clasts eroded from the adjoining surface. Evidence of size gradation in a marine conglomerate may show larger clasts proximal to the unconformity surface changing to smaller clasts that are more distal in a relationship. Joints, fissures, or other irregularities in the unconformity surface may be filled by eroded clasts called neptunian dikes. Surface irregularities may be on such a large scale as to denote buried sea stacks. The fossils preserved at or attached to eroded clasts above the unconformity surface may include marine invertebrates like barnacles from a high-energy environment typical of strong wave shock [14] or more delicate organisms like encrusting bryozoa characteristic of a sheltered setting. Fossil algae, especially the coralline red algae, may occur encrusted on an unconformity surface and clasts eroded from the unconformity surface. Overall, these criteria were applied to a compilation of 1622 extinct and extant species represented by fossils through Phanerozoic strata from the Cambrian to the Pleistocene at 361 localities identified as former rocky shores around the world [10]. A dozen examples of paleoislands related to entire archipelagos with extensive rocky shores also conform to these criteria and provide a compelling database for tracing the development of inter-tidal rocky-shore biotas across Phanerozoic time over the last 485 million years [17].

4. Results

4.1. *Review of Granite Rocky Shores and Biotas with Different Climates by Latitude*

During the late summer season in Maine (Figure 1, triangle 1), the water temperature varies between 17 °C to 20 °C, falling to 2 °C in the winter. The coastal waters around the equatorial Seychelles Islands (Figure 1, triangle 2) remain steady year-round at an average temperature of 26 °C. The seasonal change in temperature in the subtropics around Australia's Lizard Island (Figure 1, triangle 3) varies from 22 °C in the winter to a high of 29 °C in the summer. The climates supported by these temperatures at different latitudes affect the kind of marine organisms that best thrive there independently on granite rocky shores. Year-round chemical erosion of granite shores is a phenomenon peculiar to places with high, year-round humidity, as in the Seychelles archipelago.

4.1.1. Mount Desert Island (Maine, USA)

With an area of 280 km^2, Mount Desert Island is the largest island on the coast of Maine in the Atlantic Ocean's Gulf of Maine (Figure 2). Granite shores are exposed almost continually around the island's periphery. The highest elevation is on Cadillac Mountain at 466 m above sea level. French explorer Samuel de Champlain gave the island its name in 1604, observing that the granite highlands were free of vegetation and, therefore, barren. The island is home to Acadia National Park within the United States National Park system and incorporates 70% of the island's land area. The most distinctive geographic feature on Mount Desert Island is Somes Sound (Figure 2), which is considered a fjord and the only such example of its kind in the lower forty-eight states of the USA. The narrows extend for more than 9 km to the island's center with an average depth of 30 m. Like the more classic fjords of Norway and Alaska farther north, Somes Sound was excavated by glacial action when Maine was covered by ice during much of the Pleistocene Epoch. Sand Beach (Figure 2) is one of the few places on Mount Desert Island where it is possible to bathe in the ocean. A granite shore rises steeply to an elevation of 33 m on the west side of Sand Beach (Figure 3), where the typical tidal range varies between 2.5 and 3.6 m.

In her popular book *The Edge of the Sea* [3], covering the entire eastern seaboard of the USA, Rachel Carson denotes six distinct intertidal zones on the granite rocky shores of Maine. Here, the supralittoral fringe incorporates an uppermost black zone that favors lichens and tiny filaments of blue-green algae. Also, within the supralittoral fringe is the periwinkle zone dominated by the gastropod *Littorina littorea*, which feeds by grazing on the rock-covered film of blue-green algae. The zone in which they are most abundant is reached every two weeks or so by the spring tide. Native to rocky shores in northwestern Europe, the common periwinkle is an invasive species in New England. The mid-littoral zone on the coast of Maine is exposed to the air on a diurnal basis and includes two distinct sections densely crowded in the uppermost part by acorn barnacles belonging to the species *Semibalanus balanoides* and in the lower part by blue mussels belonging to the species *Mytilus edulis*. Both are filter feeders that strain seawater for plankton. Both are securely attached to the rocks in the case of barnacles by organic cement and in the case of the mussels by strong byssal threads (Figure 4a,b).

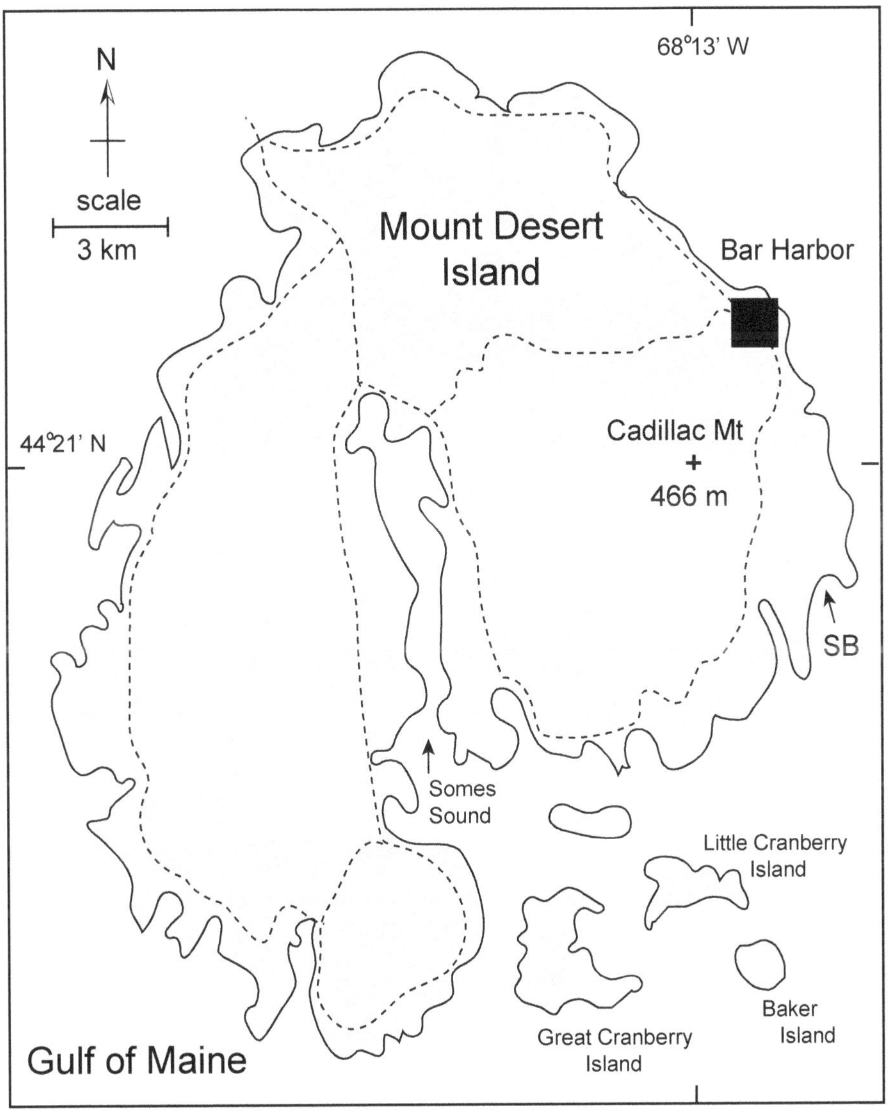

Figure 2. Map of Mount Desert Island off the coast of Maine in the Gulf of Maine on the Atlantic Ocean. The main roads around the island are shown by dashed lines. The single access onto the island (State Highway 3) is linked to the mainland by a short bridge. SB marks the location of Sand Beach east of Somes Sound.

Figure 3. View north along a granite rocky shore that intersects Sand Beach on Mount Desert Island. People can be seen congregating in the far distance on Sand Beach (right) and at the break between forest and rocks (left). Rocks densely covered by algae with a green tinge show the extent of exposure during extreme low tide.

Both species are preyed on by the carnivorous dog whelk *Nucella lapillus*, which deploys a radula mouthpiece to bore holes through barnacle and mussel shells to access the tissues within using enzymes to transform those tissues into a digestible liquid. The infralittoral fringe along the rocky coast of Maine is exposed to the air every two weeks also by the spring tide, with adjacent zones dominated above by green algae commonly called rockweed, *Fucus distichus*, and a lower band of red algae commonly called Irish moss dominated by *Chondrus crispus* (Figure 4c). During extreme low tides, the upper part of the brown algae zone dominated by kelp may be briefly exposed to the air.

A checklist maintained by the Huntsman Marine Laboratory provides a comprehensive list of marine invertebrates by habitat and depth range for Passamaquoddy Bay on the border between Maine in the USA and New Brunswick in Canada [20]. Like Mount Desert Island nearby, the shoreline around Passamaquoddy Bay is dominated by rocky shores. A list of 70 marine invertebrates that occupy the intertidal zone and infratidal fringe may be identified and separated from other species that dwell exclusively in deeper, subtidal waters. Those faunal elements in possession of calcite exoskeletons have a high likelihood of fossilization. They are dominated by mollusks that include chitons, limpets, coiled gastropods, and bivalves. A single brachiopod, four bryozoa, four barnacles, and five arthropods with a semi-durable carapace also fit this category. Together, they amount to 33 taxa. Another 33 taxa include sponges, sea anemones, flatworms, polychaete worms, and sea slugs (nudibranchs) that have little or no chance for fossilization. Sea stars and brittle stars in the phylum Echinodermata might be added to this category but have a spotty record of fossilization. Hence, the overall faunal list of rocky intertidal and littoral fringe dwellers from the coast of Maine and New Brunswick is almost evenly split between taxa that have the potential for preservation as fossils and those that do not. Marine fleshy algae from the inter-tidal fringe in temperate settings include many red, green, and brown algae, none of which have a reliable record of fossilization.

Figure 4. Images of the dominant marine invertebrates and marine algae from the mid-littoral zone and infralittoral fringe on granite rocky shores adjacent to Sand Beach on Mount Desert Island: (**a**) Densely crowded barnacles (*Semibalanus balanoides*) less than 5 mm in diameter; (**b**) Densely crowed mussels (*Mytilus edulis*) up to 9 cm in length; (**c**) Marine algae from the infralittoral fringe showing a tidal pool with rockweed, *Fucus distichus* (lower right) and concentrations of Irish Moss, *Chondrus crispus* (upper left).

4.1.2. Mahé and Praslin Islands in the Republic of Seychelles

The Seychelles archipelago consists of 115 islands spread across a relatively shallow bank covering 1,340,000 km^2 in the Indian Ocean, 1500 km off the east coast of Africa [21]. On the basis of seismic evidence, much of the bank is formed by granite, which has a thickness of 13 km draped by marine sediments [22]. Mahé and Praslin are the two largest islands among a cluster of 24 granite islands concentrated in the northern part of the bank (Figure 5). Essentially, the islands are drowned monadnocks or inselbergs that represent highland peaks of a mini-continent flooded to an average depth between 44 and 65 m with basement rocks most closely related to the nearest exposed granite on the Horn of Africa [23]. The highest peak on Mahé rises 905 m above sea level, which is twice as high as

Cadillac Mountain on Mount Desert Island. In contrast, the tidal range around Mahé is only 1.2 m, which is half that experienced on Mount Desert Island. Situated below the equator, the granite islands are affected by the southeast trade winds from April to October with maximum wind speeds on the order of 50 cm/s that favor coral reefs along the windward east side of Mahé Island and south shores of Praslin Island [24].

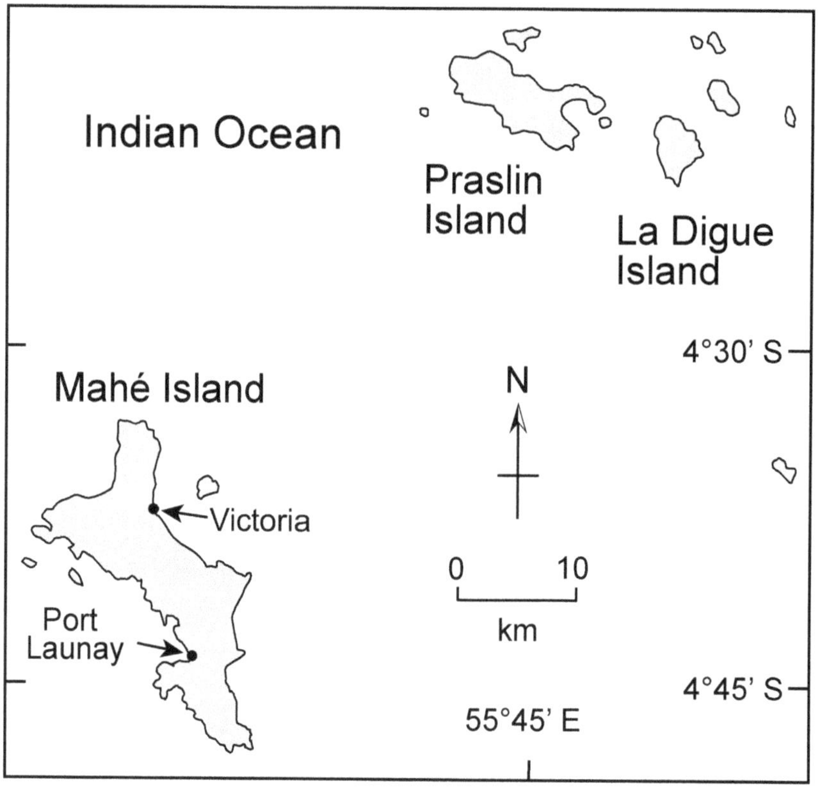

Figure 5. Map showing a cluster of a dozen granite islands in the northern, inner part of the Seychelles archipelago located in the Indian Ocean off the east coast of Africa. Arrows point to Victoria, the capital city, and Port Launay, both on Mahé the largest island.

The independent Republic of the Seychelles derives its name from the French Viscount Jean Moreau de Séchelles, Minister of Finance, during the reign of King Louis XV. The French explorer Lazare Picault landed on Mahé in 1742, and the islands remained a part of France until 1814 when they passed to British control. British influence is reflected by the name of the capital city, Victoria, named in honor of Queen Victoria in 1841. Mahé Island (Figure 5) has an area of 155 km^2 with a circumference of 120 km, 66% of which is formed by granite rocky shores [21]. The Port Launay Marine National Park is located on the southwestern, leeward side of Mahé Island, where intertidal zonation typical of a more sheltered setting is developed on granite cliffs that exhibit a superb example of chemical erosion (Figure 6).

Figure 6. Fluted development of granite rocky shores due to chemical weathering on the south side of the bay at Port Launay on Mahé Island (see Figure 5 for location). A sample grid (center) is well exposed at maximum low tide (person for scale).

A comprehensive checklist of intertidal invertebrates for the Seychelles Islands is compiled from extensive descriptions of mollusks, arthropods, echinoderms, and corals listed in published sources [24,25] from which marine gastropods and bivalves account for 80% of the fauna represented by 74 species. Some of the principal fauna from the intertidal zone at Port Launay cover the supralittoral fringe to the infralittoral fringe within the vertical space of about one meter (Figure 7). Comparable to the habitat of the gastropod *Littorina littorea* on Mount Desert Island, the species *Littorina undulata* occupies the supralittoral fringe at Port Launay (Figure 7a). The mid-littoral zone at this location is dominated by barnacles belonging to the species *Tetraclita squamosa* that are densely encrusted on the inner surface of single disarticulated oyster valves represented by the species *Crassostrea cucullate* firmly cemented to granite (Figure 7b). The limpet species *Cellana radiata* is also present in this zone. No stony corals are evident in the bay at Port Launay, but granite surfaces in the infralittoral fringe are extensively colonized by the matted growth of colonies belonging to soft corals with many different species assigned to the genus *Palythoa* (Figure 7c).

Praslin Island (Figure 5) is the second largest island in the Seychelles, with an area of 38 km^2 and a circumference of 45 km [20]. Fond Azore is the highest elevation at 367 m above sea level. The leeward side of the island occupies the north-facing shore and features extensive sand beaches, whereas the south-facing shore is in the path of the southeast trade winds and exhibits a greater concentration of granite rocky shores. The west end of the island is accessible by trails from Anse Lazio that lead to a succession of small beaches elevated behind large granite boulders that rise through the water (Figure 8).

Figure 7. Marine invertebrates from the intertidal zone at Port Launay on the leeward side of Mahé Island: (**a**) Gastropod *Littorina undulata* from the supralitltoral fringe occupying a crack in the granite (pencil for scale); (**b**) Granite surface in the mid-littoral zone initially encrusted by oysters in the species *Crassostrea cucullate* but overlain by encrusting barnacles in the species *Tetraclita squamosa* as well as mobile limpets belonging to the species *Cellana radiata* (pen cap 3.5 cm for scale); (**c**) Granite knob exposed at lowest tide in the middle of the bay densely covered by a colonial mat of soft corals belonging to the genus *Palythoa* (pen 14 cm in length for scale).

These huge boulders are positioned far enough seaward with sufficient vertical height to capture the full spectrum of intertidal zonation in a relatively rough-water setting. In contrast to the rocks at Port Launay on Mahé, there is no evidence of oysters. The smooth and rounded tops of these boulders lack crevices that provide a safe place for grazing gastropods in the supralittoral fringe. The mid-littoral zone is compressed into a space of about a half-meter, showing a gradual change from small barnacles of the species *Balanus tintinnabulum* to populations of the larger barnacle *Tetraclita squamosa* mixed together with abundant populations of the limpet species *Cellana radiata* (Figure 9a). The infralittoral fringe exposed at ebb tide is occupied by a dense cover dominated by brown algae belonging to the genus *Turbinaria*, from which pockets of encrusting coralline red algae and patches of the soft coral *Palythoa* immerge (Figure 9b). Rarely, the scarlet-colored colonial coral *Pocillopora damincornis* may be found cemented directly to the side of a granite boulder (Figure 9c). Several other coral species are known to encrust on granite surfaces but are more commonly found growing in shallow water on the upper reef slope [24]. Brittle stars also inhabit the infralittoral fringe and are known to range upward into the mid-littoral zone during high tide.

Figure 8. Line of large granite boulders seaward of the beach along the west shore of Praslin Island that exhibit intertidal zonation in a rough-water setting. Swimming person for scale (left).

Figure 9. *Cont.*

Figure 9. Elements of biota inhabiting the inter-tidal zone and infralittoral fringe in a windward, rough-water setting on Praslin Island: (**a**) Mixed populations of barnacles (*Balanus tintinnabulum* and *Tetraclita squamosa*) interspersed with the limpet *Cellana radiata* (on average 1.5 cm in diameter) in the mid-littoral zone; (**b**) Thickets of brown algae from the genus *Turbinaria* that dominate over patches with encrusting red coralline algae and soft corals in the infralittoral fringe (pen for scale); (**c**) Rare occurrence the coral *Pocillopora damincornis* attached to granite in the infralittoral fringe (20 cm across in diameter).

4.1.3. Lizard Island in the Great Barrier Reef off Queensland, Australia

Situated 43 km off the coast of Queensland (Figures 1 and 10), Lizard Island is grouped with smaller islands called Palfrey, South, and Bird amid the Great Barrier Reef centered at a latitude 14°40′ South. In total, the islands amount to an area of 10 km^{2,} with the highest elevation on the main island rising to 320 m above mean sea level. The peak is commemorated as Cook's Lookout in honor of the visit by Captain James Cook in command of the *HMS Endeavor* in 1770. It was Cook who bestowed the name Lizard Island in reference to a population of the large Yellow-spotted Monitor (*Varanus panoptes*). Aboriginal people from the Australian Dingaal tribe have a long history of foraging on the island, which, according to legend, was once attached to the mainland as a sky island. Lizard Island is home to the Marine Research Station administered by the Australia Museum in Sydney and the sponsor of major research in marine biology.

The maximum tidal range for Lizard Island is 2.70 m, as recorded against the granite shores that encircle the Blue Lagoon on the south side [26]. During the Austral winter months from June to August, the primary hydrologic influence derives from the prevailing SE trade winds. A July wind speed of 45.5 km/h is characteristic from the southeast, which influences tidal flushing into the lagoon at the entrance between the Lizard Head and the Bird islets (Figure 10). During the peak Austral summer months from December through February, monsoonal winds half as strong typically impact the island from the northwest. Lizard Island is known for its six named beaches with coarse quartz sand derived from the subtropical disintegration of granite. Collectively, the beaches around Lizard Island account for about 35% of the shoreline. The rest is dominated by granite sea cliffs that rise most abruptly to the highlands on the east side of the island.

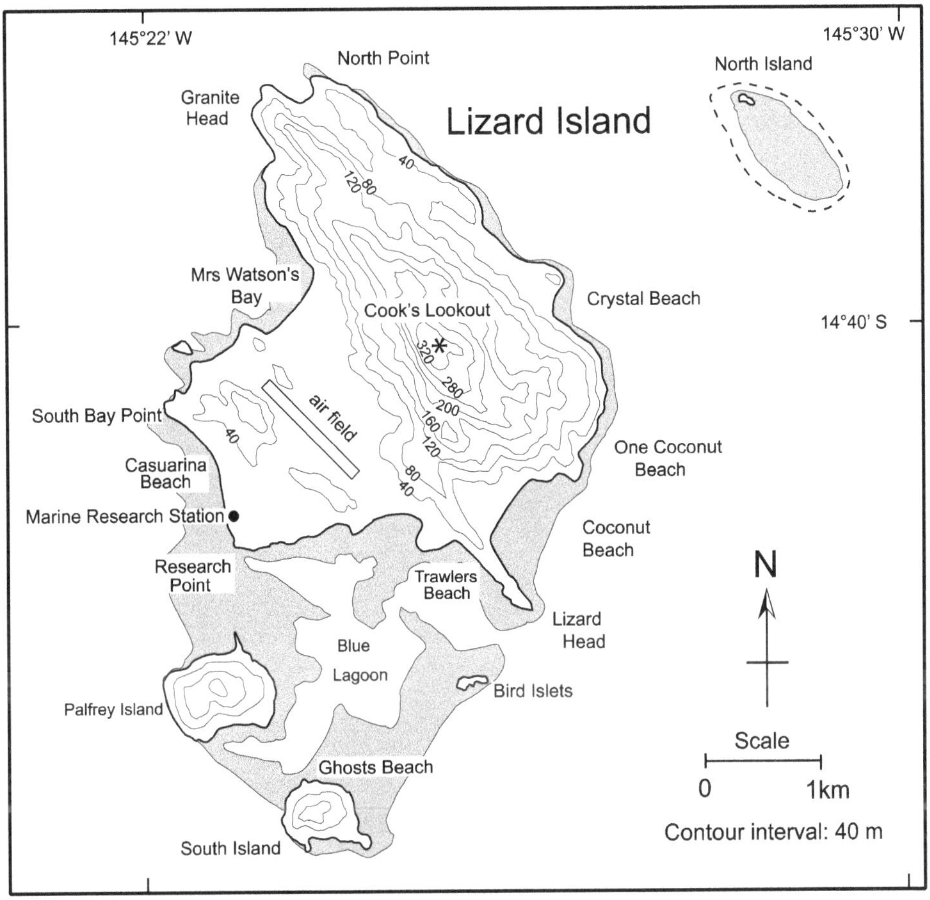

Figure 10. Topographic map covering Lizard Island and a related cluster of islets including Palfrey, South, and Bird (grey shading represents shallow-water shelf).

A checklist with 34 species for Lizard Island is dominated by marine gastropods and bivalves but also includes barnacles and some stony corals as compiled based on references in the research station's library bolstered by a photo-based guidebook [27]. The supralittoral fringe with a black zone similar to Mount Desert Island, including the gastropod *Littoraria undulata*, appears on granite rocks in a leeward setting on Mrs. Watson's Bay (Figure 10). Otherwise, observations on the island's intertidal fauna are best informed from the low rocky shores and large boulders lining Trawler's Beach on the south side of the island (Figure 10). These lack a well-defined supralittoral fringe, as described (above) from Mahé Island in the Seychelles. The mid-littoral zone on this stretch of coastline is occupied by three distinct bands of mollusks. The uppermost is dominated by the star limpet *Patelloida saccarina*. On average, eight to ten spur-shaped limpets are present within a 10-cm-by-10-cm space (Figure 11a), which also features empty depressions in the granite where the animal maintained a home base. The concept of a homing instinct is discussed by Rachel Carson [3], who states that Aristotle was the first to report that limpets go out to feed from a fixed station during high tide and return to the same spot when the tide recedes. Abandoned spots worn into the granite are about one centimeter in diameter. An intermediate band is occupied by the comb oyster, *Lopha cristagalli*. It is distinctly purple in color with a deeply crenulated commissure where the valves meet (Figure 11b). On average, 250 individuals may be counted in a sample grid covering a quarter-meter square. The ribbed gastropod, *Nerita costata*, overlaps with the upper reaches of the oyster band and is distinguished by its turban-shaped shell scored by strong ribs. Not as abundant as the oyster but unfailingly found to occupy a band directly below is another bivalve called the jewel box, *Chama limbula* (Figure 11c). Stony corals such as *Goniastrea palauensis* are rarely exposed attached to granite during the Spring tide, but soft mat corals in the genus *Palythoa* cover extensive surfaces (Figure 11d).

Figure 11. Faunal elements from the inter-tidal zone and infralittoral fringe in a windward, rough-water setting off Trawler Beach on the south side of Lizard Island: (**a**) Sparse populations of the star limpet *Patelloida saccarina* including vacant home stations (each square is 10 cm on a side); (**b**) Dense concentrations of the comb oyster, *Lopha cristagalli* occur in a middle band (pen for scale); (**c**) Scattered populations of another bivalve called the jewel box, *Chama limbula*, occupy a third and distinctly lower band (pen cap for scale); (**d**) large granite block encrusted by oysters at the top, jewel box bivalves in the middle, and a solid mat with soft corals in the genus *Palythoa* from the infralittoral fringe (author for scale).

4.2. Clues to Potential Fossilization from Beach Deposits

The mineral content of granite may reach 50% in silica (SiO_2). Adjoining beaches between granite headlands become enriched with quartz sand as a consequence of the physical weathering and slow decomposition of a granite rocky shore. Quartz itself is resistant to erosion, which means that the size of sand grains on any given beach depends on the original size of that mineral at crystallization in the parent granite. Such beach sand is likely to include eroded granite pebbles as well as the detritus of eroded exoskeletons from the marine invertebrates and certain calcareous algae that thrive on rocky shores. In tropical settings worldwide, between 35° N and S of the equator where dissolved calcium-carbonate ($CaCO_3$) is present in seawater, deposits may be lithified as beachrock. Such a process of beach lithification that has been ongoing during recent and contemporary times is evidenced by the preservation of shell casings from World War II battles in certain South Pacific islands as well as coke and rum bottles in Caribbean islands [28]. The contents of beachrock also may reflect organic materials carried landward from subtidal environments during major storms.

4.2.1. Analysis of Sand Beach on Mount Desert Island, Maine

The location of Mount Desert Island 44° north of the equator (Figure 1) is nine degrees of latitude remote from the normal production of beachrock. However, Sand Beach on Mount Desert Island (Figure 3) contains an unusually high component of calcareous materials with as much as 65% by volume derived directly from the detritus of shore-dwelling barnacles and mussels as well as echinoids and coralline red algae more characteristic of the sublittoral fringe [29,30]. Silica sand is present, but the occurrence of such a major component from a biological source is counter-intuitive because cold seawater in the Gulf of Maine drives the dissolution of calcium carbonate ($CaCO_3$). During the lifetime of a blue mussel, for example, the animal's shell is protected from dissolution by a thin outer layer of the periostracum composed of proteins and lipids impervious to leaching. Once the host animal expires, the periostracum rapidly disintegrates, and the underlying shell layers will completely dissolve, given sufficient exposure to cold seawater. An exceptionally high rate of sediment supply of shelly detritus at Sand Beach on Mount Desert Island and similar small beaches on Great Cranberry Island is due to the frequency of coastal storms that resupply those beaches on a regular basis. What is most favorable, from a paleontological perspective, is the potential preservation of a rocky-shore biota attached to a rock substrate in its original growth position. However, the burial and subsequent lithification of beach sand in high latitudes provides a reasonable alternative as a death assemblage (thanatocenosis) for the purpose of recovering census information from a past habitation.

4.2.2. Beachrock on Palfrey Island in a Subtropical Setting

Among the smaller islands clustered around Lizard Island, Palfrey Island (Figure 10) is notable for extensive beachrock deposited among the granite boulders along the south shore over a distance of 0.25 km. No systematic survey is available regarding the faunal composition and relative frequency of extant marine invertebrates entombed in the beachrock. However, general relationships are apparent to the casual visitor. Common elements represented in the Palfrey Island beachrock include large mollusk shells and various stony corals. One of the most distinctive gastropods readily identified by its conical profile in cross-section is *Trochus lineatas*, having a basal diameter of 10 cm (Figure 12a). It may occur in close proximity to stony corals in the genus *Favites* (Figure 12a) as well as *Lobophyllia*. The most abundant mollusk is represented by disarticulated valves of the large bivalve in the genus *Tridacna* (Figure 12b). Pieces of *Tridacna* shells preserved within the beachrock are relatively small, 15 cm in maximum width, but live individuals from the adjacent reef may possess shells more than a meter in width. Although the Palfrey Island beachrock incorporates granite pebbles, the larger faunal elements are characteristic of a reef habitat and not especially a rocky-shore habitat.

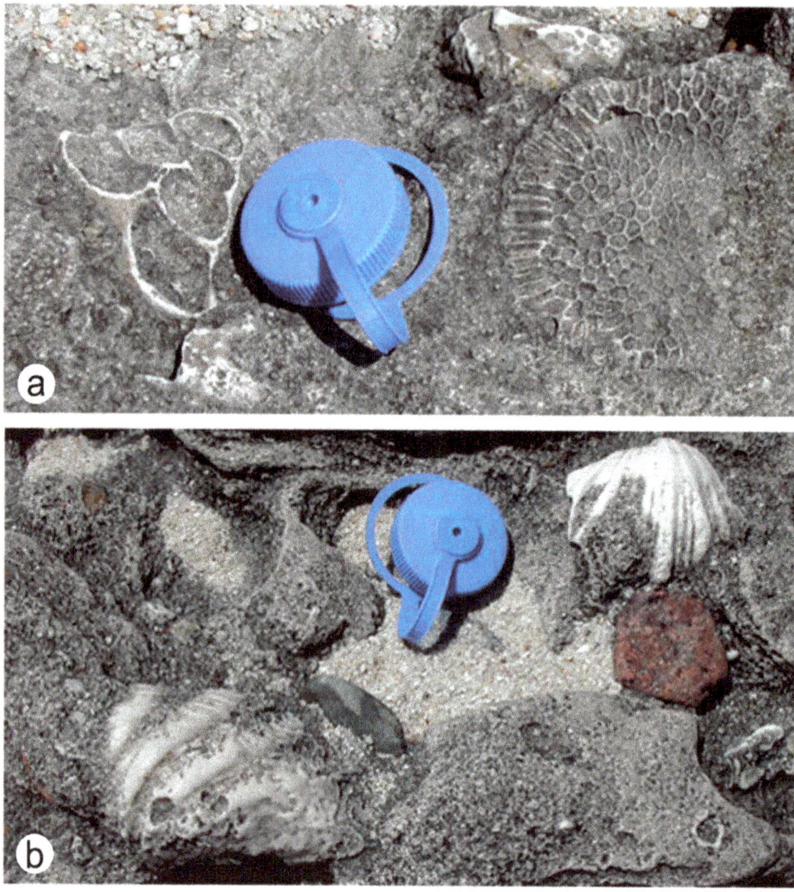

Figure 12. Faunal elements of extant marine invertebrates preserved as contemporary fossils in the Palfrey Island beachrock: (**a**) Cross-sections through the gastropod *Throchus lineatus* and stony coral from the *genus Favitres* (cap to water bottle is 7 cm in diameter); (**b**) Disarticulated and worn shells from the bivalve genus *Tridacna* together with a red granite cobble.

4.3. Review of Pre-Historic Rocky-Shore Biotas from the Fossil Record

Preferable to the worn and degraded biological elements preserved in beachrock, a superior method for the survey of rocky-shore biotas in the pre-historic fossil record relies on the availability of materials in original growth position preserved either encrusted against or in close proximity to the surface of a geological unconformity. Under the circumstances ideal for detailed analysis, the unconformity should be overlain by marine sedimentary rocks, with the underlying unconformity exposed on a scale that exceeds more than a few meters in lateral extent. It is a false assumption that all rocky shores exist in an environment that is entirely erosional in character to the detriment of fossils in anything like their original life conditions (biocenosis from the German biozönosen) [8,11]. In fact, burial of an unconformity surface will occur under conditions where a sudden storm event transfers a blanket of sediments shoreward or the ejecta from a volcano engulfs a particular coastline. The same requirement for rapid burial of any particular organism with hard parts as a fossil applies equally as well to the fossilization of a rocky shore with all those inhabitants in possession of sturdy exoskeletons. Under some circumstances, the rocky shores that encircle a small island may be fossilized intact [13]. Although there are exceptions to the rules of fossilization, the principal drawback is that animals with only soft body parts or fleshy algae, such as green and brown algae, are rarely preserved.

4.3.1. Analysis of Phanerozoic Rocky Shore Publications

A bibliography of published studies on prehistoric rocky shores was issued in 2012 [10]. It lists 151 scientific articles that describe geological unconformities and related fossil-bearing deposits diagnostic of intertidal biotas. Since 2012, the scholarly community has continued to release additional work on this topic, but the 2012 bibliography is notable for specific trends that have remained current, even with a modest increase in studies originating from additional parts of the world or as follow-up studies at the same locality. Such trends are readily apparent from a histogram showing the relative distribution of study localities arrayed according to the six time periods across the Phanerozoic Eon (Figure 13). In most cases, the 163 study localities covered in the 2012 bibliography provide detailed information that attributes a particular rocky-shore biota to a designated epoch. In theory, the fossil biotas represent a given instant in geological time during which marine invertebrates and some algae were catastrophically buried alive on a rocky shore and preserved intact in numbers and proportions faithful to their life habitat. The site may survive as a lasting marker once the former shore is uplifted and exposed.

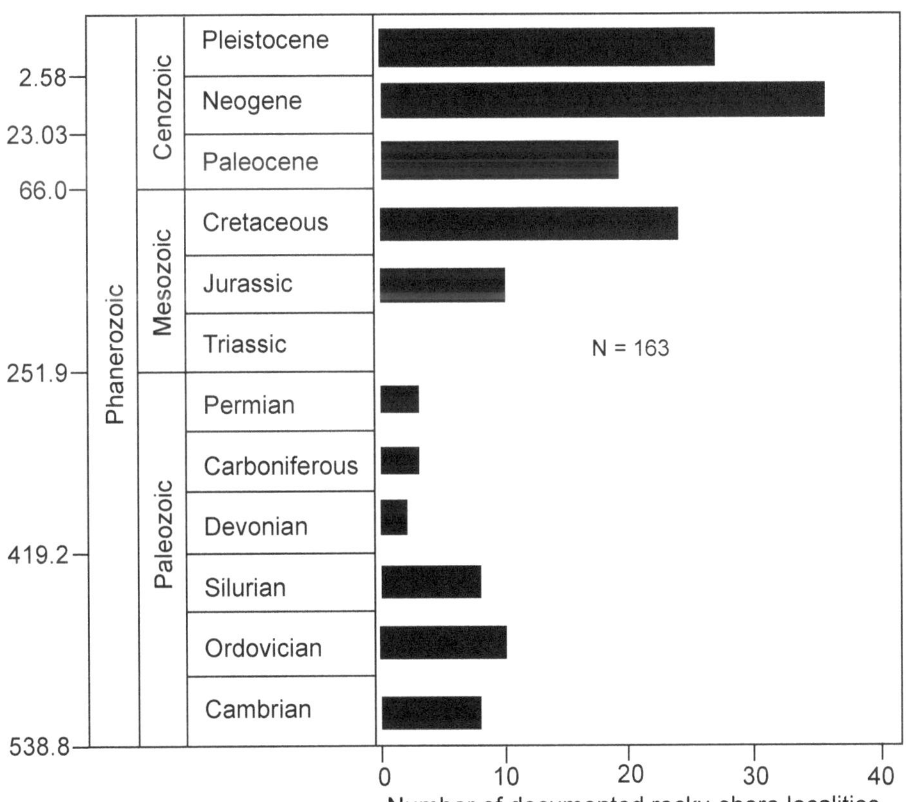

Figure 13. Histogram with the geologic time scale on the y-axis and the number of documented rocky-shore localities registered to each of twelve geologic time periods on the x-axis as of 2012. Absolute time in millions of years is also shown on the x-axis.

A notable major trend suggests that fewer well-described rocky-shore biotas derive from studies of fossils representing Paleozoic intervals of geologic time, whereas the most recent Cenozoic intervals account for almost a third of the whole. This is the case despite the fact that the Cenozoic represents only 12% of the total elapsed Phanerozoic time, whereas the Paleozoic represents more than 50% of the total. A likely explanation for the overall trend is that the Cenozoic rock record has had less time to disappear due to mass erosion and normal patterns of recycling related to plate tectonics. Another significant detail is that examples of rocky-shore biotas from the Triassic Period have yet to be recognized and described. This gap in Mesozoic data represents a major hole in our understanding of ecosystem development on rocky shores that is yet to be addressed. The lack of possible data from the Triassic Period could be related to the style of plate tectonics in operation during a time when a single land mass called Pangaea and a single great ocean called Panthalassa existed (see Discussion). More generally, however, time may have been required for the gradual evolution of the most sophisticated characteristics, allowing marine invertebrates and algae to colonize the available living space on intertidal rocky shores.

Another aspect from the 2012 data base [10] relates to the general geographic distribution of places where the research community carried out its studies. These data are presented in a pie diagram largely divided into slices by continents (Figure 14).

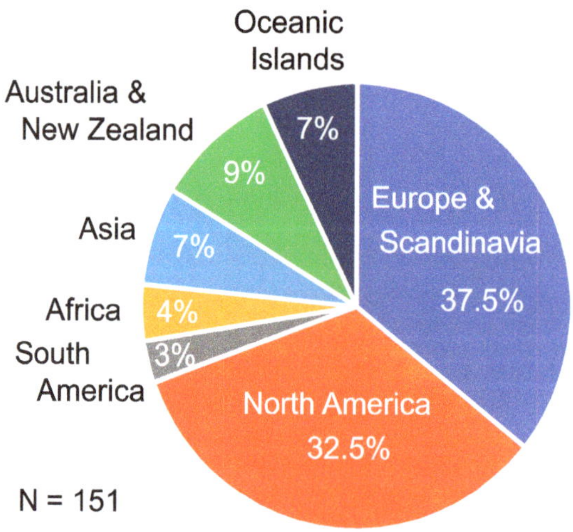

Figure 14. Pie diagram showing the geographic source distribution of places mainly by continents where researchers carried out studies eventually published on prehistoric rocky-shore biotas.

Europe (including the British Isles and Scandinavia), together with North America, account for nearly 70% of all research conducted on prehistoric rocky-shore biotas worldwide up to 2012. Australia and New Zealand combined account for a distant third place with 9% of the accessed studies. The substantially larger area of Asia as a land mass compared to any other accounts for only 7% of studies. Oceanic islands that span seas as distant from one another as the Caribbean and Mediterranean account for 7%. Lastly, the vast holdings of Africa and South America combined account for another 7% of the source areas for this kind of research. The clearest inference from these data is that researchers from lands outside of Europe and English-speaking regions, including the USA, Canada, Australia, and New Zealand, have been less aware of or perhaps less enthusiastic about contributing to this particular aspect of paleoecology and shared geoheritage. The overall paucity of data from Asia, Africa, and South America clearly represents a significant void in our understanding of the rocky shore ecosystem and how it evolved with time.

4.3.2. Review of Phanerozoic Rocky Shore Studies on Paleoislands

Although fresh discoveries have yet to be described from the greater part of the earth's surface outside Europe and North America, extraordinary examples of entire archipelagos that range in age from the late Cambrian 485 million years ago to the last inter-glacial epoch only 125,000 years ago are well known [13,17]. A dozen such island clusters are identified at their present-day locations around the world (Figure 1). They are numbered in order of age from the oldest (Cambrian) to the youngest (Pleistocene). In each example, the earth's continents were positioned in very different places than today, and the global sea level was higher than at present. This accounts for why paleoislands in North America, China, and Australia appear today as sky islands far from the nearest ocean.

The Cambrian Baraboo archipelago in south-central Wisconsin (Figure 1, locality 1) forms a cluster of about 30 individual paleoislands spread over an area of 1200 km^2 with an oval-shaped distribution around the town of Baraboo [17,31,32]. Formed by 1.7-billion-year-old Proterozoic quartzite surrounded by late Cambrian conglomerate belonging to the 485-million-year-old Tunnel City Group, the largest of the islands covers 50 km^2. More than half of the islands are much smaller at 4 km^2 in area. Although Cambrian marine sandstone deposited in between the islands is rich in fossils that include trilobites and even trace fossils representing jellyfish, there are no fossils preserved among the quartzite clasts in the eroded Cambrian conglomerate surrounding the various islands. Thus, there is no evidence of inter-tidal fauna at this location. The islands, themselves, would have been barren of land plants, the earliest fossil evidence for which would not appear for another 40 million years as moss and liverworts. Also organized around Proterozoic quartzite monoliths, Cambrian paleoislands in the Black Hills of South Dakota (Figure 1, locality 2) feature conglomerates composed of eroded quartzite clasts, some of which show evidence of small borings made by an unknown organism [33]. Additional studies of this kind are needed, but it appears that inter-tidal life during the Cambrian Period was absent or extremely limited.

The Ordovician Jens Munk archipelago in Canada's northern Manitoba (Figure 1, locality 3) forms a group of ten individual paleoislands spread over an area of 300 km^2 along the shores of Hudson Bay [17,34,35]. Like the older Cambrian islands in Wisconsin and South Dakota, the Ordovician islands are formed around a core of Proterozoic quartzite. Only 1.5 km^2 in size, one of the smaller paleoislands is flanked by a thick conglomerate deposit composed of eroded quartzite clasts, some of which are enormous. The conglomerate outcrop can be traced for 350 m along the shores of Hudson Bay but is fully exposed only during low tide. A line of individual boulders roughly 2 m in diameter forms a distinct boundary between the Ordovician shoreface and outer zones represented by layered dolostone dipping northward into the bay (Figure 15a). The quartzite boulders are aligned with the paleoshore and feature tabulate corals encrusted directly against the the rock surface (Figure 15b). Two genera of tabulate corals are known from this habitat, *Favosites* and *Ellisites*. In the seaward direction, large colonies of free-standing tabulate corals occur belonging to the genera *Palaeophyullum*, *Calapoecia*, and *Catenipora* that measure up to 46 cm in diameter. The Jens Munk archipelago is notable both for its rocky shore corals and for the preservation of offshore fossil zones that include the largest known trilobite, *Isotelus rex*.

Bater Island is a late Silurian paleoisland formed around a core of older Ordovician diorite on the grasslands of China's Inner Mongolia (Figure 1, locality 4). Measuring 610 m by 200 m, this small oval-shaped paleoisland is distinctive for its enclosing rim of sedimentary rocks that feature eroded diorite boulders in conglomerate on one side and silty limestone on the opposite side [17,36]. Large stromatoporoid colonies attributed to the genus *Plexodictyon* occur as armored sponges attached to the diorite in a leeward habitat interpreted on the basis of the prevailing silty limestone (Figure 16). Smaller stromatoporoids identified as *Clathrodicyon gotlandense* and *Actinostromella slitensis*) occur in a habitat interpreted as windward on the basis of the related diorite conglomerate. In addition, a total of seven coral species belonging to six genera are preserved as rocky shore dwellers around the island's circumference [37]. After a passage of some 25 million years post-dating the marine fauna around the Ordovician Jens Munk archipelago of Canada, it is notable that the Silurian coral fauna from Bater Island records a significant expansion in shore-dwelling corals as well as the first known occurrence of shore-dwelling stromatoporids.

Figure 15. Coastal area along Hudson Bay in northern Manitoba that overlaps with a former, late Ordovician paleoshore and paleoisland: (**a**) One of the distal quartzite boulders exposed during low tide with layers of dolostone dipping off to the north into the bay; (**b**) Example of a boulder-encrusting coral belonging to the genus *Favosities*.

The Mowanbini archipelago in Western Australia (Figure 1, locality 5) consists of two large paleoislands and some 20 satellite paleoislands formed by interbedded Paleoproterozoic quartzite and phyllite rocks in the Oscar Range surrounded by late Devonian limestone spread across an area of 175 km^2 [17,38]. The basement rocks from the Paleoproterozoic precede the surrounding Devonian strata by roughly two billion years. Exposure of the tilted quartzite and phyllite layers in Devonian time resulted in an island landscape rising 90 m above the shores of what was a set of distinctly outer and inner lagoons floored by the Pillara Limestone. A major barrier reef developed on the open seaward side of the islands (Figure 17), separated by a narrow lagoon. A larger inner lagoon separates the islands from the mainland of the granitic Kimberley Block. Tabulate corals belonging to the genus *Disphyllum* contributed to the makeup of an extensive reef structure, but the main component of reef builders was supplied by a range of species belonging to the armored sponges, or stromatoporoids.

During the 45 million years following the late Silurian development of China's Bater Island, the stromatoporoids went through a major Devonian diversification that introduced a range of new organisms to the lagoons around the islands off the Kimberley Block of Western Australia. Intricate details of off-shore zonation are preserved in the Pilara Limestone around smaller satellite islands within the inner lagoon on the present-day north side of the Oscar Range [17,39]. Exquisite preservation in a three-dimensional layout, which includes reefs, outer lagoons, paleoislands, and inner lagoons, leaves little to the imagination in exploring all parts of this former coastal region from about 375 million years in the past.

Figure 16. Leeward flank of the late Silurian Bater Island in China's Inner Mongolia: (**a**) Boundary between dark igneous diorite rocks (below) and silty limestone above with author seated for scale; (**b**) Cluster of large stromatoporoids at the changeover from igneous to sedimentary rocks (marker shows 5 cm for scale).

Occurring also in Western Australia, the usual Permian tower karst dissolved from the Callytharra Formation to expose an extensive network of intersecting galleries, much like city blocks, and was subsequently buried by Permian sandstone from the Moogooloo Formation along the Kennedy Range in the Gascoyne River district (Figure 1, locality 6). Each block is the size of a two-to-three-story, rectangular apartment building. When flooded during the deposition of the sandstone, each block would have been the equivalent of a small islet. Evidence of organic borings discovered on the top of a block near Congo Creek may have been formed anytime before the covering sandstone was deposited or long after it was stripped away [17]. The Permian topography is much like the present-day tower karst flooded by Ha Long Bay in northern Vietnam, also formed by Permian Limestone.

Figure 17. Forereef facies within the great Devonian barrier reef dipping away to the present-day south off a paleoisland in the Oscar Range (boab tree for scale).

As in the case of the Permian tower karst at Gascoyne Junction, another example of an unconformity formed by two bodies of adjoining sedimentary rock occurs along the Bristol Channel in southern Wales (Figure 1, locality 7), where Carboniferous limestone is directly followed by Jurassic limestone representing a major time gap of about 50 million years. The area has a long history of geological investigation, dating from work published in 1846 [40] by the director of the British Geological Survey, Henry De la Beche (1796–1855). His study illustrates the unconformity at Dunraven Bay (Figure 18) and describes examples of trace fossils bored into the Carboniferous limestone by Jurassic lithophagid bivalves and likely polycheates attributed to the trace-fossil genus *Trypanites* [41]. The Carboniferous limestone is also encrusted by various Jurassic bivalves like the oyster *Liostrea hisingeri* and colonial coral *Allocoeniopisis gibbose*, as well as the mobile limpet *Patella suttonenis* [17,41]. The same fauna is found in the overlying conglomerate belonging to the Sutton Stone, exposed only a short distance to the west, which is interpreted as a major storm deposit. The conglomerate fits with a catastrophic event that buried the unconformity surface and preserved the encrusting fauna in place on the rocky coast of a paleoisland in what is called the Jurassic Saint David archipelago.

Towards the close of the Cretaceous Period about 70 million years ago, a group of small islands and headlands formed by andesite lava flows were eroded along the present-day shores of the Pacific Ocean on Mexico's Baja California peninsula (Figure 1, locality 8). The uplifted strip of land north and south of Eréndira village covers an area of approximately 75 km^2 and features two distinct paleoislands, two prominent headlands, and a large sea stack formed by andesite partially enclosed by limestone belonging to the El Rosario Formation [17,42]. Cretaceous limestone in direct contact with the andesite unconformity surface on three sides of a paleoisland exhibits evidence of windward and leeward variations distinguished by the dominance of encrusting rudist bivalves in a windward, rough-water setting as opposed to encrusting oysters in a leeward, calm-water setting [17,43]. Small rhodoliths with a rind up to 0.5 cm in thickness formed by coralline red algae around andesite pebbles also characterize a distinct zone on the windward side of the paleoisland. Where pebbles are entirely coated on all sides with a uniform thickness by such algae, it signifies that the water turbulence was nearly constant due to wave activity. A greater faunal diversity occurred on the more sheltered rocky shore, with as many as 16 species of marine invertebrates.

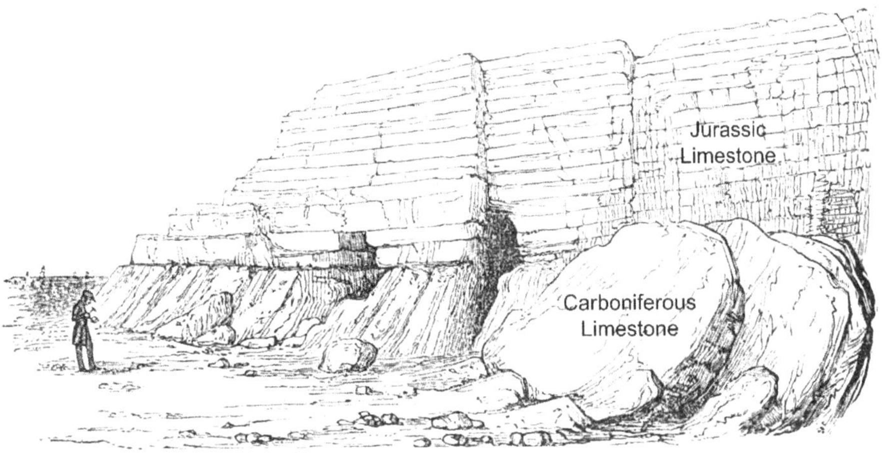

Figure 18. Reproduction of the original coastal-outcrop drawing by De la Beche from 1846 showing the geological unconformity at Dunraven Bay between tilted Carboniferous Limestone and flat Jurassic limestone. The spot where the figure stands is accessible only during extreme low tide in the Bristol Channel.

The Madeira archipelago in the northeast Atlantic Ocean (Figure 1, locality 9) includes the island of Porto Santo, formed from basalt during the middle part of the Miocene epoch between 14 and 15 million years ago. Ilhéu de Cima (the Upper Isle), off the east shore of Porto Santo, has a perimeter of only 3 km but exhibits several outstanding examples of Miocene rocky shores with marine biotas preserved intact [17]. An outcrop near the start of the trail leading to the island's lighthouse is especially interesting because it displays elements of intertidal zonation with exceptional preservation through a vertical space of 1.6 m [44]. The base of the cliff is formed by a narrow shelf that retains a fringing reef composed exclusively of the colonial coral *Tarbellastraea reussinna* directly attached to basalt in its original growth position (Figure 19).

Thousands of corals are densely crowded together over a shelf platform covering 180 m^2. Although monospecific with regard to coral composition, many show infestations by three kinds of boring bivalves in addition to borings made by barnacles and sponges. The vertical cliff that rises directly behind the platform has an uneven upper profile that is densely encrusted by coralline red algae (Figure 20). Between the top of the Miocene cliff line and the attached coral platform below, the 1.6 m cliff face is divided into separate zones with large barnacles and serpulid worm tubes limited to the upper 40 cm. Two coral species, including *Tarbellastraea reussiana* and *Isophyllastrea orbignyana*, occupy the upper half of the cliff face. A bivalve species in the genus *Spondylus* is widely cemented across the middle part. The lower part of the cliff face is dominated by borings in basalt, which is attributed to the trace fossil *Gastrochaenolites lapidicus* made by a rock-boring bivalve [44]. Additional patches of encrusted coralline red algae are scattered over the entire surface. Among all known localities with reports on intertidal biotas in the fossil record [10], this place on Ilhéu de Cima is special for its preservation of distinct vertical zones.

Figure 19. Example of a colonial coral belonging to the species *Tarbellastraea reussinna* in cross-section showing its attachment to the basalt surface of a narrow shelf on Ilhéu de Cima off the shores of Porto Santo in the Madeira archipelago.

Another island in the Northeast Atlantic Ocean is Santa Maria in the Azores archipelago, where the paleoecology of intertidal rocky shore biotas is on display (Figure 1, Locality 10). In this case, Santa Maria is much younger as a volcanic island than Porto Santo in the Madeira archipelago. The difference in geologic age is on the order of ten million years, as the volcanics on Santa Maria date to the latest Miocene epoch about 4.5 million years ago. With an area of 97 km^2 and a coastal circumference of scarcely 53 km, Santa Maria ranks seventh in size among the other Azoran islands. It is the oldest on the Azorean plateau and the only one with extensive fossil-bearing strata [17]. The south coast of Santa Maria Island extends over a distance of 15 km with high basaltic sea cliffs rising 100 m or more, separated by a single sand beach at Praia Formosa. There are several exceptional locations along the coast where Pliocene limestone may be studied as interbedded with basalt flows, but the section at Malbusca is among the best, where a 25-m thick package of sedimentary rocks has been studied in detail [45]. The unconformity that represents a Pliocene rocky shore is shaped by pillow basalt that cooled from lava under subtidal conditions to leave an undulating surface that developed as surge channels along the coast. Marine limestone that initially filled those channels was dominated by free-rolling rhodoliths formed by coralline red algae that washed landward under vigorous wave action. The channel walls are shown to be intertidal on the basis of barnacles that encrust the basalt. Succeeding limestone layers are more planar in construction and formed by a diverse group of shelly invertebrates that lived under subtidal conditions as the island underwent subsidence. Trace fossils indicate the presence of rays that left feeding depressions in an increasingly sandy sea bottom. A major storm bed consisting of thick sandstone caps off the sedimentary sequence, which in turn is sealed in place by the resumption of volcanic flows initially in deltaic cascades but finally as fully subaerial lava flows [17,46].

Figure 20. Top of a former sea cliff on Ilhéu de Cima with an irregular surface seen in profile showing a well-preserved encrustation by coralline red algae having a pustular texture on a basalt substrate (pen for scale is 12 cm long). The alga is overlain by volcanic ash and sand-size lapilli.

Santiago is the largest of ten islands in the Republic of Cape Verde, which is situated 600 km off the west coast of Africa with an area of 991 km^2 (Figure 1, locality 11). The island was visited twice by Charles Darwin (1809–1882), initially during the first landfall by the HMS *Beagle* on its circumnavigation around the world in 1832 and later in 1836 near the conclusion of the voyage on the way back to England. At home, Darwin wrote a compilation of all his geological observations regarding the many volcanic islands visited during the ship's five-year circumnavigation of the world [47]. A 10 km long stretch of Santiago's rocky coastline around the capital city of Praia was thoroughly explored by Darwin. His published work provides one of the earliest detailed descriptions of a paleo-shoreline. Better understood today as dating from between 0.7 and 1.1 million years ago [17,48], Darwin assigned a more general Tertiary age to a distinctive tri-part succession of basalt, limestone, and basalt. Especially rich in rhodolilths, the limestone was deposited over an unconformity surface of basalt with pronounced swales left by local streams that cut across the shoreface from place to place. Variable between 4.5 m and 7.5 m in thickness, the intermediate limestone is flat on the top, where it was beveled by pillow basalts from lava flows crossing the edge of the coast into shallow water. Darwin's original descriptions [47] are accurate and easy to follow along a shoreline that remains much unchanged today. Additional discoveries at Ponta das Bicudas passed over by Darwin give details on a diverse rocky-shore fauna, including the colonial coral *Siderastrea raddians*, hydrozoan *Milleora alcicornis*, solitary corals in the genera *Balanophyllia* and *Tubastreea*, the large barnacle *Megabalanus azoricus*, as well as bivalves in the genus *Spondylus*. Associated, free-dwelling mollusks include gastropods from the genus *Conus* and bivalves from the genus *Arca* [48]. The overlying pillow basalts at this locality are well exposed in cross-section, and Darwin's notes from a nearby spot show that the underlying limestone was altered to marble by the attendant heat of the magma [47].

The final paleoshore in this review is from La Digue Island in the Seychelles archipelago (Figure 1, locality 12). Contemporary rocky shores with diverse faunas are listed above. Like Mahé and Praslin islands, La Digue is a granite island and the fourth largest in the archipelago, having an area of 10 km^2 and a circumference of 16 km (Figure 5). There are 20 named beaches around the island, but rocky shores account for 53% of the whole. A late Pleistocene fauna is recorded from an outcrop at the south end of the beach at Grand Anse near the southeast end of the island [21]. Among the fossil mollusks recovered from the locality are gastropods belonging to the limpet *Cellana radiata* and gastropods *Morjula granulata, Congus frigidus*, and *Trochus virgatus*. In addition, the coral *Favia favus* occurs attached in growth position as well as vermetid gastropods in the genus *Dendropoma*. The most arresting biotic encrustations found on the vertical granite cliff face belong to sheeted coralline red algae. Large patches of bleached coatings remain in place (Figure 21). It is assumed that much of the granite surface was covered by encrusting red algae when the sea level was at least 6 m higher during the last interglacial epoch 125,000 years ago.

Figure 21. View of fluted granite cliffs at the south end of the beach at Grand Anse on La Digue Island showing large patches of limestone coatings left by sheeted coralline red algae. The coatings also occupy the underside of overhanging cliffs (person for scale).

5. Discussion

Further consideration of data from the two primary sources on the coastal geology of rocky-shore biotas [10,17] is worth exploring in the context of larger forces in the physical environment that either constrained development or controlled the degree to which they might be preserved in the rock record. Moreover, additional questions remain regarding the value of these data in regard to the registration of changes in overall biodiversity through time as well as variations at any given time between coeval biotas in support of latitudinal diversity gradients between tropical and temperate settings.

Role of External Factors in the Physical Environment

Several factors in the greater physical environment exerted a major effect on the development and preservation of rocky shore biotas across geologic time. Ranked in order of importance, they include global changes in geography through plate tectonics, global changes in sea level, and other more regional phenomena with respect to volcanos and prevailing storm tracks. It is only a crude estimate that the total length of the world's shorelines amounts to a million kilometers [2]. In theory, that value ought to increase to the extent that continents increase in number but decrease in size and circumference through dispersal brought about by plate tectonics. The converse is held to be true when the planet's seafloor-spreading zones push all the continents together into a single, massive continent. The plate-tectonics revolution was a natural outgrowth of the earlier concept of continental drift promulgated by Alfred Wegener (1880–1930), who produced and widely disseminated the earliest global maps showing changes in the position of the earth's major continents through geologic time [49]. Pangaea, the name applied to the late Paleozoic to early Mesozoic master continent, was introduced by Wegener in 1915. A reconstructed version of a Triassic Pangaea from 238 million years ago may be compared to a Paleogene reconstruction from 60 million years ago (Figure 22) to show the degree of continental dispersal between those two dates. The maximum assembly of Pangaea (Figure 22a) represents a global shoreline at a level of only 56% compared to today, whereas the dispersal of continents during a time of higher sea level near the start of the Paleogene Period signifies a 5% expansion of a global shoreline compared to today [50]. It follows that the absence of data on Triassic rocky shores in the published literature (Figure 13) may be due to the quiescence of tectonic forces around the circumference of Pangaea after the planet's principal seafloor-spreading zones exerted maximum influence. Not only was there a reduction in the amount of coastal real estate with Pangaea, but it can be argued that extensive river systems carried a huge load of sediments across the lowlands to the many deltas that ringed the single continent at that time. The fact remains that the total volume of surviving marine Triassic strata is at a nadir compared to all other periods of the Phanerozoic Eon [51]. Future discoveries on Triassic rocky shores and biotas remain possible.

As cited herein, the dozen archipelago systems with specific examples of intertidal conditions [17] are noteworthy for the preservation of island settings in three-dimensional relief due to the relative resistance to erosion of core materials like granite, basalt, and quartzite. Especially in the case of Paleozoic examples, intervals during high global sea levels resulted in the development of islands in epicontinental settings far from the periphery of continental margins. Examples from the late Cambrian of Wisconsin and South Dakota (Figure 1, localities 1, 2), the late Silurian of northern China (Figure 1, locality 4), as well as the late Devonian and early Permian of Western Australia (Figure 1, localities 5, 6) all correspond to periods of rising sea level during which paleoshores were subject to rapid burial. In such cases, the original unconformity surface that enveloped those islands became blanketed by a conglomerate consisting of clasts eroded from the core by wave action. The late Cambrian of the Baraboo area in southern Wisconsin (Figure 1, locality 1) and the early Jurassic of southern Wales on Bristol Channel (Figure 1, localities 1, 7) stand out as examples where the generation of the conglomerate was heightened by storms of hurricane intensity. Plate reconstructions for those localities during the late Cambrian and early Jurassic suggest that islands in the Baraboo and Saint David archipelagos stood in the direct path of possible hurricane tracks [52]. Finally, the catastrophic burial and preservation of rocky-shore biotas by volcanic ash and coarse tephra is exemplified by examples from islands in the Madeira, Azores, and Cape Verde archipelagos (Figure 1, localities 9, 10, 11).

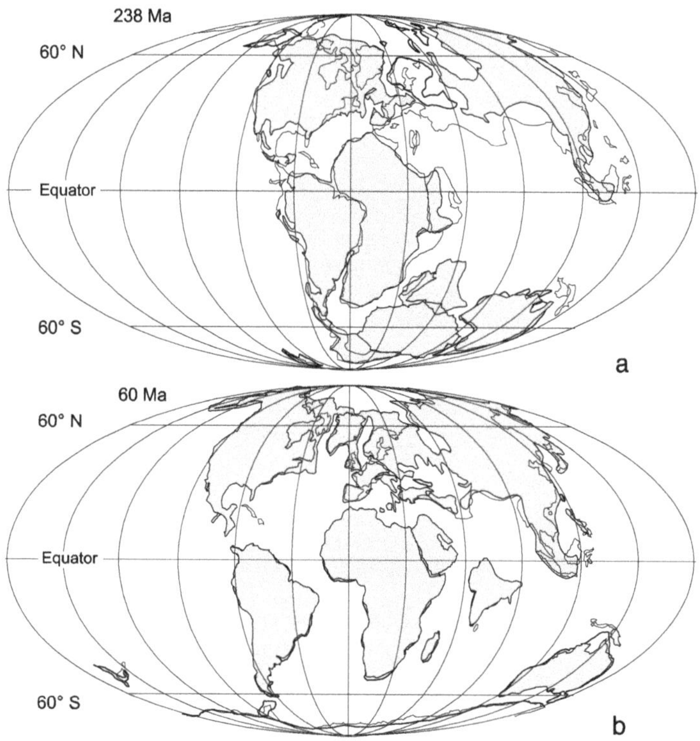

Figure 22. Global reconstructions of the continents and oceans from the early Mesozoic and early Cenozoic eras: (**a**) Pangaea during the mid-Triassic approximately 238 million years ago; (**b**) Dispersal of continents during early Paleogene Period approximately 60 million years ago. Modified from Johnson [49].

Among the oldest traces of life on a rocky shore are sedimentary-organic structures made by stromatolites preserved in growth position on ancient lava flows as old as 3.5 billion years in South Africa [8]. The stromatolites constitute an archaic element that continued to appear on rocky shores (as well as other more sheltered offshore settings) throughout Phanerozoic time (Figure 23). The search for the earliest signs of a rocky-shore biota consisting of metazoans starts with the Cambrian explosion of marine invertebrate life. No evidence of intertidal life has been detected from the extensive paleoshores surrounding the 30 islands in the Baraboo archipelago of Wisconsin [17]. Only a tantalizing clue is known for a possible rock-boring organism from the coeval Cambrian islands of the Black Hills in South Dakota [33]. From zero or near zero in the Cambrian Period, the number of intertidal rocky shore macro-invertebrates rose to 765 fossil and extant species by the last interglacial epoch of the Pleistocene [10]. At least three of the five major mass extinctions during Phanerozoic time had an effect on the continuity of diversification in rocky-shore biotas. For example, a distinctly Paleozoic biota includes the tabulated corals, which go extinct by the end of that era [10,17]. The trace-fossil genus *Trypanites* is also characteristic of the Paleozoic but continued to be preserved as borings well beyond the Paleozoic.

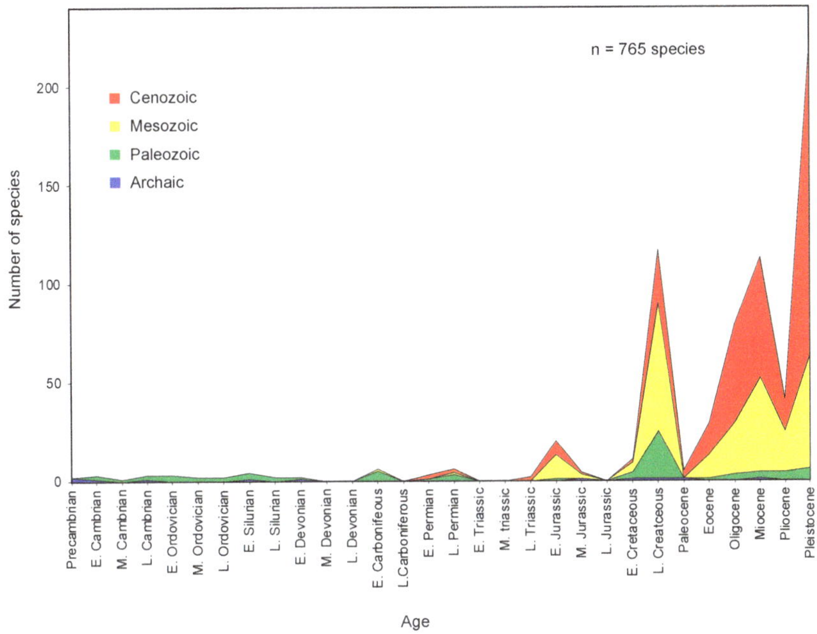

Figure 23. Changes in the diversity of the rocky-shore biota through geologic time are plotted in four overlapping clusters classified as having Archaic, Paleozoic, Mesozoic, and Cenozoic origins.

The extinction separating the Franian and Famenian epochs near the close of the Devonian had a major impact on the extermination of many stromatoporoid species recognized in the Canning Basin of Western Australia and elsewhere around the Devonian world (17). Oysters and scleractinian corals comprise two successful groups characteristic of a Mesozoic fauna, which continued to be successful through the following Cenozoic Era. However, the rudist bivalves that occupied Pacific rocky shores in western Mexico and elsewhere along the Tethys seaway were killed off during the mass extinction at the end of the Cretaceous Period. The extinction of many previously successful rocky-shore species was thus counterbalanced by the origination of other species. A highly diverse group of gastropods first appeared in the Cenozoic time and continue to thrive on rocky shores, as reflected by the diverse species recorded in the present-day tropical waters of the Seychelles Islands [25]. Despite the many extinctions in rocky-shore organisms, the clear upward trend in diversification is apparent when plotted as succeeding Archaic, Paleozoic, Mesozoic, and Cenozoic faunas (Figure 23).

The single most diverse fauna described from paleoislands in any one region derives from the Cretaceous Sachsen area in eastern Germany, with 120 fossil species of marine invertebrates recorded [53]. A detailed summary of an upper Cretaceous rocky shore from southern Sweden at Ivö Klack specifically cites zonation patterns listing close to 70 species of marine invertebrates, among which large, thick-shelled oysters are common, but brachiopods account for 27 species, gastropods for 19 species, echinoids for 18 species, and asteroids for 16 species [54]. Subsequent work at the same locality added 28 species of barnacles, including four new genera with a dozen new species [55]. Both the German and Swedish Cretaceous faunas also record abundant marine vertebrates, all of which constitute offshore faunas.

Another detailed summary of an upper Cretaceous rocky-shore fauna from the Qahlah Formation in the Oman Mountains of Arabia lists a total of 24 marine invertebrates dominated by encrusting bryozoa [56]. Given that the paleolatitude of southern Sweden in late Cretaceous time was at about 50° N and the paleolatitude of Oman was close to the

Cretaceous paleoequator, it might be expected that the Omani fauna should be more diverse than the Swedish fauna. Traditionally, latitudinal species gradients are traced within a particular group of species, for example, corals, as opposed to an entire fauna. Hence, there exist problems with any attempt to draw conclusions for an entire fauna. Seven corals are described from the upper Cretaceous hardgrounds of Oman [56], whereas a dozen corals and 28 barnacle species are identified from the roughly coeval Upper Cretaceous of southern Sweden [53,54]. The late Cretaceous Period was different from today in that no polar ice caps existed, and therefore, the latitudinal temperature gradient from the Cretaceous equator to the Cretaceous poles was moderate compared to what we understand in today's world [57]. Against this background, we find a modern-day rocky-shore fauna from the equatorial region of the Seychelles Islands that amounts to 74 species dominated by marine mollusks [24,25] compared to 24 mollusks on the rocky shores of Maine at a latitude 44° farther north under a temperate climate [20].

Thus, the present-day world provides a grounded starting point for our understanding of the multitudinous ways that life adapted to the stressful environment of intertidal rocky shores. Limited as it may be, the fossil record of life's progression in this ecosystem over the last 538 million years informs us about the challenges met in previous worlds different from our own. The endpoint of this review argues that an adequate fossil record is available for study with regard to intertidal rocky shores and that a few extraordinary examples provide a succession of nuanced purviews into the past [17].

6. Conclusions

Intact fossil assemblages preserved as distinct markers represent specific moments in time when the life relationships in a rocky shore ecosystem at a particular place were preserved in situ as time capsules. The rocky shore ecosystem represents the oldest and most widespread ecosystem on planet Earth. The perils met by a wide range of organisms in this system include challenges against waves that release great energy, exposure on a daily basis to the atmosphere with intense ultraviolet radiation, and rapid fluctuations in ambient temperature. The overall number of such known markers detected in the rock record by paleontologists remains low [10] but is expected to expand with renewed interest in the topic. Some large gaps in time, such as the entire Triassic Period, remain to be betterdocumented with distinct rocky-shore biotas. However, sufficient information has already accumulated to serve as a guide to the general evolution of the rocky shore ecosystem. A small number of outstanding examples that feature entire archipelagos across the Phanerozoic is well documented [17] and serves to encourage ongoing research. Overall, the following conclusions may be drawn from this review.

- The available habitat of rocky shores accessible by intertidal life has varied through geologic time as moderated by the mechanics of plate tectonics. A single large continent like Pangaea 250 million years ago offered a reduced shoreline compared with the wide dispersal of many and much smaller continents as today. The paucity of Triassic rocky shores that remain to be detected in the geologic record is related, in part, to this scenario in plate tectonics.
- The hard substrate of a rocky shoreline is foundational on any kind of igneous, metamorphic, or sedimentary rock. However, igneous rocks like granite or basalt with greater hardness and higher density wear better against erosion and make a more stable home for colonizing organisms, which are able to colonize and recolonize at a faster rate than rock recession.
- Changes in global sea level through geologic time have influenced the location of rocky shores, often most prominent along continental margins but also appearing in epicontinental settings far inland during intervals of high sea level.
- Marine invertebrates and many algae are capable of attachment to a rocky shore by direct cementation or other means, whereas mobile forms adhere by suction of a strong muscular "foot" or by wedging into crevices or other irregularities in the rock surface.

Some bivalves and sponges are capable of rock boring. All such groups have a distinct paleontological record.

- Available data show a steady increase in the number of such marine invertebrates and certain marine algae across geologic time, although membership was modified by the intervention of mass extinctions that affected life in other ecosystems, as well. Latitudinal temperature gradients appear to have varied through geologic time, meaning that analogous diversity gradients from tropical to more temperate settings may have been less pronounced than today.
- The preservation of rocky-shore species as fossils in their original growth position on a rock substrate is more commonly encountered than once assumed and was typically mediated by sudden catastrophic events ending in burial by rapid rates of rising sea level, storm deposits, or volcanic ejecta such as ash or coarser tephra.

The preponderance of data on which the above conclusions are based derives from the work by geologists and paleontologists in Europe, as well as English-speaking areas in the British Isles, USA, Canada, Australia, and New Zealand. It is hoped this review will attract fresh research from cohorts working elsewhere, especially in Asia, Africa, and South America.

Funding: This research received no external funding.

Acknowledgments: MEJ is grateful to B. Gudveig Baarli for assistance in editing the line drawings for this paper.

Conflicts of Interest: The author declares no conflict of interest.

References

1. National Geographic Atlas of the World, 11th ed.; National Geographic Society: Washington, DC, USA, 2019; 448p.
2. Bird, C.F.; Schwartz, M.L. *The World's Coastline*; Van Nostrand Reinhold: New York, NY, USA, 1985; 1071p.
3. Carson, R. *The Edge of the Sea*; Houghton Mifflin Co.: Boston, MA, USA, 1955; 276p.
4. Hecker, R.F. *Introduction to Paleoecology*; American Elsevier Publishing Co.: New York, NY, USA, 1966; 166p.
5. Selden, P.; Nudds, J. *Evolution of Fossil Ecosystems*; University of Chicago Press: Chicago, IL, USA, 2004; 160p.
6. Zakharov, D.O.; Bindeman, I.N. Triple oxygen and hydrogen isotopic study of hydrothermally altered rocks from the 2.43–2.41 Ga Vetreny belt, Russia: An insight into the early Paleoprterozoic seawater. *Geochem. Colsmochim. Acta* **2019**, *248*, 185–209. [CrossRef]
7. Bindeman, I.N.; O'Neil, J. Earth's earliest hydrosphere recorded by the oldest hydrothermally-altered oceanic crust: Triple oxygen and hydrogen isotopes in the 4.3–3.8 Ga Nuvvuagittuq belt, Canada. *Earth Planet. Sci. Lett.* **2022**, *586*, 117539. [CrossRef]
8. Byerly, G.R.; Lower, D.R.; Walsh, M.M. Stromatolites from the 3300–3500-Myr Swaziland Supergroup, Barberton Mountain Land, South Africa. *Nature* **1986**, *319*, 489–491. [CrossRef]
9. Logan, B.W. Cryptozoon and associated stromatolites from the recent, Shark Bay, Western Australia. *J. Geol.* **1961**, *69*, 517–533. [CrossRef]
10. Johnson, M.E.; Baarli, B.G. Development of intertidal biotas through Phanerozoic time, pp. 63–128. In *Earth and Life*; Talent, J.A., Ed.; Springer Science+Business Media: Dordrecht, The Netherlands, 2012; 1100p.
11. Emery, K.O.; Kuhn, G.G. Sea cliffs: Their processes, profiles, and classification. *Geol. Soc. Am. Bull.* **1982**, *93*, 644–654. [CrossRef]
12. Bird, E.C.F. *Coastal Geomorphology: An Introduction*; John Wiley & Sons: Chichester, UK, 2008; 448p.
13. Johnson, M.E. Why are ancient rocky shores so uncommon? *J. Geol.* **1988**, *96*, 469–480. [CrossRef]
14. Deny, M.W. *Biology and the Mechanics of the Wave-Swept Environment*; Princeton University Press: Princeton, NJ, USA, 1998; 329p.
15. Stephenson, T.A.; Stephenson, A. The universal features of zonation between tide-markes on rocky coasts. *J. Ecol.* **1949**, *37*, 289–305. [CrossRef]
16. Sunamura, T. *Geomorphologyu of Rocky Coasts*; John Wiley & Sons: Chichester, UK, 1992; 302p.
17. Johnson, M.E. *Islands in Deep Time: Ancient Landscapes Lost and Found*; Columbia University Press: New York, NY, USA, 2023; 312p.
18. Twidale, R. *Granite Landforms*; Elsevier: Amsterdam, The Netherlands, 1982; 372p.
19. Twidale, R. The research frontier and beyond: Granitic terrains. *Geomorphology* **1993**, *7*, 187–223. [CrossRef]
20. Brinkhurst, R.O.; Linkletter, L.E.; Lord, E.I.; Connors, S.A.; Dadswell, M.J. *A Preliminary Guide to the Littoral and Sublittoral Marine Invertebrates of Passamaquddy Bay*; Huntsman Marine Laboratory: St. Andrews, NB, Canada, 1976; 166p.
21. Johnson, M.E.; Baarli, B.G. Erosion and burial of granite rocky shores in the Recent and Late Pleistocene of the Seychelles Islands: Physical and biological perspectives. *J. Coast. Res.* **2005**, *21*, 867–879. [CrossRef]
22. Braithwaite, C.J.R. Geology of the Seychelles. In *Biogeography and Ecology of the Seychelles Islands*; Stoddart, D.R., Ed.; W. Junk Publishers: The Hague, The Netherlands, 1984; pp. 17–38.

23. Suwa, K.; Tokieda, K.; Hoshino, M. Palaeomagnetic and petrological reconstruction of the Seychelles. *Precambrian Res.* **1994**, *69*, 281–292. [CrossRef]
24. Taylor, J.D. Coral reef and associated invertebrate communities (mainly molluscan) around Mahé, Seychelles. *Philos. Trans. R. Soc. Lond. Ser. B Biol. Sci.* **1968**, *254*, 129–206.
25. Jerrett, A.G. *Marine Shells of the Seychelles*; Carole Green Publishing: Cambridge, UK, 2000; 149p.
26. Philipps, C.J.; Bellwood, D.R. The hydrodynamics of Lizard Island lagoon, Great Barrier Reef. *Coral Reefs* **2024**, *43*, 881–897. [CrossRef]
27. Davey, K. *Seashore Life of Australia*; New Holland Publishers: Sydney, Australia, 1998; 144p.
28. Friedman, G.M. Breachrocks record Holocene events, including natural disasters. *Carbonates Evaporites* **2011**, *26*, 97–109. [CrossRef]
29. Leonard, J.E.; Cameron, B.W. Origin of a high-latitude carbonate beach: Mt. Desert Island, Maine. *Northeast. Geol.* **1981**, *28*, 1–28.
30. Barndhardt, W.W.; Kellley, J.T. Carbonate accumulation on the inner continental shelf of Maine; a modern consequence of late Quaternary glaciation and sea-level change. *J. Sediment. Res.* **1995**, *65*, 195–207.
31. Dott, R.H., Jr. Cambrian tropical storm waves in Wisconsin. *Geology* **1974**, *2*, 243–246. [CrossRef]
32. Eoff, J.D. Sequence stratigraphy of the Upper Cambrian (Furongian, Jiangshanian and Sunwaptan) Tunnel City Group, Upper Mississippi Valley: Transgressing assumptions of cratonic flooding. *Sediment. Geol.* **2014**, *302*, 87–101. [CrossRef]
33. Johnson, M.E.; Wilson, M.A.; Redden, J.A. Borings in quartzite surf boulders from the Upper Cambrian basal Deadwood Formation, Black Hills of South Dakota. *Ichnos* **2010**, *17*, 48–55. [CrossRef]
34. Johnson, M.E.; Baarli, B.G. Encrusting corals on a latest Ordovician to earliest Silurian rocky shore, southwest Hudson Bay, Manitoba, Canada. *Geology* **1987**, *15*, 15–17. [CrossRef]
35. Nelson, S.J.; Johnson, M.E. Jens Munk Archipelago: Ordovician-Silurian islands in the Churchill area of the Hudson Bay Lowlands, Northern Manitoba. *J. Geol.* **2002**, *110*, 577–589. [CrossRef]
36. Johnson, M.E.; Rong, J.Y.; Wang, C.Y.; Wang, P. Continental island from the Upper Silurian (Ludfordian Stage) of Inner Mongolia: Implications for eustasy and paleogeography. *Geology* **2001**, *29*, 955–958. [CrossRef]
37. Rong, J.; Johnson, M.E.; Deng, Z.; Dong, D.; Yaosong, S.; Baarli, B.G.; Wang, G. Coral-stromatoporoid faunas from the shores of a late Silurian island, Inner Mongolia, North China. *Mem. Assoc. Australas. Palaeontol.* **2013**, *44*, 95–105.
38. Johnson, M.E.; Webb, G.E. Outer rocky shores of the Mowanbini Archipelago, Devonian Reef Complex, Canning Basin, Western Australia. *J. Geol.* **2007**, *115*, 583–600. [CrossRef]
39. Baarli, B.G.; Webb, G.; Johnson, M.E.; Cook, A.G.; Walsh, D.R. Shoal-water dynamics and coastal biozones in a sheltered-island setting: Upper Devonian Pillara Limestone (Western Australia). *Lethaia* **2016**, *49*, 507–523. [CrossRef]
40. De la Beche, H.T. *On the Formation of the Rocks of South Wales and South Western England*; Memoirs of the Geological Survey of Great Britain; Geological Survey: London, England, 1846; pp. 1–2167.
41. Johnson, M.E.; McKerrow, W.S. The Sutton Stone: An early Jurassic rocky shore deposit in South Wales. *Palaeontology* **1995**, *38*, 529–541.
42. Johnson, M.E.; Ledesma-Vázquez, J.; Clark, H.C.; Zwiebel, J.A. Coastal evolution of Late Cretaceous and Pleistocene rocky shores: Pacific rim of northern Baja California, Mexico. *Geol. Soc. Am. Bull.* **1996**, *108*, 708–721. [CrossRef]
43. Johnson, M.E.; Hayes, M.L. Dichotomous facies on a Late Cretaceous rocky island as related to wind and wave patterns (Baja California, Mexico). *Palaios* **1993**, *8*, 385–395. [CrossRef]
44. Santos, A.; Mayoral, E.; da Silva, C.M.; Cahão, M.; Johnson, M.E.; Baarli, B.G. Miocene intertidal zonation on a volcanically active shoreline: Porto Santo in the Madeira Archipelago, Portugal. *Lethaia* **2011**, *44*, 26–32. [CrossRef]
45. Rebelo, A.C.; Rasser, M.W.; Kroh, A.; Johnson, M.E.; Ramalho, R.S.; Melo, C.; Uchman, A.; Berning, B.; Silva, L.; Zanon, V.; et al. Rocking around a volcanic island shelf: Pliocene rhodolith beds from Malbusca, Santa Maria Island (Azores, NE Atlantic). *Facies* **2016**, *62*, 22. [CrossRef]
46. Johnson, M.E.; Uchman, A.; Costa, P.J.M.; Ramalho, R.S.; Ávila, S.P. Intense hurricane transports sand onshore: Example from the Pliocene Malbusca section on Santa Maria Isalnd (Azores, Portugal). *Mar. Geol.* **2017**, *385*, 244–249. [CrossRef]
47. Darwin, C. *Geological Observations on the Volcanic Islands Visited during the Voyage of the H.M.S. Beagle., Together with Some Brief Notices on the Geology of Australia and the Cape of Good Hope. Being the Second Part of the Geology of the Voyage of the Beagle, under the Command of Capt. FitzRoy, RN.N. during the Years 1832 to 1836*; Smith, Elder & Co.: London, UK, 1844; 175p.
48. Baarli, B.G.; Santos, A.G.; Mayoral, E.J.; Ledesma-Vázquez, J.; Johnson, M.E.; da Silva, C.M.; Cachão, M. What darwin did not see: Pleistocene fossil assemblages on a high-energy coast at Ponta das Bicudas, Santiago, Cape Verde islands. *Geol. Mag.* **2013**, *150*, 83–189. [CrossRef]
49. Wegener, A. *The Origin of Continents and Oceans*; (first English edition translated from the third German edition); Methuen & Co., Ltd.: London, UK, 1924; 246p.
50. Johnson, M.E. Uniformitarianism as a guide to rocky-shore ecosystems in the geological record. *Can. J. Earth Sci.* **2006**, *43*, 1119–1147. [CrossRef]
51. Husson, J.M.; Peters, S.E. Nature of the sedimentary rock record and its implications for Earth System evolution. *Emerg. Top. Life Sci.* **2018**, *2*, 125–136. [PubMed]
52. Marsaglia, K.M.; Klein, G.D. The paleogeography of Paleozoic and Mesozoic storm depositional systems. *J. Geol.* **1983**, *91*, 117–142. [CrossRef]
53. Pietzsch, K. *Geologie von Sahzen*; Veb Deutcher Verlag der Wissenschaften: Berlin, Germany, 1962; pp. 17–22.

54. Surlyk, F.; Sørensen, A.M. An early Campanian rocky shore at Ivö Klack, southern Sweden. *Cretac. Res.* **2010**, *31*, 567–576. [CrossRef]
55. Gale, A.S.; Sørensen, A.M. Taxonomy and palaeoecology of thoracican cirripedes (Crustacea) from a Campanian rocky shoreline at Ivö Klack, southern Sweden. *Cretac. Res.* **2015**, *54*, 212–242. [CrossRef]
56. Wilson, M.; Taylor, P.D. Palaeoecology of hard substrate faunas from the Cretaceous Qahlah Formation of the Oman Mountains. *Palaeontology* **2001**, *44*, 21–41. [CrossRef]
57. Zhang, L.; Hay, W.W.; Wang, C.; Gu, X. The evolution of latitudinal temperature gradients from the latest Cretaceous through the Presnt. *Earth-Sci. Rev.* **2019**, *189*, 147–158. [CrossRef]

Journal of
Marine Science and Engineering

MDPI

Review

Exposure of Carboniferous Granitoids on Triassic–Jurassic Seashores in the Western Caucasus: A Stratigraphical Review

Dmitry A. Ruban

Department of Organization and Technologies of Service Activities, Institute of Tourism, Service and Creative Industries, Southern Federal University, 23-ja Linija Street 43, Rostov-on-Don 344019, Russia; ruban-d@mail.ru

Abstract: Granitoids are known to crop out on ancient seashores, but the related geological evidence remains limited. The information from the Western Caucasus sheds light on the stratigraphical distribution of coarse siliciclastic beds associated with late Carboniferous granitoids of the Dakh, Rafabgo, and Sakhray crystalline massifs. For the purposes of this study, the available information was reviewed and verified against the modern stratigraphical scales. It is established that the considered coarse sisliciclastic beds occur at five stratigraphical levels of the Triassic–Jurassic succession. A rocky seashore with granitoid exposures existed for a short time around the Sakhray and probably Rufabgo massifs at the very beginning of the Triassic. The Dakh Massif possessed such a shore twice (at least), i.e., in the Norian–Rhaetian and the Early Toarcian. However, it cannot be excluded that rocky shores persisted there for >50 Ma. Generally, the Western Caucasus provides an example of granitoid exposures on Mesozoic seashores and adds knowledge of the global distribution of rocky shores in the Triassic and Jurassic periods.

Keywords: coarse siliciclastics; Greater Caucasus; Mesozoic; rocky coast; sea-level changes; stratigraphical correlation; transgression

1. Introduction

Rocky shores constitute a specific depositional environment [1–5] with a relatively poor potential for preservation in the geological record. The works by Johnson [6–8] laid a conceptual foundation for the development of the related ideas, which are important to marine geology, stratigraphy, sedimentology, palaeogeography, and palaeoecology. During decades of research, evidence of this depositional environment was collected in many regions and geological time slices [9–15]. Particularly, it became clear that some modern and ancient rocky shores are linked to exposure of granitoids [16–18]. If so, attention to coarse siliciclastic beds on margins of granitic massifs can facilitate the understanding of ancient rocky shores; the related features are interesting not only theoretically, but also practically, i.e., as objects of geoheritage and geotourism [19–23].

In a relatively small area of the Western Caucasus (western part of the Greater Caucasus orogen), three crystalline massifs representing Late Paleozoic granitoids are known [24]. The evidence of several coarse siliciclastic beds associated with these massifs accumulated and was discussed for about a century [23,25–28], but this knowledge remained limited, fragmented, and non-systematized. Moreover, a joint consideration of some old ("forgotten") and very new works can shed light on exposure and wave abrasion of granitoids in the Mesozoic. Apparently, these really spectacular geological records of ancient rocky shores were almost missed in the geological research, despite their potentially international importance. Moreover, a consideration of these records can help to fill some gaps in the understanding of the geological evolution of the Western Caucasus. The information about possible granitoid exposure on Triassic–Jurassic seashores of the Western Caucasus [23,25–28] is already significant relative to the general scarcity of this kind of evidence [8]. But this information is available in an imperfect form, and, thus, it is urgent to review it, with certain updates, improvements, and additions.

The objective of the present work is to review the stratigraphical distribution of the coarse siliciclastic beds associated with Late Paleozoic granitoids of the Western Caucasus. Particularly, the presence of these beds and granitoids in Mountainous Adygeya is considered. This review becomes possible thanks to new, original field observations in the study area and critical consideration of the available, but yet to be systematized, published evidence. Extraction and reinterpretation of the information from some old works [27,28] can facilitate the understanding of the presence of several coarse siliciclastic beds. The present work continues, but does not replicate, the discussion started in the previous work [23]; this paper differs from the latter substantially and presents more advanced interpretations. It should be stressed that this review is essentially stratigraphical and deals with rock relationships and ages. Indeed, composition of coarse siliciclastic beds is considered, but only generally; a new, fully separate research project will be necessary to investigate this composition in detail. Nonetheless, the reviewed stratigraphical information creates a basis for further, pure sedimentological investigations.

2. Geological Setting

The study area is a central part of so-called Mountainous Adygeya, which is a part of the Western Caucasus in southwestern Russia (Figure 1a). Tectonically, this area belongs to the western segment of the late Cenozoic orogen, the development of which marked the active contacts between the Eurasian lithospheric plate in the north and smaller plates (particularly, the Anatolian plate) in the south [29–31]. The ongoing orogeny was preceded by a long-term evolution of northern Neo-Tethyan island arcs and back-arc basins [32–34].

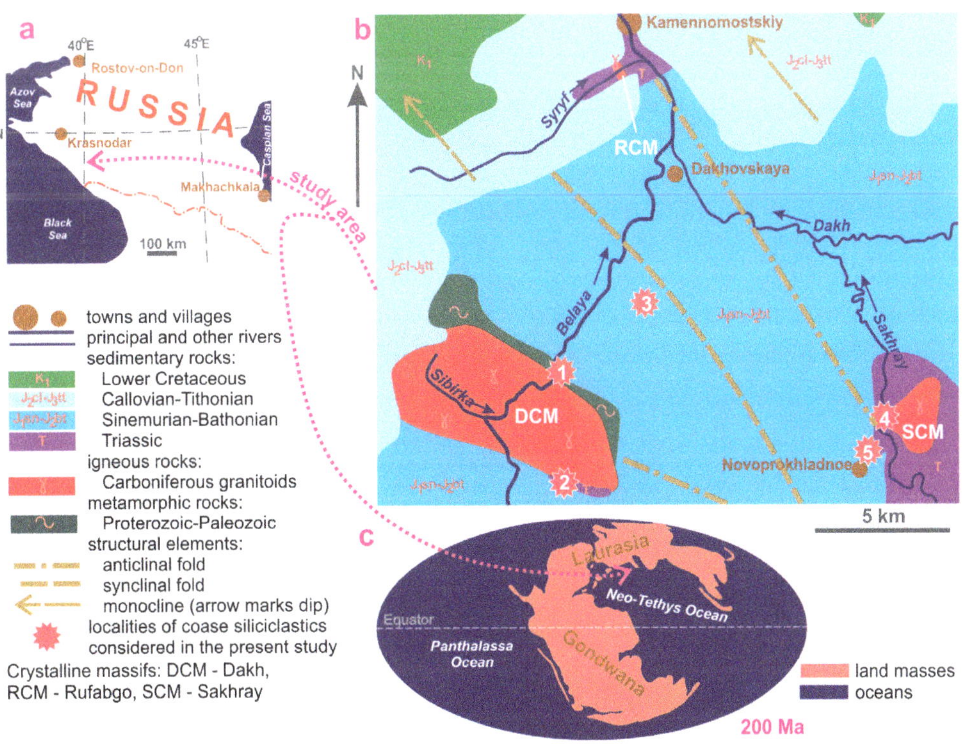

Figure 1. Position of the study area: geographical location (**a**), geological setting (**b**), and palaeogeographical location (**c**). Locality numbers are explained in the text. Structural elements are shown on the basis of the reconsidered information from Popov and Sharova [35]. The palaeogeographical scheme is modified strongly from Scotese [36].

The geology of the study area is heterogeneous (Figure 1b). It is dominated by Lower–Middle Jurassic siliciclastic rocks and Upper Jurassic carbonate and siliciclastic rocks [35]. The former are folded and faulted in a complex manner; principally, there is a northwest–southeast-trending syncline with an axis located between the Dakh and Sakhray massifs [35]. The Upper Jurassic sedimentary complex forms a kind of north-dipping monocline, and this complex covers older rocks transgressively from the north and the west.

There are three crystalline massifs (Dakh, Rufabgo, and Sakhray—Figure 1b), which consist chiefly of Late Paleozoic rocks (Figure 2a), the exact age of which is established as late Carboniferous [37]. One massif (at least) includes also Proterozoic and Paleozoic metamorphic rocks (Figure 2b). All massifs associate with Triassic sedimentary rocks (Figure 2c,d). Apparently, the crystalline massifs and Triassic complexes are represented in uplifted tectonic blocks affected by deep erosion. The boundaries of these blocks and the related major faults are yet to be known precisely.

Figure 2. Selected geological features of the study area: late Carboniferous granitoids of the Rufabgo Crystalline Massif (**a**), Precambrian metamorphic rocks and late Carboniferous granitoids of the Dakh

Crystalline Massif (**b**), limestones of the Yatyrgvarta Formation exposed near the Rufabgo Crystalline Massif (**c**) and the Sakhray Crystalline Massif (**d**), latest Triassic–earliest Jurassic conglomerate-breccias near the Sakhray Crystalline Massif (**e**).

The Mesozoic stratigraphy of the study area is known rather well [24,25,38–42], although many details and even the nomenclature of lithostratigraphical units are yet to be fixed. The local sedimentary succession is dominated by siliciclastic and carbonate packages with a total thickness measuring several thousand meters and interrupted by a series of major hiatuses (Figure 3), often marked by coarse siliciclastic beds (Figure 2e). The most notable and widely distributed packages include dark and hard limestones of the Yatyrgvarta Formation, shales of the Sakhray Group, grey-to-black shales of the Psebay and Dzhangur formations, massive limestones and dolostones of the Gerpegem Formation, and siliciclastic red beds of the Mezmay Formation. These deposits accumulated in a tropical and subtropical sea on the margin of the Neo-Tethys Ocean (Figure 1c), which embraced back-arc basins [23,43], although temperate conditions could also exist in some time slices [44]. In the Norian–Rhaetian and the Oxfordian–Kimmeridgian intervals, reefal ecosystems flourished in the study area and marked a development of vast carbonate platforms.

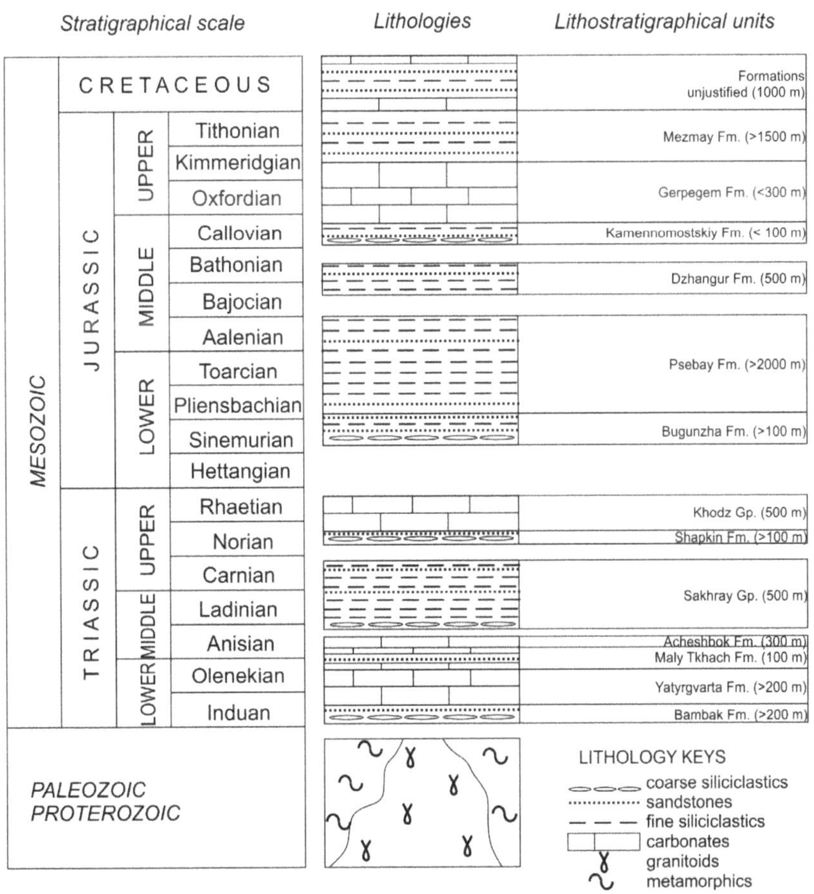

Figure 3. A composite stratigraphical section of the study area. See text for additional explanations and sources of stratigraphical information.

3. Material and Methods

The information of the presence of coarse siliciclastic beds associated with the three crystalline massifs known from the study area was collected over the course of several years of field investigations, as well as from the literature. Particularly, one should note that the geology of Mountainous Adygeya was studied intensively in the first half of the 20th century [27,28], but that information remained almost forgotten later. This creates a biased vision because the study area is relatively large, and its total geological reinvestigation requires the efforts of dozens of specialists and many years (even decades) of field works. Moreover, some places are poorly accessible due to steep slopes, dense vegetation cover, and absent trails, and finding small outcrops can be challenging there. This is why the old works should not be omitted, and they can bear a lot of precious information that makes a substantial addition to the present knowledge. Importantly, the old works were based on investigations carried out and reported carefully.

To avoid the above-mentioned biased vision, a careful search in libraries and on-line archives was carried out to find the old works dealing with the geology of Mountainous Adygeya. It was established that two of them [27,28] contain much important information about coarse siliciclastic beds associated with granitoids of the study area. Indeed, this information can be used together with the newest lines of evidence [23,25,26] and the novel field observations. Finding old sources and reinterpreting their content are important elements of the present review. Indeed, the latter also takes into account the results of new, original field investigations.

A total of five localities representing coarse siliciclastic beds associated with the three crystalline massifs of Mountainous Adygeya are considered in this review (Figure 1b). To avoid further misunderstandings, these localities are not named, but numbered. Localities 1 and 5 were detected in the field, and they were also characterized in the literature [23,26]. The other localities were detected only in the literature [25,27,28], and they were not revisited due to the limited accessibility (the appropriate quality of the published information from these localities does not require revisiting them). The information available in both the old and new sources is detailed enough to be used for the purposes of the present analysis. One should note that the considered crystalline massifs are relatively small (Figure 1b), and deposits of ancient shores are rather rare in the geological records [8]. Therefore, the considered number of the localities known in the study area seems to be appropriate for this review. Moreover, the presence of five localities representing ancient rocky shores in one area makes the area almost unique on the global scale.

The procedures of the present review are rather elementary. Nonetheless, they required much effort and complex solutions. Moreover, these procedures are essential and typical to regional lithostratigraphical investigations.

First of all, the position of each locality relative to the crystalline massifs was verified. Then, stratigraphical relationships of coarse siliciclastic beds with other rock packages and granitoids were documented and, where necessary, interpreted. Although this paper has a stratigraphical focus, the principal composition peculiarities of the beds were considered. In the considered localities, the beds do not bear fossils that would permit the establishment of their age directly. However, the noted attention to the relationships with the other rocks with known ages (contemporary understanding of the local stratigraphy (Figure 3) was followed) and some compositional peculiarities permitted judgments of the age of the beds at all five localities. On the basis of this knowledge, a stratigraphical correlation (a traditional, but powerful tool in stratigraphical, sedimentological, and palaeogeographical research [45–49]) of the coarse siliciclastic beds associated with the crystalline massifs was carried out. This permitted the proposition of a general model depicting the chronology of exposure of Carboniferous granitoids of the study area on Triassic and Jurassic seashores.

4. Results

4.1. Key Localities of Coarse Siliciclastic Beds Associated with Three Crystalline Massifs

The occurrence of coarse siliciclastic beds in the considered localities is mentioned below. Attention is paid to the general view of these localities and the contacts of principal lithologies. This information is essential for the subsequent establishment of ages of these localities and their correlation.

Locality 1 is found near the mouth of the Lipovy Stream (a small right tributary of the Belaya River) and the northern periphery of the Dakh Crystalline Massif (Figure 1b). There, a thin (~2 m) package of conglomerates crops out. These conglomerates overlay transgressively Proterozoic and Paleozoic metamorphic rocks of the noted massif (Figure 4a). Their contacts with any other sedimentary packages are not registered, but these coarse siliciclastic beds were attributed provisionally to the basal horizons of the Bagovskaya Formation (a former lithostratigraphical unit which is understood presently as a middle part of the Psebay Formation—Figure 3) [23]. These conglomerates include numerous clasts of Carboniferous granitoids of the Dakh Crystalline Massif [26]. These granitoids crop out very closely to Locality 1 (Figure 4a).

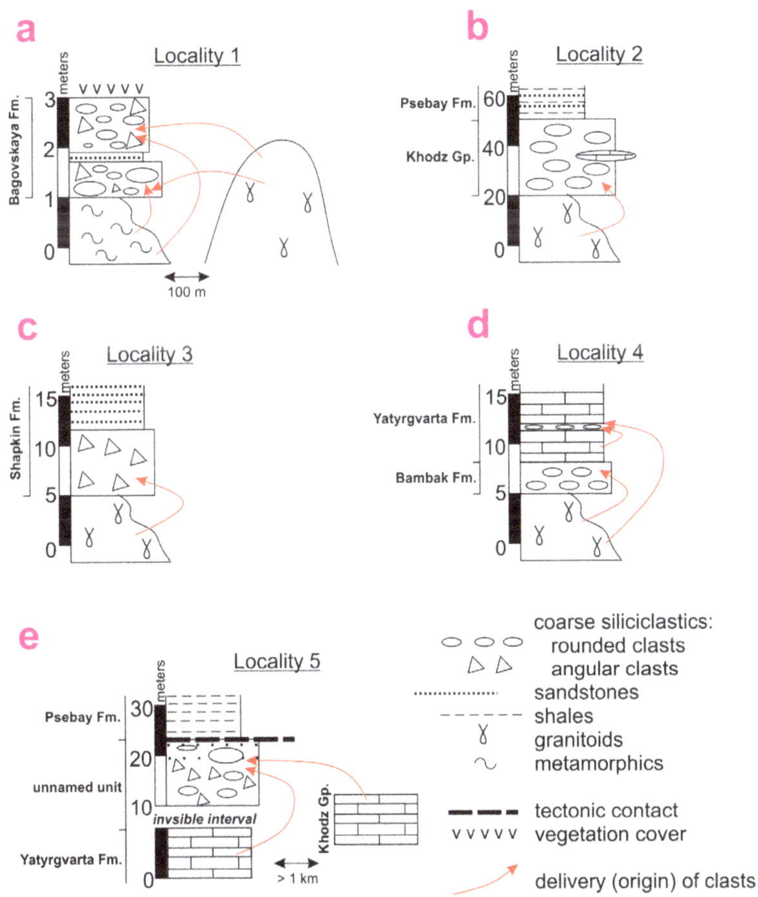

Figure 4. Stratigraphical outline of the considered localities: Locality 1 (**a**), Locality 2 (**b**), Locality 3 (**c**), Locality 4 (**d**), and Locality 5 (**e**). See text for additional explanations and sources of stratigraphical information.

Locality 2 corresponds to the watershed between the Syuk River and the Grushevaya River (both are right tributaries of the Belaya River) and the upper parts of their valleys; this is the southeastern edge of the Dakh Crystalline Massif (Figure 1b). This locality was characterized initially in the work by Vyalov and Nikshich [28], whose descriptions were later confirmed and extended by Diakonova-Savelieva [27]. According to these authors, a package of conglomerates with well-rounded granitoid clasts (up to 15 cm in size) overlays granitoids (Figure 4b). Interestingly, this package also includes small lenses of spotted, reddish limestones with a thickness of ~2.5 m. These limestones are similar to the Khodz Group, which bears reefal limestones (Figure 3). If so, it can be postulated that the considered packages of conglomerates from Locality 2 can be assigned to the Khodz Group (Figure 4b). As this can be deduced from the descriptions offered by Vyalov and Nikshich [28], this conglomerate package is overlain by Lower Jurassic sandstones and shales, which represent the Psebay Formation (Figure 4b).

Locality 3 hosts a borehole drilled south of Dakhovskaya village, where the northern periphery of the Dakh Crystalline Massif is overlain by a thick (>150 m) Upper Triassic–Middle Jurassic sedimentary cover (Figure 1b). The evidence from this locality was reported by Chaitsky et al. [25]. According to these authors, granitoids are overlain by a bed (7 m) composed of angular clasts made of these igneous rocks (Figure 4c). This bed is thought to be a result of physical weathering of granitoids [25]. It is overlain conformably by arkose and then quartz sandstones of the Shapkin Formation (Figure 3), and the bed can be considered as a basal part of this formation (Figure 4d).

Locality 4 corresponds to the middle part of the valley of the Sakhray River (a right tributary of the Dakh River, which flows to the Belaya River), where it crosses the Sakhray Crystalline Massif (Figure 1b). There, a thin (~3 m) package of conglomerates with quartz clasts and mica grains transgressively overlays granitoids (Figure 4d). These deposits were reported by Vyalov and Nikshich [28]. The upper part of the section is made of hard limestones, which are typical to the Yatyrgvarta Formation (Figure 3). Diakonova-Savelieva [27] noted the existence of conglomerate interbeds in these limestones; these conglomerates bear rare granitoid pebbles and common limestone pebbles (Figure 4d).

Locality 5 is situated in the lower part of the valley of the Gosh River (a left tributary of the Sakhray River), close to the southwestern edge of the Sakhray Crystalline Massif (Figure 1b). There, a package of conglomerate–breccias with both angular and rounded clasts of different sizes and a visible thickness of ~12 m crops out (Figure 4e). Initially, it was reported by Ruban [23] and reconsidered by Ruban et al. [24]. New investigations showed that these deposits include numerous clasts of Lower Triassic limestones (Yatyrgvarta Formation) and Upper Triassic limestones (Khodz Group). The former crop out at the same locality, and contemporary exposures of the latter are located close to the Sakhray Crystalline Massif (Figure 4e). Some clasts (e.g., quartz clasts) from the considered package of conglomerate–breccias can be interpreted as products of destruction of granitoids, although such an interpretation is highly hypothetical. Due to the local tectonic complexity, the lower and upper contacts of this package remain invisible, and it is represented in a small tectonic block exhibited among Lower Jurassic fine siliciclastics of the Psebay Formation (Figure 3). Nonetheless, the local geological setting is such that the presence of the Yatyrgvarta Formation close to the base of the package of conglomerate–breccias is clear (Figure 4e).

The information presented above implies that localities 1, 2, and 4 represent ancient granitoid seashores. Undoubtedly, Locality 5 also represents a rocky shore [23], although with exposures of limestones, not granitoids. Nonetheless, this locality is of utmost importance in the present study because it marks non-exposure of the Sakhray Crystalline Massif at the time of accumulation of conglomerate–breccias. Locality 3 does not permit the hypothesis of the presence of a rocky seashore because the basal part of the Shapkin was deposited where granitoids were exposed directly on land, and sandstones of this formation mark marine deposition. Anyway, this locality is relevant to each study when local exposures of granitoids is addressed.

4.2. Ages of Coarse Siliciclastic Beds Associated with Crystalline Massifs

The presented characteristics of the considered localities of the coarse siliciclastic beds (Figure 4a–e) facilitates establishing their ages and correlation. Indeed, the regional stratigraphy (Figure 3) is taken into account for the purpose of the related interpretations.

The age of conglomerates from Locality 1 is unclear due to the absence of visible contacts with the other sedimentary rocks (Figure 4a). However, the presence of clasts of sedimentary rocks and siderite concretions, which are typical to the Psebay Formation, indicates that the age of the conglomerates is younger than the time when the lower part of the noted formation accumulated [26]. If so, these conglomerates can be linked to the basal horizons of the Dzhangur or Kamennomostskiy formations, i.e., they can be either late Bajocian or early Callovian in age (Figure 3). However, these options should be ruled out because the noted formations cannot crop out locally, taking into account the topography and the tectonics of the study area. If so, the only acceptable option is to prove the earlier proposition [23] of the assignment of these conglomerates to the former Bagovskaya Formation [40] and, thus, an early Toarcian age.

The presence of highly specific and, thus, easy-to-distinguish limestones in conglomerates of Location 2 (Figure 4b) implies that these coarse siliciclastics were deposited synchronously with the Khodz Group consisting of such limestones (Figure 3). This group accumulated in the Late Triassic (Figure 3). Diakonova-Savelieva [27] and Vyalov and Nikshich [28] made similar conclusions. If so, the age of these conglomerates is middle Norian–early Rhaetian.

The age of breccia from Locality 3 is no younger than early Norian, when the Shapkin Formation, at the base of which they occur (Figure 4c) [25], accumulated (Figure 3). Although weathering of subaerially exposed granitoids could be a long-term process, this breccia formed, most probably, together with a local tectonic uplift at the Carnian–Norian transition, which interrupted the deposition of the Sakhray Group (Figure 3). If even some weathering products accumulated earlier, their thin cover was eroded as a result of this uplift.

The position of conglomerates of Locality 4 between granitoids and the Yatyrgvarta Formation (Figure 4d) [28] means that these conglomerates should be attributed to the Bambak Formation of an early Induan age (Figure 3). The presence of this formation consisting of coarse siliciclastics in the study area was documented by Chaitsky et al. [25].

Conglomerate–breccias from Locality 5 pose a true challenge for the stratigraphical interpretation due to unclear contacts with the other sedimentary packages (Figure 4e). Ruban [23] attributed it to the Bambak Formation, but new observations [24] imply that these deposits are no older than the Khodz Group of a Late Triassic age (Figure 3), because clasts of its characteristic limestones are found in the conglomerate–breccias (this observation was also proven in the course of new field investigations in the summer of 2024). However, these conglomerate–breccias cannot represent basal horizons of the Psebay Formation (the Bugunzha Formation is locally absent) of a Jurassic age (Figure 3) because of the presence of an angular unconformity between the package of conglomerate–breccias and the Psebay Formation. The dip direction and angle of conglomerate–breccias is similar to that of Triassic deposits [23] and differs strikingly from those of the younger Psebay Formation. If so, the only acceptable option is that this coarse siciliclastic bed is older than the Khodz Formation and younger than the Psebay Group. Apparently, its age ranges within the late Rhaetian–early Hettangian interval. A bit younger, late Hettangian–early Sinemurian age cannot be excluded, but it is rather improbable because deposits of this age (if present) should be related to the Psebay Formation and share its structural setting.

4.3. Stratigraphical Correlation of Coarse Siliciclastic Beds Associated with Crystalline Massifs

The established ages of the coarse siliciclasitc beds associated with the crystalline massifs of the study area (see above) and their justification against the local stratigraphical scheme (Figure 3) make possible a stratigraphical correlation of these beds (Figure 4). Generally, these beds occur at five stratigraphical levels, some of which correspond to the known hiatuses. At each level, the only local distribution of the beds can be documented, and it is limited to the peripheral parts of the crystalline massifs. Nonetheless, two beds can be grouped provisionally (Figure 5). Localities 2 and 3 represent southern and northern peripheries of the Dakh Crystalline Massif, respectively (Figure 1b). One can hypothesize that the deposition of breccias from Locality 3 postdated the denudation of the massif before a further transgression, and the deposition of conglomerates from Locality 2 marked this transgression, which affected remote parts of this massif (see palaeogeographical interpretations below).

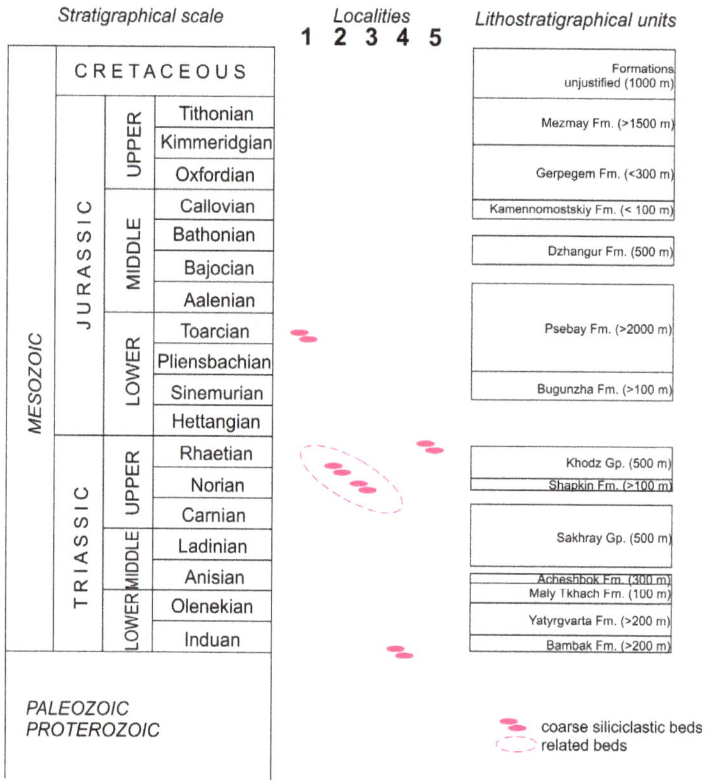

Figure 5. Stratigraphical correlation of the considered coarse siliciclastic beds.

Besides the considered localities, there is some other information related to the Triassic–Jurassic evolution of the crystalline massifs in the study area. First, Chaitsky et al. [25] used borehole data and noted the existence of the Bambak Formation in the vicinity of Kamennomostskiy town in the very north of the study area (Figure 1b), where conglomerates of this unit underlay limestones of the Yatyrgvarta Formation. Principally, this means the existence of this formation in the northern part of the study area, where the Rufabgo Crystalline Massif is located (Figure 1b). Despite several attempts, observations made in the valley of the Syryf River, which cuts this massif, do not either prove or disprove the presence of these lower Induan conglomerates above granitoids because outcrops of the latter are very small and slopes above them are covered by soil and vegetation. However, the above-mentioned information [25] means that these conglomerates can be hypothesized there.

Second, the Kamennomostskiy Formation of an early–middle Callovian age (Figure 3) underlays Late Jurassic limestones everywhere in the western and northern parts of the study area (Figure 1b). The lower part of this formation consists of conglomerates (Figure 3), which include clasts of various sandstones and Triassic limestones [40]. This formation does not and did not overlay granitoids of the Sakhray Crystalline Massif, which remained covered by the Bambak and Yatyrgvarta formations during the Triassic–Jurassic. The same can also be true for the Rufabgo Crystalline Massif, although the contacts of granitoids with sedimentary packages are not visible there. However, it appears that the Kamennomostskiy Formation may overlay directly granitoids of the Dakh Crystalline Massif (Figure 1b). The present state of the knowledge does not permit proving this idea due to the very limited accessibility of the related parts of the study area, and an exposure of granitoids of this massif at the time of the Bathonian–Callovian hiatus can only be hypothesized.

5. Discussion and Conclusions

5.1. Modelling Exposure of Crystalline Massifs on Ancient Seashores

The stratigraphical information about the coarse siliciclastic beds associated with Late Paleozoic granitoids (Figures 4a–e and 5) provides evidence of an exposure of the latter on rocky seashores in the Triassic–Jurassic history of the study area (Figure 6). The proposed model and the related palaeogeographical interpretations are given below. This pioneering reconstruction offers a new vision of the local geological history, but is tentative and, thus, can be detailed and even corrected in the light of future investigations.

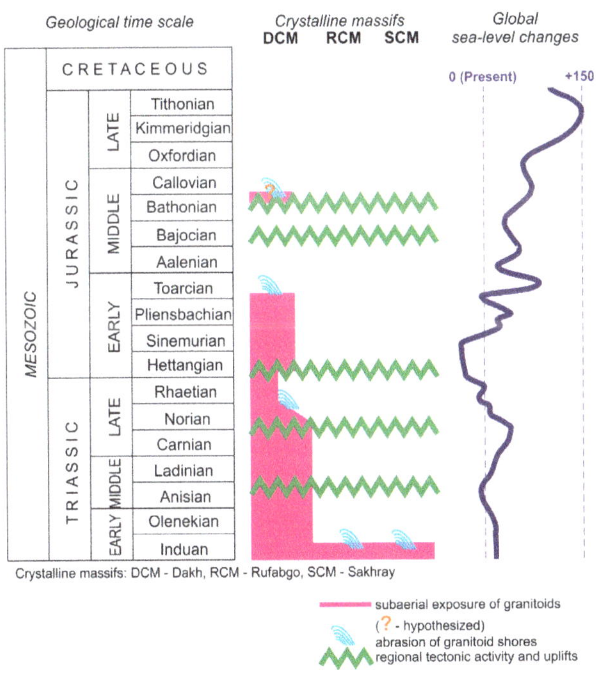

Figure 6. A proposed model of the chronology of exposure of granitoids on Triassic–Jurassic seashores in the study area. Global sea-level changes are shown for reference according to Haq [50,51].

In the beginning of the Triassic, all three crystalline massifs were exposed, and a wave abrasion of the related rocky shores resulted in a widespread accumulation of the Bambak Formation on a northeastern periphery of a large land mass (Figure 7a). The presence of megaclasts [25] makes realistic the idea of possible severe storms and even tsunamis [23]. Although the global sea level did not rise in the Early Triassic [50], a wide regional transgression led to the drowning of the Rufabgo and Sakhray massifs, which, probably, were not exposed again during the Triassic–Jurassic interval (Figure 6). Despite multiple uplifts during the phases of tectonic activity (Figure 6), erosion was not deep enough to reach granitoids of these two massifs beneath limestones of the Yatyrgvarta Formation.

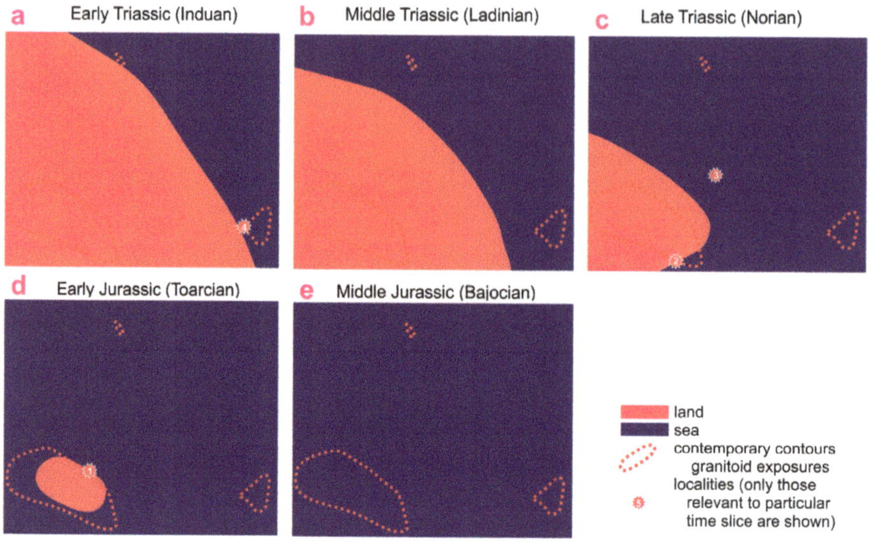

Figure 7. Tentative local palaeogeographical schemes for five geological time slices: Induan (**a**), Ladinian (**b**), Rhaetian (**c**), Toarcian (**d**), and Bajocian (**e**). Each scheme corresponds exactly to the area shown on Figure 1b, and contours of the massifs and the localities are shown accordingly.

In contrast, it appears that the Dakh Crystalline Massif remained a landmass throughout the Triassic. At least, any Lower–Middle Triassic deposits are not known from its periphery in contrast to two other massifs, which are surrounded by vast "fields" of Triassic rocks (Figure 1b). Moreover, the information provided by Chaitsky et al. [25] implies that the Shapkin Formation comprises the first marine deposits formed on the northern periphery of this massif. An intensified tectonic activity at the Anisian–Ladinian transition [38] did not change the situation (Figure 7b), as well as the perturbations at the Carnian–Norian transition, although the size of the land mass centered around the Dakh Crystalline Massif could increase temporary (it cannot be excluded that the entire study area became a land mass at the mentioned transitions). Moreover, the regional uplift at the Carnian–Norian transition protected the study area from any transgression caused by a major global sea-level rise [50] (Figure 6).

A significant transgression started in the Norian (Figure 7c) and culminated in the Rhaetian; apparently, it was triggered by a regional subsidence because the global sea level tended to fall stepwise [50] (Figure 6). The Dakh Crystalline Massif submerged gradually, but to a significant degree. In the north, the sea covered weathering-related breccias of the Shapkin Formation. Apparently, rocky shore did not exist there. In the south, a true rocky shore established. Deposition of conglomerates from Locality 2 marked the maximum incursion of the sea into the land corresponding to this massif. The presence of large rounded granitoid clasts [28] reflects wave abrasion. As these conglomerates accumulated during the entire interval of the Late Triassic–earliest Jurassic deposition on the southern margin of the massif, it appears that the latter was not drowned entirely. Conglomerate–breccias reported from Locality 5 are found close to the outcrops of granitoids of the Sakhray Crystalline Massif (Figure 1b). Their composition implies the existence of a shore affected by severe storms and/or tsunamis [23], but granitoids were not exposed there. These deposits formed when the study area experienced an intensification of tectonic activity and overall regression (Figure 6). Coarse siliciclastic beds formed in more or less similar conditions are known from some other places of the world [52–54].

During the Hettangian–Sinemurian, a strong deformation phase occurred (Figure 6). Most probably, the land mass grew, and granitoids of the Dakh Crystalline Massif were exposed more widely than in the end-Triassic. However, the two other massifs of the study area were not exposed. Marine sedimentation in the study area restarted widely in the mid-Early Jurassic (the locally distributed Bugunzha Formation and the widely distributed Psebay Formation—Figure 3). Taking into account the age of conglomerates from Locality 1 (see above), it is possible that the Dakh Crystalline Massif remained a land mass until the beginning of the Toarcian, when a global sea-level rise [51] coupled with a regional subsidence caused full drowning of the granitoids (Figure 6). A rocky shore existed in the early Toarcian (Figure 7d), when granitoids and older metamorphic rocks were abraded by waves. A subsequent basin subsidence and voluminous deposition of fine siliciclastics of the Psebay Formation (Figure 3) kept granitoids unexposed (Figure 7e). Their short-term, ephemeral exposure was possible in the late Bathonian due to regional tectonic activity [33,38,55,56] (Figure 6), but the evidence in support of this interpretation is unknown (see above).

5.2. Additional Inferences

The international importance of the outcomes of the present study is linked to the insufficiency of information about Triassic and Jurassic rocky shores [8]. This review implies that three localities (1, 2, and 4) represent granitoid seashores of three different time slices (Induan, Norian–Rhaetian, and Toarcian), and one locality (5) represents a non-granitoid rocky shore of the Triassic–Jurassic transition. Moreover, the reviewed knowledge of the exposure of granitoids on the Triassic–Jurassic seashores of the study area permits various, supplementary interpretations. These interpretations are linked to the time span of seashores, local palaeogeography, and lithostratigraphy. The related thoughts are given below.

The present stratigraphical review implies that rocky seashores with exposed granitoids appeared several times in the Triassic–Jurassic history of the study area (Figure 6). But how long did they exist? The Bambak Formation accumulated during the only part of Induan (Figure 3). This stage was very short (~0.7 Ma) [57], which implies an exposure of granitoids on a rocky shore for <0.2–0.3 Ma. Nonetheless, this shore could be rather long (Figure 7a).

The situation with the younger rocky shores developed on the margins of the Dakh Crystalline Massif could differ. Localities 1 (early Toarcian) and 2 (Norian–Rhaetian) can represent the only short episodes when the preservation of coarse siliciclastics was possible. However, the massif remained a land mass since the beginning of the Triassic and until the mid-Toarcian (Figure 6), i.e., during >50 Ma (according to the numerical time scale [57]). During long time intervals, seas existed in the neighboring parts of the study area, and, thus, rocky shores could evolve for millions of years. If so, their deposits were eroded together with uplifts at the Anisian–Ladinian, Carnian–Norian, and Triassic–Jurassic transitions.

The established chronology of exposure of Carboniferous granitoids on Triassic–Jurassic seashores of the study area (Figure 6) facilitates the understanding of the local palaeogeography (Figure 7a–e). First, it appears that the three massifs formed a "core" of an ancient landmass (a possible island), which existed at the Permian–Triassic transition and diminished in size in the Early Triassic when the Rufabgo and Sakhray massifs were drowned. Indeed, this island was larger in size in the Late Triassic than in the Early Jurassic because Norian weathering-related breccias from Locality 3 were located far northward than Toarcian marine conglomerates from Locality 1 (Figure 1b). Second, the sea transgressed more actively from the east than from the north because Locality 3 remained land until the mid-Norian. Third, although all three massifs can be parts of a single igneous body emplaced in the late Carboniferous, they had already separated in the palaeogeographical sense in the Triassic. For instance, the distance between the Dakh Crystalline Massif and Locality 5 is very short (<10 km), but a denudation of granitoids of this massif in the Late Triassic did not result in massive accumulation of the related clasts at Locality 5.

Conglomerate–breccias exposed at Locality 5 need justification in the terms of regional lithostratigraphy. The composite scheme indicates a late Rhaetian–early Sinemurian hiatus

(Figure 3), although the age of the package of these conglomerate–breccias is established as late Rhaetian–early Hettangian. Two alternative options are possible. First, this package can be related to the upper horizons of the conglomerate package from Locality 2. The latter belongs to the Khodz Group (see above), which means the uppermost part of the latter exists at Locality 5, and the age of this group should be extended into the early Hettangian. Second, this package can be related to a new local lithostratigraphic unit that occurs in the only east of the study area (it is logical to name this unit as the Gosh Formation, with Locality 5 as its reference section). The second option seems to be a bit more realistic regarding the different evolution of the heterogeneous palaeogeographical development of the study area noted above. Anyway, the major hiatus at the Triassic–Jurassic transition should be dated by the middle Hettangian–early Sinemurian, irrespective of which option is correct. Moreover, the information from the works by Diakonova-Savelieva [27] and Vyalov and Nikshich [28] makes reasonable a hypothesis that marine sedimentation continued locally through the Triassic–Jurassic transition in the vicinities of the Dakh Crystalline Massif. In such a case, the local lithostratigraphy will need even deeper revision. Verification of this hypothesis is a task for future investigations.

5.3. Conclusive Remarks

The present review of the stratigraphical distribution of the coarse siliciclastic beds associated with Carboniferous granitoids of the study area in the Western Caucasus permits the making of three general conclusions, as follows. First, the existence of these beds on five stratigraphical levels is established. Second, the early Induan, Norian–Rhaetian, and early Toarcian conglomerates mark ancient rocky shores where granitoids were exposed. Third, granitoids of the Rufabgo and Sakhray massifs were exposed on rocky seashores only once, i.e., in the earliest Triassic, whereas the Dakh Crystalline Massif could have rocky shores for >50 Ma.

Methodologically, this study stresses the utility of the works, which are old, but not outdated, as sources of the precious information about difficult-to-access and small-sized geological features. Geologists who worked many decades (even a century) ago made observations and descriptions which can easily be reinterpreted against modern stratigraphical schemes and extend a stratigraphical vision of highly complex areas. The international significance of this study is linked to the stratigraphical interpretations of ancient granitoid shores, the evidence of which remains demanded, but scarce, on the global scale. This review is chiefly stratigraphical, which means that future investigations should pay attention to the compositional peculiarities of the reported coarse siliciclastic beds. Of course, new field investigations in the remote parts of the study area (e.g., along small streams crossing crystalline massifs or along new roads and trails constructed for the purposes of tourism) will be essential for the comprehensive understanding of the locally existing ancient granitoid shores and islands. Moreover, an application of remote sensing techniques and geophysical approaches may help to outline the true contours of the crystalline massifs and their relationships with younger rocks.

Funding: This research received no external funding.

Data Availability Statement: Data are contained within the article.

Acknowledgments: The author gratefully thanks S.V. Platonova (Sankt-Petersburg, Russia) for literature support and N.V. Ruban (Rostov-on-Don, Russia) for field assistance.

Conflicts of Interest: The author declares no conflicts of interest.

References

1. Erdmann, W.; Scheffers, A.M.; Kelletat, D.H. Holocene Coastal Sedimentation in a Rocky Environment: Geomorphological Evidence from the Aran Islands and Galway Bay (Western Ireland). *J. Coast. Res.* **2018**, *34*, 772–792. [CrossRef]
2. Felton, E.A. Sedimentology of rocky shorelines: 1. A review of the problem, with analytical methods, and insights gained from the Hulopoe Gravel and the modern rocky shoreline of Lanai, Hawaii. *Sediment. Geol.* **2002**, *152*, 221–245. [CrossRef]

3. Manikam, S.L.; Green, A.N.; Cooper, A.; Deacon, G.; Flemming, B. Development and preservation of transgressive sandy versus rocky shorelines: Observations from the SE African shelf. *Geomorphology* **2022**, *419*, 108485. [CrossRef]
4. Trenhaile, A.S. Climate change and its impact on rock coasts. *Geol. Soc. Mem.* **2014**, *40*, 7–17. [CrossRef]
5. Violante, C. Rocky coast: Geological constraints for hazard assessment. *Geol. Soc. Spec. Publ.* **2009**, *322*, 1–31. [CrossRef]
6. Johnson, M.E. Why are ancient rocky shores so uncommon? *J. Geol.* **1988**, *96*, 469–480. [CrossRef]
7. Johnson, M.E. Uniformitarianism as a guide to rocky-shore ecosystems in the geological record. *Can. J. Earth Sci.* **2006**, *43*, 1119–1147. [CrossRef]
8. Johnson, M.E. Ecology of Intertidal Rocky Shores Related to Examples of Coastal Geology across Phanerozoic Time. *J. Mar. Sci. Eng.* **2024**, *12*, 1399. [CrossRef]
9. Bizzarri, R. Early Pleistocene rocky coasts (Orvieto area, Western Umbria, Central Italy): Facies analysis and sedimentation models. *Ital. J. Geosci.* **2010**, *129*, 251–268.
10. Chatalov, A.; Ivanova, D.; Bonev, N. Transgressive Eocene clastic-carbonate sediments from the Circum-Rhodope belt, northeastern Greece: Implications for a rocky shore palaeoenvironment. *Geol. J.* **2015**, *50*, 799–810. [CrossRef]
11. Desrochers, A. Rocky shoreline deposits in the Lower Silurian (upper Llandovery, Telychian) Chicotte Formation, Anticosti Island, Quebec. *Can. J. Earth Sci.* **2006**, *43*, 1205–1214. [CrossRef]
12. Foix, N.; Ocampo, S.M.; Paredes, J.M.; Allard, J.O.; Giacosa, R.E.; González, P.D.; Olazábal, S.X. Maastrichtian-Danian Northpatagonian rocky shore, Argentina. *Sediment. Geol.* **2023**, *454*, 106463. [CrossRef]
13. Harland, T.L.; Pickerill, R.K. Ordovician rocky shoreline deposits–the basal Trenton Group around Quebec City, Canada. *Geol. J.* **1984**, *19*, 271–298. [CrossRef]
14. Rousse, S.; Duringer, P.; Stapf, K.R.G. An exceptional rocky shore preserved during Oligocene (Late Rupelian) transgression in the Upper Rhine Graben (Mainz Basin, Germany). *Geol. J.* **2012**, *47*, 388–408. [CrossRef]
15. Surlyk, F.; Sørensen, A.M. An early Campanian rocky shore at Ivö Klack, southern Sweden. *Cretac. Res.* **2010**, *31*, 567–576. [CrossRef]
16. Kennedy, D.M.; Ierodiaconou, D.; Schimel, A. Granitic coastal geomorphology: Applying integrated terrestrial and bathymetric LiDAR with multibeam sonar to examine coastal landscape evolution. *Earth Surf. Process. Landf.* **2014**, *39*, 1663–1674. [CrossRef]
17. Nehyba, S.; Roetzel, R. High-energy, microtidal nearshore deposits and their provenance (Lower Miocene, Burdigalian/Eggenburgian, Alpine-Carpathian Foredeep, Lower Austria). *Geol. Q.* **2022**, *66*, 33. [CrossRef]
18. Puig López, J.M.; Howell, J.; Roetzel, R.; Poyatos-Moré, M. Transgressive rocky coasts in the geological record: Insights from Miocene granitic rocky shorelines and modern examples. *Sediment. Geol.* **2023**, *446*, 106344. [CrossRef]
19. Brocx, M.; Semeniuk, V. The '8Gs'—A blueprint for Geoheritage, Geoconservation, Geo-education and Geotourism. *Aust. J. Earth Sci.* **2019**, *66*, 803–821. [CrossRef]
20. Mazzucato, E.; de La Corte Bacci, D.; de Gouveia Souza, C.R. Geomorphological Heritage on the North Coast of the State of São Paulo: A Perspective About Current and Past Climate Changes. *Geoheritage* **2022**, *14*, 121. [CrossRef]
21. Nazaruddin, D.A. Granite landforms of Samui Island (southern Thailand) from geoheritage, geoconservation and geotourism perspectives. *Int. J. Geoheritage Parks* **2020**, *8*, 75–86. [CrossRef]
22. Roig-Munar, F.X.; Martín-Prieto, J.Á.; Rodríguez-Perea, A.; Gelabert, B.; Vilaplana, J.M. Proposed geosites for tsunamitic blocks in the rocky coasts of Formentera (Balearic Islands). *Rev. Soc. Geol. Esp.* **2018**, *31*, 35–48.
23. Ruban, D.A. Islands in the Caucasian Sea in Three Mesozoic Time Slices: Novel Dimension of Geoheritage and Geotourism. *J. Mar. Sci. Eng.* **2022**, *10*, 1300. [CrossRef]
24. Ruban, D.A.; Mikhailenko, A.V.; Ermolaev, V.A. Temporal outline of geological heritage sites in the Western Caucasus. *Int. J. Geoheritage Parks* **2024**, *12*, 295–310. [CrossRef]
25. Chaitsky, V.P.; Popkov, V.I.; Popkov, I.V.; Pinchuk, T.N. Triassic of the Northern Caucasus. *Geol. Geogr. Glob. Energiya* **2020**, *77*, 11–21. (In Russian)
26. Chepurnoy, E.A. Sedimentary rocks of the northern contact of the Dakh Crystalline Uplift (Adygeya). In *Praktika Geologov na Proizvodstve*; Granovskaya, N.V., Ed.; SFU: Rostov-on-Don, Russia, 2021; pp. 132–134. (In Russian)
27. Diakonova-Savalieva, E.S. Granitoids of the Dakh and Sakhray intrusions. *Uchyonye Zap. Leningr. Gos. Univ. Seriya Geol. Pochvennykh Nauk.* **1940**, *45*, 190–223. (In Russian)
28. Vyalov, O.S.; Nikshich, I.I. Age of Dakh and Shibaba granite intrusion in the Northern Caucasus. *Vestn. Geol. Kom.* **1928**, *9–10*, 13–15. (In Russian)
29. Ismail-Zadeh, A.; Adamia, S.; Chabukiani, A.; Chelidze, T.; Cloetingh, S.; Floyd, M.; Gorshkov, A.; Gvishiani, A.; Ismail-Zadeh, T.; Kaban, M.A.; et al. Geodynamics, seismicity, and seismic hazards of the Caucasus. *Earth-Sci. Rev.* **2020**, *207*, 103222. [CrossRef]
30. Pierce, I.; Guliyev, I.; Yetirmishli, G.; Muradov, R.; Kazimova, S.; Javanshir, R.; De Pascale, G.P.; Johsnon, B.; Marshall, N.; Walker, R.; et al. Surface Rupturing Earthquakes of the Greater Caucasus Frontal Thrusts, Azerbaijan. *Tectonics* **2024**, *43*, e2023TC007758. [CrossRef]
31. Van Hinsbergen, D.J.J.; Torsvik, T.H.; Schmid, S.M.; Matenco, L.C.; Maffione, M.; Vissers, R.L.M.; Gürer, D.; Spakman, W. Orogenic architecture of the Mediterranean region and kinematic reconstruction of its tectonic evolution since the Triassic. *Gondwana Res.* **2020**, *81*, 79–229.
32. Adamia, S.; Alania, V.; Chabukiani, A.; Kutelia, Z.; Sadradze, N. Great Caucasus (Cavcasioni): A long-lived North-Tethyan back-arc basin. *Turk. J. Earth Sci.* **2011**, *20*, 611–628. [CrossRef]

33. Saintot, A.; Brunet, M.-F.; Yakovlev, F.; Sebrier, M.; Stephenson, R.; Ershov, A.; Chalot-Prat, F.; McCann, T. The Mesozoic-Cenozoic tectonic evolution of the Greater Caucasus. *Geol. Soc. Mem.* **2006**, *32*, 277–289. [CrossRef]
34. Scotese, C.R. An Atlas of Phanerozoic Paleogeographic Maps: The Seas Come in and the Seas Go out. *Annu. Rev. Earth Planet. Sci.* **2021**, *49*, 679–728. [CrossRef]
35. Vasey, D.A.; Garcia, L.; Cowgill, E.; Trexler, C.C.; Godoladze, T. Episodic evolution of a protracted convergent margin revealed by detrital zircon geochronology in the Greater Caucasus. *Basin Res.* **2024**, *36*, e12825. [CrossRef]
36. Popov, Y.V.; Sharova, T.V. *Formations and Structure-Formation Zoning of Geological Complexes in the Territory of the Mountainous Part of Adygeya (Greater Caucasus)*; Southern Federal University: Rostov-on-Don, Russia, 2022. (In Russian)
37. Somin, M.L.; Levchenkov, O.A.; Kotov, A.B.; Makeev, A.F.; Komarov, A.N.; Ro, N.I.; Lavrishchev, V.A.; Lebedev, V.A. The Paleozoic age of high-pressure metamorphic rocks in the Dakhov Salient, Northwestern Caucasus: Results of U-Pb geochronological investigations. *Dokl. Earth Sci.* **2007**, *416*, 1018–1021. [CrossRef]
38. Gaetani, M.; Garzanti, E.; Poline, R.; Kiricko, Y.; Korsakhov, S.; Cirilli, S.; Nicora, A.; Rettori, R.; Larghi, C.; Bucefalo Palliani, R. Stratigraphic evidence for Cimmerian events in NW Caucasus (Russia). *Bull. Société Géologique Fr.* **2005**, *176*, 283–299. [CrossRef]
39. Rostovtsev, K.O.; Savel'eva, L.M.; Efimova, N.A.; Shvemberger, Y.N. *Decision of the 2nd Interdepartmental Regional Stratigraphical Maeeting on the Mesozoic of the Caucasus (Triassic)*; VSEGEI: Leningrad, Russia, 1979. (In Russian)
40. Rostovtsev, K.O.; Agaev, V.B.; Azarian, N.R.; Babaev, R.G.; Besnosov, N.V.; Hassanov, N.A.; Zesashvili, V.I.; Lomize, M.G.; Paitschadze, T.A.; Panov, D.I.; et al. *Jurassic of the Caucasus*; Nauka: St. Petersburg, Russia, 1992. (In Russian)
41. Ruban, D.A.; Zerfass, H.; Pugatchev, V.I. Triassic synthems of southern South America (southwestern Gondwana) and the Western Caucasus (the northern Neotethys), and global tracing of their boundaries. *J. S. Am. Earth Sci.* **2009**, *28*, 155–167. [CrossRef]
42. Vuks, V.J. Lower and Middle Jurassic Stratigraphic Scheme of the Western Caucasus: Problems of Correlation and Division. In *STRATI 2013*; Rocha, R., Pais, J., Kullberg, J.C., Finney, S., Eds.; Springer: Cham, Switzerland, 2013; pp. 609–618. (In Russian)
43. Yasamanov, N.A. *Landscape-Climatic Conditions of the Jurassic, the Cretaceous and the Paleogene of the South of the USSR.*; Nedra: Moscow, Russia, 1978. (In Russian)
44. Gale, L.; Rigaud, S.; Gennari, V.; Rettori, R.; Martini, R.; Gaetani, M. Recognition of upper Triassic temperate foraminiferal assemblages: Insights from the Khodz Group (NW Caucasus, Russia). *Glob. Planet. Change* **2020**, *188*, 103152. [CrossRef]
45. Akulov, N.I.; Melnikov, A.I.; Shtelmakh, S.I.; Akulova, V.V.; Hearn, P.P. A geochemical and lithological correlation of lower Jurassic conglomerates in the area surrounding the Lake Baikal rift zone: An improved reconstruction of the region's palaeogeographic and tectonic evolution. *Int. Geol. Rev.* **2022**, *64*, 1–16. [CrossRef]
46. Alçiçek, H. Stratigraphic correlation of the Neogene basins in southwestern Anatolia: Regional palaeogeographical, palaeoclimatic and tectonic implications. *Palaeogeogr. Palaeoclimatol. Palaeoecol.* **2010**, *291*, 297–318. [CrossRef]
47. Michelsen, O.; Clausen, O.R. Detailed stratigraphic subdivision and regional correlation of the southern Danish Triassic succession. *Mar. Pet. Geol.* **2002**, *19*, 563–587. [CrossRef]
48. Sartori, M.; Gouffon, Y.; Marthaler, M. Harmonization and definition of the Briançonnais lithostratigraphic units in the Penninic nappes of the Valais region. *Eclogae Geol. Helv.* **2006**, *99*, 363–407. [CrossRef]
49. Smith, T.F.; Waterman, M.S. New stratigraphic correlation techniques. *J. Geol.* **1980**, *88*, 451–457. [CrossRef]
50. Haq, B.U. Triassic Eustatic Variations Reexamined. *GSA Today* **2018**, *28*, 4–9. [CrossRef]
51. Haq, B.U. Jurassic sea-level variations: A reappraisal. *GSA Today* **2018**, *28*, 4–10. [CrossRef]
52. Armitage, I.A.; Pemberton, S.G.; Moslow, T.F. Facies succession, stratigraphic occurrence, and paleogeographic context of conglomeratic shorelines within the Father "C", Spirit River Formation, Deep Basin, west-central Alberta. *Bull. Can. Pet. Geol.* **2004**, *52*, 39–56. [CrossRef]
53. Massari, F. High-frequency cycles within Pleistocene forced-regressive conglomerate wedges (Bradanic area, southern Italy) filling collapse scars. *Sedimentology* **1997**, *44*, 939–958. [CrossRef]
54. Üner, S.; Dirik, K.; Çiner, A. Late Miocene evolution of Kargı fan delta (Aksu basin, Antalya). *Earth Sci.* **2011**, *32*, 121–138.
55. Giorgobiani, T.V. Conditions of formation of the Alpine folded system of the Greater Caucasus and unique features of its structure. *Geol. Geofiz. Yuga Ross.* **2019**, *9*, 43–57.
56. International Commission on Stratigraphy (ICS). International Chronostratigraphic Chart, v2023/09. Available online: https://stratigraphy.org (accessed on 16 May 2024).
57. Kushcheva, Y.V.; Latysheva, I.V.; Golovin, D.I.; Gavrilov, Y.O. Textural-structural, mineralogical, isotopic, and age characteristics of Jurassic terrigenous rocks of the northwestern Caucasus (the Belaya River section). *Lithol. Miner. Resour.* **2007**, *42*, 257–267. [CrossRef]

Journal of
Marine Science
and Engineering

MDPI

Review

Indication of Deep-Water Gravity Flow Types by Shelf-Edge Trajectory Migration Patterns: A Case Study of the Quaternary Qiongdongnan Basin, South China Sea

Chang Ma, Hongjun Qu * and Xian Liu

State Key Laboratory of Continental Dynamics, Department of Geology, Northwest University, Xi'an 710069, China; machang0604@foxmail.com (C.M.); liu_xian1996@163.com (X.L.)
* Correspondence: hongjun@nwu.edu.cn

Abstract: The shelf-edge trajectory is comprehensively controlled by tectonics, sediment supply, sea level, and climate fluctuations; its migration and evolution have a strong influence on what happens in the deep-water depositional system during the Quaternary. The shelf-edge trajectory pattern, sediment-budget partitioning into deep-water areas, and reservoir evaluations are focused topics in international geosciences. In this paper, the Qiongdongnan Basin (QDNB) in the northern South China Sea is taken as an example to study how shelf-edge trajectory migration patterns can influence the types of deep-water gravity flow which are triggered there. Through quantitatively delineating the Quaternary shelf-edge trajectory in the QDNB, four types of shelf-edge trajectory are identified, including low angle slow rising type, medium angle rising type, high angle sharp rising type, and retrogradation-slump type. A new sequence stratigraphic framework based on the migration pattern of shelf-edge trajectory is established. There are four (third-order) sequences in the Quaternary, and several systems tracts named lowstand systems tract (LST), transgressive systems tract (TST), and highstand system tract (HST) are identified. This study indicates that the type of deep-water gravity flow can be dominated by the shelf-edge trajectory migration patterns. When the shelf-edge trajectory angle (α) ranged between 0° and 4°, the continental canyons were mostly small-scaled and shallowly incised, with multiple large-scale sandy submarine fan deposits with few MTDs found in the deep-water area. When the angle (α) ranged from $4° < \alpha < 35°$, the size and incision depth of the continental slope canyons increased, relating to frequently interbedded sandy submarine fan deposits and MTDs. When angle (α) ranged from $35° < \alpha < 90°$, only a few deeply-incised canyons were present in the continental slope; in this condition, large-scaled and long-distance MTDs frequently developed, with fewer submarine fans deposits. When angle (α) ranged from $90° < \alpha < 150°$, the valley in the slope area was virtually undeveloped, sediments in the deep-sea plain area consisted mainly of large mass transport deposits, and submarine fan development was minimal. Since the Quaternary, the temperature has been decreasing, the sea level has shown a downward trend, and the East Asian winter monsoon has significantly enhanced, resulting in an overall increase in sediment supply in the study area. However, due to the numerous rivers and rich provenance systems in the west of Hainan Island, a growing continental shelf-edge slope has developed. In the eastern part of Hainan Island, due to fewer rivers, weak provenance sources, strong tectonic activity, and the subsidence center, a type of destructive shelf-edge slope has developed. The above results have certain theoretical significance for the study of shelf-edge systems and the prediction of deep-water gravity flow deposition type.

Keywords: shelf-edge trajectory; deep-water sediment gravity flow; the northern South China Sea; Qiongdongnan Basin; Quaternary

1. Introduction

The shelf-edge slope connects the shallow water area of the shelf with the deep-water area of the seabed. The analysis of shelf-edge trajectory involves studying the vertical and lateral migration process of the shelf-edge slope break point (Figure 1). The shelf-edge trajectory is divided into three types: negative-angle descending type, low-angle slowly ascending type, and high-angle ascending type [1,2]. In recent years, with the continuous development and improvement of the migration theory of shelf edge and slope, the study of shelf-edge trajectory has become an international frontier field of sedimentology and sequence stratigraphy [3–6]. Through the study on the migration types of shelf-edge trajectory, it is believed that the deep-water area of the Porcupine Basin in Ireland has two completely different sedimentary models: one is guided by the formation of canyons and MTDs, and the other is guided by the shelf-edge delta and submarine fans. Pointing out that there is a good response relationship between relative sea level change and shelf-edge migration trajectory [7,8]. Three factors, namely climatic conditions, provenance supply, and accommodative space, are considered to jointly control the shelf margin system and its deep-water response model [5,6,9]. By calculating sediment flux, the relationship between shelf-edge migration trajectory and stratigraphic superposition pattern can be established. A negative-angle descending pattern corresponds to a progradational superposition pattern, a low-angle slowly rising pattern corresponds to a mix between progradative and aggradation superposition patterns, and a high-angle rising pattern corresponds to an aggradational superposition pattern [10,11]. Although sequence stratigraphy itself was also aimed to correlate coastal systems with basins, the analysis of shelf boundary migration trajectory focuses on the continuous sedimentary sequence [12]. By studying the migration trajectory of the shelf edge, the corresponding relationship between it and the deep-sea sedimentary system can be established, and the dynamic dispersion process of sediment and sand bodies can be predicted and analyzed.

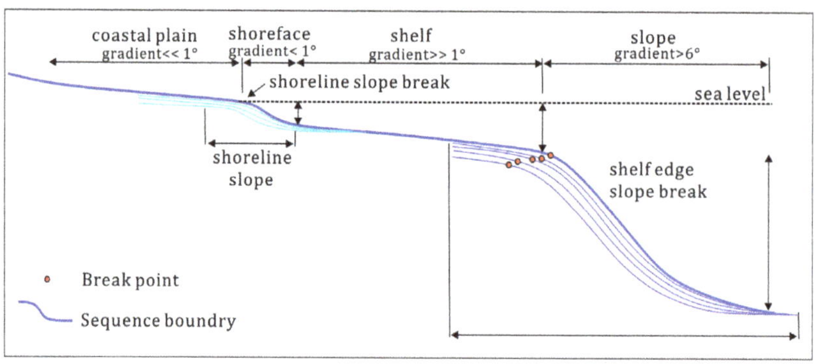

Figure 1. Schematic diagram of shelf-edge slope.

In recent years, the shelf-edge trajectory and connected deep-water system in the QDNB have gradually attracted the attention of scholars [13]. The continental slope of the QDNB can be subdivided into four types: progradation type, slump type, channelized type, and gentle type, with each exhibiting a sequence stacking pattern [14,15]. Based on the quantitative characterization of the types of shelf-edge trajectories and the thickness of sand bodies in each sedimentary unit in the QDNB. It is confirmed that under the condition of constant source supply, the ratio of sand to mudstone in the deep-water plain area corresponding to different types of shelf-edge trajectories is also different. And the angle of the shelf-edge trajectory is inversely proportional to it [5,16]. It is considered that the shelf-edge trajectory of the Quaternary Ledong Formation is mainly a high-angle sharp rising type by using quantitative characterization methods to study the shelf-edge trajectory during the Quaternary in the QDNB [10]. At the same time, it is pointed out that river input, large accommodation, and narrow shelf are the key elements for the development of large-scale sandy submarine fans during this period.

Recently, the broad prospect of natural gas hydrate resources in the deep-water area of the QDNB has been widely discussed [17]. But up to now, there is still a lack of research examples of the shallow Ledong Formation in the QDNB. The lateral difference and evolution of the shelf-edge migration pattern and its relationship with the sedimentary system types in the deep-water area are still unclear. So, we believe that systematic analysis of the migration patterns and evolution of the shelf-edge plays an important role in the prediction of deep-water systems.

2. Geological Setting

2.1. Regional Structural Division

The QDNB, in the northern South China Sea, covers approximately 4.5×10^4 km^2 with a maximum water depth of the QDNB > 3000 m. With an arc-shaped spreading around Hainan Island with an NE-SW orientation (Figure 2). This sedimentary basin has good hydrocarbon reservoir formation conditions and promising exploration prospects [18]. The QDNB is bounded by the Shenhu Uplift to the east, the Yinggehai Basin and No. 1 Fault to the west, the Yongle Uplift to the south, and Hainan Island to the north [19–21]. From north to south, the basin can be divided into four primary tectonic units, including the Northern Depression Belt, Northern Uplift Belt, Central Depression Belt, and Southern Uplift Belt (Figure 2). The Northern Depression Belt is composed of the Yabei Sag, Songdong Sag, and Songxi Sag. The Central Depressional Belt consists of the Yanan Sag, the Ledong Sag, the Lingshui Sag, the Baodao Sag, the Songnan Sag, and the Beijiao Sag [16,22].

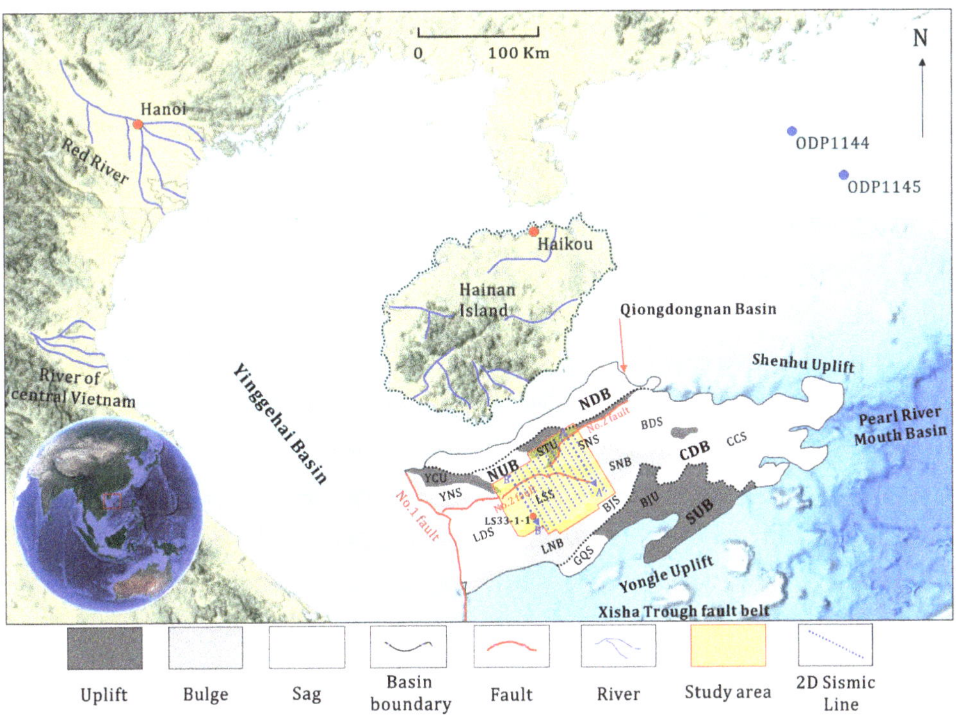

Figure 2. Location of the study area (red polygon). QDNB: Qiongdongnan Basin. SCS: South China Sea. NDB: Northern Depression Belt. CDB: Central Depression Belt. SUB: Southern Uplift Belt YCU: Yacheng Uplift. STU: Songtao Uplift. LNB: Lingnan Bulge. SNB: Songnan Bulge. LDS: Ledong Sag. LSS: Lingshui Sag. YNS: Yanan Sag. BJS: Beijiao Sag. SNS: Songnan Sag. BDS: Baodao Sag. CCS: Changchang Sag. BJU: Beijiao Uplift. GQS: Ganquan Sag.

The study area is located in the Central Depressional Belt, covering approximately 8000 km^2 (Figure 1). The study area spans the Lingshui Sag and Songtao Uplift to the north, the Lingnan Bulge and Beijiao Sag to the south, the Ledong Sag to the west, and the Songnan Sag and Songnan Bulge to the east. The study area includes a shelf that degrades to a slope and to deep-water bottomsets, has large topographic variations, and the depth of seawater gradually increases from north to south, reaching 2000 m (Figure 2).

2.2. Structural Evolution and Sedimentary Filling

The formation of the QDNB is closely related to the complicated interactions among the Indo-Australian Plate, Eurasia Plate, Pacific Plate, and the opening of the SCS [23]. Four regional tectonic movements in the Cenozoic period influenced the stages of tectonic evolution of the basin [24]: (1) the Shenhu movement in the Late Cretaceous–Paleocene, (2) the South China Sea movement in the Eocene–early Oligocene, (3) the South China Sea seabed spreading in the late Oligocene–middle Miocene, and (4) the Dongsha movement in the late Miocene to Quaternary. According to previous studies, the rifting of the QDNB may have started in the Eocene [19]. Since the Eocene, the basin has undergone two major stages of tectonic evolution: rifting and post-rifting depression, leading to a two-layer structure characterized by an "upper depression and lower fault" [21]. Seismic horizon T60, interpreted as a major tectonic event (Figure 3), is used to divide the basin into two major tectonic layers, namely, upper and lower, which represent the products of the rifting and post-rifting subsidence stages, respectively. Combined with the characteristics of extensional fracturing in the basin, they can be further divided into the pre-rifting stage, the syn-rifting stage, the thermal subsidence stage, and the accelerated subsidence stage [24,25].

High sedimentation rates result in Cenozoic sediment thicknesses ranging from 3000 to 16,000 m [21,26,27]. The stratigraphic sequence is subdivided into eight formations based on the drilling and seismic horizons (Figure 3). The Lingtou Formation is undrilled and is interpreted to represent Eocene lacustrine sedimentary facies; this formation is an important hydrocarbon source rock based on seismic data interpretation [28]. The Yacheng Formation was deposited in the early Oligocene and contains marsh–coastal plain facies [29]. This formation is widely considered to be the main hydrocarbon source rock of the Lingshui 17-2 gas field [30]. The Lingshui Formation contains coastal facies that were deposited in the late Oligocene [18]. The T60 unconformity is the boundary between the Lingshui Formation and the Sanya Formation (coastal shallow marine facies). The Meishan and Huangliu Formations were deposited in the middle and late Miocene and contain shallow to deep coastal facies [31]. The Huangliu Formation sands may be the hydrocarbon reservoir of the QDNB [18]. T20 is the boundary between the Yinggehai Formation and the Ledong Formation (Quaternary), which contains bathyal to abyssal sedimentary facies dominated by fine-grained sediments [18]. Since the Quaternary, there have been three major source areas in the QDNB, namely, the Red River provenance, the Hainan Island provenance, and the central Vietnam provenance. In the early Quaternary, the supply of sediments of the Red River was weak, which only affected the northwestern shelf area of the QDNB. The provenance of central Vietnam has always affected the middle part of the basin, and almost the whole basin is under the influence of the provenance of Hainan Island. During the late Ledong Formation (0.8 My~present), the source of the Red River was greatly strengthened, and the influence range could reach the middle part of the basin. At the same time, the intensity of supply in central Vietnam was slightly reduced, and the influence range was smaller than that of the early Quaternary. The sediment supply from Hainan Island still occupies a dominant position [20,32,33] (Figure 3).

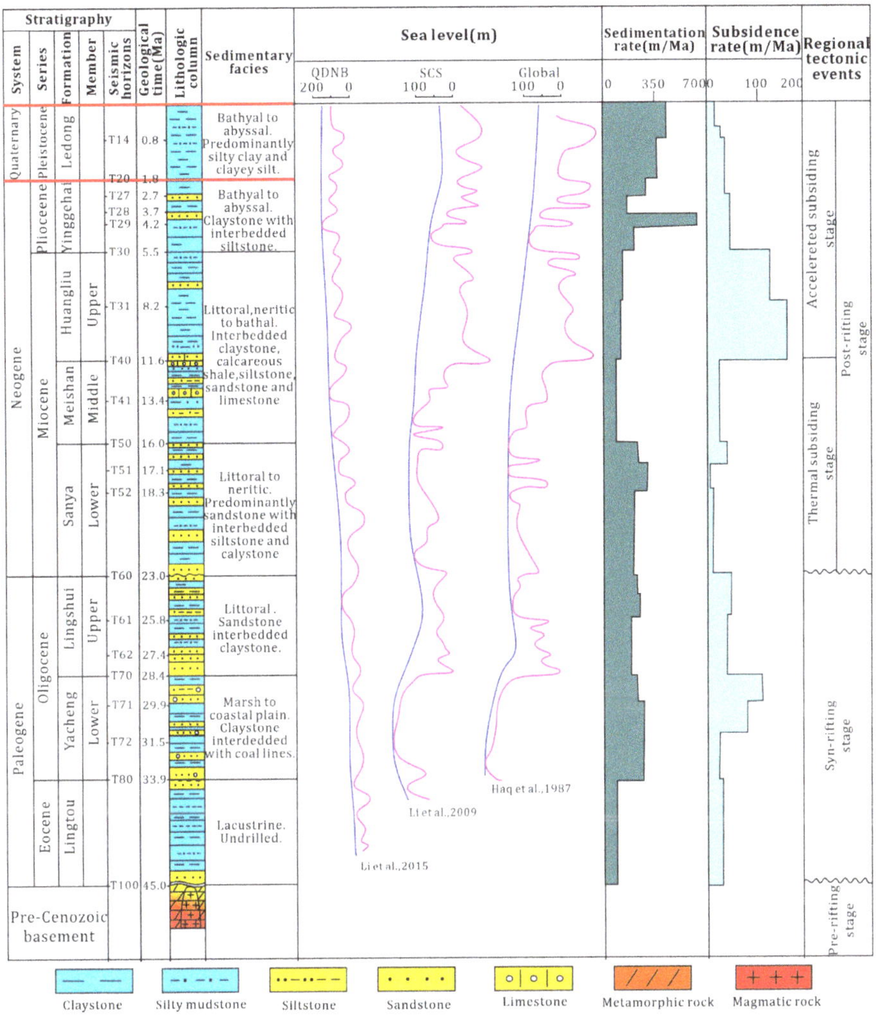

Figure 3. Comprehensive stratigraphic column of the QDNB (modified from [34,35]. The sedimentation rate and subsidence rate are redrawn after [27,36]. Sea level curves are derived from [37–39]. QDNB: Qiongdongnan Basin. SCS: South China Sea.

3. Data and Methods

3.1. Data

In this study, seismic stratigraphy was applied to analyze the shape and trajectory angle of the continental slope of the QDNB. The study area was crossed by 2D seismic data from 113 seismic lines and covered approximately 10,000 km^2 with high density (Figure 2). The frequency was nearly 38 Hz in the interval of interest, providing a vertical resolution of 20 to 30 m, and the seismic wave velocity was 1700 m/s.

Twelve along shelf-slope (north-west/south-east) seismic lines were selected for data statistics and analysis, and two seismic lines (AA′ and BB′) were used for detailed sequence stratigraphic analysis integrated with either well data (LS33-1-1). Line L1 (AA′) is located in the western part of the basin, and Line L2 (BB′) is located in the eastern one (Figure 2).

These lines represent the characteristics of two different types of continental slopes in the study area.

3.2. Methods

The shelf-edge trajectory migration is composed of a series of slope breakpoints, which record the development and evolution of the shelf edge (Figure 4). Its morphology can be characterized by the following parameters [7]: (1) accretion distance and progradation distance between the shelf edge, (2) accretion rate and progradation rate, (3) sediment flux, (4) shelf-edge trajectory angle, (5) horizontal distance and vertical distance between slope break point and slope toe point, and (6) slope topography angle.

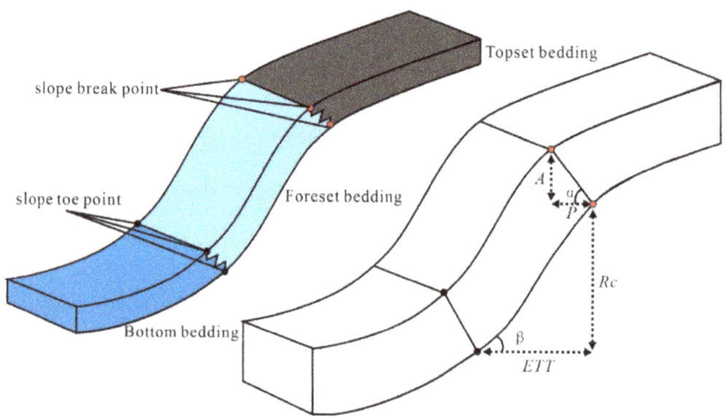

Figure 4. A model for calculating the shelf-edge trajectory (α), the shelf-edge migration (P), the vertical distance between the successive shelf edges (A), the height (in m) between the toe and the shelf edge (Rc), the edge-to-toe distance (ETT), and the slope angle (β).

The shelf-edge trajectory angle (α) can be calculated as:

$$\alpha \text{ (degree)} = \tan^{-1} (A/P)$$

where α is the shelf-edge trajectory angle (in degrees), A is the height between successive shelf edges (in m) and P is the length of the shelf-edge migration (in m).

The slope angles are calculated as:

$$\beta \text{ (degree)} = \tan^{-1} (Rc/ETT)$$

where β is the slope angle (in degrees), Rc is the height (in m) between the toe and the shelf edge and ETT is the edge-to-toe distance (in m).

Accretion rate (Ra), progradation rate (Rp), and sediment flux (Fc) cross sections at each stage can be calculated as:

$$Ra = A/T$$

$$Rp = P/T$$

$$Fc = Rp \times A \times 10^{-3}$$

Ra is the accretion rate (in m/My), A is the height between successive shelf edges (in m), and T is the time of the shelf-edge migration (in My). Additionally, Rp is the progradation rate (in m/My), P is the length of the shelf-edge migration (in m), T is the time of the shelf-edge migration (in My), Fc is the sediment flux (in m^3/My), Rp is the progradation rate (in m/My), and A is the height between successive shelf edges (in m) (Tables S1–S3). Although the current trajectory angle measurement method of shelf edge has high accuracy, how to correctly statistical accretion and progradation rate is still a challenge.

Based on the comparison of previous research data, the coupling relationship between sea and land development, relative sea level fluctuation, and sediment supply was established. Thus, how these mechanisms control various types of stacking patterns and parameters could also be determined.

4. Results

4.1. Sequence Stratigraphic Framework

According to classical sequence stratigraphy combined with the relative change of sea level in QDNB [40,41], the sequence boundary is identified by using typical seismic termination relations including upper exceedance, top exceedance, lower exceedance, and local cutting. By using this method, and combining gamma-ray logs and lithology, five sequence boundaries and four third-order sequences (S1, S2, S3, and S4) were identified (Figure 5).

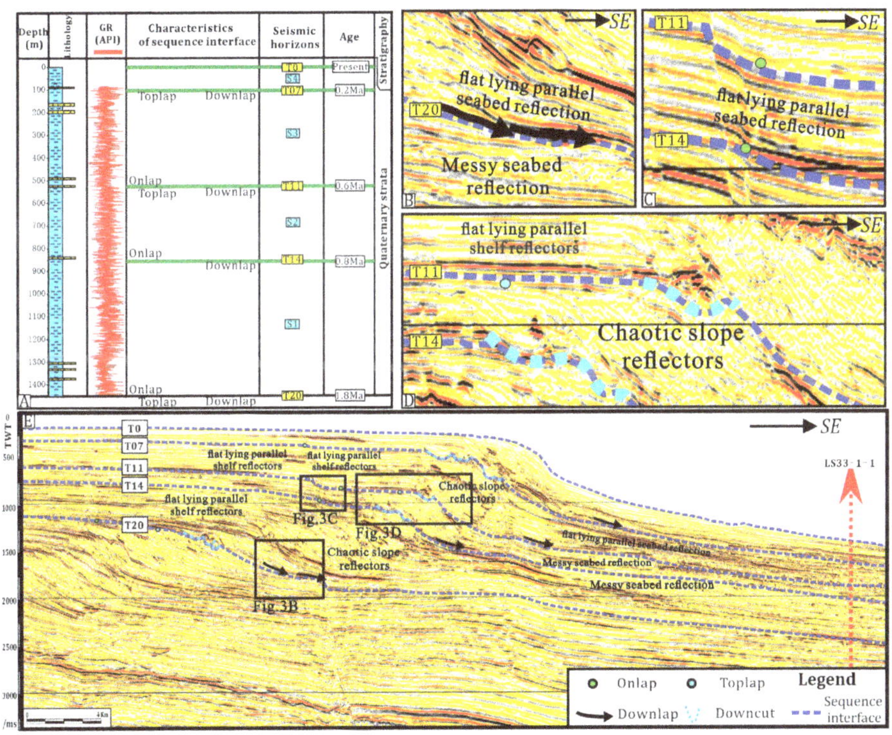

Figure 5. Sequence stratigraphic framework established from the deep-water well LS33-1-1 in the QDNB. (**A**) Lithology and well-logging data of LS33-1-1. (**B–D**) Typical seismic contact relationship. See (**E**) for the locations of (**B–D**). (**E**) 2D seismic profile crossed the study area. See Figure 1 for the locations of LS33-1-1 and (**E**) QDNB: Qiongdongnan Basin.

The lithology changes above and below the T20 sequence boundary, with sandstone above the sequence boundary and mudstone below the sequence boundary. GR log curve (Figure 5) is abnormally small at the sequence boundary, showing a zigzag bell shape above the sequence boundary and a zigzag box shape below the sequence boundary. The lithology changes above and below the T14 sequence boundary, with fine sandstone above the sequence boundary and mudstone below the sequence boundary. The GR curve is abnormally small at the sequence boundary (Figure 5), showing a zigzag bell shape above the sequence boundary and a zigzag box shape below the sequence boundary. The lithology changes above and below the T11 sequence boundary, with mudstone above the sequence boundary and thin sandstone below the sequence boundary. The GR curve is abnormally small at the sequence boundary (Figure 5), and the upper and lower parts of the sequence boundary are sawtooth bell-shaped. The T07 sequence boundary is dominated by medium-weak amplitude and medium-high continuous seismic reflection in the shelf area, and a typical onlap can be seen. The shelf-edge deposits are mainly progradated, and the formation thickness is small (40~90 m). In the slope area, the main axis is the medium-high amplitude and low-continuous seismic reflection, and local top superimposed and truncated structures can be seen. The deep-water area shows a meso-strong amplitude and medium-continuous type of sheet seismic reflection, with a wide range of lower superstructures above the sequence boundary, and large-scale seismic in-phase axes of strong amplitude, high continuity, and medium frequency can be seen. T0 is the seabed (Figure 5).

The shelf edge of Sequence S1 (T20-T14) is characterized by parallel progradation and accretion. The progradation distance and the strata thickness are both large in the northwest of the basin, and the accretion and progradation are both clearly recognizable. The thickness is small in the northeast of the basin, with accretion and progradation occurring mainly in small amounts. Strong amplitude and high continuous seismic in-phase axis can be seen in the continental shelf area. The continental slope area is dominated by a set of thick shelf-edge delta. In the deep-water area, chaotic reflection structures can be seen with massive medium-strong amplitude reflections, which are MTDs; on the other hand, parallel reflections of weak amplitude and high continuity are widely seen, mainly in the mudstone deposits (Figure 5).

The eastern shelf edge and shelf edge of sequence S2 (T14-T11) are characterized by strong accretion. The strata in the northwest of the basin are thick and the progradation is obvious. In the northeast, the stratum thickness is thin and accretion is dominant. In the continental shelf area, strong amplitude and high continuous seismic in-phase axis can be seen, and a large number of continental shelf waterways are developed. The continental slope area is dominated by the marginal delta of the shelf and the valley of the slope. In the deep-water area, multi-stage MTDs can be seen stacked on top of one another, with parallel reflections of weak amplitude and high continuity widely developed, mainly in the mudstone deposits (Figure 5).

The shelf edge of Sequence S3 (T11-T07) is characterized by strong accretion, and the strata are thick in the west and thin in the east. The continental shelf area is dominated by strong amplitude and high continuous seismic in-phase axis, and the continental shelf channel is more developed. The continental slope area is dominated by the MTDs formed by the delta collapse at the edge of the shelf and the MTDs from the inner wall of the slope canyon. Multi-stage large-scale MTDs can also be seen in the deep-water area; the flow direction is mainly southeast, and a large number of weak amplitude, high continuous type of parallel reflections are also developed, which are mainly mudstone deposits (Figure 5).

The shelf edge of Sequence S4 (T07-T0) is characterized by strong accretion, and the strata inherit the characteristics of thick west and thin east. The continental shelf delta plain is widely distributed in the continental shelf area. The northwestern continental slope area is dominated by shelf-edge delta deposits with a large number of MTDs, and the northeast is dominated by slope canyon development with a large number of MTDs. The deep-water plain area is dominated by chaotic reflection structures, mainly large MTDs (Figure 5).

4.2. *Types of Shelf-Edge Migration*

Based on the statistical analysis of the parameters of the shelf-edge trajectory of 12 typical seismic profiles in the study area, the Quaternary shelf-edge trajectory in QDNB is classified into the following four types:

(1) Low-angle slow rising type shelf-edge trajectory: The accretion distance of this type of shelf-edge trajectory ranges from −50 to 100 m, and the progradation distance of the four types of shelf-edge trajectory is the smallest. The accretion distance ranges from 0 to 8000 m, the ratio of accretion distance to progradation distance is −0.01 to 0.07 (Figure 6A and Table S1), and the angle of the shelf-edge trajectory is −1 to 4° (Figure 6B). The shelf-edge slope height is 350 to 680 m and the length is 3800–11,000 m, which is the longest among the four shelf-edge trajectory types, and the slope toe angle is 2.5 to 9° (Table S2).

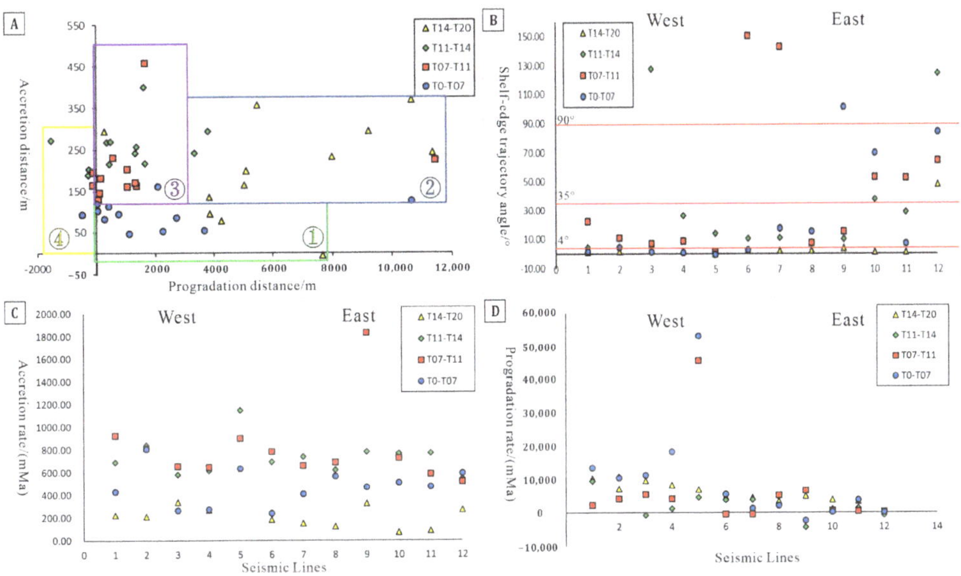

Figure 6. (**A**) The intersection between accretion distance and progradation distance at the edge of the Quaternary shelf in QDNB (The numbered colored boxes represent the parameter range (accretion and progradation distance) of the shelf-edge trajectory); (**B**) Shelf-edge trajectory angle statistics of each section of the Quaternary in QDNB; (**C**) Accretion rate statistics of each section of the Quaternary in QDNB; (**D**) Progradation rate statistics of each section of the Quaternary in QDNB.

(2) Medium-angle rising type shelf-edge trajectory: The accretion distance of this type of shelf-edge trajectory is between 100 to 400 m, and the progradation distance is about 300 to 12,000 m (Figure 6A and Table S1). Compared with type (1), the accretion distance and progradation distance are larger, the ratio of accretion distance to progradation distance is 0.07 to 0.55, and the angle of the shelf-edge trajectory is 4 to 35° (Figure 6B). The height of the slope at the edge of the shelf is 200–570 m, the length is 200–10,500 m, and the angle of the slope toe is 2 to 7° (Table S2).

(3) High-angle sharp rising type shelf-edge trajectory: The accretion distance of this type of shelf-edge trajectory is between 100 to 500 m, which is the largest among the three types of shelf-edge trajectory, and the progradation distance is 0 to 3000 m (Figure 6A and Table S1). The ratio of accretion distance to progradation distance is 0.76 to 9.45, the shelf-edge trajectory angle is between 35 to 90° (Figure 6B), the slope height of the shelf edge is between 200 to 600 m, the length varies between 3000 to 8000 m, and the slope toe angle is 3 to 7° (Table S2).

(4) Retrogradation-slump type shelf-edge trajectory: The accretion distance of this type of shelf-edge trajectory is between 0–300 m, and the progradation distance is −2000-0 m (Figure 6A and Table S1), which is the smallest among the four types. The ratio of accretion distance to progradation distance is less than 0, the shelf-edge trajectory angle is between

90–150° (Figure 6B), the slope height of the shelf edge is in the range of 350–550 m, the length varies little in the range of 3000–8000 m, and the slope toe angle is 3–6° (Table S2).

4.3. Migration and Evolution of Shelf-Edge Trajectory

4.3.1. S1 (T20-T14)

S1 is characterized by strong accretion, progradation, and high sediment flux (Table S3), and most of shelf-edge trajectories are low-angle slow rising types. The accretion distance of the shelf edge ranged from −4 to 369.5 m, with an average value of 204.85 m. The shelf-edge progradation distance is in the range of 262.5 to 11,362.5 m, with an average of 6221.88 m (Figure 6A). Most of the shelf-edge trajectory angles are between 0 and 4°. The slope height ranged from 282.36 to 575.2 m, with an average value of 455.9 m. The slope length is in the range of 3850 to 8625 m, with an average value of 6457.5 m. The slope angle is between 2 and 6° (Table S2).

The accretion rate of the northwestern shelf edge ranges from −4 to 350 m/My, and the accretion rate ranges from 6900 to 10,500 m/My. The accretion rate of the northeastern shelf edge is between 70 and 270 m/My, and the accretion rate is between 230 and 3800 m/My. The sediment flux in the northwest is in the range of −33 to 3500 m^3/My, the sediment flux in the middle is in the range of 450 to 1800 m^3/My, and the sediment flux in the northeast is in the range of 70 to 330 m^3/My (Table S3). The sediment flux decreased gradually from west to east during this period, indicating a decline in source activity from west to east (Figure 7).

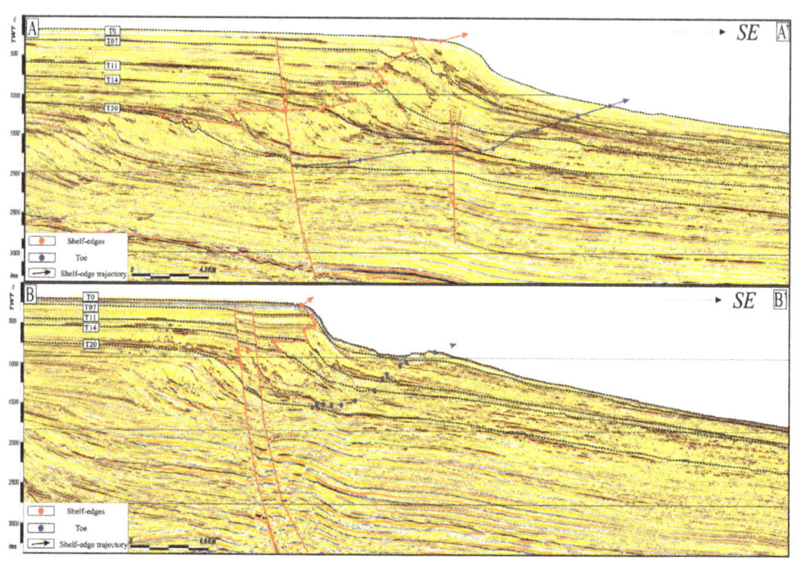

Figure 7. Two seismic sections (AA′ and BB′) which shown the shelf-edge migration trajectories, see locations in Figure 1. Shelf edges and toes are marked by red and blue dots, respectively.

4.3.2. S2 (T14-T11)

S2 is characterized by strong progradation and high sediment flux (Table S3), and most of the shelf-edge trajectories on the northwestern are low-angle slow rising types, or medium-angle rising types. Most of the trajectories of the northeastern shelf edge are of high-angle sharp rising type.

The accretion distance of the shelf edge ranges from 400.5 to 188.96 m, with an average value of 255.72 m. The shelf-edge progradation distance is in the range of −1537 to 3800 m, with an average of 1023.96 m (Figure 6A). In the northwestern region, the trajectory angle of the shelf edge is between 4 and 35°. In the northeastern region, the trajectory angle of the shelf edge increases rapidly, with a variation range of 30 to 40°. There are also a few retrogradation-slump type shelf-edge trajectories, with angles between 100 and 130°. The slope height ranges from 270.88 m to 520.92 m, with an average value of 391.57 m. The slope length is in the range of 2125 to 8300 m, with an average of 5387.5 m. The slope angle is between 2 and 8° (Table S2).

The accretion rate of the northwestern shelf edge ranges from 579.09 to 1144.11 m/My, and the progradation rate ranges from −750 to 10,857.14 m/My. The accretion rate of the middle shelf edge ranges from 600 to 800 m/My, and the progradation rate ranges from −4392.86 to 4750 m/My. The accretion rate of the northeastern shelf edge is between 500 and 800 m/My, and the progradation rate is between −785.71 and 1392.86 m/My. The sediment flux in the northwest is in the range of −152.01 to 3188.96 m^3/My, and the sediment flux in the northeast is in the range of −148.47 to 374.12 m^3/My (Table S3). The progradation rate gradually decreased from west to east during this stage. The sediment flux decreased gradually from west to east, and the sediment flux increased slightly in S2 compared with that in S1, indicating that the provenance supply in this stage inherited the characteristics of "strong in west and weak in east" in S1 (Figure 7).

4.3.3. S3 (T11-T07)

S3 is characterized by strong accretion, weak progradation, and sediment flux (Table S3). Most of the shelf-edge trajectories in the northwest are a medium-angle rising type, and few retrogradation-slump types are caused by slumps. Most of the northeastern shelf-edge trajectories are high-angle sharp rising types.

The accretion distance of the shelf edge ranges from 129.2 to 457.6 m, with an average value of 202.72 m. The shelf-edge progradation distance is in the range of −125 to 11,437.5 m, with an average value of 1541.67 m. In the northwestern region, the shelf-edge trajectory angle is between 1 and 23°. In the northeastern region, the shelf-edge trajectory angle increases rapidly, with a variation range of 52 to 65°. There are also a few retrogradation-slump type shelf-edge trajectories, with angles between 100 and 130°. The slope height ranges from 381.2 to 585.28 m, with an average value of 458.59 m. The slope length is in the range of 3200 to 8800 m, with an average value of 7073.96 m. The slope angle is between 2 and 8° (Table S2).

The accretion rate of the northern shelf edge ranges from 646.24–924.8 m/My, and the progradation rate ranges from 2250–45,750 m/My. The accretion rate of the northeastern shelf edge is between 516.8–725.6 m/My, and the progradation rate is between 250–550 m/My. The sediment flux in the northwest is in the range of 520.2–10,266.3 m^3/My, and the sediment flux in the northeast is in the range of 32.3–99.77 m^3/My (Table S3). The accretion rate decreased from west to east during this stage. The rate of accumulation gradually decreased from west to east. Sediment flux decreased significantly from west to east. However, compared with S2, the sediment flux decreased slightly. It shows that although the source supply of S2 in the west is more powerful than east, it has a decreasing trend in general (Figure 7).

4.3.4. S4 (T07-T0)

The northwestern part of S4 is characterized by strong progradation and high sediment flux. The northeastern part is characterized by strong accretion, weak progradation, and low sediment flux (Table S3), while the northwestern part of the shelf edge is mostly low-angle slow rising type, and most of the trajectories of the northeastern shelf edge are of high-angle sharp rising type. There also the retrogradation-slump type caused by the slump.

The accretion distance of the shelf edge ranged from 47.4–126.68 m, with an average value of 94.12 m. The shelf-edge progradation distance ranges from −475–10,650 m, with an average value of 1960.42 m. In the northwestern region, the shelf-edge trajectory angle is between 0 and 4°. In the northeastern region, the shelf-edge trajectory angle increases rapidly, with a range of 60–90°. There are also a few retrogradation-slump type shelf-edge trajectories with an angle of 101°. The slope height ranged from 390.28 to 647.96 m, with an average value of 543.67 m. The slope length was in the range of 3500–10,537.5 m, with an average value of 7410.42 m. The slope angle is between 3–9° (Table S2).

The accretion rate of the northwestern shelf margin ranges from 263.4–805.6 m/My, and the progradation rate ranges from −10,500–53,250 m/My. The accretion rate of the northeastern shelf edge is between 472.2–590.6 m /My, and the progradation rate is between 62.5–3812.5 m /My. The sediment flux in the northwest is in the range of 595.94–6745.71 m^3/My, and that in the northeast is in the range of 7.38–360.05 m^3/My (Table S3). The deposition rate and sediment flux decreased significantly from west to east during this stage, with the sediment flux showing a decreasing trend compared with S3. It shows that the provenance supply in this stage inherited the characteristics of "strong in the west and weak in the east" in the S3 stage and continued to decrease (Figure 7).

4.4. Types of Gravity Flow Deposits in Deep Water

4.4.1. Submarine Fan

The submarine fan is an element of the basin architecture that is composed of the varied types of deposits [42]. In the plane, it is foliated, and in the section, it is a flat hill with bidirectional downward superposition; in the fan body, it is a parallel and sub-parallel structure, showing the reflection characteristics of high continuity and strong amplitude [43].

In the northwestern part of the study area, the submarine fan develops on a large scale, with a width of 8 to 12 km and a thin thickness, fluctuating between 25 to 30 m. The northeast submarine fan has a small distribution scale, with a width between 6 and 10 km and a thickness between 25 and 30 m (Figure 8).

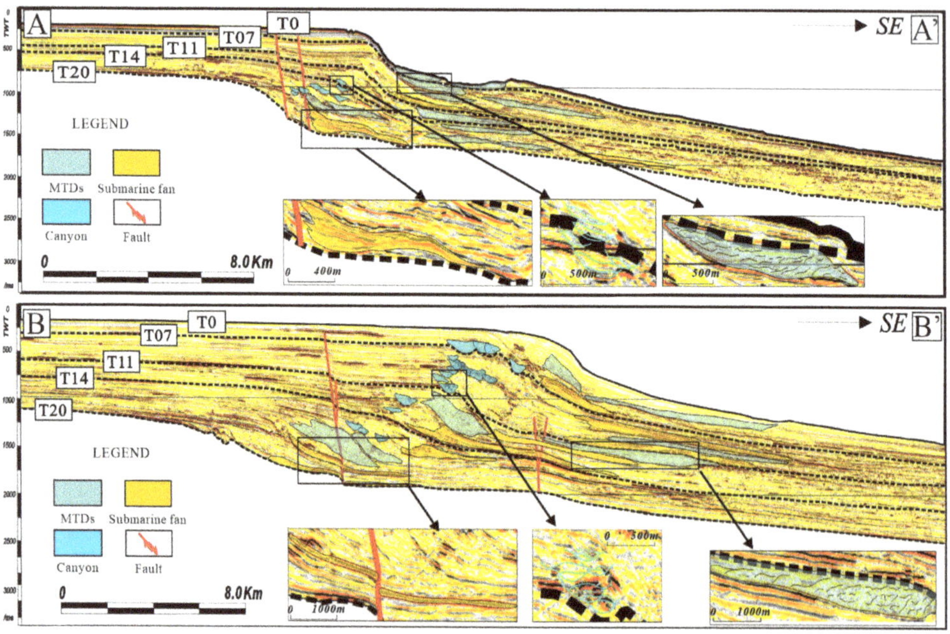

Figure 8. Sedimentary types of gravity flow are shown in seismic sections (AA′ and BB′) in the QDNB. See locations in Figure 1.

4.4.2. Mass Transport Deposits

Mass transport deposits (MTDs) are a kind of event deposits formed by the instability and collapse of continental slope sediments [44,45]. There is a chaotic reflection on the profile, and residual blocks retaining the original formation information can be seen inside. The thrust fault can be seen on the toe, with poor continuity and weak amplitude reflection characteristics (Figure 8).

The orientation of MTDs in the study area is mainly from the northwest to the southeast, and the primary block is the main component near the continental slope, while the clastic flow is the main component near the submarine plain. They are superimposed on each other, the overlying MTDs erode the underlying layer evidently, the thrust fault can be seen, and the shear plane between MTDs is evident (Figure 8).

4.4.3. Canyon Filled Deposit

The canyon is the conduit through which sediment is transported from the shelf to the deep-sea basin [46,47]. Its cross-section is a "U" or "V" shape; the "U" shape is mainly filled by debris flow deposits, and the "V" shape is mainly filled by muddy deposits, and its internal structure is complex and diverse due to different sediments. Most of the canyons in the study area have chaotic internal structures with weak continuity and moderate-weak amplitude reflection characteristics, and the interior is filled with debris flow deposits or deep-sea mud deposits, and the block of canyon wall collapses. Therefore, it is considered that the canyon-fill deposits are mainly gravity flow or deep-sea mud deposits (Figure 8).

Since 0.8 My, a large number of canyons have developed in the study area, with an average cutting depth of 60 m and an average width of 90 m. The northwestern canyons are mostly characterized by lateral tangential overlapping relationships, while the northeastern canyons are characterized by vertical tangential overlapping relationships (Figure 8).

5. Discussion

5.1. The Evolution of Systems Tract Indicating by the Shelf-Edge Trajectory

The relationship between relative sea-level change and sediment supply may lead to different sequence accumulation patterns and shelf margin trajectories [48–52]. Due to the response of the shelf edge to the changes in sediment supply and relative sea level change, it presents a descending, rising, or flat trajectory [6,7,9]. This suggests that the sedimentary environment can be distinguished based on the shelf-edge trajectory, and thus corresponding system tract types can be inferred. The observed patterns are discussed below [49,53].

5.1.1. Lowstand Systems Tract (LST)

During this time, the relative sea level began to rise from the lowest point, and the shelf-edge trajectory was medium-angle rising type, which had the characteristics of medium accretion and weak accretion. The topset bedding is medium thick, the foreset bedding is thin, and the bottom bedding is thin, indicating that sediments are still transported to the basin, but submarine fan deposits are easily formed near the slope of the shelf edge due to rising sea level (Figure 9).

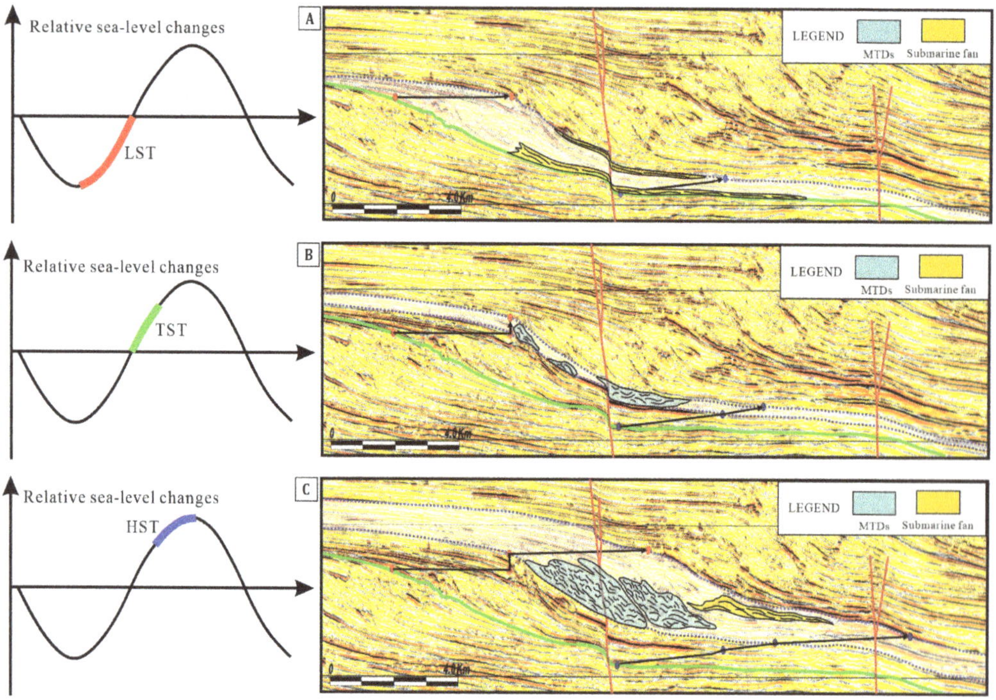

Figure 9. Relationship between relative sea level and shelf-edge trajectories. Three combinations of the shelf-edge trajectories with the corresponding stages of sea-level fluctuation. (**A**) LST has more submarine fans, (**B**) TST has more MTDs, and (**C**) HST has both of submarine fans and MTDs.

5.1.2. Transgressive Systems Tract (TST)

During this time, the relative sea level rose rapidly, and the shelf-edge trajectory angle was retrogradation-slump type, which was characterized by strong regression and forced accumulation. The thickness of the topset bedding is medium, the foreset bedding is thin, and the bottom bedding is thin, indicating that sediments accumulate rapidly at the shelf edge due to the rapid rise of sea level and retreat of the source. It could easily become instable and collapse, forming MTDs (Figure 9).

5.1.3. Highstand System Tract (HST)

During this time, the relative sea level rise rate slowed down, and the shelf-edge trajectory angle was low-angle slow rising type or high-angle sharp rising type, which had the characteristics of strong prograde and forced prograde. The thickness of the topset bedding, foreset bedding, and bottom set bedding are large, indicating that the sediment supply is strong, and the submarine fan and MTDs are superimposed near the slope of the shelf edge (Figure 9).

Based on the research results, in seismic data interpretation, we should pay more attention to finding favorable reservoir groups at the sequence interface (HST) or under the sequence interface. Fans developed above the sequence interface (LST) tend to have longer transport distances and thinner thicknesses, while fans cultivated under the sequence interface (HST) have shorter transport distances and larger thicknesses.

5.2. Relationship Between Shelf-Edge Migration and Deep-Water Gravity Flow Deposition

During the S1 (1.8 Ma to 0.8 Ma), the overall trajectory of the shelf edge was dominated by a low-angle and slow-rising type (Figure 10), and the sediment supply rate in the northwest was greater than that in the northeast. A large volume of submarine fan deposits with low thickness and the short transport distance was present in the northwest. In contrast, the submarine fan is less developed in the northeast, with thin thickness, a small width, and a long transport distance. There were fewer canyons in the northwest and more in the northeast. Some MTDs are developed in the slope area (Figure 10).

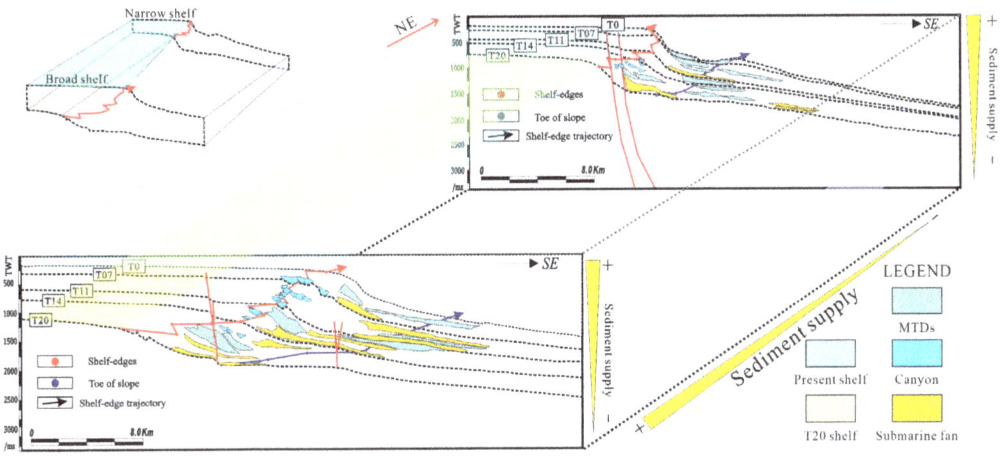

Figure 10. Response mode of deep-water deposition to shelf-edge trajectory in the Quaternary Qiongdongnan Basin. The western shelf margin is dominated by progradation, while the eastern shelf margin is dominated by accretion.

During S2 (0.8 Ma to 0.6 Ma), the shelf-edge trajectory in the northwest was dominated by the low-angle slow rising type, while that in the northeast was dominated by the medium-angle rising type (Figure 7). The sediment supply rate in the northwest was still higher than that in the northeast. There were more submarine fan deposits in the northwest, which had a certain inheritance, but less in the northeast. Canyons were more developed in the northwest but less in the northeast. A small number of MTDs developed in the northwest and thin layers developed in the northeast (Figure 10).

During S3 (0.6 Ma to 0.2 Ma), the northwest shelf-edge trajectory was dominated by medium-angle rising type, and the northeast shelf-edge trajectory was mainly of the high-angle sharp rising type. The development scale of submarine fans in the northwest have a long transport distance, while only a few submarine fans developed in the northeast. MTDs and submarine fan deposits overlapped and more MTDs developed in the northeast at this time (Figure 10). During this time, with the increase of the shelf-edge trajectory angle, the development of submarine fan sediments began to decrease, while the development of MTDs increased.

During the S4 (0.2 Ma to present), the northwest shelf-edge trajectory was dominated by the medium-angle rising type, and the northeast shelf-edge trajectory was dominated by the high-angle sharp rising type. The development scale of submarine fan in the northwest was the smallest, canyons developed extensively, and MTDs and submarine fan overlapped. In the northeast, there was almost no submarine fan deposition, no canyon development, and large-scale development of MTDs (Figure 10). During this time, with the increase of the shelf-edge trajectory angle, the development of submarine fans began to decrease, and MTDs began to develop on a large scale.

Based on what has been mentioned above, it is concluded that the shelf-edge trajectory is a record of the changes in sea level, which also can indicate the development of the depositional system in the basin. When the shelf-edge trajectory angle is $-1^\circ < \alpha < 4^\circ$, the canyon size in the slope area is small and the downcutting is shallow, and the deep-sea plain area has multi-stage large-scale submarine fan deposits with fewer MTDs. When $4^{\circ} < \alpha < 35^{\circ}$, the canyon size increases in the continental slope area, and both submarine fan deposits and MTDs appear in the deep-sea plain area. When $35^{\circ} < \alpha < 90^{\circ}$, the valley in the continental slope area is less developed but deep cut, the sediments in the deep-sea plain area are mainly large MTDs, and the submarine fan is rarely developed. When $90^{\circ} < \alpha < 150^{\circ}$, the valley in the continental slope area is almost undeveloped, the sediments in the deep-sea plain area are mainly large MTDs, and the submarine fan is nearly undeveloped (Figure 10).

5.3. *Shelf-Edge Trajectory Driven by Multiple Factors*

5.3.1. Influence of Relative Sea Level Change on Shelf-Edge Trajectory

(1) Tectonic subsidence

At 10.5 Ma, the subduction of the Pacific plate and the Eurasian plate, the right-lateral strike-slip action of the Red River fault, and the regional activity of Fault No. 2 caused the subsidence center to shift eastward, and the topography of the basin showed the characteristics of high in the west and low in the east, which has continued to the present day [20,54]. Since the Quaternary, the sedimentation rate of the QDNB peaked [55], and the difference between the sedimentation rates of the east and west part of the QDNB affected the sediment accumulation rate, resulting in the lateral difference of the shelf-edge trajectory in the study area (Figures 10 and 11).

Qiongdongnan Basin is located between the Red River fault zone, Xisha Trough fault zone, and west margin fault zone of the South China Sea, with a complex structure and faults. Since Quaternary, the basin has been in a period of accelerated thermal subsidence, and the tectonic movement has weakened. Most of the faults have stopped activities at this time, but some faults still remain active in the slope area [26]. The No. 2 fault zone runs from east to west of QDNB, with a length of about 300 km (Figure 1). Since Quaternary, the northeastern region of the No. 2 fault zone is still active [21]. The regional activity of the No. 2 fault zone also led to the high-angle rising and recession-slump migration patterns of the shelf edge in the northeast of the study area, and the large-scale development of MTDs. Therefore, the regional activities of fault zone No. 2 not only control the trajectory migration of the shelf edge, but also control the type of sedimentary system in the northeastern part of the basin (Figures 10 and 11).

(2) Global sea level change

Since 1.9 Ma, due to the influence of the ice age, the global temperature has dropped, and the formation of the North and South polar ice sheet has led to sea level fall. During this time, the sea level of QDNB underwent fluctuations [56,57]. The vertical evolution of the shelf-edge trajectory in QDNB show strong relation with sea level changes. When the sea level rises, the progradation rate of the shelf-edge trajectory is small, and when the sea level falls, the progradation rate is large (Figures 10 and 11). Since 0.8 Ma, the eccentricity period at the scale of 100 Ka has increased significantly, the fluctuation amplitude of sea level has increased significantly (up to 120 m), the shelf-edge system has repeatedly collapsed on a large scale, and the shelf-edge trajectory is dominated by the high-angle rising type or the recession-slump type.

5.3.2. Source Supply

There are two main provenance supply systems in QDNB, namely the Red River provenance system and the Hainan Island provenance system. The Red River provenance system plays an important role in controlling the shelf-edge trajectory migration in the northwest QDNB. According to the calculation of sedimentation rate and sediment transport of the Red River since the formation of QDNB [55], since the Quaternary, the maximum sediment transport capacity of the Red River is 38 km^3/My. Since Quaternary, due to the decreasing sedimentation rate of Yinggehai Basin, it has been gradually filled, and the influence range of Red River provenance has gradually increased, with the northwestern part of the basin as a whole being under the influence of the Red River provenance system [33]. Studies on the distribution patterns of rare earth elements in drilling samples from different regions of QDNB show that the distribution characteristics of rare earth elements in the sediments of northeastern QDNB since Quaternary are almost identical with those in the source region of Hainan Island [24,58]. Therefore, it is believed that the provenance system of Hainan Island dominates the shelf-edge trajectory in the northeast QDNB. Since the Quaternary, the sediment supply of Hainan Island has been increasing, and the sediment transport is within the range of 15–45 km^3/My [59].

Since 1.9 Ma, the Red River provenance has entered through the northwest of QDNB [60], and while there are rivers in the southwest of Hainan Island, there are fewer rivers in the southeast. As a result, the sedimentary center of the basin has always been located in the northwest, and the amount of sediment received in the northwest of the study area is much larger than that in the northeast [55,61], resulting in the extreme advance of the shelf-edge system in the northwest of the study area, whose maximum distance is up to 24 km. Due to the small supply of sediment, the shelf-edge trajectory of the northeastern basin is dominated by the high-angle rising type, resulting in little advance of the shelf-edge system (25–500 m), and the instability of sediment accumulation is prone to collapse. It is concluded that provenance supply plays a leading role in the differential evolution of the east and west sides of QDNB [11].

5.3.3. Climate Change

According to the study of Cenozoic palynological assemblages in Qiongdongnan, the climate in the northern part of the South China Sea has gradually changed from hot and humid to cool and dry, and from tropical to subtropical since Quaternary [62,63]. Based on the analysis of sediment records at ODP1145 and ODP1144, it is concluded that the carbonate content at ODP1145 showed a significant decrease at around 1.9 Ma, which may indicate the significant enhancement of the East Asian winter monsoon during this time [31,64]. After about 0.8 Ma, the 100 Ka scale eccentricity period was significantly enhanced in response to the Mid Pleistocene Transition (MPT). Before the Pleistocene climate change, the climate in QDNB was a typical tropical climate [36]. After the transition, the QDNB was affected by the strengthening of the East Asian winter monsoon, the temperature dropped, and the climate changed to subtropical (Figure 11).

The change in the overall climate pattern determines the strength of the provenance supply and indirectly controls the shelf-edge trajectory [36]. Through the comparison of climate change and the shelf-edge trajectory in the study area, it can be found that the accumulation rate of the shelf-edge trajectory will increase with the decrease of temperature and the intensification of the East Asian winter monsoon. After the Mid Pleistocene Transition (MPT) occurred, the sediment flux in the study area increased significantly, resulting in a significant increase in the formation thickness in the study area since 0.8 Ma (Figure 11).

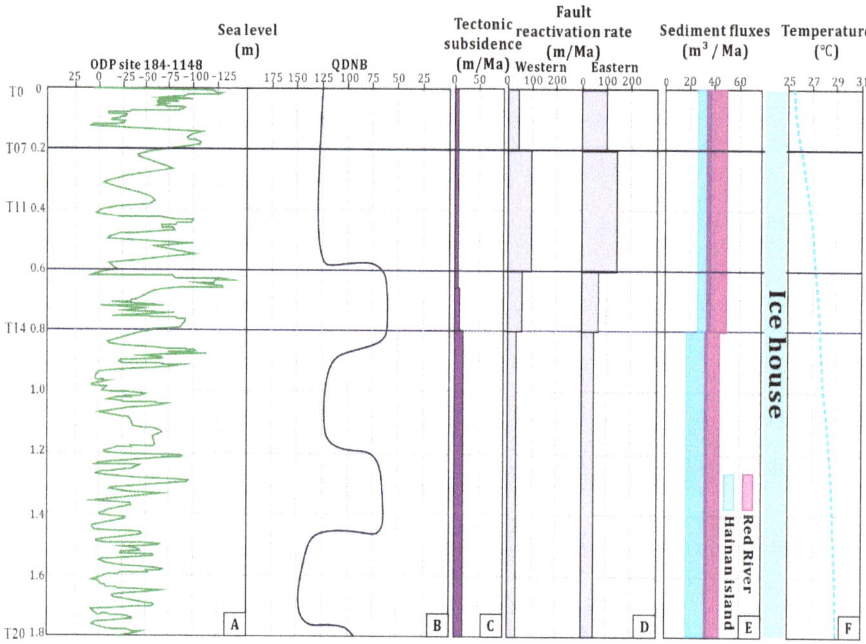

Figure 11. (**A**) Sea level records in the northern South China Sea (ODP site 184–1148) (green line) from 0 to 1.8 Ma derived from [65,66] (**B**) Sea level records in the QDNB (deep blue line) from 0 to 1.8 Ma derived from [27]. (**C**) Tectonic subsidence records in the QDNB (purple rectangle) from 0 to 1.8 Ma derived from [54]. (**D**) Fault reactivation rate records in the QDNB (brown rectangle) from 0 to 1.8 Ma. (**E**) Sediment flux of QNDB (blue rectangle represents Hainan Island sources; pink rectangle represents Red River sources) from 0 to 1.8 Ma derived from [20,32–34,36]; (**F**) temperature curve changes from 0 to 1.8 Ma derived from [16].

6. Conclusions

(1) Through quantitative characterization of Quaternary shelf-edge trajectory migration, four types of shelf-edge trajectory migration have been recognized, including low-angle slow ascending type, medium-angle ascending type, high-angle sharp ascending type, and recession-slump type.

(2) In this study, a new sequence stratigraphic framework based on shelf-edge trajectory migration was established. There are four third-order sequences in the Quaternary, and several system domains (LST, TST, HST) have been identified. During the LST period, the shelf-edge trajectory was mainly low-angle type, the progradation rate was higher than the accretion rate. In the TST period, the shelf-edge trajectory was mainly regression-slump type, the sea level rose rapidly, leading to shelf-edge sediment instability and collapse, forming MTDs. During the HST period, the shelf-edge trajectory was mainly medium-angle rising type or high-angle sharp rising type, the accretion rate was almost the same as the progradation rate, the sea level rising rate slowed down, and a large amount of sediment accumulated on the shelf edge.

(3) When the canyon in the slope area is small and the downcutting is shallow, and the deep-sea plain area has multi-stage large-scale submarine fan deposits and fewer MTDs, the trajectory angle is $-1^{\circ} < \alpha < 4^{\circ}$. When the canyon increases in the continental slope area, and both submarine fan deposits and MTDs appear in the deep-sea plain area, the trajectory angle is $4^{\circ} < \alpha < 35^{\circ}$. When the canyon in the continental slope area is less developed but downcutting is deep, the sediments in the deep-sea plain area are mainly large MTDs, and the submarine fan is poorly developed. In this case, the trajectory angle

is 35° < α < 90°. When the canyon in the continental slope area is almost undeveloped, the sediments in the deep-sea plain area are mainly large MTDs, and the submarine fan is almost undeveloped, the trajectory angle is 90° < α < 150°.

(4) Since the Quaternary, the temperature in QDNB has been decreasing, the sea level has been falling, and the East Asian winter monsoon has been significantly enhanced, resulting in an overall increase in sediment supply in the study area. However, due to the numerous rivers and rich sources of sediment in the west of Hainan Island, which is not in the subsidence center, a growing shelf-edge slope has developed. In the eastern part, due to fewer rivers, weak provenance system, strong tectonic activity, and subsidence center, a type of destructive shelf-edge slope has developed in the western part of basin.

Supplementary Materials: The following supporting information can be downloaded at: https://www.mdpi.com/article/10.3390/jmse12112051/s1, Table S1: Statistics of shelf-edge trajectory parameters in QDNB. A: Accretion, P: Progradation, α: Shelf-edge trajectory. Table S2: Statistics of slope angle parameters in QDNB. ETT: Edge to toe, Rc: Height between edge and toe. Table S3: Statistics of sediment supply parameters in QDNB. A: Accretion, P: Progradation, Ra: Accretion rate, Rp: Progradation rate, Fc: Sediment flux.

Author Contributions: Conceptualization, H.Q.; methodology, C.M.; software, C.M.; validation, C.M.; formal analysis, C.M.; investigation, X.L.; resources, H.Q.; data curation, C.M.; writing—original draft preparation, C.M. and X.L.; writing—review and editing, H.Q.; visualization, C.M.; supervision, H.Q.; project administration, H.Q.; funding acquisition, H.Q. All authors have read and agreed to the published version of the manuscript.

Funding: This research received no external funding. And The APC was funded by Hongjun Qu.

Data Availability Statement: The original contributions presented in this study are included in the article/Supplementary Material. Further inquiries can be directed to the corresponding author.

Conflicts of Interest: The authors declare no conflict of interest.

References

1. Helland-Hansen, W.; Martinsen, O.J. Shoreline Trajectories and Sequences; Description of Variable Depositional-Dip Scenarios. *J. Sediment. Res.* **1996**, *66*, 670–688.
2. Løseth, T.M.; Helland-Hansen, W. Predicting the Pinchout Distance of Shoreline Tongues. *Terra Nova* **2001**, *13*, 241–248. [CrossRef]
3. Payton, C.E. *Seismic Stratigraphy—Applications to Hydrocarbon Exploration*; American Association of Petroleum Geologists: Tulsa, OK, USA, 1977.
4. Steel, R.; Olsen, T. Clinoforms, Clinoform Trajectories and Deepwater Sands. In *Sequence Stratigraphic Models for Exploration and Production: Evolving Methodology, Emerging Models, and Application Histories: 22nd Annual*; SEPM Society for Sedimentary Geology: Tulsa, OK, USA, 2002; pp. 367–380.
5. Chenglin, G.; Wang, Y.; Steel, R.J.; Olariu, C.; Xu, Q.; Liu, X.; Zhao, Q. Growth Styles of Shelf-Margin Clinoforms: Prediction of Sand-and Sediment-Budget Partitioning into and across the Shelf. *J. Sediment. Res.* **2015**, *85*, 209–229.
6. Tugdual, G.; Rovere, M.; Pellegrini, C.; Cattaneo, A.; Campiani, E.; Trincardi, F. Factors Controlling Margin Instability During the Plio-Quaternary in the Gela Basin (Strait of Sicily, Mediterranean Sea). *Mar. Pet. Geol.* **2021**, *123*, 104767.
7. Helland-Hansen, W.; Hampson, G.J. Trajectory Analysis: Concepts and Applications. *Basin Res.* **2009**, *21*, 454–483. [CrossRef]
8. Gemma, E.; Casas, D.; Estrada, F.; Vázquez, J.T.; Iglesias, J.; García, M.; Gómez, M.; Acosta, J.; Gallart, J.; Maestro-González, A. Morphosedimentary Features and Recent Depositional Architectural Model of the Cantabrian Continental Margin. *Mar. Geol.* **2008**, *247*, 61–83.
9. Stefano, P.; Scisciani, V.; Helland-Hansen, W.; D'Intino, N.; Reid, W.; Pellegrini, C. Upslope-Climbing Shelf-Edge Clinoforms and the Stepwise Evolution of the Northern European Glaciation (Lower Pleistocene Eridanos Delta System, Uk North Sea): When Sediment Supply Overwhelms Accommodation. *Basin Res.* **2020**, *32*, 224–239.
10. Si, C.; Steel, R.; Olariu, C.; Li, S. Growth of the Paleo-Orinoco Shelf-Margin Prism: Process Regimes, Delta Evolution, and Sediment Budget Beyond the Shelf Edge. *Bulletin* **2018**, *130*, 35–63.
11. Mengnan, L.; Liu, H.; van Loon, A.J.; Xu, J.; Hao, S.; Zhang, Y. Analysis of the Geometric Characteristics of Clinothems and the Relationship with Shelf-Edge Trajectories of the Plio-Pleistocene Continental Slope in the Qiongdongnan Basin, South China Sea. *Sedimentology* **2023**, *70*, 5–30.
12. Chenglin, G.; Steel, R.J.; Wang, Y.; Lin, C.; Olariu, C. Shelf-Margin Architecture Variability and Its Role in Sediment-Budget Partitioning into Deep-Water Areas. *Earth-Sci. Rev.* **2016**, *154*, 72–101.
13. Chenglin, G.; Wang, Y.; Zhu, W.; Li, W.; Xu, Q.; Zhang, J. The Central Submarine Canyon in the Qiongdongnan Basin, Northwestern South China Sea: Architecture, Sequence Stratigraphy, and Depositional Processes. *Mar. Pet. Geol.* **2011**, *28*, 1690–1702.

14. Chen Percy, P.H.; Chen, Z.Y.; Zhang, Q.M. Sequence Stratigraphy and Continental Margin Development of the Northwestern Shelf of the South China Sea. *AAPG Bull.* **1993**, *77*, 842–862.
15. Ming, S.; Xinong, X.; Junliang, L.; Tao, J.; Cheng, Z.; Yunlong, H.; Shanshan, T.; Cuimei, Z. Gravity Flow on Slope and Abyssal Systems in the Qiongdongnan Basin, Northern South China Sea. *Acta Geol. Sin. Engl. Ed.* **2011**, *85*, 243–253. [CrossRef]
16. Si, C.; Steel, R.; Wang, H.; Zhao, R.; Olariu, C. Clinoform Growth and Sediment Flux into Late Cenozoic Qiongdongnan Shelf Margin, South China Sea. *Basin Res.* **2020**, *32*, 302–319.
17. Dawei, W.; Wu, S.; Yao, G.; Lü, F.; Strasser, M. Analysis of Quaternary Mass Transport Deposits Based on Seismic Data in Southern Deep-Water Region of Qiongdongnan Basin, South China Sea. In *Landslide Science for a Safer Geoenvironment*; Springer: Cham, Switzerland, 2014; Volume 3.
18. Yuhong, X.; Zhang, G.; Sun, Z.; Zeng, Q.; Zhao, Z.; Guo, S. Reservoir Forming Conditions and Key Exploration Technologies of Lingshui 17-2 Giant Gas Field in Deepwater Area of Qiongdongnan Basin. *Pet. Res.* **2019**, *4*, 1–18.
19. Tao, J.; Xie, X.; Wang, Z.; Li, X.; Zhang, Y.; Sun, H. Seismic Features and Origin of Sediment Waves in the Qiongdongnan Basin, Northern South China Sea. *Mar. Geophys. Res.* **2013**, *34*, 281–294.
20. Shi, X.; Kohn, B.; Spencer, S.; Guo, X.; Li, Y.; Yang, X.; Shi, H.; Gleadow, A. Cenozoic Denudation History of Southern Hainan Island, South China Sea: Constraints from Low Temperature Thermochronology. *Tectonophysics* **2011**, *504*, 100–115. [CrossRef]
21. Jianye, R.; Zhang, D.; Tong, D.; Huang, A.; Wang, Y.; Lei, C.; Zuo, Q.; Zhao, Y.; He, W.; Yang, L. Characterising the Nature, Evolution and Origin of Detachment Fault in Central Depression Belt, Qiongdongnan Basin of South China Sea: Evidence from Seismic Reflection Data. *Acta Oceanol. Sin.* **2014**, *33*, 118–126.
22. Cao, L.; Jiang, T.; Wang, Z.; Zhang, Y.; Sun, H. Provenance of Upper Miocene Sediments in the Yinggehai and Qiongdongnan Basins, Northwestern South China Sea: Evidence from Ree, Heavy Minerals and Zircon U–Pb Ages. *Mar. Geol.* **2015**, *361*, 136–146. [CrossRef]
23. Stefan, N.; Granjeon, D.; Willett, S.; Lin, A.T.-S.; Castelltort, S. Stratigraphic Modeling of the Western Taiwan Foreland Basin: Sediment Flux from a Growing Mountain Range and Tectonic Implications. *Mar. Pet. Geol.* **2018**, *96*, 331–347.
24. Lei, C.; Ren, J. Hyper-Extended Rift Systems in the Xisha Trough, Northwestern South China Sea: Implications for Extreme Crustal Thinning Ahead of a Propagating Ocean. *Mar. Pet. Geol.* **2016**, *77*, 846–864. [CrossRef]
25. Zhao, Z.; Sun, Z.; Sun, L.; Wang, Z.; Sun, Z. Cenozoic Tectonic Subsidence in the Qiongdongnan Basin, Northern South China Sea. *Basin Res.* **2018**, *30*, 269–288. [CrossRef]
26. Mao, K.; Xie, X.; Xie, Y.; Ren, J.; Chen, H. Post-Rift Tectonic Reactivation and Its Effect on Deep-Water Deposits in the Qiongdongnan Basin, Northwestern South China Sea. *Mar. Geophys. Res.* **2015**, *36*, 227–242. [CrossRef]
27. Zhao, Z.; Sun, Z.; Wang, Z.; Sun, Z.; Liu, J.; Zhang, C. The High Resolution Sedimentary Filling in Qiongdongnan Basin, Northern South China Sea. *Mar. Geol.* **2015**, *361*, 11–24. [CrossRef]
28. Jie, H.; Wan, S.; Xiong, Z.; Zhao, D.; Liu, X.; Li, A.; Li, T. Geochemical Records of Taiwan-Sourced Sediments in the South China Sea Linked to Holocene Climate Changes. *Palaeogeogr. Palaeoclimatol. Palaeoecol.* **2016**, *441*, 871–881.
29. Xiao, X.M.; Xiong, M.; Tian, H.; Wilkins, R.W.; Huang, B.; Tang, Y. Determination of the Source Area of the Ya13-1 Gas Pool in the Qiongdongnan Basin, South China Sea. *Org. Geochem.* **2006**, *37*, 990–1002. [CrossRef]
30. Huang, H.; Huang, B.; Huang, Y.; Xing, L.; Hui, T. Condensate Origin and Hydrocarbon Accumulation Mechanism of the Deepwater Giant Gas Field in Western South China Sea: A Case Study of Lingshui 17-2 Gas Field in Qiongdongnan Basin. *Pet. Explor. Dev.* **2017**, *44*, 409–417. [CrossRef]
31. Huang, B.; Tian, H.; Li, X.; Wang, Z.; Xiao, X. Geochemistry, Origin and Accumulation of Natural Gases in the Deepwater Area of the Qiongdongnan Basin, South China Sea. *Mar. Pet. Geol.* **2016**, *72*, 254–267. [CrossRef]
32. Wan, S.; Li, A.; Clift, P.D.; Jiang, H. Development of the East Asian Summer Monsoon: Evidence from the Sediment Record in the South China Sea since 8.5 Ma. *Palaeogeogr. Palaeoclimatol. Palaeoecol.* **2006**, *241*, 139–159. [CrossRef]
33. Xiaoming, Z.; Qi, K.; Patacci, M.; Tan, C.; Xie, T. Submarine Channel Network Evolution above an Extensive Mass-Transport Complex: A 3d Seismic Case Study from the Niger Delta Continental Slope. *Mar. Pet. Geol.* **2019**, *104*, 231–248.
34. Cheng, C.; Jiang, T.; Kuang, Z.; Ren, J.; Liang, J.; Lai, H.; Xiong, P. Seismic Characteristics and Distributions of Quaternary Mass Transport Deposits in the Qiongdongnan Basin, Northern South China Sea. *Mar. Pet. Geol.* **2021**, *129*, 105118. [CrossRef]
35. Wang, Z.; Jiang, T.; Zhang, D.; Wang, Y.; Zuo, Q.; He, W. Evolution of Deepwater Sedimentary Environments and Its Implication for Hydrocarbon Exploration in Qiongdongnan Basin, Northwestern South China Sea. *Acta Oceanol. Sin.* **2015**, *34*, 1–10. [CrossRef]
36. Zhuo, H.; Wang, Y.; Shi, H.; He, M.; Chen, W.; Li, H.; Wang, Y.; Yan, W. Contrasting Fluvial Styles across the Mid-Pleistocene Climate Transition in the Northern Shelf of the South China Sea: Evidence from 3d Seismic Data. *Quat. Sci. Rev.* **2015**, *129*, 128–146. [CrossRef]
37. Bilal U, H.; Hardenbol, J.A.; Vail, P.R. Chronology of Fluctuating Sea Levels since the Triassic. *Science* **1987**, *235*, 1156–1167.
38. Li, Q.; Zhong, G.; Tian, J. Stratigraphy and Sea Level Changes. In *The South China Sea: Paleoceanography and Sedimentology*; Springer: Berlin/Heidelberg, Germany, 2009; pp. 75–170.
39. Wei, L.; Alves, T.M.; Wu, S.; Völker, D.; Zhao, F.; Mi, L.; Kopf, A. Recurrent Slope Failure and Submarine Channel Incision as Key Factors Controlling Reservoir Potential in the South China Sea (Qiongdongnan Basin, South Hainan Island). *Mar. Pet. Geol.* **2015**, *64*, 17–30.
40. Vall, P.R. The Stratigraphic Signatures of Tectonics, Eustacy and Sedimentology-an Overview. In *Cycles Events Stratigraphy*; Springer: Berlin/Heidelberg, Germany, 1991; pp. 617–659.

41. Pyles, D.R.; Syvitski, J.P.; Slatt, R.M. Defining the Concept of Stratigraphic Grade and Applying It to Stratal (Reservoir) Architecture and Evolution of the Slope-to-Basin Profile: An Outcrop Perspective. *Mar. Pet. Geol.* **2011**, *28*, 675–697. [CrossRef]
42. Rotzien, J.R.; Caldwell, R.L.; Goggin, L.R. Deepwater Depositional Environments. In *Deepwater Sedimentary Systems*; Elsevier: Amsterdam, The Netherlands, 2022; pp. 251–300.
43. Xiaoming, Z.; Li, M.; Qi, K.; Liu, L.; Hu, G.; Wu, W.; Zhang, W.; Ge, J. Development of a Distinct Submarine Depositional System on a Topographically Complex Niger Delta Slope. *Geol. J.* **2020**, *55*, 3732–3747.
44. Rui, Z.; Chen, S.; Olariu, C.; Steel, R.; Zhang, J.; Wang, H. A Model for Oblique Accretion on the South China Sea Margin; Red River (Song Hong) Sediment Transport into Qiongdongnan Basin since Upper Miocene. *Mar. Geol.* **2019**, *416*, 106001.
45. Jinhua, L.; Peimin, Z.H. Gravity Induced Deposits in the Continental Slope of Qiongdongnan Basin Based on Ultrahigh Resolution Auv Data. *Geol. Sci. Technol. Inf.* **2019**, *38*, 42–50.
46. Pickering, K.T.; Hiscott, R.N. *Deep Marine Systems: Processes, Deposits, Environments, Tectonics and Sedimentation*; John Wiley & Sons: Hoboken, NJ, USA, 2015.
47. Jon R, R.; Hernández-Molina, F.J.; Fonnesu, M.; Thieblemont, A. Deepwater Sedimentation Units. In *Deepwater Sedimentary Systems*; Elsevier: Amsterdam, The Netherlands, 2022; pp. 203–249.
48. Hampson, G.J.; Storms, J.E.A. Geomorphological and Sequence Stratigraphic Variability in Wave-Dominated, Shoreface-Shelf Parasequences. *Sedimentology* **2003**, *50*, 667–701. [CrossRef]
49. Claudio, P.; Asioli, A.; Bohacs, K.M.; Drexler, T.M.; Feldman, H.R.; Sweet, M.L.; Maselli, V.; Rovere, M.; Gamberi, F.; Valle, G.D. The Late Pleistocene Po River Lowstand Wedge in the Adriatic Sea: Controls on Architecture Variability and Sediment Partitioning. *Mar. Pet. Geol.* **2018**, *96*, 16–50.
50. Ramon-Duenas, C.; Rudolph, K.W.; Emmet, P.A.; Wellner, J.S. Quantitative Analysis of Siliciclastic Clinoforms: An Example from the North Slope, Alaska. *Mar. Pet. Geol.* **2018**, *93*, 127–134. [CrossRef]
51. Pszonka, J.; Žecová, K.; Wendorff, M. Oligocene Turbidite Fans of the Dukla Basin: New Age Data from the Calcareous Nannofossils and Paleoenvironmental Conditions (Cergowa Beds, Polish–Slovakian Borderland). *Geol. Carpathica* **2019**, *70*, 311–324. [CrossRef]
52. Pszonka, J.; Wendorff, M.; Godlewski, P. Sensitivity of Marginal Basins in Recording Global Icehouse and Regional Tectonic Controls on Sedimentation. Example of the Cergowa Basin, (Oligocene) Outer Carpathians. *Sediment. Geol.* **2023**, *444*, 106326. [CrossRef]
53. Anell, I.; Midtkandal, I.; Braathen, A. Trajectory Analysis and Inferences on Geometric Relationships of an Early Triassic Prograding Clinoform Succession on the Northern Barents Shelf. *Mar. Pet. Geol.* **2014**, *54*, 167–179. [CrossRef]
54. Xi, X.; Müller, R.D.; Ren, J.; Jiang, T.; Zhang, C. Stratigraphic Architecture and Evolution of the Continental Slope System in Offshore Hainan, Northern South China Sea. *Mar. Geol.* **2008**, *247*, 129–144. [CrossRef]
55. Clift, P.D.; Sun, Z. The Sedimentary and Tectonic Evolution of the Yinggehai–Song Hong Basin and the Southern Hainan Margin, South China Sea: Implications for Tibetan Uplift and Monsoon Intensification. *J. Geophys. Res. Solid Earth* **2006**, *111*, B06405. [CrossRef]
56. Stow, D.A.V.; Howell, D.G.; Nelson, C.H. Sedimentary, Tectonic, and Sea-Level Controls. In *Submarine Fans and Related Turbidite Systems*; Springer: Berlin/Heidelberg, Germany, 1985; pp. 15–22.
57. Miller, K.G.; Kominz, M.A.; Browning, J.V.; Wright, J.D.; Mountain, G.S.; Katz, M.E.; Sugarman, P.J.; Cramer, B.S.; Christie-Blick, N.; Pekar, S.F. The Phanerozoic Record of Global Sea-Level Change. *Science* **2005**, *310*, 1293–1298. [CrossRef]
58. Hao, L.; Meng, J.; Zhang, Y.; Yang, L. Pliocene Seismic Stratigraphy and Deep-Water Sedimentation in the Qiongdongnan Basin, South China Sea: Source-to-Sink Systems and Hydrocarbon Accumulation Significance. *Geol. J.* **2019**, *54*, 392–408.
59. Xubiao, L.; Ge, J.; Zhao, X.; Qi, K.; Jones, B.G.; Fang, X. Geochemistry of Quaternary Sediments in the Northwestern South China Sea: Sediment Provenance and Mid-Pleistocene Transition. *Mar. Geol.* **2024**, *477*, 107371.
60. Xu, D.R.; Xia, B.; Li, P.C.; Chen, G.H.; Ma, C.; Zhang, Y.Q. Protolith Natures and U-Pb Sensitive High Mass-Resolution Ion Microprobe (Shrimp) Zircon Ages of the Metabasites in Hainan Island, South China: Implications for Geodynamic Evolution since the Late Precambrian. *Isl. Arc* **2007**, *16*, 575–597. [CrossRef]
61. Long, V.H.; Clift, P.D.; Schwab, A.M.; Huuse, M.; Nguyen, D.A.; Zhen, S. Large-Scale Erosional Response of Se Asia to Monsoon Evolution Reconstructed from Sedimentary Records of the Song Hong-Yinggehai and Qiongdongnan Basins, South China Sea. *Geol. Soc. Lond. Spec. Publ.* **2010**, *342*, 219–244.
62. Alley, R.B. Wally Was Right: Predictive Ability of the North Atlantic "Conveyor Belt" Hypothesis for Abrupt Climate Change. *Annu. Rev. Earth Planet. Sci.* **2007**, *35*, 241–272. [CrossRef]
63. Jie, H.; Li, A.; Wan, S. Sensitive Grain-Size Records of Holocene East Asian Summer Monsoon in Sediments of Northern South China Sea Slope. *Quat. Res.* **2011**, *75*, 734–744.
64. Mei, S.; Wang, X.; Zhang, S.; Chu, G.; Su, Y.; Yang, Z. A 20,000-Year High-Resolution Pollen Record from Huguangyan Maar Lake in Tropical–Subtropical South China. *Palaeogeogr. Palaeoclimatol. Palaeoecol.* **2017**, *472*, 83–92.

65. Tian, J.; Zhao, Q.; Wang, P.; Li, Q.; Cheng, X. Astronomically Modulated Neogene Sediment Records from the South China Sea. *Paleoceanography* **2008**, *23*. [CrossRef]
66. Bates, S.L.; Siddall, M.; Waelbroeck, C. Hydrographic Variations in Deep Ocean Temperature over the Mid-Pleistocene Transition. *Quat. Sci. Rev.* **2014**, *88*, 147–158. [CrossRef]

Journal of
Marine Science and Engineering

MDPI

Review

Stromatolites and Their "Kin" as Living Microbialites in Contemporary Settings Linked to a Long Fossil Record

Ed Landing [1] and Markes E. Johnson [2,*]

1 New York State Museum, Albany, NY 12230, USA; ed.landing@nysed.gov
2 Department of Geosciences, Williams College, Williamstown, MA 01267, USA
* Correspondence: mjohnson@williams.edu; Tel.: +1-(413)-597-2329

Abstract: Organo-sedimentary deposits that result from fine-grained sediment trapping, binding, and likely precipitation (of carbonate) by microbes in flat-mat, branching, and dome-shaped constructions are termed microbialites. They were first identified as stromatolites by paleontologists well before the discovery of cyanobacteria that build the same kinds of structures in contemporary settings around the world. Earth's earliest life forms were prokaryotes (bacteria and bacteria-like forms) that reproduced under anaerobic conditions and later produced increasingly aerobic conditions. Stromatolites persisted through later Archean and Proterozoic times through the subsequent Phanerozoic to the present. At the start of the Cambrian Period 538 million years ago, stromatolites continued alongside rapidly diversifying plant and animal phyla during the Cambrian explosion of eukaryotic life, which have complex cells with internal structures and tissue-grade organization in multicellular taxa. The type locality exhibiting clear examples of stromatolite structures is conserved at Lester Park near Saratoga Springs in northeastern New York State. Paleontologist James Hall (1811–1898) was the first in 1884 to assign a Latin binomen (*Cryptozoon proliferum*) to stromatolite fossils from Lester Park. Thereafter, reports on formally named stromatolites proliferated, as did examples from virtually all subsequent geological time intervals including the Pleistocene Epoch. However, recognition that living cyanobacteria formed stromatolites identified as *Cryptozoon* took place much later in 1961 with the announcement by geologist Brian W. Logan (1933–2008) who described modern constructions in Hamlin Pool, Shark Bay, Western Australia. Initially, Shark Bay was regarded as a one-of-a-kind sanctuary for stromatolites living under restricted conditions with elevated levels of salinity that prohibited competition or grazing by eukaryotes. Most notably, among other settings with living stromatolites discovered and described since then are the Bahamas, East African rift lakes, Mexico's Baja California, and saline lakes in Argentina. This report reviews the history of discoveries of modern-day stromatolites, more commonly called microbialites by biologists. All are predicated on the ground-breaking efforts of geologists and paleontologists who first described fossil stromatolites but were unaware of their living counterparts. The Lester Park locality is highlighted together with a master list of other North American localities that feature purported Cryptozoons.

Keywords: prokaryotes; eukaryotes; cyanobacteria; organo-sedimentary structures; oligotrophs; extremophiles

1. Introduction and Perspective

Dominant groups of larger organisms preserved as fossils are used to name popular divisions of geological time even though those groups have much earlier or later occurrences in the rock record. Thus, an "Age of Dinosaurs" is a casual term for the Mesozoic Era ca. 251–66 Ma (where Ma is "million years ago"), even though dinosaurs have a late, ca. 233 Ma, oldest occurrence and their feathered cousins (birds) persisted through the subsequent "Age of Mammals". Earlier still, an "Age of Invertebrates" characterizes the Paleozoic Era (ca. 538–251 Ma) though marine invertebrates persist to the present and had a somewhat earlier first appearance. Similarly, the record of bioconstructed rocks at a macro-scale could be termed the "Age of Stromatolites", with macroscopic planar, domal, and columnar build-ups formed by microbial microorganisms appearing at ca. 3.45 Ga (where Ga is "billion years ago"), which contributed to the photosynthetic production and accumulation of oxygen [1,2]. There is no convenient upper bracket to this fundamentally important "Age of Stromatolites" as stromatolites persisted through the origin and diversification of the Cambrian Evolutionary Radiation at ca. 538 Ma [3] and, as reviewed herein, occur in modern aqueous environments.

Humans perceive the living world to be dominated by relatively large (macroscopic) organisms referable to three biological kingdoms: multicellular higher plants and animals, with a passing awareness of fungi. These groups are eukaryotes with complex cell structures characterized by a nucleus and other organelles enclosed by intracellular membranes and capable of asexual cell division by mitosis and production of sexual cells (gametes) by meiosis. Organisms classified in these kingdoms thrive under aerobic conditions and multiply by sexual reproduction but also by asexual reproduction in certain cases. Biologists recognize five kingdoms, which include unicellular protoctists (commonly termed "protists") as a fourth kingdom that share the same cellular and reproductive characteristics as animals and plants [4].

Unseen by the human eye, the fifth kingdom is yet more diverse and divided between two types of bacteria: the eubacteria (commonly termed "bacteria") and the archaea. They are exclusively unicellular with a prokaryotic organization lacking a separately enclosed nucleus and with cell reproduction reliant on binary fission. During the first four billion years of Earth's development, the two groups reigned supreme with the archaea living under anaerobic conditions before sufficient oxygen became available by the activity of photosynthetic eubacteria [5]. Both kinds of prokaryotes continued to exist with the other four kingdoms that exploded in diversity at the beginning of the Phanerozoic (or "visible life") Eon at 538 Ma, coincident with the start of the Cambrian Period. Exactly when the earliest prokaryotes appeared during the Archean Eon remains under debate, but evidence from the 3.4 Ga Strelley Pool Chert in Western Australia supports multiple physical settings in which the archaea proliferated in zones interpreted as peritidal, rocky-shore, and reef-like environments [6]. The term "stromatolite" is applied to these early forms in a general way to emphasize a shared organization based on laminate structures.

Meaning "layered rock", the word "stromatolith", anglicized to "stromatolite", was introduced to the scientific literature in 1908 [7] by the German geologist Ernst L. Kalkowsky (1851–1938) to define laminar structures previously described in the literature as having possible non-biologic and biologic origins, including animal relationships. Shortly after, the role of microbes in sediment trapping, binding, and perhaps cementation in producing late Proterozoic stromatolites was proposed [8,9]. Confirmation of fossilized microbes associated with stromatolites was first recognized in 1965 on the basis of evidence from the two-billion-year-old Gunflint Chert of Ontario, Canada [10]. Subsequently, additional evidence of fossil microbes from stromatolites preserved in Precambrian rocks in Australia and Siberia became widely publicized [11].

Fossilized microbes from the Cambrian of northwestern China's Tarim Basin as well as He'nan Province are also well documented [12,13]. Fossil and present-day microbialites are comparable in having filaments that encase a string of individual bacterial cells within a larger network of layered films or mats. Cell damage by ultraviolet light in shallow water is limited in living stromatolites by filaments coated with mucus. Strands with a thickness of only 10 μm elongate as neighboring cells encased within multiply through binary fission. The exposed surface of mat-forming and domed structures is thus able to fix and bind fine sediment from the water column until a new layer of microbial growth is added above. Oxygenated water in this environment provides for aerobic respiration of bacterial cells, but archaea colonies may thrive below the surface where anaerobic conditions prevail within a given stromatolite structure. Thus, modern stromatolites and their functional biology can be linked to analogs by a shared design based on fossilized microbes that persisted as procaryotes from the remote geologic past.

The goals of this review are two-fold: (1) to summarize the history of ongoing discoveries on the life properties and environments of modern stromatolites in varied settings around the world and (2) to review the history of discoveries made by paleontologists that outline the success and persistence of stromatolites through geologic time. A particular emphasis is given to the recognition and conservation of the site at Lester Park in eastern New York, which is relevant to ancient stromatolites in terms of what may be regarded as the "type locality" for stromatolites. Additional background material is provided to illustrate how the descriptive terminology has advanced by a growing knowledge of the complexities in stromatolite morphology. Examples of stromatolites as recognized in Cambrian strata are reviewed to show the current status of studies, particularly in North America.

2. Background Geography and Geological Oceanography

On the basis of the original definition [4], stromatolites and their past and present kinship are almost exclusively aqueous in habitat, although rare exceptions are recognized with the production of layered (i.e., stromatolitic) structures by lichens in the formation of caliche in subaerial (albeit wet) environments [14]. Examples from the geologic record include common occurrences of stromatolites in shallow marine (i.e., well-illuminated) conditions during intervals characterized by globally high sea levels and extensive shallow marine shelves. The type reference locality for fossil stromatolites is at Lester Park in northeastern New York State on a Late Cambrian (Furongian Epoch, ca. 490 Ma) tropical marine shelf marginal to ancient (i.e., 1.0 + Ga, Mesoproterozoic) basement rock of the Adirondack massif on the Laurentia ("ancestral North America") paleocontinent (Figure 1, dot 1). The ancestral North American continent was in a different position during that time and the Adirondacks and its surrounding marine shelf were located at ca. 35° S in the southern hemisphere in a fully tropical setting [15].

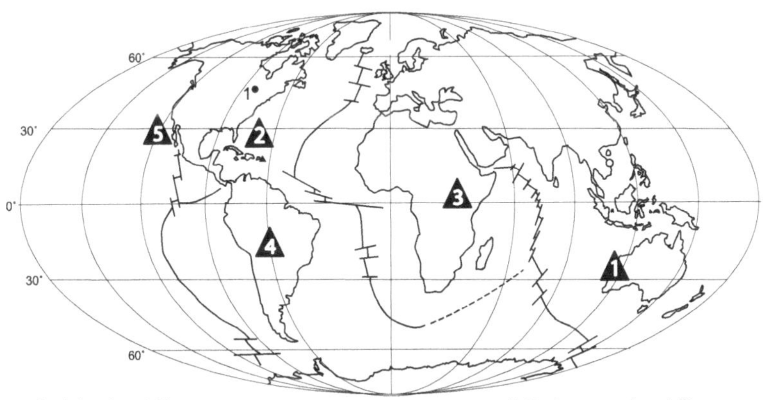

Figure 1. Global representation of today's continents and oceans on a Mollweide projection showing major ocean spreading zones. Numerals on black triangles denote the locations of oligotrophic stromatolites living today in (1) Shark Bay, Western Australia; (2) Bahamas; (3) Lake Tanganyika, Africa; (4) saline lakes in northwestern Argentina; and (5) saline ponds in Mexico's Baja California. The numbered black dot marks the location of fossil stromatolites from present-day Upper New York State (1).

In the context of paleoceanography and paleoclimatology, stromatolites and kin are disproportionately reported from carbonate platform successions, and thus are primarily tropical to subtropical in ancient and modern occurrences. However, siliciclastic stromatolites are known in mixed carbonate and siliciclastic facies, and this suggests that early cementation by carbonate mineral (calcite and aragonite) cementation in mats was prominent in lending rigidity to the form of stromatolites and kin [16]. Locally, rhodochrosite (manganese carbonate) and typical carbonate stromatolites occur in high temperate latitude Middle Cambrian deposits in Maritime Canada and North Wales on the Avalonia paleocontinent. These Avalonian stromatolites mantle depositional early Middle Cambrian depositional sequence unconformities where their development as the first deposits during epeirogenic onlaps may reflect a sort of sedimentary condensation with restriction of siliciclastic sediments to adjacent on-shore facies [17,18]. A middle latitude occurrence at 43° N is shown by calcareous thrombolites (discussed in Section 3.2) and large (baseball-size) spherical microbialites at Green Lake State Park near Syracuse, New York [19]. Thrombolites were defined for the first time by Aitken in 1967 [20] as distinctly non-laminated bodies characterized by a clotted fabric.

Living stromatolites and related microbial build-ups from Hamlin Pool in Shark Bay occur in an intertidal to shallow subtidal setting in Western Australia on the Indian Ocean (Figure 1, triangle 1). Somewhat deeper habitats for microbial build-ups also appear off the Exuma Cays in the Bahamas in the eastern Atlantic Ocean (Figure 1, triangle 2). In contrast, large domal microbial build-ups are found in lakes in the East African Rift System, including Lake Tanganyika (Figure 1, triangle 3), as well as in interior western Argentina in South America (Figure 1, triangle 4). Finally, mat-forming stromatolites are restricted to high-salinity coastal and island lagoons along Mexico's Gulf of California (Figure 1, triangle 5). All represent contemporary settings that entail elevated salinity with the exception of the Exuma Cays in the Bahamas. In effect, the geographical spread of known living stromatolites is reliant on settings where competing eukaryotic life is limited or excluded.

3. Definitions and Review Methods

3.1. *Growth of Stromatolites and Kin*

Studies on the morphogenesis of living and fossil stromatolites that followed after the first introduction of the term "stromatolite" [4] typically focus on intrinsic or internal biological factors that affect the growth of an organo-sedimentary structure as opposed to competing extrinsic or environmental factors [9]. Internal factors entail the production of extra-cellular polymeric secretions (EPSs) that make mucus coatings on the filaments housing microbial cells but also the biomineralization that cements laminae.

3.2. *Types of Microbialites*

Additional terms that differentiate several types of microbialites from densely laminated stromatolites are important in the discussion of these ancient and modern build-ups and are reviewed in [2,17]. Although the terminology for microbialite build-ups is sometimes vague and contradictory [2], the consistent confusion of stromatolites with "thrombolites" should be corrected [16]. As first described in 1967 by Aitken [20], thrombolites differ significantly from stromatolites due to reticulopodial protists that disrupt the lamination of these marine and freshwater microbialites and create voids [21]. What this means is that stromatolites are the earliest appearing microbialite build-ups, while thrombolites appear only later in the Neoproterozoic, seemingly with the appearance of eumetazoans, as the earliest indication of the Cambrian Evolutionary Radiation. Technically, it should not be said that a thrombolite is a specific sub-category of a generalized stromatolite. Along with stromatolites and thrombolites, three additional types of late Proterozoic–modern microbialite build-ups are part of the lexicon, as shown in [2,16]. Dendrolites feature columnar build-ups that are shrub-like but lack laminae. Leiolites, commonly called mud mounds, are internally structureless, very fine-grained, and commonly dome-like, but include irregularly formed shapes sometimes at a different scale. Finally, MISS (microbially induced sedimentary structures) are typically siliciclastic, laminated structures that can often be shown to be cohesive and relatively "plastic" during formation.

3.3. Macroscopic Form of Stromatolites and Kin

Extrinsic factors affecting stromatolite morphology include the nature of suspended sediment in the water column, water salinity, relative availability of nutrients, sunlight intensity, and wave or current activity that may abrade build-ups or focus growth and are responsible for the shape of the build-up. The combination of these factors is responsive to simulations based on a stochastic model to illustrate different structural shapes that range from flat mats to simple domes and to branching forms with variable morphology described as stubby, tuberous, divergent branches, or ragged columns [22].

Yet another consideration relates to the more formal naming of stromatolites based on the Linnean binomial system of nomenclature. Now outmoded and essentially of historical interest in the opinion of many workers, this has meant that genera and species names have been proposed in the literature based primarily on the morphological shape of a stromatolite build-up and other structural details (e.g., presence/absence of columnar build-ups and their shapes). A Linnean nomenclature has primarily been the practice in Precambrian build-ups where the detailed morphology is thought to be useful in biostratigraphic correlations. Modern microbialite build-ups occur worldwide (Figure 1), and can be compared with fossil examples based on the authors' experience.

4. Results

4.1. Review of Contemporary Regions with Stromatolites and Kin

In this section, five regions widely spread around the globe with living stromatolites (Figure 1) are described as a sample of key localities based on published materials and from personal experience in two of those locations.

4.1.1. Hamelin Pool, Shark Bay, Western Australia

The Western Australia coast along the Indian Ocean (Figure 2a) includes the Shark Bay World Heritage conservation area that was designated in 1991 by the United Nations Educational, Scientific and Cultural Organization (UNESCO). Located at the inner southeast corner of Shark Bay within the World Heritage zone, (Figure 2b), an over-lapping conservation zone is maintained by the Western Australia Department of Parks and Wildlife. This zone is designated as the Hamelin Pool Marine Nature Reserve. Hamelin Pool is famous for its living stromatolites (Figure 1, triangle 1). It is protected within an adjoining land and marine district covering 127,000 hectares that includes a nature center and boardwalk, allowing visitors to explore the intertidal to shallow subtidal zone without damaging wildlife. A scientific paper from 1961 by geologist Brian W. Logan [23] described the stromatolites at Hamelin Pool, which at that time was the only place in the world where living stromatolites were recognized. The average salinity of marine water at Hamelin Pool is 65 parts per thousand, which is nearly twice the salinity of normal seawater. Few eukaryotic organisms are able to tolerate such a high salinity and it is argued that the lack of competition, as discussed below, by encrusting eukaryotes or by eukaryote grazing with disruption of the microbial mats is the principal reason why the stromatolites are able to thrive there. The Latin binomen *Cryptozoon proliferum*, meaning proliferating hidden animal (with the frequently coalesced domes suggesting the tubers of the common onion, *Alium proliferum*), was initially assigned to the Australian stromatolites, following the usage first applied to fossil stromatolites in New York State, USA, by paleontologist James Hall in 1884 [24]. In this section, the Hamelin Pool microbialites are loosely designated "stromatolites", though their structure has long been known to be transitional into thrombolites [25], with intertidal forms in nearby Lake Clinton dominantly thrombolites and stromatolites restricted to the upper intertidal zone [26].

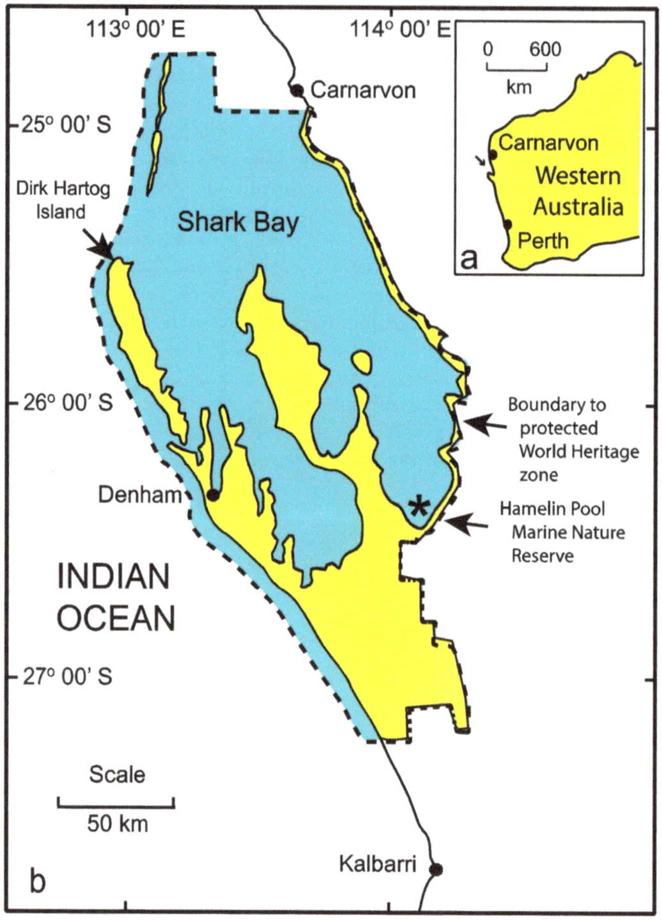

Figure 2. Maps showing the coast of Western Australia on the Indian Ocean: (**a**) Western Australia with a small arrow pointing to Shark Bay; and (**b**) enlargement showing the location of the Hamelin Pool Marine Nature Reserve (asterisk) within the UNESCO World Heritage zone protecting the greater Shark Bay.

Variations in stromatolite morphology are readily observable at Hamelin Pool [27]. The dominant forms are represented by thin club-shaped or stockier stool-shaped stromatolites that rise from intertidal to very shallow subtidal waters to a height of 50 cm or more. The visitor to the Hamelin Pool Marine Nature Reserve will observe that stromatolites exposed along the inner shore during low tide are stony hard. These are not living stromatolites but the semi-fossilized remains of stromatolites that lived during the recent past before the coastal zone underwent exposure due to epeirogenic rise 1000 years ago [28]. Exposure to subaerial conditions at the shore over time caused the surface layers with delicate microbial films to dry out. A distinctly local variation in morphology is discernable among the stony stromatolites close to shore. These are markedly loaf-shaped or elongated stromatolites oriented with the long axis perpendicular to the shore. Structures that are individually 12 to 15 cm high, 9 to 15 cm wide, and 40 to 50 cm long are fully exposed at low tide (Figure 3a). They are barely covered by water during the returning tide (Figure 3b). These stromatolites are oriented in a uniform direction with roughly 10 to 20 cm of open space in between each "bread loaf". The logical explanation for this shape is due to the influence of tidal currents washing back and forth along the shore. The shapes are well preserved in three dimensions, with all surfaces being stony hard. The upper surfaces are smooth to the touch and exhibit patches of iron-brown discoloration, which is likely due to oxidation. Clearly, there was a failure of extra-cellular polymeric secretions (EPSs) necessary to keep exposed surfaces sticky with mucus.

Figure 3. Examples of stromatolites shaped like large bread loaves oriented perpendicular to the shore at the Hamelin Pool Marine Nature Reserve, Shark Bay: (**a**) stromatolites exposed during low tide (pocket knife 9 cm long for scale); (**b**) stromatolites barely awash at high tide with narrow, open galleries from 10 to 20 cm wide.

Living stromatolites occur offshore, where club-shaped and dome-shaped structures are prevalent in the subtidal. Samples on display from this zone in the nature center are cut vertically to exhibit interior layering typical of the club-shaped or stubby morphotype [27] and have an internal fabric indicative of thrombolites.

During a 1968 visit by the junior author to Carbla Point on the mainland a short distance north of Hamelin Pool, the opportunity arose to swim into subtidal waters where stromatolites the size of large footstools rise from the seafloor (Figure 4). Structures roughly 50 cm high with an upper-surface diameter of 50 cm to 100 cm are closely spaced with narrow open passages in between. These were found to be living stromatolites. A forefinger could be poked through the surface to a depth of 5 cm to feel mucus characteristic of extra-cellular polymeric secretions (EPSs). Below that depth, the structure was dense and hard. Presumably, the inner core of these structures is stony and dead. A more nuanced overview of eight different morphotypes around the margins of the Hamelin Pool Marine Nature Reserve provides further insight into the spatial relationship of different stromatolite morphotypes through the mapping of bathymetrical zones. These zones separate stromatolites with an elongated loaf shape from those with a club shape or footstool shape, and those characterized by a massively tabular structure from the merger of multiple columnar heads [27].

4.1.2. Exuma Cays in the Bahamas of the Western Atlantic Ocean

Following the growing interest in living stromatolites in Western Australia's Shark Bay, the next major discovery was in the Bahamas in the western Atlantic Ocean (Figure 1, triangle 2). Columnar stromatolites up to 2 m in height were found to be growing at depths as much as 8 m in current-swept channels off the Exuma Cays on the eastern bank of the Bahamas [28,29]. The stromatolites occur in rows. The laminae in the columns are produced by microbes that trap carbonate sand, including ooids and pelletal material stirred up by waves and currents into the water column. The adjacent Exuma Sound marks a precipitous drop-off into deeper water open to the Atlantic Ocean. Seawater carried by the tides onto the bank is within the range of normal salinity (37 to 38 parts per thousand). However, water salinity increases during ebb tide to reach as much as 40 parts per thousand [29].

Tidal flow rates between 60 and 100 cm s^{-1} were concluded to be the primary reason why the stromatolites remain free from disturbance by potential grazers. Thereafter, expanded research on stromatolite distribution in the Bahamas revealed that localities within the Exuma Cays are far more extensive, amounting to more than a dozen sites that exhibit tabular, ridge-forming, and columnar examples in waters mostly between 2 and 3 m in depth but as much as 10 m deep [30]. The upshot of this research was the pronouncement that Bahamian stromatolites are "uncommonly common".

Figure 4. A field of living stromatolites in permanently subtidal seawater offshore Carbla Point near Hamelin Pool in Shark Bay. Junior author for scale.

4.1.3. Lake Tanganyika in the East African Rift System

Living and fossil stromatolites are known from a freshwater setting in Lake Tanganyika (Figure 1, triangle 3), which occupies a branch of the great East African Rift System [31]. Occupying a series of structural grabens, the lake is the largest of the African rift lakes. It has a maximum depth of 1470 m and is considered the second-deepest lake in the world after Russia's Lake Baikal. As a result of investigations by SCUBA divers in 1997, at least 22 stromatolite-bearing localities are known mostly along the eastern side of the 800 km-long lake [31]. The depth of water in which the freshwater stromatolites were found to thrive ranges between 4 and 46 m, with 46 m being only the limit that the divers were able to reach.

The stromatolites occur in isolated patches, taking the form of biohermal accumulations with a diameter of tens of meters with the intergrowth of domes and columns. Variations in shape include stratiform encrustations that cover boulders to vertical walls, dome-shaped structures, and branching columns. The saturation of lake water by carbonates including calcium and magnesium is elevated, more so during the dry season. Microbes are described as responsible for producing fine-scale laminations that fit with the traditional description of stromatolites as opposed to thrombolites. The principal fauna associated with the stromatolites is dominated by the freshwater gastropod *Paramelania imperialis*, which is unusually large and thickly shelled; these characteristics are perhaps coincidental with the lake's uniquely calcium-rich water chemistry [32]. The gastropods are equipped with mouth parts that function for rasping algae, but the stromatolites apparently are not overgrazed. In turn, the gastropods are preyed on by crabs.

4.1.4. Saline Lakes of Northwestern Argentina

Preliminary information indicates giant stromatolites in hypersaline pools surrounded by salt flats on the arid plateau in the Puna de Atacama district of northwestern Argentina in the Salta Province [33,34]. Geologist Brian Hynek from the University of Colorado at Boulder and microbiologist Maria Farías from a consulting firm in Argentina first identified a region with a system of salt lagoons by satellite imagery in April 2022 and subsequently visited the remote region bordered by northern Chile (Figure 1, triangle 4). Not easily accessed, a dozen salt lagoons are spread over an area of 10 hectares located 1880 km northwest of Buenos Aires. The coalescing stromatolites in the Argentinian salt lagoons are dome-shaped, up to 4.5 m in diameter, and a meter in height. Extremely dry and at an elevation of 3650 m above sea level, the salt flats are essentially devoid of plant and animal life. This represents one of the most exciting recent discoveries related to living stromatolites. Details on the microbial biology of the Argentine stromatolites have yet to be published but will be of much interest.

4.1.5. Hypersaline Pools from Islands in Mexico's Gulf of California

Stromatolites living in a sheltered lagoon on the Pacific shores of Mexico's Baja California peninsula were first reported in 1975 [35]. Their discovery was fortuitous, much as in the case of the sighting by geologist B.W. Logan of living stromatolites fourteen years earlier in Western Australia's Hamelin Pool [23]. Geologists who landed on Isla Ángel de la Guarda in Mexico's Gulf of California made the unexpected discovery of living stromatolites in hypersaline pools on that island [36]. The junior author (of this review) was part of the 2007 expedition. Strategically located 32 km east of Bahía de los Angeles on the peninsular mainland (Figure 5a), the island was named Guardian Angel because the settlement is sheltered from major storms in the lee of the large island. With an area of 936 km^2, it is the second-largest island in the Gulf of California (Figure 5b). The island is uninhabited except for populations of large lizards (*Sauromalus hispildus*) and the endemic rattlesnake (*Crotalus angelencis*). Local fishermen generally avoid the island's east-facing windward side due to highly changeable weather conditions. The 2007 expedition went to the south end of the island with the goal of mapping a small paleoisland that rises above the coastal plain as a volcano previously spotted by aerial reconnaissance (Figure 5b).

The paleoisland on Isla Ángel de la Guarda (Figure 5b) was first investigated in 2007. In carrying out that project, however, living stromatolites were discovered on the shores of two hypersaline lagoons (Figure 5c) [36]. The small lagoon (Figure 5c) is 375 m long and covers an area of approximately 2250 m^2, whereas the big lagoon extends for a distance of 1250 m and covers an area of approximately 125,000 m^2. Thrombolites having the shape of small cauliflower heads (Figure 6a) occur around the shores of the small lagoon. Regular stromatolites with thinly matted layers of black, mucus-rich matter occur in a wide band around the shores of the big lagoon, where the surface mud is encrusted with lime and deeply cut by desiccation cracks (Figure 6b).

For the team on Isla Ángel de la Guarda in 2007, the objective was quickly prioritized to consider how the lagoons formed in the first place. It is theorized that each pond began as an embayment open to seawater with normal salinity in the Gulf of California. Through time, cobbles and boulders transported across the opening by strong longshore currents during episodic storms closed off the bays, turning them into isolated ponds. Once fully cut off from the open gulf, intense evaporation of seawater isolated in the ponds resulted in hypersaline conditions. Salinity was measured at 148 parts per thousand, more than four times the salinity of normal seawater [36]. Erosional features that scar the outer berms enclosing the ponds show that spill-over by storm waves adds normal seawater from time to time and thereby modulates lagoonal salinity. Subsequent exploration farther north on both sides of Isla Ángel de la Guarda confirmed that living stromatolite mats also occur in other closed lagoons [37]. Moreover, Isla Ángel de la Guarda is not the only island in the Gulf of California with closed lagoons where stromatolites thrive under conditions of high salinity that qualify the inhabitants as extremophiles. The geographical range of microbialites in the region raises unanswered questions about how they are seeded from one island to another.

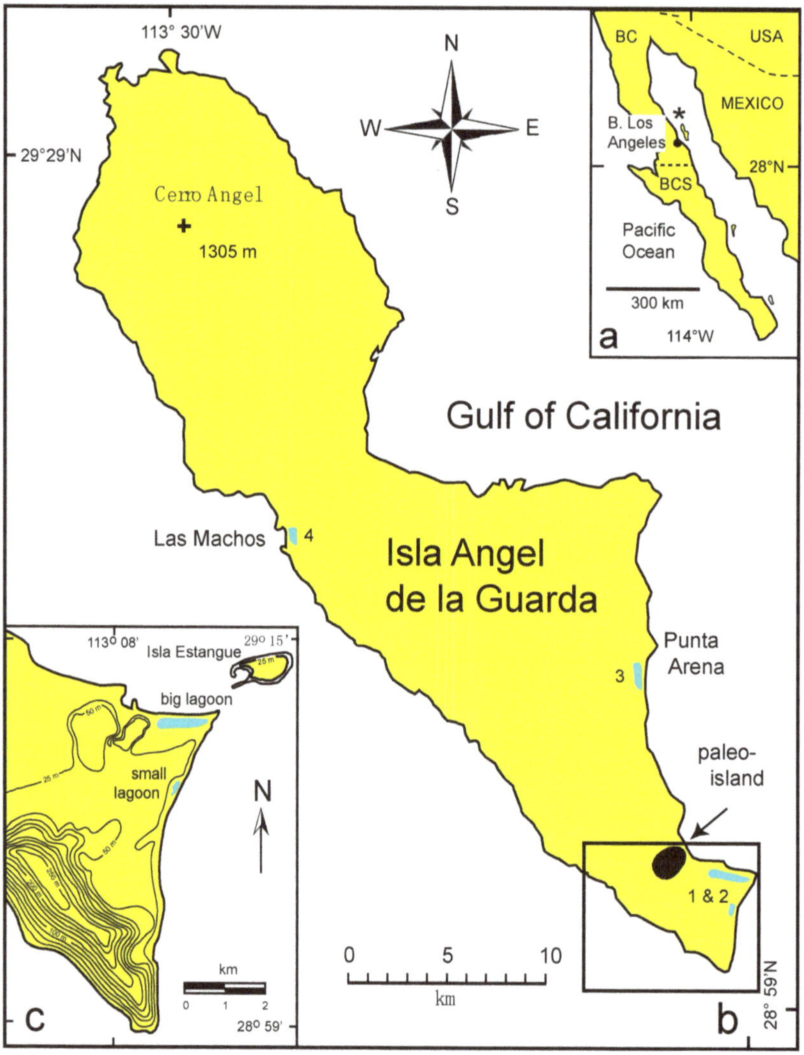

Figure 5. Maps showing Mexico's Isla Ángel de la Guarda in relation to the Baja California peninsula: (**a**) the full peninsula adjacent to the Gulf of California off the Mexican mainland with the island's location (asterisk) near the head of the gulf; (**b**) map of Isla Ángel de la Guarda marking the four localities (blue) where stromatolites occur in closed lagoons; and (**c**) topographic map enlarged from box in (**b**) showing the island's southeast end, where thrombolites and mat-forming stromatolites were discovered in 2007.

Figure 6. Photos showing living stromatolites from closed lagoons on the shores of southeast Isla Ángel de la Guarda (see Figure 5c for location): (**a**) thrombolite assemblage of branched forms the size of small cauliflower heads from the small lagoon (coin 2.4 cm in diameter for scale); (**b**) matted microbialites dissected by desiccation polygons along the shore of the big lagoon (compass case 10 cm across for scale).

4.2. Stromatolites from the Archean–Phanerozoic Eons

Comprehensive reviews of the temporal distribution of stromatolites and thrombolites [2,16] show their persistence in aqueous environments from their earliest known occurrence in the earlier part of the Archean (4.0–2.5 Ga) through to the present. Whether or not the oldest known (ca. 3.45 Ga) coniform build-ups with continuous laminae and calcite spar fabrics from the Pilbara craton, Western Australia, are stromatolites [38] was resolved with the recognition of microbial fossils in the build-ups and analysis that showed the presence of organic molecules [39]. Stromatolites are relatively rare in the Archean but abruptly appear as giant, elongate, decameter domes on the terminal Archean (2.55 Ga) carbonate platform in South Africa [40] and show mm- to cm-sized areas of calcareous cement [41], which may suggest the fabric of thrombolites.

Stromatolites show limited morphologic diversity in the earlier part of the Proterozoic Eon (2.5 Ga to 438 Ma). A purported abrupt increase in morphologic diversity characterizes the onset of the Mesoproterozoic (1.6–1.0 Ga) with diversity then falling progressively through the later Mesoproterozoic and reaching its lowest relative values at the end of the Proterozoic [16,42]. A significant development includes the appearance of calcified cyanobacteria at ca. 1.2 Ga [43], which led to the development of thrombolites by at least ca. 775 Ma [44]. A fall in pCO_2, and consequently higher marine pH and lower carbonate solubility, has been proposed as the environmental control that allowed calcification [45].

This 1.2 Ga change in oceanic geochemistry is, however, at odds with an interpretation that a purported fall in later Meso- and Neoproterozoic stromatolite morphologic diversity records the opposite change in oceanic geochemistry, i.e., a fall in carbonate saturation [16,45]. However, there does not seem to be a consistently defined trend in stromatolite diversity from the later Proterozoic into the early Phanerozoic in Western (i.e., English language) and Russian reports. As opposed to Western workers who emphasized verbal descriptions and comparisons of stromatolite morphology, the tradition of stromatolite workers in the USSR and Russia has been the use of a binomial nomenclature to distinguish stromatolites in the Riphean (ca. 1650–600 Ma, roughly equivalent to the Mesoproterozoic and most of the Neoproterozoic) and Vendian (ca. 600 Ma to base of Cambrian, roughly equivalent to the Ediacaran), e.g., [46]. Based on work on carbonate platform successions largely in Siberia, long-ranging "groups" (i.e., form genera largely based on the morphology of stromatolite columns) sequentially appeared and persisted through the Riphean, with a fall in diversity only in the Vendian and Cambrian, which show the origin of new "groups" [47].

An interesting development in some Proterozoic stromatolites is an alternation of calcite spar and micritic layers [16,48] that form dark–light layers interpreted as alternations of microbial mat and abiogenic crust and are also seen in the Late Cambrian *Cryptozoon proliferum* (discussed below). Spar-filled tubes that form a vermiform microstructure in the dark layers of *C. proliferum* were reported as first appearing in the Ediacaran [49] but are known in far older Proterozoic stromatolites [2].

The Phanerozoic featured a transient, abrupt increase in stromatolite morphologic diversity and high abundance of thrombolites in the Late Cambrian and Early Ordovician, while the younger Phanerozoic features a generalized stromatolite decline [16,43,50]. We are unaware of any explanation(s) of the efflorescence of Late Cambrian–Early Ordovician stromatolites. Indeed, no single factor may be involved. Their abundance may not reflect perturbations in pCO_2 (discussed above) but may be, in part, essentially an artifact of very high sea levels in this time interval [51]. Thus, very high eustatic levels would mean broad submergence of the large paleocontinents in this time interval. This would have led to extensive shallow shelf facies and widespread microbialite build-ups on large paleocontinents that were largely tropical [52] and in which carbonate supersaturation in shallow tropical seas would have promoted microbial and abiotic cementation of stromatolites and thrombolites. This explanation as a diversity sample area artifact may explain high Cambrian and Ordovician stromatolite and thrombolite diversity and abundance. However, the gradual diminishment of these litho/biofacies in the later Phanerozoic with resurgences in the Silurian, late Devonian–early Mississippian, early Triassic, and locally (in China) in the Jurassic and Miocene (Spain) is problematic [53].

A common explanation of this later Phanerozoic decrease and the common limitation of modern stromatolites to highly saline or freshwater environments (see modern localities, above) features the origin and diversification of animal metazoans, particularly mollusks. These animals would graze and destroy microbial films and, thus, preclude the onset and growth of microbialites; however, this interpretation is countered [54] by the presence of microbialite build-ups, as those noted above, in areas of normal marine salinity in tidal channels in the Bahamas. Examples such as the classic Late Cambrian stromatolite *Cryptozoon proliferum* and higher giant thrombolites at Lester Park, eastern New York (Figure 7), are also in unrestricted marine facies [15]. However, this association does not contribute to understanding the later Paleozoic decrease in stromatolite and thrombolite abundance because possible grazing metazoans (gastropods and polyplacophorans) are known from very rare fossil remains both locally and elsewhere. More importantly, encrusting metazoans are abundant after the Late Cambrian [55,56].

Thus, an alternative explanation of the decline in stromatolites, as well as thrombolites, after the Early Ordovician is metazoan competition and apparent changes in carbonate saturation in marine waters [55]. The origin and diversification of colonial reef-building corals, bryozoans, and other groups [57] and increased abundance of calcareous algae led to microbialite build-ups being out-competed for habitat space in shallow marine areas through the later Phanerozoic. After the Early Ordovician, a generalized long-term fall in pCO_2 took place [57], which should have promoted calcareous microbialite growth with higher pH and increasing marine carbonate saturation. However, for the resurgences listed above, higher stromatolite and thrombolite abundances in marine facies with increased carbonate saturation are not recorded. This further emphasizes that long-term eumetazoan competition for habitat area limits their abundance and occurrences generally to inhospitable (at least to grazing eumetazoans) habitats of highly saline water and freshwater. The alternative explanation of resurgences in microbialite abundance resurgences in the Phanerozoic may simply be the same as that suggested above for that in the Late Cambrian–Early Ordovician. The middle Paleozoic and Mesozoic resurgences appear to track eustatic levels and reflect increased habitat areas during times of highest sea levels but may also entail recovery from post-biological crises.

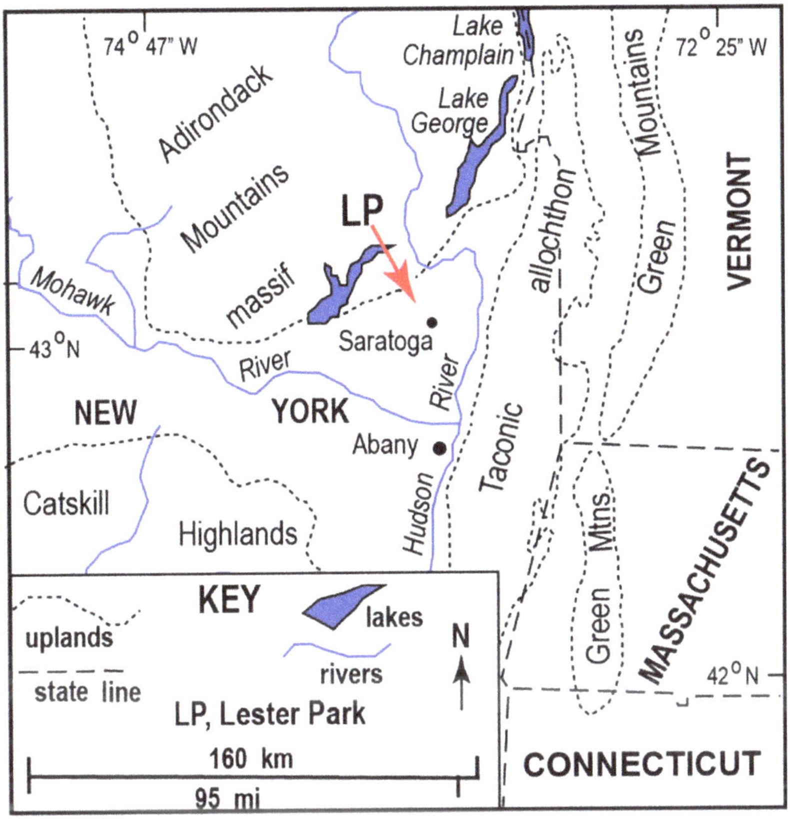

Figure 7. Location of Lester Park in the Upper Cambrian–Upper Ordovician lowlands southeast of the Mesoproterozoic Adirondack Mountains massif, northeast of the upper Middle Devonian Catskill Highlands, and west from the terminal Ediacaran–lower Upper Ordovician Taconic Allochthon.

4.2.1. Cambrian Stromatolites from Lester Park in New York State

The earliest named stromatolite received the binomial designation of *Cryptozoon proliferum* Hall, 1881 [24]. Almost all reports incorrectly report an 1883 naming of the species [2,58–66], as do such on-line international taxonomic databases as AlgaeBase and others [67,68]. The proposal for the genus and species is in the 1883 Thirty-sixth Annual Report of the State Museum of Natural History and was written in 1882. However, the report was printed and released with an 1884 cover. Hall's report [24] built on Steele's early description [69] of hundreds of domal structures that he regarded as inorganic concretions on a bedding surface in the Town of Greenfield just NE of Saratoga Springs, eastern New York. This extensive surface (ca. 500 m^2) immediately east of Lester Park Road (43°05′32.14″ N, 73°50′5342″) was part of the old Hoyt family farm and is now part of Lester Park Scientific Reservation (Figure 7). Lester Park was deeded to the New York State Museum in 1915 in honor of Willard Lester, Esq. [15,16,70]. The bedding surface extends ca. 1.0 km south to the Petrified Sea Gardens, a now-closed tourist attraction owned by the Pompa Bros. stone company, with this long exposure traditionally termed the "*Cryptozoon* ledge" [71].

The Lester Park surface (Figure 8a) is a ca. 490 Ma exposure of the Late Cambrian Laurentian (ancestral North America) shallow tropical platform [16,72]. The gently west-dipping (ca. 5°) surface is low in the Hoyt Member of the Little Falls Formation (here ca. 11.35 m exposed) and underlies the thrombolite-bearing strata in the Hoyt quarry ca. 75 m west (Figure 8b). Minor hydrothermal calcite and dolomite veins (Figure 9) that cut the stromatolite domes and surrounding limestone matrix were likely produced during the Taconic orogeny (ca. 460 Ma). The recovery of thermally blackened euconodont elements (color alteration index of 4.0) in the Lester Park area is consistent with burial temperatures in excess of 200 °C [73] and a ca. 7.5 km burial depth [74].

Figure 8. Coalesced topotype specimens of the first-named stromatolite *Cryptozoon proliferum* Hall, 1884, at Lester Park, Saratoga County, eastern New York: (**a**) View of the top of a shoaling cycle abraded and truncated to show growth laminae by the movement of coarse quartz sand that weathers brownish; narrower, lower parts of domes (upper part of the figure) were exposed by glacial (Pleistocene) plucking of the upper part of domes. (**b**) Detail of clotted thrombolite structure surrounded by bedded limestone from Hoyt quarry, ca. 5 m above the *C. proliferum* surface at Lester Park. Hammer (30 cm) for scale in both pictures.

Figure 9. *Cryptozoon proliferum* from older Galway Formation (lower Upper Cambrian *Elvina* Zone) NE of Lester Park shows two generations of fracturing: (1) brownish carbonate mud-filled cracks in the lower part of the specimen that separate and also run transverse to growth laminae are syndepositional fractures (ca. 490 Ma) and reflect continued extension of the rifted margin of NE Laurentia; (2) thin white calcite veins parallel to growth laminae produced during the Taconic orogeny (ca. 460 Ma) [74]. Hypotype NYSM 19512 from the middle of the Galway Formation railroad cut above U.S. Route 9 just N of the intersection of U.S. Route 9 with Daniels Road [74], with a USD 25 cent coin (23 mm) for scale.

The northern end of the bedding surface at Lester Park shows broad, coalesced, light grey weathering (i.e., limestone) *Cryptozoon* build-ups that have been truncated and show fine growth laminae (Figure 8a, lower part). The bedding surface underwent glacial plucking in the Pleistocene, which removed the tops of the build-ups and exposed the lower parts of the build-ups in the southern part of the Lester Park surface. The bases are embedded in buff weathering (i.e., dolomitic) fossiliferous ooid–fossil hash limestone and show a roughly E–W arrangement (Figure 8a, upper part of photograph). The upper parts of the domes were not truncated by glacial activity, as repeatedly reported [25,58–63], but were beveled during a shoaling cycle and are overlain by a brownish-weathering, coarse-grained sandstone (0–10 cm thick) [15,72].

An alternative explanation for the flat tops of the *Cryptozoon proliferum* build-ups at Lester Park is that they are analogous to modern, flat-topped thrombolites in hypersaline facies in Lake Clifton in Western Australia. These thrombolites can grow only as high as high tide levels and then expand laterally [26]. However, the specimen figured by Burne and Moore [26] as a "microatoll" shows erosional truncation of its top and lateral margins and is buried in a calcareous sandstone with clasts of *C. proliferum* (Figure 10).

Figure 10. Small *Crytozoon proliferum* dome at the north end of the Lester Park surface. This specimen was termed a "microatoll" [26], but it has an erosion-truncated top and lateral margins and is surrounded by a coarse-grained sandstone with light grey-colored *C. proliferum* clasts (yellow arrows).

Hall initially accepted Steele's interpretation of the build-ups as non-biologic [75], but later, without explanation, followed Mather [76] who regarded them as "organic" and later referred to them as "sea plants" [77,78]. Hall's [24,79] next taxonomic assignment shifted to some sort of marine animal (*Cryptozoon*, i.e., "hidden animal"). He regarded the form as similar to the much younger *Stromatopora* as understood then (now regarded as either a sponge or a hydrozoan). However, *Cryptozoon* differed, based on Hall's quite early use of thin sections, by lacking tabulae and by having build-up shapes that did not have a broad base as in *Stromatopora* but, rather, broadened upward.

Steele [69] reported that the Lester Park concretions had "compact" dark blue or black laminae and a lighter-colored "stalactical" or "granular" layer. He did not note any of the fenestrae associated with thrombolites. Similarly, Hall [78] also distinguished two types of lamellae, light and dark, noting that "the substance between the concentric lines [i.e., the dense laminae] is traversed by numerous, minute, irregular canaliculi which branch and anastomose without regularity". In terms of more recent petrographic work on fossil stromatolites, these correspond, respectively, to spongiostromate and porostromate textures [80]. The essential difference is that dense laminations are typical of spongiostromate development and Goldring [57] specifically states that calcite-filled tubes in the porostromate texture are "vesicular" in assigning *Cryptozoon proliferum* to the red algae.

4.2.2. Distribution of Cryptozoon

Hall's publications set a standard for late nineteenth-century paleontology, and *Cryptozoon* was quickly recognized across North America as a readily recognizable build-up characterized by domal shapes that could coalesce and, with thin section analysis, distinctive alternations of light laminae and dark laminae having a vesicular microfabric of carbonate-filled, straight to curving tubes. Subsequent reports suggest that build-ups attributed to *Cryptozoon proliferum* or which are likely synonyms of the type material of the "*Cryptozoon* ledge" are confidently limited to Upper Cambrian–Lower Ordovician tropical successions in Laurentia. The uniform morphology (macro and micro) and apparent limited stratigraphic range (discussed below) suggest the form is more than a form taxon and is a biological taxon. In the following review of *Cryptozoon* occurrences, an incorrect 1883 date for *C. proliferum* is regularly cited.

Midwestern United States occurrences include two form species of *Cryptozoon*-like build-ups that were named from the Shakopee Formation in Minnesota (Lower Ordovician, middle Tremadocian, Stairsian Stage [81]). Winchell [82] distinguished the broadly domal *C. minnesotense* from the type material of the Upper Cambrian *C. proliferum* (he did not include a publication year for *Cryptozoon*) based on its growth up and outward from a flat base. The flat base of a series of *C. minnesotense* build-ups is figured in outcrop drawings by Chaney [83], who proposed a nomen nudum (*C. giganteum*) for specimens larger than 40 cm in diameter. Winchell [82] also named *C. minnesotense libertis* for fragments of an elongate, cylindrical build-up that are up to ca. 9 cm in diameter. These are likely fragments of closely packed columns that Stauffer [84] named *C. rosemontensis*. Other than fine, mm-scale growth lamination illustrated from broken or slabbed specimens [82,84], little is known about the microhistology of the laminae, and these form species might be questionably attributed to *Cryptozoon*.

Eastern North American occurrences include the Upper Cambrian report of the genus in western Vermont [85] from their Beekmantown Division B, which is the Little Falls Formation [72]. The genus was early reported in the Upper Cambrian of eastern Pennsylvania [9,86]. The quality of the topotype Upper Cambrian Lester Park material led to early redescriptions by Walcott [87] and as part of Rothpletz's review of Proterozoic and Cambrian Cryptozoons [88].

Dawson [89] redescribed the topotype *Cryptozoon proliferum* and extended the range of the genus into the Proterozoic and Late Ordovician. However, his figure of *Cryptozoon boreale* from the Upper Ordovician of Quebec is a branching *Cyathophyllum*-like coral. He [89] also named the then-oldest *Cryptozoon* as *C. occidentale*, whose spartan description (i.e., both build-up shape or size are undescribed) and locality information seemingly brought *Cryptozoon* into "strata which may be regarded as Pre-cambrian" in the Grand Canyon, Arizona, succession. However, his description and supposed illustration of "ragged cells, giving off on all sides numerous small tortuous and branching canals" are not distinguishable in Figure 3, which shows euhedral calcite (?) crystals in a black field. In short, *C. occidentale* should be regarded simply as a rounded stromatolite. The youngest apparent *C. proliferum* in Laurentia is *C. saxiroseum*, which Seely [90] described from domal build-ups with spongiostromate and porostromate laminae from the middle Lower Ordovician (Floian) "Division D" (i.e., Fort Cassin Formation) at Shoreham, Vermont.

An exceptionally early *Cryptozoon* was reported in the middle Mesoproterozoic Belt Supergroup of Montana by Walcott and others [8,91,92]. These reports illustrate and differentiate *Cryptozoon* from taxa but do not provide useful thin sections of the laminae, and thus do not establish whether *Cryptozoon* is actually present in the Mesoproterozoic. Rezak [92] recognized Dawson's [89] problematic *C. occidentale* (discussed above) in the Belt Supergroup but his description and illustration of a thin section are also inadequate in resolving lamellae microstructure. Even older reports of purported *Cryptozoon* include Rothpletz's [88] proposal of *Cryptozoon walcotti* from the Steep Rock Lake area of northwestern Ontario. This taxon from the lowest Neoarchaen (ca. 2.8 Ga) Mosher carbonate platform would seemingly be the oldest occurrence of *Cryptozoon*, [93]. Fralick and Riding [94] detailed the stromatolite-rich interval in the Hogarth Member that yielded *C. walcotti*. Although many of the Hogarth's small, coalesced columnar build-ups are similar to lower Phanerozoic *Cryptozoon*, their microfabric is essentially undocumented. Rothpletz [88], pl. 2, his Figure 6), showed a chaotic (tectonized?) fabric of alternating coarser carbonate and darker, fine-grained carbonate that lacks the filaments and threads of the porostromate layers diagnostic of *C. proliferum*. Thus, the several reports of purported lowest Neoarchaen and Mesoproterozoic *Cryptozoon* do not establish a Precambrian record of *Cryptozoon* s.s. with its characteristic spongiostromate and porostromate laminae.

The geologically youngest reports of purported *Cryptozoon* in North American successions are from the Carboniferous and Permian. Two species were proposed by J.H. Johnson [95] of *Cryptozoon* from the upper Carboniferous (*C. kansasensis*) and Permian (*C. wrefordensis*) of Kansas. They are described as small domal build-ups but their shapes are unillustrated. Neither species can be assigned to the form genus. *Cryptozoon kansasensis* seems to be poorly laminated and consists of a dense fabric of anastomosing, carbonate-filled, *Girvanella*-like threads while *C. wrefordensis* is completely composed of stacked, very lenticular laminae with finer calcite-filled threads.

We are unaware of younger reports of *Cryptozoon* except for the early reports of the genus from modern environments at Shark Bay, Western Australia, e.g., [23]. However, these build-ups are not stromatolites and should not be attributed to *Cryptozoon*. Even if coarsely laminated, they show the small to large voids of thrombolites [25,26]. The designation of "stromatolites" for the Western Australian and other modern domal build-ups is, unfortunately, incorrectly embedded in the literature, e.g., [96]. The fundamental biological difference by which reticulopodial protists are associated with the production of thrombolite porosity in marine to lacustrine habitats [21] should be recognized in the distinction of stromatolites from thrombolites [16].

5. Discussion of Modern *Cryptozoon* Research

A remaining question about *Cryptozoon* and *C. proliferum* is whether these terms are appropriate only to form taxa that reflect environmental factors or are actually biological entities as claimed for many Precambrian stromatolites. The definitive criteria of Hall's [24] diagnosis of the taxon are not only a domal build-up form but the alternation of stromatostromate and porostromate laminae as determinable by thin/polished section analysis. The absence of this information on many purported occurrences of *Cryptozoon*, which have emphasized only a laminated domal/coalescing build-up form, has created a number of binomial/Linnean taxa that are distinct from the genus. The demonstration of spongiostromate and porostromate microstructure, perhaps reflecting a consortium of sediment grain-binding and carbonate-secreting organisms, limits the demonstrated occurrence of *Cryptozoon* to lower Paleozoic reports.

At the very least, the application of a new non-traditional Linnean taxonomy to Phanerozoic build-ups [97], even if used to compare build-ups that are characterized/ distinguished by overall morphology in an individual study, would simplify literature searches and make them more efficient and thorough. There is undoubtedly any number of Phanerozoic or older *Cryptozoon*-like build-ups with the distinctive lamellae microsculpture of the type "species", but they are hidden under informal morphological descriptions used in individual reports.

The first-named "classic" stromatolite, *Cryptozoon proliferum*, is biologically enigmatic. Does the alternation of spongiostromate and porostromate laminae reflect the possibly environmentally controlled activity of a microbial consortium or the alternation of the activities of much different organisms? These laminae in topotype *C. proliferum* from Lester Park are detailed by Lee and Riding [63,64] and Neuweiler et al. [66], who reach contrary interpretations, respectively, that the porostromate texture shows that the "classic stromatolite" is formed by alternations of keratose (non-spiculate) demosponges and cyanobacteria [63,64] or that the porostromate layer is merely an accumulation of diagenetically altered micrite [66]. The problems in assigning the porostromate microtexture to keratose sponges or interpreting it as microbially mediated micrite are reviewed by Kershaw et al. [65], who conclude that current techniques are inadequate to provide a conclusive interpretation for each and every porostromate texture. In yet another interpretation by Neuweiler et al. [98], the porostromate laminae reflect the effects of deep burial metamorphism and hydrothermal activity of the Lester Park area, i.e., [75].

This ambiguity in the interpretation of porostromate textures makes the proposal of "keratolite" [63,64] for the "classic stromatolite" *Cryptozoon* and other stromatolites very problematic. "Keratolite" is doubly problematic as it concludes formation by a taxonomic group, which is at least uncertain or questioned, i.e., [65,66], and also creates a synonymous new designation (keratolite) for build-ups that have long been termed "stromatolites". Furthermore, "keratolite" leads to ambiguity as to whether stromatolites such as *Cryptozoon* are ecologically a metazoan–microbial (cyanobacterial) or simply a microbial consortium. This has consequences for understanding the early occurrence of eumetazoans (i.e., keratose sponges) in the Cambrian Evolutionary Radiation.

Indeed, porostromate-like fabrics have been proclaimed as evidence for early sponges in the early Neoproterozoic (ca. 890 Ma) by Turner [99] and are relevant to understanding the origin of metazoans. However, this interpreted early existence drives the origin of simple metazoans and sponges into an interval well before the Cambrian Evolutionary Radiation as traditionally documented in the Ediacaran–Cambrian transition. Alternatively, porostromate laminae might show the persistence of a non-spiculose sponge–microbial (cyanobacterial) consortium for ca. 400 Ma into the early Paleozoic as recorded by *Cryptozoon*. However, associations of tangled calcareous filaments may have any number of origins, such as microbial or diagenetic alterations in micrite. Indeed, the presence of filamentous fabric in the Mesoproterozoic (e.g., [100], their Figure 3e), and well before most estimated earliest origins of sponges may be only a taphonomic artifact that does not support the participation of keratose sponges in the construction of *Cryptozoon*-like build-ups in the Precambrian.

6. Conclusions

The "Age of Stromatolites" persists into the present with the occurrence of such charismatic build-ups as stromatolites and thrombolites in a variety of habitats. Examples of such modern habitats described in this report include normal marine (Exuma Cays, Bahamas) and high-salinity marine facies (Hamlin Pool, Shark Bay, Western Australia, with enclosed lagoons in the Gulf of California) more indicative of extremophiles. In addition, microbialite build-ups occur in freshwater (Lake Tanganyika; Green Lakes State Park, central New York), hypersaline lacustrine environments (saline lakes in Argentina), and waterlogged subaerial habitats where a type of caliche is formed.

In many of these modern habitats, a large domal microbial build-up has come to be casually, and incorrectly, termed a stromatolite [100]. However, ancient and modern microbial build-ups with mm- to cm-sized voids should be termed a "thrombolite". This is a significant difference, not merely a terminological distinction, as colonization by reticulopodial protists controls the development of voids in lacustrine and marine microbialites. In short, although the shape of the columnar build-ups in lateral profile in Shark Bay is that of *Cryptozoon*, i.e., [101], and many studies report "stromatolites" from Shark Bay, Western Australia, repeated studies over several decades of the diverse fabrics of the Hamlin Pool build-ups show that they are best regarded as thrombolites, commonly with fenestrae, e.g., [102]. As they lack the diagnostic spongiostromate and porostromate laminae of *C. proliferum*, there is no modern record of *Cryptozoon*.

The abundance and morphologic diversity of stromatolites and their "kin" beginning in the Archean into the present reflects both global changes in the environment and the evolution of eumetazoans. Thus, stromatolite diversity and abundance, after reaching a maximum, fall through much of the Precambrian. This change tracks an apparent fall in pCO_2 associated with a rise in pO_2 and burial of carbon, due at least in part to photosynthesis by microbialite communities. After a second rise in diversity and abundance in the

Late Cambrian–Early Ordovician, stromatolites and thrombolites again become rare, with resurgences in the Silurian, Devonian, and Carboniferous. This suppression and their abundance in some high-salinity and freshwater habitats have frequently been explained by the origin and diversification of eumetazoan grazers that would have restricted the origination and growth of microbialites. Alternatively, a combination of substrate competition with the origin and diversification of encrusting eumetazoan plants and animals, along with a generalized trend to lower pCO_2 levels in the Phanerozoic, would have the effect of limiting their occurrence and abundance. Ultimately, however, abundance resurgences in the later Phanerozoic, at times with long-established grazing and competition for habitat space, may simply reflect protracted high eustatic levels and greater areas of shallow marine habitats.

The earliest named and, thus, "classic stromatolite" is *Cryptozoon proliferum*, which has a range bracketed to the Late Cambrian–Early Ordovician based on reports that illustrate not only a build-up shape but the presence of alternating spongiostromate and porostromate growth laminae. Recent work on topotype *C. proliferum* shows that this early named fossil remains the subject of controversy, with its alternating laminae variably seen as alternations in microbial communities, simple diagenetic changes in microbial mud, the presence of keratose sponges in a microbial–sponge consortium, or the result of alteration with deep burial.

Author Contributions: The junior author (M.E.J.) was responsible for the conceptualization of the project and wrote the front sections on living stromatolites. The senior author (E.L.) was responsible for the validation of the literature and wrote the later sections on fossil stromatolites and thromobolites. All authors have read and agreed to the published version of the manuscript.

Funding: This research received no external funding.

Acknowledgments: Much of E.L.'s field and lab work was financed by U.S. National Science Foundation grants and the New York State Museum. Much of M.E.J.'s field studies in Mexico's Baja California and Western Australia were financed by the Petroleum Research Fund administered by the American Chemical Society. As a note of explanation, capitalized Lower/Early, Middle/Middle, and Upper/Late Cambrian follow the proposal of formally defined, unambiguous Cambrian subsystems/subperiods [103].

Conflicts of Interest: The authors declare no conflicts of interest.

References

1. Heard, A.W.; Bekker, A.; Kovalick, A.; Evans, H.; Ireland, T.; Dauphas, N. Oxygen production and rapid iron oxidation in stromatolites immediately preceding the Great Oxidation Event. *Earth Planet. Sci. Lett.* **2022**, *282*, 117416. [CrossRef]
2. Grey, K.; Awramik, S.M. *Handbook for the Study and Description of Microbialites*; Geological Survey of Western Australia, Bulletin No. 147; Geological Survey, Western Australia: Perth, WA, Australia, 2020; 278p.
3. Landing, E.; Geyer, G.; Brasier, M.D.; Bowring, S.A. Cambrian Evolutionary Radiation: Context, correlations, and chronostratigraphy—Overcoming deficiencies of the first appearance datum (FAD) concept. *Earth-Sci. Rev.* **2013**, *123*, 133–177. [CrossRef]
4. Margulis, L.; Schwartz, K.V. *Five Kingdoms—An Illustrated Guide to the Phyla of Life on Earth*, 2nd ed.; W.H. Freeman & Co.: New York, NY, USA, 1982; 375p.
5. Margulis, L.; Sagan, D. *Microcosmos—Four Billion Years of Microbial Evolution*; Allen & Unwin: Boston, MA, USA, 1987; 301p.
6. Allwood, A.C.; Walter, M.R.; Kamber, B.S.; Marshall, C.P.; Burch, I.W. Stromatolite reef from the Early Archaean era of Australia. *Nature* **2006**, *441*, 04764. [CrossRef] [PubMed]
7. Kalkowsky, E. Oölith und Stromatolith im norddeutchen Bundsandstein. *Z. Deutshen Geol. Ges.* **1908**, *60*, 68–125.
8. Walcott, C.D. Cambrian geology and paleontology. III. No. 2.—Pre-Cambrian Algonkian algal flora. *Smithson. Misc. Collect.* **1914**, *65*, 79–156.
9. Wieland, G.R. Further notes on Ozarkian seaweeds and oolites. *Bull. Am. Mus. Nat. Hist.* **1914**, *33*, 237–260.
10. Barghoorn, E.S.; Stanley, A.T. Microorganisms from the Gunflint Chert. *Science* **1965**, *147*, 563–577. [CrossRef]
11. Schopf, J.W. *Cradle of Life—The Discovery of Earth's Earliest Fossils*; Princeton University Press: Princeton, NJ, USA, 1999; 367p.
12. You, X.L.; Sun, S.; Zhu, J.Q. Significance of fossilized microbes from the Cambrian stromatolites in the Tarim Basin, Northwest China. *Sci. China Earth Sci.* **2014**, *57*, 2901–2913. [CrossRef]
13. Chang, Y.; Bai, W.; Qi, Y.; Sun, F.; Wang, M. Microfossil assemblage and its sedimentary environment in Cambrian stromatolites, western He'nan. *Adv. Earth Sci.* **2014**, *29*, 456–463. (In Chinese with English Summary)

14. Klappa, C.F. Lichen stromatolites: Criterion for subaerial exposure and a mechanism for formation of lamellar calcretes (caliche). *J. Sediment. Petrol.* **1979**, *49*, 387–410.
15. Landing, E. Tropical seas and volcanic fire in eastern New York. *Nat. Hist.* **2022**, *130*, 36–41.
16. Riding, R. Microbialites, stromatolites, and thrombolites. In *Encyclopedia of Geobiology*; Reiter, J., Thiel, V., Eds.; Springer: Dordrecht, The Netherlands, 2011; pp. 635–654.
17. Landing, E.; Westrop, S.R. Upper Lower Cambrian depositional sequence in Avalonian New Brunswick. *Can. J. Earth Sci.* **1996**, *33*, 404–417. [CrossRef]
18. Landing, E.; Westrop, S.R.; Geyer, G. Trans-Avalonian green–black boundary (early Middle Cambrian): Transform fault-driven epeirogeny and onset of 26 m.y. of shallow-marine, black mudstone in Avalonia (Rhode Island–Belgium) and Baltica. *Can. J. Earth Sci.* **2023**, *60*, 133–171. [CrossRef]
19. Vanuxem, L. *Geology of New York. Part III*; Survey of the Third Geological District of New York: Albany, NY, USA, 1842; 306p.
20. Aitken, J.D. Classification and environmental significance of cryptalgal limestones and dolomites with illustrations from the Cambrian and Ordovician of southwest Alberta. *J. Sediment. Petrol.* **1967**, *37*, 1163–1178. [CrossRef]
21. Bernhard, J.M.; Fisher, L.A.; Murphy, Q.; Sen, L.; Yeh, H.D.; Louyakis, A.; Gomaa, F.; Reilly, M.; Batta-Lona, P.G.; Bucklin, A.; et al. Transition from stromatolite to thrombolite fabric: Potential role for reticulopodial protists in lake microbialites of a Proterozoic ecosystem analog. *Front. Microbiol.* **2023**, *14*, 1210781. [CrossRef]
22. Dupraz, C.; Pattisina, R.; Verrecchia, E.P. Translation of energy into morphology: Simulation of stromatolite morphospace using a stochastic model. *Sediment. Geol.* **2006**, *185*, 185–203. [CrossRef]
23. Logan, B.W. *Cryptozoon* and associated stromatolites from the Recent, Shark Bay, Western Australia. *J. Geol.* **1961**, *69*, 517–533. [CrossRef]
24. Hall, J. *Cryptozoön*, n.g., *Cryptozoön proliferum*. n.sp. In *Thirty-Sixth Annual Report of the Trustees of the State Museum of Natural History to the Legislature 1883. Transmitted January 10, 1883. New York State Senate Paper 53*; Pierson, H.R., Ed.; (Two covers, same material, inner printed January 10, 1883, and outer January 10, 1884), unnumbered end page and Pl. 6 with caption; Weed, Parsons & Company: Albany, NY, USA, 1884. [CrossRef]
25. Playford, P.E.; Cockbain, A.E. Modern algal stromatolites at Hamelin Pool, a hypersaline basin in Shark Bay, Western Australia. In *Developments in Sedimentology, 20, Stromatolites*; Walter, M.R., Ed.; Elsevier: Amsterdam, The Netherlands; Oxford, UK; New York, NY, USA, 1976; pp. 389–411.
26. Burne, R.V.; Moore, L.S. Microatoll microbialites of Lake Clifton, Western Australia: Morphological analogues of *Cryptozoon proliferum*, the first formally named stromatolite. *Facies* **1993**, *29*, 149–178. [CrossRef]
27. Suosaari, E.P.; Reid, R.P.; Playford, P.E.; Foster, J.S.; Stolz, J.F.; Casaburi, G.; Hagan, P.D.; Chirayath, V.; Macintyre, I.G.; Planavsky, N.J.; et al. New multi-scale perspectives on the stromatolites of Shark Bay, Western Australia. *Sci. Rep.* **2016**, *1*, 20557. [CrossRef]
28. Dravis, J.J. Hardened subtidal stromatolites Bahamas. *Science* **1983**, *219*, 385–386. [CrossRef]
29. Dill, R.F.; Shinn, E.A.; Jones, A.T.; Kelly, K.; Steinen, R.P. Giant subtidal stromatolites forming in normal salinity waters. *Nature* **1986**, *324*, 57–58. [CrossRef]
30. Reid, R.P.; Macintyre, I.G.; Browne, K.M.; Steneck, R.S.; Miller, T. Modern marine stromatolites in the Exuma Cays, Bahamas: Uncommonly common. *Facies* **1995**, *33*, 1–17. [CrossRef]
31. Cohen, A.S.; Talbot, M.R.; Awramik, S.M.; Dettman, D.L.; Abell, P. Lake level and paleoenvironmental history of Lake Tanganyika, Africa, as inferred from late Holocene and modern stromatolites. *Geol. Soc. Am. Bull.* **1997**, *109*, 444–460. [CrossRef]
32. West, K.; Cohen, A. Morphology and behavior of crabs and gastropods from Lake Tanganyika, Africa: Implications for lacustrine predator-prey coevolution. *Evolution* **1991**, *45*, 589–607.
33. Argentinian Stromatolites. Available online: https://www.cnn.com/2023/12/16/americas/stromatolites-argentina-discovery-lagoons-ecosystem-scn/index.html (accessed on 14 November 2024).
34. CU Boulder Today. Available online: https://www.colorado.edu/today/2023/12/06/deep-within-inhospitable-desert-window-first-life-earth (accessed on 14 November 2024).
35. Horodyski, R.J.; Vonder Haar, S.P. Recent calcareous stromatolites from Laguna Mormona (Baja California) Mexico. *J. Sed. Petrol.* **1975**, *45*, 894–906.
36. Johnson, M.E.; Ledesma-Vázquez, J.; Backus, D.H.; González, M.R. Lagoon microbialites on Isla Angel de la Guarda and associated peninsular shores, Gulf of California (Mexico). *Sediment. Geol.* **2012**, *263–264*, 76–84. [CrossRef]
37. Bowen, T. *On Desert Shores—Archaeology & History of the Western Midriff Islands in the Gulf of California*; University of Utah Press: Salt Lake City, UT, USA, 2022; 313p.
38. Baumgartner, R.J.; Van Kranendonk, M.J.; Wacey, D.; Fiorentini, M.L.; Saunders, M.; Caruso, S.; Pages, A.; Homann, M.; Guagliardo, P. Nano-porous pyrite and OM in 3.5-billion-year-old stromatolites record primordial life. *Geology* **2019**, *47*, 1039–1043. [CrossRef]
39. Hickman-Lewis, K.; Cavalazzi, B.; Giannoukos, K.; D'Amico, L.; Vrbaski, S.; Saccomano, G.; Dreossi, D.; Tromba, G.; Foucher, F.; Brownscombe, W.; et al. Advanced two- and three-dimensional insights into Earth's oldest stromatolites (ca. 3.5 Ga): Prospects for life on Mars. *Geology* **2022**, *51*, 33–38. [CrossRef]

40. Beukes, N.J. Facies relations, depositional environments and diagenesis in a major early Proterozoic stromatolitic carbonate platform to basinal sequence, Campbellrand Subgroup, Transvaal Supergroup, Southern Africa. *Sediment. Geol.* **1987**, *54*, 1–46. [CrossRef]
41. Sumner, D.Y.; Grotzinger, J.P. Implications for Neoarchaean ocean chemistry from primary carbonate mineralogy of the Campbellrand-Malmani platform, South Africa. *Sedimentology* **2004**, *51*, 1273–1299. [CrossRef]
42. Awramik, S.M.; Sprinkle, J. Proterozoic stromatolites: The first marine evolutionary biota. *Hist. Biol.* **1999**, *13*, 241–253. [CrossRef]
43. Kah, L.C.; Riding, R. Mesoproterozoic carbon dioxide levels inferred from calcified cyanobacteria. *Geology* **2007**, *35*, 799–802. [CrossRef]
44. Aitken, J.D.; Narbonne, G.M. Two occurrences of Precambrian thrombolites from the Mackenzie Mountains, northwestern Canada. *Palaios* **1989**, *4*, 384–388. [CrossRef]
45. Riding, R. Cyanobacterial calcification, carbon dioxide concentrating mechanisms, and Proterozoic-Cambrian changes in atmospheric composition. *Geobiology* **2006**, *4*, 299–316. [CrossRef]
46. Grotzinger, J.P. Geochemical model for Proterozoic stromatolite decline. *Am. J. Sci.* **1990**, *290-A*, 80–103.
47. Chumakov, N.M.; Semikhatov, M.A. Riphean and Vendian of the USSR. *Precambrian Res.* **1981**, *15*, 229–253. [CrossRef]
48. Sami, T.T.; James, N.P. Synsedimentary cements as Paleoproterozoic platform building blocks, Pethei Group, northwestern Canada. *J. Sediment. Res.* **1996**, *66*, 209–222.
49. Riding, R. Abiogenic, microbial and hybrid authigenic carbonate crusts: Components of Precambrian stromatolites. *Geol. Croat.* **2008**, *61*, 73–103. [CrossRef]
50. Walter, M.R. Stromatolites and the biostratigraphy of the Australian Precambrian and Cambrian. *Spec. Pap. Palaeontol.* **1972**, *11*, 190.
51. Landing, E. Time-specific black mudstones and global hyperwarming on the Cambrian-Ordovician slope and shelf of the Laurentia palaeocontinent. *Palaeogeogr. Palaeoclimatol. Palaeoecol.* **2012**, *367–368*, 256–272. [CrossRef]
52. Landing, E.; Keppie, J.D.; Keppie, F.D.; Geyer, G.; Westrop, S.R. Greater Avalonia—Latest Ediacaran–Ordovician "peribaltic" terrane bounded by continental margin prisms ("Ganderia," Harlech Dome, Meguma): Review, tectonic implications, and paleogeography. *Earth-Sci. Rev.* **2022**, *255*, 103863. [CrossRef]
53. Kiessling, W. Secular variations in the Phanerozoic reef ecosystem. In *Phanerozoic Reef Patterns*; Kiessling, W., Flügel, E., Golonka, J., Eds.; Society of Economic Paleontologists and Mineralogists—SEPM Special Publication; SEPM: Tulsa, OK, USA, 2002; Volume 72, pp. 625–690.
54. Pratt, B.R. Stromatolite decline—A reconsideration. *Geology* **1982**, *10*, 512–515. [CrossRef]
55. Fischer, A.G. Fossils, early life, and atmospheric history. *Proc. Natl. Acad. Sci. USA* **1965**, *53*, 1205–1215. [CrossRef]
56. Landing, E.; Antcliffe, J.B.; Geyer, G.; Kouchinsky, A.; Bowser, S.S.; Andreas, A. Early evolution of colonial animals (Ediacaran Evolutionary Revolution-Cambrian Evolutionary Radiation-Great Ordovician Diversification Interval). *Earth-Sci. Rev.* **2018**, *178*, 105–135. [CrossRef]
57. Foster, G.; Royer, D.; Lunt, D. Future climate forcing potentially without precedent in the last 420 million years. *Nat. Commun.* **2017**, *8*, 14845. [CrossRef]
58. Goldring, W. Algal barrier reefs in the Lower Ozarkian of New York with a chapter on the importance of coralline algae as reef builders through the ages. *N. Y. State Mus. Bull.* **1938**, *315*, 5–75. [CrossRef]
59. Friedman, G.M. Hoyt Limestone of Late Cambrian, eastern New York State: Spectacular domed stromatolites as Lester Park and Petrified Sea Gardens. *Northeast. Geol. Environ. Sci.* **2000**, *22*, 336–349.
60. Lindemann, R.H. The historic significance of a petrified cabbage patch at Greenfield, New York. *Northeast. Geol. Environ. Sci.* **2006**, *28*, 101–109.
61. Friedman, G.M. Late Cambrian cabbage-head stromatolites from Saratoga Springs, New York, USA. *Carbonates Evaporites* **2012**, *15*, 37–44. [CrossRef]
62. Lee, J.-H. Stromatolite. In *Encyclopedia of Geology*, 2nd ed.; Alderton, D., Elias, S.A., Eds.; Elsevier: Rotterdam, The Netherlands, 2019; pp. 75–388.
63. Lee, J.-H.; Riding, R. The 'classic stromatolite' *Cryptozoon* is a keratose sponge-microbial consortium. *Geobiology* **2021**, *9*, 189–198. [CrossRef]
64. Lee, J.-H.; Riding, R. Keratolite–stromatolite consortia mimic domical and branched columnar stromatolites. Palaeogeography, Palaeoclimatology. *Palaeoecology* **2021**, *571*, 110288. [CrossRef]
65. Kershaw, S.; Li, Q.; Li, Y. Phanerozoic carbonate facies conundrum—Sponges or clotted fabric? Evidence form Early Silurian reefs. *Stratigr. Rec.* **2021**, *19*, 3–10.
66. Neuweiler, F.; Kershaw, S.; Boulva, F.; Matysik, M.; Sendino, C.; McMenamin, M.; Munnecke, A. Keratose sponges in ancient carbonates—A problem of interpretation. *Sedimentology* **2023**, *70*, 927–969. [CrossRef]
67. Guiry, M.D.; Guiry, G.M. AlgaeBase. World-Wide Electronic Publication. National University of Ireland, Galway. Available online: https://algaebase.org (accessed on 1 October 2024).
68. World Register of Marine Species. Available online: https://www.marinespecies.org/ (accessed on 19 November 2024).
69. Steele, J.H. A description of the oolite formation lately described in the County of Saratoga and in the State of New York. *Am. J. Sci.* **1825**, *9*, 16.

70. Flick, A.C. Suggestions of a state policy relating to historic and scientific reservations. Twenty-third Report of the Director of the Division of Science and the State Museum. *N. Y. State Mus. Bull.* **1929**, *284*, 68–71.
71. Cushing, H.P.; Ruedemann, R. Geology of Saratoga Springs and vicinity. *N. Y. State Mus. Bull.* **1914**, *169*, 177.
72. Landing, E. Ediacaran–Ordovician of east Laurentia—geologic setting and controls on deposition along the New York Promontory. In *Ediacaran–Ordovician of East Laurentia—S. W. Ford Memorial Volume. New York State Museum Bulletin 510*; Landing, E., Ed.; The NewYork State Education Department: Albany, NY, USA, 2007; pp. 5–24.
73. Landing, E.; Westrop, S.R.; Kröger, B.; English, A.M. Left behind—Extinction and a relict trilobite fauna in the Cambrian–Ordovician boundary succession, (northeast Laurentian platform, New York). *Geol. Mag.* **2010**, *148*, 529–557. [CrossRef]
74. Friedman, G.M.; Sanders, J.E. Time-temperature-burial significance of Devonian anthracite implies former great (~6.5 km) depth of burial of Catskill Mountains, New York. *Geology* **1982**, *10*, 93–96. [CrossRef]
75. Landing, E.; Webster, M.; Bowser, S.S. Terminal Ediacaran–Late Ordovician of the NE Laurentia palaeocontinent: Rift-drift–onset of Taconic orogeny, sea-level change, and "Hawke Bay" onlap (not offlap). In *Supercontinents, Orogenesis, and Magmatism*; Nance, R.D., Strachan, R.A., Quesada, Q., Lin, S., Eds.; Geological Society of London, Special Publication; Geological Society Publications: London, UK, 2023; Volume 542. [CrossRef]
76. Hall, J. *Natural History of New York. Geology of New-York. Part 4—Comprising the Geology of the Fourth Geological District*; Carroll and Cook: Albany, NY, USA, 1843; 683p.
77. Mather, W.W. *Geology of the First Geologic District of New York*; Carroll and Cook: Albany, NY, USA, 1843; 832p.
78. Hall, J. *Palaeontology of New York. 1. Organic Remains of the New York System*; C. van Benthuysen: Albany, NY, USA, 1847; 338p.
79. Hall, J. Report of the Director. Albany December 23, 1882, p. 1–17. In *Thirty-Sixth Annual Report of the Trustees of the State Museum of Natural History to the Legislature. Transmitted January 10, 1883: New York State Senate Paper, 53*; Pierson, H.R., Ed.; Weed, Parsons & Company: Albany, NY, USA, 1884; printed 10 January 1884.
80. Monty, C. Spongiostromate vs. porostromate stromatolites and oncolites. In *Phanerozoic Stromatolites. Case Histories 1–4*; Monty, C., Ed.; Springer: Berlin/Heidelberg, Germany, 1981.
81. Smith, G.L.; Clark, D.L. Conodonts of the Lower Ordovician Prairie du Chien Group of Wisconsin and Minnesota. *Micropaleontology* **1996**, *42*, 363–373. [CrossRef]
82. Winchell, N.H. VII New species of fossils. In *The Fourteenth Annual Report for the Year 1885. Geological and Natural Historical Survey of Minnesota*; Winchell, N.H., Ed.; Minnesota Geological Survey; Pioneer Press Company: St. Paul, MN, USA, 1886; pp. 313–318.
83. Chaney, L.W., Jr. *Cryptozoon minnesotense* in the Shakopee Limestone at Northfield, Minnesota. *Minn. Acad. Nat. Sci. Bull.* **1889**, *3*, 280–284.
84. Stauffer, C.R. Cryptozoons of the Shakopee Dolomite. *J. Paleontol.* **1945**, *19*, 376–379.
85. Brainerd, E.; Seely, H.M. The Calciferous Formation in the Champlain valley. *Geol. Soc. Am. Bull.* **1890**, *1*, 501–516. [CrossRef]
86. Bassler, R.S. Systematic paleontology of the Cambrian and Ordovician deposits of Maryland. In *The Cambrian and Ordovician deposits of Maryland*; Bassler, R.S., Ed.; Maryland Geological Survey Report; The Johns Hopkins Press: Baltimore, MD, USA, 1919; pp. 184–373.
87. Walcott, C.D. Cambrian geology and paleontology II No. 9. New York Potsdam–Hoyt fauna. *Smithson. Misc. Collect.* **1912**, *57*, 251–304.
88. Rothpletz, A. Über die systematische Deutung und die stratigraphishe Stellung der ältesten Versteinerungen Europas und Nordamerika mit besonderer Berücksichtigung der *Cryptozoon* und Oolithe. Teil II: Über *Cryptozoon*, *Eozoon* und *Artikokania*. *Abh. Könglich Bayer. Akad. Wiss. Matimatische-Phys. Kl.* **1916**, *28*, 91.
89. Dawson, J.W. Notes on *Cryptozoon* and other ancient fossils. *Can. Rec. Sci.* **1896**, *7*, 203–209.
90. Seely, H.M. Beekmantown and Chazy formations in the Champlain Valley. Contribution to their geology and paleontology. In *Fifth Report of the Vermont State Geologist*; Vermont Department of Environmental Conservation: Montpelier, VT, USA, 1906; pp. 115–145.
91. Fenton, C.L.; Fenton, M.A. Algae and algal beds in the Belt Series of Glacier National Park. *J. Geol.* **1931**, *39*, 670–688. [CrossRef]
92. Rezak, R. Stromatolites of the Belt Series in Glacier National Park, Montana. *U. S. Geol. Surv. Prof. Pap.* **1957**, *294*, 127–150.
93. Wilks, M.E.; Nisbet, E.G. Archaean stromatolites from the Steep Rock Group, northwestern Ontario, Canada. *Can. J. Earth Sci.* **1985**, *22*, 792–797. [CrossRef]
94. Fralick, P.; Riding, R. Steep Rock Lake: Sedimentology and geochemistry of an Archean carbonate platform. *Earth-Sci. Rev.* **2015**, *151*, 132–175. [CrossRef]
95. Johnson, J.H. Lime-secreting algae from the Pennsylvanian and Permian of Kansas. *Geol. Soc. Am. Bull.* **1946**, *57*, 1087–1120. [CrossRef]
96. Vahrenkamp, V.; Chandra, V.; Garuglieri, E.; Marasco, R.; Hachmann, K.; Khanna, P.; Daffonchio, D.; Petrovic, D. Discovery of modern living intertidal stromatolites on Sheybarah Island, Red Sea, Saudi Arabia. *Geology* **2024**, *52*, 347–351. [CrossRef]
97. Shapiro, R.S. Stromatolites: A 3.5 billion year ichnologic record. In *Trace Fossils: Concepts, Problems, Prospects*; Ajello, M., Brotton, S.J., Eds.; Elsevier: Amsterdam, The Netherlands, 2007; pp. 381–389.
98. Neuweiler, F.; Mueller, M.; Walter, B.; Landing, E.; Beranoaguirre, A.; Sendino, C.; Amati, L.; Kershaw, S. Fossil record misconstrued: Sponge-like fabric reflect incipient carbonate metamorphism. *Res. Sq.* **2024**. [CrossRef]
99. Turner, E.C. Possible poriferan body fossils in early Neoproterozoic microbial reefs. *Nature* **2021**, *596*, 87–91. [CrossRef] [PubMed]

100. Grotzinger, J.P.; Knoll, A.H. Stromatolites in Precambrian carbonates: Evolutionary mileposts or environmental dipsticks? *Annu. Rev. Earth Planet. Sci.* **1999**, *27*, 313–358. [CrossRef]
101. Hoffman, H.J. Stromatoid morphometrics. In *Stromatolites. Developments in Sedimentology 20*; Walter, M.R., Ed.; Elsevier: Amsterdam, The Netherlands, 1976; pp. 45–54.
102. Reid, P.R.; James, N.P.; Macintyre, I.G.; Dupraz, C.P.; Burne, R.V. Shark Bay stromatolites: Microfabrics and reinterpretation of origins. *Facies* **2013**, *49*, 299–324. [CrossRef]
103. Landing, E.; Geyer, G.; Schmitz, M.D.; Wotte, T.; Kouchinsky, A. (Re)proposal of three Cambrian Subsystems and their geochronology. *Episodes* **2021**, *44*, 273–283. [CrossRef]

Journal of
*Marine Science
and Engineering*

MDPI

Review

The Bakken Model: Deposition of Organic-Rich Mudstones and Petroleum Source Rocks as Shallow-Marine Facies Through the Phanerozoic

Ed Landing

New York State Museum, Albany, NY 12230, USA; ed.landing@nysed.gov; Tel.: +1-(518)-438-8728

Abstract: Many organic-rich marine mudstones, which are key hydrocarbon sources, were deposited on continent margins in mid-water oxygen-minimum zones (OMZs) that expanded and intensified during oceanic anoxic events (OAEs). Other marine hydrocarbon sources include platform and forearc black shales that record trans-continental, long-erm anoxic/dysoxic environments with no modern analog. Their explanation as recording deep-water, Black Sea-type basins or low-oxygen upwelling is not satisfactory for occurrences on shelves that lack significant epeirogenic activity, while modern studies show that upwellings do not cross the shelf break. The alternative is the Bakken model, which concludes that regionally extensive shelves and forearc organic-rich mudstones are shallow-water facies. These Bakken facies reflect hyper-warming conditions with high sea-levels, high water temperatures with increased insolation and low oxygen solubility, turbid water due to algal blooms and mud eroded from orogenic highlands, and possible LIP activity. Early Paleozoic black shales indicate that increased nutrients presumed to accompany the Devonian appearance of forests with deep roots that enhanced weathering simply cannot explain older Cambrian–Ordovician shelf anoxia/dysoxia. Shallow-marine deposition by the Bakken model is mandated by black shales deposited on subaerial unconformities that show high-energy facies (wave cross beds, HCS) and common bioturbation. The Bakken model explains shallow anoxia/dysoxia with high Paleozoic sea levels and tropical distribution of large continents. It is based on the Upper Devonian–lower Mississippian Bakken Formation (western U.S. and adjacent Canada). Rising temperatures, diminished oxygen solubility, and eustatic rise with deglaciation accompany modern climate change and mean that near-future platform seas will feature the reappearance of low-oxygen Bakken facies and environments.

Keywords: Bakken model; petroleum source rocks; Black Sea model; anoxia facies; dysoxic facies; marine oxygenation; Paleozoic

1. Introduction

Oil and natural gas exploration and production focus on organic-rich marine mudstones. These facies, with 5–65% total organic carbon (TOC) [1], are the main economic source rock for petroleum and natural gas, which may be fracked (hydraulically fractured) or have migrated into more porous units. These fine-grained units can also form relatively impermeable top seals for hydrocarbon accumulation. No single explanation exists for organic-rich mudstones, and their origin, other than being the product of low-oxygen, dysoxic/anoxic (d/a) deposition, is contentious. Organic-rich marine mudstones occur in three types of settings. These include the giant fields on continental margins in break-up or collisional settings. Break-up settings include the northern Gulf of Mexico (Figure 1, upper Jurassic and middle Cretaceous [2]) and the North Sea (Middle Jurassic [3]), and collisional (and barred) basins include the Permian Basin of New Mexico and West Texas [4] and the Jurassic of the Arabian Gulf [5].

The third type of giant (or potentially giant) oil field occurs within continents and comprises widespread organic-rich mudstones. Some examples have undergone little thermal maturation, such as the terminal Devonian–lower Mississippian heavy oils of the Bakken Formation in the Williston Basin, western US and adjacent Canada (Figure 1, intervals 17, 18) [6]. An older organic-rich mudstone is the Upper Ordovician Utica Shale (and related units) that covers much of the eastern United States west of the Appalachian Mountains [7]. The uppermost Jurassic of the extensive West Siberian Basin is the source of much exported Russian oil (Figure 1, horizon 25) [8,9]. Note that in this report, Lower/Early, Middle/Middle, Upper/Late Cambrian refer to the precisely defined proposed subsystems/subperiods that replaced the undefined subsystem-level terms "lower"/"early", "middle"/"middle", and "upper"/"late" of many reports [10].

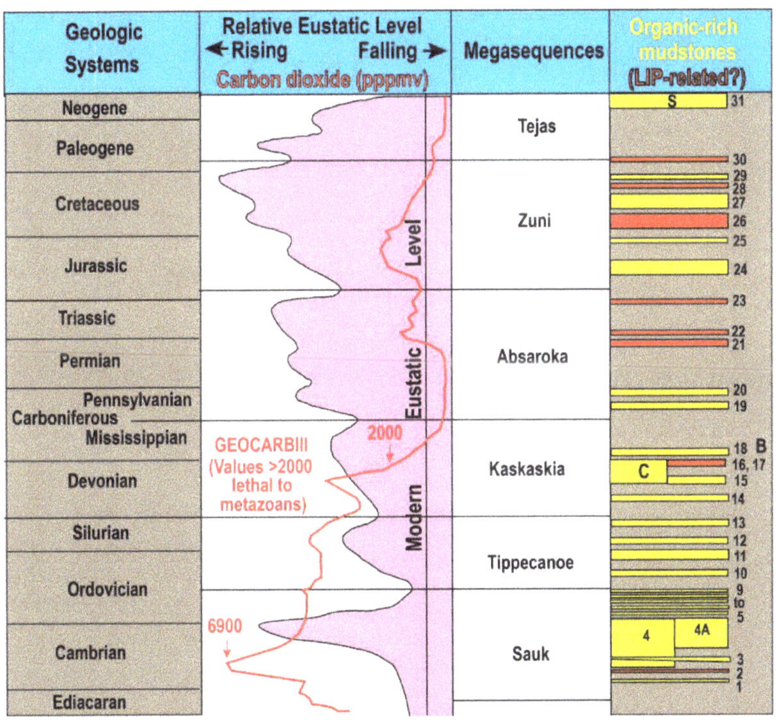

Figure 1. Phanerozoic organic mudstones. Sections 2.2 and 4 discuss this figure. Most organic-rich intervals in [11,12]; eustatic curves of [13,14], mega-sequences in [15].

2. Regionally Extensive Phanerozoic Organic-Rich Marine Mudstones

2.1. Succession of Phanerozoic Black Mudstones on Continents and in the Ocean

Black shale intervals representing variable time intervals occur through the Phanerozoic, particularly in the Paleozoic through the lower Tertiary (Figure 1). Most have been listed in earlier syntheses [11,12,16], but they have been supplemented by additional intervals documented on several paleocontinents in the Cambrian–Lower Ordovician [17–22]. Almost all of the black shale intervals in Figure 1 were deposited on, or marginal to, paleocontinents, with only the later Tertiary black sapropels deposited as deep-water facies of the Mediterranean and Black Sea (Figure 1, "S"). These sapropels correspond to the traditional deep-water/silled basin "Black Sea model" for ancient black mudstone deposition used in many reports [23] (Figure 2a).

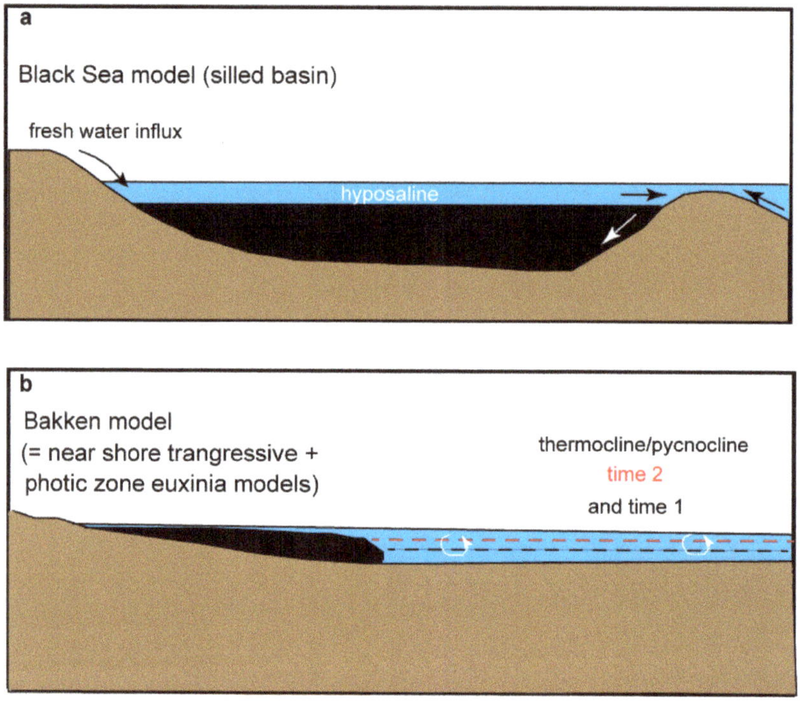

Figure 2. Models for marine organic-rich mudstone deposition: (**a**) Black Sea (silled basin model); (**b**) Bakken model featuring origin of strongly dysoxic/anoxic facies in shallow water and transgression with sea-level rise, either eustatic or epeirogenic; the model corresponds to the near-shore transgressive model [16] and with photic zone euxinia [17].

This review is largely limited to the early Paleozoic. Some of the black shale intervals described below, beginning as early as the Cambrian, can be shown to occur on a number of paleocontinents. Similarly, a number of marginal black shale oxygen-minimum zones (OMZs) can be temporally linked to on-platform (Bakken) black shales or to coeval, unconformity-bounded depositional sequences on tropical paleocontinents [12–18,24].

2.2. Bakken Model for Shallow-Marine Black Shale Deposition

The designation "Bakken" is derived from the shallow-marine Late Devonian and lower Carboniferous (lower Tournaisian) Bakken Formation in the Williston Basin or the western US and adjacent Canada ([6,22], Figure 1, intervals 17, 18).

Organic-rich black mudstones occur in a wide variety of marine depositional environments. These include marginal facies deposited under an oxygen-minimum zone (OMZ) known to develop with or without upwelling in marginal and upper slope environments (Figure 2b). OMZ black shales may be coeval with shallow, anoxic/dysoxic Bakken-type shale facies on the platform as eustatic rise is accompanied by the thickening and intensification of the OMZ. However, expansion of the OMZ is not regarded as leading to the onlap of OMZ water onto the shelf of a paleocontinent (Figure 3b). The distinction is that the Bakken model has been proposed for shallow, often peritidal organic-rich shale deposition, with Bakken organic-rich facies arising via the de novo origin of low-oxygen facies with eustatic or epeirogenic onlap of shallow sea water. The model likely features the proposed heating of onlapping/transgressing waters during regional onlaps that replace higher-albedo land surfaces with higher insolation, heat-trapping shallow waters with lowered oxygen solubility due to their high temperatures, and likely with photic zone algal proliferation and the sinking of organic detritus (Figure 2b [6,17–24]).

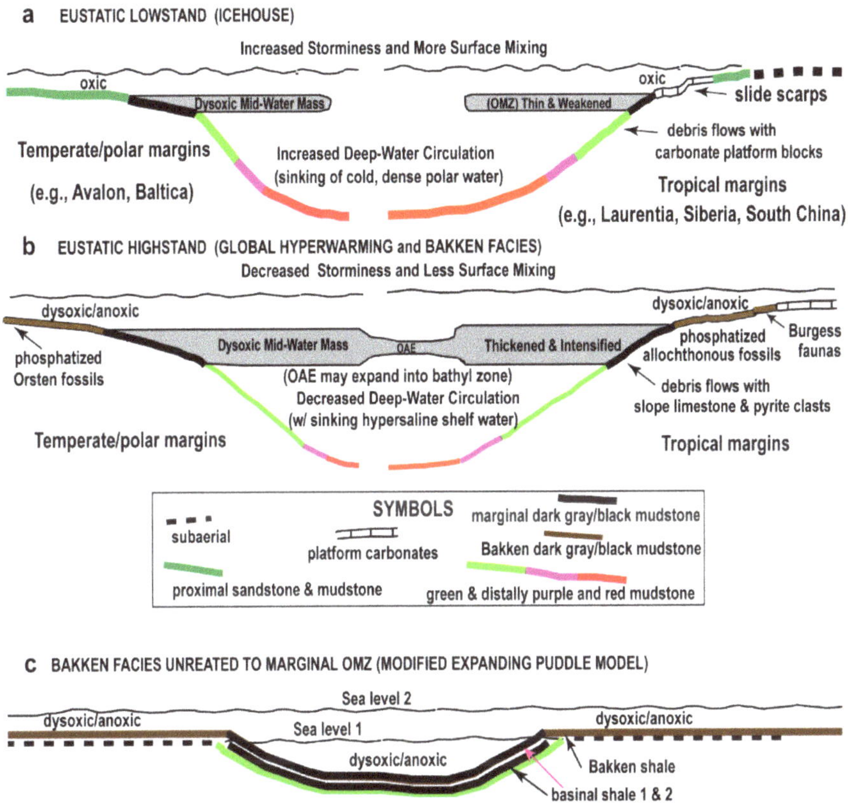

Figure 3. Depositional models for marine organic-rich mudstone: Global hyper-warming model for eustatic lowstand (**a**) and highstand (**b**) showing litho- and biofacies patterns and changes in mid-water oxygen-minimum zones, with Bakken facies unrelated to OMZ transgression in the Cambrian–Ordovician and developing as Bakken facies with the hyper-warming of the shelf; (**c**) development of Bakken facies on shelf unrelated to the expansion of anoxic puddle limited to on-shelf basin.

The Bakken model resembles the expanding anoxic "puddle" for the origin of organic-rich facies on continental platforms with sea-level rise [25,26], but with the "puddle" not originating with the transgression of OMZ water from the shelf margin (Figures 2b and 3a–c).

The Bakken model for the hallow-marine deposition of organic-rich mudstones is an alternative to several existing models for the origin of these facies. As noted above, the Bakken facies is only coeval with an expanded, thickened OMZ at times of high sea-level; it does not represent the onlap of the OMZ, as in frequently cited traditional models [27]. Indeed, modern shallow-shelf onlaps of marginal OMZs are unknown and do not seem physically possible (Section 2.5).

A second model for the origin of organic-rich mudstones proposes the deep subsidence of parts of paleocontinents that otherwise are not involved in orogeny or known epeirogenic activity (Figure 2a). This includes the Cambrian Alum Shale deposition described below (Figure 1, interval 4a).

Finally, anoxic black mudstone facies have appeared, particularly in the Mesozoic–early Tertiary (Figure 1), with major oceanic changes (i.e., OAEs, oxygen anoxic events). These ocean-wide intervals result in the deposition of black muds in pelagic facies to bathyal depths and are correlated with the origin of dysoxic/anoxic (Bakken) intervals in paleocontinental successions [27].

2.3. Cold Water Upwelling and Deep-Water Organic-Rich Mudstones in Early Paleozoic Forearc Troughs

The Bakken model is an alternative to proposals of black shale deposition during orogeny that have proposed the upwelling of deep, poorly oxygenated marine water into Paleozoic forearc troughs that led to the production and preservation of organic-rich mud. As much as a kilometer of rapid subsidence in Paleozoic fore-arc troughs in eastern North America has been presumed in these models.

The upwelling has been regarded as possibly wind-driven, which entrained cold, low-oxygen marginal waters into elongate (1000+ km) forearc troughs during the Taconian (Late Ordovician) [28,29] and Acadian (Middle–Late Devonian) [30–32] orogenies.

An alternative explanation to deep-water fore-arc black shale deposition is needed as no modern forearc regime features deep-water, low-oxygen deposition. In some cases, this may be due to the rapid accumulation and aggradation of orogen-derived sediment that keeps modern fore arc troughs relatively shallow. Indeed, an ancient example involving forearc syndepositional mud- and sandstones immediately in front of and under the leading edge of the Taconian fore-arc thrusts suggests depths of only tens of meters [33].

2.4. Pre-Latest Ordovician Cold-Water and Glaciations

Wind-driven upwelling of deep- and cool-water low-oxygen OMZ facies in synorogenic fore-arc facies has been suggested to play a part in the formation of Late Ordovician and Late Devonian black shales across the eastern U.S. at the margin of, or in-board of, the Appalachian Mountains [31,34]. This model proposes the appearance of supposed cool-water carbonates, as shown by phosphate lag and manganese deposits and the replacement of coral by bryozoan biofacies [30,31,34]. An alternative explanation of these "cool water" developments is that they reflect the replacement of coral by bryozoan biofacies in turbid water sedimentary condensation and the development of sediment-laden water that led to surface water algal blooms, to the development of benthic cyanobacterial mats, and to the accumulation of organic-rich mud [35–37]. Indeed, the origin of carbon from cyanobacterial or algal sources probably explains the low $\delta\ ^{13}C_{org}$ values of Phanerozoic black shales [12].

One key suggested ingredient of onlap models, such as in the late Middle–Late Ordovician, is cold, low-oxygen OMZ water. This water was purportedly drawn into and became the bottom water mass of the successive synorogenic, late Middle–middle Late Ordovician Sevier and Sebree fore arc troughs in eastern Laurentia. This cold water is said to have been produced by the possible cooling or glaciations before the brief (ca. < 1 Ma) terminal Ordovician Hirnantian glaciations across Gondwana [38].

Pre-latest Ordovician glaciations have been discussed based on rapid, presumed eustatic changes in Jordan [39]. However, no physical evidence for these glaciations has been found [40]. A caveat to proposing glaciation is that rapid third-and fourth-order eustatic changes cannot be assumed to be glacioeustatic. Indeed, 100 Ka, 405 Ka, 1.2 Ma, 2.4 Ma, and 3.0 Ma sea level changes correspond to obliquity and thus humid–arid cycles in a time with no glaciation (the Cretaceous). As a result, short-term eustatic changes actually track storage of water in global aquifers [41]. The takeaway is that the production and onlap of cold dysoxic/anoxic waters cannot be imagined to be an element of the extensive transcontinental Utica Shale Bakken facies (discussed below) or onlaps of other shales during times with no or limited polar glaciation (e.g., Cambrian, most of Ordovician, later Silurian, Mesozoic and lower Paleogene) (Figure 1).

The one key ingredient in the explanation of shallow-water (Bakken facies) or marginal black shale facies is obviously the depositional significance of siliciclastic mud, which can be moved by waves and currents for greater or smaller distances (depending on steep vs gentle basin paleoslopes) [39]. Subaerial erosion was the obvious source of mud in the widespread Late Ordovician and Middle–Late Devonian black shales in Laurentia (Figure 1, intervals 10, 11) [32,42–44]. An often overlooked additional consequence of muddy, turbid water would be that it would reduce the depth and intensity of light in the euphotic zone and would promote low oxygenation in extensive seas that already had lowered oxygen solubility at times of high global temperatures [18].

2.5. Is Upwelling Plausible in Formation of Platform or Forearc Trough Black Shales?

The equation of this thin black shales on continents (Section 3) with upwelling that drove water from an OMZ across tens of kilometres of shallow carbonate shelf seems unlikely [42]. Indeed, upwelling is also invoked in many reports [29,45,46] to explain low-oxygen black mudstones and supposed "high productivity" in very elongate (>1000 km) forearc troughs that traversed the Laurentia paleocontinent in the later Ordovician.

However, no modern or Pleistocene examples of upwelling across very wide shelves exist. All examples of upwelling driven by trade or long-shore winds do not show OMZ water onlapping far across the continental shelf margin, and no examples of such nutrient-rich water reaching into near-shore environments exist. Thus, the powerful Gulf Stream only episodically crosses the northern Florida–North Carolinas shelf margin but remains 70–100 k offshore, with upwelling limited to shelf depths of 30–70 m [47]. Similarly, a review around all of North America showed upwelling water masses crossing the continental shelf only to depths of 40–120 m [48]. Theoretical calculations of upwelling [49] do not show upwelling crossing the shelf margin, with most upwelling actually being 50–250 km offshore and seaward of the shelf break [47,50].

If an OMZ were actually to cross a shelf break and enter the Ordovician Sevier and Sebree forearc troughs (discussed below), any significant along-trough flow would be precluded. Physical limitations (e.g., internal friction with the shelf floor and between water masses; thermohaline density contrasts of shelf and onlap water) would seem to preclude such regional upwelling models [45]. The locations of the mouths of the Sevier and Sebree troughs were at low southern latitudes at the margin of Iapetus [32,34]. These locations would have allowed for westerly movement of the surface water under the action of trade winds, as well as for southerly transport via Ekman transport and the Coriolis effect, and would thus be the "perfect" conditions for upwelling. However, both the physical limitations imposed on upwelling into the troughs [45] and the likely shallow depths of the Sevier and Sebreee troughs, with likely frictional and density constraints on the movement of wind-driven Eckman transport of upwelling water into a forearc trough (discussed below), limit the possibility of any deep offshore water (OMZ or otherwise) entered the forearc troughs.

Early Paleozoic Laurentia, Baltica, Avalonia, and South China had black marine shales (Figure 1, intervals 1–9, discussed below) as onlapping platform deposits on older oxic facies. These platform Bakken shales either developed following the "anoxic puddle" model (Figure 3c) or reflect the development of hyper-warming as a consequence of regional marine onlaps. Black shale facies in Laurentian forearc troughs (Figure 1, intervals 10, 11) also involved development on a tropical paleocontinent. In all cases, the low-oxygen black muds were deposited under turbid water due to phytoplankton growth, with the forearc trough water infused with mud derived from the eroding mountain belts [32]. In all these cases, black mudstone deposition was likely associated with the heating of shallow water and the resulting diminished oxygen solubility in lower latitudes [18].

2.6. Role of Global Temperatures in Black Mudstone Deposition

Figure 1 follows Hallam's [13] eustatic curve and shows the long-appreciated association of black shale deposition with sea-level rises and onlap [14]. The major craton-wide transgressive intervals of North America are Sloss' [15] (mega)sequences. The history of pCO_2 is taken from the GEOCARBIII proposal [51]. Black d/a mud deposition is said in many reports to be regularly associated with high eustatic level, high global temperatures, and high pCO_2. However, some aspects of these syntheses have to be questioned.

Indeed, a *negative* correlation exists between global temperatures and pCO_2 [52], and the sensitivity of proxies for published levels of temperature and pCO_2 should be regarded as uncertain. An unquestioned factor in paleoclimate reconstruction is the very high pCO_2 levels in many Phanerozoic reports. It is significant that even at the modern ca. 420 PPM pCO_2, Arctic and tropical oceans are losing their bicarbonate buffering capability. This means that aragonite is becoming soluble, with plants and animals losing the ability to precipitate it. At the very questionable 1000–7000 PPM levels reconstructed by GEOCARB III (Figure 1), bacteria would undergo DNA damage, while higher levels are associated with kidney damage and calcification and bone degradation in vertebrates [53,54].

Thus, significantly lower pCO_2 levels through the Phanerozoic, with maximum levels of 2000 PPM [55,56], may be more accurate, with the proviso noted above that there may be no systematic correspondence of pCO_2 with climate temperature variations [52,55]. This agreement of Jackson Davis [52] and Rothman [55] actually demonstrates what has been known from the "dawn" of the role of atmospheric pCO_2 in temperature regulation in 1896; i.e., that levels above 1000 PPM do not progressively or significantly increase atmospheric temperatures [57]. This limited control of global temperatures with the rise of pCO_2 leads to a necessary questioning of the presumed consequence of increased pCO_2 levels with large igneous province (LIP) activity.

Higher global temperatures have traditionally been associated with large igneous province (LIP) activity and consequent high atmospheric pCO_2 levels [12,58,59]. However, in addition to the plausible negative association of temperature and pCO_2, this presumed association of LIP activity and temperatures is problematic given the limited temporal resolution of individual $^{40}Ar/^{39}Ar$ dates (commonly ± 4 Ma) and long duration of LIP events [60]. In short, the role of LIPs in the occurrence of black shale intervals (see organic-rich shales shown in red in Figure 1) is questionable as most LIPs cannot be unquestionably associated with a particular black mudstone interval.

3. Black Shales and Bakken Facies Through the Early Paleozoic

Almost all early Paleozoic marine successions deposited off early Paleozoic paleocontinents have been lost to subduction or are highly metamorphosed in orogenic belts. The best-preserved early Paleozoic continental slope successions occur in the Taconian overthrust belt in NE North America, i.e., from western Pennsylvania into eastern New York State and western Vermont and then extending from southern Quebec to western Newfoundland. These slope successions preserve a record of the oxygenation of the Iapetus Ocean off the margin of Laurentia and provide a standard of comparison for oceanic oxygenation of much of the early Paleozoic.

In this region, upper–middle-slope OMZ intervals are best known from the leading edge of the Taconic overthrust belt in eastern New York and southern Quebec, south of the St. Lawrence River. In eastern New York and Vermont, they are exposed in the narrow (ca. 20–60 km-wide), outermost thrust slices (i.e., Sunset Point and Giddings Brook slices) [22,61]. A brief summary of Taconian successions is that coarse grained, feldspathic sandy, rift deposition continued into the late Early Cambrian (estimated ca. 510 Ma). The rift-related facies is followed by six third-order alternations of green and black mudstone-dominated intervals. All known black mudstone OMZ localities from eastern Pennsylvania to western Newfoundland, with their correlations into Laurentian carbonate platform units and into all early Paleozoic paleocontinents, have been discussed [18].

The black OMZ intervals may be slightly calcareous, are strongly dysoxic to anoxic with small and limited burrows, and they commonly have thin, nodular-to-bedded turbiditic and contourite limestones and debris flows composed of slope-derived limestones in the Cambrian–lower Middle Ordovician (Figure 1, intervals 3–9, 3a, b). The onset of the Taconian orogeny in NE Laurentian is marked by a lowest Upper Ordovician (*Nemagraptus gracilis* Zone) black mudstone and then by a middle Upper Ordovician OMZ, with black mudstone and abundant sandstones being derived from the encroaching Taconian accretionary prism (Figure 1, horizons 10, 11). Relatively thin black mudstones (Figure 1, OMZ intervals 5–9) are intercalated with the Lower–lower Middle Cambrian green mudstone alteration.

3.1. *Cambrian Black Shales and Bakken Facies*

3.1.1. Lowest Cambrian

Deposition of black mudstones of the Lower Cambrian Shuijingtuo Formation (Figure 1, interval 1) began at ca. 526 Ma at about the beginning of Stage 2 [62]. This black mudstone is older than the earliest black mudstone in the Taconic overthrusts. The low-energy shelf facies of the formation blankets a huge area of the SE part of the South China Platform, but its possible lateral continuation into slope facies and into an OMZ is not recorded. Relatively little data are available to allow for the determination of whether this is a shallow (i.e., Bakken facies) or relatively deep water unit. No coeval black mudstones on other paleocontinents seem to have been described.

3.1.2. Browns Pond Dysaerobic/Anoxic Interval and Sinsk Event

Three Lower-Cambrian black mudstones (Figure 1, interval 2) that may be coeval include the OMZ black shale of the Browns Pond d/a interval in the Taconic allochthon of eastern New York [18,22]. The Browns Pond interval, with the Moosamaloo Shale, correlates with Cheshire Quartzite and overlying Dunham Dolostone onlap of the east New York–west Vermont shelf and is referable to Cambrian Series 2, upper Stage 3–4 [61].

A tentative correlation of Stage 3–4 is with organic-rich mudstones of the Sinsk Formation [63], which forms an extensive, relatively shallow Bakken facies on the SE Siberia paleocontinent [64,65]. This black shale facies was deposited in the photic zone [66].

A ca. 507.7 U-Pb volcanic zircon date on the *Callavia broeggeri* Zone (Stage 3–4) in Avalonian SE Newfoundland [67] may be younger than the Sinsk Formation, which records the Sinsk event and the extirpations of archaeocyaths and trilobites. A black shale unit occurs immediately below molasse of the Ross-Delamerian orogeny in SW Australia and in the Ross Peninsula of Antarctica. Based on associated trilobites, this East Gondwanan shale is correlated with the Sinsk Formation [68]. Dates on volcanics of the Kalkarindji LIP of northern and central Australia were used to propose an age of ca. 510–511 Ma. This date is older than the ca. 507.7 Ma on the *C. broeggeri* Zone, but it may be accurate based on the relatively low-resolution $^{40}Ar/^{39}Ar$ ages that cluster around 510 Ma, with a resolution of

±4 Ma [61]. The possible association of the Sinsk-Ross/Delamerian black mudstones with an LIP event is indicated by the red color proposed in Figure 1, horizon 2.

3.1.3. Lower–Middle Cambrian Boundary Interval

Black mudstones occur in the Lower–Middle Cambrian boundary interval (Figure 1, interval 3). These mudstones yield the lowest Middle Cambrian trilobites, including what is cautiously regarded as the long-ranging, dysoxia-associated *Oryctocephala indicus*-plexus, used to recognize the base of the Middle Cambrian/"Miaolingian Stage", and they have been discussed by Landing et al. [67]. In South China, the mudstone occurs at the base of a meter-scale shaly, calcareous redox alternation termed a Logan cycle in early Paleozoic ramp and slope successions [18,22]. This succession was deposited in what appears to be a dysoxic upper slope succession [67]. In SW Laurentia, the *Oryctocephala indicus*-plexus and a low-diversity trilobite assemblage have a lowest occurrence with the onlap of dark mudstones across platform sandstones in SE California and SW Nevada. Limited details of the facies are available, but the Laurentian occurrences may record the development of a shallow-marine Bakken facies. Finally, the type of locality of the lowest Middle Cambrian *Oryctocephala indicus*-plexus is in dark mudstones that abruptly overlie a subaerial unconformity surface [68,69], which is best interpreted as the abrupt onlap of shallow-water of a low-oxygen Bakken facies. Via their interpretation, and with no supporting biostratigraphy, Jiang et al. [60] interpreted their ca. 511–510 Ma age on the Kalkarindji LIP as the Lower–Middle Cambrian boundary age. A more precise bracket on the boundary is 507.7–506.6 Ma [67], with the ±4 Ma resolution of the Kalkarindji date overlapping a possible age for the Sinsk organic-rich mudstones.

3.1.4. Later Cambrian–Lowest Ordovician Boundary Interval Hatch Hill d/a and Alum Shale Intervals

Two distinct major organic-rich mudstones characterize the later Cambrian–lowest Ordovician (lower Tremadocian) on a number of paleocontinents. The Hatch Hill d/a interval (Figure 1, interval 4) in the Taconian allochthons of eastern New York, along the St. Lawrence River of southern Quebec, and western Newfoundland, is a long-term (late Early Cambrian–early Tremadocian) composite OMZ [18]. The designation "composite" is used for this heterogeneous association of black shales with slope-derived, carbonate clast debris flows, turbiditic sandstones, and local thin-bedded ("ribbon") limestones. This "composite" interval likely consists of a succession of macroscale sequences linked to local epeirogenic sea-level and eustatic changes. Thus, evidence for possible eustatic changes in the Lower–Middle Cambrian boundary interval, traditionally termed the global "Hawke Bay regression" but re-interpreted as a local epeirogenic slowdown in the epeirogenic subsidence rate in its "type" area of NE Laurentia [22], remains unknown in the Hatch Hill interval.

The Hatch Hill d/a interval is recognized in the parautochthonous Franklin Basin in NW Vermont, which is located north of the Taconic allochthon in eastern New York. The Franklin Basin originated in the late Early Cambrian via the abrupt foundering of the ENE-trending Ottawa aulacogen, which is the passive arm of the triple junction that defined the eastern NY–St. Lawrence River embayment of NE Laurentia [70]. This foundering also extended along strike into the Vermont passive margin shelf, which rapidly subsided to form the Franklin Basin. Indeed, Early Cambrian peritidal carbonate platform carbonates (Dunham Formation) are abruptly overlain by strongly anoxic/dysoxic black shales [61]. The late Early Cambrian–Upper Cambrian succession of black slate-dominated units in the Franklin Basin (i.e., Parker-St. Albans, Skeels Corners, Humgerford, Gorge formations [71]) is likely an upper-slope sequence of the Browns Pond dysoxic-anoxic interval OMZ, with the Gorge Formation likely deposited at a depth of only ca. 100 m [72].

The second later Cambrian–lowest Ordovician interval of organic-rich shale deposition (Figure 1, interval 4a) is a Bakken facies known on several paleocontinents. It has a later onset (middle Middle Cambrian) than the Laurentian Hatch Hill OMZ but also persists into the early Tremadocian for a ca. 26 Ma duration [21]. Interval 4a includes the Alum Shale in southern Scandinavia, the Baltic states, and adjacent Russia and Poland. The coeval interval on the Avalonia paleocontinent in NE North America and southern Britain is a regionally extensive succession of three unconformably bounded units. In North American Avalonia, these are the successive Manuels River Formation (black mudstones), the McLean Brook Group (black mudstones and thin shelf sandstones), and the black mudstone-dominated Chesley Drive Group [21,69,73]. The Alum Shale TOC in southern Norway is high and was used as a petroleum source in World War II.

The onset of interval 4a deposition is essentially simultaneous in Avalonia and Baltica (i.e., within the seemingly short upper *Ptychagnostus gibbus* Zone as defined in Baltica) [21]. Initial deposition of 4a was often initiated on an unconformity surface with initial shoreline sediments composed of black muds and even highly pyritic remané sandstone (Figure 4a,b). Although a number of reports have traditionally regarded the Alum Shale and other organic-rich black shales as a deep shelf facies (150–900 m) [74,75], a shallow-marine origin is suggested by evidence for photosynthetic activity, i.e., both benthic algal mats [76,77] and stromatolites with cyanobacterial filaments, as well as possible algal mats on large calcite crystals that grew on the sea floor [78,79].

The Alum Shale is a superb example of the "expanding puddle model" by which the areal extent of a low-oxygen, organic-rich water mass expands laterally across the shelf with the rise in sea level [25] (Figure 3c). Indeed, the Alum Shale progressively onlaps to the north and northeast through the Late Cambrian [80]. This cumulative onlap was not continuous and the Alum Shale shows numerous offlaps, often marked by calcareous nodules and nodular limestones prior to its further onlap directly across these unconformity surfaces [81].

A surprising development in the Alum Shale is a number of erosional surfaces with karstic cavities that record subaerial exposure [82]. As noted above (Figure 5), a similar development of subaerial exposure and weathering occurs in the black mudstones of the Upper Cambrian Chesley Drive Group of Cape Breton Island, Nova Scotia [83]. The Nova Scotia occurrence shows the development of an unconformity surface mantled with eroded limestone nodules, black mudstone clasts, phosphate nodules, and brachiopod and trilobite debris cut by caliche-lined fractures (Figure 5).

In short, the upper part of the Cambrian–lowest Ordovician, lower Tremadocian not only features a black mudstone-dominated OMZ (Hatch Hill d/a interval) off NE Laurentia but also a shallow-shelf black shale Bakken facies in Baltica and Avalonia. These persistent, major black shale successions roughly correspond to high eustatic levels (Figure 1, intervals 4, 4a) which, at their maximum, covered all of the Avalonian platform and much of southern Scandinavia and the East European platform in shallow-water black mud. Coeval Laurentia, except for shallow sandstones of the northern Mississippi River valley region, was blanketed by shelf carbonates.

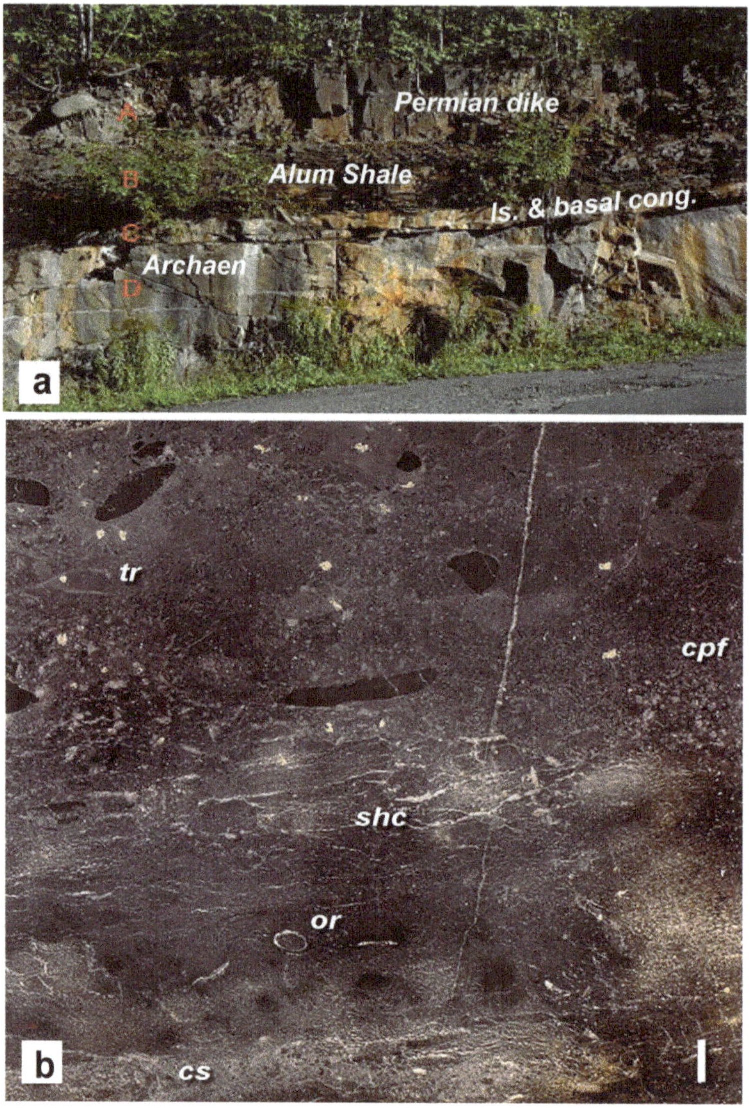

Figure 4. Unconformable subaerial onlaps of Cambrian Bakken black shales: (**a**) Alum Shale on Archaen at Slemmmertad, Oslofjord, southern Norway (from Creative Commons Attribution 2.5 Generic License, by J.O.R. Ebbestad and E. Sundring); (**b**) Basal pyritic remané sandstone, feather edge of upper Middle Cambrian Cavendish Formation, bed 125 of Howell [79], on Manuels River Formation, Manuels River, southeastern Newfoundland. Sandstone has a basal thin layer of coarse quartz sand and caliche fabrics (cpf, undifferentiated crystic plasmic fabric; shc, subhorizontal cracks). The golden area at the lower right is a mixture of caliche and fine-grained, euhedral pyrite. Orthothecid hyolith conch (or) and trilobite (tr) fragments. Scale bar = 1 cm.

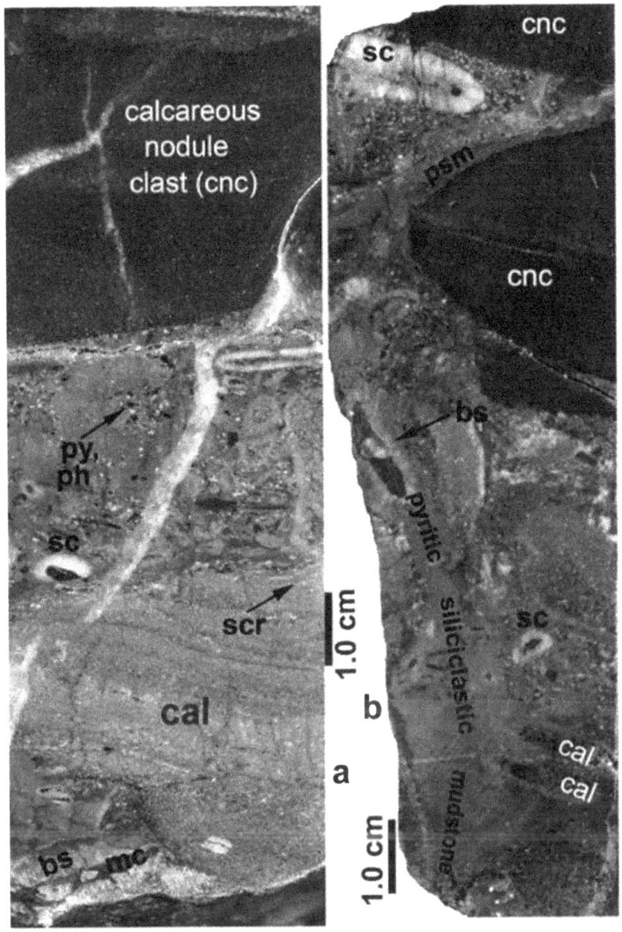

Figure 5. Caliche development in subaerially exposed, shallow-marine (Bakkan facies) black mudstone with carbonate and dark mudstone clasts and dyke of pyritic (psm) filling intraclast space in the Upper Cambrian Chesley Drive Group, Cape Breton Island, eastern Canada [82]. (a) and (b) show large, dark grey, fine-grained, diagenetic calcareous nodules at the top of the bed (cns) and lenses of phosphatic sand and granules (ph) and pyrite sand (ps) that yielded *Leptoplastus* cf. *ovatus* trilobite sclerites. The lower part of the figure depicts clasts of non-fossiliferous, light-to-medium grey, laminar caliche hardpan caliche (cal) composed of microsparitic carbonate laminae with quartz sand (small white grains), scattered darker grey glaebules (gl), and sub-horizontal, caliche calcite spar (scr). Small pebbles include rounded white clasts dominated by syntaxial calcite rims (sc) and sparry (bladed) calcite (bs) that envelope fragments of black siliciclastic mudstone and caliche. Pyritic siliciclastic mudstone (psm) dyke fills interclast space.

3.2. Lower–Lower Middle Ordovician OMZ Black Shales and Coeval Bakken Facies

A basis for the correlation of Lower–lower Middle Ordovician OMZ black shales and shallow-water Bakken facies is provided via the conodont-based biostratigraphy of successions in the Taconic allochthons from eastern Pennsylvania and eastern New York through western Newfoundland. As noted above, the Taconian allochthons preserve the only remaining, unambiguous continental slope sequences of the early Paleozoic. This is a greenish mudstone-dominated stratigraphic interval with five named black mudstone OMZ intervals [18,22,61] (Figure 1, intervals 5–9).

This interval ends with the deposition of a condensed, upper Middle Ordovician red mudstone with radiolarian cherts and volcanic ashes from the Taconic arc. This red mudstone (Indian River Formation in the Taconian allochthon of eastern New York and western Vermont) reflects passage of the peripheral bulge in front of the advancing Taconic arc and accretionary prism [25,61,84]. Volcanic ashes give U-Pb zircon dates of ca. 466.2–464.2 Ma on the Indian River Formation [85] and allow for correlation into the middle Darriwillian Stage, which comprises the unconformity bounded Chazy Group carbonates (base of Tippeecanoe Megasequence, Figure 1) on the eastern New York platform.

The five successive mesoscale OMZ black shales (Figure 1, intervals 5–9) in the greenish mudstone-dominated, Lower–lower Middle Ordovician of the Taconian overthrust belt are geographically limited. They occur at or within several km inboard of the leading edge of the outermost thrust sheet of the Taconian allochthon in eastern NY–western Vermont and along the south shore of St. Lawrence River in Quebec. These OMZs cannot be recognized further east or SE even within the outermost thrust and seem to lens out away from the upper slope. Three of the OMZs (Figure 1, intervals 5–7) are known to occur on the upper slope, such as at Levis, Quebec, where wave ripples occur on the top of carbonate clast debris flows [86].

Almost every one of the relatively thin (1–15 m) black shale OMZs can be correlated biostratigraphically with a eustatic onlap that defines one of the unconformity-bounded formations of the Beekmantown Group on the eastern New York–western Vermont shelf. These carbonate-dominated formations were also the sources for thin-bedded ("ribbon") turbiditic (easterly transport) and contourite limestones that occur in the upper part of the five OMZ black shales [18,22,61]. As each of the OMZ black shales can be correlated onto another paleocontinent, they are regarded as eustatic records through the earlier Ordovician [18,61].

The succession and global correlation of Lower–lower Middle Ordovician mesoscale black OMZs (Figure 1, intervals 5–9), as earlier detailed [15,61], include the following.

3.2.1. Schaghticoke d/a Interval

The upper lower Tremadocian Schaghticoke d/a interval (Figure 1, interval 5) is correlated with the trans-Laurentian Stonehenge transgression [87]. The Schaghticoke d/a OMZ correlates across Laurentia with carbonate platform units that unconformably overlie the Upper Cambrian such as the Tribes Hill Formation in eastern New York, Oneota Dolostone in the upper Mississippi River valley, and Manitou Formation in the Colorado Rocky Mountains. The middle of the Tribes Hill Formation is a 1–3 m thick mudstone (Van Wie Member) that changes from black to light grey over a 60 km E–W distance in eastern New York. The presence of a mollusk-dominated limestone without deeper-marine faunal elements (i.e., trilobites) under the Van Wie; the 25 cm thick, wave-rippled carbonate sand at its base; and the massive thrombolite cap all suggest a shallow-marine, Bakken facies for the Van Wie Member.

That this thin black shale with upwelling drove water from an OMZ across tens of kilometers of shallow carbonate shelf seems unlikely (discussed above [45]).

3.2.2. Rte. 299 d/a Interval

A second OMZ black mudstone (Figure 1, interval 6) associated with sea level rise is the late (but not latest) Tremadocian Rte. 299 d/a interval (Laurentian *Macerodus dianae* Zone, conodonts). This OMZ is known only from the Taconian allochthon in southern Quebec [18,61]. On the NE Laurentian platform, it includes carbonate platform units with basal unconformities in western Newfoundland (middle Boat Harbour Formation), the Ottawa aulacogen (Theresa Formation), eastern New York (Rochdale Formation and its junior synonym the "Fort Ann Formation"), and an unnamed black shale and limestone interval in the outer shelf of the Frederick valley, central Maryland [88]. The Rte. 299 d/a interval could also mark a transgression driven by strong epeirogenic subsidence in NE Laurentia as it does not seem to be present on other paleocontinents.

3.2.3. Levis d/a Interval

The Levis d/a OMZ (Figure 1, interval 7) is widespread in the Taconian allochthons of eastern Pennsylvania, New York, and Quebec, but coeval shelf black shales are not known. This OMZ brackets the Tremadocian–Floian boundary interval (Laurentian upper *Acodus deltatus* Zone, conodonts). There is no shelf equivalent in eastern New York, but this interval includes "cycle 3" in the middle Rockdale Run Formation, central Maryland shelf [88]; the unconformity bounded middle Boat Harbour Formation, western Newfoundland shelf; the Kinblade Formation on the Arkansas shelf; and the Billingen onlap (,) in Baltica [18,61,89].

3.2.4. Laignet Point d/a Interval

The Laignet Point d/a OMZ (Figure 1, interval 8) records a strong sea level rise in the upper Lower Ordovician (Floian) across eastern Laurentia from northern Greenland and western Newfoundland to eastern New York (Fort Cassin Formation). This strong sea level rise includes the stratigraphically abrupt appearance of Bakken facies-type black shale deposit (*Didymograptus protobifidus* Zone, graptolites) on the Laurentian platform and outer shelf in Arkansas, Oklahoma, Nevada, and western Utah. It is correlated with the *Oepikodus evae* drowning event in Baltica [89], with the occurrence on two paleocontinents showing it to be eustatic [18,61].

3.2.5. Raceville d/a Interval

The Raceville d/a interval is the highest OMZ in the Taconian allochthons before the onset of the Taconian orogeny in the Appalachians SW of western Newfoundland (Figure 1, interval 9). This OMZ and its carbonate platform equivalents are lower—but not the lowest—Middle Ordovician units (middle Dapingian–lower Darriwillian stages), as shown by the bracketing by graptolites that go as low as the *Isograptus gibberulus* Zone, with *Undulograptus austrodentatus* Zone graptolites at its top. The Raceville OMZ correlates with onlaps of carbonate platform units with basal unconformities in western Newfoundland (upper Aguathuna Formation); New York, Vermont, and adjacent Quebec (Providence Island Formation and junior synonyms Carillon and Bridport formations); and Joins Formation in Oklahoma [22,61]. The Raceville OMZ was likely associated with earliest Middle Ordovician eustatic rise given the appearance of the slightly older bases of the petroliferous Kanosh Shale in the Ibex area, west-central Utah [90], and the correlative "basal Llanvirnian drowning event" in Baltica [89].

3.3. *Upper Ordovician Black Shales*

Two Late Ordovician black shale OMZs occur in the Taconian allochthon of eastern New York. These OMZs occur above the upper Middle Ordovician red mudstones of the Indian River Formation (discussed above) and record synorogenic developments on the east Laurentian slope during the onset of the Taconian orogeny [19,61].

3.3.1. Glenmont d/a Interval

The lower unit is the Glenmont d/a (Figure 1, interval 10), which is a black shale interval in the dominantly greenish mudstones of the Mount Merino Formation. The Glenmont d/a has the classic graptolites of the Normanskill Group in eastern New York [91]. These graptolites (*Climacograptus bicornis* Zone) indicate a correlation into the unconformity-bounded Black River Group carbonates on the New York shelf and coeval carbonates across Laurentia [18,22,61].

3.3.2. Mohawkian d/a Interval

The higher Mohawkian d/a interval (Figure 1, interval 11) [18,61] is a long interval in the Upper Ordovician. Its lower part includes the black mudstone and turbiditic sandstones in synorogenic flysch. This trench fill is the highest unit (Austin Glen Formation, upper *Climacograptus bicornis*—at least *Corynoides americanus* zones) in the Taconic allochthon of eastern New York [92]. The Austin Glen reflects rapid burial and deposition under a low-oxygen water mass in the early Upper Ordovician (Sandbian and lower Katian stages). These d/a conditions spread onto the eastern margin of Laurentia with the foundering of the forearc region as the allochthonous Taconian orogen (i.e., accretionary prism and volcanic arc) moved on to eastern New York.

Erosion of the moving and subaerially emergent allochthon led to the deposition of synorogenic fore arc sandstones and mudstones (Snake Hill Formation) that are overrun by the allochthon and are difficult to distinguish from the Austin Glen Formation in the allochthon. The Snake Hill has yielded lower Katian graptolites (*Orthograptus ruedemanni* Zone) and likely overlaps the upper Austin Glen in terms of age [33]. In eastern New York, the synorogenic flysch rapidly changes westward into the black and brown mudstones of the Utica Shale that prograde west through the middle Katian before. With aggradation, they are replaced by overlying shallow-marine shales and sandstones of the Queenston delta that persist to the Ordovician–Silurian boundary unconformity (North American Cherokee unconformity) [92].

The Utica Shale and synonymous units overlie a large area east of the Mississippi River valley and west of the folded and thrust-faulted Taconian belt of the western Appalachian Mountains. Utica Shale facies are the oldest synorogenic siliciclastics in the Martinsburg forearc trough in front of the overthrust belt in southeastern New York– northern Virginia. As a result of erosion with the westward movement and the uplift of the Taconian peripheral bulge, the Utica Shale overlies units as low as Lower Ordovician carbonates, with its western onlap across New York State (Figure 6). With regional extension later in the Taconian orogeny, the NNE striking Sebree trough subsided along basement faults, with the Utica deposited being the lowest synorogenic unit in the trough.

Figure 6. Utica shale contacts: (**a**) Succession at Floral Park on Canajoharie Creek, Canajoharie, eastern New York, showing the Upper Ordovician Trenton Group unconformity (dashed white line) on the Lower Ordovician Tribes Hill Formation and Utica Shale unconformity on Trenton; (**b**) Utica Shale (upper) unconformity on Trenton Group, Bay City core, eastern Michigan, USA (from [93]; Figures 2 and 3, master's thesis—open access).

4. Upper Ordovician Black Mudstones—Not Deep Water

4.1. Sevier Belt

In eastern North America, two major regional developments of black mudstone deposition took place in the forearcs of two successive stages of the Taconian orogeny. The earlier, or Blountian stage (late Middle Ordovician [late Darriwillian Stage] to early Late Ordovician [Sandbian Stage]), features black mudstones (Athens, Blockhouse, Paperville, Rich Valley shales). These shales were deposited in a long (ca. 1200 km), NNE-trending

Sevier belt that extends from northern Alabama and Georgia to southern Virginia [28]. The Sevier belt is a remarkably narrow (to 50 km) belt in the Valley and Ridge province of the western Appalachians.

The stratigraphic and sedimentologic data indicate the very rapid deposition of ca. 2.5 km of black mudstones (>1 m/Ka) with the foundering of a forearc belt. What proves to be an idealized depositional motif is a lateral transition from east to west from dysoxic/anoxic black mudstone to somewhat calcareous shales with limestone turbidites, then to carbonate clast debris flows from a deep shelf, and then a peritidal carbonate shelf [28,94].

The model of an abrupt peritidal–bathyal transition along the Sevier belt is problematic and has no modern analog. To begin with, what is presumed in these stratigraphic and depositional environment summaries [94–97], as well as the subsequent commentaries on the Sevier belt [28,31], is a Black Sea model with a purported 800–1000 m mudstone deposition depth.

This water depth is problematic because the dominant facies motif along the Sevier trough [25] is actually not a lateral transition from the black shales–turbidites and debris flows–deep shelf to the peritidal limestone; rather, it is generally an upward transition from a peritidal carbonate (Lenoir Limestone) abruptly into black shale at a bedding plane, with this transition being diachronous over several upper Middle and lower Upper Ordovician graptolite and conodont zones along the Sevier belt [28]. In short, the proposed depositional model cannot be a *local* sudden foundering at sites along the Sevier belt of "~250 to 800 m" [25]; it records a cumulative south-to-north and east-to-west peritidal carbonate–black mudstone facies transition. Again, the designation "local" is used because the peritidal carbonate–black mudstone transition is diachronous and changes by ca. 8 Ma northward along the Sevier belt (see Ordovician geochronology in [98]). The tectonic mechanism for cumulative foundering along a thousand kilometers of the Sevier belt, with no evidence of a down-to-basin faulting of the Lenoir or a turbulent deposition on the Lenoir, is unexplained by a deep-water Black Sea model for black mud deposition.

4.2. Upper Ordovician Utica Shale Black and Brown Mudstones and Sebree Trough

In eastern New York, d/a conditions are shown right in front of the Taconic overthrust by the younger Snake Hill Formation (*Corynoides americanus–Orthograptus ruedemanni* zones) [91]. One of the few studies that have examined depositional conditions in front of the advancing Taconian overthrust [33] concluded that anoxic/dysoxic black mud deposition likely took place at water depths of a few tens of meters, with black muds rapidly replaced by a shoaling-up sequence that includes turbidite sands and sandstone clast debris flows, higher HCS sandstones, and then shoreface sandstones deposited in front of, and then overrun by, the Taconic allochthon.

Thus, *O. ruedemanni* Zone graptolites from the black shale of the lower Snake Hill Formation mark the onset of a shoaling-up succession [33]. This shoaling-up succession correlates ca. 120 km west in the Mohawk River valley into an offlap interval where the middle Utica Shale is laterally transitional into turbiditic limestones of the Dolgeville facies and then into the peritidal, shell-hash rich Denley and Steuben limestones [99].

Even younger forearc shales ("Utica Shale" and equivalents) record subsequent onlap and dysoxic/anoxic mud deposition that spread across the Laurentian Midcontinent into the Midwest (i.e., Ohio, Indiana, upper Michigan and north into central Quebec, Canada) [99,100]. The designation "Utica Shale" is effectively a group-level designation in its type-area in New York, where it is divided into the black shale-dominated lower Flat Creek and upper Indian Castle formations [99].

The Utica Shale and synonymous units consequently overlie a large area east of the Mississippi River valley and west of the folded and thrust-faulted Taconian orogenic belt of the western Appalachian Mountains. Utica Shale facies are the oldest synorogenic siliciclastics in the Martinsburg forearc trough in front of the overthrust belt in southeastern New York–northern Virginia. As a result of erosion following the movement of the Taconian peripheral bulge, Utica black shales overlie units as low as Lower Ordovician carbonates, with their onlap across New York State [31].

Simultaneously with the Utica Shale onlap in northerly areas, regional extension in the Taconian orogeny later opened a second fore arc trough called the Sebree trough. The 1000 km long Sebree trough extends from the southern Laurentian margin with the Iapetus Ocean in the Ouachita Mountains region of Oklahoma and persists into Ohio. It parallels the older Sevier trough (discussed above) but is ca. 500 km to the west and has a smaller thickness of organic-rich black shales. Both the Sevier and Sebree troughs opened along old fault lines in the Mesoproterozoic basement of the Laurentia paleocontinent [31].

The Sebree trough records two successive intervals of extension. The older featured organic-rich sediment deposited somewhat earlier in the Late Ordovician (lower Katian Stage) and coeval with shelf carbonates of the Lexington Limestone to the east, the Galena Group to the west, and within and on the Trenton Group on the axis of the trough and to the north. Subsequent uplift and erosion removed this earlier episode of organic-rich deposition, but subsequent subsidence due to regional extension allowed the Utica and coeval units to be deposited in the Sebree trough [34,37].

Studies of depositional environments of the Utica Shale are incomplete by comparison with those on the Middle Paleozoic black shales, which led to the proposal of their shallow-water deposition. The latter include work on the Middle and Upper Devonian in New York State [32,101] (Figure 1, intervals 15–17); the Upper Devonian part of the Chattanooga Shale in the Midcontinent of the U.S. [42,102–104] (Figure 1, interval C); and the uppermost Upper Devonian and lowest Mississippian Bakken Shale [6] (Figure 1, intervals 17, 18).

A brief survey of the shallow-marine deposition of the Utica Shale includes features such as unconformities on subaerially exposed older units as low as the Lower Ordovician. This basal Utica Shale unconformity extends laterally east and west into local subaerial unconformity on the lower Upper Ordovician Trenton Group (Figure 6). Similarly, a younger (and second) sub-Utica Shale unconformity is with the uppermost Trenton Group [99].

These unconformities show an abrupt transition into black shale at a bedding plane with transgression of the Utica muddy facies (Figure 6). These contacts may show the mm thicknesses of remané sediment (e.g., phosphatic grains) on eroded surfaces, but these contacts do not feature significant accumulations of reworked/lithified sediment or significant mineralization. If present, such features might suggest that the basal Utica contact represents a long time interval, but their absence is evidence that the contact marks a brief transgressive event featuring dysoxic/anoxic water onlap at the paleoshoreline.

This Bakken-type unconformity, marked by the onlap of anoxic/dysoxic water across a subaerial unconformity, is discussed above for the Cambrian Manuels River and Alum Shale formations (Figures 4b and 5). Similarly, the lowest beds of the Utica locally feature wave-generated cross bedding (Figure 7), ripples, burrowing (indicating at least periodic oxygenation), wave-disarticulated brachiopods, and evidence of algal blooms (i.e., presence of amorphinite). These features are consistent with shallow-marine deposition [43,104,105]. Somewhat surprisingly, ripples, HCS-like beds, and large cross beds are known in black shales, but these are present as the mud formed floccules that behaved similarly and produced bed forms similar to sand grains responding to wave and current control [102,103].

Figure 7. Lower Whitehouse formation shales (upper Utica Shale equivalent) with wave-cross beds formed on mud floccules. Whitehouse unconformably overlies karstic limestone with corals of the eroded lower Saluda Formation. A 25 mm diameter coin is depicted at contact. Rte. 421 road cut at Madison, Indiana. Photograph courtesy of C.E. Brett.

Once an inundation had taken place, little evidence is available to determine the depth of Utica Shale deposition. Although depths > 150 m are proposed in many reports [29], this is based on the presumed modern analog, i.e., the modern Black Sea [106]. However, Utica Shale thicknesses are rarely more than 100 m and underlie open- shelf, wave-dominated sandstones. This stratigraphic context suggests that Utica facies were relatively shallow. Lateral facies change from Utica Shale mudstones into peritidal carbonates; the Denley and Steuben limestones in central New York State [99] are consistent with, but do not prove, relatively shallow Utica deposition. A suggestion of shallow Utica Shale deposition comes from drill cores along a 300 km, NW–SE transect across the northern Appalachian basin from southern Michigan to western West Virginia [104,105]. The cores show a relatively uniform ca. 400 m thickness of the Trenton Group carbonate shelf and Utica Shale, with the upper Utica onlapping the Trenton. As the Trenton and Utica are bracketed by underlying shallow-water Black River Group carbonates and overlying Cincinnatian limestones, there is no possibility of a Utica depocenter that would accommodate significantly deeper Utica facies [107].

5. Review of Phanerozoic Black Mudstones

Most of the Phanerozoic black shales (i.e., OMZ (marginal oxygen-minimum zones); platform Bakken facies; and OAE (ocean anoxic events)) of Figure 1 have been outlined previously [11,16]. The red bars suggest equation with large igneous province (LIP) activity. The limited temporal resolution of LIP events is limited, and their equation with organic-rich mudstones is tentative [60]. Sections 3 and 4 of this report describe early Paleozoic dysoxic/anoxic black shale intervals (Figure 1, intervals 1–11).

Post-early Paleozoic black shales include intervals: 12, Llandoverian; 13, Pridolian; 14, Emsian; 15 Givetian, with Bakken facies black mudstones being unconformable on somewhat older limestones as an initial forearc unit in Acadian orogeny, NE Laurentia

(Figure 8); 16–17, Frasnian–Famennian black shales [32]; 15–17, Bakken facies [104] of Chattanooga Shale [108]; 16 may correspond to Yakutsk-Viliya LIP; 18, lower Mississippian (lower Tournaisian); 17 and 18, Bakken Formation [6]; 19, core black shales (Bakken facies) of Pennsylvanian cyclothems, U.S. Midcontinent; 20, Gzeliaan (latest Carboniferous) Bakken facies; 21 and 22, Changhsingian (latest Permian) and lower Triassic Bakken facies, likely corresponding to Siberian traps; 23, upper Triassic (Rhaetian)–lower Jurassic correspond to Central Atlantic Magmatic Province; 24, upper lower Jurassic (Toaracian) with T-OAE (e.g., Posidonienschiefer); 25, Upper Jurassic (Oxfordian) OAE and Bakken facies [49]; 26, lowest Cretaceous (Valanginian–Hauterivian) OEA-1a may correspond to Pararà-Etendeka LIP; 27, lower Cretaceous (Albian), including OEA 1b and 1c; 28, Cenomanian–Turonian boundary, where OAE-2 may correspond to High Arctic LIP; 29, middle Cretaceous Coniacian–Santonian OAE-3; 30, Paleocene–Eocene boundary (PETM thermal maximum anoxic events), which may correspond to North Atlantic Igneous Province activity; and 31, middle Miocene (Langian)–Modern Mediterranean sapropels (S), Black Sea facies.

Figure 8. Middle Devonian (early Givetian) unconformity of synorogenic black Marcellus Shale (Figure 1, interval 15) on carbonate platform Onondaga Limestone (grey field in foreground with red arrow). Kaaterskill Creek, Catskill village, eastern New York State. Figures 8 and 9 photographs compliments of C.E. Brett.

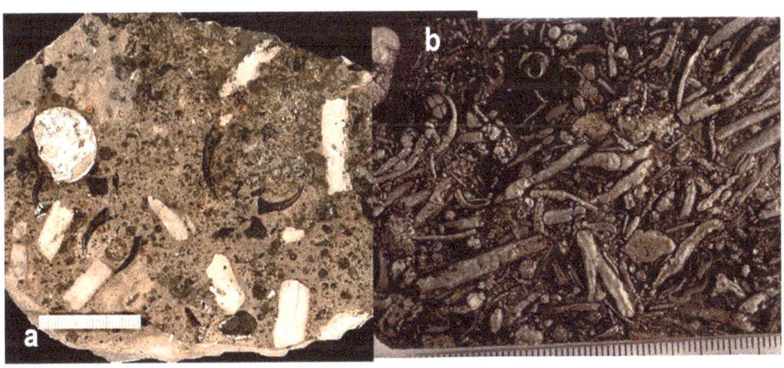

Figure 9. Remané sediment at base of black shale onlaps, note mm-scales: (**a**) Middle Devonian (Givetian) Onondaga Limestone–Marcellus Shale contact with phosphatized crinoid fragments, Syracuse, central New York State; (**b**) Givetian reworked, current-aligned pyritized burrows, Leicester pyrite, basal Genessee Formation and Geneseo Member, Erie County, western New York.

6. Discussion

Shallow-marine organic-rich, commonly black mudstones that overlie large tracts of paleocontinents and forearc basins, appear through the Paleozoic and Mesozoic. A consensus does not exist on the change in the Earth systems responsible for the occurrence of these economically important hydrocarbon source rocks. However, the frequently referenced equation of forestation of the land as driving organic-rich Devonian mudstone deposition due to increased nutrient run-off with root penetration and greater soil weathering [109] has to be questioned. Indeed, trans-continental and forearc, organic-rich shale deposition took place in the Cambrian and Ordovician of Baltica, Avalonian, Laurentia, and South China long before the appearance of significant land plants (Figure 1, intervals 1–3, 4a, 5, 8–11).

Trans-continental organic-rich mudstones such as the Late Ordovician Utica Shale and Devonian Chattanooga Shale record sea-level highs (Figure 1), which have been associated with the thickening and expansion of mid-water OMZs on paleocontinent margins. However, the origin of trans-continental organic-rich shales, unlike a number of proposals [1,58,61], likely did not involve the expansion of an OMZ onto shallow shelves (Figure 3b). Indeed, upwellings of marginal water do not cross the shelf break, even with persistent long-shore or offshore winds. Thus, there is a fundamental impossibility of proposals that feature the wind-driven Ekman transport of deeper, low-oxygen, marginal water and the deposition of organic-rich mudstone in forearc basins. A different mechanism for the appearance of black shales as the oldest synorogenic deposits in the east Laurentian late Middle and Late Ordovician Sevier and Sebree troughs is required. In addition, other physical limitations (thermal and salinity density contrast with shelf water; friction during transport of water masses) would likely pose problems for the movement of offshore OMZ water masses onto a paleocontinent's shelf [45].

The origin of these black shales on a number of paleocontinents (Laurentia, Baltica, Avalonia) and in the Ordovician forearc basins of Laurentia includes the onset of early Paleozoic black shale deposition, sometimes directly on a subaerial unconformity. This unexpected depositional history has no modern analog but is also present in later inundations of unconformity surfaces. A later example is the onlap of the synorogenic, Middle Devonian (Givetian) Marcellus Formation black shale (Figure 1, interval 14) on the somewhat older carbonate platform deposits of the Onondaga Limestone (Figure 8). The duration of the hiatus and initial deposition of the black shale may have been relatively protracted as the mineralization of some unconformity surfaces occurs (Figure 6b). In addition, some of these unconformities may have involved the relatively protracted development of an interval of oxic on-shore facies represented by thin lags of skeletal fragments that were cemented prior to black mud deposition (Figure 9). Another unexpected development is the presence of karstic unconformities and cavities in Late Cambrian Bakken black mudstones in Baltica and Avaonina (Figure 5). The continued shallow water deposition of at least the lower part of these Bakken facies mudstones is shown by HCS and the wave trough cross-bedded structures developed in beds of clay floccules, the photic zone algal mats, and the burrows and shell hash laminae (Figure 7).

Trans-continental Bakken facies comprising black and organic-rich mudstones require low-oxygen water masses that are proposed to have developed on continents via the solar heating (hyper-warming) of onlap waters, which reduced oxygen solubility in lower water layers, while photic algal mats developed in the water column as a source of organic carbon (Figures 2b and 3). An additional factor in the development of synorogenic black mudstones in forearc troughs would have been the increased turbidity due to suspended mud eroded from emergent orogens, thus lowering the oxygen in the bottom waters without photic activity.

The Bakken model has predictive value in this time of rapid climatic change. With modern eustatic rise with the melting of glaciers and the marine onlap of low-relief continental margins, the increased areal extent of low-latitude transgressive seas would be expected to transform land areas with high albedo into shallow seas with high insolation and high temperatures. These low-oxygen-shelf seas would feature the onset of a new interval of Bakken facies deposition.

Funding: This research received no external funding.

Data Availability Statement: Not applicable.

Acknowledgments: Much of E.L.'s field and lab work was financed by US National Science Foundation grants and the New York State Museum.

Conflicts of Interest: The author declares no conflicts of interest.

References

1. Duncan, D.C.; Swanson, V.E. Organic-rich shale of the United States and world land areas. *U.S. Geol. Surv. Circ.* **1966**, *523*, 30.
2. Snedden, J.W.; Cunningham, R.C.; Virdell, J.W. The northern Gulf of Mexico offshore super basin: Reservoirs, source rocks, seals, traps, and successes. *AAPG Bull.* **2020**, *104*, 2603–2642. [CrossRef]
3. Underhill, J.R.; Richardson, N. Geological controls on petroleum plays and future opportunities in the North Sea rift super basin. *AAPG Bull.* **2022**, *106*, 573–631. [CrossRef]
4. Fairhurst, E.; Ewing, T.; Lindsay, B. West Texas (Permian) Super Basin, United States: Tectonics, structural development, sedimentation, petroleum systems, and hydrocarbon reserves. *AAPG Bull.* **2021**, *105*, 1099–1147. [CrossRef]
5. Kendall, C.G.S.C.; Chiarenzelli, J.; Hassan, S. World source rock potential through geological time: A function of basin restriction, nutrient level, sedimentation rate, and sea-level rise. In Proceedings of the AAPG Annual Convention, Denver, CO, USA, 7–10 June 2009; 46p.
6. Petty, D.M. Shallow-water versus deep-water: Stratigraphic geometries in the organic-rich shale/mudstone debate. *GSA Today* **2022**, *32*, 4–10. [CrossRef]
7. Ryder, R.T. Assessment of Appalachian basin oil and gas resources; Utica-Lower Paleozoic Total Petroleum System, Chapter G.10. In *Coal and Petroleum Resources in the Appalachian Basin; Distribution, Geologic Framework, and Geochemical Character*; Ruppert, L.F., Ryder, R.T., Eds.; Chapter G.10 supersedes USGS Open-File Report 2008–1287; U.S. Geological Survey: Reston, VA, USA; p. 44.
8. Ulmishk, G.F. Petroleum geology and reservoirs of the West Siberian Basin, Russia. In *U.S. Geological Survey Bulletin 2201-G*; U.S. Geological Survey: Reston, VA, USA, 2004.
9. Khafizov, S.; Syngaevsky, P.; Dolson, J.C. The West Siberian Super Basin: The largest and most prolific hydrocarbon basin in the world. *AAPG Bull.* **2022**, *106*, 517–572. [CrossRef]
10. Landing, E.; Geyer, G.; Schmitz, M.D.; Wotte, T.; Kouchinsky, A. (Re)proposal of three Cambrian subsystems and their geochronology. *Episodes* **2021**, *44*, 273–283. [CrossRef]
11. Negri, A.; Wagner, T.; Meyers, P.A. Introduction to "Causes and consequences of organic carbon burial through time". *Palaeogeogr. Palaeoclimatol. Palaeoecol.* **2006**, *235*, 1–7. [CrossRef]
12. Meyers, P.A. Why are the $\delta^{13}C_{org}$ values in Phanerozoic black shales more negative than in modern marine organic matter? *Geochem. Geophys. Geosyst.* **2014**, *15*, 3085–3106. [CrossRef]
13. Hallam, A. *Phanerozoic Sea-Level Changes*; Columbia University Press: New York, NY, USA, 1992; 266p.
14. Hallam, A.; Bradshaw, M.J. Bituminous shales and oolitic ironstones as indicators of transgressions and regressions. *J. Geol. Sci.* **1979**, *136*, 157–164. [CrossRef]
15. Sloss, L.L. Sequences in cratonic interior of America. *GSA Bull.* **1963**, *74*, 93–114. [CrossRef]
16. Wignall, P.B.; Newton, R. Black shales on the basin margin: A model based on examples from the Upper Jurassic of the Boulonnais, northern France. *Sediment. Geol.* **2001**, *144*, 335–356. [CrossRef]
17. Słowakiewicz, M.; Tucker, M.E.; Perri, E.; Pancost, R.D. Nearshore euxinia in the photic zone of an ancient sea. *Palaeogeogr. Palaeoclimatol. Palaeoecol.* **2015**, *426*, 242–259. [CrossRef]
18. Landing, E. Time-specific black mudstones and global hyperwarming on the Cambrian–Ordovician slope and shelf of the Laurentia palaeocontinent. *Palaeogeog. Palaeoclim. Palaeoecol.* **2012**, *367–368*, 256–272. [CrossRef]
19. Landing, E. Extended Abstract—The Great American Carbonate Bank in northeast Laurentia: Its births, deaths, and linkage to continental slope oxygenation (Early Cambrian–Late Ordovician). In *The Great American Carbonate Bank, Essays in Honor of James Lee Wilson*; Derby, J.R., Fritz, R.D., Longacre, S.A., Morgan, W.A., Sternbach, C.A., Eds.; American Association of Petroleum Geologists: Tulsa, OK, USA, 2012; pp. 253a–260a.

20. Landing, E. Great American Carbonate Bank in northeast Laurentia: Its births, deaths, and linkage to continental slope oxygenation (Early Cambrian–Late Ordovician). In *The Great American Carbonate Bank, Essays in Honor of James Lee Wilson*; Derby, J.R., Fritz, R.D., Longacre, S.A., Morgan, W.A., Sternbach, C.A., Eds.; American Association of Petroleum Geologists: Tulsa, OK, USA, 2012; pp. 451–492.
21. Landing, E.; Westrop, S.R.; Geyer, G. Trans-Avalonian green–black boundary (early Middle Cambrian): Transform fault-driven epeirogeny and onset of 26 m.y. of shallow-marine, black mudstone in Avalonia (Rhode Island–Belgium) and Baltica. *Can. J. Earth Sci.* **2022**, *60*, 133–171. [CrossRef]
22. Landing, E.; Webster, M.; Bowser, S.S. Terminal Ediacaran–Late Ordovician of the NE Laurentia palaeocontinent: Rift–drift–onset of Taconic orogeny, sea-level change, and "Hawke Bay" onlap (not offlap). In *Supercontinents, Orogenesis, and Magmatism*; Nance, R.D., Strachan, R.A., Quesada, C., Lin, S., Eds.; The Geological Society of London: London, UK, 2023; p. 542. [CrossRef]
23. Demaison, G.J.; Moore, G.T. Anoxic environments and oil source bed genesis. *Amer. Assoc. Petrol. Geol. Bull.* **1980**, *64*, 1179–1209. [CrossRef]
24. Landing, E.; Keppie, J.D.; Keppie, F.D.; Geyer, G.; Westrop, S.R. Greater Avalonia—Latest Ediacaran–Ordovician "peribaltic" terrane bounded by continental margin prisms ("Ganderia," Harlech Dome, Meguma): Review, tectonic implications, and paleogeography. *Earth-Sci. Rev.* **2022**, *224*, 103863. [CrossRef]
25. Wignall, P.B. Model for transgressive black shales. *Geology* **1991**, *19*, 167–170. [CrossRef]
26. Wignall, P.B. *Black Shales*; Oxford University Press: New York, NY, USA, 1994.
27. Schlanger, S.O.; Jenkyns, H.C. Cretaceous oceanic anoxic events: Causes and consequences. *Geol. Mijnb.* **1976**, *55*, 179–184.
28. Finney, S.C.; Grubb, B.J.; Hatcher, R.D., Jr. Graphic correlation of Middle Ordovician graptolite shale, southern Appalachians: An approach for examining the subsidence and migration of a Taconic foreland basin. *GSA Bull.* **1996**, *108*, 355–371. [CrossRef]
29. Ettensohn, F.R.; Hohman, J.C.; Kulp, M.A.; Rast, N. Evidence and implications of possible far-field responses to Taconian orogeny: Middle–Late Ordovician Lexington Platform and Sebree Trough, east-central United States. *Southeast. Geol.* **2002**, *41*, 1–36.
30. Kolata, D.R.; Huff, W.D.; Bergström, S.M. The Ordovician Sebree Trough: An oceanic passage to the Midcontinent United States. *GSA Bull.* **2001**, *113*, 1067–1078. [CrossRef]
31. Ettensohn, F.R.; Lierman, R.T. Large-scale tectonic controls on the origin of Paleozoic dark-shale source-rock basins: Examples from the Appalachian Foreland Basin, eastern United States. In *Tectonics and Sedimentation: Implications for Petroleum Systems*; Gao, D., Ed.; American Association of Petroleum Geologists: Tulsa, OK, USA, 2012; pp. 95–124.
32. Smith, L.B.; Schieber, J.; Wilson, R. Shallow-water onlap model for the deposition of Devonian black shales in New York, USA. *Geology* **2019**, *47*, 279–283. [CrossRef]
33. English, A.M.; Landing, E.; Baird, G.C. Snake Hill—Reconstructing eastern Taconic foreland basin litho- and biofacies from a giant mélange block in eastern New York, USA. *Palaeogeog. Palaeoclim. Palaeoecol.* **2006**, *242*, 201–213. [CrossRef]
34. Ettensohn, F.R. Origin of Late Ordovician (mid-Mohawkian) temperate-water conditions on southeastern Laurentia: Glacial or tectonic? In *The Ordovician Earth System*; Finney, S.C., Berry, W.B.N., Eds.; The Geological Society of America: Boulder, CO, USA, 2010; pp. 163–175.
35. Kazmierczak, J.; Kremer, B.; Racki, G. Late Devonian marine anoxia challenged by benthic cyanobacterial mats. *Geobiology* **2012**, *10*, 371–383. [CrossRef]
36. Hints, R.; Hade, S.; Soesoo, A.; Voolma, M. Depositional framework of the East Baltic Tremadocian black shale revisited. *GFF* **2014**, *34*, 464–482. [CrossRef]
37. Leonowicz, P. Nearshore transgressive black shale from the Middle Jurassic shallow-marine succession from southern Poland. *Facies* **2016**, *62*, 16. [CrossRef]
38. Pohl, A.; Donnadieu, Y.; Le Hir, G.; Ladant, J.-B.; Dumas, C.; Alvarez-Solas, J.; Vandenbroucke, T.R.A. Glacial onset predated Late Ordovician climate cooling. *Paleoceanography* **2016**, *31*, 800–821. [CrossRef]
39. Turner, B.R.; Armstrong, H.A.; Wilson, C.R.; Makhlouf, I.M. High frequency eustatic sea-level changes during Middle and early Late Ordovician of southern Jordan: Indirect evidence for a Darriwilian ice age in Gondwana. *Sediment. Geol.* **2012**, *251–252*, 34–48. [CrossRef]
40. Torsvik, T.H.; Cocks, L.R.M. *Earth History and Palaeogeography*; Cambridge University Press: Cambridge, UK, 2016; 317p.
41. Sames, B.M.; Wagreich, M.; Wendler, J.E.; Haq, B.U.; Conrad, C.P.; Melinte-Dobrinescu, M.C.; Hu, X.; Wendler, I.E.; Wolfgring, E.; Yilmaz, I.Ö.; et al. Short-term sea-level changes in a greenhouse world—A view from the Cretaceous. *Palaeogeog. Palaeoclim. Palaeoecol.* **2016**, *441*, 393–411. [CrossRef]
42. Schieber, J. Mud re-distribution in epicontinental basins: Exploring likely processes. *Mar. Pet. Geol.* **2016**, *71*, 119–133. [CrossRef]
43. Smith, L. Shallow transgressive onlap model for Ordovician and Devonian organic-rich shales, New York State. In Proceedings of the Unconventional Resources Technology Conference, Denver, CO, USA, 12–14 August 2013. [CrossRef]
44. Landing, E.; Smith, L. Global hyperwarming and shallow marine deposition of hydrocarbon source rocks. *Geol. Soc. Am. Abstr. Programs* **2012**, *44*, 125.

45. Landing, E.; Geyer, G. Comment: Carbonate production and reef building under ferruginous seawater conditions in the Cambrian rift branches of the Avalon Zone, Newfoundland by J.J. Álvaro and A. Mills. *Sedimentology* **2025**, *71*, 1245–1269.
46. Pope, M.C.; Steffen, J.B. Widespread, prolonged late Middle to Late Ordovician upwelling in North America: A proxy record of glaciation? *Geology* **2015**, *31*, 63–66. [CrossRef]
47. Atkinson, L.P.; Yoder, J.A.; Lee, T.N. Review of upwelling off the southeastern United States and its effect on continental-shelf nutrient concentrations and primary productivity. *Rapp. Procès-Verbaux Réunions* **1984**, *183*, 70–78.
48. Freely, R.A.; Sabine, C.L.; Hernandez-Ayon, J.M.; Ianson, D.; Hales, B. Evidence for upwelling of corrosive "acidified" water onto the continental shelf. *Science* **2008**, *320*, 1490–1492. [CrossRef]
49. Hill, R.B.; Johnson, J.A. A three-dimensional theory of coastal currents and upwelling over a continental shelf. *Tellus* **1975**, *23*, 249–258. [CrossRef]
50. Mohrholz, V.; Eggert, A.; Junker, T.; Nausch, G.; Ohde, T.; Schmidt, M. Cross shelf hydrographic and hydrochemical conditions and their short term variability at the northern Benguela during a normal upwelling season. *J. Mar. Syst.* **2019**, *140*, 92–110. [CrossRef]
51. Berner, R.A.; Kothavala, Z. Geocarb III: A revised model of atmospheric CO_2 over Phanerozoic time. *Am. J. Sci.* **2001**, *304*, 397–437. [CrossRef]
52. Jackson Davis, W. The relationship between atmospheric carbon dioxide and global temperatures for the last 425 million years. *Climate* **2017**, *5*, 76. [CrossRef]
53. Schaefer, K.E. Effects of increased ambient CO_2 levels on human and animal health. *Experentia* **1982**, *38*, 1163–1168. [CrossRef] [PubMed]
54. Ezraty, B.; Chabalier, M.; Maisonneive, E.; Dukan, S. CO_2 exacerbates oxygen toxicity. *EMBO Rep.* **2011**, *12*, 321–326. [CrossRef]
55. Rothman, D.H. Atmospheric carbon dioxide levels for the last 500 million years. *Proc. Natl. Acad. Sci. USA* **2002**, *99*, 4167–4171. [CrossRef] [PubMed]
56. Judd, E.J.; Tierney, J.E.; Lunt, D.J.; Montañez, I.P.; Huber, B.T.; Wing, S.L.; Valdes, P.J. A 485-million-year history of Earth's surface temperature. *Science* **2024**, *385*, eadk3705. [CrossRef] [PubMed]
57. Arrhenius, S. On the influence of carbonic acid in the air upon the temperature on the ground. *Philos. Mag. J. Sci.* **1896**, *41*, 237–276. [CrossRef]
58. Jenkyns, H.C. Geochemistry of oceanic anoxic events. *Geochem. Geophys. Geosyst.* **2010**, *11*, Q03004. [CrossRef]
59. Bergman, S.C.; Eldrett, J.S.; Minisin, D. Phanerozoic large igneous province, petroleum system, and source rock links. In *Large Igneous Provinces: A Driver of Global Environmental and Biotic Changes*; Dickson, A.J., Bekker, A., Eds.; American Geophysical Union; Wiley: Washington, DC, USA, 2021; pp. 191–228. [CrossRef]
60. Jiang, Q.; Jourdan, F.; Olierook, H.K.H.; Renaud, E.M. An appraisal of the ages of Phanerozoic large igneous provinces. *Earth-Sci. Rev.* **2023**, *237*, 104314. [CrossRef]
61. Landing, S. Ediacaran–Ordovician of east Laurentia—geologic setting and controls on deposition along the New York Promontory. In *Ediacaran–Ordovician of East Laurentia—S. W. Ford Memorial Volume*; Landing, E., Ed.; New York State Museum: Albany, NY, USA, 2007; Volume 510, pp. 5–24.
62. Yang, C.; Bowyer, F.T.; Condon, D.J.; Li, X.-H.; Zhu, M. New U-Pb age from the Shuijingtuo Formation (Yangtze Gorges area) and its implications for the Cambrian timescale. *Palaeogeogr. Palaeoclimatol. Palaeoecol.* **2023**, *616*, 111477. [CrossRef]
63. Geyer, G. A comprehensive Cambrian correlation chart. *Epis. J. Int. Geosci.* **2019**, *42*, 321–332. [CrossRef]
64. Varlamov, A.I.; Rozanov, A.Y.; Khomentovsky, V.V.; Shabanov, Y.Y.; Abaimova, G.P.; Demidenko, Y.E.; Karlova, G.A.; Korovnikov, I.V.; Luchinina, V.A.; Malakhovskaya, Y.E.; et al. The Cambrian System of the Siberian Platform. Part 1: The Aldan-Lena region. In *XIII Field Conference of the Cambrian Stage Subcommission Working Group*; PIN Russian Academy of Science: Moscow, Russia, 2008; p. 300.
65. Bowyer, F.T.; Zhuravlev, A.Y.; Wood, R.; Zhao, F.; Sukhov, S.S.; Alexander, R.D.; Poulton, S.W.; Zhu, M. Implications of an integrated late Ediacaran to early Cambrian stratigraphy of the Siberian Platform, Russia. *GSA Bull.* **2023**, *135*, 2428–2450. [CrossRef]
66. Ivantsov, A.Y.; Zhuravlev, A.Y.; Leguta, A.V.; Krassilova, A.V.; Melnikova, L.M.; Ushatinskaya, G.T. Palaeoecology of the Early Cambrian Sinsk biota from the Siberian Platform. *Palaeogeogr. Palaeoclimatol. Palaeoecol.* **2005**, *205*, 69–88. [CrossRef]
67. Landing, E.; Schmitz, M.D.; Westrop, S.R.; Geyer, G. U-Pb zircon dates from North American and British Avalonia bracket the Early–Middle Cambrian boundary interval, with evaluation of the Miaolingian Series as a global unit. *Geol. Mag.* **2023**, *160*, 1790–1816. [CrossRef]
68. Myrow, P.M.; Goodge, J.W.; Glenn ABrock, G.A.; Betts, M.J.; Park, T.-Y.; Hughes, N.C.; Gaines, R.R. Tectonic trigger to the first major extinction of the Phanerozoic: The early Cambrian Sinsk event. *Sci. Adv.* **2024**, *10*, eadl3452. [PubMed]
69. Singh, I.B.; Singla, G.; Bhargava, O.N.; Kaur, R.; Stopden, S. Miaolingian transgression and the *Oryctocephalus indicus* Biozone in the Sumna Valley (Spiti), Himalaya, India. *Comptes Rendus. Géosci.* **2020**, *352*, 157–168. [CrossRef]

70. Landing, E.; Amati, L.; Franzi, D.A. Epeirogenic transgression near a triple junction: The oldest (latest Early–Middle Cambrian) marine onlap of cratonic New York and Quebec. *Geol. Mag.* **2009**, *146*, 552–566. [CrossRef]
71. Shaw, A.B. Stratigraphy and structure of the St. Albans area, northwestern Vermont. *Geol. Soc. Am. Bull.* **1957**, *69*, 519–568.
72. Landing, E. Highgate gorge: Upper Cambrian and Lower Ordovician continental slope deposition and biostratigraphy, northwestern Vermont. *J. Paleontol.* **1983**, *57*, 1149–1187.
73. Landing, E. Avalon—Insular continent by the latest Precambrian. In *Avalonian and Related Peri-Gondwanan Terranes of the Circum-North Atlantic*; Nance, R.D., Thompson, M., Eds.; The Geological Society of America: Boulder, CO, USA, 1996; pp. 27–64. [CrossRef]
74. Kohl, D.; Slingerland, R.L.; Arthur, M.A.; Bracht, R.; Engelder, T. Sequence stratigraphy and depositional environments of the Shamokin (Union Springs) Member, Marcellus Formation and associated strata in the middle Appalachian Basin. *AAPG. Bull.* **2014**, *98*, 483–513. [CrossRef]
75. Egenhoff, S.O.; Fishman, N.S.; Ahlberg, P.; Maletz, J.; Jackson, A.; Kolte, K.; Lowers, H.; Mackie, J.; Newby, W.; Petrowsky, M. Sedimentology of SPICE (Steptoean positive carbon isotope excursion); a high-resolution trace fossil and microfabric analysis of the Middle to Late Cambrian Alum Shale Formation, southern Sweden. In *Paying Attention to Mudrocks: Priceless*; Larsen, D., Egenhoff, S.O., Fishman, N.S., Eds.; the Geological Society of America: Boulder, CO, USA, 2014; pp. 87–102.
76. Dworatzek, M. Sedimentology and petrology of carbonate intercalations in the Upper Cambrian Olenid Shale facies of southern Sweden. *Sver. Geol. Undersökning C* **1987**, *81*, 73.
77. Buchardt, B.; Nielsen, A.T.; Schovsbo, N.H. Alun Skiferen i Skandinavien. *Geol. Tidskift* **1997**, *3*, 1–30.
78. Majchrzyk, A.; Kozłowski, W.; Żylińska, A. Authigenic calcium carbonate precipitation in the "bathtub ring" around the anoxic Alum Shale Basin during the Furongian SPICE event (Baltic Basin, northern Poland). *GFF* **2022**, *144*, 41–58. [CrossRef]
79. Howell, B.F. *The Faunas of the Cambrian Paradoxides Beds at Manuels, Newfoundland*; Harris Company; Cornell University: Ithaca, NY, USA, 1925; Volume 11, 140p.
80. Schovsbo, N.H. Uranium enrichment shorewards in black shales: A case study from the Scandinavian Alum Shale. *GFF* **2002**, *124*, 107–111. [CrossRef]
81. Nielsen, A.T.; Schovsbo, N.H. Cambrian to basal Ordovician lithostratigraphy in southern Scandinavia. *Bull. Geol. Soc. Den.* **2007**, *53*, 47–92. [CrossRef]
82. Lehnert, O.; Caner, M.; Ahlberg, P.; Ebbestad, J.O.; Harper, D.A.T.; Meinhold, J. Palaeokarst formation in the early Palaeozoic of Baltoscandia—evidence for significant sea-level changes in a shallow epicontinental sea. In Proceedings of the e 3rd IGCP 591 Annual Meeting, Lund, Sweden, 9–19 June 2013.
83. Landing, E.; Westrop, S.R. Late Cambrian (middle Furongian) shallow-marine dysoxic mudstone with calcrete and brachiopod-olenid-*Lotagnostus* faunas in Avalonian Cape Breton Island, Nova Scotia. *Geol. Mag.* **2015**, *152*, 973–992. [CrossRef]
84. Landing, E. Depositional tectonics and biostratigraphy of the western portion of the Taconic allochthon, eastern New York State. In *The Canadian Paleontology and Biostratigraphy Seminar, Proceedings*; Landing, E., Ed.; New York State Museum: Albany, NY, USA, 1988; pp. 96–110.
85. Macdonald, F.A.; Karabinos, P.M.; Crowley, J.L.; Hogbin, E.B.; Crockford, P.W.; Delano, J.W. Bridging the gap between the foreland and hinterland II: Geochronology and tectonic setting of Ordovician magmatism and basin formation on the Laurentian margin of New England and Newfoundland. *Am. J. Sci.* **2017**, *317*, 555–596. [CrossRef]
86. Landing, E.; Benus, A.P. The Levis Formation: Passive margin slope processes and dynamic stratigraphy in the western area. In *Field Trips Guidebook. Canadian Paleontology and Biostratigraphy Seminar, Ste. Foy, Quebec*; Riva, J.F., Ed.; Université Laval Press: Québec, QC, Canada, 1985; pp. 1–11.
87. Taylor, J.F.; Repetski, J.E.; Orndoff, R.C. The Stonehenge transgression: A rapid submergence of the central Appalachian platform in the Early Ordovician. In *Global Perspectives on Ordovician Geology*; Webby, B.D., Laurie, J.R., Eds.; CRC Press: Rotterdam, The Netherlands, 1992; pp. 409–418.
88. Brezinski, D.K.; Repetski, J.E.; Taylor, J.F. Stratigraphic and paleontologic record of the Sauk III regression in the central Appalachians. In *National Park Service Paleontological Research Volume 3*; Santucci, V.L., McClelland, L., Eds.; Geological Resources Division Technical Report NPS/NRGDRD/GRDTR-99/o3; National Park Service: Washington, DC, USA, 1999; pp. 32–41.
89. Nielsen, A.T. Ordovician sea level changes: A Baltoscandian perspective. In *The Great Ordovician Biodiversification Event*; Webby, B.D., Paris, F., Droser, M.L., Percival, I.G., Eds.; Columbia University Press: New York, NY, USA, 2004; pp. 189–201.
90. Miller, J.F.; Evans, K.R.; Dattilo, B.F. The great American carbonate bank in the miogeocline of western central Utah: Tectonic influences on sedimentation. In *The Great American Carbonate Bank: The Geology and Economic Resources of the Cambrian–Ordovician Sauk Megasequence of Laurentia*; Derby, J.R., Fritz, R.D., Longacre, S.A., Morgan, W.A., Steinbach, C.A., Eds.; AAPG Mem: Tulsa, OK, USA, 2012; pp. 769–854. [CrossRef]
91. Berry, W.B.N. Stratigraphy, zonation, and age of Schaghticoke, Deepkill, and Normanskill shales, eastern New York. *Geol. Soc. Am. Bull.* **1962**, *73*, 695–718. [CrossRef]

92. Ross, R.J.; Adler, F.J., Jr.; Amsden, T.W.; Bergstrom, D.; Bergström, S.M.; Carter, C.; Chrukin, M.; Cressman, E.A.; Derby, J.R.; Dutro, J.T.; et al. The Ordovician System in the United States. Correlation Chart and Explanatory Notes. *Inter. Union Geol. Sci.* **1982**, *12*, 73.
93. Sattler, F.R. Lithologic Properties of the Upper Ordovician Utica Formation, Michigan Basin, USA: A Geological Characterization and Assessment of Carbon Dioxide Confinement Potential. Master's Thesis, Michigan State University, East Lansing, MI, USA, 2015; 94p. Available online: https://scholarworks.wmich.edu/masters_theses/608 (accessed on 1 February 2024).
94. Walker, K.R. *A brief introduction to the ecostratigraphy of the Middle Ordovician of Tennessee (southern Appalachians, U.S.A.), In The Ecostratigraphy of the Middle Ordovician of the Southern Appalachians (Kentucky, Tennessee, and Virginia) U.S.A.*; Ruppel, S.C., Walker, K.R., Eds.; Studies in Geology University of Tennessee: Tennessee, TN, USA, 1977; pp. 12–17.
95. Shanmugam, G.; Walker, K.R. Tectonic significance of distal turbidites in the Middle Ordovician Blockhouse and lower Sevier formations in east Tennessee. *Am. J. Sci.* **1978**, *278*, 551–578. [CrossRef]
96. Shanmugam, G.; Walker, K.R. Sedimentation, subsidence, and evolution of a foredeep basin in the Middle Ordovician, southern Appalachians. *Am. J. Sci.* **1980**, *280*, 479–496. [CrossRef]
97. Shanmugam, G.; Walker, K.R. Anatomy of the Middle Ordovician Sevier Shale basin, eastern Tennessee. *Sediment. Geol.* **1983**, *34*, 315–337. [CrossRef]
98. Ogg, J.G.; Ogg, G.; Gradstein, F.M. *The Concise Geologic Time Scale*; Cambridge University Press: Cambridge, UK, 2008; 177p.
99. Goldman, D.; Mitchell, C.E.; Bergstrom, S.M.; Delano, J.W.; Tice, S. K-Bentonite and graptolite biostratigaraphy in the Middle Ordovician of New York State and Quebec: A new chronstratigraphic model. *Palaios* **1994**, *9*, 124–143. [CrossRef]
100. Bergstrom, S.M.; Goldman, D. Conodont biostratigraphy and biofacies of the Jacques Cartier River Ordovician section, Quebec. In *Studies in Stratigraphy and Paleontology in Honor of Donald W. Fisher*; Landing, E., Ed.; New York State Museum: Albany, NY, USA, 1994; Volume 481, pp. 1–4.
101. Ver Straeten, C.; Brett, C.E.; Baird, G.; Boyer, D.; Lindemann, R.; Ivany, L.; Over, J.; Witzke, B. Forum Comment Shallow-water onlap model for the deposition of Devonian black shales in New York. *Geology* **2019**, *47*, e495. [CrossRef]
102. Schieber, J.; Lazar, R.O. Devonian black shales of the eastern U.S.: New insights into sedimentology and stratigraphy from the subsurface and outcrops in the Illinois and Appalachian basins. In *Field Guide for the 2004 Great Lakes Section SEPM Annual Field Conference*; SEPM: Tulsa, OK, USA, 2004; 90p.
103. Schieber, J.; Southard, J.; Thiesen, K. Accretion of mudstone beds from migrating floccule ripples. *Science* **2007**, *318*, 1760–1767. [CrossRef]
104. Schieber, J. Evidence for high-energy events and shallow water deposition in the Chattanooga Shale, Devonian, central Tennessee, USA. *Sediment. Geol.* **2009**, *93*, 193–208. [CrossRef]
105. Smith, L.B. Core description, petrography, sedimentology, stratigraphy, TOC and depositional environment. In *A Geologic Play Book for Utica Shale*; Patchen, D.G., Carter, K.M., Eds.; Final report of the Utica Shale Appalachian Basin Exploration Consortium; Darrin West: Houston, TX, USA, 2015; pp. 49–73. Available online: https://www.wvgs.wvnet.edu/utica (accessed on 28 April 2025).
106. Rhoads, D.C.; Morse, J.W. Evolutionary and ecological significance of oxygen-deficient marine basins. *Lethaia* **1971**, *4*, 413–438. [CrossRef]
107. Popova, O. *Utica Shale Play—Geology Review*; U.S. Energy Information Administration: Washington, DC, USA, 2017; Volume 21.
108. Over, J.D. Conodont biostratigraphy of the Chattanooga Shale, Middle and Upper Devonian, southern Appalachian Basin, eastern United States. *J. Paleontol.* **2007**, *81*, 1194–1217. [CrossRef]
109. Algeo, T.J.; Berner, R.A.; Maynard, J.B.; Scheckler, S.E. Late Devonian oceanic anoxic events and biotic crises: "rooted" in the evolution of vascular land plants? *GSA Today* **1995**, *5*, 64–67.

Journal of
Marine Science and Engineering

MDPI

Review

Contemporary Issues and Advancements in Coastal Eolianite Research

David H. Backus

Department of Environmental Studies, Haverford College, Haverford, PA 19041, USA; dbackus@haverford.edu

Abstract: This review of eolianite research—following a comprehensive overview earlier this century—looks at several areas of overlapping research in coastal eolianite deposition, including the following: tectonic setting; carbonate–sediment type and source areas (carbonate factories); the relationship between relative sea-level change and eolianite deposition; and the evolution of depositional models. Several locations are highlighted in order to emphasize the variety in the eolianite depositional record based on more recent research. In particular, a historical review of eolianite depositional models developed over the last 100-plus years indicates that purely high-stand models of the last century—developed on steep-sided carbonate platforms—do not adequately reflect the complex depositional histories found on other coastal shelves. More recent depositional models emphasize high-stand carbonate production followed by the mobilization of shelf sediments during subsequent relative drops in sea level.

Keywords: carbonates; depositional model; eolianite; rhodolith; relative sea level

1. Introduction

This review article is not intended as an exhaustive compilation of the literature devoted to eolianites since the research of Brooke [1] but is a presentation of the issues that continue to inform the directions that eolianite research has taken since the early 2000s. Eolianites reflect the environmental conditions in which they formed as uncemented dunes. Carbonate dunes are of scientific interest as they track with changes in sea level, while the composition of carbonate clasts reflects the offshore carbonate system/factory. Changes in clast composition through time also provide information related to changing climatic or oceanographic conditions. Pleistocene-age eolianites are of particular interest as they help constrain the magnitude and timing of environmental changes associated with the glacial–interglacial cycles over the past 800k+ years, helpful in predicting future changes to coastal systems. Eolianite cements also have the potential to track shifts in local-to-regional environmental conditions [2].

This review also highlights several localities that illustrate eolianite and dune deposition: (1) in an unusually high latitude area (the Outer Hebrides, UK); (2) in a transtentional tectonic setting (the Gulf of California) and; (3) in an area where the dominant bioclastic component is derived from rhodoliths (Cabo Verde). As modern dune systems provide a ready comparison or contrast to adjacent eolianite deposits, comparisons are made when possible. Locations discussed in the text are shown using a global Mollweide projection (Figure 1). Satellite images as well as coordinates for localities under discussion—determined using Google Earth Pro (7.3.6.10201)—will also be provided as an additional aid to visualization.

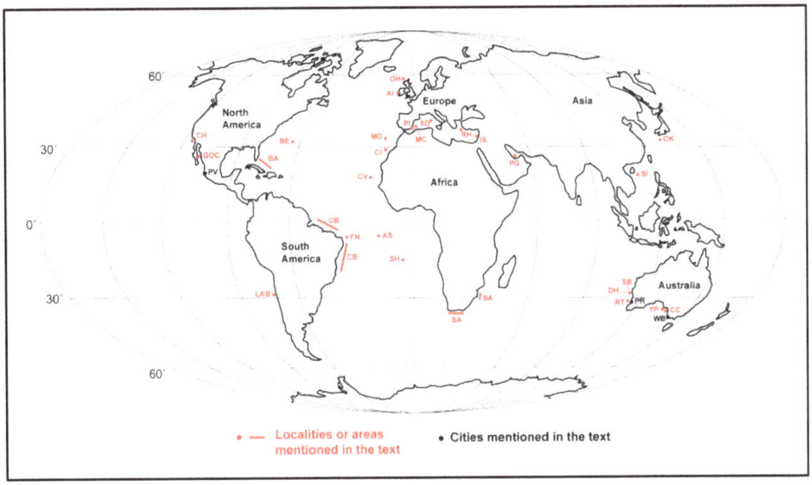

Figure 1. Locations mentioned in the text: AI, the Aran Islands; AS, Ascension Island; BA, the Bahamas; BE, Bermuda; CB, Coastal Brazil; CC, Coorong Coast; CH, the Channel Islands; CI, the Canary Islands; CV, the Cape Verde Islands; DH, Dirk Hartog Island; FN, Fernando de Noronha Island; GOC, the Gulf of California; IS, Israel; LAB, Llano Agua de los Burros, Chile; MC, Mallorca; MD, the Madeira Islands; OH, the Outer Hebrides; OK, Okinawan Archipelago; PG, the Persian Gulf; PI, the Pityusic Islands; RH, Rhodes Island; RT, Rottnest Island; SA, the South African Coast; SB, Shark Bay; SD, Sardinia; SH, St. Helena; SI, Shidao Island; YP, Yorke Peninsula; PR, Perth, AUS; WB, City of Warrnambool, AUS; PV, City of Puerto Vallarta, MX.

The terms eolianite and calcareous eolianite were introduced by Sayles in 1929 to refer to, respectively, any carbonate-cemented dune and those composed of largely calcareous material [3]. By common usage, the term eolianite now only refers to carbonate-rich dunes and is equated with dune calcarenite, a genetic term describing any detrital rock composed of >50% carbonate grains of sand size [4,5]. Regional synonyms for eolianite include aeolianite (UK), dunerock (South Africa), grès dunaire (Mediterranean), kukar (Israel), and miliolite (India and the Middle East) [1,6].

Carbonate-rich dunes form primarily along shorelines where offshore carbonate production (bioclastic or inorganic) contributes significantly to the aeolian sediments blown onshore. Factors that influence the development of eolianite dunes include the following: wind energy, offshore carbonate production, low siliciclastic input, climate (rate of cementation), tectonic setting, shelf profile (platform vs. ramp), and changes in sea level [2,7]. Eolianite deposits have been primarily identified from Quaternary rocks deposited in subtropical to tropical conditions, although there are exceptions, such as the carbonate-rich dunes of Scotland and Ireland. Table 1 provides information (climate, tectonic setting, and type of clast) for localities with eolianites and modern dune systems that are described and discussed further in this review.

A comprehensive list of Quaternary eolianite localities is available, as are inventories of more ancient deposits [1,2,39]. The recognition of eolianites in rock records older than the Quaternary period has been a focus of several studies [40–42] as eolianites may be confused with adjacent near-shore or subtidal calcarenites in outcrop or drill core [39]. No single feature of eolian carbonates distinguishes them from subtidal carbonates [41]. Features used to distinguish eolian structures from their subtidal counterparts include relatively large-scale (>1 m) cross beds and well to very-well sorted allochems (<4 mm in size) that are also well-rounded and abraded (Table 1 in [41]; Table 2 in [42]). When present in dunes or sand sheets, low-amplitude wind-ripples (with a high ratio of wavelength to height) inversely grades pinstripe laminations, and adhesion ripples are diagnostic of eolian deposition [41,42]. The presence of rhizoliths and vadose cements that occur below exposure surfaces are also helpful in distinguishing eolian strata from subtidal carbonates [41,42].

Table 1. Climate, tectonic setting, and clast type for eolianite/dune localities in Figure 1.

Locality	Climate	Tectonic Setting	Clast	References *
Ascension Island (AS)	Subtropical	Intraplate	bioclasts	[8]
The Bahamas (BA)	Sub- to Semi-Tropical	Passive	ooids/peloids	[9,10]
Bermuda (BE)	Subtropical	Intraplate	bioclasts	[11]
Coorong Coast (CC)	Warm Temperate Dry	Passive	bioclasts	[12,13]
The Channel Islands (CH)	Warm Temperate Dry	Strike-slip	bioclasts	[14]
The Canary Islands (CI)	Warm Temperate Dry to Subtropical	Intraplate	bioclasts	[15,16]
The Cape Verde Islands (CV)	Subtropical	Intraplate	bioclasts	[15,17,18]
Dirk Hartog Island (DH)	Subtropical	Passive	bioclasts	[19]
Fernando de Noronha Island (FN)	Tropical	Intraplate	bioclasts	[20]
The Gulf of California (GOC)	Subtropical	Transtensional	bioclasts	[21,22]
Israel (IS)	Warm Temperate Dry	Divergent Rift	bioclasts	[23,24]
Llano Agua de los Burros (LAB)	Warm Temperate Dry	Convergent	bioclasts	[25]
Mallorca (MC)	Warm Temperate Dry	Extensional	bioclasts	[26,27]
The Madeira Islands (MD)	Warm Temperate Dry	Passive	bioclasts	[15,28]
The Outer Hebrides (OH)	Cool Temperate	Passive	bioclasts	[29]
The Persian Gulf (PG)	Tropical	Foreland Basin	ooids/peloids	[30]
The Pityusic Islands (PI)	Warm Temperate Dry	Extensional	bioclasts	[31]
Rhodes Island (RH)	Warm Temperate Dry	Convergent	bioclasts	[32]
Rottnest Island (RT)	Subtropical	Passive	bioclasts	[33]
The South African Coast (SA)	Warm Temperate Dry	Passive	bioclasts	[34]
Shark Bay (SB)	Subtropical	Passive	bioclasts	[35]
Sardinia (SD)	Warm Temperate Dry	Convergent	bioclasts	[36]
St. Helena (SH)	Subtropical	Intraplate	bioclasts/ooids	[8]
Shidao Island (SI)	Subtropical	Intraplate	bioclasts	[37]
Yorke Peninsula (YP)	Warm Temperate Dry	Passive	bioclasts	[38]

* References identify clast type in eolianites or dunes at the locality.

The most favorable conditions for the accumulation of carbonate eolianites are generally found between 50° N and 45° S, and more commonly where warm conditions support carbonate production and the trade winds blow [39,42]. With sufficient wind velocity, subtidal carbonate sediments exposed during tidal or higher-order cycles in sea level are winnowed and deposited as coastal dunes behind the beach. Under higher velocity conditions and a relatively low sediment budget, dunes may form inland of the beach with sediments sourced through the deflation of the beach or adjacent dunes [41]. With a sufficiently high sediment budget, a coastal dune system can develop into an oceanward prograding strand plain (a series of transverse sand dunes that pile up against the shore) [43]. Sand plains may develop where there is a high water table, coarser sediments, or vegetation that prevents dune formation [41].

A review of the distribution of eolianites [1] found that somewhat greater than 80% of eolianite deposits occur between 20 and 40° in both hemispheres, with the greatest extent and thickness of eolianite found along the southern coast of Australia and the southern and eastern coasts of South Africa, adjacent to major areas of carbonate production [1]. Preservation of carbonate dune deposits is enhanced in arid regions where there are seasonal rains that promote lithification by secondary cements in the form of crusts, calcrete horizons, and through pedogenic processes—the development of clay-rich soils that act as resistant caps (e.g., the alternating eolianite/paleosol allostratigraphy of Bermuda [3,44]).

2. Eolianite Cements and Climate

Eolianite cements have the potential to reflect changes in regional climatic and oceanographic systems [37]. As mentioned above, the cements that consolidate dune sands in the formation of eolianites depend on seasonal or intermittent rains that facilitate the diagenetic transfer of calcium carbonate from sediment to binding cements [45]. Seasonal changes in the water table also promote the formation of cements in eolianites (e.g., machair sand plains of the Outer Hebrides [46]). Variations in the dissolution of sediments within an eolianite sequence (alteration of high-Mg calcite to low-Mg calcite and dissolution of aragonite) should, in theory, reflect the volume of water that passed through the sediment in the process of cement formation (increase in void-filling calcite spar) as outlined in a study of vadose diagenesis in aeolianites near Warrnambool, SE Australia [47]. A more recent study illustrates potential issues with accepting the evidence of vadose diagenesis with an uncritical eye, showing that variation in diagenetic effects can occur across a single outcrop [48]. Multiple samples of eolianites from arid, semi-arid, Mediterranean, and subtropical settings were compared from more-exposed and less-exposed parts of individual outcrops with variations depending on particle size, percent cement, and sea spray [48]. It was concluded that local (meso-scale) factors commonly have stronger effects on outcrop diagenesis than regional (macro-scale) factors, such as climate [48].

An example of how eolianite cements may reflect changing climatic conditions is an analysis of drill cores extracted on Shidao Island (16°50′ N; 112°20′ E), Xisha Ids, China, which sits within a region affected by seasonal monsoons [37]. The drill core included five eolianite–paleosol cycles, deposited during MIS 3 and the transition to MIS 2. Analyses of the core, including stable isotope and major element geochemistry, particularly showed variations in the δ18O curve and Sr, Mg, Fe, and Al concentrations. In conjunction with the observation of changes in the bulk composition of the eolianite bioclasts, it was shown that more extreme diagenetic alteration of one of the eolianites (Unit 2) resulted in the bulk loss of bioclasts, lower Sr and Mg concentrations in the bulk rock analysis, and more negative δ18O values. It is also proposed that analysis of the core reflects dune formation occurred in a dry-cold climate with higher velocity winds deflating the upwind shelf during relatively low-stands in sea level. In addition, an increase in Fe and Al concentrations was linked to an increase in terrigenous dust. It was proposed that paleosols, interbedded with the eolianites, were formed in more humid environmental conditions [37].

3. Eolianite Dating Methods

Depending on the geological setting and potential age of an eolianite deposit, a variety of dating techniques have been used. An excellent example of how multiple geochronologic methods can be used to investigate the timing of eolianite deposition and consolidation is a study in the Channel Islands of California (USA) [14]. In conjunction with the mapping of stratigraphic relations with underlying marine terraces and related geosols, a combination of radiocarbon dating (14C), luminescence chronologic methods—including optically stimulated luminescence (OSL) and infrared stimulated luminescence (IRSL)—amino-acid geochronology, and previously reported uranium series (U/Th) datasets are used to constrain the ages of eolianite deposits (~80,000 to ~120,000 yr) and correlate them with the marine isotope stages [MIS] 4, 3, and 2 [14,49].

The increasing number of studies using luminescence techniques (OSL, TL, and IRSL) for dating coastal dune deposits, including eolianites, is illustrated by the review of ~200 studies from around the world and the extensive use of OSL measurements (n = 51) in a recent investigation of Lower to Upper Pleistocene coastal deposits (MIS 22 to MIS 2) on the Pityusic Islands of the Balearic Archipelago in the Western Mediterranean [31,50].

When eolianites do not include a siliciclastic component (e.g., Bahamas (32°18′ N; 64°46′), Western Australian coast), amino-acid racemization (AAR) analyses can be used to determine relative ages on whole-rock samples [51,52]. This method is particularly helpful if used in conjunction with (U/Th) analysis of suitably sized bioclasts from adjacent stratigraphic units [53–55]. Generating age data for eolianites can be difficult [55,56] but any model for eolianite deposition suffers if not supported by appropriately precise age data [56], particularly as models are becoming more nuanced with respect to variations in the fifth-order sea-level curve and more subtle stadial/interstadial variations [26,57–60].

4. Eolianites and Tectonic Setting

Eolianite sediments accrete and are preserved where accommodation space is available downwind of the source area for carbonate sand. Differences in coastal landscapes are, in part, reflective of a region's tectonic history, as are the repose and spatial dimensions of offshore source areas where carbonates are produced [2,43]. The volumetrically largest eolianite deposits are found on tectonically passive margins such as coastal south-central Australia [1,12] and coastal South Africa [34,57,61,62] where complex shore–parallel coastal dune complexes are preserved adjacent to broad continental shelves with highly productive carbonate systems.

Though of much smaller volumes, dune sequences are also preserved in island areas that sit in stable settings such as Bermuda, where stacks of Pleistocene eolianites sit upon a slowly residing volcanic core astride the Bermuda Pedestal (Figure 2)—an extinct shield volcano [63]—or the Bahamas, where islands and smaller cays largely composed of oolitic eolianites are perched upon on an extensive carbonate platform, near the base of which are Upper Jurassic bioturbated limestones (Figure 3) [64].

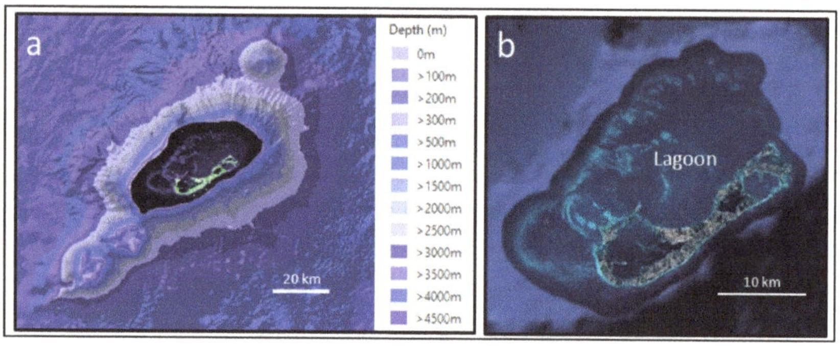

Figure 2. Images showing the bathymetry, island, reef, and lagoonal areas of the Bermuda Rise: (**a**) bathymetric map of the Bermuda Rise, overlain by a false-color composite satellite image of the Bermuda Islands, reef, and lagoon; (**b**) close-up image of the Bermuda Island area.

An interesting counterpoint to stable or slowly subsiding passive margins is the Coroong to Mount Gambier Coastal Plain of Australia where Quaternary emplacement of volcanics has uplifted the region and preserved a series of shore-parallel belts of eolianite that reach 80–100 km inland along 300 km of coastline and represent almost 1 million years of sea-level high-stands (Figure 4) [13,56,65,66].

A region with a complex tectonic history where eolianites have been extensively studied is the Mediterranean Basin (see [1,31,36] and others listed in Table 1). As the basin touches or includes parts of the Eurasian, African, and Arabian Plate boundaries, eolianite localities within the region can be found in both extensional and convergent settings (Table 1) with eolianites preserved in both coastal plains (e.g., [58]) and rugged, cliff-sided coastlines (e.g., [27,36]).

An example from a tectonic region where eolianites have not been extensively studied is the Pleistocene eolianites (MIS 5e-5a) from a convergent tectonic boundary (subduction zone) described from the coastal region of the Atacama Desert (Llano Agua de los Burros) in Chile (27°55′ S; 71°05′ W), where the shelf drops precipitously toward the Peru–Chile

Trench and wind velocities were sufficient to transport carbonate sediments over 3.5 km inland and over 250 m in elevation above the modern coastline [25].

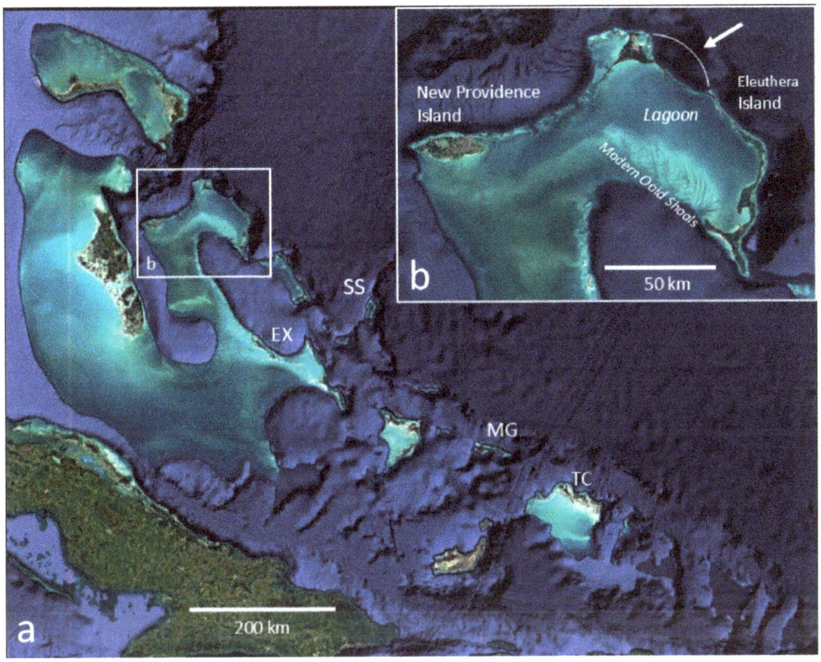

Figure 3. Images showing the location of the Bahamas, bathymetric features, and islands or features mentioned in the text: (**a**) Islands and banks that make up the Bahamas (EX—Exuma Islands, SS—San Salvador Island, MG—Mayaguana Island, TC—Turks and Caicos); (**b**) inset showing Eleuthera Island's lagoon and modern oolite shoals. The white arrow points to the area proposed to have been an oolite shoal in the past (outlined in white) but subsequently eroded into the adjacent abyssal deep.

Additional examples of eolianites from a less-common tectonic setting are found along the coast of the Gulf of California (GOC), an extensional/transtensional tectonic basin [21,22]. Initially, the Gulf of California region formed as part of a generally E-W extensional tectonic system in the Late Miocene (~12 to 6 mya). By 5 mya, the extension within the "Proto-Gulf" had rotated 30–50° to become a NW-SE transtensional system with the northern and southern parts of the Gulf of California opening at different rates. Eventually, the Baja California Peninsula became part of the Pacific Plate as the boundary with the North American Plate was transferred into the Gulf of California [67,68]. As a result of the shift to a transtensional tectonic system within the Gulf of California, fault blocks of Miocene-age volcanics along the Baja California coast have been uplifted along with Plio-Pleistocene rocky shore, eolianite, and other related deposits [22,67,69].

As the Gulf of California now extends for more than 1000 km, those parts of the coast that project into the Gulf are subject to high wind velocities due to fetch alone (Figure 5) [21,70]. Winter winds can sustain velocities in the range of 5–7 m/s over multiple days, with gusts of 8+ m/s not unusual [21,70]. As a result, a wide variety of modern dune field morphologies with varying percentages of carbonate material are also found perched on, ramped against, or in pockets along much of the western edge of the Gulf of California where accommodation space is available [71]. These modern dune systems provide a benchmark model for interpreting eolianites preserved along the coast of the Gulf [21,22,71]. A survey of 84 individual dune fields along the peninsular coast of the GOC indicates an average migration direction of S8° E within the modern wind field that blows directly down the axis of the GOC with an average azimuth of S24.5° E [70,71]. Plio-Pleistocene coastal and inland eolianites are best preserved at Punta Chivato and Isla Carmen [21,22]. These localities from the central region of the Gulf of California provide interesting examples of eolianite deposition in this rugged and seasonally windy environment.

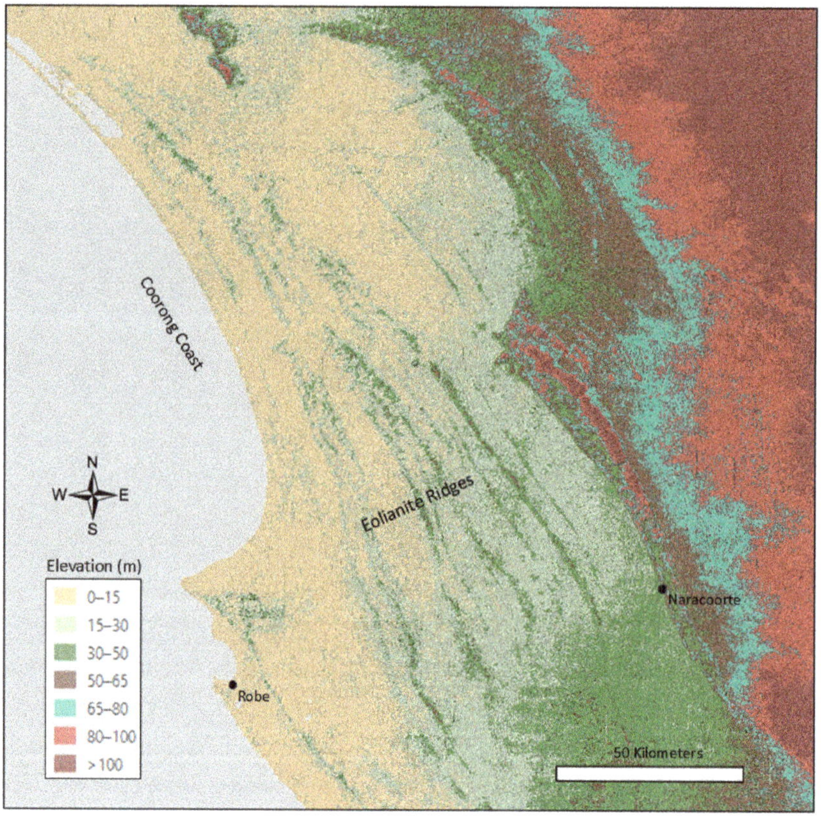

Figure 4. Part of the Coorong—Mt Gambier region of Australia as represented by a Digital Elevation Model (DEM). Multiple fossil strandline ridges are shown between the towns of Robe and Naracoorte.

At Punta Chivato, a narrow Plio-Pleistocene valley extends behind the north-facing beach of Ensenada el Muerto (Figure 6) [21,22]. Behind a cobble beach, the eroded remnants of an eolianite deposit extend 400 m to the south, in places crossing the 40 m contour while also rising 4 to 7.5 m above the arroyo stream bed. Cross-bedded calcarenite dipping SE at 32° is found along the eastern wall of the arroyo, while higher on the western wall, cross-beds are found dipping SW at 28° where wind-blown sediment collected against the arroyo walls (Figure 6b) [21]. Point counts of rock samples show 44% of the clasts as volcanic lithics with bioclasts (mollusks, coralline red algae, and echinoid) accounting for 56% of the sample, though dissolution of the clasts (resembling mollusk grain) suggests an undercount of the molluscan component [21]. Stratigraphic relationships with a presumed MIS 5e terrace outside of the canyon suggest a similar age for Punta Chivato eolianite [21].

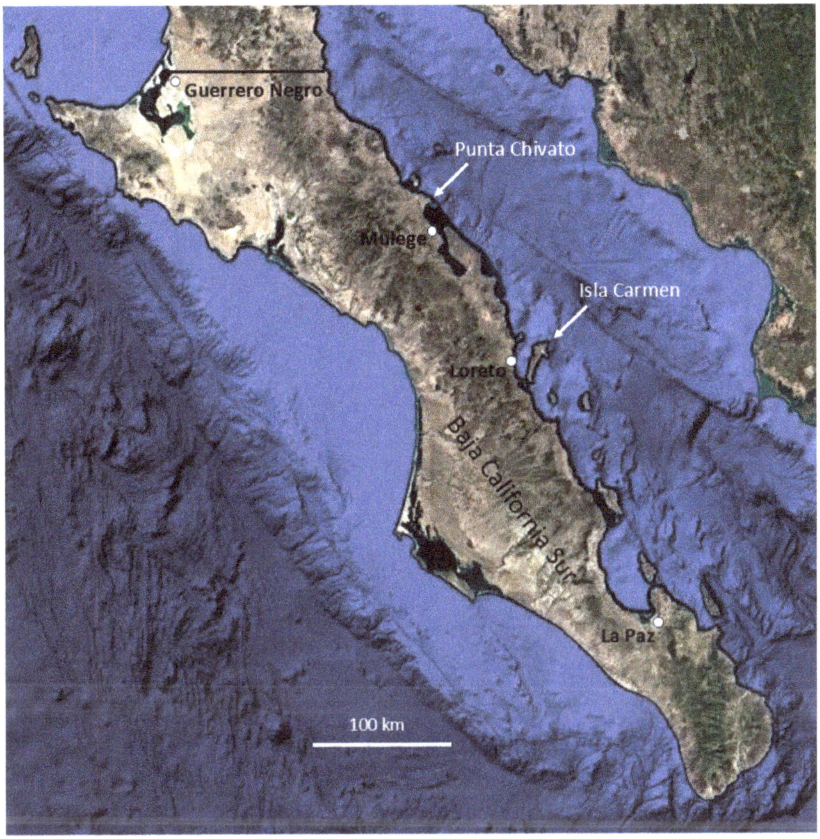

Figure 5. Image showing the location of Punta Chivato and Isla Carmen along the coast of Baja California Sur, Mexico, as well as the bathymetric features of the Southern Gulf of California. Notice the NW-SE rift basins within the GOC.

The modern dune and beach at Punta Chivato are representative of many north-facing rocky promontories or island coasts in the Gulf of California and provide a modern example of dune deposition during the current high-stand in sea level. The sediments on the shore and in dunes plastered along the cliff fronts at Punta Chivato are primarily composed of mollusk fragments (50–86%) derived primarily from clams, such as the locally abundant infaunal bivalve, *Megapitaria squalida*. Along the coast behind Ensenada el Muerto at Punta Chivato, modern dune sediments are commonly found blown up and over the 20 m contour and in places over and around the 40 m and 60 m contours as well (Figure 6c) [21]. However, the modern coastal dune system is relatively starved of sediment and does not extend inland as extensively when compared with the Pleistocene eolianites preserved nearby, suggesting that the eolianite was deposited under a different set of conditions (e.g., sediment budget, relative sea level, coastal morphology, and wind velocity) [21,22].

To the south of Punta Chivato, on the north end of Isla Carmen, dune rocks are found stacked or draped over andesite ridges on the NW side of the island at Bahia Oto and the NE side of the island at Puerto de la Lancha (Figures 5 and 7a–c) [22]. At low tide, the depositional sequence at Oto Bay can be seen to consist of a 2-m-thick conglomerate of andesite boulders eroded from the underlying Miocene basement volcanics, followed by 1.5 m of pink, coarse molluscan sands (beach deposit), which is in turn overlain by 4 m of beige-colored eolianite pock-marked by dissolution cavities and composed of 0.5 mm fragments of molluscan bioclasts and volcanic lithics. A Pleistocene faunal component of the basal conglomerate provides the only age limit on the overlying eolianite [22].

Figure 6. Satellite image and photographs showing modern dune and Pleistocene eolianite features at Punta Chivato: (**a**) Locations on the northern coast of the Punta Chivato promontory mentioned in the text; (**b**) looking south-east within the Plio-Pleistocene valley at the remnants of eolianite slip-face bedding. Notice other remnant deposits both in the foreground and background. Figures standing on the cross-beds for scale; (**c**) just west of the NE corner of the promontory, looking east at dune sand being removed from the deflating beach and blown inland over and around the coastal heights.

To the east at Bahia de la Lancha (Figure 7b), a succession of five fossil carbonate dunes stack to a height of ~70 m are exposed across 300 m of coastline (Figure 7d,e). Sand samples from an active dune deposited within a slot canyon cut into the north face of the eolianite suggest a bioclast component of about 68 percent, dominantly composed of mollusk fragments [22,67]. The base of the eolianite sequence drapes over faulted blocks of Miocene volcanics, which likely controlled the shape of the valley during deposition [22]. An Upper Pliocene-age estimate is assigned to the lowest dune rock unit [72].

An additional outcrop of interest is a karstic, eroded eolianite found approximately 2.5 km south of the Bahia Oto beach and banked behind and against the back of a 200-m-high, E-W trending ridge of andesite (Figure 7c,f). Only 16 m thick but extending 450 m across the valley, an extensive talus pile below the outcrop suggests that this falling dune was originally much more substantial in the past [22]. The relatively large volume of calcarenite preserved at Bahia de la Lancha and the inland eolianite deposit suggest a much larger accommodation space for beach and dune deposits than presently exists along the northern coast of Isla Carmen.

Figure 7. Images showing the locations of the Pleistocene-age dune rocks on the north end of Isla Carmen: (**a**) False-color satellite image showing the location of fossil dune deposits displayed in 7b,c; (**b**) aerial photograph showing the dirty-white Plio-Pleistocene limestone and eolianite deposits at the Puerto de la Lancha locality, black arrow points to a modern carbonate-rich slot dune. There is glare off the water in the image; (**c**) aerial photograph of the dirty-white Plio-Pleistocene limestone and eolianite deposits at the Bahia Oto locality and the location of an inland, karst-eroded, falling dune deposit. The black arrow points to a modern, carbonate-rich slot dune; (**d**) looking east across the elevated dune-rock deposits at Puerto de la Lancha; (**e**) close-up of a section topped by eolianite along the Puerto de la Lancha coast; (**f**) looking north at the karst-eroded remnants of falling dune deposit, 0.5 km from the nearest Pleistocene beach source.

5. Carbonate Sediments Track Source Environments

The study of carbonate-rich dune systems has an additional advantage over siliciclastic-dominated eolian sandstones in that the bioclasts of the offshore marine environment are sampled with each successive depositional event. For example, a study documenting changes in the cool-water carbonate depositional system along the southern coast of Australia found that the coastal eolianites of the Yorke Peninsula dependably represented the upwind marine environment [38]. A similar analysis on the island of Bermuda documented a shift in the bioclastic composition of eolianites (coralline-rich to Halimeda-rich sediments) that reflected oceanographic controls (cool to warm conditions) consistent with other palaeozoological studies of the MIS 5e sea-level inflection from across the globe [11]. An interesting development in this relationship of eolianite to the upwind source area is the conundrum presented by the comparison of pre-Sangamon eolianite sediments that make up the windward islands and cays of the Bahamas with the modern bioclastic-rich carbonates found on the narrow shelf to windward. No oolites are presently found along the edge of the Blake-Bahama Escarpment while, the composition of dune ridges on perimeter islands or cays (e.g., Eleuthera, Cat, or Joulters) are primarily composed of oolites or oolitic-peloidal grainstones (Figure 3a,b) [9,10].

Suggested models for the origin of the eolianite ooids include the transportation of ooids from the western basin side of the Bahamas Bank, while theoretical calculations suggest that the narrow shelf along the windward edge of the bank could possibly produce the necessary volume of ooids in the proper conditions. A plausible alternative narrative, related to shelf size and therefore ooid production volumes, is the proposal that scallop-shaped areas along the windward escarpment are evidence for the erosional collapse of what would have been a wider shelf in the past (Figure 3b) [9].

Connecting bioclasts and inorganic carbonate constituents of eolianites with their upwind, offshore carbonate "factories" leads to the evolving models for the classification of oceanic carbonate systems on a global scale. In the previous review of the global distribution of eolianites, Brooke [1] followed the works of James [73] who classified carbonate sediments as either heterozoan or photozoan associations. A new ecosystem approach to carbonate production introduced a tripartite division (tropical shallow-water, cool-water, and mud mound) of carbonate factories [74,75]. This conceptual framework is, in turn, modified further by splitting the Tropical (T) factory into a Photozoan-T factory (roughly 30° N–30° S)—bounded by the 18 °C winter limit—and a Biochemical factory (Persian Gulf), while the Cool-water (C) factory was subdivided into the Heterozoan-C (poleward of the Photozoan-T factory limits) and Photo-C (Mediterranean Sea) factories [76]. A predictive model for the distribution of the carbonate factories across the globe was then created using the main environmental controls of light, temperature, nutrients, salinity, substrate, and carbonate saturation [77] (Figure 8).

A more explicit description of the environmental controls on the carbonate factories that contribute to eolianite sediments is presented as part of a major synthesis of marine carbonate systems [7]. Not unexpectedly, the main environmental controls on the Photozoan-T factory are light and temperature (e.g., tropical reef areas and the Great Bahama Bank), while the Persian Gulf is considered a special case of the T-factory (Biochemical) where salinity and temperature are the controlling factors, with corals and other bioclast producers absent or limited to deep water areas [7]. The main environmental control on the Heterozoan-C factory is nutrients (e.g., southern coast of Australia), while the Photo-C (Mediterranean) is a variant controlled by light and temperature [7]. Local environmental controls that may have an impact on carbonate production include ocean currents, upwelling, shallow-water dynamics, and the input of terrestrial sediment and water [7].

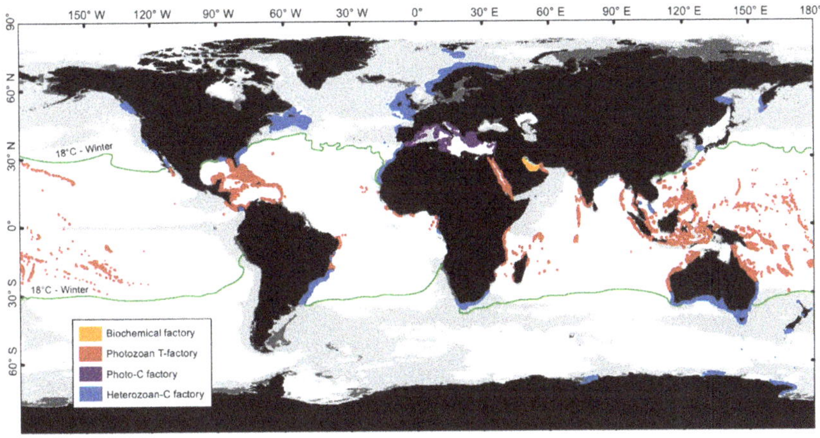

Figure 8. Map of the distribution of the carbonate factories, adapted from [77]. The map is based on the environmental model described in [77] and expanded on in [7].

6. Rhodoliths as Eolianite Bioclasts

Reflected in previous reviews and illustrated by many of the depositional environments described above, a majority of studies focus on the role marine invertebrates (mollusks, foraminifera, coral, echinoderms, and bryozoans) play in the production of bioclasts or the contribution of inorganic carbonate (e.g., ooids/peloids) to eolianite sediments [1,17]. The contribution to aeolian sediment by the Rhodophyta (coralline red algae) in the form of rhodoliths has become more apparent as there have been multiple studies since the last major review of eolianites [1,15,32,78]. Rhodoliths are unattached spherical to subspherical structures created by the concentric growth of precipitated calcium carbonate around a central core, often a small pebble or shell fragment. As algae are dependent on light for photosynthesis, rhodoliths thrive in relatively clear water [79,80]. A dependable rolling motion—as rhodoliths grow concentrically outward—means rhodolite beds are found below fair-weather base and above storm-weather base or within areas where currents are tidal or driven by surface winds [79,80]. For reference purposes, see studies of rhodolith-rich eolianite deposits that include images of rhodoliths in the field (see Figure 9.4 in [15]) and of rhodolith-derived bioclasts in thin-section (see Figure 6 in [17]).

Rhodolith beds play an important role in the ecosystems where they are found [79,80]. They are habitat modifiers, with hundreds of species (cryptic, soft-bodied, and hard-bodied) living within the beds or in the natural spaces created by the branches of individual rhodoliths. A habitat for polychaete worms, crustaceans, mollusks, and echinoderms, the hard surfaces of individual rhodoliths also support a variety of sessile organisms and borers [79,80]. Therefore, abundant rhodolith clasts in a dune or eolianite deposit indicate not just rhodoliths offshore but the entire ecosystem that rhodolith beds support [79,80]. In areas where rhodolith-rich deposits can be found representing significant time intervals (e.g., Pliocene and Pleistocene deposits in the Gulf of California), analysis of preserved rhodoliths can provide information on changes in community diversity or stability through time—as has been the case with corals [79]. Modern rhodolith beds have been reported from diverse areas such as, but not exclusively from, the following: the east coast of Brazil (2° N to 27° S); the shelf SE of Okinawa, Japan; the Western Australian coast centered on Shark Bay; the shelf outboard of the windward islands of the Caribbean; along much of the Gulf of California coastline and south to Puerto Vallarta in Mexico; along the Atlantic coastal areas of Northern Europe; as well as in the Mediterranean [79–82].

Thus, modern rhodolith beds are found within all the carbonate factory zones (Figure 8), with the exception of the Persian Gulf. Though species of algae that construct rhodoliths are found in tropical-to-polar latitudes, the low-latitude areas within the trade-wind zones of both hemispheres are where rhodoliths contribute most to beach and dune sediments [15,17,18,79,81]. Dunes or eolianites with high concentrations of rhodolith debris are found within the Gulf of California [65]; on the Macaronesian Islands of the North Atlantic [15,17,18,28,83]; Dirk Hartog Island, Shark Bay, Western Australia [19]; and Middle-Pleistocene eolianite on the island of Fernando de Noronha (3°51′ S; 32°25′ W) off the NE coast of Brazil [20] (Figure 1). A weakly cemented Late-Pleistocene mixed siliciclastic–algal sand dune is reported from the eastern Mediterranean island of Rhodes as well [32]. The paucity of eolianites reported with high concentrations of rhodolith debris suggests either that rhodoliths and rhodolith-rich dunes are in environments with low preservation potential or they have not been the focus of a sufficient number of studies to reflect their true presence in the geologic record.

Examples from the Cape Verde Islands off the Atlantic Coast of Africa (Figure 9) suggest that rhodoliths can be found as a major component in modern dune systems and fossil dune deposits, and may be present as an overlooked component of eolianites in other parts of the globe [15,17,18]. In the Cape Verde Archipelago, an extensive eolianite deposit is found on Maio, the easternmost island of the leeward islands, about 600 km west of Dakar, Senegal (Figures 1 and 9). The Cape Verde Islands are a group of shield volcanoes with a varied geologic history related to the opening of the Atlantic Ocean. On Maio, a sizable portion of the landscape is composed of uplifted Jurassic- and Cretaceous-age deep-ocean deposits of black shale and pelagic limestone [17,84].

The deposit at Lomba Greija is a north–south-oriented prominence on the eastern side of the island, 2.25 km inland from the shore, near the village of Pilão Cão (Figure 9b) [17]. Approximately 800+ m, 40 m wide, and 127 m high, this Pleistocene-age deposit includes a basal marine conglomerate overlying Cretaceous basement rocks, back beach deposits, and a 35-m-thick eolianite that caps the sequence. Preserved slip faces dip 28–32° SW and SE with stoss slope beds dipping 8–10° NE indicating the local wind field that is the result of the interaction between the Pleistocene landscape and the regional aspect of the trade winds (Figure 10a,b) [17]. Sedimentological analysis of the Lomba Greija dune rock shows significant loss (29%) due to bioclast dissolution. Micritic matrix accounts for 38% of the sample, while only 5% of the eolianite is composed of basaltic and lithic grains. The remaining component of bioclasts has coralline red algae as the largest contributor at 74%, followed by 11% for bivalve mollusks and 6% for corals [17].

To the north and east of Maio, the island of Boa Vista has a relatively unstudied set of modern and fossil dunes (Figure 9a,c). The largest modern dune field falls within a wind corridor that originates on the NW edge of Boa Vista Island (NE of Sal Rei) and extends from the Boa Esparança coast south-west toward the Praia do Curralinho beach where aeolian sands re-enter the Atlantic Ocean [85]. In Figure 9c, the modern dune field can be seen to split around the airport with the barchan dunes of the eastern part of the dune field coalescing into barchanoid ridges as they are somewhat stabilized locally by vegetation, while the western lobe of the dune field somewhat skirts the bay at Cabeçadas. Barchnoid dunes that originate at Praia de Chaves beach, SW of Cabeçadas (Figure 10c), eventually exit along the SW corner of the island (Figure 10e). Other smaller dune fields can be seen perched on the more rugged coastline of the north-eastern, eastern, and southern side of the island [85]. A recent study of beach sands from across the Cabo Verde Archipelago, including beaches (n = 7) from the perimeter of Boa Vista, indicates that the modern carbonate fraction (dominantly bioclasts) composes roughly 70–90% of the Boa Vista sands analyzed [18]. Abundance data reported for the carbonate fraction of the sands show a relative predominance of red algae clasts (48–87%) with bivalve (11–44%), echinoderm (0–14%), and minor amounts of bryozoan plus foraminifera clasts (0–5%) accounting for the remainder of the bioclasts across all samples analyzed [18].

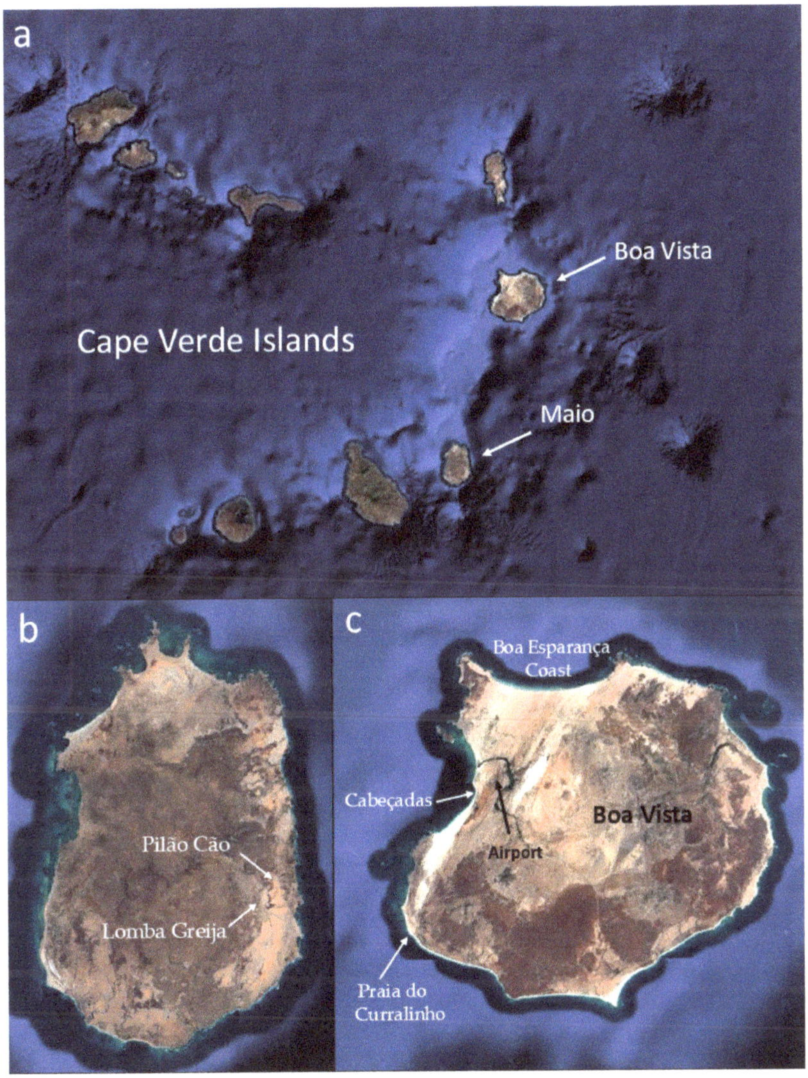

Figure 9. Images showing the distribution of the Cape Verde Islands and the locations of eolianite deposits on the islands of Maio and Boa Vista: (**a**) the location of Maio and Boa Vista along the eastern edge of the Cape Verde Island chain; (**b**) the location of the Lomba Greija rhodolith-rich, dune-rock deposits on Maio Island; (**c**) the location of rhodolith-rich dune carbonate-rich deposits and other localities on Boa Vista mentioned in the text.

A recent preliminary study of dune systems on Boa Vista by a field party that included Dr. Markes Johnson of Williams College (Personal Communication, 6 September 2024) suggests that extensive eolianites underlay the modern dune deposits found across the island (Figure 10d,f).

Figure 10. Images of eolinites or carbonate-rich dunes on Maio and Boa Vista Islands: (**a**) View of multiple cross-beds delimiting overlapping sets of dunes along the flanks of the Lomba Greija locality on Maio. Human figures on the ledge in the middle of the photograph for scale; (**b**) looking north at interacting dune sets at Lomba Greija. Goat standing in the upper-right of the image for scale. These deposits are dominantly composed of rhodolith debris; (**c**) on Boa Vista, looking west across barchan dunes, just south of Cabeçadas with the photograph taken while standing on an eolianite ledge; (**d**) looking south at eolianite exposed along a roadcut, just south of the airport on Boa Vista; (**e**) looking east across the front of a barchanoid ridge at Praia do Curralinho on the SW corner of Boa Vista; (**f**) outcrop of eroded eolianite being overtopped by modern dune ridge on the SW corner of Boa Vista Island. All images courtesy of Markes E. Johnson.

7. Unusual Carbonate Dunes/Eolianites of Scotland and Ireland

An exceptional case of carbonate dune and sand plain deposition appears along the cool temperate Atlantic coasts of Scotland and Ireland (Figure 11a), well outside the usual subtropical-to-tropical distribution of eolianite and possibly the only carbonate-rich sediment facies within the zone of the last glaciation (MIS 2) [1,29]. In this region, carbonate-rich dunes form the predominantly western edge of a fertile, sandy, and carbonate-rich coastal plain ecosystem (machair in Scotland) and are found in the NW coastal Highlands and island areas of Scotland, as well as the NW coast of Ireland from Malin Head south to the Aran Islands and the northern coastline of Galway Bay [86–89].

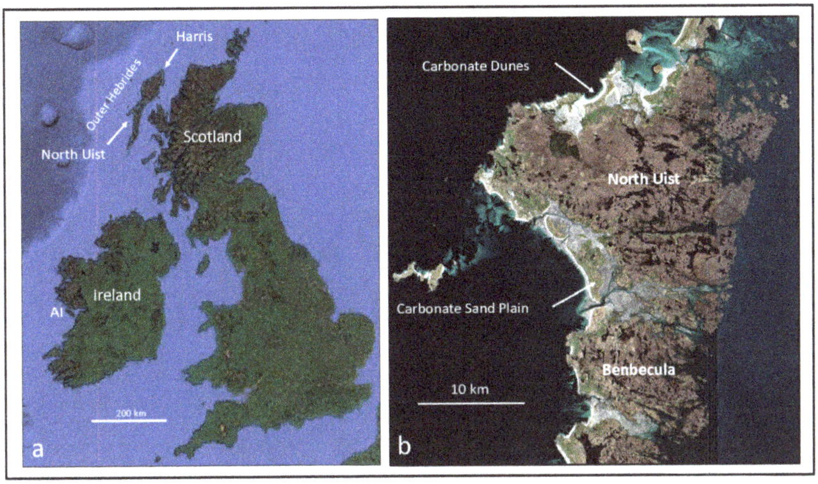

Figure 11. Images showing the region and examples of where the machair ecosystem is found: (**a**) image showing the location of Scotland, Ireland, the Outer Hebrides (Harris and North Uist Ids.), and the Aran Islands, AI; (**b**) image of the North Uist Island area showing modern carbonate dunes and carbonate sand plains associated with the machair ecosystem.

From beach to hillside, the machair ecosystem typically includes carbonate-rich coastal dunes behind which is a sand plain stabilized by vegetation (non-grass) that has adapted to carbonate–sand environments [88]. In almost all cases, the carbonate dune and sand ramp portions of the machair ecosystems are underlain by eolianite cemented by seasonal fluctuations in the water table [88]. In Scotland, machair dunes and sand plains are found in the Hebrides from Islay north of Shetland. Along the Atlantic coast of the Outer Hebrides, the most extensive expression of the machair ecosystem is found along the western coasts of the islands SW of Harris (Figure 11b) [29,87,88,90].

The carbonate-rich dunes of the Outer Hebrides are considered to be part of an unusual barrier/lagoon system whose distribution is strongly controlled by the basal bedrock of the region (Proterozoic Gneiss) [90]. As the region was covered by a kilometers-thick ice sheet during the last continental glaciation, the system is relatively starved of siliciclastic sediments as there are no streams (historically or presently) that would carry post-glaciation sediments across the low-lying, low-angle, west dipping landscape of the region [29]. The continental shelf to the west of the Outer Hebrides is shallow dipping (0.2%) and broad in dimension (>50 km to ≈130 km wide) [29]. Glacially derived sediments deposited on the continental shelf are presumed to have been redistributed with the Holocene rise in sea level. Winds are highly energetic in the region, primarily out of the south-west, and averaging 6–8 m/s year-round [90]. Modern dune sediments are predominantly molluscan in origin with greater volumes of fragmented shell material pushed ashore during storm events [29]. Fragmented molluscan shell is usually reported as the dominant sediment type in the carbonate dunes of both Scotland and Ireland [29,89].

A study utilizing radiocarbon and optically stimulated luminescence (OSL) dating methods outlined in broad strokes a chronology of sand movement onto areas of the Uist Islands (Figure 11b) [91]. Initial accumulations were of quartzose sands (≈13,700 to 10,500 BP-cal) with significant accumulations of carbonate sand starting ≈ 9000 ago. Multiple episodes of sand/dune deposition occurred during the remainder of the Holocene with the last significant episode (600–200 BP-cal) associated with the latest Medieval Warm Period (MWP) and earliest Little Ice Age (LIA) [91]. Subsequent work from additional sites along the Uist Island Atlantic coasts suggests substantial variation in the sand incursion chronology but emphasized that the bulk of the machair (carbonate-rich) sand inventory was deposited between ≈5800–4200 (calendar years), a period known for coastal sand movement across NW Europe [29]. It is likely that the machair dune and sandplain ecosystem will disappear due to coastal erosion as the sea level continues to rise [88].

8. Development of Depositional Models

Though glacio-eustatic changes in sea level are not considered a requirement for eolianite deposition in much of the geologic record, it has become a primary focus in the development of Pleistocene eolianite depositional models [2]. In the first attempt at modeling the development of a carbonate dune, Darwin (Part 2, pp. 86–88, Superficial Deposits [8]) proposed that very fine material (bioclastic and oolitic) on the island of St. Helena (15°58′ S; 5°42′ W)—found between 600 and 700 ft above sea level—must have accumulated at a previous time when there was a marine shelf, "like that of Ascension" (7°57′ S; 14°22′ W). Darwin [8] recognizes that wind is the mechanism of deposition after reading about the violence of the trade winds and the description of similar deposits on the Canary Islands [16]. He also surmised that the relationship between land and water must have been different, though Darwin proposed "elevation of the land", as the concept of eustatic changes in sea level was not yet conceived [8,92]. That percolating rainwater was the agent for the lithification of the calcareous material into rock was the final touch in his analysis [8]. Thus, almost all the components for a model for the origin, deposition, and lithification of carbonate-rich aeolian sediments were in place.

As part of a subsequent study on the geology of St. Helena, Daly [93] noticed offshore patches of calcareous debris, which led him to propose that a drop in sea level of a moderate speed would allow for the mobilization of the sand deposits beachward with the wind moving them further landward. He also makes the case for a 5-m eustatic sea-level drop related to the last glaciation, compiling and discussing data from around the globe and suggesting that this type of analysis will be useful in the study of "earlier oscillations" [93]. Thus, the case for sea-level drop and the deflation of shelf sediments in the formation of carbonate-rich dune sands is conceptually realized.

However, it is the subsequent work by Sayles [3] on the islands of Bermuda where the case for the link between the alternating sequence of five eolianite formations and intervening paleosols with Pleistocene glacio-eustatic sea-level cycles is made. He proposed that dune development was triggered by exposure of lagoon sediments as global sea level dropped due to continental glaciation, coupled with increased winds and a loss of vegetation as cooler climatic conditions prevailed. Intervening paleosols were proposed to have developed as relict weathering features during interglacial periods (Figure 2b) [3]. Observations made by Sayles [3] that have implications on subsequent models proposed for the relationship between eolianite genesis and the recession of relative sea levels are: (1) that modern dunes are rare on Bermuda (interglacial period) and; (2) there is a relatively older group of eolianites that form the core of Bermuda, while younger dune sets form the perimeter of the island. In a more recent review of the models proposed for eolianite genesis on Bermuda, it is noted that the unstated but implied aspect of Sayles' low-stand model for eolianite development is that the extensive lagoon (~12 km wide) along the north-western border of the island group (see Figure 2), as well as other perimeter shelf areas, act as a sink for calcareous sediments generated during high-stand conditions prior to becoming the source material for Pleistocene dunes during subsequent sea-level fall [94].

Contrary to the model proposed by Sayles [3], a rise in relative sea level (RSL) is proposed by Bretz [95] and endorsed by others [96,97]. Bretz proposed that younger, perimeter eolianites along Bermuda's north shore were developed as rising sea levels progressively moved sediments out of the lagoon, onto the beach, and landward no further than behind the beach. In this proposed sequence of events, lagoon sediments are washed to the beach, winnowed by winds as they are blown into dunes, and cemented while they sit just behind the beach. In this way, the dunes are preserved as eolianites sitting above or adjacent to their source beach, preserved as a subtidal calcarenite.

Key to this model is overriding the energy-dampening effect of the reef tract outboard of the lagoon [95]. Sea levels must rise quickly enough to outpace reef catch-up so that waves capable of transporting sand from the lagoon to the beach are not impeded by the reef [95,97]. Vacher [97] suggests different modes of eolianite formation for the north and south coasts of Bermuda as the reef structures associated with two sides of the island are very different. Dune-building episodes along the north side of the island—which faces the lagoon (see Figure 2b)—occur only during sea-level rise, while eolianite deposition along the high-energy south coast occurs only when relative sea level exposes the slow-growing boiler reefs (composed of algae and vermetids, not corals) and associated sediments along the comparatively narrow (~1 km wide) shelf. North- and south-shore eolianite deposition would, therefore, be active or stalled depending on the RSL and not at the same time (Figure 19 in [97]). A further requisite criterion for this depositional model is the prompt cementation of the carbonate dunes by rainwater, preventing further landward movement and preservation of the foredunes adjacent to their beach source [97].

The expectation that younger eolianites could not overtop or impinge on the older, smaller core of eolianites that make up the backbone of Bermuda is an underlying assumption for both the Vacher depositional model and a preferred stratigraphy for Bermuda's eolianite sequences [97]. In support of the idea that younger eolianites were cemented near the beach, Vacher [97] illustrates what is described as a typical cross-section for a dune on Bermuda: a foreset wedge draped or rolled over by a windward topset wedge with the intervening sub-horizontal surface merging with the foresets to landward. Vacher implies that this structural configuration shows vertical dune growth by aggradation and not the recycling of older sediments, as would be expected if the dune were moving landward (Figure 10 in [97]).

In laying out a preferred stratigraphy for Bermuda, Vacher [97] rejects the proposed previous stratigraphic schemes [3,96], instead arguing that analytical error allows for the inverting of radiometric ages and, therefore, the reordering of associated marine units and eolianites into a "preferred" stratigraphic order [97]. Additionally, this stratigraphic scheme allows for "missing" eolianites to be attributed to removal by subsequent high-stand events as opposed to being emplaced and preserved inland after overtopping the older eolianite core of the island [97]. It should be noted that age determinations for Bermudian eolianites are difficult to achieve as there is no suitable quartz fraction available for analysis. Therefore, radiometric age determinations from bioclasts preserved the marine units (beach calcarenites) that underlay eolianites must be used in determining the time-stratigraphic ordering of the island's dunes. As a consequence, organizing the eolianites of Bermuda in a refined time-stratigraphic order has been an evolving process, complicated by the fact that only a portion of Bermuda Island's geological record is found at any single locality. For example, the idea that "terra rossa" paleosols might be island-wide depositional markers [96] was later shown to be erroneous [98].

As part of their thorough and extensive structural analysis of Bermuda's eolianites, Rowe and Bristow [59] show that the typical dune of Vacher [97] actually mirrors the architecture of a sub-critical climbing transverse dune [99]. In addition, their structural analysis of Bermuda's eolianites using ground-penetrating radar provides ample evidence that younger dunes did "transgress" or advance inland over the older eolianitic core of the island [59]. If Bermuda's Pleistocene dunes were not emplaced as transverse coastal dune sets, a re-evaluation of the stratigraphic and age data used to assemble the geologic story of Bermuda's eolianites with respect to changes in Pleistocene sea levels becomes necessary.

Building on an earlier, detailed facies analysis of the interbedded coastal dune, beach, and paleosol deposits of the Belmont Formation (MIS 7) [53], Rowe and Bristow [94] review previous U-series age determinations for coral fragments collected from shoreface and foreshore calcarenites at the base of the upper three (Belmont, Rocky Bay, and Southampton) of Bermuda's formations and provide new age determinations of their own. One of the primary results of the re-evaluation of facies relationships and age data is the adjustment/positioning of the three formations and their member eolianites in relation to the [MIS] 7.1, 5e, and 5a peaks of the Pleistocene eustatic sea-level curve [94] and a consequential depositional model. Rowe and Bristow [94] conclude that, in all three cases, eolianite deposition occurred during the drawdown in sea level following an RSL rise and high-stand, during which carbonate sediments accumulated on the lagoon and shelf, a refinement of Sayles' original model.

The model developed for Bermuda by Vacher [97] was also applied to the extensive eolianite deposits of the Bahamas by Carew and Mylroie [100,101], who subdivided eolianite deposition into three phases (transgressive, stillstand, and regressive) corresponding to the runup, maximum flooding, and regression of water from the Bahamian carbonate banks during high-stands in sea level (Figure 3). As both Bermuda and the Bahamas occur on steep-sided platforms, carbonate production would be restricted to times when advancing, stillstand, or regressing sea levels occur without exposure of the bank [2,100]. Eolianites developed in early phases of the runup to sea-level high-stands would have low preservation potential as the continually encroaching beach would cannibalize these dunes as well as potentially erode eolianites deposited during earlier high-stand events [100]. Paleosols would have been developed during low-stands when the banks were exposed. In addition, source beach and eolianite would move in concert and be cemented relatively quickly, as with the "Vacher Model" proposed for Bermuda [97,100,101].

Given the 1000+ km distance from Little Bahama Bank to the Caicos Bank to the south-east, it is not surprising that the geologic record for the Bahama Banks is complex. As noted by Kindler and Hearty [55], linking depositional events to specific sea-level events is challenging. As such, the stratigraphic column for the Bahamas has evolved since the initial depositional history proposed by Garret and Gould on New Providence Island [102]. A more recent synthetic column that reflects more recent studies on Mayaguana Island is provided by Kindler and Hearty, who also review key studies—focused primarily on outcrops from San Salvador, Eleuthera, and Great Inagua—that have advanced the physical stratigraphic understanding of the Bahamas up to now [55].

Representative of the multi-method approach needed to unravel the complex sequence of depositional events across the Bahamas is the work of Kindler and Hearty establishing the Whale Point Formation (WPF) [55]. Utilizing morphostratigraphic, stratigraphic, sedimentological, and petrographic methods anchored by amino-acid racemization (AAR) analyses from 11 islands across the island banks, Kindler and Hearty reposition eolianites and related paleosols—previously thought to be related to the (MIS) 5e maximum in sea level—to the younger (MIS) 5a high-stand. Adding to the evolving thinking around depositional models for the emplacement of eolianites, they found that the WPF eolianites advanced landward and progressively away from their beach source areas, similar to what was found on Bermuda by Rowe and Bristow [55,94]. In addition, Kindler and Hearty proposed that the two eolianites and the scant intervening protosol of the WPF represent deposition during phases of RSL drop, related to two sub-peaks in the MIS 5a high-stand record, with carbonate sediments produced during the preceding, sequential high-stands [55].

In contrast to the advancing eolianites described from Bermuda and the Bahamas, seaward prograding dunes are less commonly recognized and only from the Late Holocene. In the Bahamas, strand plains of Late Holocene age (MIS) 1, created by multiple beach ridges that sequentially accrete as the system progrades oceanward, are described on San Salvador, Eleuthera, and both the Great and Little Exuma Islands [100,103,104]. Grouped together in the Hanna Bay Member of the Rice Bay Formation, eolianites are found above or inboard and laterally adjacent to foreshore and backshore deposits, evidence that dunes either prograded oceanward over beach deposits or that the beach deposits were the source area for the ridges forming the strand plain [104].

The range of whole-rock radiocarbon ages obtained from the Hanna Bay Member suggests deposition during and somewhat beyond the Medieval Warm Period (MWP) [105], similar to the time frame for the deposition of sands in the Outer Hebrides of Scotland [93]. Rejecting the idea that deposition of the intertidal and eolian deposits of the Hannah Bay Member was driven by a decelerating rise in eustatic sea level, it is proposed that more ephemeral sea-level high-stands, related to increased storminess during the MWP, resulted in the oceanward progradation of these strand plains [103–106]. It is also surmised that these strand plain deposits will not survive continued sea-level rise [104]. This prediction reflects the generally accepted proposition that many interstadial or interglacial variations in the past may have resulted in eolianite formations that were eroded away during subsequent and more significant sea-level high-stands [97,100,104].

9. Depositional Models on Coastal Shelf vs. Rimmed Platform

As described above, depositional models for eolianite deposition have been—since the initial work of Darwin [8] and Daly [93]—developed in reference to steep-sided and rimmed platforms (Bermuda and the Bahamas), where deposition of sediments can only occur when the RSL overtops the edge of the platform [96,97,100]. Therefore, it is necessary to compare eolianite deposition on coastal ramps since the high-stand depositional model—developed on rimmed and steep-sided platforms—in some cases has been unilaterally applied to eolianites deposited on coastal shelves [107,108].

In contrast to carbonate platforms, deposition can occur on a coastal ramp with sea-level regression as the carbonate factory is able to track with the shoreline [2]. The limiting factors in this depositional system become the width/profile of the downwind shelf area as this delimits the volume of sediment that can be stored in the offshore environment and the extent of the RSL drop relative to the edge of the local shelf. Examples of broad coastal shelves with substantial carbonate production include the south-west corner of the Persian Gulf [30], the southern tip of South Africa (Figure 12) [34], and along the southern and western coasts of Australia [33,66].

Strand plains, composed of multiple shore-parallel beach ridges that accumulate sequentially oceanward from the coast, have commonly been linked to high-stand sea levels along these broad shelves [1]. For example, the low-angle, broad (150+ km) coastal shelf off of the Coorong Coast Plain of Australia (Figure 4) has a cool-water carbonate factory with a high rate of bioclast production [56,108]. The extensive eolianite ridges produced over the last 800k+ years are interpreted as being shoreline barriers correlated with successive Pleistocene interglacial sea-level highs [13,56,65]. Uplift of the region due to Quaternary neotectonism has led to the elevation and spatial separation of these paleoshorelines after deposition, though several of the ridges are compound structures of multiple, stacked dune sets that complicate deciphering the depositional chronology [13].

In comparison, the more tectonically stable coastal shelf of Western Australia, near Perth and Rottnest Island, gently deepens to around 75 m before any significant shelf break. As along the southern coast of Australia, the onshore eolianite ridges have very complex depositional histories as earlier ridges have been eroded, reworked, or overstepped by younger preserved dune systems (see Figure 7 in [33]). These composite deposits are mirrored by offshore structures that appear to be drowned eolianite ridges representing relict coastal barriers. Comparable eolianite ridges with similarly complex depositional histories are also found on the coastal plains of Israel [109] and South Africa [110,111], while analogous submerged aeolianite structures are found along other parts of the Western Australian, Southeastern Australian, and South African shelves [111–114]. Taken as a whole, there is the suggestion that stable, gently to modestly deepening shelves will record a more sensitive eolianite depositional record, with the formation of carbonate dunes occurring at times other than peak high-stands [33].

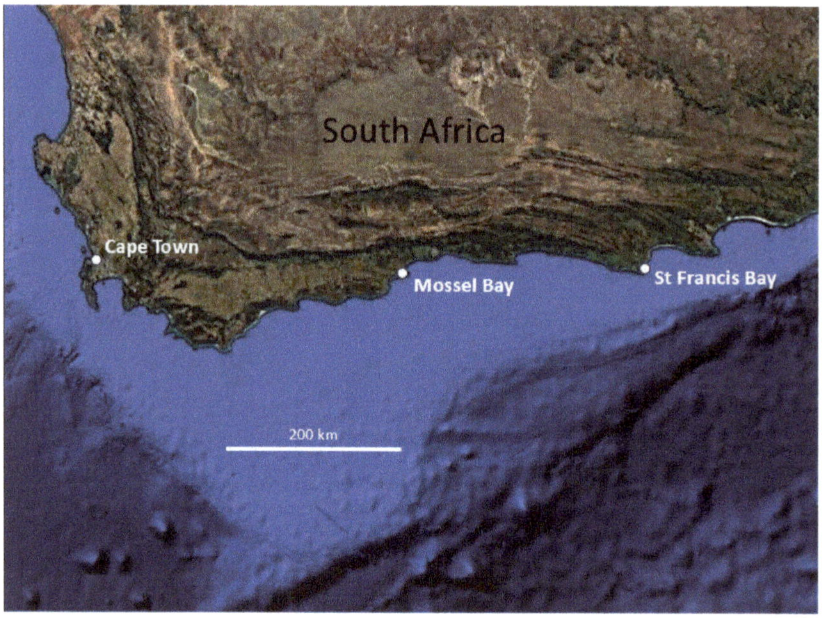

Figure 12. Image of a representative broad coastal shelf along the southern coast of South Africa. Modern carbonate-rich dunes and eolianites and Pleistocene eolianites are found along all coastal areas shown in this image.

Another example that illustrates, albeit obliquely, the contrast in depositional models for coastal shelf systems as compared to rimmed platforms is the contrasting approaches taken in establishing the timing of eolianite accumulation represented in the outcrops of the Tamala Limestone along the south-western coast of Western Australia. Hearty and O'Leary use U/Th age determinations of corals to calibrate a series of whole-rock amino-acid racemization (AAR) ratios [107]. AAR is a technique developed in the Bahamas that requires another dating method to anchor any progression of the AAR data [115]. In evaluating their U/Th results, Hearty and O'Leary question the veracity of the age data that did not fall within the age range (too old and too young) indicating the 5e high-stand as they support an "a priori" assumption that eolianites only develop during sea-level high-stands [97,107]. A similar approach is taken by Lipar and Webb in support of the "high-stand Model" for eolianite deposition of the Tamala Limestone [35]. U/Th age data are reported ($\pm$20 to 50 ka) with the measurements not falling in the range of high-stand peaks described as potentially problematic [35]. Similar to the high-stand model of Vacher for Bermuda, early cementation of the eolianites due to the corresponding warm, wet climate is a component of the model proposed for the Carnavon shelf eolianites [107].

In contrast, Vimpere and others use detailed sedimentological and stratigraphic analyses to support the idea that the eolianites of the Tamala Limestone represent landward-advancing dunes [60]. They use a climate model to support the idea that the Tamala Limestone eolianites developed during glacial periods, when the climate would have been drier and windier. Also, they plot the available age data against the Lisiecki and Raymo MIS stack [116], illustrating that the preponderance of the data does not support the "high-stand" model for eolianite deposition [60]. In addition, they note that sea levels control sediment production with generally lower production rates on ramps compared to rimmed platforms [117], while sediment availability is controlled by climate [2,30]. As the Tamala Limestone eolianites developed adjacent to the 50 to 100 km wide Carnavon shelf, carbonate production is not limited by sea level compared to a rimmed platform [60].

The process of refining depositional models requires placing eolianites within a stratigraphic framework where the sequence and relationships between lithofacies can be understood so that the relative sea-level position can be established [2]. In an effort to standardize the language used to describe the relationship between eolianite deposition and RSL, more than twenty studies were interpreted and categorized in terms of sequence stratigraphy nomenclature [58]. Of the studies from coastal ramp or shelf areas, about three-quarters of the studies connected eolianite accumulation to high-stand systems tract (HST) conditions or related variations in timing (e.g., approaching or exiting high-stand conditions). The remainder could be categorized as being emplaced during either the falling-stage systems tract (FSST), approaching a low-stand, or low-stand systems tract (LST) conditions.

Overall, there is the suggestion that where sediments accumulate in the offshore shelf-to-beach system, the transect may influence the perceived timing of dune/eolianite emplacement onshore when evaluating age data. If the source sediments for eolianite accumulate predominantly in a seaward prograding beach face, then the distribution of the age data recovered from the dune system will reflect emplacement during high-stand conditions. In contrast, if sediments are primarily stored in offshore accommodation space, the mobilization of this source material during an RSL drop will overwhelm any signal from a preceding high-stand beach.

For example, in their study of eolianites along the southern coast of the Persian Gulf, Wiliams and Walkden found that modern carbonate-rich sediments are sequestered outboard of a coastal barrier complex with no active dune accumulation behind the beach [30]. Eolianites found 85 km downwind from the modern beach were interpreted as sourced from the Gulf during sea-level fall after the MIS 5e high-stand. In this case, it is worth noting that the eolianites are thought to have been decoupled from their source area due to high aridity preventing quick cementation [30].

Another example is from the cliff-lined coast of Pollenca Bay, Mallorca, where eolianites (MIS 5c/b, 4, 3) are preserved along the southern edge of the bay. Sediments preserved in the eolianite are sourced from the adjacent bowl-bottomed bay, exposed during relative drops in sea level—not from the coastal beach [26,27]. Studies of similarly rugged coastal areas of the Western Mediterranean show that high-volume storage of sediments also does not occur in the near-shore but on the adjacent shelf during high-stand conditions, with sediment then mobilized, deposited, and preserved as eolianite after being exposed during the subsequent sea-level drop [31,36,118].

10. Conclusions

Since the publication of the extensive review on eolianites by Brooke [1], the study of eolianites has become increasingly multi-disciplinary with stratigraphic, petrologic, and sedimentological analyses enhanced by the application of ground-penetrating radar, the use of climate models, and the creative application of dating techniques.

There has been an increase in the use of luminescence methods of dating eolianites in mixed carbonate–siliciclastic environments, while U/Th, the amino-acid racemization, and radiocarbon dating techniques predominate in depositional settings where minor to no siliciclastics are present. The dating of Plio-Pleistocene and particularly post-Sangamon (Eemian) eolianite deposits will require more precise and accurate analysis in teasing apart the complex depositional record related to interstadial/interglacial fluctuations and the concomitant cyclical generation of sediment and eolianite.

There has been a re-evaluation of the "high-stand" model, which is developed on the steep-sided carbonate platforms of Bermuda and is a dominant model in the literature. Many of the key precepts of the "high-stand" model, such as the early cementation of dunes next to their beach sources, have not been upheld by more recent studies that point to regression exposing sediments accumulated during the preceding sea-level rise. On broad coastal shelves, the depositional histories are more complex as both carbonate factory and shoreline positions shift with changes in relative sea levels, leading to complex depositional histories where younger eolianites have overstepped, cannibalized, or coalesced with older eolianites. The link between regression and exposure of offshore sediment reservoirs is also asserted for rugged, cliff-walled coastlines where accommodation space on the beach is limited. Establishing eolianites in a sequence stratigraphic framework allows for a better understanding of the role that relative sea level plays in depositional models.

The most extensive eolianite deposits of the Pleistocene age (volumetric and time span) are found on tectonically passive margins associated with highly productive offshore carbonate factories. Eolianites preserved in the Mediterranean Basin are found in more varied coastal environments, reflecting the region's complex tectonic history. More recently, eolianites have been identified in the transtensional Gulf of California basin and along the convergent margin of coastal Chile.

The recognition, refined definition, and geographic distribution of global carbonate factories have been a significant advancement in the understanding of where carbonate bioclastic and inorganic sediments (eolianite source sediments) are produced. Primarily controlled by light and temperature, the Photozoan-T factory includes tropical (~30 N to 30 S) localities (e.g., the Great Bahama Bank) with the Persian Gulf subregion (Biochemical Zone) an area of high salinity and temperature. The volumetrically large eolianite deposits along the southern coasts of Australia and South Africa fall within the Heterozoan-C factory, which lies poleward of the Photozoan-T factory and is primarily controlled by nutrients. The Mediterranean Sea (Photo-C factory) is a variant of the Heterozoan-C factory, with the primary controls of light and temperature. Eolianite clasts continue to be a source of information concerning offshore depositional systems of the past, tracking shifts in and within carbonate factories. In depositional settings with sufficient wave action and relatively clear light conditions, rhodolith clasts are a substantial or dominant component in coastal eolianites. Rhodoliths are found in all regions with carbonate deposition, with the exception of the Persian Gulf (Biochemical factory).

There have been few studies that focus primarily on eolianite cements; however, alterations in eolianite cements may reflect environmental changes, particularly in regions where seasonal rain patterns are affected by shifts in climate.

Funding: This research received no external funding.

Acknowledgments: The author is grateful to Rob Haley of the Haverford College Library for assistance in tracking down references. Haverford College also provided the ArcGIS Pro (3.1.0) software used to generate images for this paper. Field images shown in Figure 10a–f were provided by Markes

E. Johnson (Williams College). The author thanks the anonymous reviewers who provided helpful comments that improved this article. The author is responsible for any errors in the text. Multiple images were used in the creation of figures for this article. The citations for the image data are as follows: (1) Figure 2a includes bathymetric data from the GEBCO 2024 dataset, while the image data are from the ASTER sensor (AST_L1T_00304112021151443_20220322185348_552), accessed 13 November 2024 through the NASA EARTHDATA portal. Figure 2b is an image from Google Earth Pro that includes data from Landsat, Copernicus, SIO, NOAA, the U.S. Navy, NGA, LDEO-Columbia, and GEBCO; (2) Figure 3 is an image from Google Earth Pro with data provided by Landsat, Copernicus, SIO, NOAA, the U.S. Navy, NGA, and GEBCO; (3) The digital elevation data used to create Figure 4 are from the ASTER DEM 003 dataset ASTGTM_003-20241108_144052, accessed 13 November 2024 through the NASA EARTHDATA portal; (4) Figure 5 is an image from Google Earth Pro that includes data from Landsat, Copernicus, SIO, NOAA, the U.S. Navy, NGA, LDEO-Columbia, and GEBCO; (5) Figure 6a is a multiband image from the Copernicus Browser, Sentinel 2A dataset, granule L2A_T12RUQAO48515_20241005T181316, accessed on 24 October 2024; (6) Figure 7 includes a satellite image from the ASTER sensor (L1B_003_04082001182114_10312003072215), accessed 14 March 2005 from the UMD-GLCF data repository. The aerial photographs were downloaded from the INEGI digital data archive; (7) Figure 9 is from Google Earth Pro, with data provided by Landsat, Copernicus, SIO, NOAA, the U.S. Navy, NGA, and GEBCO; (8) Figure 11a is from Google Earth Pro, with data provided by Landsat, Copernicus, SIO, NOAA, the U.S. Navy, NGA, and GEBCO. Figure 11b is a composite multiband image from Sentinel 2A data, granules T29VND_A040875_20230420T114345 and T29VPD_A040875_20230420T114345, downloaded from the Copernicus Browser on 14 October 2024; (9) Imagery for Figure 12 is from Google Earth Pro, with data provided by Landsat, Copernicus, SIO, NOAA, the U.S. Navy, NGA, and GEBCO.

Conflicts of Interest: The author declares no conflicts of interest.

References

1. Brooke, B. The distribution of carbonate eolianite. *Earth Sci. Rev.* **2001**, *55*, 135–164. [CrossRef]
2. Abegg, F.E.; Loope, D.; Harris, P.M. Carbonate eolianites: Depositional models and diagenesis. In *Modern and Ancient Carbonate Eolianites: Sedimentology, Sequence Stratigraphy, and Diagenesis*; Abegg, F.E., Harris, P.M., Loope, D.B., Eds.; SEPM Special Publication: Claremore, OK, USA, 2001; Volume 71, pp. 17–30.
3. Sayles, R.W. Bermuda during the ice age. *Proc. Am. Acad. Arts Sci.* **1931**, *66*, 381–467. [CrossRef]
4. Pettijohn, F.J. *Sedimentary Rocks*, 2nd ed.; Harper and Row Publishers: New York, NY, USA, 1975; 628p.
5. Fairbridge, R.W.; Johnson, D.L. Eolianite. In *The Encyclopedia of Sedimentology*; Fairbridge, R.W., Bourgeois, J., Eds.; Dowden, Hutchinson and Ross: Stroudsburg, PA, USA, 1978; pp. 279–282.
6. McLaren, S.J. Aeolianites. In *Geochemical Sediments and Landscapes*; Nash, D.J., McLaren, S.J., Eds.; Blackwell Publishing: Oxford, UK, 2007; pp. 144–172.
7. Reijmer, J.J.G. Marine Carbonate Factories: Review and Update. *Sedimentology* **2021**, *68*, 1729–1796. [CrossRef]
8. Darwin, C. *Geological Observations on Coral Reefs, Volcanic Islands, and on South America*; Smith, Elder & Co.: London, UK, 1851; 700p.
9. Kindler, P.; Hine, A.C. The paradoxical occurrence of oolitic limestone on the eastern islands of Great Bahama Bank: Where do the ooids come from? *Int. Assoc. Sedimentol. Spec. Publ.* **2009**, *41*, 113–122.
10. Harris, P.M.; Purkis, S.J.; Ellis, J.; Swart, P.K.; Reijmer, J.G. Mapping bathymetry and depositional facies on Great Bahama Bank. *Sedimentology* **2015**, *62*, 566–589. [CrossRef]
11. Rouse, J.; James, N.P.; Kyser, T.K. Aeolianites reveal Pleistocene marine history of Bermuda. *Sedimentology* **2019**, *66*, 1580–1599. [CrossRef]
12. Belperio, A.P.; Murray-Wallace, C.V.; Cann, J.H. The Last Interglacial shoreline in southern Australia: Morphostratigraphic variations in a temperate carbonate setting. *Quat. Int.* **1995**, *26*, 7–19. [CrossRef]
13. Murray-Wallace, C.V.; Belperio, A.P.; Cann, J.H. Quaternary neotectonism and intra-plate volcanism: The Coorong to Mount Gambier Coastal Plain, Southeastern Australia: A review. *Geol. Soc. Lond. Spec. Publ.* **1998**, *146*, 255–267. [CrossRef]
14. Muhs, D.R.; Pigati, J.S.; Schumann, R.R.; Skipp, G.L.; Porat, N.; DeVogel, S.B. Quaternary sea-level history and the origin of the northernmost coastal aeolianites in the Americas: Channel Islands National Park, California, USA. *Palaeogeogr. Palaeoclim. Palaeoecol.* **2018**, *491*, 38–76. [CrossRef]
15. Johnson, M.E.; Ledesma-Vázquez, J.; Ramalho, R.S.; Da Silva, C.M.; Rebelo, A.C.; Santos, A.; Baarli, B.G.; Mayoral, E.; Cachão, M. Taphonomic range and sedimentary dynamics of modern and fossil rhodolith beds: Macaronesian Realm (North Atlantic Ocean). In *Rhodolith/Maërl Beds: A Global Perspective*; Riosmena-Rodríguez, R., Nelson, W., Aguirre, J., Eds.; Coastal Research Library, Springer International Publishing: Cham, Switzerland, 2017; pp. 221–261.

16. von Buch, L.; Boulanger, C. *Description Physique des Îles Canaries, Suivie d'une Indication des Principaux Volcans du Globe*; F.G. Levrault: Paris, France, 1836; 525p.
17. Johnson, M.E.; Baarli, B.G.; da Silva, C.M.; Cachão, M.; Ramalho, R.S.; Ledesma-Vázquez, J.; Mayoral, E.D.; Santos, A. Coastal dunes with high content of rhodolith (coralline red algae) bioclasts: Pleistocene formations on Maio and São Nicolau in the Cape Verde archipelago. *Aeolian Res.* **2013**, *8*, 1–9. [CrossRef]
18. Le Pera, E.; Morrone, C.; Arribas, J.; Arribas, M.E.; Ancochea, E.; Huertas, M.J. Petrography and provenance of beach sands from volcanic oceanic islands: Cabo Verde, Atlantic Ocean. *J. Sediment. Res.* **2021**, *91*, 92–115. [CrossRef]
19. Harvey, A.; Johnson, M.E.; Harvey, R. Heterozoan carbonate-enriched beach sand and coastal dunes—With particular reference to rhodoliths, Dirk Hartog Island, Shark Bay, Western Australia. *Facies* **2018**, *64*, 23. [CrossRef]
20. Angulo, R.J.; de Souza, M.C.; Fernandes, L.A.; Disaró, S.T. Quaternary sea-level changes and aeolianites in the Fernando de Noronha archipelago, northeastern Brazil. *Quat. Intern.* **2013**, *305*, 15–30. [CrossRef]
21. Russell, P.; Johnson, M.E. Influence of seasonal winds on coastal carbonate dunes from the Recent and Plio-Pleistocene at Punta Chivato (Baja California Sur, Mexico). *J. Coast. Res.* **2000**, *16*, 709–723.
22. Johnson, M.E.; Backus, D.H. Beach Deflation and Accrual of Pliocene-Pleistocene Coastal Dunes of the Gulf of California Region. In *Atlas of Coastal Ecosystems in the Western Gulf of California: Tracking Limestone Deposits on the Margin of a Young Sea*; Johnson, M.E., Ledesma-Vazquez, J., Eds.; University of Arizona Press: Tucson, AZ, USA, 2009; pp. 134–144.
23. Frechen, M.; Neber, A.; Dermann, B.; Tsatskin, A.; Boenigk, W.; Ronen, A. Chronostratigraphy of aeolianites from the Sharon coastal plain of Israel. *Quat. Intern.* **2002**, *89*, 31–44. [CrossRef]
24. Frechen, M.; Neber, A.; Tsatskin, A.; Boenigk, W.; Ronen, A. Chronology of Pleistocene sedimentary cycles in the Carmel Coastal Plain of Israel. *Quat. Intern.* **2004**, *121*, 41–52. [CrossRef]
25. Nash, D.J.; Bateman, M.D.; Bullard, J.E.; Latorre, C. Late Quaternary coastal evolution and aeolian sedimentation in the tectonically-active southern Atacama Desert, Chile. *Palaeogeogr. Palaeoclimatol. Palaeoecol.* **2018**, *490*, 546–562. [CrossRef]
26. Fornós, J.; Clemmensen, L.B.; Gómez-Pujol, L.; Murray, A.S. Late Pleistocene carbonate aeolianite deposits on Mallorca, western Mediterranean: A luminescence chronology. *Quat. Sci. Rev.* **2009**, *28*, 2697–2709. [CrossRef]
27. Fornós, J.J.; Clemmensen, L.B.; Gómez-Pujol, L.; Ginés, A.; Gines, J. Pleistocene eolianites and low sea levels. *Monogr. De La Soc. D'historia Nat. De Les Balears.* **2012**, *18*, 85–110.
28. Goodfriend, G.A.; Cameron, R.A.D.; Cook, L.M.; Courty, M.-A.; Fedoroff, N.; Livett, E.; Tallis, J. The Quaternary eolian sequence of Madeira: Stratigraphy, chronology, and paleoenvironmental interpretation. *Palaeogeogr. Palaeoclimatol. Palaeoecol.* **1996**, *120*, 195–234. [CrossRef]
29. Ritchie, W.; Whittington, G.; Edwards, K.J. Holocene changes in the physiography and vegetation of the Atlantic littoral of the Uists, Outer Hebrides, Scotland. *Trans. R. Soc. Edinb. Earth Sci.* **2001**, *92*, 121–136. [CrossRef]
30. Williams, A.H.; Walkden, G.M. Carbonate Eolianites from a Eustatically Influenced Ramp-Like Setting: The Quaternary of the Southern Arabian Gulf. In *Modern and Ancient Carbonate Eolianites: Sedimentology, Sequence Stratigraphy, and Diagenesis*; Abegg, F.E., Harris, P.M., Loope, D.B., Eds.; SEPM Special Publication: Claremore, OK, USA, 2001; Volume 71, pp. 77–92.
31. del Valle, L.; Timar-Gabor, A.; Fornós, J.J. Chronology of Pleistocene sedimentary cycles in the western Mediterranean. *Quat. Sci. Rev.* **2024**, *330*, 108451. [CrossRef]
32. Titschack, J.; Nelson, C.S.; Beck, T.; Freiwald, A.; Radtke, U. Sedimentary evolution of a Late Pleistocene temperate red algal reef (Coralligéne) on Rhodes, Greece. correlation with global sea-level fluctuations. *Sedimentology* **2008**, *55*, 1747–1776. [CrossRef]
33. Brooke, B.P.; Olley, J.M.; Playford, P.E.; Haines, P.W.; Murray-Wallace, C.; Woodroffe, C. Chronology of Quaternary coastal aeolianite deposition and the drowned shorelines of southwestern Western Australia—A reappraisal. *Quat. Sci. Rev.* **2014**, *93*, 106–124. [CrossRef]
34. Bateman, M.D.; Holmes, P.J.; Carr, A.S.; Horton, B.P.; Jaiswal, M.K. Aeolianite and barrier dune construction spanning the last two glacial–interglacial cycles from the southern Cape coast, South Africa. *Quat. Sci. Rev.* **2004**, *23*, 1681–1698. [CrossRef]
35. Lipar, M.; Webb, J.A. Middle–late Pleistocene and Holocene chronostratigraphy and climate history of the Tamala Limestone, Cooloongup and Safety Bay Sands, Nambung National Park, southwestern Western Australia. *Aust. J. Earth Sci.* **2014**, *61*, 1023–1039. [CrossRef]
36. Andreucci, S.; Clemmensen, L.B.; Murray, A.S.; Pascucci, V. Middle to late Pleistocene coastal deposits of Alghero, Northwest Sardinia (Italy): Chronology and evolution. *Quatern. Internat.* **2010**, *222*, 3–16. [CrossRef]
37. Li, R.; Qiao, P.; Cui, Y.; Zhang, D.; Liu, X.; Shao, L. Composition and diagenesis of Pleistocene aeolianites at Shidao, Xisha Islands: Implications for palaeoceanography and palaeoclimate during the last glacial period. *Palaeogeogr. Palaeoclimatol. Palaeoecol.* **2018**, *49*, 604–616. [CrossRef]
38. James, N.P.; Bone, Y. Pleistocene aeolianites at Cape Spencer, South Australia; record of a vanished inner neritic cool-water carbonate factory. *Sedimentology* **2015**, *62*, 2038–2059. [CrossRef]
39. McKee, E.D.; Ward, W.C. Eolian environment. In *Carbonate Depositional Environments*; Scholle, P.A., Bedout, D.G., Moore, C.H., Eds.; AAPG Memoir; American Association of Petroleum Geologists: Tulsa, OK, USA, 1983; Volume 33, pp. 132–169.
40. Kindler, P.; Davaud, E. Recognizing eolianites in thin-section. Review and case study: The Lower Cretaceous Chambotte Formation, Salève Chain, southeastern France. In *Modern and Ancient Carbonate Eolianites: Sedimentology, Sequence Stratigraphy, and Diagenesis*; Abegg, F.E., Harris, P.M., Loope, D.B., Eds.; SEPM Special Publication: Claremore, OK, USA, 2001; Volume 71, pp. 141–150.

41. Loope, D.; Abegg, F.E. Recognition and geological preservation of ancient carbonate eolianites. In *Modern and Ancient Carbonate Eolianites: Sedimentology, Sequence Stratigraphy, and Diagenesis*; Abegg, F.E., Harris, P.M., Loope, D.B., Eds.; SEPM Special Publication: Claremore, OK, USA, 2001; Volume 71, pp. 3–16.
42. Frébourg, G.; Hasler, C.A.; Le Guern, P.; Davaud, E. Facies characteristics and diversity in carbonate eolianites. *Facies* **2008**, *54*, 175–191. [CrossRef]
43. Woodroffe, C.D. *Coasts: Form, Processes and Evolution*; University Press: Cambridge, UK, 2002; 623p.
44. Vacher, H.L.; Rowe, M.P.; Garrett, P. *The Geological Map of Bermuda*; Public Works Department Publication, The Government of Bermuda: Hamilton, Bermuda, 1989.
45. Gardner, R.A.M. Aeolianite. In *Chemical Sediments and Geomorphology*; Goudie, A.S., Pye, K., Eds.; Academic Press: London, UK, 1983; pp. 265–300.
46. Ritchie, W. Machair development and chronology in the Uists and adjacent islands. *Proc. Roy. Soc. Edin.* **1979**, *77B*, 107–122. [CrossRef]
47. Reeckmann, S.A.; Gill, E.D. Rates of vadose diagenesis in quaternary dune and shallow marine calcarenites, Warrnambool, Victoria, Australia. *Sediment. Geol.* **1981**, *30*, 157–172. [CrossRef]
48. McLaren, S.; Gardner, R. Late Quaternary vadose carbonate diagenesis in coastal and desert dune and beach sands: Is there a paleoclimatic signal? *Earth Surf. Process. Landf.* **2004**, *29*, 1441–1458. [CrossRef]
49. Muhs, D.R.; Simmons, K.R.; Groves, L.T.; McGeehin, J.P.; Schumann, R.R.; Agenbroad, L.D. Late Quaternary sea-level history and the antiquity of mammoths (*Mammuthus exilis* and *Mammuthus columbi*), Channel Islands National Park, California, USA. *Quat. Res.* **2015**, *83*, 502–521. [CrossRef]
50. Jacobs, Z. Luminescence chronologies for coastal and marine sediments. *Boreas* **2008**, *37*, 508–535. [CrossRef]
51. Hearty, P.J.; Vacher, H.L.; Mitterer, R.M. Aminostratigraphy and ages of Pleistocene limestones of Bermuda. *Geol. Soc. Am. Bull.* **1992**, *104*, 471–480. [CrossRef]
52. Rutter, N.W.; Blackwell, B. Amino acid racemization dating. In *Dating Methods for Quaternary Deposits*; Rutter, N.W., Catto, N.R., Eds.; Geological Association of Canada: Newfoundland, NL, Canada, 1995; pp. 125–167.
53. Rowe, M.; Wainer, K.; Bristow, C.; Thomas, A. Anomalous MIS 7 sea level recorded on Bermuda. *Quat. Sci. Revs.* **2014**, *90*, 47–59.
54. Godefroid, F.; Kindler, P.; Chiaradia, M.; Fischer, G. The Misery Point cliff, Mayaguana Island, SE Bahamas: A unique record of sea-level highstands since the Early Pleistocene. *Swiss J. Geosci.* **2019**, *112*, 287–305. [CrossRef]
55. Kindler, P.; Hearty, P.J. The Whale Point Formation: A stratigraphic record of high-frequency climate and sea-level fluctuations in the Bahamas during Marine Isotope Stage 5a (ca. 80 ka BP). *Sediment. Geol.* **2022**, *432*, 106107. [CrossRef]
56. Murray-Wallace, C.V.; Bourman, R.P.; Prescott, J.R.; Williams, S.F.; Price, D.M.; Belperio, A.P. Aminostratigraphy and thermoluminescence dating of coastal aeolianites and the later Quaternary history of a failed delta: The River Murray mouth region, South Australia. *Quat. Geochron.* **2010**, *5*, 28–49. [CrossRef]
57. Carr, A.S.; Bateman, M.D.; Cawthra, H.C.; Sealy, J. First evidence for onshore marine isotope stage 3 aeolianite formation on the southern Cape coastline of South Africa. *Mar. Geol.* **2019**, *407*, 1–15. [CrossRef]
58. Mauz, B.; Hijma, M.P.; Amorosi, A.; Porat, D.; Galili, E.; and Bloemendal, E.J. Aeolian beach ridges and their significance for climate and sea level: Concept and insight from the Levant coast (East Mediterranean). *Earth-Sci. Rev.* **2013**, *121*, 31–54. [CrossRef]
59. Rowe, M.; Bristow, C. Landward-advancing Quaternary eolianites of Bermuda. *Aeolian Res.* **2015**, *19*, 235–249. [CrossRef]
60. Vimpere, L.; Del Piero, N.; Le Cotonnec, A.; Kindler, P.; Castelltort, S. Depositional timing and palaeoclimate interpretation of the Tamala Limestone aeolianites in Shark Bay, Western Australia. *Aeolian Res.* **2022**, *54*, 100700. [CrossRef]
61. Carr, A.S.; Bateman, M.D.; Holmes, P.J. Developing a 150ka luminescence chronology for the coastal dunes of the southern Cape, South Africa. *Quat. Geochron.* **2007**, *2*, 110–116. [CrossRef]
62. Carr, A.S.; Bateman, M.D.; Roberts, D.L.; Murray-Wallace, C.V.; Jacobs, Z.; Holmes, P.J. The last interglacial sea-level high stand on the southern Cape coastline of South Africa. *Quat. Res.* **2010**, *73*, 351–363. [CrossRef]
63. Foreman, F. Study of some Bermuda rock. *Geol. Soc. Am. Bull.* **1951**, *62*, 1297–1330. [CrossRef]
64. Initial Core Descriptions DSDP Legs 44 and 44A, Western North Atlantic. *Deep Sea Drilling Project: Archive of Core and Site/Hole Data and Photographs from the Deep-Sea Drilling Project (DSDP)*; National Geophysical Data Center, NOAA: Boulder, CO, USA, 1989. [CrossRef]
65. Murray-Wallace, C.V. Pleistocene coastal stratigraphy, sea-level, highstands and neotectonism of the southern Australian passive continental margin—A review. *J. Quat. Sci.* **2002**, *17*, 469–489. [CrossRef]
66. Blakemore, A.G.; Murray-Wallace, C.V.; Westaway, K.E.; Lachlan, T.J. Aminostratigraphy and sea-level history of the Pleistocene Bridgewater Formation, Mount Gambier region, southern Australia. *Aust. J. Earth Sci.* **2015**, *62*, 151–169. [CrossRef]
67. Ledesma-Vázquez, J.; Johnson, M.E.; Gonzalez-Yajimovich, O.; Santamaría-del-Angel, E. Gulf of California Geography, Geological Origins, Oceanography, and Sedimentation Patterns. In *Atlas of Coastal Ecosystems in the Western Gulf of California: Tracking Limestone Deposits on the Margin of a Young Sea*; Johnson, M.E., Ledesma-Vazquez, J., Eds.; University of Arizona Press: Tucson, AZ, USA, 2009; pp. 1–10.

68. Helenes, J.; Carreno, A.L.; Marcos-Giron, R.Y. Miocene-Pliocene biochronology of the Obregon Basin and it's bearing on the evolution of the Proto-Gulf of California. *J. S. Am. Earth Sci.* **2020**, *104*, 102758. [CrossRef]
69. Johnson, M.E.; Ledesma-Vazquez, J. Pliocene and Pleistocene Development of Peninsular and Island Rocky Shores in the Gulf of California. In *Atlas of Coastal Ecosystems in the Western Gulf of California: Tracking Limestone Deposits on the Margin of a Young Sea*; Johnson, M.E., Ledesma-Vazquez, J., Eds.; University of Arizona Press: Tucson, AZ, USA, 2009; pp. 28–44.
70. Parés-Sierra, A.; Mascarenhas, A.; Marinone, S.G.; Castro, R. Temporal and spatial variation of the surface winds in the Gulf of California. *Geophys. Res. Lett.* **2003**, *30*, 1312–1316. [CrossRef]
71. Backus, D.H.; Johnson, M.E. Sand dunes on peninsular and island shores in the Gulf of California. In *Atlas of Coastal Ecosystems in the Western Gulf of California: Tracking Limestone Deposits on the Margin of a Young Sea*; Johnson, M.E., Ledesma-Vazquez, J., Eds.; University of Arizona Press: Tucson, A, USA, 2009; pp. 117–133.
72. Anderson, C.A. Geology of islands and neighboring land areas. In *1940 E. W. Scripps Cruise to the Gulf of California*; Part, 1, Anderson, C.A., Durham, J.W., Shepard, F.P., Natland, M.L., Revelle, R., Eds.; Geological Society of America Memoir; Geological Society of America: Boulder, CO, USA, 1950; Volume 43, p. 53.
73. James, N.P. The cool-water carbonate depositional realm. In *Cool-Water Carbonates*; James, N.P., Clarke, A.D., Eds.; SEPM Special Publication: Claremore, OK, USA, 1997; Volume 56, pp. 1–20.
74. Schlager, W. Benthic carbonate factories of the Phanerozoic. *J. Earth. Sci.* **2003**, *92*, 445–464. [CrossRef]
75. Schlager, W. *Carbonate Sedimentology and Sequence Stratigraphy*; SEPM Concepts in Sedimentology and Paleontology: Claremore, OK, USA, 2005; Volume 8, pp. 1–200.
76. Michel, J.; Borgomano, J.; Reijmer, J.J.G. Heterozoan carbonates: When, where and why? A synthesis on parameters controlling carbonate production and occurrences. *Earth Sci. Rev.* **2018**, *182*, 50–67. [CrossRef]
77. Laugié, M.; Michel, J.; Pohl, A.; Poli, E.; Borgomano, J. Global distribution of modern shallow-water marine carbonate factories: A spatial model based on environmental parameters. *Sci. Rep.* **2019**, *9*, 16432. [CrossRef] [PubMed]
78. Sewell, A.A.; Johnson, M.E.; Backus, D.H.; Ledesma-Vázquez, J. Rhodolith detritus impounded by a coastal dune on Isla Coronados, Gulf of California. *Cienc. Mar.* **2007**, *33*, 483–494. [CrossRef]
79. Foster, M.S. Rhodoliths: Between rock and soft places. *J. Phycol.* **2001**, *37*, 659–667. [CrossRef]
80. Steller, D.L.; Riosmena-Rodríguez, R.; Foster, M.S. Living Rhodolith Bed Ecosystems in the Gulf of California. In *Atlas of Coastal Ecosystems in the Western Gulf of California: Tracking Limestone Deposits on the Margin of a Young Sea*; Johnson, M.E., Ledesma-Vazquez, J., Eds.; University of Arizona Press: Tucson, AZ, USA, 2009; pp. 72–82.
81. Amado-Filho, G.M.; Moura, R.; Basto, A.C.; Salgado, L.T.; Sumida, P.; Guth, A.Z.; Francini-Filho, R.B.; Pereira-Filho, G.H.; Abrantes, D.P.; Brasileiro, P.S.; et al. Rhodolith beds are major $CaCO_3$ biofactories in the tropical South West Atlantic. *PLoS ONE* **2012**, *7*, e35171. [CrossRef]
82. Teichert, S.; Woelkerling, W.J.; Rüggeberg, A.; Wisshak, M.; Piepenburg, D.; Meyerhöfer, M.; Form, A.; Freiwald, A. Arctic rhodolith beds and their environmental controls. *Facies* **2014**, *60*, 15–37. [CrossRef]
83. Soares, A.F. A formação eolianítica da ilha de Porto Santo. *Memórias e Notícias, Publ. Museu Lab Min. Geológico Univ. Coimbra Cent. Estud Geol.* **1973**, *75*, 47–88.
84. Robertson, A.H.F. Mesozoic deep-water and Tertiary volcaniclastic deposition of Maio, Cape Verde Islands: Implications for Atlantic paleoenvironments and ocean island volcanism. *Geol. Soc. Am. Bull.* **1984**, *94*, 433–453. [CrossRef]
85. Hernández, L.; Suárez, C. Characterization of the contemporary aeolian sediment dynamics of Boa Vista (Cape Verde). *J. Coast. Res.* **2008**, *48*, 64–68.
86. Ritchie, W. The meaning and definition of machair. *Trans. Botan. Soc. Edinb.* **1976**, *42*, 431–440. [CrossRef]
87. Bassett, J.A.; Curtis, T.G.F. The Nature and Occurrence of Sand-Dune Machair in Ireland. *Proc. Roy. Irish Acad. Sect. B Biol. Geol. Chem. Sci.* **1985**, *85B*, 1–20.
88. Angus, S. The conservation importance of machair systems of the Scottish Islands, with particular reference to the Outer Hebrides. In *The Islands of Scotland: A Living Marine Heritage*; Baxter, J.M., Usher, M.B., Eds.; HMSO: Edinburgh, UK, 1994; pp. 95–120.
89. Gaynor, K. The vegetation of Irish Machair. *Biol. Environ. Proc. R. Ir. Acad.* **2006**, *106B*, 311–321. [CrossRef]
90. Pile, J.; Cooper, J.A.G.; Jackson, D.W.T. Stratigraphy and internal structure of wind-dominated barrier islands (dune and machair) of the Outer Hebrides, Scotland. *Earth Surf. Process. Landf.* **2019**, *44*, 1482–1493. [CrossRef]
91. Gilbertson, D.D.; Schwenninger, J.L.; Kemp, R.A.; Rhodes, E.J. Sand-drift and Soil Formation Along an Exposed North Atlantic Coastline: 14,000 Years of Diverse Geomorphological, Climatic and Human Impacts. *J. Arch. Sci.* **1999**, *26*, 439–469. [CrossRef]
92. Fairbridge, R.W. Eolianites and Eustacy: Early Concepts on Darwin's Voyage of HMS Beagle. *Carbonates Evaporites* **1995**, *10*, 92–101. [CrossRef]
93. Daly, R.A. The Geology of Saint Helena Island. *Proc. Am. Acad. Arts Sci.* **1927**, *62*, 31–92. [CrossRef]
94. Rowe, M.; Bristow, C. Sea-level controls on carbonate beaches and coastal dunes (eolianite): Lessons from Pleistocene Bermuda. *Geol. Soc. Am. Bull.* **2015**, *127*, 1645–1665. [CrossRef]
95. Bretz, J.H. Bermuda: A partially drowned, late mature, Pleistocene Karst. *Geol. Soc. Am. Bull.* **1960**, *71*, 1729–1754. [CrossRef]

96. Land, L.S.; Mackenzie, F.T.; Gould, S.J. Pleistocene history of Bermuda. *Geol. Soc. Am. Bull.* **1967**, *78*, 993–1006. [CrossRef]
97. Vacher, H.L. Coastal dunes of younger Bermuda. In *Coastal Geomorphology*; Coates, D.R., Ed.; State University of New York: Binghampton, NY, USA, 1973; pp. 355–391.
98. Vacher, H.L.; Hearty, P.J.; Rowe, M.P. Stratigraphy of Bermuda: Nomenclature, concepts, and status of multiple systems of classification. In *Terrestrial and Shallow Marine Geology of the Bahamas and Bermuda*; Curran, H.A., White, B., Eds.; GSA Special Paper; Geological Society of America: Boulder, CO, USA, 1995; Volume 300, pp. 271–288.
99. Rubin, D.M.; Hunter, R.E. Bedform climbing in theory and nature. *Sedimentology* **1982**, *29*, 121–138. [CrossRef]
100. Carew, J.L.; Mylroie, J.E. Depositional model and stratigraphy for the Quaternary geology of the Bahama Islands. In *Terrestrial and Shallow Marine Geology of the Bahamas and Bermuda*; Curran, H.A., White, B., Eds.; GSA Special Paper; Geological Society of America: Boulder, CO, USA, 1995; Volume 300, pp. 5–32.
101. Carew, J.L.; Mylroie, J.E. Quaternary carbonate eolianites of the Bahamas: Useful analogues for the interpretation of ancient rocks? In *Modern and Ancient Carbonate Eolianites: Sedimentology, Sequence Stratigraphy, and Diagenesis*; Abegg, F.E., Harris, P.M., Loope, D.B., Eds.; SEPM Special Publication: Claremore, OK, USA, 2001; Volume 71, pp. 33–45.
102. Garrett, P.; Gould, S.J. Geology of New Providence Island, Bahamas. *Geol. Soc. Am. Bull* **1984**, *95*, 209–220. [CrossRef]
103. Savarese, M.; Hoeflein, F.J. Sea level and the paleoenvironmental interpretation of the middle to late Holocene Hanna Bay Limestone, San Salvador, Bahamas: A high foreshore setting without a higher-than-present eustatic highstand. In Proceedings of the Fifteenth Symposium on the Geology of the Bahamas and Other Carbonate Regions; Gamble, D.W., Kindler, P., Eds.; Gerace Research Centre: San Salvador, Bahamas, 2012; pp. 163–183.
104. Savarese, M.; Curran, H.A. Origin of Late Holocene Strandplains in the Southern Exuma Islands, Bahamas: Progradation, Ephemeral Highstands, and Storminess. In Proceedings of the Sixteenth Symposium on the Geology of the Bahamas and Other Carbonate Regions; Gamble, D.W., Kindler, P., Eds.; Gerace Research Centre: San Salvador, Bahamas, 2016; pp. 39–59.
105. Mann, M.E.; Woodruff, J.D.; Donnelly, J.P.; Zhang, Z. Atlantic hurricanes and climate over the past 1500 years. *Nature* **2009**, *460*, 880–883. [CrossRef]
106. Graham, N.E.; Ammann, C.M.; Fleitmann, D.; Cobb, K.M.; Luterbacher, J. Support for global climate reorganization during the "Medieval Climate Anomaly". *Clim. Dyn.* **2011**, *37*, 1217–1245. [CrossRef]
107. Hearty, P.J.; O'Leary, M.J. Carbonate eolianites, quartz sands, and Quaternary sea-level cycles, Western Australia: A chronostratigraphic approach. *Quat. Geochronol.* **2008**, *3*, 26–55. [CrossRef]
108. James, N.P.; Bone, Y. Quaternary aeolianites in south-east Australia—A conceptual linkage between marine source and terrestrial deposition. *Sedimentology* **2017**, *64*, 1005–1043. [CrossRef]
109. Sivan, D.; Porat, N. Evidence from luminescence for Late Pleistocene formation of calcareous aeolianite (kurkar) and paleosol (hamra) in the Carmel Coast, Israel. *Palaeogeogr. Palaeoclim. Palaeoecol.* **2004**, *211*, 95–106. [CrossRef]
110. Porat, N.; Botha, G. The luminescence chronology of dune development on the Maputaland coastal plain, southeast Africa. *Quat. Sci. Rev.* **2008**, *27*, 1024–1046. [CrossRef]
111. Bateman, M.D.; Carr, A.S.; Dunajko, A.C.; Holmes, P.J.; Roberts, D.L.; McLaren, S.J.; Bryant, R.G.; Marker, M.E.; Murray-Wallace, C.V. The evolution of coastal barrier systems: A case study of the Middle-Late Pleistocene Wilderness barriers, South Africa. *Quat. Sci. Rev.* **2011**, *30*, 63–81. [CrossRef]
112. Ramsay, P.J. Marine geology of the Sodwana Bay shelf, southeast Africa. *Mar. Geol.* **1994**, *120*, 225–247. [CrossRef]
113. Nichol, S.L.; Brooke, B.P. Shelf habitat distribution as a legacy of Late Quaternary marine transgressions: A case study from a tropical carbonate province. *Cont. Shelf Res.* **2011**, *31*, 1845–1857. [CrossRef]
114. Sprigg, R.C. Stranded and submerged sea-beach systems of southeast south Australia and the aeolian desert cycle. *Sediment. Geol.* **1979**, *22*, 5396. [CrossRef]
115. Hearty, P.J.; Kaufman, D.S. Whole-rock aminostratigraphy and Quaternary sea-level history of the Bahamas. *Quat. Res.* **2000**, *54*, 163–173. [CrossRef]
116. Lisiecki, L.E.; Raymo, M.E. A Pliocene-Pleistocene stack of 57 globally distributed benthic δ 18O records. *Paleoceanography* **2005**, *20*, PA1003. [CrossRef]
117. Sultana, D.; Burgess, P.; Bosence, D. How do carbonate factories influence carbonate platform morphology? Exploring production-transport interactions with numerical forward modelling. *Sedimentology* **2021**, *69*, 372–393. [CrossRef]
118. Andreucci, S.; Clemmensen, L.B.; Pascucci, V. Transgressive dune formation along a cliffed coast at 75 ka in Sardinia, Western Mediterranean: A record of sea-level fall and increased windiness. *Terra Nova* **2010**, *22*, 424–433. [CrossRef]

Journal of
Marine Science and Engineering

MDPI

Review

Australia's Two Great Barrier Reefs: What Can ~360 Million Years of Change Teach Us?

Gregory E. Webb

School of the Environment, The University of Queensland, Brisbane, QLD 4072, Australia; g.webb@uq.edu.au; Tel.: +61-7-3365-2181

Abstract

Coral reefs are among the most important marine habitats but face significant threats from anthropogenic sources, including climate change. This paper reviews and compares the modern Great Barrier Reef Province and the 360-million-year-old Devonian Great Barrier Reef of western Australia. Despite occurring at times with different climates, biota (both marine and terrestrial), weathering processes and marine chemistry, similar reefs were constructed under certain circumstances. Major differences in global temperature, marine carbonate saturation, sea level behavior and reef community constituents were evaluated. The comparison highlights the integration of, and interdependencies within, reef communities and the need for both carbonate producers and significant binders, whether skeletal or microbial, to construct a reef in a high-energy setting. Devonian communities with abundant corals and skeletal sponges were incapable of making modern reef types without competent binders to unify framework into rigid substrate. The current strong focus on corals and bleaching in modern reef conservation may be obscuring the equally significant issue of ocean acidification, which impacts on equally crucial framework unification, i.e., hard binding by coralline algae and microbialites and early cementation. The comparison also supports the idea that 'empty bucket' carbonate platform morphologies require increased accommodation from high-amplitude icehouse sea level oscillations.

Keywords: coral reef; Great Barrier Reef; Canning Basin; Devonian; framework; carbonate factory

1. Introduction

Coral reefs, despite their small area of the globe, are among the most biodiverse environments on Earth, potentially harboring as many as 25–33% of all marine taxa [1]. At the same time, coral reefs provide many important ecological services (ESs), including fisheries/food security, building materials, recreational centers, opportunities for economic development, and direct coastal protection, for as many as 500 million people in tropical regions of the world [2–5]. Significant economic value accrues from these services (e.g., [6,7]). Reefs and their habitable islands also serve as, and support, important cultural heritage for many indigenous populations (e.g., [8–12]). The great value of, and now payment for, reef ESs have been the focus of growing attention owing to the many threats faced by coral reefs from predicted climate change [13,14]. Coral bleaching from marine heat waves [15], increased storminess [16,17] and ocean acidification [18,19] threaten reef growth along with the services they provide. Previous anthropogenic changes to reef environments ranging from indigenous fishing and gathering [20] to regional industrialization and declining water quality from proximal land use change already have impacted significantly on many of the world's reefs [21–23].

A major contributor to coral reef ESs, especially where coastal protection is concerned, is the characteristic of reef communities (including ancient reef communities that lacked corals) to be perhaps the world's best natural engineers. These communities interact directly with local geography to modify proximal environments and construct rigid structures at sea level that directly support and benefit their constituent communities. These rigid, relief-bearing structures, defined by reef geomorphology, protect adjoining lagoons and coasts and thus uniquely support a variety of other marine and coastal ecosystems. Reef growth dynamics have been of great interest to ecologists and geologists since Darwin [24] and have been studied in a comparative sense (modern reefs as analogs for ancient reefs, and ancient reefs as analogs for the longer-term drivers of, constraints on, and anthropogenic effects on modern reef formation) since that time [25–30]. As builders, reef communities were the first life on Earth to create elevated structures large enough to be visible from space. That said, no two reefs today are exactly alike owing to differences in latitude, local climate, bathymetry, tectonic setting, basement substrate, local oceanographic settings, etc., and, when geological time is added, even marine and proximal terrestrial biological constituents and marine chemistry have changed significantly. Regardless, reefs in the ancient past commonly bear striking similarities to modern reefs in terms of geomorphology and, where studied in detail, even their local community structures and the distributions of morphological types and functional roles of organisms across different reef environments share many similarities [31–39].

As reefs are very sensitive to both climate and sea level, understanding them is even more important now with those factors projected to change owing to anthropogenic forcing. Additionally, as short-term emitters and long-term carbon sinks in the modern shallow ocean, reefs themselves play a role in the carbon cycle and their carbon dynamics are important to understand in that regard alone [40–43]. Understanding the developmental history of Quaternary reefs is one of the best ways to predict potential future effects of environmental changes that are now happening or are projected to happen (e.g., [25,28,30,44–46]). However, the broader constraints on, and patterns in, reef dynamics are relatively poorly understood and the immediate precursors of modern reefs are generally difficult to study. In areas of subsidence like the Great Barrier Reef Province (GBRP) in northeastern Australia previous (penultimate) reefs are buried beneath living reefs making them inaccessible without scientific drilling [45,47–51]. Elsewhere in the world where precursor reefs are not overlain by modern reefs, they commonly are uplifted and exposed to meteoric (freshwater) systems that destroy or alter original reef materials making them difficult or impossible to date and disturbing their environmental geochemical archives [52–56].

As reefs of the more distant past appear to have experienced many of the same types of oceanographic and climatic issues that face modern reefs, understanding how such reefs handled previous changes provides important information for forward modeling potential future changes in reefs [57,58] and possibly for predicting outcomes of some types of environmental change [26,27,29,59–62]. Although the need to understand broader reef dynamics through time is recognized, functional approaches to reef ecology (e.g., reef guilds as functional roles [32,33,63]; functional groups of herbivorous fish [64]; functional ecology of coralline algal crusts [65–67], coral community dynamics [68–70]) have lagged behind many other types of reef ecology and paleoecology studies [71,72]. Thus, the broader principles that guide reef communities through times of change are not adequately understood despite increasing comparative data being available [37,73].

Even the GBRP of eastern Australia, the largest reef complex in the modern world (Figure 1a), faces serious challenges [74–79]. The >3000 reefs of the GBRP contain many different reef types in different settings (onshore to offshore, higher or lower tidal ranges (meso- to macrotidal), greater or lower current velocities, differing water quality, etc. [80,81]). The latitudinal range of the GBRP is so large that the reef encompasses different climatic zones and the history of the reef itself differs significantly from north to south owing to the northward movement of the Australian Plate through those different climatic zones [82,83] (Figure 2). Few opportunities exist to investigate fossil reefs over such large and complex geographic distributions, but extensive reef growth during the Devonian Period provides possibilities [84,85]. Major Devonian reef provinces extended for more than 1000 km in many regions, peaking in the Middle Devonian [85]. Late Devonian reefs were less abundant but extensive reefs continued in western Canada (Alberta), NW Europe, NE Laurussia and eastern slopes of the Urals, South China and western Australia [84–86] (Figure 1b). Perhaps the best exposed Devonian reef for detailed study is in the Canning Basin, western Australia where the 'Devonian Great Barrier Reef' (DGBR) currently crops out over a lateral distance of 350 km [87–90] (Figure 1b). This ~360-million-year-old reef occurred mostly after peak global Devonian reef development (Givetian) but survived for more than 20 million years (Figure 2), through major changes in sea level and even one of the large mass extinction intervals that affected reef builders in particular [84,91]. The purpose of this paper is to review and compare these two major Australian reef provinces (the GBRP and DGBR) with their differing reef builders and settings to allow inferences to be made about reef building in general.

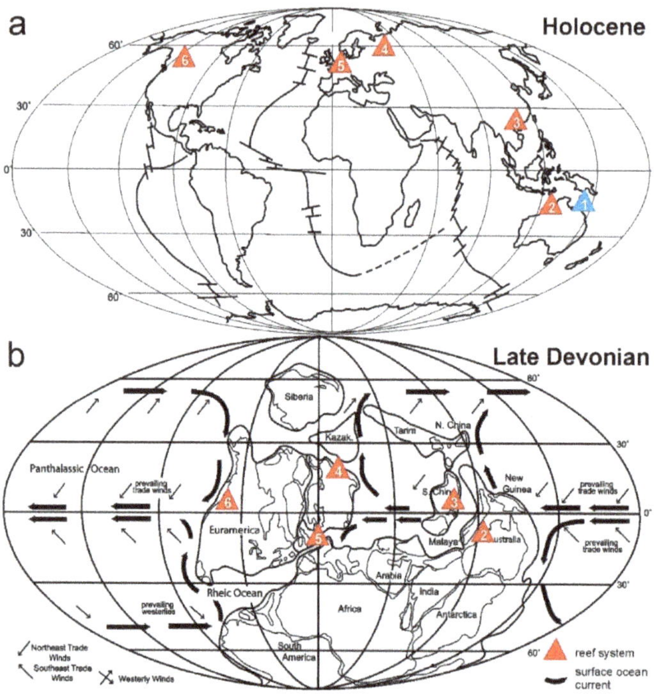

Figure 1. Holocene (**a**) and Late Devonian (**b**) Mollweide projections showing locations of Great Barrier Reef Province (blue triangle 1) and major Late Devonian reef provinces (red triangles): (2) Devonian Great Barrier Reef of Western Australia; (3) South China; (4) NE Laurussia—Russian platform, from Timan-Pechora to the pre-Caspian of Kazakhstan; (5) SE Laurentia—Belgium and (6) Alberta, Canada. Base maps provided by Markes Johnson and (**b**) modified from [92].

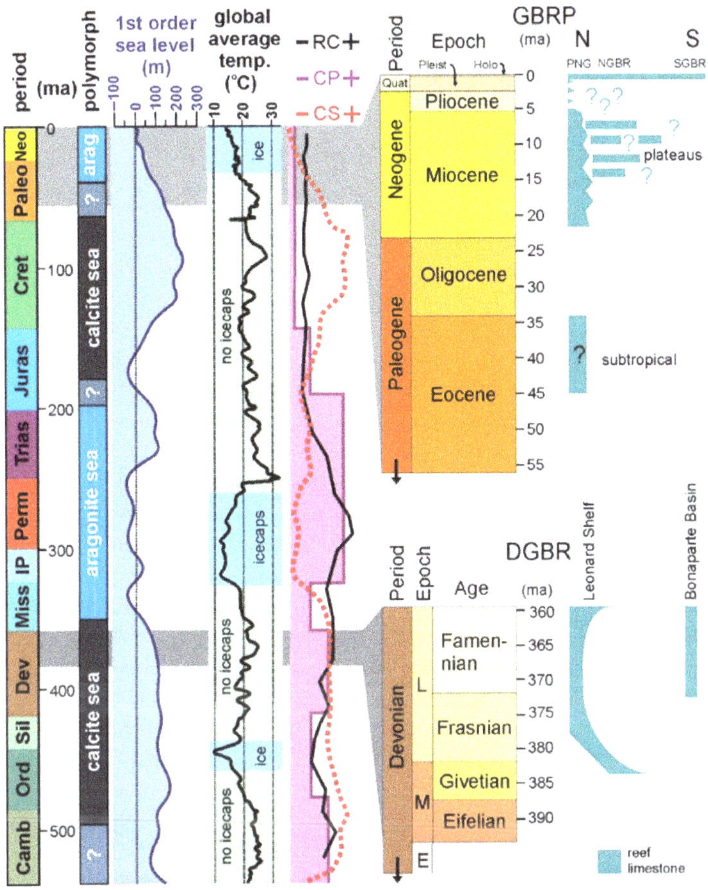

Figure 2. Stratigraphic distribution of the Great Barrier Reef Province and the Devonian Great Barrier Reef. Calcite–aragonite seas modified from [93], there after [94]. Sea level from [95]; global average temperature and ice caps versus no ice caps from [96]; reef cement (RC) from [97]; cement potential (CP) from [98]; modeled calcite saturation (CS) from [99]; geological time [100].

2. Operational Definitions—Methods, Perspectives, Concepts and Semantics

The present review is based on the literature and field work carried out on the Leonard Shelf between 1988 and 2010 and in the GBR between 1984 and 2024. The goal is to make the discussion accessible both to modern reef workers and paleontologists–geologists working on ancient reefs, but reef concepts vary widely between modern ecologists and geologists and even the definition of what constitutes a 'reef' or 'reef framework' varies between different practitioners [25,31,98,101–103]. Different reef semantics make better sense to different workers based on their background, experience, aims and scale of study [72,104,105]. The important thing, though, is to be clear about the semantics being used in an analysis so that the concepts can be directly compared, even if the terminology differs. For this review, a general definition is used [72,106] where reefs are defined as autochthonous structures that (1) have relief above the surrounding sea floor, (2) have adequate synsedimentary rigidity to withstand local wave action or potential wave action, and (3) were constructed by a specialized biological community that differs from that on the surrounding level (non-reef) sea floor. Although reef ecosystems today are dominated by scleractinian corals and crustose coralline algae (CCA), and corals are commonly considered the dominant and/or most significant feature of a reef, almost to the exclusion of other organisms [107], the living biology is but a thin veneer situated upon the substantial structure of the reef edifice itself. It was these massive structures that loomed up from deeper waters to provide hazards to shipping just below the ocean's surface that first brought reefs to the attention of Western mariners (and no doubt many earlier Indigenous mariners) and then to modern science. The term 'reef' was most likely developed from an old Norse or German word 'rif', meaning a ridge or 'Er Rif', an Arabic term for submerged hills [102]. Using the present definition, a reef is a physical, self-supported edifice that includes both the so-called 'framework' and the particulate sediment that is protected behind it from wave energy (i.e., the shallow back-reef) along with the attached sediment that is gravitationally deposited on the reef slopes that helps create and maintain the relief. This definition avoids the difficult semantic task of differentiating a 'reef core' from the rest of the edifice, but regardless, there must be a demarcation between what is and is not part of the reef. Reef sediments transported away from the reef by currents or gravity (e.g., turbidites) to end up in different depositional settings, such as the deep sea, would not be considered part of the reef. In the case of modern linear, 'ribbon' reefs, reef sediments slope off the back of the windward reef margin and eventually intergrade with normal level bottom, non-reef sediments on the shelf. The demarcation could be chosen based on relief or by sediment source depending on the semantics chosen. For this paper, elevation above the level bottom is the key factor, but members of the reef-building community may intergrade continuously with those of the non-reef, level bottom community. In shore-attached platforms and fringing reefs, reef-derived sediments may grade into continental sediments at the same elevation. In that case composition would provide a necessary demarcation, but the elevation of even the continental sediments is supported by the reef rim and thus could be considered part of the overall edifice. Platform reefs are perhaps easiest to conceive, as they are elevations that are isolated on all sides by deeper waters, but every reef and reef system is unique in some way.

The thin veneer of living reef organisms makes these truly huge and impressive structures (and navigational hazards) given time. However, the nature of that 'time' has led to different perspectives being held by different reef workers [104] (Figure 3). In general, biologists–ecologists focus mostly on modern observations and instrumental records, but increasingly, written historical records are used to understand ecological processes on the surface of the reef. Geologists, on the other hand, more commonly study reefs internally as larger structures with longer histories. However, as our understanding of reefs has increased, the interface between 'biological' and 'geological' processes and time scales has merged [108]. The dynamics of reef environments through ecological time (e.g., life of a coral as demonstrated by annual banding [109], to intermediate disturbance regimes—years, decades or centuries [110]) have been well studied by modern ecologists. With modern radiometric dating [111–113] and geochemical proxies [30,114–118], such studies are being pushed back to multi-centennial and millennial scales and beyond. While whole reefs record the positions of sea level through time, coral skeletons and reefal microbialites can record geochemical archives of their local environments (temperature, salinity, pH, water quality, nutrients, etc.) from daily increments [119], to seasonal, yearly, decadal and millennial scales (Figure 3). Coral-based geochemical archives provide data on Holocene climate dynamics, such as El Niño Southern Oscillation (ENSO), Pacific Decadal Oscillation (PDO), and Indian Ocean Dipole, that are required for better climate modeling [120–122]. At longer time scales (10s to 100s of thousands of years; e.g., Milankovitch orbital forcing frequencies and icehouse–greenhouse climate phases), reefs have been studied more by geologists [25,26,39,46,123]. That work and interest in ancient reefs as hydrocarbon reservoirs in the subsurface [124] have focused most geological attention on older reefs. As a result, important reef-building processes, such as 'lithification' (turning reef products into rocks), are commonly considered the realm only of geologists and 'geological time.' However, construction of a rigid structure must occur on ecological timescales to provide adequate substrate for coral recruitment after disturbance [125–128]. Increasingly, the processes that control early lithification in reefs have been demonstrated to be biologically controlled or mediated over these shorter ecological time scales [129–133].

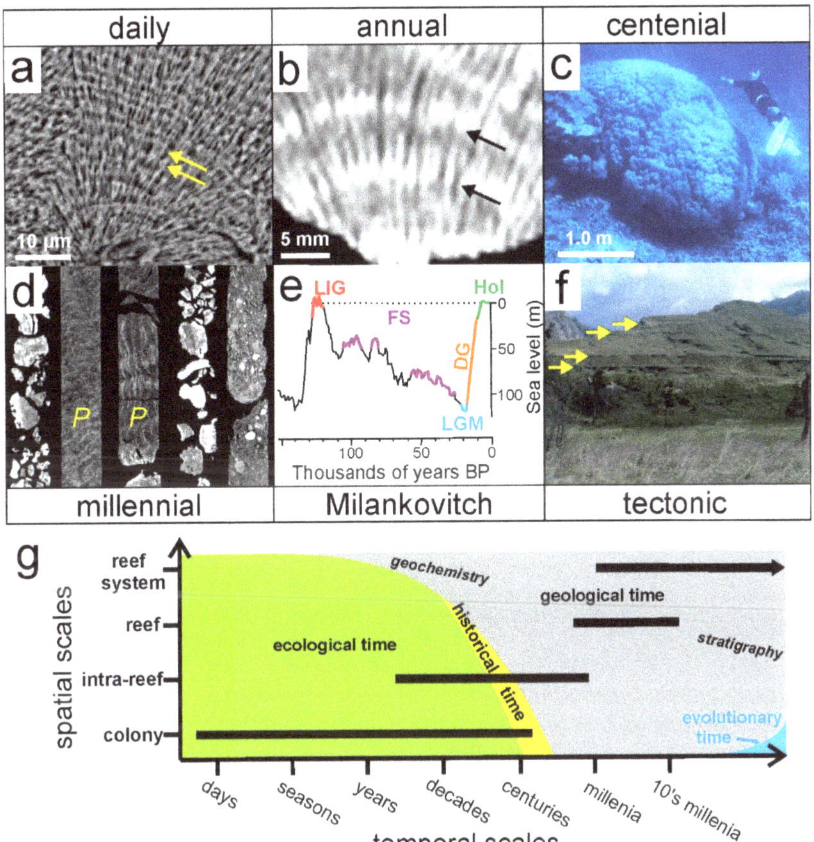

Figure 3. Temporal scales of processes preserved by corals and coral reefs. (**a**) Scanning electron photomicrograph of *Porites* coral showing daily growth increments in aragonite skeleton (yellow arrows), Heron Reef, Great Barrier Reef (GBR); (**b**) computed tomography (CT) X-ray scan slice, *Porites* coral showing annual density bands (black arrows), Heron Reef, GBR; (**c**) massive *Porites* colony, Heron Reef, GBR—such massive colonies record continuous decadal to centennial records of coral growth; (**d**) CT scan slice of core box showing *Porites* (*P*) in mid-Holocene reef core, Heron Reef, GBR—older recovered corals aid reconstruction of records over thousands of years; (**e**) representative sea level curve that highlights the upper limits of vertical coral reef growth through the Holocene (Hol), Deglacial (DG), Last Glacial Maximum (LGM), falling stage (FS) and Last Interglacial (LIG)—modern reefs are constrained by sea level fluctuations associated with Milankovitch scale (20–400 ka) climate cycles; (**f**) tectonically uplifted Pleistocene reef terraces (yellow arrows), Huon Peninsula, Papua New Guinea—sea level and climate constraints make coral reefs useful recorders of global change over very long tectonic scales of 10 s to 100 s of millions of years; (**g**) schematic diagram showing time frames over which ecological, historical, geological and evolutionary processes operate on reefs relative to spatial and temporal scales.

Although modern reefs are dominated by corals, CCA are crucial [72,134] and many other organisms contribute quantitatively to carbonate production [135–138] (Figure 4). Scleractinian corals are 'keystone' taxa in modern reefs and the diversity and range of morphology of corals directly supports much of the non-coral biodiversity for which coral reefs are noted. However, modern reefs generally are not composed mostly of corals preserved in growth position. Most reef corals are ephemeral and destined for destruction to form sediment, which, with other reef organisms, is transported and deposited to make up the vast bulk of a given reef edifice [41,139–143]. Ancient corals played major roles in reef building in some time intervals, but major reef systems also occurred in the complete absence of corals or even skeletons of any kind when microbialites (rocks formed by microbial biofilms) dominated [98,144]. Combined, carbonate producers, both skeletal and microbial, in a given reef community can be thought of usefully as a carbonate factory [145]. Members of a given carbonate factory are constrained by a variety of physicochemical factors (e.g., light, temperature, energy levels, pH, alkalinity, oxygen, etc.) in the local environment as well as nutrients [146] and have changed through time as skeletal and non-skeletal organisms have evolved to keep pace with changing marine chemistry, climate and each other [31,35,37,98,99,147–149]. No reef can form without a carbonate factory to make an adequate volume of carbonate sediment. However, for the volume of sediment to make an elevated structure in high-energy settings at sea level, the sediment must be well stabilized, and this is traditionally considered a function of 'reef framework.' In modern coral reefs, in situ coral skeletons provide a degree of strength, but they are essentially fragile (Figure 4), especially in the face of bioerosion. Some modern coral reefs are notable mainly for their lack of coral skeletons in growth position [141–143,150,151]. These rubble-rich reefs have gained considerable recent attention as coral bleaching and large storms have increased rubble production on some modern reef flats. Understanding how reef rubble becomes stabilized has become a conservation imperative [152–154].

Although every reef is unique in some ways [31] key ongoing processes (Figure 5) that generate and constrain reefs typically include:

- **Construction** of calcium carbonate by biomineralization (e.g., coral skeletons, foraminifer tests) or by induction of carbonate precipitation by microbial biofilms (microbialites); in situ carbonate production is the major role of the carbonate factory, and importantly, includes production of both skeletons in growth position and particulate sediment;
- **Destruction** of in situ calcium carbonate (e.g., today, coral skeletons) both by physicochemical processes (waves, storms, chemical dissolution) and by bioerosion (destruction by other organisms) create mobile sediment from initially immobile substrates;
- **Transportation** of sediment by gravity, waves, currents and wind;
- **Deposition** of sediment by settling, baffling and/or trapping;
- **Unification**, where skeletons in growth position, sediment (including coarse rubble) and microbial deposits are bound into rigid, hard substrate by (1) direct encrustation by other skeletal organisms, generally termed binders (e.g., CCA), (2) microbial precipitates within larger and smaller cavities and pore spaces and in some cases (3) early marine cement that precipitates naturally from seawater.

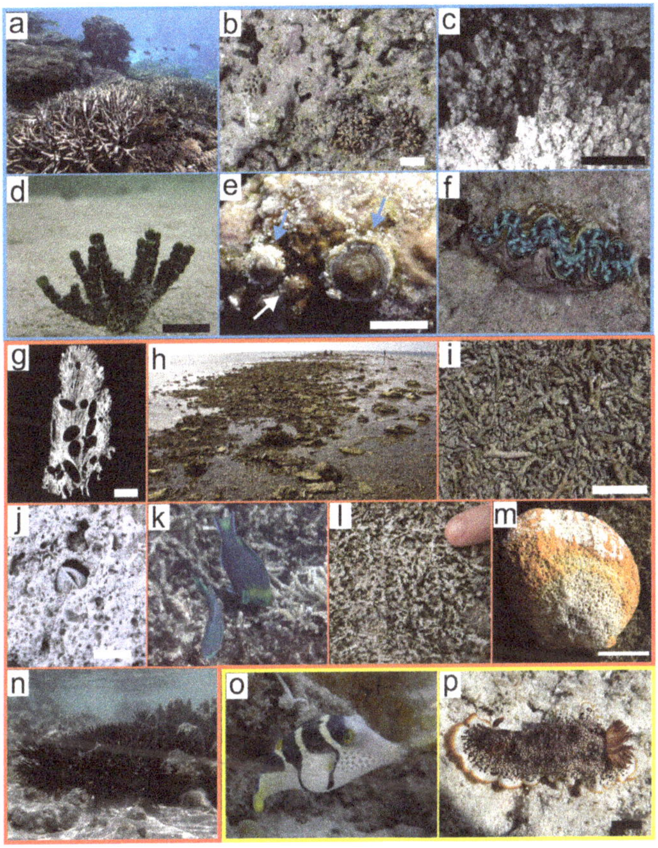

Figure 4. Key modern skeletal reef carbonate producers and examples of main functional guilds [31]. (**a–f**) Key carbonate producers in the Great Barrier Reef (GBR), blue outline. (**a**) Scleractinian corals, key constructor guild; (**b**) crustose coralline algae (CCA), dominant binder guild, scale = 2 cm; (**c**) cryptic microbialite, secondary binder guild, scale = 1 cm; (**d**) *Halimeda* (calcareous green alga), baffling guild along with branching corals and non-calcareous algae, scale = 1 cm; (**e**) large benthic foraminifers—LBFs (blue arrows *Marginopora*, white arrow *Calcarina*) on algal turf with CCA, dweller guild, but carbonate producer, scale = 5 mm; (**f**) mollusk (live *Tridacna*), mainly dweller guild but carbonate producer; (**g–m**) bioerosion products and members of the destroyer guild, red outline; (**g**) computer-aided tomography scan of live collected *Isopora* coral showing bivalve borings (*Gastrochaenolites* ichnotaxon)—such borings weaken corals so that they break easier to become coral rubble (i.e., shingle), scale = 2 cm; (**h**) coral shingle accumulated on the rubble subzone of the reef margin; (**i**) enlargement of branching coral rubble from rubble subzone, scale = 15 cm; (**j**) bivalve occupying boring in heavily bored coral skeleton, scale = 1 cm; (**k**) parrot fish (Scaridae), important raspers in the GBR; (**l**) bite marks where filamentous algae have been removed from beachrock by parrot fish; (**m**) *Spondylus* bivalve shell riddled with boring sponge galleries, scale = 2 cm; (**n**) Crown of Thorns Starfish (COTS—*Acanthaster planci*), a significant corallivore on modern reefs, but it does not actually erode coral skeleton it just kills areas of coral; (**o**,**p**) example dweller guild members (yellow outline) that represent high biodiversity on reefs, but do not make carbonate nor destroy it; their roles in the trophic structure, such as fish grazing on fleshy algae can still be critical to reef growth, but they may leave no trace in ancient reefs; (**o**) *Canthigaster* (Blacksaddle Toby Fish); (**p**) *Hexabranchus sanguineus* nudibranch (Spanish Dancer shell-less sea slug), scale = ~3 cm. All photos from Heron Reef, except (**m**) from One Tree Reef, southern GBR.

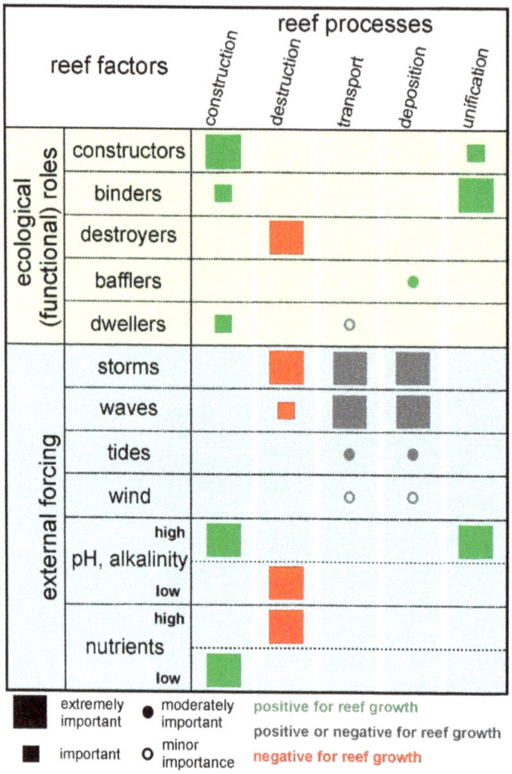

Figure 5. The five key reef growth processes correlated with roles of functional guilds of reef organisms and external forcing.

Carbonate factories have three key ecological functions in producing a reef and these roles can be mapped to ecological or community structures using the functional 'guilds' of Fagerstrom [31–33] (Figures 4 and 5). Notwithstanding that the term guild also has different meanings and contexts for different users, a functional approach is particularly useful for comparing reefs of different ages. However, it is important to distinguish the broader reef 'construction process' outlined above, from the functional 'constructor guild' [31] (Figures 4 and 5). The constructor guild includes members of the carbonate factory that create calcium carbonate in growth position, thus initiating framework development where they remain preserved in growth position (e.g., modern corals on the margin: upper reef slope, crest or rim). The construction process, on the other hand, includes all quantitatively important members of the carbonate factory, whether they are preserved in growth position or not (e.g., large benthic foraminifers—LBFs). Corals dominate the constructor guild in modern reefs (i.e., they initiate most framework construction), but they also play a major role, along with other organisms, in the broader process of constructing the bulk calcium carbonate that makes up the reef edifice. Other organisms may contribute to the constructor guild, including CCA, which can be volumetrically important framework constructors in some high-energy settings [155,156], and even encrusting foraminifers, vermetid gastropods and bryozoans may play a role. Different organisms have dominated the constructor guild in previous time intervals and the nature of reef growth and substrate depend significantly on the types of organisms in that role. As members of the constructor guild, many scleractinian corals have significant size and relief above the seafloor and thus enhance vertical growth (which leads to reef accretion). Such morphology is termed suprastral [157] as the constructors rise 'above' the sea floor stratum (Figure 6). Other organisms, such as CCA, some corals and sponges in the past, typically grew as thinner crusts on and parallel to the substrate, thus being termed constratal—'with' the sea floor stratum [157]. The ability to build vertical relief above the sea floor is in part a function of these constructor morphologies along with their growth and unification rates relative to ambient energy levels.

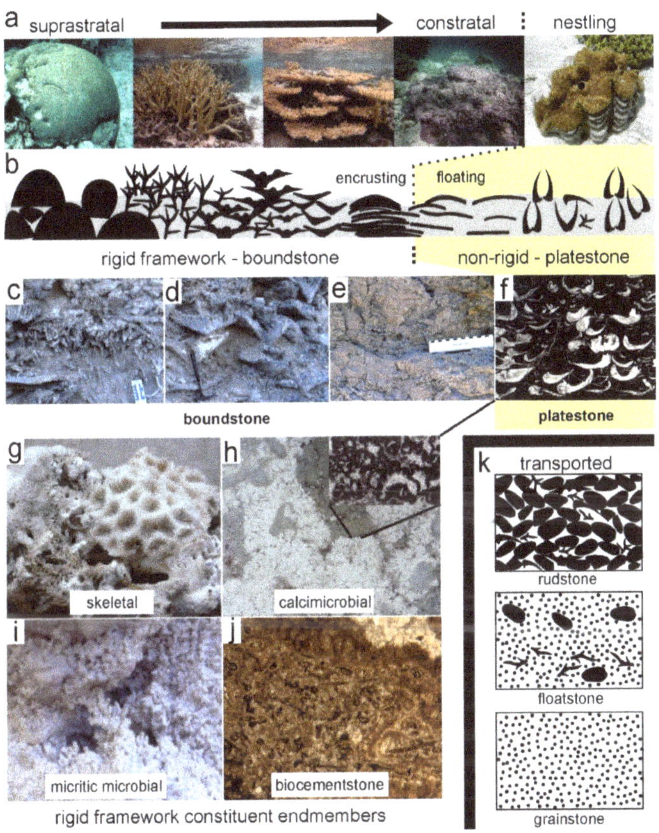

Figure 6. Reef-building organisms, substrate relationships and terminology for framework facies and common reef rock types. (**a**) Modern reef builders classified using the scheme of [157] wherein skeletons projecting above the sediment are suprastratal whereas skeletons growing along the substrate are constratal; skeletons sitting on the sediment or partly buried are nestlers; (**b**) silhouette showing growth morphologies in (**a**) reef rock; (**c–f**) reef–bioherm rock types; (**c**,**d**) Holocene suprastratal coral frameworks, uplifted reef terrace, Huon Peninsula, PNG; (**e**) Late Devonian constratal stromatoporoid sponge framework, Leonard Shelf, W.A.; (**f**) Middle Devonian megalodontid bivalve platestone (shells nestling in or laying on loose sediment), Chillagoe Limestone, Queensland); (**g–j**) four key components of reef framework from [158]; (**g**) Skeletal framework of self-encrusting scleractinian corals, Heron Reef, southern GBR; (**h**) *Renalcis* calcimicrobial framework, Late Devonian, Leonard Shelf, W.A.; (**i**) micritic microbialite framework (cryptic), Heron Reef, southern GBR; (**j**) biocementstone framework consisting of early marine cement (brown color) on delicate fenestrate bryozoans, Mississippian, Muleshoe Mound, New Mexico, U.S.A.; (**k**) silhouettes of common rock types made from transported reef sediments, rudstone is composed of coarse reef rubble, floatstone equates to coarse rubble floating in sand, grainstone is composed of sand [159,160].

The binding guild has a critical functional role in reef frameworks and today is dominated by CCA at the surface (Figure 4), but microbialite-producing biofilms occur in cryptic settings (Figure 4). A given organism can overlap in functional guild (CCA are dominate binders but also may be volumetrically significant constructors in growth position [156] and dwellers producing abundant mobile sediment [134]). Binders are generally constratal in morphology and encrusting corals play a minor role as modern binders. A variety of non-skeletal organisms, such as sponges and algae, stabilize mobile rubble prior to hard binding (e.g., by CCA) [152–154]. Hence, they are binders, and potentially important contributors to the broader unification process [150,161], but they are not part of the carbonate factory. They may leave little or no record in ancient reefs. Regardless, initial binding performed by CCA and encrusting corals is a major step in reef unification in most modern reefs. The formation of microbialites within cryptic settings or in exposed benthic positions also gives biofilms a critical binder role, making them major agents of unification (Figure 4). The baffling guild today includes branching corals and algae, such as *Halimeda* (calcareous alga) and *Caulerpa* (non-calcareous alga) (Figure 4), which reduce proximal current velocities to allow settling of finer sediment than otherwise would be deposited. These same organisms help stabilize mobile sediment once it has been deposited, thus supporting the generation of relief by the carbonate factory. Bafflers do not play a dominant role in reef formation today, but were important at other times in Earth history, e.g., the Pennsylvanian when fragile phylloid algae helped construct large buildups [162].

The destroyer guild includes borers (today, bivalves, sponges and worms) and raspers (today, fish, mollusks and echinoids) and corallivores (today, gastropods, fish and crown of thorns starfish) (Figure 4n). Borers may directly create sediment in excavating their borings (e.g., silt sized 'sponge chips' made by boring clionid sponges [163]), but more importantly they decrease the physical strength of in situ substrates, thus enhancing the destruction process caused by non-biological agents (e.g., waves, storms) (Figure 4g). Microborers, such as cyanobacteria and filamentous green algae, also weaken substrates, but can increase local alkalinity causing localized cementation in some environments [164]. Raspers can turn a significant volume of rigid substrate into mobile sand [165,166]. Importantly, raspers, such as parrot fish, are not part of the carbonate factory. Although their feeding behavior produces sand, the sand consists of carbonate made by a different organism (coral, CCA, etc.). They produce no carbonate themselves other than perhaps their otoliths. The final ecological role includes a great variety of community 'dwellers', which have many other types of ecological functions and roles in reef trophic structure and reef services. Some of these functions are critical to modern coral reef development, such as the grazers that keep levels of fleshy algae down so that the slower growing calcifiers typical of the carbonate factory can accumulate [167], but these organisms may never be recorded directly in a reef's preserved structure (Figures 4 and 5). Filamentous algae and cyanobacteria also contribute to trapping and stabilizing sediments in 'turf algae'. Other dwellers are very important members of the carbonate factory as they produce significant volumes of sediment. The most important dweller members of the carbonate factory in modern coral reefs are the baffler *Halimeda*, whose modular calcified thallus becomes sediment upon death, LBFs and mollusks [135,138] (Figure 4).

Concepts and semantics of, and perspectives on, reef 'framework' may vary even more than for the term reef [104,168]. For many, perhaps most, authors reef 'framework' consists only of skeletons in growth position [31,169], although Fagerstrom [31] treated ancient calcimicrobes as calcareous alga skeletons, rather than microbialites. Today, such a definition applies only to scleractinian coral skeletons for some workers. However, to discuss and compare reefs of the ancient past, a skeletal definition of framework is inadequate because reefs occurred literally billions of years before skeletons did [98,144]. The approach used below is to define the framework as the reef material that is unified, both skeletons in growth position and bound rubble, to make rigid reef substrate on ecological time scales [98,158]. Modern reef framework includes not only the corals, but the binding CCA and microbialites, and deeper stromatolites, where they occur [170]. Some authors [104,171] used a hybrid definition with skeletons in growth position being 'primary framework' and unifying organisms (e.g., microbial biofilms and CCA) providing a 'secondary framework.' However, the distinction becomes unnecessary from a functional perspective, because there is little difference between the rigid carbonate dome made by a biofilm on a hard, calcium carbonate stromatolite, for example, and the rigid calcium carbonate dome made beneath a thin veneer of living coral tissue in a *Porites* skeleton [98]. Both are rigid, relief-bearing structures made by biological members of a carbonate factory. Using a skeleton-in-growth-position-only framework concept, even many modern reefs would lack framework entirely. Some reefs occupy positions where hydrodynamically deposited rubble controls the position of the carbonate edifice more than the biological community itself [104,140,142,150,151]. These structures could be considered 'reefs' that lack framework or hydrodynamic sediment piles and not reefs at all. However, where the rubble is unified by binders in ecological time [140], which also occurs in reefs that have skeletons in growth position, it can be considered reef framework. Additionally, reefs dominated by microbialites would be considered as lacking any framework using a skeletal definition, despite meeting all criteria to be reefs, and that includes parts of modern reefs, such as the deglacial reefs of Tahiti, where microbialites make up as much as 80% of 'framework' volume [172]. The very different nature of carbonate factories and marine alkalinity in the past adds considerable complexity with the formation of 'hybrid carbonates' [169]. Morphological and ecological differences between reef-building corals and less well-understood Paleozoic stromatoporoid sponges and microbes, for example, led to four major types of reef margin framework being recognized in Late Devonian reefs in Canada: metazoan–microbial; sediment-laden metazoan; metazoan-dominated, and metazoan–marine cement types [168]. Thus, not only corals and microbialites had to be accounted for, but also marine cements [173]. Broader framework concepts [158] are used below to facilitate comparison of modern and ancient reefs that have different carbonate factories (Figure 6). This classification system applies to rigid boundstones [159] whether framestones, bindstones or bafflestones, but there has been no attempt to classify different types of skeletal morphology within the framework types (e.g., framestone, pillar stone, etc. [157]). All boundstones (reef frameworks) discussed below were adequately rigid in ecological time to make a hard substrate.

Reef framework is an example of a carbonate 'facies', which is a term (both singular and plural) that is applied in different ways and at different spatial scales to groupings of sediment or limestone based on a specific set of attributes [174]. For this paper, facies refers to a specified collection of sediments, be they particulate or in situ skeletons, along with cements, that are being, or have been, deposited together under a set of conditions that reflect a specific environment or sub-environment, in this case, on or around a reef. Each 'geological' reef facies generally has an analogous environmental–ecological connotation typically with a geographic or geomorphological distribution and Devonian and modern reef facies have previously been found to be comparable [34]. Hence, reef framework facies is part of a broader reef margin facies, which is part of a broader reef facies that also contains lagoon, reef flat, and back-reef facies, etc. The facies connotations are the same for modern and fossil reefs and terminology for reef zonation used in this paper is a combination of that devised for the DGBR [88] and GBRP [175,176] (Figure 7).

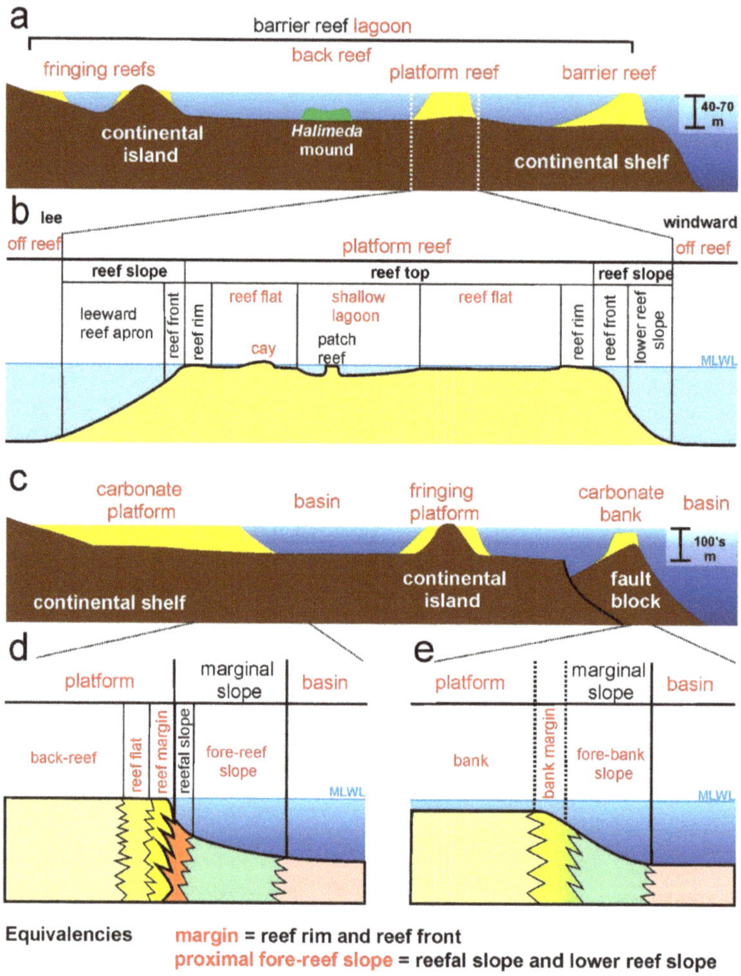

Figure 7. Reef zonation terminology. (**a**,**b**) Terminology for the Great Barrier Reef; (**b**) modified from [176]; (**c**–**e**) terminology for the Devonian Great Barrier Reef (**d**,**e**) modified from [177]. Terminology in red is adopted for this paper.

Another key concept required for comparing reefs of different settings or ages relates to stratigraphy—the layers of reefs built up through time (Figure 8). 'Accommodation' is a term for the vertical space available for deposition of sediment. For reefs, accommodation is the vertical space between the seafloor and ~mean low sea level. Modern reef growth is confined to shallow bathymetry owing to the need for light of the dominant photoautotrophs that drive the carbonate factory, but reef organisms cannot grow significantly above mean sea level owing to desiccation. Changes in local relative sea level (RSL) affect the accommodation for a given reef and can result from either eustatic (global) sea level rise or fall or to local subsidence or uplift. Major increases in eustatic sea level, such as the last deglacial, provided increasing accommodation for existing coral reefs such that they had to accrete vertically (aggrade) so as to 'keep up' with the rising sea surface [178,179]. Where reefs could not keep up with increasing accommodation, their carbonate factory could shut down and they could eventually 'give up' and become drowned with new reefs forming up-slope in shallower settings (i.e., back-stepping) [46,180]. Where sea level rise slowed to a halt, some reefs that were not able to keep up were able to 'catch up' to the new sea level [178,179]. The key control on a given reef's behavior through time then is the ratio between the rate of change in accommodation versus the rate of vertical reef growth produced by its carbonate factory. Once caught up to sea level, accommodation has been filled, so reefs cannot continue to aggrade. They then grow laterally, causing the margin to 'prograde' at sea level over older slope deposits causing the areal footprint of the reef to increase. Where the slope in front of the reef is very deep and steep, such as on oceanic atolls or on continental shelf edges, seaward progradation may be severely limited [181].

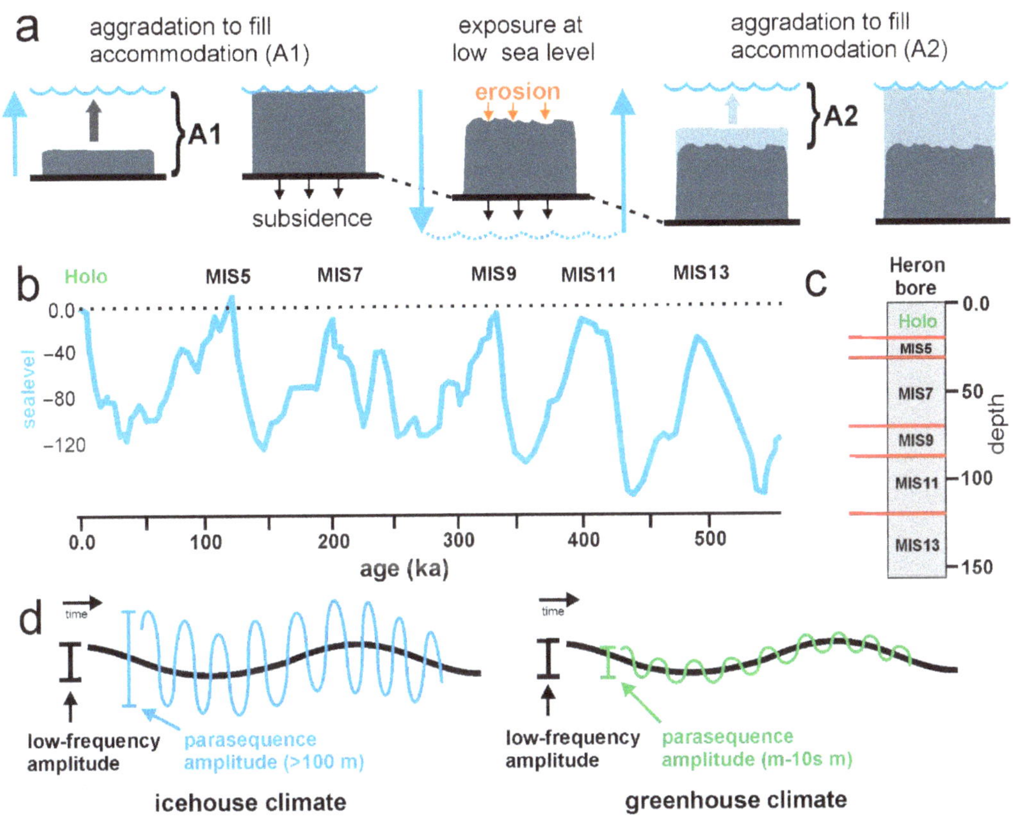

Figure 8. Schematic diagram showing how highstand reefs are stacked one above the other in the Great Barrier Reef Province. (**a**) Upon flooding of a platform with sea level rise new reef growth fills vertical accommodation (A1, A2) to sea level, followed by sea level fall and erosion with ongoing subsidence lowering the platform prior to new inundation when the cycle is repeated. (**b**) Late Pleistocene sea level curve (based on foraminifer stable isotope curve from ODP Site 823A [182]) showing time intervals relative to typical shelf depths where highstand reefs can grow (light blue band). Highstands are labeled by Marine Isotope Stages (MIS). (**c**) Stack of highstand reefs penetrated by the Heron Reef bore of 1937. MIS assignments are based on counting unconformities [47] and have not been verified by radiometric dating. (**d**) Comparison of amplitude of parasequences during icehouse climates (like today) and greenhouse climates (like the Devonian); icehouse parasequences provide much more accommodation than greenhouse parasequences (modified from [183]).

Over long time frames, major cyclic changes in accommodation are caused by Milankovitch forcing of continental ice volumes, and thus eustatic sea levels, and these changes dominate reef behavior over scales of 10s to 100s of thousands of years [184].

During icehouse climates, such as today, large changes in ice volume at the poles create high-amplitude (~100 m+) changes in sea level that have resulted in a 'layer cake' stacking of reefs in regions with adequate ongoing subsidence, like the GBRP (Figure 8). Where there is little or no subsidence, the reefs are 'stacked' laterally between each glacial cycle. In southern Florida, for example, little to no recent subsidence has led to Holocene reefs growing laterally, seaward of the Last Interglacial (LIG) maximum sea level Marine Isotope Stage (MIS) 5e reef on older reefs that tracked lowering sea levels into the glacial maximum (substages 5a–c) [185,186]. The MIS 5e LIG reef currently is exposed above modern sea level to form the Florida Keys because LIG sea level was higher than today [187,188]. However, in times of greenhouse climate, sea level fluctuation has much smaller amplitude (m to 10 s of m) as there is less continental ice storage [183]. These lower-amplitude changes in sea level create less accommodation, so thinner layers result. These externally forced, high-frequency (in geological time) cycles dominate the stratigraphy of most ancient carbonate platforms and are called parasequences and their amplitudes vary between icehouse and greenhouse climates [189]. As renewed reef growth seeks to fill newly created accommodation with each new sea level rise, each parasequence generally shows a shallowing-up trend as the reef community grows up to, or progrades over deeper deposits at, the newly established maximum sea level.

3. Results

3.1. Location, Age and Tectonic Settings

3.1.1. Great Barrier Reef Province

The GBRP in northeastern Australia stretches from Lady Elliot Reef in the south to the Gulf of Papua in the north, a distance of 2300 km and ~14°55′ of latitude (Figure 9). The province's NNW to SSE orientation gives it a latitudinal gradient from the subtropics (south of the Tropic of Capricorn) to the tropics north of 10° S latitude. It is bordered by the Coral Sea to the east and the Australian continent to the west. Although inshore reefs occur, the vast majority of the >3000 individual reefs occur within a 25–65 km wide zone on the outer margin of the continental shelf [181]. The zone extends as much as 300 km offshore in the Swains Reef complex where it is separated from shore by the Capricorn Channel (Figure 9e), but it is closer to shore in the north [181]. In the central GBR, south of Cairns, the reef zone lies mostly east of a NNW–SSE oriented structural 'hinge line' where structural basement deepens abruptly to the east (Figure 9a) [190]. As a mixed carbonate–siliciclastic domain, inshore areas commonly host significant siliciclastic sediment delivered by local catchments, which also helps restrict the most abundant reef growth farther offshore in a pure carbonate zone [181,191]. However, inshore reef communities have adapted to more turbid conditions [192,193]. Circulation in the Coral Sea is dominated by the southern part of the west flowing South Equatorial Current, which splits against the northern part of the GBR, flowing largely south to form the East Australian Current, but also to the north (North Queensland Coastal Current) where it forms complex eddies in the Gulf of Papua (Figure 9) [81]. The currents are complicated seasonally, especially in the north where they are affected by the Indonesian Australian Summer Monsoon. Dominant winds are from the SE for most of the reef for most of the year with a change to more NNW winds during the summer monsoon [81]. Tides are mostly semi-diurnal to diurnal and mesotidal with offshore Spring tidal ranges mostly >2.5 m, but tidal ranges increase towards the coast and are macrotidal in some coastal areas, such as Broad Sound (to 8 m) [80,81]. Throughout the province marine circulation has been heavily modified, or 'bioengineered' [81], by reef growth.

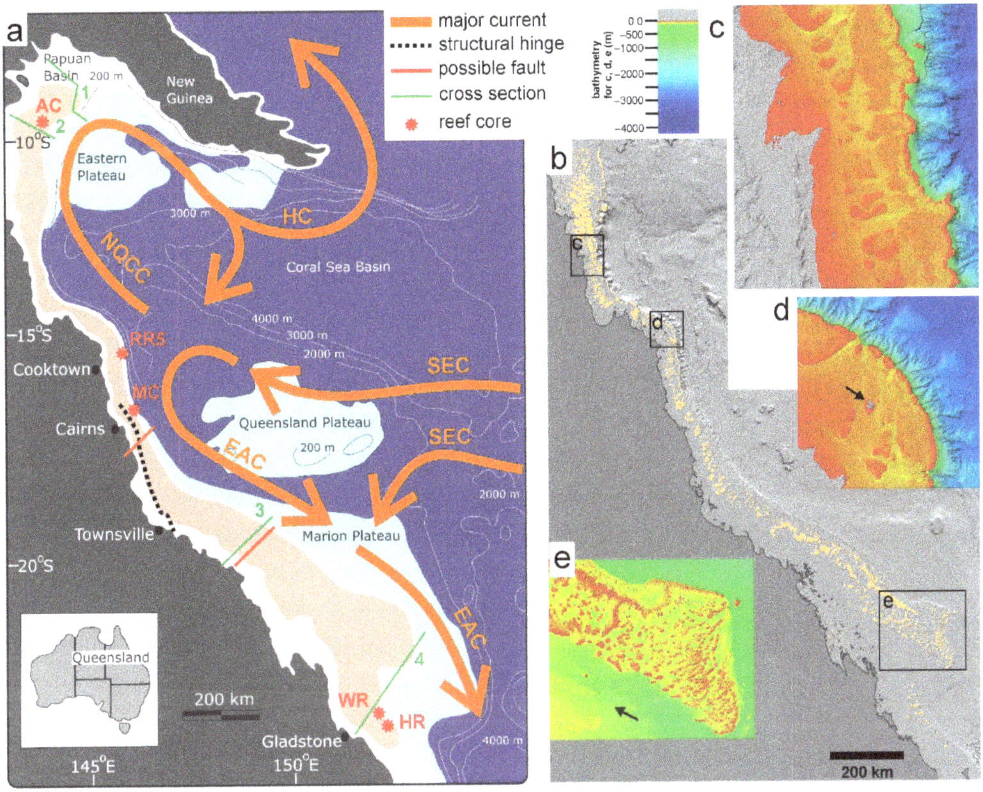

Figure 9. (**a**) Map of the Great Barrier Reef Province showing the main zone occupied by reefs (pink) [181], major current systems (South Equatorial Current: SEC, East Australian Current: EAC, North Queensland Coastal Current: NQCC, Hiri Current: HC), structural hinge with deepening to the east [190], positions of possible reactivated transform faults (red) that isolate the more rapidly subsiding central zone of the reef [194], positions of cross-sections (green) shown in a subsequent figure, and positions of deep reef cores: Anchor Cay—AC, Ribbon Reef 5—RR5, Michaelmas Cay—MC, Wreck Reef—WR, and Heron Reef—HR. (**b**) Map showing Late Pleistocene platforms and reefs (light orange) with positions of insets (**c**–**e**) shown; (**c**–**e**) bathymetry [195]; (**c**) ribbon reefs with narrow shelf; (**d**) shelf area with Lizard Island (arrow), a mid-shelf continental island complex with fringing and barrier reefs surrounding a shallow lagoon; (**e**) Swains Reefs separated from the continent by the Capricorn Channel (arrow).

Modern reefs in the GBR are Holocene in age (less than ~10 kyr old) but offshore reefs grow directly on top of older reefs in a 'layer cake' fashion where shallow reef facies formed during interglacial high sea levels, like today, are separated by unconformities that represent low sea levels at glacial maxima [47,190,196,197] (Figure 10). The existing Holocene reef represents a single parasequence. Older, subsurface reefs have been difficult to date owing to their burial beneath the modern reef (i.e., poor access) and because the rare available carbonate materials obtained from cores have generally been altered by meteoric water, thus disturbing the geochemistry required for radiometric dating. There are many more dated rocks from the moon than from the base of the GBR [51]. The first deep GBR reef cores were recovered in 1926 (Michaelmas Cay [198]) and 1937 (Heron Island [198]) but the first geochemical and geophysical dating was done on a core recovered from Ribbon Reef 5 (RR5) only in 1995 [49,182] (Figure 9a). Although the Michaelmas and Heron cores sampled Pleistocene rock from beneath the Holocene reef [198], the cores could not be dated. A petroleum exploration well drilled on Wreck Reef, southern GBR in 1959 also penetrated the entire reef sequence, but also provided no radiometric dates [182,199] (Figure 9a). An exploration well drilled at the northern margin of the GBRP off Anchor Cay (Gulf of Papua) did not penetrate adjacent Holocene or Pleistocene reef rock [80] (Figure 9a). Prior to the 1995 RR5 core, seismic data and shallow reef sediments shed into deeper settings on the slope near Cairns recovered by IODP Leg 133 drilling suggested an initiation age as late as ~500 kyr ago [200]. Direct dating of the initial reef phases of RR5 in the GBR showed that shelf carbonate deposition initiated since ~770 kyr ago as it all represents the Brunhes geomagnetic chron (a time of consistent global magnetic polarity) [182]. Reef facies appear to have initiated around 400–600 kyr BP years ago, during sea level highstands at Marine Isotope Stages (MIS) 11, 13 or 15 based on packages of reef facies separated by unconformities [197] (Figure 10b) and poorly constrained strontium isotope ($^{87}Sr/^{86}Sr$) ages and limited open-system U-Th ages [182]. Based on current evidence, the reefs at the base of the modern GBR 'layer cake' initiated around 600 kyr BP and will remain relatively poorly constrained until it is possible to directly date reefal carbonates from the earliest reef package.

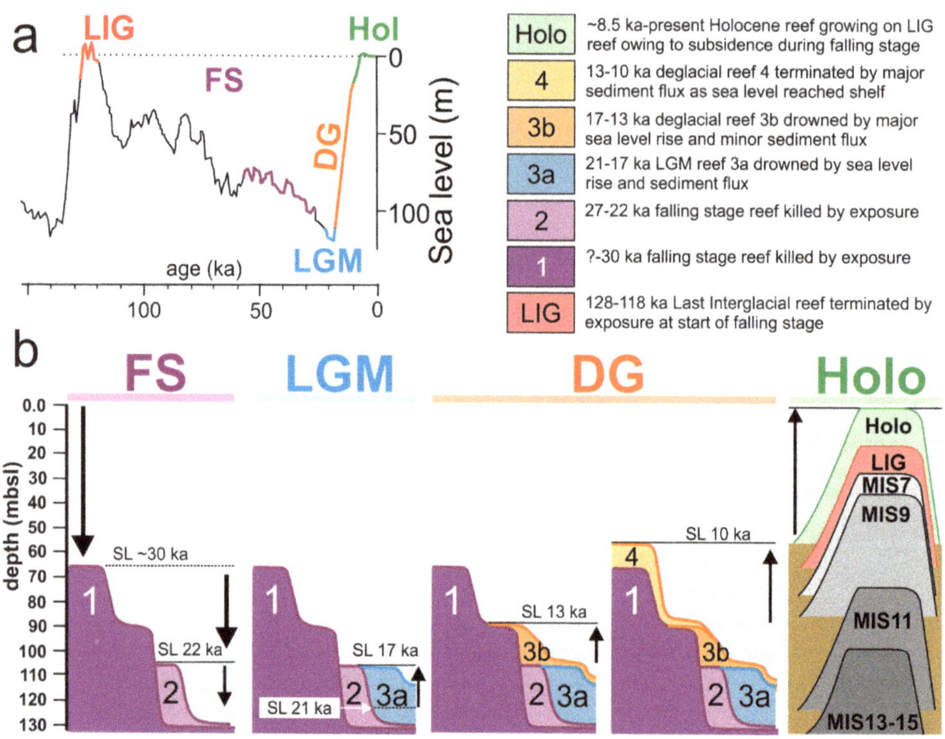

Figure 10. History of the modern Great Barrier Reef. (**a**) Sea level curve showing times of reef growth in the Great Barrier Reef: Holocene (Hol), Deglacial (DG), Last Glacial Maximum (LGM), Falling stage (FS) and Last Interglacial (LIG). (**b**) Sequence of reef growth since the LIG in relation to sea level change, modified from [46]; Holocene stacked reef section is based on the Ribbon Reef 5 core, depths and interpreted packages from [182].

While the age of the 'layer cake' underlying the current GBR has been constrained to the Late Pleistocene–Holocene, the great latitudinal extent of the GBRP provides an additional complication. The northernmost part of the GBRP reached the tropics, and thus warmer SSTs, >30 My earlier than the southern part because of the slow northern movement of the Australian Plate and the nearly 15° latitudinal range of the reef [82,83] (Figure 11). Thus, the interplay of tectonic movement and global climate has controlled the development of the GBRP and the two factors are inextricably linked. The southern-most zone of the GBR currently straddles the Tropic of Capricorn, so reached that latitude only recently, much later than in the north. Although the northern GBRP reached the tropics during the Oligocene, the very warm conditions in the Early Cenozoic 'Greenhouse' meant that SSTs were adequately warm at higher, subtropical latitudes for carbonate development in the Eocene [83,201] (Figures 2, 11 and 12). However, very cool global temperatures with the onset of Antarctic glaciation at the end of the Eocene suppressed carbonate deposition through most of the Oligocene. With Australia's continued movement north, subtropical and finally tropical carbonates formed again in the Miocene. Increased SSTs in the SW Pacific allowed reef facies to thrive on the Queensland and Marion plateaus in the Coral Sea from the Early to Middle Miocene [202–204] consistent with warm seawater surface temperature (SST) proxies (e.g., $\delta^{18}O$ from foraminifers [205], foraminifer Mg/Ca ratios and organic molecule evolution, such as $TEX_{86}{}^{H}$ [206]) and Sr isotope-dated shallow tropical carbonate sediments [200] from IODP cores (Figure 12). Late Miocene changes in oceanic currents [203], sea level [204] and/or cooling [207] led to platform demise or shift to carbonate ramps [202,204]. That and the onset of northern hemisphere ice sheets in the Pliocene led to a 'Pliocene Reef Gap' and continued low carbonate production in the offshore Coral Sea [207]. The plateaus only began to support shallow carbonate production again in the Late Pleistocene as temperatures warmed [205,206], presumably at the same time as tropical–subtropical reefs began growing on the continental shelf in the central [182] and southern GBR [208] (Figures 2 and 12).

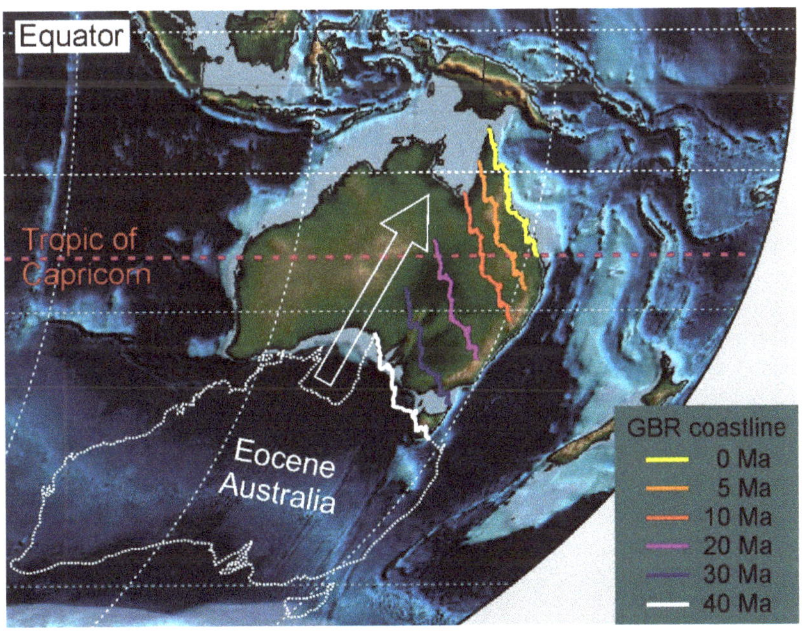

Figure 11. Australia's movement through time from the Eocene to Holocene on a Robinson projection. Outlines of Great Barrier Reef: 40 Ma (white), 30 (blue), 20 (purple), 10 (red), 5 (orange) on zero (yellow). Base map and movement data from C. R. Scotese [209,210].

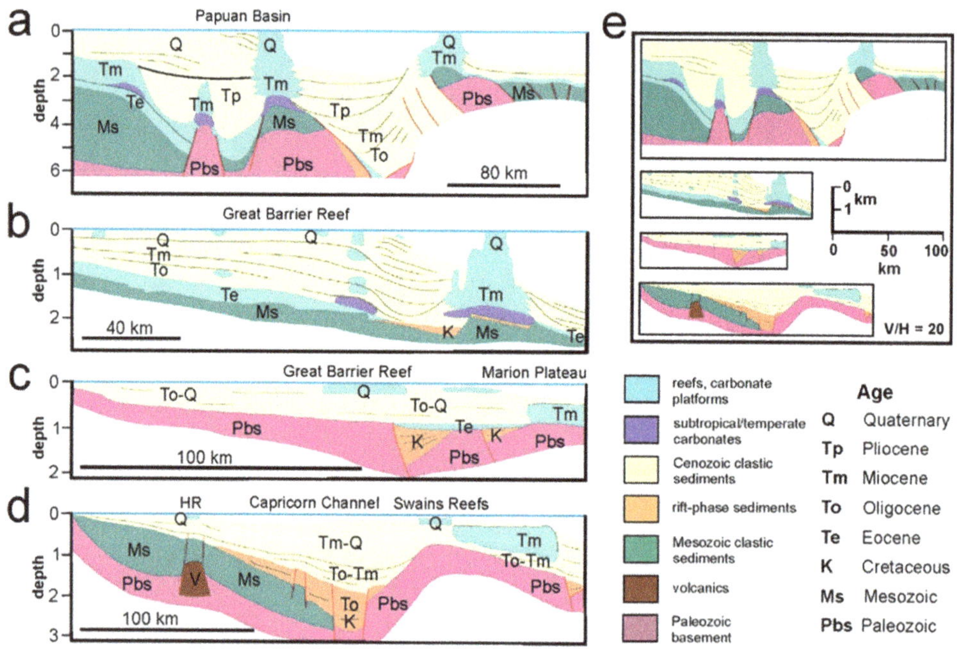

Figure 12. Reconstructed cross-sections based on seismic data and limited drilling across the Great Barrier Reef Province from the Papuan Basin in the north to the southern GBR; positions of sections are shown in green on Figure 9; (**a**) section 1; (**b**) section 2; (**c**) section 3; (**d**) section 4; from [201] as modified by [211]; (**e**) simplified versions plotted with the same scales relative to each other; vertical exaggeration: V/H = 20. Note the much thicker stratigraphy to the north owing to subsidence caused by crustal loading from sediments delivered to the back-arc basin.

In the northernmost parts of the GBRP, Miocene reef carbonates also flourished on the continental shelf based on seismic data [83,201]. At the northern end of the GBRP shelf edge reefs east of the Torres Strait overlie a 1.5 km thick sequence of stacked, but intermittent, carbonate reefs–platforms that date from the Miocene to Pleistocene [201] but these older platforms reach thicknesses of nearly 3 km father north in the Bay of Papua [211,212] (Figure 12). Those reefs have been penetrated by few exploration wells and remain poorly known. Farther north in the Gulf of Papua Early Miocene carbonate shelf deposition appears to have initiated above Eocene carbonate ramps but may not have been continuous to the Pleistocene owing to changes in climate and sea level [211–214]. Broad older carbonate platforms are tilted to the NE owing to foreland loading from the Papuan Fold and Thrust Belt but younger carbonates remain near horizontal. Extensive siliciclastic sediments were delivered from the New Guinea highlands from the Late Miocene and reefs and platforms at the northern end of the GBRP were progressively buried by continental sediments shed from the New Guinea highlands [201,211,212] (Figure 12a). Current seismic exploration and drilling in the Gulf of Papua will eventually shed much light on the older parts of the GBRP that are now buried beneath thick siliciclastic sediments.

A recent review of earlier tectonic development of the GBRP and offshore Queensland and Marion plateaus in the western Coral Sea [215] showed opening of the Coral Sea began in the Late Cretaceous and terminated in the Eocene. Intracratonic rifting on the continental margin produced the complex basement relief that controls the position of the plateaus, coastal basins and modern shelf edge. Aside from spanning climate zones, the northward movement of the Australian plate through the Cenozoic led to collision and active convergence on the northern margin of New Guinea [215] with major implications for carbonate environments in the associated foreland basin [213]. The positions of most current reefs and carbonate platforms are controlled by basement highs, such as the edges of fault blocks, that resulted from that tectonic history [190,201] (Figure 12).

Today, northeastern Queensland is generally considered tectonically quiescent and an example of a passive continental margin in an intraplate setting [201,216]. However, much of the GBRP has experienced significant subsidence during the time of reef growth, thus differentiating it from modern passive margin reefs, such as those in Florida. Offshore Holocene reefs in the GBR represent only the current parasequence at the top of the Late Pleistocene stack [47,51,190,196,197]. Where cores that penetrate beneath Holocene reefs on the Queensland continental shelf have been dated using U-Th techniques, the immediately underlying reef is LIG in age [182,217,218]. The depth to the Pleistocene–Holocene boundary varies significantly around the GBRP [80]. This partly reflects localized differences in elevation on the underlying LIG substrates themselves [196,218,219] but substrates also deepen seaward of a structural hinge line [190] and along latitude, where they mostly deepen from north to the south [80], unlike older reefs farther north near Papua New Guinea. Some abrupt differences in depth to the LIG reef may be associated with active faults [220] and ongoing increased subsidence in the central region of the GBR is partitioned, north and south, by major NNE–SSW trending faults that presumably represent reactivation of older transform faults that formed much earlier during the opening of the Coral Sea [194] (Figure 9a). These NNE–SSW trending structures are associated with modern seismicity [194,221] although northeastern Queensland is generally thought to have relatively low seismic risk in general. In the northern part of the GBR near the Torres Strait, individual Quaternary reefs have been offset laterally suggesting active right-lateral strike slip movement since initiation of the Pleistocene platforms. The current right-lateral strike slip movement presumably occurs along reactivated NNW–SSE trending normal faults, also probably associated with the opening of the Coral Sea [222]. Movement on these faults is consistent with the current stress field caused by active convergence on the northern margin of Papua New Guinea, the northern edge of the Australian Plate [222]. Hence, although a passive continental margin, a degree of active tectonism still affects the GBR, but it is based on reactivation of older structures formed during the rifting that gave rise to the Coral Sea.

Although a degree of neotectonic (i.e., recent to current) activity in the GBR has been demonstrated relatively recently, the cause(s) of the significant subsidence that allows the Pleistocene reefs to be stacked into a 'layer cake' (a thickness of 260 m [190]) remains enigmatic. Minor (a few meters) but systematic differences in the modern elevation of LIG shoreline indicators around Australia suggested continent-scale subsidence to the north [223], which has subsequently been acknowledged as a major ongoing tilting of the Australian continent of ~250 m up in the SW to ~250–300 m down in the N since the mid-Miocene [224]. The rate of tilt was calculated at ~15–20 m/myr and was attributed primarily to vertical movements related to underlying mantle flow (dynamic topography) associated mainly with subduction beneath Papua New Guinea to the north [224]. However, that subsidence rate cannot explain the depth to LIG reef substrates in the central GBR [80], which, all other things being equal, would require ~10× the rate. Even where the LIG substrate is shallower (e.g., 5 to 8 m depth [80,217,219]) it would require > 3× the continent-scale subsidence rate. Significant subsidence occurred in northern Australia (Gulf of Papua, Arafura Sea to the Indian Ocean) in the Oligocene and Miocene owing to thrust and sediment loading as part of the foreland basin south of the Papuan subduction zone [213]. There, it accumulated more than 2000 m of carbonate shelf sediments (Darai Limestone) during the Miocene [213]. However, that subsidence was too early, oriented in the wrong direction (i.e., more E–W) and was too distal to account for ongoing subsidence through most of the offshore GBRP. However, it has implications for the northernmost part of the GBRP, which are discussed below.

Although tectonic subsidence occurred (and is occurring) in northern Australia since at least the Miocene, it is important to remember that the 260 m reef thickness in the GBR [190] formed only within the last ~700,000 years. This is particularly enigmatic, because it reflects a major difference in subsidence between onshore and offshore positions across the reef. Onshore LIG sea level indicators are above present sea level owing to the high sea levels of the LIG [188]. However, offshore LIG reef top elevations beneath modern reefs are significantly lower [80,225,226]. Karst removal of LIG reef was suggested as a possible reason for the difference in elevation [196] but was rejected owing to inadequate rates of karst removal to account for many of the depths to LIG substrates [227]. Karst was rejected in the southern GBR where coring has allowed the best study, because: (1) shallow water facies occur at the top of the LIG reef succession; (2) calcrete–caliche at the unconformity suggests a more arid climate (i.e., less dissolution); and (3) the depth does not systematically increase northward as would be expected for reefs transitioning into the wetter tropics [218,219]. Karst removal could not account for 30 m-thick Holocene reef successions resting on LIG reefs in any case. Sediment compaction is unlikely to explain the difference either. Although coastal LIG shore indicators occur farther onto the ~stable basement rocks of the continent, and have less overlying loading than the reef substrates, reef limestones, at least at the margins, are not as compactable as fine siliciclastic sediments owing to their early rigidity and are unlikely to account for so much difference. Dynamic topography, glacial isostacy and sediment loading also have been rejected as the cause [225,226]. Hence, minor neotectonic reactivations of normal faults formed during the opening of the Coral Sea are possible and are consistent with deepening east of the NW–SE 'hinge' line [190] (Figure 9). Unfortunately, adequate modern seismic data to test such a hypothesis are not available as appropriate surveys have not been allowed since the GBR Marine Park was declared in 1975 and few older seismic lines exist of adequate quality.

3.1.2. Devonian Great Barrier Reef

The DGBR in western Australia crops out over a length of 350 km with a width of as much as 50 km on the Lennard Shelf along the southwestern margin of the Kimberly Block, an ancient shield area of Proterozoic metamorphic and igneous rocks that make up the North Australian Craton [88] (Figure 13). However, the reef extends into the subsurface to the northwest and originally may have bordered the Kimberly Block for 1000 km turning back to the north and east to the Bonaparte Basin [88,90,228,229]. The reef also may have extended to the east and south along the edges of the Fitzroy Trough and Gregory Sub-basin and then turned west again along the shallower Crossland and Broome platforms [90] where similar Late Devonian reef facies have been recognized in seismic data and intersected in boreholes [230–233]. Hence, it may stand as one of the larger Devonian reef complexes, not the smallest of twenty [85]. Playford and others [90] provided a detailed history of study of the DGBR. Although first studied geologically in 1883 [234], the limestones were not interpreted as fossil reefs until 1924 [235]. They were investigated for their resource potential from the 1930s to 1950s culminating in major regional and integrative studies by the Geological Survey of Western Australia since the 1960s [87,88,90,236]. They are associated with minor hydrocarbon development [233,237] but host significant base metal sulfide deposits [238] and many hundreds of papers have been published on various aspects of the reefs over the last century.

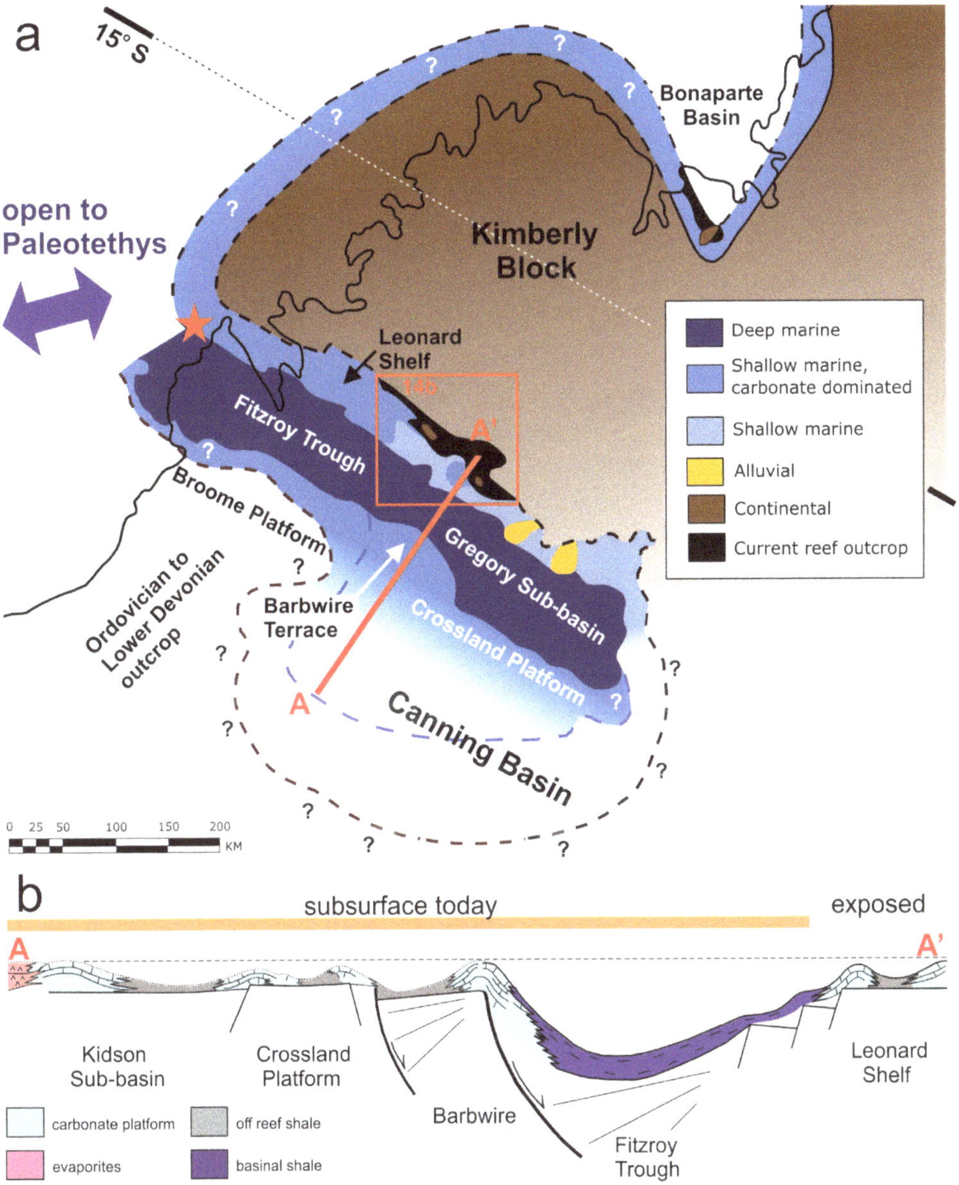

Figure 13. Paleogeographic reconstruction of the Devonian Great Barrier Reef (DGBR); paleolatitude for 375 ma [210]. (**a**) DGBR potentially borders the Kimberley land mass to the north and east to the exposed Famennian reefs in the Bonaparte Basin, and to the south along the Barbwire Terrace, where it has been penetrated in drill core, modified from [90,230,233], red star marks DGBR in seismic data [229]; transect A-A′ is shown in (**b**), inset (14b) is shown in a subsequent figure; (**b**) hypothetical transect across Leonard Shelf to south across Fitzroy Trough to the Kidson Sub-basin (after [233]).

The Fitzroy Trough was an active, extensional intracratonic basin during the Devonian and now forms the northernmost region of the broader Canning Basin [239,240]. The northern part of the Late Devonian, western Australia occupied the southern tropics located in the northeastern part of Pangea, thus representing similar latitudes to the current

GBRP [210] (Figure 1b). However, the orientation of the exposed portion of the reef complex on the Lennard Shelf was near parallel to latitude, so represented a tighter latitudinal range at a given time than does the GBRP. Although a variety of rifted tectonic terranes occurred to the north and west of Australia at the time, faunal evidence suggests that the Canning Basin was open to Tethyan Ocean circulation to the west [241], but the basin was partially restricted [233] (Figure 13). The eastern coast of northern Australia also occupied the tropics during the Devonian and hosted shallow carbonate environments, but they were mostly Early and early Middle Devonian in age [85,242,243]. Givetian (upper Middle Devonian) shelf limestones occur in Queensland in the Broken River [243] and small patch reefs with corals and stromatoporoids occur in the Fanning River [69,244,245] groups, but major reef complexes similar to those in the Caning Basin have not been recognized [242,246]. However, stromatoporoid–coral buildups to 30 m thick and 2 km long occur in the lower Devonian part of the Broken River Group [247]. Older Lower Devonian limestones near Chillagoe to the north also contain stromatoporoids and corals, but no major reef complexes have been described [248]. Limestones were particularly rare in Queensland during the Frasnian and Famennian, the main age of the DGBR, and reefs of that age are unknown there [84].

The DGBR reef facies range in age from late Middle Devonian (late Givetian) to Late Devonian (Frasnian to late Famennian) (~385–355 million years ago) [100] with a total thickness of more than 2.5 km [88]. The Late Devonian was an interval of major global change [249] wherein 'greenhouse climate' temperatures (Figure 2) decreased following an end-Givetian thermal peak only to rise to a late Frasnian peak before shifting to somewhat cooler conditions through the Famennian [96,250,251]; terrestrial floras diversified with the introduction of forests with deep rooting trees and deeper soil development with implications for terrestrial weathering and coastal sedimentary environments [252–255]; fish underwent major diversification [256,257]; the first tetrapods invaded the land [258]; and major extinction events occurred: Givetian–Frasnian, Frasnian–Famennian, and end-Famennian [259–262].

Representing more than 20 million years of active growth, the DGBR reef complex is divided into a stratigraphic 'supersequence' or cycle consisting of two major depositional phases: the Givetian to late Frasnian transgressive back-stepping Pillara phase and the late Frasnian to late Famennian regressive prograding Nullara phase [263–266] (Figure 14). The dominant transgressive regressive supersequence most likely represents a global sea level signal (i.e., eustasy) as it has been recognized in other regions (Alberta, Canada and South China) that have differing local tectonic settings [265,267,268]. Currently, two Givetian, seven Frasnian and more than three Famennian (the largest part being undifferentiated) 3rd order sequences have been recognized through the reef complex [269–271] with ages based on biostratigraphy (primarily conodonts and ammonoids [272,273] but also microvertebrates [274]) and chemostratigraphy [269]. Higher-frequency, meter-scale parasequences dominate stratigraphy in parts of platforms [275], primarily being attributed to Milankovitch orbital forcing [276–278] but differentiating cycles is complex [183] and many lower-frequency cycles are associated with tectonic forcing, especially apparent in coastal sections where siliciclastic and carbonate sediments interfinger [278–280]. Interfingering of siliciclastic and carbonate facies in proximal basin deposits also has been attributed to sea level cycles with carbonate sediments delivered past the shelf edge during highstands and siliciclastic sediments bypassing the reefal carbonates or overtopping them during lowstands (reciprocal sedimentation) [281].

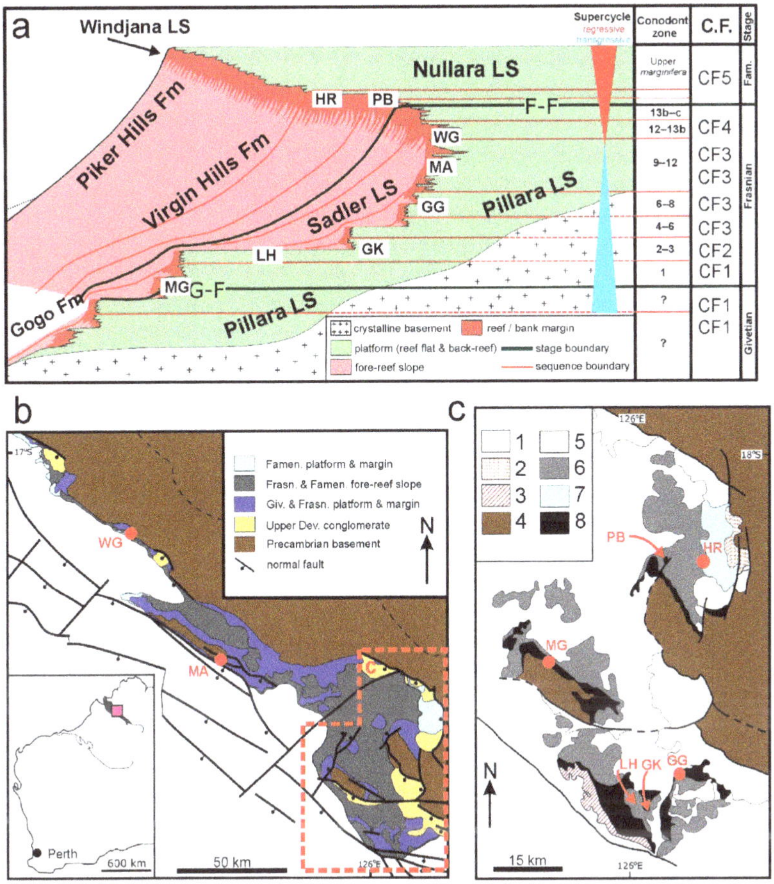

Figure 14. Devonian Great Barrier Reef stratigraphy and localities. (**a**) Simplified cross-section of Leonard Shelf stratigraphy (modified from [267,269]) with main formation names (black letters) and proposed carbonate factory distribution (C.F.); key localities are plotted at approximate stratigraphic levels: Menyous Gap—MG, Lloyd Hill—LH, Glenister Knolls—GK, Galeru Gorge—GG, Mowanbini Archipelago—MA, Windjana Gorge Classic Face—WG, Pothole Bore—PB, Horseshoe Range—HR; (**b**) Leonard Shelf locality map (inset 14b in Figure 13) (modified from [265] there after [89]) with localities shown in red; (**c**) inset map (modified from [270]), legend numbers: 1—Quaternary cover; 2—Mississippian Fairfield Group; 3—Ordovician Prices Creek Group; 4—Precambrian basement; 5—Devonian siliciclastic strata; 6—Frasnian–Famennian fore-reef slope to basin; 7—Famennian platform facies; 8—Frasnian platform facies.

Reef growth in the DGBR was syntectonic [87] (i.e., occurred at the same time as major basin extension) and tectonic movements controlled RSL, and thus accommodation, in parts of many depositional sequences [90,278,280,282–285]. Some shelf edge collapse events that led to the deposition of very large olistoliths composed of shallow reef margin facies into deeper basinal settings (e.g., McIntrye Knolls) may in part be the result of coeval seismic activity [286,287]. However, margin collapse in general was associated with oversteepening and changes in relative sea level, that also were controlled partly by active subsidence along faults [270]. Neptunian dykes (fractures that ~parallel platform margins) are common in the reef complex owing to early cementation/unification and high rates of progradation of margins over less well-indurated slopes [177,287–289]. Since deposition, the reefs have experienced relatively little subsequent structural deformation and no metamorphism and many whole platforms have consistent low structural dips [89]. Currently, parts of the reef complex have been exhumed with near original topographic expression so that local bathymetric relationships have been preserved [87,290]. This in part reflects the position of the Leonard Shelf on the edge of the North Australian Craton where it has not been subject to as much deep burial as the basin to the south, but also to the removal of some overlying sediments during the Late Paleozoic glaciation events when the reefs were in some cases exhumed by continental glaciation from the south [291]. Paleogeographic relationships are also enhanced by the great disparity in relative erodibility of both underlying and overlying rocks. The readily eroded shales and sandstones of the overlying Mississippian Fairfield Group have allowed reef limestones to stand in high relief relative to their contemporaneous basins allowing one to walk up the exhumed reef slopes in some areas and onto the elevated reef margins and into back-reef facies and lagoons. At the same time, the very durable quartz-rich metavolcanics and metasediments of the underlying Proterozoic rocks still retain much of their original Devonian relief so that the relationship of the reefs with contemporaneous continental paleoislands and rocky shores are well preserved [92,290,292,293].

3.2. Global Constraints—Climate and Water Chemistry

3.2.1. Great Barrier Reef Province

The closely linked climatic–tectonic interactions through most of the Cenozoic [83,201] (Figures 11 and 12), including the Eocene 'Greenhouse' climate, are mentioned above with the tectonic setting (Section 3.1.1). Today's GBRP experiences an 'icehouse' climate spanning nearly 15° of latitude from fully tropical conditions with a well-developed Indonesian–Australian Summer Monsoon (IASM) in the north and more arid subtropics at the southern end. Thus, it experiences both spatial and temporal (Milankovitch) changes in climate. Both terrestrial and marine temperatures generally decrease from north to south but are partly ameliorated by the south-flowing EAC, which carries warm tropical waters south along the Queensland coast (Figure 9). Rainfall and fluvial discharge are significantly higher and more consistent in the north. South of the Wet Tropics, and encompassing the two largest catchments in the GBR (Burdekin and Fitzroy river systems), discharge is lower, but much more flashy and flood-prone [294] with disproportionate impacts on water quality in the GBR lagoon [295]. Cyclones (hurricanes) have major impacts on the GBR in terms of significant physical damage to the reef, but also with increased terrestrial runoff and water quality declines inshore and increased sedimentation and re-suspension of sediments in the lagoon [75]. Cyclone frequency increases eightfold in a near linear latitudinal trend from around ten cyclones in the south to eighty in the north at Torres Strait based on data collected since 1906 [296].

Although the increase in marine heatwaves and attendant coral bleaching is of major concern for coral reefs today, the greatest climatic effect on the modern GBR is the larger scale temporal oscillation of sea level with each glacial cycle that has resulted in the layer cake stacking of highstand reefs. During each glacial interval, the reefs of the GBR migrated seaward where appropriate substrates were available (currently subsea terraces) and then grew until sea level again lowered, exposing them [46] (Figure 10). The reefs of the Last Glacial Maximum (LGM) grew farthest seaward and then aggraded so as to keep up or catch up with sea level during deglacial time [297]. Then as sea level continued to rise, reefs drowned, and communities back-stepped onto the previously abandoned MIS3 reef terraces where not hampered by siliciclastic sediment flux, until finally reestablishing themselves on the previous highstand platforms during the Early Holocene [46,297] (Figure 10). Overlain on these parasequence level changes are smaller changes in climate [30]. Unlike some areas of the world (e.g., the Caribbean) the Holocene local sea level curve (RSL) is complex for the GBR with sea level initially reaching more than a meter higher than present by 8–7 kyr BP and then falling to the present sea level through the Holocene [298,299]. However, the fall was not gradual, and small but synchronous sea level oscillations at 5.5, 4.6, 2.8 and 1.2 kyr BP regulated various reef turn-on and turn-off events in both inshore and offshore reefs [300–305]. Some of the hiatuses in reef growth lasted more than 1000 years and were coeval with reef hiatus events in reefs in other parts of the world [306], but their exact relationship to global climate is not known. Inshore reef behavior is complicated by sedimentation, water quality, and potentially by localized subsidence from compaction. The high tidal ranges for GBR reefs make measurements of previous sea levels more difficult, but even offshore reefs show well-dated hiatuses in growth. The broad shut down of reef flat growth between 3.6 and 1.6 kyr BP occurred in offshore GBR reefs throughout its latitudinal range with the only exception being the central region that was undergoing greater ongoing subsidence [194,305]. Hence, eustatic sea level may be implicated in that case, suggesting a more global driver.

Although latitudinal gradients in climate existed throughout a given parasequence, the temperature gradient for SSTs along the GBR was steeper during the LGM and deglacial than it is today [307]. The change was attributed to the expansion of cooler subtropical waters northward associated with changes in oceanic circulation, but core data are not available to test broader changes to biodiversity that may or may not have accompanied the changes. SSTs increased dramatically between the LGM and Holocene but there were also smaller temporal perturbations. Immediately before the Holocene the Younger Dryas cool event did not cause cooling where it has been tested in deglacial GBR cores (IODP Expedition 324) [308]. Modeling based on associated coral $\delta^{13}C$ also showed that rapid reef accretion at that time had negligible effect on atmospheric CO_2 [42]. Deglacial reef accumulation was heavily dependent on local bathymetry slope angles and proximity to fresh terrestrial or re-suspended particulate sediments [288], thus complicating broad carbonate budget modeling.

Within the Holocene the GBR experienced a mid-Holocene thermal optimum as did the rest of the western Pacific [309–312]. Data from the southern GBR suggests SSTs were cooler than present during the Holocene both before and after the optimum when they were as much as 1–2° higher between ~6 and 7 kyr BP [309]. Snapshots of mid-Holocene climate (~4–6 kyr BP) have been documented from geochemical archives in corals showing differences in monsoon patterns, seasonality and terrestrial flooding based on coral luminescence bands and $\delta^{18}O$ [313,314]. However, longer records of secular change in Holocene climate of the GBR are scarce [30]. Major changes in climate and Pacific Ocean circulation potentially associated with a shift in El Niño Southern Oscillation (ENSO) are recorded in changing ^{14}C reservoir corrections [315,316]. Major frequent changes in the correction occurred prior to stabilization at ~5.5 kyr BP, when more modern ENSO behavior engaged, altering the degree of upwelling of deep older waters that reached the GBR. *Halimeda* bioherms in the GBR lagoon record changes in $\delta^{15}N$ that reflect changes in terrestrial versus upwelled water sources that also correlate well with ENSO frequency and intensified El Niño activity after ~5 kyr BP [317]. The data also correlate with monsoon (IASM) reconstructions based on independent speleothem data [318]. Longer secular records of trace element geochemistry preserved in reefal microbialites record similar differences in Holocene terrestrial runoff in the GBR [319] with major controls in precipitation being the IASM and ENSO. Limited analyses of coral $\delta^{15}N$ and Ba/Ca ratios also show much higher terrestrial influence on offshore GBR waters from ~8 to 6 kyr BP, with peak terrestrial nutrients associated with significantly lower coral extension rates and deeper, more turbidity-resilient taxa [320]. Cyclone abundance and intensity varied through the Holocene with the GBR experiencing very active times (>5 to ~3.5 kyr; ~2.5 to

~1.6 kyr; ~600 yr to today) interspersed with periods of lower activity [321]. It is not clear how the cyclone record relates to global or regional climate systems [321], but the intervals of activity and inactivity vary regionally. The record is not adequately long in the GBR to compare to other water quality data, but one reef hiatus in the central GBR was interpreted to reflect intense cyclone activity that did not allow reef accretion in some inshore areas for 5000 years between ~6.5 and 1.5 kyr BP [322].

The GBR today occurs during an interval of 'aragonite seas' with high Mg/Ca ratios, low pCO_2 and low temperatures [94,323,324] (Figure 2) thus favoring aragonite precipitation by organisms such as scleractinian corals and *Halimeda* [325]. Most of the GBR is a classic example of an oligotrophic system, favoring low-nutrient, open ocean seawater [146]. Modern tropical seawater pH and alkalinity promote an adequate carbonate saturation state for abundant carbonate precipitation. However, lower atmospheric CO_2 during the LGM and deglacial, relative to the current interglacial (i.e., the Holocene), appear to have increased pH and saturation, and thus, the precipitation of reefal microbialites [148], which as non-obligate calcifiers are greatly affected by ambient water chemistry [98,132,170].

3.2.2. Devonian Great Barrier Reef

The Middle to Late Devonian was a time of global greenhouse conditions [96] with average global temperatures mostly 5–10 °C higher than today [326], but these high global temperatures fluctuated through the Middle and Late Devonian. Cooling during the Famennian is associated with minor continental glaciation at very high paleolatitudes in South America [327,328], but there is little evidence for large-scale glaciation at that time [329], and it occurs against a backdrop of very high temperatures based on most lines of evidence [96,330,331]. Continued cooling into the Mississippian eventually led into the Late Paleozoic icehouse phase. Local paleoclimate in the DGBR was consistent with the tropical to subtropical latitude with arid conditions in the Early Devonian [332], and this continued into the late Givetian where the oldest part of the exposed reef complex, the Cadjebut Formation, contains abundant evaporites [90,333]. That is consistent with the shallow intracratonic setting within the southeast trade wind belt with evaporation greater than rainfall at that time. However, modeling suggests higher humidity in the Late Devonian Canning Basin [250] with tropical monsoons [326] and cyclones tracking from the northeast [334]. Storm deposits occur in inner and mid-ramp settings in the Hull Range region [285], but it is unclear to what extent event beds represent storms, seismic activity or stochastic platform collapse. However, Late Devonian paleoclimate data are complex over the entire Canning Basin. Although evaporites are not known from Frasnian or Famennian platforms on the Lennard Shelf [90], they did occur, at least in the Famennian, in the southeastern part of the basin on the Barbwire Terrace just across the Fitzroy Trough [230,232] (Figure 13).

The common, high-frequency parasequences, including alternating siliciclastic and carbonate deposition, throughout DGBR stratigraphy are generally attributed to Milankovitch orbital forcing of minor sea level fluctuations [277], although the driving forces for cyclicity are complex with other climatic and tectonic forcing potentially dominating some sequences [278]. Late Devonian southern hemisphere monsoons were modeled to peak during precession maxima, thus leading to intervals of higher terrestrial runoff at those times [335]. Weaker monsoons during precession minima would have led to lower continental runoff and thus, might have supported greater carbonate production at those times. The alternation of carbonate and siliciclastic sediments would be enhanced during more arid times in general [285], and thus climate may have modulated aspects of the carbonate–siliciclastic contribution to parasequences beyond regulating eustatic sea level.

3.3. Carbonate Factories

Carbonate factories are intimately related to local conditions, such as bathymetry, energy levels, oxygen and nutrients and typically define the spatial distribution and productivity of particular facies and their zonation across carbonate environments [174,336,337]. Maximum carbonate production typically occurs on reef margins facing the open ocean (Figure 15). That is also where reef framework constructors are concentrated. Where those constructors are suprastratal and adequately unified, they can support much greater aggradation than the factories in other environments leading to barrier reef morphologies and back-reef lagoons where adequate accommodation is available. The degree of aggradation also can be limited by the rate and amplitude of accommodation generation (Figure 15b). Inadequate accommodation can dampen aggradation, thus equalizing the potential across different factories. Although generalities exist in carbonate production potential between carbonate factories of different ages, their constituent members have changed significantly over time and in some cases, the aggradation potential is more even across an entire platform (Figure 15c).

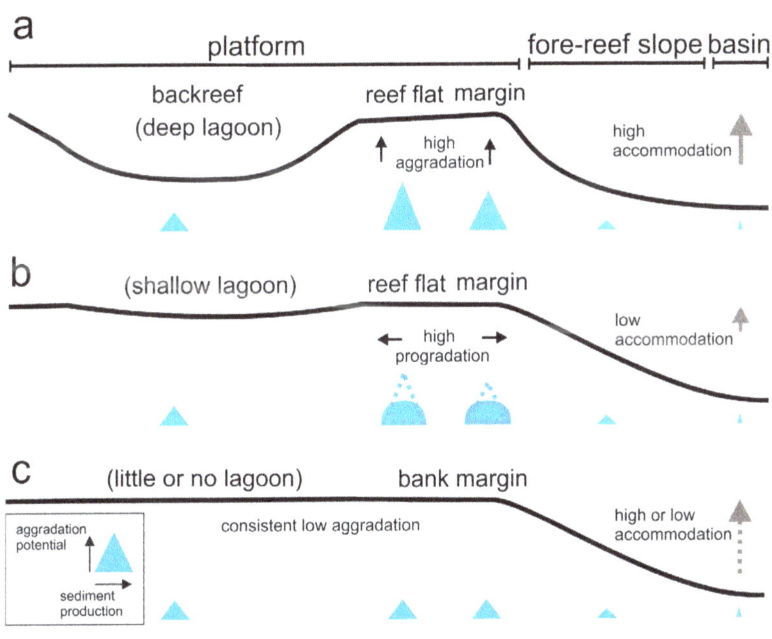

Figure 15. Differential sediment production and potential aggradation rates (blue triangles) of carbonate factories as distributed across a platform and basin. The relative width of triangles represents sediment production rate and height represents potential aggradation (modified after [336]). (**a**) Example similar to the Great Barrier Reef today where margin and reef flats have the highest sediment production and aggradation, so can take advantage of high accommodation during each high-amplitude sea level rise. The adjoining back-reef carbonate factory cannot produce adequate sediment to fill accommodation leading to a deep lagoon. (**b**) Example more like the Devonian Great Barrier Reef (CF3) wherein aggradation potential is limited by low accommodation from low-amplitude sea level fluctuations. Thus, even though margin aggradation potential is higher, actual aggradation is limited to a level more similar to the back-reef. Sediment from the highly productive margin is shifted laterally causing progradation and filling of back-reef to limit depth of the lagoon. (**c**) In the case of a bank (CF1,2), aggradation potential may be nearly the same across the entire platform.

3.3.1. Reef Builders of the Great Barrier Reef Province

The GBR is known for its vast biodiversity [338,339] but the dominant, and 'namesake,' reef builders of the GBR are scleractinian corals. As keystone taxa they have been the key focus of most recent conservation studies [15,22,72,75,79] and have been considered the main or only constituent of framework by many workers [31,340]. What we know about the broader GBRP carbonate factory is based almost entirely on surface sediments [135,175,341] and current living substrates [217,342–345], supplemented only in places by underlying Holocene reef cores [80,196,219,346]. Broader biodiversity has been sampled and modeled using a variety of techniques [338,347–350], but the large scale of the province has put a major focus on satellite proxy data [351–355]. Very few cores penetrate the immediately underlying Late Pleistocene sections [80,197,217,356]. Data are much more limited for older (late Eocene to early Pleistocene) carbonate platforms of the northern GBRP, which are known primarily from seismic stratigraphy [190,201,212]. Where recovered from exploration wells or dredging from exposures on the sea floor typical older back-reef platform facies have been identified with, for example, abundant LBFs (e.g., nummulitids) and CCA rhodoliths in Miocene strata [211]. Reef margin facies have never been adequately described and may not have been encountered in core. Extensive drowned Miocene carbonate platforms also occur offshore from the GBRP on the Marion and Queensland plateaus [202–204], but of necessity this review focuses on the Late Pleistocene sections, i.e., the modern GBR, where data are better constrained. Modern (i.e., Late Pleistocene) carbonate factories differ greatly from those of their poorly known Tertiary and early Quaternary progenitors based on scant existing evidence. However, based on that evidence, older carbonate factories of the GBRP are consistent with better known, similarly aged equivalents on the northwest shelf of Australia where there has been more subsurface exploration [357–359].

On the Queensland continental shelf proper, the deepest reef cores (Michaelmas—1926; Heron Reef—1937; Wreck Reef—1959; Ribbon Reef 5—1995) (Figure 9a) did not encounter reef facies older than the Late Pleistocene [182]. The succession at RR5 was divided into three major sections. The lowest section consists of mudstones, packstones and grainstones interpreted at the base as a deep slope or ramp environment [182]. Carbonate sand and gravel were transported downslope from shallower settings and include coral, CCA, LBFs, *Halimeda* and bryozoans, but the muds represent deeper facies and contain planktic foraminifers. Higher in the section an in situ carbonate factory of corals, CCA, bivalves, bryozoan and echinoderms was interpreted to represent a deeper and cooler, non-reefal carbonate setting [182,197,356]. The middle section is dominated by in situ deeper CCA rhodolith beds amid wackestone and packstone. It contains a few thin intervals of shallow reef facies but is mostly interpreted as a non-reefal, shallow shelf. Only the upper 96 m of core, the youngest section, consists of true late Pleistocene reef facies [182,197,356]. The core did not reach crystalline basement, but shows a general shallowing-up trend [182], with CCA assemblages showing a pronounced shift from deeper melobesioid assemblages at the base to shallower, cool-water lithophylloid assemblages at the top of the middle section to shallow tropical mastophoroid assemblages in the true reef section at the top [356]. Six cycles of reef development separated by exposure surfaces occur in the uppermost section, each typically having lower-energy, massive coral assemblages at the base grading into shallower, higher-energy assemblages at the tops [197] (Figure 10b). Each cycle is interpreted as a parasequence.

Deep cores farther south at Heron and Wreck reefs have not been so well studied. The Wreck Reef petroleum exploration well reached silicified conglomerate basement at 547 m deep, but was not fully cored and did not sample any Holocene reef facies [199]. The top 121 m originally interpreted as 'Recent' is likely Pleistocene/Holocene reef facies overlying Pleistocene quartz sandstone. The underlying sequence primarily represents deeper calcareous marine sediments thought to be Pliocene in age down to 288 m with additional marine carbonates and siliciclastics of Miocene age below that to 547 m depth based on the occurrence of LBFs including *Cycloclypeus* and *Lepidocyclina* [199]. No reef facies were encountered lower in the core and the upper section of reef facies was not described. The Heron bore consists of an upper 156 m of stacked reef facies, similar to those at RR5, underlain by calcareous quartz sandstone and foraminiferal facies [198]. It did not reach hard basement at 223 m depth where drilling stopped. The 154 m thick Late Pleistocene reef sequence at the top contains six cycles, as at RR5, similarly interpreted as parasequences associated with glacial—interglacial cycles [47,360]. Although it was reported previously that the lower parts of the carbonate succession were Pliocene in age, based on foraminifers [201,361], the entire reef sequence is now considered Late Pleistocene, but radiometric dates are lacking.

Quantification of the carbonate factory is very difficult for a variety of reasons. Although some authors have suggested that corals alone are responsible for as much as 95.5% of carbonate production on the GBR (see case study 2 in [107]), such analyses are based primarily on modeled coral growth rates and coral cover data from reef slopes. They do not take into account the broader carbonate factory, including cryptic communities, and or sediment producers from other parts of the reef, like reef flats. They also generally fail to measure the carbonates that make up the reef edifice as preserved in reef cores. The broader modern GBR carbonate factory is dominated by scleractinian corals, CCA, *Halimeda*, LBFs, mollusks [134,135,137,362–364] and microbial biofilms (that make microbialites) [129,132,158,170] (Figure 4; Table 1). Coral diversity decreases from north to south along the GBRP [365,366] but has had some medium scale (multi-decadal) monitoring in places [344,355]. Assemblages vary from lower-energy, more turbid, nutrient-rich, inshore to more exposed, open marine, offshore settings [347,367] with specialized coral communities adapted to more turbid inshore environments [192,193]. Corals in mesophotic settings (30–150 m depth) are becoming better known as these environments are evaluated as potential refugia in the face of increased coral bleaching [368,369]. In more open, clean settings, bathymetry and energy levels are the key determinants of coral ecology [197]. Algal assemblages also vary inshore to offshore [370] and non-calcareous macroalgae assemblages show complex patterns of distribution based on a variety of substrate and local environmental conditions [371]. Distributions of CCA have been poorly studied across the GBR as a whole [370,372], but bathymetric trends are well established for shallow reef-building forms [217,356] and CCA death assemblages correlate well with those of corals [46,217,288]. For the southern GBR, windward and leeward coralgal assemblages were defined for 0–6 m, 6–20 m, and 20–30 m depth [217]. The shallowest habitats are dominated by robust branching and corymbose *Acropora*, *Isopora* and *Pocillopora* corals with *Porolithon onkodes* and a variety of other CCA giving way to platy acroporids and massive *Porites* in deeper settings with more variable CCA [217]. Importantly, CCA thickness decreases significantly below 6 m depth, as does the abundance of vermetid gastropods encrusted within the thalli [217]. In the back-reef lagoon, extensive *Halimeda* bioherms occur in water depths of 20 to 50 m and with an area of >6000 km^2, they occupy more than twice the area covered by coral reefs [317,373]. The bioherms reach as much as 20 m in height and initiated ~10,000 yr ago, when the shelf was first flooded during the deglacial [374]. Other non-coral members of the carbonate factory have been studied relatively little in terms of distribution, e.g., [107,342], production rates or volumes through time.

Table 1. Main carbonate factories and their functional roles for the Great Barrier Reef Province (GBRP) and exposed Devonian Great Barrier Reef (DGBR).

	Modern GBRP	DGBR CF1–2	DGBR CF3	DGBR CF4–5
dominant carbonate factory	scleractinian coral, crustose coralline algae, *Halimeda*, large benthic foraminifer, mollusk, microbial biofilm	massive, branching stromatoporoid, rugose coral, tabulate coral, brachiopod, crinoid, microbial biofilm	massive, branching stromatoporoid, calcimicrobes, microbial biofilm, rugose coral, brachiopod, crinoid mollusk	calcimicrobes, microbial biofilm mollusk
functional constructors *	**scleractinian coral,** crustose coralline algae	massive, branching **stromatoporoid, rugose coral, tabulate coral**	**massive, branching stromatoporoid, calcimicrobes**	**calcimicrobes, micritic microbialite, tabular stromatoporoid**
binders *	**crustose coralline algae,** scleractinian coral, micritic microbialite	none	**calcimicrobes, marine cement** micritic microbialite	**calcimicrobes, marine cement** micritic microbialite
bafflers *	**branching *Acropora*** *Halimeda*	branching stromatoporoids and corals	branching **stromatoporoids** and corals	None (?)

* Dominant forms boldface.

Interest in reef sand has increased owing to its significance in carbonate budgets [41,364,375,376] and as evidence of bioerosion rates [377,378], but its constituents have rarely been quantified directly [135], commonly being estimated using hydrochemical, census-based or bulk accumulation estimation techniques [378]. Abundance of particular carbonate producers in sand varies such that less mature reefs have greater proportions of coral and CCA, whereas development of broader reef flats leads to a greater proportion of the *Halimeda*, mollusks and LBFs that live in those shallow environments [175]. Some mature reef flat sands are dominated by LBFs [137,175,379]. Hence, composition data from reef sands may be biased to a particular stage of reef development or sub-environment (e.g., mature versus immature reef flat), in a similar way that coral cover analyses on reef slopes may bias carbonate production data towards coral. However, sands do provide a means of sampling a broad range of the carbonate factory owing to today's high bioerosion rates, where many larger carbonate producers are reduced to sand, e.g., [377], and the sheer volume that sands make up in the reef edifice. Reefal microbialites are a commonly neglected aspect of carbonate production on modern coral reefs, primarily because they are mostly cryptic in occurrence and only became widely recognized in the mid-1990s [172,380]. However, where quantified in Holocene sections in the GBR they account for 7.2% of total framework volume (including cavities) and ~13.6% of preserved carbonate in shallow, leeward reef framework, compared to ~32.4% corals in growth position, ~30% CCA and ~21% rubble (mostly coral, CCA and mollusk) [176] (Figure 16). Although generally not quantified, even where recognized, microbialites have been recovered in many if not most GBR Holocene reef cores [319,346,381]. During the LGM and deglacial, GBR microbialites were much more abundant, including both typical shallow-water cryptic forms and deeper benthic forms that lacked coral substrates and were not restricted to crypts [132,170]. Where quantified as percentage area on cut core surfaces microbialites account for ~3.1% to 81.6% of LGM and deglacial reef frameworks, with microbialite dominating mainly deeper slope settings [170]. Although some microbialite formed at the tops of successions as individual reefs drowned, most of it was coeval with the coralgal framework facies with which it occurs [170].

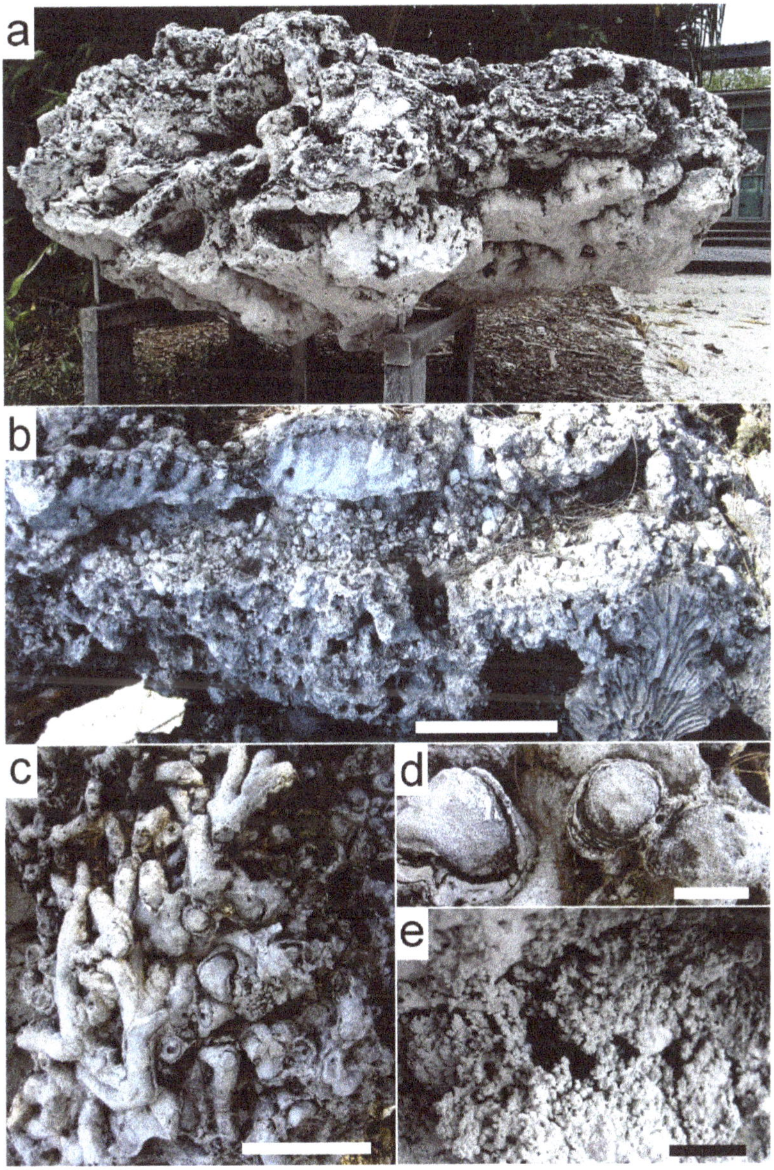

Figure 16. Excavated blocks of typical reef rock, leeward end, Heron Reef; (**a**) large block of rigid vase-shaped corymbose *Acropora* framework with crustose coralline algae (CCA) and cryptic thrombolites, diameter ~1.5 m; (**b**) platy and massive corals bound in situ with rubble, scale bar = 15 cm; (**c**) branching *Acropora* framework with abundant CCA, scale bar = 15 cm; (**d**) enlargement of (**c**) showing layers of CCA on eroding Acropora branches scale = 3 cm; (**e**) cryptic thrombolite (microbialite) growing cavities between corals, scale = 1 cm.

Typical high-energy GBR reef frameworks prograding at sea level are classified as skeletal–microbial frameworks consisting of robust branching, corymbose and digitate corals encrusted by CCA at the exposed surface with growth cavities and hollows filled by coral rubble, itself encrusted by CCA, and in some cases sand before encrustation within larger cavities and penetration of finer pore spaces in skeletal remains by cryptic microbialites (Figure 16). However, the percentages of corals in growth position, coarse rubble and microbialites vary [176,197,217,346]. Frameworks aggrading from greater depths, or prograding at greater depths on slopes, contain more massive and platy corals with coarse rubble and less CCA encrustation [197,217,346]. Additionally, the abundance of microbialites in high-energy windward settings is less well constrained, and although they occur in younger framework sections (e.g., 1–2 kya BP), they have not been adequately quantified in the rare cores of those ages. During the LGM and deglacial, reef frameworks were mixed skeletal–microbial or microbial, containing abundant microbialites (to >50% volume) with corals in growth position and some deeper slope settings were dominated by massive or laminated microbialite frameworks [46,132,170,288].

Over the last six interglacials, coralgal assemblages in RR5 reassembled after each successive lowstand [197] (Figure 10), in a similar way to coeval reef communities in northern PNG [59]. Deeper or cooler assemblages at the base of each cycle were dominated by massive *Porites*, faviids and *Montipora*, which gave way to robust *Acropora, Isopora, Stylophora, Pocillopora, Goniastea* and *Platygyra* in shallower, higher-energy settings at the top of each cycle. During glacial intervals, reefs tracked the lower sea levels down to the glacial maxima and then tracked deglacial sea levels back up, at least for MIS5 to MIS1 [46] (Figure 10). Similar coral assemblages occurred in the glacial reefs [288], but cycles commonly went from deeper at the base to shallower and then deeper again before individual reefs drowned where aggradation did not keep up with deglacial sea level rise [46]. However, in areas with abundant terrigenous flux, the reef facies were less well developed and had more turbidity tolerant species [288]. Drowning of reefs was associated both with rapid sea level rise and intervals of higher turbidity and nutrient flux from adjacent continental runoff (Figure 10). Hence, the Pleistocene reef carbonate factory has been very resilient through the last glacial cycle. However, detailed comparison of Holocene and underlying warmer LGM reef facies shows some significant differences. Although the carbonate factories are basically the same, and the LGM reef reached sea level with a similar catch-up succession to that in the Holocene [197,218,346], initial phases of growth suggest that once the platform was colonized by corals a rapid deepening up sequence occurred resulting in dominance of *Halimeda* and deeper-turbid coralgal assemblages [217]. Timing of the event suggests that the rapid sea level rise was associated with meltwater pulse 2B (MIS 6) of the penultimate deglacial [382]. That, and/or increased nutrient supply from upwelling may have led to the near drowning of the reef before the carbonate factories recovered adequately to catch up with the stabilizing sea level [217]. Hence, each parasequence was not identical and the warmer, higher-sea-level LIG differed from the current Holocene based on limited existing cores.

Aside from the carbonate factory itself, many other dwellers including vertebrates (e.g., fish, cetaceans, turtles) and invertebrates (e.g., arthropods, annelids, shell-less mollusks) occur in GBRP waters. More than 16,500 species occur across the habitats, including 450 scleractinian corals, at least 6000 mollusks and 630 echinoderms [339], and that does not include calcareous or silicious tests of LBFs and plankton. Biodiversity for many groups (e.g., corals, some fish) increases from south to north with temperature [339]. The understood biodiversity, in particular for corals, is probably underestimated, as modern genetic studies identify cryptic species in what were once thought to be very well-known and well-studied taxa (e.g., *Stylophora pistillata* [383]). Regardless, the overall habitat is highly differentiated and some taxa prefer more inshore, turbid or open shelf, inter-reef settings [338,347] whereas others are restricted to shallow reefs or mesophotic depths depending on individual requirements and along known trends (e.g., nutrients [146]). The destroyer guild is very well developed with scarid fish being the dominant raspers [377] but polyplacophorans are also abundant. Rasping echinoids are not generally acknowledged as major destroyers in the GBR but boring bivalves and worms are abundant [384]. Excavating sponges are ubiquitous [385] as are endolithic cyanobacteria, green algae and fungi [384]. Corallivores, such as *Acanthaster planci* (Crown of Thorns Starfish), do not actually damage carbonate substrates, but kill significant amounts of live coral during outbreaks, leaving the exposed skeleton more susceptible to bioeroders.

3.3.2. Reef Builders of the Devonian Great Barrier Reef

Perhaps the most comprehensive general descriptions of carbonate petrology and reef facies from the DGBR remain those of Kerans [236] and Copp [386], despite many more recent works focused on more regionally or temporally constrained parts of the complex, e.g., [265,266,270,285,387,388] and the major review by the Geological Survey of Western Australia [90]. Key members of the DGBR carbonate factory were summarized by Playford [88] and include stromatoporoid sponges [389], corals [390,391] (Figure 17), calcimicrobes [236,392], microbial biofilms (makers of stromatolites, thrombolites, many oncoids and probably ooids) [87] (Figure 18), receptaculitids, calcareous algae [392], brachiopods [393,394], crinoids, gastropods and bivalves [99,236,395] (Figure 19). Stromatolites [282,396,397] and siliceous lithistid [398] and soft sponges [236] also were significant on reef slopes and in deeper settings (Figure 19h) and encrusted broad event surfaces in the middle to late Frasnian [283]. Soft and siliceous sponges, although not carbonate producers, enhanced carbonate production by promoting carbonate precipitation through microbial metabolic processes [399], thus localizing microbialite formation as in Mississippian and Jurassic reefs [98]. Carbonate factories changed significantly through the Late Devonian, with the most drastic changes associated with the Frasnian–Famennian extinction event [84], but also the earlier end-Givetian event [70], which coincided with the end of the peak of Devonian reef growth globally [85]. In general, three major sets of organisms dominated the carbonate factories through DGBR time: Givetian–early Frasnian stromatoporoids and corals; middle–late Frasnian microbes and stromatoporoids; and Famennian microbes [90] (Figures 17 and 18). However, as with the GBRP, the reef complex was extensive with complex geography and significant and abrupt changes in accommodation due to active tectonism leading to significant temporal and/or regional environmental differences in factories [236,267,400]. These major differences allow more or less distinct carbonate factories to be recognized (Figure 14). The carbonate factories specified below represent a much larger spectrum and combine some of the previously defined lithofacies and lithotopes [236] but are intended to represent the generalized temporal succession. However, difficulties in correlating shallow facies in the DGBR render differentiation of exact temporal versus spatial–geographic relationships difficult [269,271,401]. The exact relation of margin framework facies, for example, to coeval back-reef facies is hampered by incomplete exposure and subsequent erosion. Some margin facies are known primarily from debris on associated slopes, where more complete exposures may occur, e.g., [265,266,270,402].

Figure 17. Devonian carbonate factory skeletal constructors. (**a**) Domal stromatoporoid, CF1, Limestone Billy Hills, scale = 10 cm; (**b**) 'cathedral' complex stromatoporoid, CF3 back-reef, Napier Range, scale = 10 cm; (**c**) tabular stromatoporoid sloping on winnowed CF2 margin with abundant brachiopods, Loyd Hill, Bugle Gap, scale = 5 cm, note cement-filled shelter cavity beneath stromatoporoid formed by winnowing; (**d**) branching stromatoporoid *Stachyodes* showing astrorhizae on branch, CF3 back-reef, Oscar Range, scale = 1 cm; (**e**) branching stromatoporoid *Amphipora*, CF3 back-reef, Oscar Range, scale = 1 cm; (**f**) massive rugose coral, *Argutastrea*, deep platform interior, CF1, Pillara Range, scale = 20 cm; (**g**) non-encrusting *Argutastrea*-dominated facies, CF1, southern Hull range, scale = 10 cm; (**h**) branching rugose coral, *Disphyllum*, CF3 reef flat, Mowanbini (Oscar Range), scale bar = 2 cm; (**i**) tabular tabulate corals, *Alveolites* (arrows), sloping on winnowed CF1 margin, Menyous Gap, scale = 2 cm; (**j**) branching tabulate coral, *Thamnopora*, deeper CF1 back-reef facies, scale bar = 2 cm.

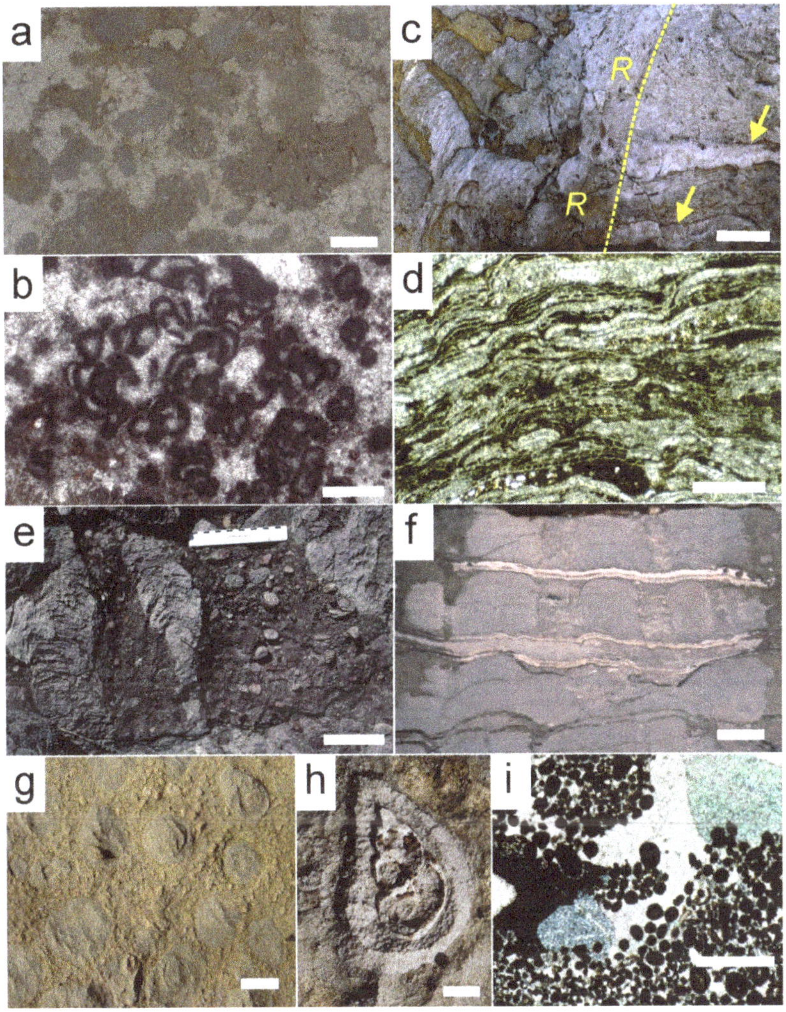

Figure 18. Calcimicrobes and biofilm members of the Devonian carbonate factory. (**a**) CF3 *Renalcis* crusts (white) on *Stachyodes* with cavities accumulating sediment, calcimicrobial–skeletal framework, Galeru Gorge, scale = 1 cm; (**b**) thin section photomicrograph of *Renalcis*, scale = 250 µm; (**c**) *Rothpletzella* (*R*) crust (calcimicrobial framework) becoming upward-directed columns to left on a collapse scar that has cut into CF1-2 reef flat facies with tabulate stromatoporoid (arrows) platestone, scale = 5 cm; (**d**) thin section photomicrograph of *Rothpletzella*, scale = 5 mm; (**e**) stromatolites (micritic microbialite framework) growing within siliciclastic sediment in CF5 back-reef, Horseshoe Range, scale = 10 cm; (**f**) stromatolites (micritic microbialite framework) on Famennian slope, Ngumban Cliff, scale = 5 cm; (**g**) CF3 microbial oncoids in fenestral oolitic grainstone, Mowanbini, scale = 1 cm; (**h**) CF3 microbial oncoid with gastropod as nucleus, Mowanbini lagoon, scale = 1 cm; (**i**) thin section photomicrograph of oolitic grainstone with fenestral fabric cemented by meteoric poikilotopic calcite, back-reef Mowanbini, scale = 1 mm.

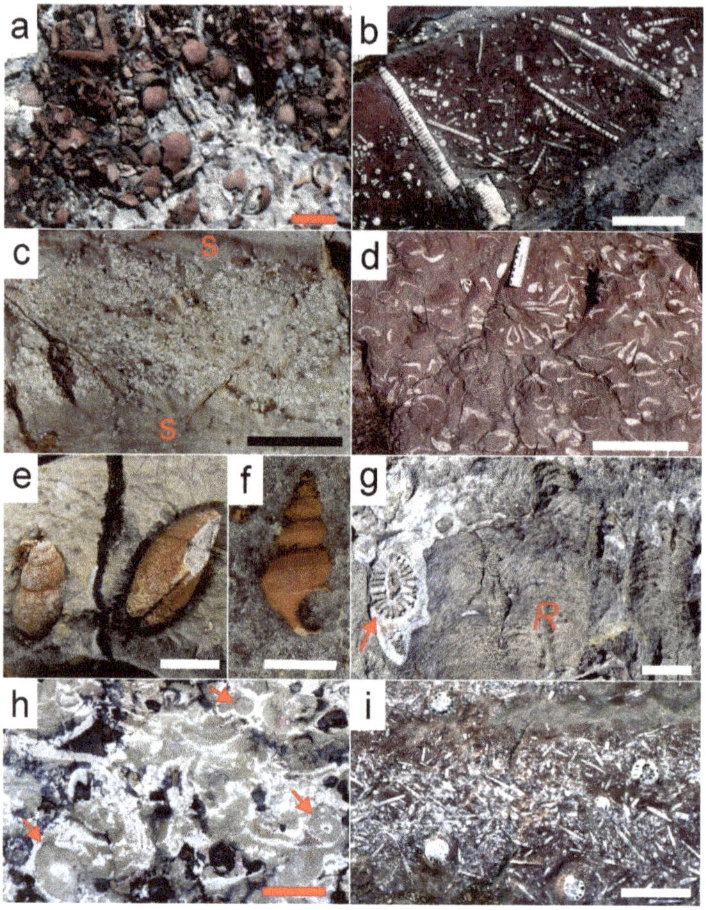

Figure 19. Functional dweller members of the Devonian carbonate factory (**a**) silicified brachiopods and tabulate corals in CF1–2 back-reef, Lloyd Hill, scale = 2 cm; (**b**) CF5 distal slope crinoid columns, Casey Falls, scale = 5 cm; (**c**) CF3 platestone on proximal fore-reef slope with tabular stromatoporoids (S) resting in crinoidal grainstone, Mowanbini fringing reef, scale = 2 cm; (**d**) megalodontid bivalves in situ in CF5 back-reef, Pothole Bore, scale = 30 cm; (**e**,**f**) silicified gastropods and bivalve from protected lagoon around Christopher Bore Island, scales = 1 cm; (**g**) receptaculitid (arrow) with *Rothpletzella* (*R*), fore-reef slope, Glenister Knolls, scale = 2 cm; (**h**) sponges (arrows) in distal slope mound, Glenister Knolls, scale = 5 cm; (**i**) CF5 pelagic cephalopods (nautiloids and goniatites) in distal slope facies, Kelly's Pass, Lawford Range, scale = 10 cm.

Initial carbonate deposits (Carbonate Factory 1: CF1) were confined mostly to low-relief platforms and banks that initially may have had ramps, rather than steeper shelf edges, on the seaward sides [280,403]. These mid-late Givetian to early Frasnian platforms typically display m-scale, mostly subaqueous parasequences in their interiors [90,275–277] (Figures 20 and 21) and in some places, at their margins [267] (Figure 22). Higher in section, and/or possibly in more open settings, Early Frasnian CF2 margins of low- to steeper-relief platforms became more stromatoporoid-rich with more abrupt, but still low-relief laminar stromatoporoid buildups at their margins. Back-reef areas had similar variable carbonate producers to CF1. In the early middle to late Frasnian, CF3 margin facies changed considerably with a major increase in the calcimicrobe *Renalcis* (used in the broadest sense for similar renalcid-like microfossils that can be divided into separate taxa, e.g., *Renalcis*,

Shuguria, *Izhella*, etc. [404–406]) (Figure 18a,b) and different stromatoporoids, including much more abundant branching *Stachyodes* on the reef flat and proximal fore-reef slope and laminar *S. australe* on the margin [88,236,407]. Early marine cement became much more abundant at this time and increased aggradation led to larger framework cavity systems and higher relief with steep, upright shelf edges [88,236,267]. These three factories all occurred within the transgressive Pillara phase of the supersequence (Figure 14). Towards the end of the Frasnian, the abundance of stromatoporoids decreased in the CF4 margin to be replaced increasingly by low-domal stromatolites [88,236,267,402]. This may represent the beginning of the regressive Nullara phase. Stromatoporoids and colonial corals were almost entirely removed at the Frasnian–Famennian boundary extinction event following the previous major collapse at the end of the Givetian [84]. Subsequent early Famennian CF5 margins, where studied, were constructed primarily by microbes [88,236,388,408]. Key members for the different carbonate factories are shown in Table 1 with dominant ecological roles of reef margin and reef flat organisms.

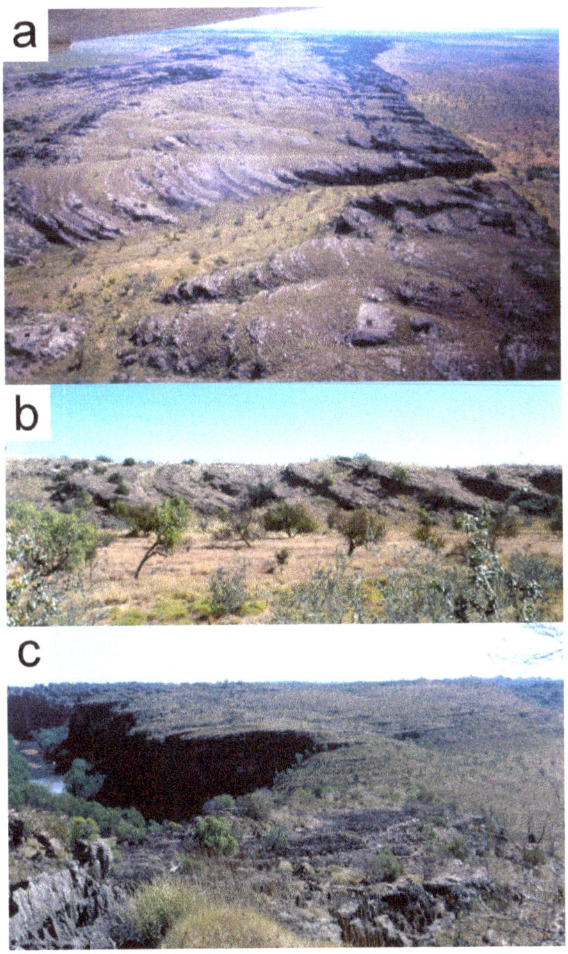

Figure 20. Late Devonian parasequence-scale cyclicity; (**a**) CF1 cyclicity of Pillara Range at Menyous Gap; (**b**) cyclicity within Menyous Gap showing resistant shallow stromatoporoid layers alternating with recessive deeper coral-dominated layers, note ~20 degrees structural dip to north (right side); (**c**) CF4 to CF5 cyclicity in the Napier Range behind Windjana Gorge.

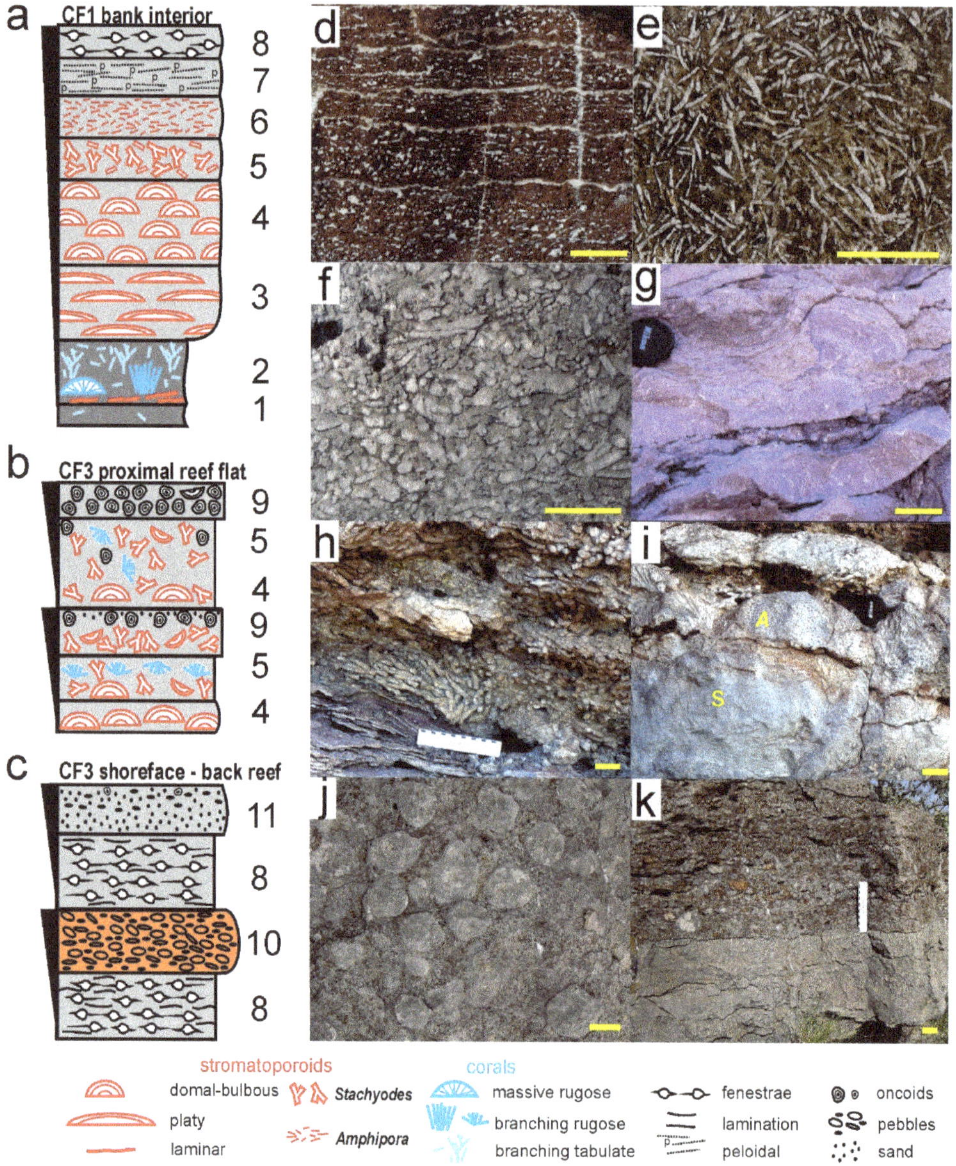

Figure 21. Typical parasequence cycles in parts of the Devonian Great Barrier Reef. Numbers by columns represent facies described in more detail in the test. (**a**) Typical composite CF1 parasequence [276,277]; (**b**,**c**) CF3 back-reef parasequences form Mowanbini [292]; (**d**) fenestral laminate, proximal back-reef, Mowanbini, scale = 5 cm; (**e**) *Amphipora* rudstone, back-reef, Mowanbini, scale = 5 cm scale; (**f**) *Stachyodes* rudstone, CF1, McWhae Ridge, scale = 5 cm (**g**) platy constratal stromatoporoids, CF1, Pillara Range, scale = 5 cm; (**h**) crushed thamnoporoid corals overlying laminar stromatoporoids, CF1 Menyous Gap, scale = 5 cm; (**i**) massive *Argutastrea* (*A*) coral at base of recessive facies 2 directly on domal and constratal stromatoporoid (S) facies 4, CF1, Menyous Gap, scale = 5 cm; (**j**) very large oncoids, proximal reef flat, CF3, Mowanbini, scale = 5 cm; (**k**) facies 11 on 8, shore proximal back-reef, Mowanbini, scale = 5 cm.

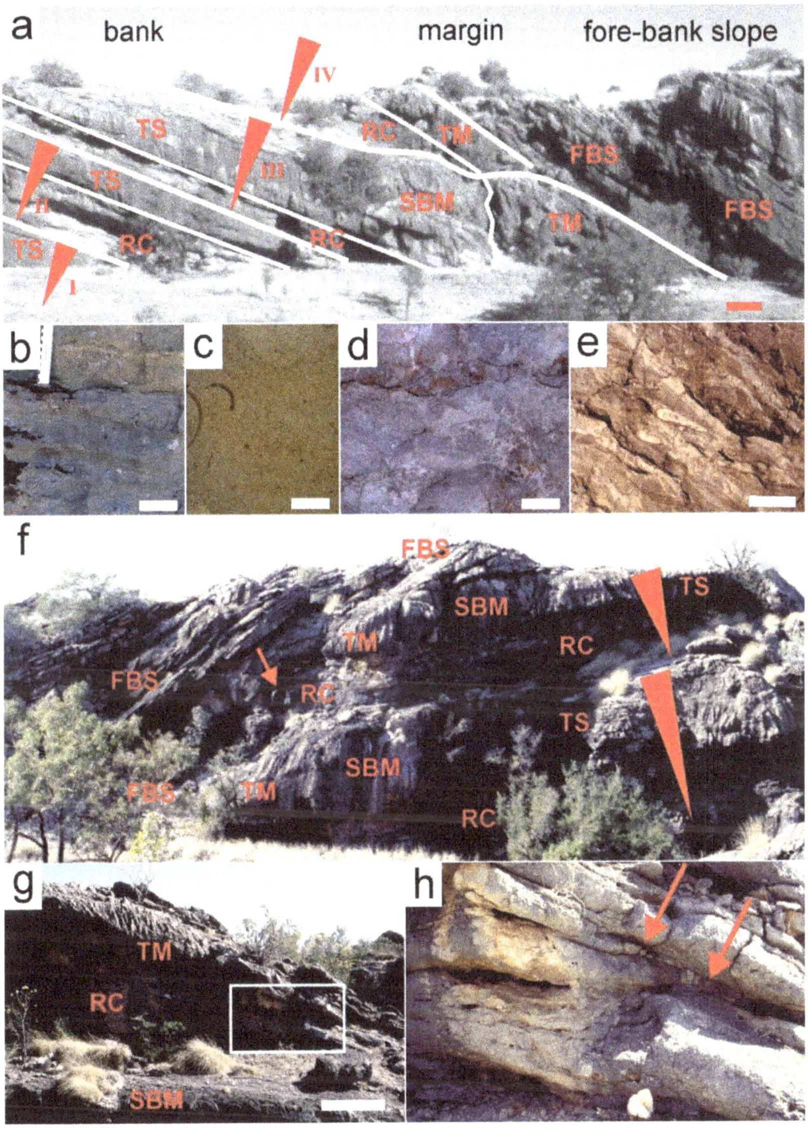

Figure 22. Pillara Range northern CF1 bank margins. (**a**) Bank margin at Menyous Gap (MG), note additional 14° degrees of original dip on slope relative to ~20° degrees of structural dip to N, scale = 5 m, modified from [267], red triangles show parasequence shallowing-up cycles, recessive coral facies—RC, facies interpretations: tabular stromatoporoid facies—TS, mounding stromatoporoid bank margin facies—SBM, tabular tabulate mound facies—TM; fore-bank slope—FBS; (**b**) protected bedded mudstones–wackestones from lagoon behind mounded margin near MG, scale = 10 cm; (**c**) polished slab from lagoon mudstone, note recrystallized mollusk fragments, scale = 1 cm; (**d**) SBM facies, scale = 10 cm; (**e**) TM facies, note sloping thin alveolitid corals, scale = 6 cm; (**f**) bank margin ~1 km east of Menyous Gap, note people at arrow for scale; (**g**) bank margin in opposite exposure shown in (**f**), scale = 40 cm; (**h**) erosive truncation (arrows) of RC facies at margin beneath SBM facies is consistent with increasing water energy as sea level fell to deposit the shallower facies SBM facies at the top of the parasequence, scale = 2 m.

Carbonate Factory 1 is highly variable with significant ecological shifts, e.g., [70,407]. Its age is late Givetian probably to early middle Frasnian (conodont zone 5) [270], but it is very difficult to correlate exposures even within the Pillara Range [409]. Basal CF1 carbonates are Frasnian to the north in the Hull Range [270,280] reflecting transgression and coastal onlap. Although the CF1 carbonate factories were highly varied they are included together here as they built similar low-relief carbonate banks. The earliest parts, to the south in the Pillara Range, were dominated in the east by the delicate branching stromatoporoid *Amphipora* (Figure 17e), which occurred mostly in more protected platform interiors (back-reef settings) [403,409]. That transgressive unit occurred non-conformably over Precambrian basement relief with feldspathic sandstone flooring paleovalleys and abundant siltstone units in the lower parts. The unit is less dominated by *Amphipora* to the west [407] and is absent farther to the north and landward in the Hull Range [285]. Succeeding bank facies are cyclic with stromatoporoid (*Actinostroma*) and fenestral facies or just fenestral facies to the west that then become more coral-rich with abundant massive *Argutastrea* (Figure 17f,g) before passing into more stromatoporoid-dominated cycles up section [409]. In offshore platforms (e.g., Pillara Range), the cycles represent low-relief biostromes in the bank interior that typically alternate between shallower, more resistant stromatoporoid and intertidal facies and recessive, deeper coral facies [70,267,275,276] (Figure 20). A composite cycle at Menyous Gap (Figure 21a) includes eight typical facies from base to top: (1) rare recessive carbonate mudstone–wackestone; (2) recessive coral facies dominated by the corals *Disphyllum*, *Argutastrea*, *Alveolites*, and/or *Thamnopora* in growth position, but crushed, with minor thin stromatoporoids in a muddy matrix; (3) more resistant tabular stromatoporoid facies characterized by tabular to low-domal constratal stromatoporoids in growth position in muddy matrix (platestone—equivalent to floatstone of [236]); (4) much more resistant domal stromatoporoid facies with more abundant, large (60 cm) stromatoporoids in growth position in variable wackestone–packstone matrix; (5) resistant *Stachyodes* facies consisting of abundant *Stachyodes* (stromatoporoid) floatstone with peloidal packstone–grainstone matrix; (6) *Amphipora* (stromatoporoid) facies consisting of *Amphipora* floatstone in peloidal wackestone–packstone matrix; (7) resistant peloidal facies consisting of well-sorted peloidal packstone–grainstone; and (8) resistant fenestral limestone facies consisting of peloidal packstone–grainstone containing abundant fenestrae (see [276] for more detail). Elsewhere cycles proximal to the margin may be capped by a high-energy oncoid facies (Figure 21b, facies 9), whereas shore proximal cycles can contain coastal siliciclastic units (facies 11) at lowstands and even fan deltas (facies 10) (Figure 21c) [292]. Exposure surfaces are very rare at the tops of Pillara cycles [275] and many cycles are incomplete, never having had the shallowest facies deposited [267,276]. Cycles generally shallow to the north with those in the Guppy Hills and parts of the southern Hull Range [280] being deeper to shallow subtidal, similar to those in the Pillara Range [276], but with fewer deeper coral facies and many more intertidal facies at cycle tops in the northern Hull Range [280]. Bank margins range from carbonate ramps [403] to more abrupt roll-over margins with high-energy oncoid facies and rare lagoonal facies protected behind thicker margin buildups [267] (Figure 22). Cycles are interpreted as parasequences resulting from Milankovitch forced minor eustatic sea level oscillations [277,278]. Shore attached cycles with major siliciclastic content are more complicated to interpret and have additional tectonic and climatic controls superposed on eustacy [90,278,280].

In perhaps the most detailed biological study of the cyclic successions [70] biotic membership in the back-reef platform interior was quantified for as many as 70 successive parasequences at Menyous Gap, Pillara Range (Figure 20a,b). Those data show how variable the biostromes were, but cluster analysis supported the major differences between deeper and shallower communities (i.e., typical parasequences). Overall, they found that cycles tended to have relatively stable community memberships over short time scales, but that over longer time scales taxonomic shifts occurred. In particular, they noted three major temporal shifts in broader associations of shallow biostromes from base to top: (1) branching rugose coral (*Disphyllum*) massive stromatoporoid (domes to >1 m in diameter); (2) stromatoporoids largely lacking corals, but with greater diversity of stromatoporoid morphology (massive, irregular and tabular); and (3) branching tabulate (alveolitids) and massive rugose (*Argutastrea* to >1 m diameter) coral and massive stromatoporoid. They [70] provisionally attributed the shifts in coral communities to changing sea surface temperatures near the end of the Givetian [410] but noted that the succession is not adequately dated for firm conclusions. Owing to the focus on corals and stromatoporoids in growth position, it is difficult to quantify the contributions to the carbonate factory of more fragile branching taxa, such as *Stachyodes, Amphipora* and *Disphyllum*, but the data uniquely highlight the long-term biostrome community dynamics while demonstrating some of the spatial and temporal heterogeneity (i.e., complexity) within the setting.

A retrograding CF1 platform margin is exposed at the northern end of Menyous Gap for the youngest few cycles [267]. Cycles at this margin near Menyous Gap may contain some of the highest energy facies (e.g., the oncoid facies), although subaqueous throughout, and entire cycles roll over the margin intact (Figure 22a,f). Although the current topographic expression of the northern margin of the limestone is eroded into a relatively abrupt, more or less straight outcrop (Figure 20a), the platform margin was irregular, with seaward projecting extensions potentially like buttresses exposed far to the east at Teichert Hills (Figure 23). Those structures were interpreted as spur and groove morphology [411] and contain a constructional margin community dominated by sloping tabular tabulate corals with minor low-domal stromatoporoids (Figure 23). The proximal fore-bank slope tabular tabulate facies at Menyous Gap [267] is very similar (Figure 22e). Shallowest intertidal and oncoid facies were not developed at the exposure within Menyous Gap itself [267], but nearby the deeper recessive facies underwent erosion above wave base beneath overlying shallower margin deposits when sea level fell (Figure 22a,h). This erosional truncation of recessive beds at the margin provides evidence that the cycles were indeed driven by changes in sea level [267] and were not autocyclic (formed by progradation alone—the Ginsburg model [183]). The shallow margin thickened to varying degrees to have some relief relative to the back-reef facies and consisted of constratal tabular stromatoporoids floating in a matrix of peloidal grainstone forming a platestone facies. This facies is not to be confused with the recessive 'tabular stromatoporid' facies of other authors [70]. It is not clear to what degree it may have been 'wave-resistant,' as the observed examples were poorly winnowed and clearly subtidal. At the margin the facies passes laterally into massive proximal slope facies of carbonate mudstone (microbial?) to packstone containing isolated, sloping tabular tabulate corals (alveolitids) in growth position (Figure 22e). Cement-filled cavities occur beneath overhanging alveolitids on the slope and coarse debris from higher on the platform is abundant. Recessive (deeper) facies continue over the margin largely unchanged except where pinched from below by higher-relief shallow margin facies (Figure 22a) and in places eroded by the succeeding shallower environment (Figure 22g,h) [267]. Other members of the carbonate factory, such as crinoids, bryozoans, brachiopods and the microbes that made peloids (and potentially ooids [412–414]) were also active, but their contributions have not been quantified. Where mounded margins accreted higher than platform interiors, microbes may have been significant, as skeletal frameworks were not developed and the isolated, non-encrusting stromatoporoids were constratal. It is unclear if they developed any significant rigidity, but in some cases protected deeper muddy (lagoonal) facies containing a mollusk fauna behind them (Figure 22b,c). Deeper facies at the toes of the banks include microbial mounds containing debris from up slope in clotted mudstone with subaqueous fenestrae [285], and sponges and stromatolites were likely present.

Figure 23. (**a**) Bank margin at Teichert Hills showing chute and buttress structure similar to (**b**) spur and groove structures on modern reefs, this example Heron Reef.

Carbonate Factory 2 represents a shift to greater tabular stromatoporoid abundance near the margin. Within the Pillara Range it may represent part of the lower-stromatoporoid unit, may be equivalent to the lower-reef-cap member in the western Pillara Range or may extend into the upper-stromatoporoid members [407]. Difficulty in correlation renders its spatial versus temporal distribution unclear and it may be distinctive only at the margin. Regardless, it occurs early in the prominent early Frasnian back-stepping phase, but before the major shift to aggradation associated with the CF3 association (Frasnian platform phases 3 or 4; conodont zones 4–5) [270] (Figure 14). CF2 margins are dominated by stacked constratal tabular stromatoporoids (generally less than 10 cm thick), in some cases reaching total thicknesses >2 m (Figure 24). These low relief frameworks are similar to the laminar stromatoporoid framestone of low-relief reef margins in the middle Frasnian Alexandra Reef Complex #2 of the Northwest Territories, Canada [168] and are classified as skeletal frameworks. Reef flat facies are difficult to define owing in part to the lack of abundant *Renalcis*, which is an indicator of reef flat facies higher in section (CF3) [236], but proximal back-reef to reef flat facies include grainstones with scattered constratal tabular stromatoporoid platestones and abundant brachiopod beds, crinoids and peloids. At Glenister Knolls exposures of the margin are commonly missing owing to margin collapse. The arcuate nature of the present margin of the knolls in map view combined with abundant debris in associated slope deposits and clear failure surfaces that cut into back-reef facies prior to encrustation by *Rothpletzella* (Figure 18c) suggest that some margins were quite rigid and even back-reef facies were well indurated prior to collapse events. However, the stratigraphy is very complex in three dimensions with younger, deeper facies overlaying older, shallower facies [415]. Some apparent in situ margin facies represent steep collapse scarps and it can be difficult to discriminate original margin slopes from winnowed or collapsed slopes. They contain very little *Renalcis* where margins can be traced directly into coeval adjoining slope and where a laminar stromatoporoid skeletal framework facies does not occur at the margin, non-encrusting marginal tabular stromatoporoids rarely have original dips greater than 22° relative to associated geopetal horizons. In other places, the substrate was not well indurated at these margins and sediment eroded from beneath tabular stromatoporoids and tabulate corals causing them to collapse seaward, steepening them, and forming shelter cavities beneath (Figure 17c). The nature of the outcrop makes it unclear as to where in a parasequence the less indurated facies is in respect to the thick laminated stromatoporoid facies but the two margin types may represent the deeper and shallower parts of a parasequence, respectively. An additional complication comes from the back-stepping itself, as it superposed the deeper facies over older, relief forming shallow facies formed earlier during the transgression [415]. Regardless, CF2 is associated with major reef drowning and back-stepping events in the early Frasnian. The skeletal stromatoporoid margins were not able to aggrade and keep up or catch up to sea level. Although *Rothpletzella* occurs (especially on slopes) throughout the Late Devonian succession, drowned reef margins were commonly encrusted by laterally persistent columnar *Rothpletzella* crusts as much as 50 cm thick, consistent with its abundance in deeper slope settings [88,90]. Some layers may mark individual drowning events.

Figure 24. (**a**,**b**) Eroded CF2 margin facies composed of stacked constratal tabular stromatoporoid skeletal framework, Glenister Knolls.

Carbonate Factory 3 is entirely Frasnian in age (conodont zones 4?–11 or 12 [269], Frasnian sequences 2 to 5 [265]) (Figure 14) and differs from previous factories most prominently at the margin where suprastral branching stromatoporoids were more abundant and commonly supported by the calcimicrobe *Renalcis* and abundant early cements. This allowed them to aggrade and keep up better with increasing RSLs leading to a shift from the earlier low-relief bank facies to higher-energy, upright reef margins [236,267]. Margin facies tend to be massive without obvious bedding but may be thin zones laterally of just a few to tens of meters width (Figure 25a). Generalities are difficult to make as margins frequently experienced collapse. Key constructors at the margin were low-domal to complex stromatoporoids or more laminar forms, such as *Stachyodes australis* (Figure 25c,d), supported and suspended above the sediment–water interface by *Renalcis* and early marine cements. In other areas thicker tabular stromatoporoids were well bound by *Renalcis* and cements commonly holding open very large framework cavities, to be filled subsequently by cycles of internal sediment and cement (Figure 25f). Spatial or temporal trends in dominant CF3 margin constructors are difficult to constrain, but they include the constructors of the following lithotopes [236]: coral–stromatoporoid framestone; laminar stromatoporoid framestone; and *Renalcis–Spharocodium* (i.e., *Rothpletzella*)–stromatoporoid bindstone. Cemented oolitic grainstone shoal margins also occur in places (e.g., at Nardji Cave [265]). Proximal reef flats could contain similar facies but with framework cavities partially filled by peloidal microbial fabrics and minor bioclasts (crinoid, brachiopod and mollusk debris). More commonly, branching *Stachyodes* encrusted by *Renalcis* dominated the proximal reef flat (Figure 25e). The strength of the framework allowed subsequent filling of framework cavities by a variety of platform sediments, including siliciclastic sands from the mainland [270], and where pieces of reef margin and reef flat facies collapsed into the adjacent basin (e.g., McIntyre Knolls) the final internal sediment fillings were commonly basinal muds. Such collapse was particularly frequent in the earliest phases of Frasnian sequence 5 as highstand deposits initially built over basin facies that could not support progradation of new margins [265]. Some high-energy windward settings also had large oncoids and round, potentially mobile bulbous stromatoporoids floating in mud-lump (micritized grain) grainstones just behind the reef crest [290,292]. In other areas, probably later in the Frasnian, large *Renalcis*-bound peloidal sediment masses occurred (Figure 25g), but it is unclear exactly where they fit within the stratigraphy, as they mostly represent debris collapsed from the margin. Similar facies become more abundant in CF4 and CF5.

Cycles are not easy to discern in massive margin facies, but back-reef areas alternate between shallower and deeper facies and with increasingly prominent bedding, the reef flat passes into back-reef facies with the loss of abundant *Renalcis* [236]. Reef flat facies may pass into *Amphipora* rudstones and in deeper parts of cycles the branching coral (*Disphyllum*) [292]. Typical high-energy back-reef cycles have scattered massive stromatoporoid biostromes at the base and shoal up through mud-lump and intraclastic grainstones containing oncoids with common gastropods to terminate in shoaling oolitic grainstones with fenestral fabrics at cycle tops [290]. Where ooids do not occur, fenestral algal laminites made of mud-lump grainstones with linked fenestrae and exposure surfaces with desiccation cracks occur [292]. More protected distal reef flat facies contain peloidal and mud-lump grainstones with domal stromatoporoids, branching *Stachyodes* and *Amphipora* and a diverse fauna of mollusks with abundant gastropods [293].

Proximal fore-reef slopes contain a variety of boundstones, platestones, rudstones, floatstones, grainstones and packstones containing predominantly branching stromatoporoid debris [236]. *Stachyodes* is very common in the fore-reef slope where it occurs encrusted by *Renalcis* in places or as open, well-winnowed and cemented rudstones providing significant rigidity to proximal fore-reef slope environments (Figure 25). Where the *Stachyodes* was cemented adequately, these facies may be considered biocementstones (Figure 25h). Although tabular–laminar stromatoporoid platestones occur in places within grainstone (Figure 19c), *Renalcis* and *Rothpletzella* were important on some slopes and receptaculitids and lithistid sponges may be important [236] (Figure 19g). Farther down slope stromatolites and soft sponges became more abundant as well (Figure 19h). Debris flows of shallower shelf facies are common in many horizons on the slope [88,265,270].

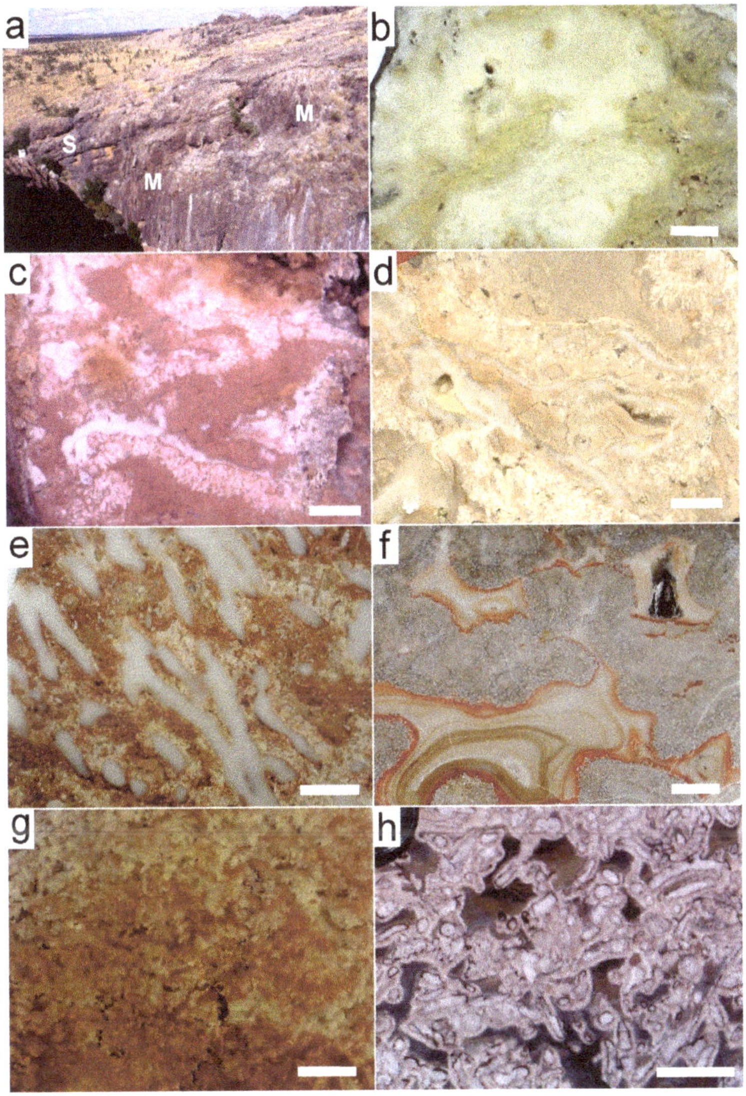

Figure 25. CF3 reef margin and framework constructors; (**a**) CF3 reef margin at Galeru Gorge, massive margin facies (M), proximal fore-reef slope (S); (**b**) irregular to massive stromatoporoids in skeletal platestone, scale = 1 cm; (**c**) laminar stromatoporoid *Stachyodes australe* supported by thick *Renalcis* crusts mostly underneath, subsequently filled with finer sediment (pink) from succeeding slope, caclimicrobial–skeletal framework, Lawford Range, scale = 2 cm; (**d**) as (**c**), but from Galeru Gorge with minor tabular stromatoporoids, scale = 2 cm; (**e**) branching *Stachyodes* in growth position encrusted by *Renalcis*, skeletal–calcimicrobial framework, scale = 2 cm; (**f**) thick *Renalcis* crusts on scattered *Stachyodes* branches with abundant early marine cement (red and white), well-cemented calcimicrobial–skeletal framework, scale = 2 cm; (**g**) *Renalcis* intergrown with sediment in calcimicrobial framework, scale = 1 cm; (**h**) *Stachyodes* debris with isopachous early marine cement making a biocementstone framework where open cavities were subsequently filled by darker muds on the slope, scale = 5 cm, this framework was made rigid by early marine cement, as were the open calcimicrobial frameworks in (**c**,**d**), and has no modern analog.

Carbonate Factory 4 is most abundant towards the end of the Frasnian (Frasnian sequences (?)5–top, 6 and 7 [265,269]) and has increased microbial input and decreased skeletal reef builders [416], particularly missing massive and tabular stromatoporoids [402], but also with fewer branching rugose and tabulate corals and branching sponges [236]. At steep margins (Figure 26), microbe-dominated frameworks include microlenticular–cyanobacterial and *Renalcis*–fenestral bindstones [236]. Early cements still provided significant rigidity for margin facies, but although margin collapse was not as common as earlier in the Frasnian, margins are commonly missing or difficult to access, so are poorly known [265]. Dominant margin frameworks are low-relief calcimicrobial and peloidal cementstones.

Figure 26. CF4–5 reef margins and framework constructors; (**a**) Late Frasnian 'classic face' at Windjana Gorge (WG), note that platform facies are flat-lying and slope is near original dip; (**b**) abrupt, upright margin with slope possibly abutting a collapse scar, near WG; (**c**) distinctive platy–columnar stromatoporoid *Actinostroma windjanicum* Cockbain in CF4 back-reef at WG, scale = 10 cm; (**d**) more tabular *Actinostroma* showing *Renalcis* crust (*R*) on lower surface, scale = 3 cm; (**e**) Famennian margin at Horseshoe Range (HR), massive margin (M), fore-reef slope (S); (**f**) large microbial mound in Famennian proximal fore-reef slope at Pothole Bore; (**g**) low-growing columnar stromatolites amidst terrigenous siliciclastic sediment, Famennian HR, scale = 3 cm; (**h**) robust *Renalcis* crusts (*R*) on irregular underlying surface with abundant siliciclastic sediment, reef flat, Famennian, HR, scale = 3 cm.

CF4 reef flat and back-reef facies are dominated by microbial deposits, but they are poorly known owing to erosion [402] and their exact relationship to margin facies is challenging. Reef flat and back-reef facies developed at the top of Windjana Gorge may represent the end of CF3 or CF4 (Figure 26a,b). The *Actinostroma* community [387] contains a variety of very large (to 5 m diameter), but isolated stromatoporoids (Figure 26c,d), commonly encrusted by *Renalcis* and other microbialites in largely constratal settings in bedded grainstones to packstones. They supported some cavities containing early marine cements, microbes and internal sediments. The previously described *Stachyodes–Shuguria* (i.e., *Renalcis*) community [387] most likely represents the younger CF3 factory, but the fenestral 'microbialite-forming community' and '*Rothpletzella* stromatoporoid community' [387] are characteristic of CF4 reef flats. Cyclicity becomes more obvious in back-reef facies [89,90] with more isolated stromatoporoids in the deeper parts of cycles shoaling up to fenestral microbialites in the shallowest parts (Figure 20c). At Windjana Gorge, abundant siliciclastic sediments from the proximal coast to the north interfinger with the microbe-dominated facies. The proximal fore-reef slope community is similar to that of CF3 with some remaining laminar stromatoporoids and lithistid sponges and abundant calcimicrobes in stromatactis-rich microbial facies [236]. Small sponge–rugose coral mounds occur in the fore-reef slope in the Napier Range [282] and some of these small mounds continued into the Famennian [397,402].

Carbonate Factory 5 represents the prograding Famennian Nullara cycle [88] after the Frasnian–Famennian extinction 'event' where most, but not all, stromatoporoids were removed from the DGBR [88,389]. Stromatoporoid abundance was already decreasing in CF4 [402,416] and CF5 intergrades with CF4 to a large degree, rather than representing a sharp boundary. CF5 margins were still abrupt (Figure 26e) and dominated almost entirely by microbes with diverse calcimicrobes, but calcareous algae, encrusting foraminifers and problematica also occurred [236,388,392,408]. Microbialites include micritic (i.e., non-skeletal microbialites [158]) microdigitate thrombolites (clot-shaped) and stromatolites (laminated) forming bedded and mounded structures and calcimicrobes are dominated by *Renalcis*, *Rothpletzella* and *Girvanella* (Figure 26h). Calcareous algae include both solenoporoids and *Parachaetetes*. Minor laminar stromatoporoids still occurred in the margin, at least in the lower Famennian [236,408], but their role in the framework was incidental. Lithistid sponges were more abundant. Minor bryozoans, encrusting foraminifers and problematica contributed to very diverse local factories, and account for at least four major lithotopes, some of which were similar to those in CF4 and the most dominant of which is a *Uralinella–Renalcis–Rothpleztella* bindstone [236]. Most frameworks would be classed as micritic microbialite or mixed calcimicrobial–micritic microbial frameworks. In all cases frameworks were massive, very porous and early marine cements were important and helped maintain relief on the high-energy margins [236]. Individual surfaces were low-relief mounds and lenses (Figure 26f), in many cases interspersed with lenses of siliciclastic sand [236] and low-relief, elongated mounds–buttresses probably extended from the margin onto the proximal fore-reef slope [266,408].

Reef flat facies graded back from the margin with decreasing cemented framework, but a columnar–fenestral stromatolitic bindstone facies is common [236] (Figure 26g). These columnar fine-grained stromatolites contain abundant carbonate or siliciclastic sand, but only minor calcimicrobes. Similar, but larger mounds occurred in Frasnian back-reef settings where there was abundant siliciclastic influence [285]. The discrete stromatolites also graded in places into more irregular thrombolitic textures, but cavities were still abundant and could be partly filled by *Renalcis* [388]. Other major members of the carbonate factory include the microbes that formed peloids and micritized grains and possibly the abundant ooids. Large megalodontid (Figure 19d) and lithiotid bivalves and gastropods also occurred [236]. The coarse, well-winnowed grainstones were dominated by microbial peloids, ooids, grapestones (grains made of several attached ooids), mud lumps, rounded intraclasts and minor skeletal debris and occur in discrete erosive channels that traverse the margin facies [282] or fill topography between stromatolite heads [267,388]. Back-reef facies are dominated by oolitic and intraclastic grainstones with many intraclasts themselves being oolitic grainstone, and agglutinated stromatolites have been observed [417]. Landward, a variety of peloidal, pisolitic and microbial grainstones and packstones occur with minor skeletal contributions from bivalves and gastropods. They commonly contain fenestral fabrics and interfinger with terrigenous siliciclastic sediments at the coast [236]. High-frequency Famennian cycles are recognized in slope deposits [266] and back-reef (Figure 20c) [267] but remain poorly known. Near the coast the carbonates interfinger with siliciclastic sediments [236], and siliciclastic sediments commonly extend to the margin [388], but the degree to which a given instance represents climatic, tectonic or eustatic forcing is poorly constrained. Regardless, the development of shoaling grainstones and microbial fenestral fabrics along with desiccation cracks observed in back-reef facies [417] are consistent with Milankovitch band cycles as in the Frasnian.

Fore-reef slope facies are basically continuations of the margin facies [236,266]. They consist of mounded microbial boundstones [408] (Figure 26f) with grainstones derived from the platform top and coarse debris generated from occasional margin collapse and smaller scale brecciation on the slope itself [266]. Some of the grainstones are turbidites [282]. The dominant in situ carbonate factory consists of the marginal *Uralinella–Renalcis–Rothpleztella* bindstone lithotope [236]), but brachiopods, gastropods and crinoids occur as in situ dwellers [402]. Cycles of platform collapse occurred throughout the sequence [90]. Distal slope boundstones and grainstones interfinger increasingly with well-bedded silty peloidal wackestones to packstones [266].

The destroyer guild is poorly represented in the DGBR [31], but massive stromatoporoids contain scattered *Trypanites* and *Gastrochaenolites*-like borings. The abundant micritized mud-lump grains in back-reef settings also presumably represent endolithic cyanobacteria and green algae, but micrite rims are not prominent on coarser skeletal allochems. Apart from the carbonate factories, the DGBR had extensive biodiversity amongst its non-carbonate producing dweller guild. Although many of these types of organisms are rarely preserved in shallow reef facies, a significant Frasnian lagerstätte developed in the deeper inter-reef environment, especially well-exposed in the Gogo Formation in Bugle Gap [418,419]. This inter-reef basin deposit provides significant insights into the co-occurring DGBR dweller guilds. Early diagenetic carbonate concretions grew around organic remains just below the seafloor in deep shelf pelagic facies preserving everything from conodonts, radiolarians [420] and arthropods [421], including eurypterids [422], to fish [256]. Like the GBRP, the DGBR was host to a diverse range of fish with more than seventy species identified in the Gogo Formation [418,423]. Fish teeth and scales also occur in the shallow platform facies [274], but pale in comparison to the lagerstätte. More typical pelagic carbonate fossils (ammonoids, nautiloids) (Figure 19i) are also preserved, but the vertebrate preservation is especially impressive, including three-dimensional preservation of delicate skeletons and preservation of many different types of soft tissues [424,425]. The deposit has revolutionized our understanding of early fish biology and evolution [418]. It is probable that the deposits formed during transgression-related Early Frasnian basinal anoxic episodes [423] that also preserved the organic matter responsible for oil generation from the Gogo Formation [233]. Localized anoxia also explains the lack of benthic scavenger activity that allowed articulated skeletons and soft part preservation. Devonian source rocks in the Gogo and younger Fairfield formations formed owing to eutrophication during transgressions associated with very high primary productivity from planktic algae [233].

Although it is unclear whether the nutrients that drove the productivity were derived from terrestrial runoff or more from upwelling from the deepening Fitzroy Trough over the entire age range of the Gogo Formation, biomarker studies are consistent with terrestrial pulses leading to photic zone euxinia at times and lower-water-column anoxia [426]. Hence, few of the larger preserved organisms in the Gogo nodules were local benthic fauna as the bottom waters were anoxic [421]. Most of the preserved fish were pelagic taxa and some other fauna may have been washed into the setting from shallower benthic settings, such as the eurypterid [422]. There are also ten genera of lungfish [423], which were adapted to stirring up shallow muds in a well-oxygenated setting [419]. Thus, the deposit is in some ways similar to the Jurassic Solnhofen Lagerstätte in Germany where shallow and even terrestrial taxa (e.g., the vertebrate *Archaeopteryx* and plants) were washed into deeper anoxic lagoons protected by sponge–microbial and coral reefs [427,428]. This setting allowed sampling from a much broader geography than the depositional environment itself [429].

3.4. Comparing Reef Geomorphology and Palaeogeography

Both the GBRP and exposed portion of the DGBR developed on continental shelves proximal to terrestrial terrigenous sources but with some reefs isolated on offshore fault-controlled basement highs (either exposed or subsurface) that formed during rifting of continental crust. Hence, they both represent mixed carbonate–siliciclastic settings with shore attached and more isolated, cleaner carbonate environments. Both reef provinces are confined to continental crust and true atolls on oceanic crust are absent. A few modern reefs occur on oceanic seamounts in the Coral Sea just east of the GBRP as part of the Tasmantid Seamount chain (e.g., Cato, Wreck, Kenn reefs) [430], but they are not part of the GBRP itself. Fringing reefs occur in both settings attached to the mainland and around continental islands composed of crystalline basement; barrier reefs isolate lagoons behind them, with isolated platform reefs occurring amidst deeper shelf settings. Both complexes faced deeper basins seaward, the Coral Sea, a true ocean, for the GBRP and the Fitzroy Trough, an epicontinental sea, for the DGBR. Owing to the size and complexity of the two provinces comparisons must be made at different scales. At the broadest scale (i.e., carbonate platforms) the two provinces vary considerably. The carbonate banks of CF1 and possibly CF2 in the DGBR do not have close analogs in the modern GBRP but may be similar to some of the Miocene platforms developed on the Marion and Queensland Plateaus or attached to the continent farther north (Figure 12). Equally, the vast prograding CF5 Famennian platform of the DGBR has no close analog in the GBRP. Regardless of those difference, where platforms occur at the edge of major relief, such as the ribbon reefs of the GBR facing the Coral Sea or the Frasnian and Famennian platforms of the Napier Range facing the Fitzroy Trough, great linear reef belts were formed (Figure 27a). However, the best DGBR analogs for modern GBRP reef geomorphology are the Frasnian CF3 shelf edge reefs (Figure 27b) and isolated platform (i.e., 'pinnacle') reefs (Figure 27c) surrounded by deeper shelf depths that formed during back-stepping. These platform-scale similarities and differences relate to the abilities of the carbonate factories to fill accommodation through time and are discussed below.

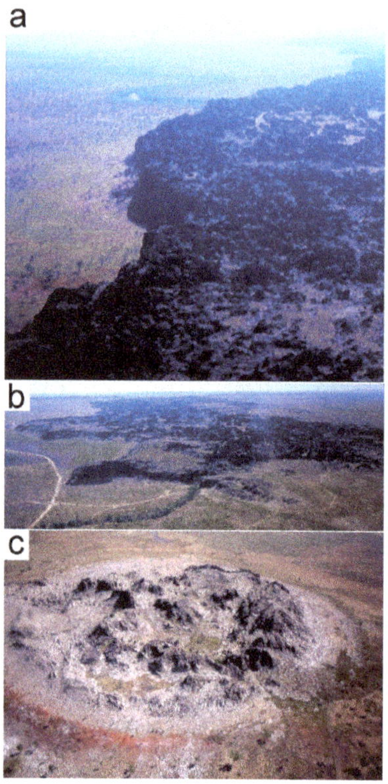

Figure 27. Aerial views of reef geomorphology in the Devonian Great Barrier Reef. (**a**) View of reef looking west along Napier Range near Windjana Gorge formed by the Lennard River (seen in distance); (**b**) view looking north at southern Lawford Range showing irregular shape of reef platforms; (**c**) view of isolated platform or pinnacle reef of Lloyd Hill, reef is ~600 m in diameter.

At smaller spatial and temporal scales, intra-reef geomorphology is far more comparable as the very different carbonate factories reacted in similar ways to the same types of coastal processes. Sedimentologic evidence documents many of the same types of coastal processes in the DGBR that are well known in the GBRP. The DGBR includes evidence for coastal erosion, waves, storms, tides and currents. The Middle to Late Frasnian (CF3) Mowanbini Archipelago (Oscar Range) provides a good example of Devonian reef palaeogeography [92,290,292,293] (Figure 28). The current Oscar Range stands in relief on the modern landscape as it is composed of very resistant siliceous Proterozoic metasediments and metavolcanics. During the Late Devonian virtually identical relief (90 m) formed islands that were buried progressively through time by the Frasnian reef complex [89]. Today, the upper part of the reef complex (Famennian and latest Frasnian) and overlying strata have been removed by erosion in places to expose a nearly coeval Frasnian paleogeography from the basin to the south across the reef fringed continental island and carbonate shelf to the mainland continent to the north. Examples of fringing and barrier reefs, high-energy and more protected, lower-energy lagoons, and high- and low-energy rocky shores are all well exposed. Although the archipelago was larger than Lizard Island on the GBR (Figure 9d), that too represents a group of continental islands connected by fringing and barrier reefs to enclose a shallow lagoon [431] (Figure 29). Lizard Island is surrounded by shelf depths within the broader GBR lagoon. The Mowanbini Archipelago, o the other hand, occurred on the edge of, and facing, a deep basin, the Fitzroy Trough to the south

(Figure 28a–c). The western part of the largest island is encased by a fringing reef where reef limestone occurs directly against the metamorphic basement (phyllite) that makes up the low-lying southern part of the island (Figure 28d). Just to the east the reef separated from the island to become a barrier reef and a shallow, high-energy lagoon opened up behind the margin. A drainage flowing off the elevated island reached the coast near this point and fluvial sediments were reworked into a small shoal-water delta [292] (R1 in Figure 28a,e). Modern sediments being delivered by a stream most likely occupying the same channel in the very hard basement is compositionally the same as the sediment in the ancient delta, but there the sediment is better sorted, more rounded and occurs with grain-oriented fabric sloping seaward consistent with deposition on a beach (Figure 28f). The reef margin itself was not very wide and was dominated by typical CF3 *Renalcis* and stromatoporoids. The reef flat to back-reef shows well-developed cycles wherein the delta and other coastal siliciclastic sediments interfinger with tidal flat facies near shore, but pass into high-energy cycles just behind the margin with domal stromatoporoids and *Disphyllum* biostromes in the deeper parts of cycles overlain by reworked stromatoporoids, oncoids and ooid shoals in the highest energy sections (Figures 17h and 21b,c) [292].

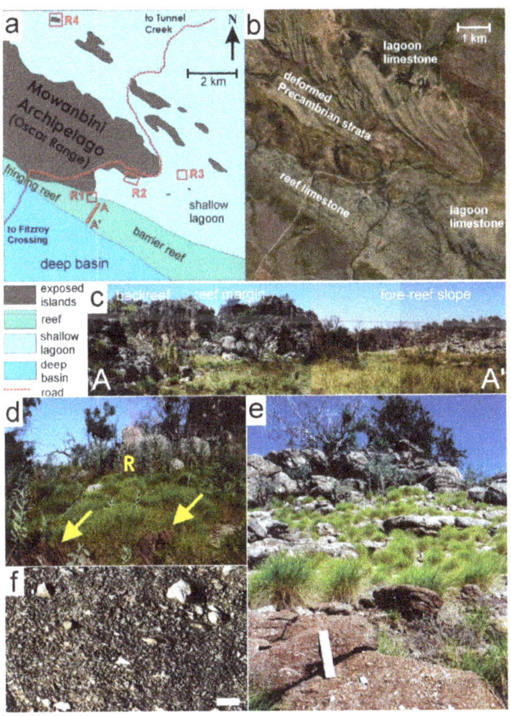

Figure 28. Geomorphology of the CF3 Mowanbini Archipelago, Oscar Range. (**a**) Paleogeographic reconstruction showing location of section A–A′ shown in (**c**) and areas R1—shoal-water delta in back-reef [292], R2—well-exposed rocky shore in lagoon [290], R3—shallow lagoon [290], R4—Christopher Bore Island [293]; (**b**) aerial photograph, Nearmap © 2025, showing part of area in (**a**); (**c**) photo montage across line A–A′ showing transition from back-reef (north) to fore-reef slope (south) with near original dips; (**d**) position in western side of R1 where reef limestone (R) occurs directly on Precambrian basement (phyllite at arrows) making it a fringing reef; (**e**) view in R1 looking south across shore-attached shoal-water delta and encroaching reef limestone, scale in foreground is 15 cm; (**f**) shore-face conglomerate of shoal-water delta showing seaward-sloping grain-oriented fabric, scale = 2 cm.

Figure 29. (**a**–**e**) Rocky shore environment, Mowanbini Archipelago, area R2 from Figure 28; (**a**) air photograph, Google Earth, Digital Globe 2010, showing steeply dipping metamorphic rocks of island to north and ~flat-lying lagoonal limestone beds to south, inset (white box) is shown in (**b**), scale = 100 m; (**b**) reconstruction of rocky shore, after [290] where lagoon sediments lap onto quartzite sea stacks (orange), longshore current reconstructed from clast orientation along shore, red zone near sea stacks is a wave-worked shore-face scree breccia; (**c**) view of large quartzite ridge and small sea stacks (Q) with limestone (L) lapping up from south; (**d**) polished slab of shore-face breccia showing grain-oriented fabric, scale = 1 cm; (**e**) offshore biostrome with in situ irregular stromatoporoids, scale = 15 cm; (**f**,**g**) comparative photographs from Lizard Island, Great Barrier Reef (sourced from M.E. Johnson); (**f**) view of reef connecting islands south of Lizard Island around a central lagoon; (**g**) view of granite sea stacks on shore looking north towards Lizard Island from Palfrey Island.

To the north along the edge of the lagoon the island has higher elevation held up by quartzite and metasediments. The southernmost quartzite ridges formed a prominent rocky shore to the lagoon [290] (Figure 29a–d). The well-bedded, nearly flat-lying lagoon sediments lapped up around individual sea stacks and immediately proximal beach sediments consist of seaward sloping grain-oriented beach breccia made up mostly of proximal quartzite clasts (Figure 29c,d). Although wave energy was not adequate to round the very hard quartzite clasts, its distribution and the orientations of toppled elongate stromatoporoids offshore showed consistent longshore current to the east, consistent with waves breaking over the margin to the south and onshore directed winds expected from heating of the continental mass to the north [290]. Smaller islands within the lagoon generally provided smaller sediment input to the shore-face, but longshore current directions were consistently to the east on the south sides of the islands. The shallow lagoon supported domal stromatoporoids as well as *Stachyodes* and *Amphipora*, but the last two taxa occur only as transported debris. Oncoids also occur in some horizons as well as abundant gastropods, which themselves became the nuclei for oncoid growth (Figures 18h and 30a). Around the eastern end of the large island (Figure 28a, R3), flat bedded cyclic lagoonal facies continue with some high-energy shoals (ooids and oncoids) suggesting high hydrodynamic energy was funneled between the island groups (Figure 30b).

Figure 30. (**a**,**b**) Mowanbini lagoon deposits, from area R3; (**a**) lagoonal facies with abundant gastropods and oncoids suggestive of algal turfs and high energy, scale bar = 2 cm; view to east above flat bedded grainstone shoals in foreground towards paleoisland rising in the background; (**c**–**k**) protected paleoisland in protected shallow lagoon north of major island near Christopher Bore [293], area R4, with common fauna; (**c**) near flat lying lagoonal limestone beds leading up to low island in background (yellow arrows); (**d**) quartzose metaconglomerate ridge near edge of small island; (**e**) abundant, well-preserved but wave-worked *Amphipora* from near shore environment, scale bar = 2 cm; well-preserved, but broken *Stachyodes*, scale bar 1 cm; (**g**,**h**), common gastropods from near-shore settings, scale bars = 5 mm; (**i**) scaphopod from near shore setting, scale bar = 1 cm; (**j**) bivalve-dominated 'shell hash', western margin of paleoisland, scale bar = 3 cm; (**k**) isolated domal to columnar stromatoporoids in shallow lagoon with abundant siliciclastic sand from paleoisland, scale = 15 cm.

Across the large island to the north a small island (Figure 28a R4) crops out near Christopher Bore [293]. The island is composed of Precambrian metasediments, including metaconglomerate and phyllite, similar to the larger island to the south (Figure 30c,d). The shallow protected lagoon had a diverse fauna of *Amphipora*, *Stachyodes*, gastropods and bivalves and even scaphopods (Figure 30e–j). Although wave-worked and distributed, many of the *Amphipora* are well preserved. Farther offshore domal stromatoporoids occur in growth position within bioclastic grainstones, floatstones and fenestral algal laminites (Figure 30k). The sediments are commonly partially dolomitized. Although waves from the southeast dominated the protected environment, storms and larger waves occurred sporadically as indicated by toppled stromatoporoids and intraclasts in the sediment. Siliciclastic sediment also eroded from the island to be intermixed with the shore-face carbonates. Although detailed biofacies analysis has been done [293], the nature of parasequences cannot be explored owing to the very low dip of the strata and flat outcrop.

4. Discussion

Comparisons between modern and ancient reef systems suffer from biases inherent in studying the two different types of deposits (i.e., living communities with poorly accessible stratigraphy versus scattered sections of stratigraphy and paleogeography with varying degrees of preservation bias). Modern reefs are much better known in terms of everyday processes, both biological and geological, from seasonal to longer-term, stochastic event scales that drive intermediate disturbance [110]. The knowledge represents a very limited time span in the history of a carbonate system (i.e., stratigraphy). Almost all we know about modern coral reefs represents only the surface of a single ~10 kyr interglacial episode within the broader 100 kyr+ cyclicity that has driven modern reef growth over the last 500+ kyr. That knowledge has been supplemented by coring, and seismic surveys, but we had little understanding about what happened to the GBR in the majority of the time between interglacials until IODP Expedition 325 drilling in 2010 [46,288,432] (Figure 10). That said, the extent of the modern GBRP across climate zones causes the biodiversity and reef constraining factors (temperature, cyclone frequency, terrestrial runoff, etc.) to vary considerably along its length. Generalizing about the modern GBR as a single entity carries significant risk [30]. Although we now understand that the modern GBR formed over the last several Late Pleistocene glacial cycles, we know surprisingly little about the older Tertiary and Quaternary reef systems in the subsurface farther north. However, isolated modern reefs are growing on carbonate platforms that initiated in the Miocene if not earlier in the northern part of the province [201,211] (Figure 11). There may have been breaks, but the >20 million year history of the GBRP is of the same scale as the >20 million year history of the DGBR. We just have very little knowledge about most of it.

In studying fossil reefs, even the incredibly well-exposed DGBR, we have the inverse problem. We have detailed knowledge about aspects of its variation through stratigraphy (e.g., for fore-reef slopes [265,266]) with only scattered snapshots of a very complicated paleogeography for some intervals and places as allowed by existing exposure and exploration drilling (e.g., the Mowanbini Archipelago [292]). More detailed paleogeographic and paleobiological investigations are hampered by the vagaries of exposure and erosion, difficulty in correlating between different exposures and the sheer amount and age range of existing outcrop. Most biostratigraphic correlation for the DGBR is based on conodonts and ammonoids [272,273], which are restricted mainly to deeper slope and basin environments. Correlation of individual sections of shallow platform facies is still difficult, although microvertebrates [274] and chemostratigraphy [269] will help in time. Even assuming adequate correlation, the range of geographic settings provides similar problems as in the GBRP and makes it unwise to overly 'lump' the DGBR into a single 'reef' [283]. Sequence stratigraphy is very useful at different scales. The supercycle concept [88] and attempts to break the succession into shorter temporal units with different carbonate facies or factories [236,267,283] as attempted above, still leaves uncertainties as to what represents a purely temporal or spatial change in some cases (e.g., CF1 and CF2 margins). Matching to higher-frequency cycles—parasequences, e.g., [275,277] could provide very high temporal resolution but such cycles are notoriously difficult to correlate between platforms. Additionally, the exposed DGBR represents but a small part of a much larger reef system, of which we know very little as most of it is in the subsurface (Figure 12). Current exploration on the Barbwire Terrace may elucidate the reefs on the other side of the Fitzroy Trough, but existing cores appear to intersect less informative back-reef environments [233]. Despite those issues, the reef provinces described above can be compared meaningfully using the concepts of key processes, functional guilds and unified jargon (see introduction). Devonian reefs have things to tell us about modern reefs and vice versa.

4.1. Comparing GBRP and DGBR Reef Builders

Little if any overlap exists in the organisms that built these two reef systems (Table 1). Corals occurred in both provinces, but the scleractinian corals of the GBRP are most likely no more closely related to the rugose and tabulate corals of the DGBR than to modern non-skeletal anemones, although discussion continues, e.g., [433]. However, it is a testament to evolutionary convergence how similar many of the morphologies are between the different orders of corals as the skeletal needs of relatively simple cnidarian polyps are highly constrained [434]. The different coral orders also played similar functional roles in many cases in their respective carbonate environments [31]. Some of their morphologies, especially for the tabulates, include features suggestive of photosymbiosis, as with modern scleractinian corals [72,410,435]. However, it is likely that the trophic structure was quite different between the two reef provinces (more on that below). Hence, corals filled constructor roles in both provinces with massive forms, such as modern faviids and ancient *Argutastrea* (Figure 31a,b), and branching forms, such as living *Acropora* and the extinct tabulate *Thamnopora* (Figure 31d,e) occupying similar roles in some cases. Corals were important constructors in older Middle Devonian reefs although, there too, they were generally, but not invariably [436], subordinate to stromatoporoids [84]. However, corals in the Late Devonian DGBR were not major reef builders and appear never to have dominated shallow margins as do modern scleractinian corals. Rather, they were best developed on platforms in the deeper phases of subtidal parasequences (CF1–3), at platform margins in some places (CF1), or less abundantly in more protected back-reef settings and lagoons where margins had aggraded to sea level (CF3). They became rare or absent in CF4 and CF5. Platy stromatoporoids and alveolitid tabulate corals formed overhangs on fore-reef slopes in CF1 to CF3, so provide morphological analogs for some of the deeper platy corals in the GBRP fore-reef (Figure 31h,i), but they did not reach the level of benthic cover that modern corals do and generally were not self-encrusting although limited encrusting stromatoporoids and tabulate corals occurred.

Figure 31. Functional analogs between modern Great Barrier Reef (GBR) constructors and those of the Devonian Great Barrier Reef (DGBR). (**a**) Modern *Goniastrea* colony, ~20 cm in diameter; (**b**) small *Argutastrea* corallum with similar corallite size and structure; (**c**) columnar stromatoporoid from Christopher Bore paleoisland, which with other stromatoporoids are functional analogs for massive *Porites* colonies in the GBR, scale bar = 5 cm; (**d**) modern branching *Acropora* in shallow ponded reef flat, Heron Reef, GBR; (**e**) in situ *Thamnopora* rugose coral in reef flat setting, Lloyd Hill, scale bar = 10 cm; (**f**) modern crustose coralline alga encrusting live *Acropora* branch where it is actively overgrowing and killing the polyps, Heron Reef, GBR, scale bar = ~5 mm; (**g**) in situ branching *Stachyodes* encrusted by layers of *Renalcis*, proximal reef flat, Lawford Range, scale bar = 1 cm; (**h**) modern seaward-sloping, platy acroporids, fore-reef slope, Heron Reef, GBR; (**i**) seaward-sloping tabular stromatoporoids, fore-reef slope, Lloyd Hill, scale bar = 10 cm.

The best functional analogs for scleractinian corals in the DGBR are the stromatoporoids, with encrusting, laminar, tabular, massive and branching morphologies (Figure 17a–e) that align closely with those of modern scleractinians and to a lesser extent with their ecological roles. However, even there, there are key differences. The vast majority of massive stromatoporoids in the DGBR were constratal tabular to low-domal forms and many if not most lived free on the sediment substrate without encrusting it or each other. They do not seem to have required a hard substrate to encrust for long-term recruitment as do modern reef-building scleractinian corals [437]. Platestone facies with loose stromatoporoids in growing position are abundant on some margins, reef flats and slopes that were not well cemented (CF1,2) and occurred on grainstone-dominated slopes of CF3 in places as well. Although some stromatoporoids reached very large dimensions [388], they were still basically constratal and no analogs for the very large multi-meter, free-standing *Porites* colonies of modern reefs have been noted. Branching *Stachyodes* was suprastratal and very similar in morphology to some of the less robust branching *Acropora* in the GBR; they generally did not reach the same size as many *Acropora* colonies but occurred in very similar fore-reef and back-reef settings and were responsible for similar beds of debris in those settings (Figure 32e,h). Hence, apart from CF4 and CF5 in the DGBR, the skeletal constructor guild was well developed in both provinces. The micritic microbial constructor guild was well developed in the DGBR, with micritic stromatolites abundant on deeper slopes throughout the Late Devonian but also building frameworks in shallow margin settings and reef flats in CF4 and CF5 (Table 1). Although less well-known, the benthic micritic microbialites (some stromatolitic) that formed in deeper reef slope environments of the LGM and deglacial reefs of the GBR [170] are analogous but appear not to have made benthic framework in shallow settings. Massive benthic structures built by calcimicrobes (*Renalcis*) in a constructor functional role were also abundant in the DGBR through CH3 and CH4 but were lower in relief in CF5 (Figure 26h). In most cases, they occurred with significant amounts of early marine cement. Such calcimicrobes have no real analog as constructors in modern reefs. Modern CCA (the dominant functional binder today) cross over as constructors on some high-energy oceanic reef margins [156]. Those constratal CCA crusts that reach great thicknesses in those settings (e.g., >5 m in Martinique [438]) may be analogous to laterally extensive calcimicrobial crusts like *Rothpletzella* (Figure 18c) or constratal stromatoporoids on margins in CF2 (Figure 24), but such thick CCA crusts are more common in open ocean reefs than in the GBR.

Regardless, CCA are the dominant skeletal binders in Holocene reefs, but have no skeletal analog in the DGBR, where encrusting types of tabulate corals, stromatoporoids and calcareous algae were uncommon. DGBR calcareous algae [392] generally did not form lateral crusts, occurring rather as small, commonly isolated, thalli (e.g., various solenoporoids) in more complex microbial frameworks [236,408]. They are best known from CF4 and CF5, but that could partly reflect their small size or scarcity in earlier factories, or the difficulty in recognizing them in frameworks containing other abundant skeletons and cement. They stand out better in more micritic frameworks. The dominant binders in the DGBR were the calcimicrobes, especially *Renalcis* (CF3–5) and micritic microbialites, both stromatolitic and thrombolitic (CF4,5). *Rothpletzella* was also common in shallow settings in CF5. In both cases, early marine cements also were very abundant [88,236], enhancing the binding role and providing significant strength to the framework. Whereas modern highstand reef frameworks in the GBRP are basically skeletal with a minor, but important component of cryptic microbialite [176], the lowstand and deglacial frameworks have much greater volumes of microbialite [132,170]. Although still largely cryptic in occurrence, in those cases, it overlapped with the constructor guild. Frameworks were variable in the DGBR [236], but were basically lacking in CF1, which primarily built low-relief banks and ramp margins. Thickened margins occur (e.g., Menyous Gap), but were formed from microbial sediments with abundant, but non-encrusting stromatoporoids in platestones (stromatoporoid bank margin facies) (Figure 22). CF2 constructed fully skeletal, but constratal, frameworks consisting almost entirely of tabular stromatoporoids (Figure 24). CF3 has the greatest diversity of framework types ranging from skeletal frameworks composed of tabular stromatoporoids with minor *Renalcis* and cement to a variety of mixed skeletal–calcimicrobial frameworks with stromatoporoids, *Renalcis* and abundant cement. These frameworks vary between laminar and tabular stromatoporoids with *Renalcis* and cement on margins to encrusted branching *Stachyodes*, more common on the slope and reef flat (Figures 25e and 31g) (and see [236]). Fully microbial frameworks with varying percentages of micritic and calcimicrobial components and minor but locally important skeletal builders like laminar stromatoporoids or calcareous algae round out the framework types in CF4 and CF5.

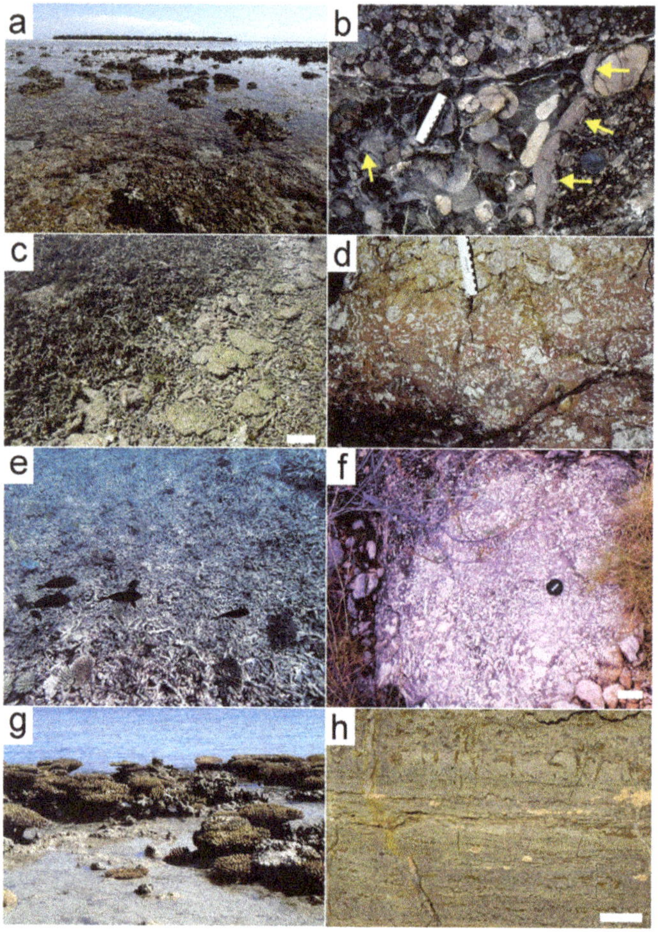

Figure 32. Evidence for similar responses to hydrodynamic energy and tides within carbonate facies of both reef provinces. (**a**) Large blocks of reefrock (to > 1 m diameter) thrown up by storm on leeward reef flat, Heron Reef, GBR c. 2013, Heron Island in background; (**b**) well-rounded boulders in Late Devonian fanglomerate, Painted Rocks, DGBR, scale bar = 15 cm, note stromatolites growing on and between boulders (yellow arrows) showing degree of coastal winning associated with common high-energy coastal events; (**c**) rubble subzone of reef margin juxtaposed against live corals on reef flat, Heron Reef, scale bar = 15 cm; (**d**) coarse reef rubble in proximal reef flat at Mowanbini reef margin (Figure 28a, area R1); (**e**) branching *Acropora* rubble on fore-reef slope after storm, c. 2013, Heron Reef; (**f**) branching *Stachyodes* debris (rudstone) on fore-reef slope, Wagon Pass, scale bar = 10 cm; (**g**) exposed corymbose and tabular *Acropora* on reef margin at spring low tide, Heron Reef (corals only withstand so much exposure during spring low tides and at accompanying high tides are submerged by more than 2 m of water); (**h**) back-reef tidal flat facies at area R1 showing fenestrae and desiccation cracks consistent with low tide exposure at Mowanbini, consistent with high micro- to mesotidal ranges as in the GBR, scale bar = 2 cm.

The baffling guild includes branching taxa, such as *Acropora* and *Stachyodes*, which presumably baffled sediment in their respective provinces, but bafflers similar to modern *Halimeda* are difficult to identify in the DGBR. A possible candidate is the delicate branching stromatoporoid *Amphipora*, which is typically preserved as fragments a few mm in diameter and a few cm long. Although not known to be modular, like *Halimeda*, the skeleton was fragile and when erect, would have baffled currents in protected back-reef areas but became a major sediment source when broken. It is assumed that fleshy brown and red algae, such as are common in the GBRP, would have been present in back-reef settings of the DGBR, but there is little evidence to confirm it as such taxa do not fossilize well. From a theoretical basis [146], settings with high terrestrial sediment and nutrient flux should have favored such algae, and some lagoons had abundant grazing gastropod faunas and oncoids (Figure 30). Fleshy algae and other soft algal turfs may have occurred in inshore settings throughout the Late Devonian but may have been more important in the terrigenous-rich, regressive CF4 and CF5 facies, such as in the Napier Range. It is also probable that high nutrient loads would have favored micritic microbialite deposition at those places and times [439] and calcimicrobes also may have required elevated nutrient flux [440] (see below).

Easier to constrain are the bioeroders. Microborers (e.g., cyanobacteria, fungi) are known since the Archean and larger metazoan borers in carbonate substrates already existed in the Paleozoic, but increased dramatically in the Mesozoic [35,97,441,442]. Lithophagid clams appeared in the Eocene, but large *Gastrochaenolites* (bivalve borings) were already abundant in rocky shore settings by the earliest Pennsylvanian [443,444] and similar, but smaller, examples have been noted in the DGBR. Acrothoracican barnacle borings already occurred in the Devonian as well as *Trypanites* [444] and as many as eleven different ichnogenera were documented in Upper Devonian reefs [445]. Of those, *Trypanites* is the most abundant form in the DGBR [387] but is not responsible for significant carbonate removal where it occurs. Although boring sponges existed in the Devonian [446], they were not pervasive and the earliest definitive clionid sponge style borings (*Entobia*), which are pervasive in modern reefs, did not occur until the Jurassic [444]. Raspers are more difficult to evaluate. The abundant gastropods in back-reef settings in the DGBR presumably grazed on algal turfs, but their ability to rasp and excavate hard substrate has not been demonstrated. Polyplacophorans (chitons) were present in the Cambrian, but their rasping abilities through the Paleozoic also remain unknown and they have not been documented in the DGBR. The abundant modern rasping groups that reduce hard substrate to sand (scarid fishes and echinoids) expanded or appeared only in the Mesozoic [35]. The increased grazing pressure that they then applied may have disfavored the less well-calcified solenoporoid algae, such as those in the DGBR, leading to the rise of the better calcified CCA that dominate binding in reefs today [447].

The dweller guild was well developed in the carbonate factories of both provinces, but with major differences. Dwellers of the modern carbonate factory are dominated by mollusks, *Halimeda* and LBFs. Among mollusks, gastropods and bivalves were well represented in the DGBR back-reef, but benthic mollusks were subordinate to brachiopods in most settings. Nektic nautiloids and goniatites were abundant, commonly preserved in deeper inter-reef settings throughout the DGBR factories (Figure 19i). Of those, only nautiloids (the chambered *Nautilus*) remain today in the GBR where they occupy deeper off-reef settings. Bryopsidales like *Halimeda* were not abundant in the DGBR, but small solenoporoid algae were abundant in CF4 and CF5. Significant LBFs had not yet evolved and only small encrusting foraminifers occur in DGBR frameworks [236], where they were far less abundant than are the common cryptic encrusting foraminifers in the GBR today (e.g., *Homotrema*). Among echinoderms, crinoids were more abundant and major sediment producers in the DGBR, whereas they are very minor sediment producers in the GBR today, overtaken among the echinoderms by asteroids and echinoids. Non-carbonate producing dwellers are relatively poorly known in the DGBR, except for the great diversity of fish and arthropods preserved in the Gogo Formation Lagerstätte [419]. Although many preserved taxa were pelagic, some clearly came from well-oxygenated benthic settings on the reef margins and platforms [419]. The microvertebrate record also suggests diverse fish, including chondrichthyans, in shallow reef settings [274]. The in situ bottom fauna of the Gogo Formation was restricted, owing to low oxygen, with most preserved taxa introduced from shallower reef settings [421]. Regardless of the differences, a swim along any of the Frasnian reefs of the DGBR would probably have provided a recognizably 'reefy' experience for a modern reef goer. It is less clear how similar the experience would have been swimming along a Famennian microbe-dominated reef margin.

Modern tropical reefs are dominated by 'photozoan' assemblages where photosymbionts enhance biomineralization in typically oligotrophic settings [146,448]. As both reef provinces described above were constructed by tropical factories [145] it is easy to assume that they are both driven by photoautotrophs restricted to relatively shallow settings. However, the trophic structure of the Late Devonian factory is distinct from the modern one. For example, tabulate and rugose corals may, or may not, have had photosymbionts based on photoresponsive morphology, phenotypic plasticity and ecological distributions [410,449–452] but there is no clear consensus, and they were not major frame-builders in the DGBR in any case. Stromatoporoids, on the other hand, were important in DGBR frameworks but were filter feeders with no clear evidence of photosymbionts [453,454]. Reef-building calcimicrobes may have been mixotrophs rather than obligate photoautotrophs [455] and many occur in both exposed and clearly cryptic settings. Hence, the effective depth range of binding was greater in the Late Devonian owing to the greater bathymetric ranges of calcimicrobes and micritic microbialites. Although modern red algae can live deeper than 200 m [456], CCA are not abundant in GBR reef frameworks below ~6 m [217]). Devonian calcimicrobes, especially *Rothpletzella*, extended to depths where they occurred with toe-of-slope 'mudmound' sponge–microbialite factories that are less dependent on light [149], but they did display phototropic behavior, commonly growing vertically upward even from steep slopes (Figure 18c). Hence, although the DGBR was a tropical reef system, we cannot make standard assumptions based on modern reefs as the trophic structure was certainly different, even if not well understood. Some Cenozoic tropical carbonates also had different trophic structures than classical photozoan communities, such as more nutrient-rich SE Asian reefs that have more heterozoan content [457]. Those types of communities may have been more typical of the Miocene carbonates of the GBRP.

4.2. Tectonic Constraints on Reefs

Although both reef provinces formed in intraplate settings with geologic structures controlled the distribution of reef growth, there are major differences in tectonic forcing. Northward movement of the Australian Plate restricted the oldest growth in the GBRP to the north, where it currently rests in the subsurface (Figures 11 and 12). The latitude-parallel orientation of the currently exposed portion of the DGBR changed little due to plate movement through the Late Devonian [210]. That latitudinal stability through time provided it with a more uniform stratigraphy along its exposure, but where it extends into the subsurface far to the north and south it would have encountered different climate zones. An example is the more arid and restricted setting to the south across the Fitzroy Trough on the Barbwire Terrace (Figure 13), which may have been affected more by the subtropical high-pressure zone. At shorter temporal scales, the two provinces differ greatly in the degree of synsedimentary tectonism. Although the GBRP experiences minor reactivation of old structures with little seismic activity the Fitzroy trough was actively rifting during deposition of the DGBR and this caused both subsidence, tilting and uplift of active carbonate platforms with major effects on local stratigraphy [280,283–285]. Earthquakes associated with movement on major faults also likely initiated at least some of the significant margin collapse episodes made so evident by debris in the slopes [265,402].

4.3. Climatic Constraints on Reefs

Global climate (i.e., greenhouse versus icehouse) constrained both reef provinces but also provides an overarching factor to differentiate them. The earliest carbonate deposits in the GBRP, high-latitude Eocene carbonates (Figures 11 and 12), were able to form only owing to the very warm 'Paleocene–Eocene Hothouse' climate [96] near the end of the last global greenhouse climate phase (Figure 2). The shift to icehouse climate in the Oligocene shut down most carbonate production [83] before Miocene warming helped re-establish it at more tropical latitudes. Major late Miocene to Pliocene cooling then greatly restricted Pliocene carbonate development and the current GBR is fully the product of icehouse climate at mostly tropical latitudes. The DGBR on the other hand formed entirely during a very warm 'super greenhouse' [85] 'Late Devonian Hothouse' [96] world (Figure 2). Although cooling occurred with minor polar glaciation in the Famennian [85], it was not equivalent to Ordovician or Late Paleozoic glacial phases and temperatures were very warm by 'icehouse' (e.g., modern) standards [96].

Apart from the effects of global temperature on the carbonate factories themselves (e.g., potential breakdown of photosymbiont relationships as in modern coral bleaching [72,410,458]), which remain speculative for DGBR reef builders for reasons discussed above, the amplitude of high-frequency (parasequence) sea level change was much lower in the Devonian owing to the absence of major polar ice caps at that time (Figure 8d). Whereas modern parasequences in the GBRP reflect >120 m of sea level change for each glacial–interglacial cycle [459], those in the DGBR may have been only meters to tens of meters in amplitude [276]. Difference in sea level cycle amplitude between icehouse and greenhouse climates place fundamental constraints on reefal carbonate factories, sedimentary dynamics and resulting preserved limestones and parasequences [336,460]. Much greater sea level falls in the Late Pleistocene caused each parasequence beneath existing reefs to be bounded by an unconformity with some degree of exposure-related karstification–calcretization of the underlying reef. Associated offshore lowstand reef cycles are also stacked, with deglacial (transgressive) reefs having formed and drowned on top of the reefs that formed during the falling sea level (regressive) stage [46,288] (Figure 14). Many Devonian parasequences were entirely subtidal, meaning that sea level never fell below the edge of the platform [267] (e.g., Figure 22). Those cycles represent a relatively small difference in bathymetry, although some settings were deeper overall and some shallower (Figure 21). In any case, deeper and shallower deposits are stacked together one above the other. Although intertidal, and even supratidal facies are not uncommon at cycle tops in some intervals, especially near paleoshore, karst surfaces are rare at parasequence tops in the DGBR [275], and in some cases are restricted to areas with synsedimentary uplift [284,461]. Thus, much deeper karstification occurred in the GBRP, as the entire existing reef edifice on the continental shelf was exposed to meteoric water and diagenesis during each lowstand. The result is significant carbonate alteration and the formation of caves in parts of the GBR, where collapse has led to sinkhole formation with the surface expression being, for example, the 'blue holes' of the Pompey Reefs [462]. Although the southern part of the reef was in a more arid climate setting during glacial times, which limited karstification [217], the aragonite and high-Mg calcite (HMC) skeletons of the carbonate factory are particularly subject to dissolution and recrystallization. Hence, preservation of the Pleistocene GBRP is relatively poor, and one of the reasons that early radiometric dates from cores remain so elusive [51]. Only some LBFs (e.g., *Amphistegina*) and mollusks (e.g., oysters) in the GBR carbonate factory construct more stable low-Mg calcite (LMC) skeletons.

Parasequence-related karstification in the DGBR, where exposure even occurred, would have been much shallower, affecting fewer underlying parasequences, and important parts of the carbonate factory initially made skeletons of the more stable LMC (e.g., rugose and tabulate corals, brachiopods). Calcimicrobes, other microbialites and early marine cements also would have consisted of LMC. The original mineralogy of Paleozoic stromatoporoids remains enigmatic. They appear not to have been aragonite or LMC, but may have been HMC wherein they converted readily to LMC during early diagenesis [463]. Deep rooting trees and other soil-developing terrestrial flora, although increasing in the Late Devonian [253], were far less advanced than in the Pleistocene, meaning that karst-enhancing organic soil acids also were less abundant during DGBR lowstands. Thus, DGBR frameworks were less susceptible to dissolution in general and DGBR limestones are better preserved in many cases than those of the pre-Holocene GBR despite the latter experiencing significantly less time and burial. There are exceptions. Some DGBR limestone was subsequently or synsedimentarily dolomitized, and some of that was later de-dolomitized [232,236,464]. Dolomitization is not abundant on the Lennard Shelf but is associated with the economic base metal sulfide mineralization [236,464]. It is also common at non-conformities [290] and some reef margins [267]. However, Famennian dolomitization was more prevalent to the south on the Barbwire Terrace where it partly represents synsedimentary formation within an arid sabkha environment [232]. Deeper burial also caused preservation issues for some DGBR limestones. Well-cemented margins commonly did not undergo intense compaction, but initially less well-cemented back-reef facies underwent significant pressure solution (chemical compaction) during burial, losing as much as 25% of their initial thickness [88]. This resulted in relatively poor preservation. Regardless, only limited karst horizons occur in DGBR stratigraphy although significant unconformities have been recognized [90,283,284,461] and are not to be confused with significant post-depositional sub-ice dissolution that occurred following exhumation during late Paleozoic glaciation [291]. Certain DGBR reef flat features called 'sand tubes' appear very similar to subsoil karst solution pipes, but are interpreted as growth features made by microbialites and are considered pseudokarst [236,291]. As a complicating factor, microbialites could easily form within true solution pipes once re-inundated. Calcimicrobes like *Renalcis* were common in crypts and stromatolites even grew in intergranular spaces within siliciclastic conglomerates [90] (Figure 32b). Microbial deposits also grew in Neptunian dykes in the DGBR [289] as in Mississippian reefs in eastern Australia [465]. Regardless, karstification is a much less significant feature of DGBR stratigraphy than in the modern GBRP.

Higher parasequence amplitude combined with adequate subsidence (i.e., increasing RSL) also led to smaller, more restricted, particularly barrier, carbonate platforms in the GBRP. There, carbonate factories managed to fill the entire accommodation only at the margins or on pre-existing highs formed by previous reefs. Back-reef carbonate factories, even where isolated by distance and longshore currents from terrigenous siliciclastic flux, could not fill all accommodation provided by each cycle (Figure 15a). This resulted in a deeper back-reef lagoon on the shelf, i.e., an 'empty bucket' morphology [466,467]. As an example, the DGBR CF3 back-reef lagoon at Mowanbini [290] (described above) contains cycles with shallow subtidal stromatoporoids at the deepest and ooid shoals or intertidal fenestral algal flats at the top. Most available accommodation was filled to very shallow shoaling or intertidal depths for each cycle (Figure 15b). Deeper back-reef settings in the

GBR contain discrete carbonate factories, like the *Halimeda* mounds that can reach 20 m in height [468], but none were able to aggrade to sea level along with their associated reef margins. The comparison supports the contention [460] that 'empty bucket' reef morphologies may better represent icehouse than greenhouse climates.

The best partial geomorphological analogs in the DGBR for many modern GBR reef types are the back-stepping and upright margin platforms made by CF2 and CF3 factories that struggled to keep up with increasing accommodation during the early part of Pillara supercycle transgression. That eustatic rise provided adequate accommodation to isolate individual reefs on the deeper drowned shelf (e.g., Lloyd Hill in Bugle Gap) (Figure 27c). However, even there, back-reef settings appear to have kept up better with the margins so that deeper back-reef lagoons ('empty buckets') are not the norm. Kerans [236] used the abundance of *Renalcis* behind the reef margin as an arbitrary, but definable, means of differentiating reef flat from back-reef facies owing to the gradual nature of the facies changes without major bathymetric change.

Over longer time scales parasequence amplitude also may have affected the evolution of the carbonate factories. The great accommodation imposed on reef growth (keep-up or catch-up) by Late Pleistocene parasequences put increased pressure on newly established reef communities to aggrade, or they might give up and drown, as may nearly have occurred in the southern GBR during the recent deglacial [46] and the previous LIG, e.g., [217]. This 'pressure' to aggrade may have favored the very fast growing and disturbance resistant branching acroporids that dominate most Indo-Pacific reefs today [27]. Suprastratal branching acroporids were present in reefs throughout the Cenozoic but took on their present dominance only in the Late Pleistocene when the amplitude of sea level change for each parasequence reached its very high current levels [27]. Morphological analogs in the DGBR, branching tabulate corals and *Stachyodes*, were not associated with major aggradation until the introduction of abundant *Renalcis* and early cement in CF3, which provided branching (and laminar) *Stachyodes* adequate strength to make rigid framework on high-energy platform margins, slopes and back-reefs. Branching tabulate corals, although prolific at CF1–CF2 highstands, made only thin biostromes, but may have lacked the high extension rates of modern branching *Acropora*. Prior to the provision of adequate binding and cement to cause framework unification, the late Givetian–early Frasnian (CF1–CF2) margin communities were incapable of aggrading rapidly enough to keep up with sea level rise during the Pillara transgression and drowned and/or retreated and back-stepped.

The potential enhancement of branching *Acropora* in Late Pleistocene parasequences of modern reefs occurred against a backdrop of general biological stability between cycles. As shown in other regions [59,68], highstand reef communities in the GBR reassembled in a very similar way after each glacial exposure interval, despite every new reef having to initiate independently on the previously exposed, dead substrate [197,217,219,356]. The same basic carbonate factory even re-established in the reefs through the glacial times as the reef tracked sea level down and back up [288] (Figure 14). Similar ecological stability occurred in CF1–CF2 communities in the DGBR [70]. Over short time scales—a few consecutive parasequences, communities remained stable, but larger changes occurred over longer time spans, presumably in reaction to broader environmental trends [70]. Although such quantification has not been accomplished on younger cycles in the DGBR, the recognition of cycles there in part reflects the repeatable ecological membership in similar geological circumstances through a longer succession. Hence, it is likely that such ecological stability was the norm over similar limited time spans.

This highlights the comparability issues between modern and fossil reefs imposed by differing temporal and spatial scales of study (Figure 3). At small spatial and temporal scales modern and DGBR reef communities reacted in similar ways producing similar facies zonation in similar settings with comparable geographical expressions resulting from external forcing, such as waves, tides and currents (Figure 32). However, the modern GBRP that we know well represents only a single parasequence highstand. Much larger changes may occur in reefs and reef communities at the longer temporal scales of supersequences, and these affect larger scale reef geomorphology (e.g., the difference between CF1–CF2 banks, CF3 upright margins and CF4–CF5 prograding platforms). Much of this depends on the differing rates of accommodation creation relative to the aggradation potential of the carbonate factories. The major driver of long-term DGBR stratigraphy was eustatic sea level change [266] with similar supercycles in Upper Devonian reef systems in Alberta [469] and South China [267,268] although each region has local tectonic differences [262,400,402]. Parasequences too would have been global if driven by Milankovitch dynamics. Aggradation of CF3 communities in the DGBR during a time of rising sea level generated geomorphological and facies similarities to the GBRP, because significant aggradation occurred with each parasequence. The main difference was that the GBRP experiences relatively slow subsidence (slow RSL rise) overlain by high-frequency, high-amplitude sea level fluctuations, whereas the DGBR experienced a low-frequency, high-amplitude sea level change (supercycle) overlain by high-frequency, low-amplitude parasequences (Figure 8d). Although the framework types differed significantly in the two provinces (specifically, Late Devonian mixed calcimicrobial–biocementstone–skeletal frameworks compared to primarily skeletal frameworks today), the key factor is the ability to aggrade adequately to keep up with accommodation at the margin. However, this same factor led to a major difference between the two reef systems, in that deeper back-reef lagoons of the GBRP hosted carbonate factories that could not fill required accommodation for each parasequence (Figure 15a). The following question then arises: Did increased accommodation of the Pillara transgression in the DGBR promote calcimicrobes like *Renalcis*, which then allowed CF3 reef margins to aggrade and keep up with accommodation in an analogous way to high-amplitude Late Pleistocene parasequences favoring fast-growing, branching *Acropora*? The answer to that question requires investigation of other types of external forcing.

4.4. Marine Chemistry Constraints on Reefs

Most high-energy DGBR reef margin facies (at least CF3–CF5) are dominated by calcimicrobes and early marine cement [88,236]. The abundance of microbial carbonates and marine cement has varied through the Phanerozoic, particularly in reefs [98,99,158,470,471]. Calcimicrobes are non-obligate calcifiers, meaning that they do not intentionally precipitate $CaCO_3$ skeletons. Rather, they may promote, localize and/or ease nucleation of carbonate minerals and/or locally increase alkalinity as a byproduct of their metabolism, thus incidentally increasing precipitation [98,472]. Hence, their presence/absence partly represents taphonomic processes (i.e., preservation potential) [404,473–475], and these depend on ambient water chemistry (alkalinity, cation availability, pH and temperature). Although concerns about ocean acidification (OA) causing reduced biomineralization in modern carbonate factories are valid [18,19], many reef-building organisms can still calcify in lower-pH waters at an additional metabolic cost [476,477]; there is a limit [478]. Both micritic microbialites and calcimicrobes, on the other hand, are completely dependent on advantageous water chemistry to form and cannot form widely, if at all, in waters with adverse chemistry (i.e., low alkalinity, low pH, etc.) unless partially isolated in a microenvironment [130,479]. Simple filamentous calcimicrobes, such as *Girvanella*, occur through most of the Phanerozoic, but their distribution partly reflects adequate preservation [475]. Where rapidly calcified in waters with abundant dissolved carbonate, identifiable calcimicrobes were formed, but where calcification was slower or less complete, micritic microbialites or mud were formed. Hence, microbialite occurrence and type should provide an independent proxy for marine calcification potential [98]. However, critical interrelationships between the metabolic inputs, influences and constraints of skeletal and microbial communities on each other during reef building are poorly understood [480].

Changing microbialite distribution through the Phanerozoic was attributed to the combination of fluctuating marine carbonate saturation and competition with skeletal metazoans as reflected by biodiversity curves [99,481], the latter partly regulated by atmospheric oxygen levels [482]. Marine alkalinity and carbonate saturation should be primary controls on calcimicrobe abundance, but major anomalies exist in their correlations [482] (Figure 2). Marine alkalinity and buffering capacity are highly complex, even in the modern ocean [478,483], but depend on basic chemical equilibria that should allow modeling through time [484]. This modeling is increasingly complicated with less secure proxies the farther back it is attempted [485]. Although modeling Phanerozoic saturation states is beyond the scope of this paper, several observations can be made relative to the comparison of Late Devonian and modern reefs. The general decreasing trend in Phanerozoic microbialites [471] is undoubtedly correct, but the initial data analyzed include earlier studies that commonly overlooked microbial carbonates in reef frameworks owing to a greater focus on, and easy recognition of, associated skeletal organisms (see [158,465,486,487]). Microbialites are still not considered part of the reef framework by many workers despite the large role and volumes of micritic microbialites in deglacial and Holocene reefs [132,439]. Hence, existing microbialite abundance curves [37,471] may not take micritic or cryptic reefal microbialites adequately into account. However, that cannot account for the major differences between modeled saturation states and abundance of reefal microbialite and especially reef cement [97].

Through time, modeled equilibrium saturation states may not adequately consider carbonate removal rates from changes in biomineralizers [482]. Differing biomineralization processes affect how organisms respond to changes in pH and carbonate removal [478]. A temporal mismatch between the rate of anthropogenic CO_{2atm} increase and natural removal by chemical weathering leads to the current OA problem [488], thus representing a temporal disequilibrium. Major anomalies between modeled carbonate saturation state and empirical proxies for ease of marine carbonate precipitation (e.g., marine cement) could in part represent temporal disequilibria in the broader carbon cycle or fundamental changes to mass balance. The largest anomaly in modeled saturation state versus microbialite abundance occurs from the mid-Mesozoic, when very high calculated saturation states were not associated with abundant microbialites or cement (Figure 2). This major difference was previously attributed to the evolution of calcareous plankton [98], which were able to expand carbonate precipitation into open oceanic settings, where nutrients allowed, and remove and sequester alkalinity to deep ocean waters and deep-sea sediment. Prior to calcifying plankton, most marine carbonate biomineralization occurred only on continental shelves. Such major changes in the marine carbonate system could have fundamentally, and unidirectionally, altered the dynamics of the marine carbonate system. Hence, the diversity of organisms through the Phanerozoic may be less critical than the source, volume and rate of carbonate removal they accomplished. Increased efficiency of calcification amongst some skeletal organisms (e.g., photosymbiosis in scleractinian corals, LBFs and mollusks) and expansion of biomineralization into new open ocean habitats by plankton could have caused major shifts in shallow marine carbonate saturation dynamics. Something must account for the massive decrease in biocementstones in reef frameworks since the Early Mesozoic [97,98] (Figure 2). Regardless, that decrease represents a major empirical difference in calcification potential between the DGBR and GBRP. Thus, the lack of evidence that OA was a problem in Late Devonian reefs despite the very high pCO_{2atm} [489] provides no security for modern reefs.

Regardless of broad Phanerozoic calcification patterns, the rise of abundant calcimicrobes and cements to dominate CF3 reef frameworks in the DGBR requires explanation, as modeled carbonate saturation states were high throughout the Devonian [482]. So why did *Renalcis* and associated calcimicrobes bloom when they did in the Late Devonian reefs of the Canning Basin? It is unlikely to represent the occurrence of new microbes. Calcimicrobes like *Renalcis*, as with *Girvanella*, occurred both before and after the Late Devonian (Cambrian [490]; Ordovician [491]; Mississippian [492]) but were commonly missing in intervening stratigraphic intervals where adequate substrates were available and other members of reef carbonate factories flourished (e.g., corals and stromatoporoids in CF1 in the DGBR). Calcimicrobes became abundant in reefs around the world during the Frasnian transgression, but were not as abundant everywhere as in the DGBR. They occurred in transgressive Frasnian reefs in Alberta, and the Northern Territories, Canada [168,493,494], in some cases as cryptic binders with abundant cement in aggrading and prograding margins as in the DGBR [469]. Frasnian reefs in South China had abundant calcimicrobes, but they were less pervasive, although more diverse, than in the DGBR [267,268]. The abundance of calcified algae, cyanobacteria and microproblematica increased by 34% in South China from the Givetian to the Frasnian, but then declined by 63% in the Famennian, as micritic microbialites dominated [406]. Local paleogeography controlled carbonate factory distribution in Late Devonian South China with microbial reefs primarily developing on isolated platforms, not attached to the continental mass [268]. The steep aggrading stromatoporoid–*Renalcis* frameworks of the DGBR are not common in South China and stromatoporoid–coral communities back-stepped during Frasnian transgression [268]. Calcimicrobes, including *Renalcis* also occurred in Frasnian reefs in Belgium [495] and in the Urals from Timan-Pechora [496–498] to the pre-Caspian of Kazakhstan [84,86,499]. Some of the larger (300–700 m thick) Frasnian buildups in Timan have stromatoporoid–microbial frameworks with *Renalcis* and abundant cements [496]. There, a general trend of decreasing skeletal frame-builders with increasing microbialite occurred through the Frasnian, but biocementstones were even more abundant during the Early Devonian, being somewhat reduced through the Frasnian before dropping off in the Late Famennian [498]. In all regions, margins were dominated by micritic microbialites (stromatolites and thrombolites) at or prior to the end of the Frasnian with major carbonate platforms continuing through the Famennian.

The Late Devonian increase in micritic microbialites in Alberta was attributed to increased nutrients associated with lowstands, but the aggrading calcimicrobial–skeletal frameworks of highstand margins were considered to represent oligotrophic conditions [469]. Other authors considered only coral–stromatoporoid frameworks to represent oligotrophic settings, with calcimicrobe-rich facies and micritic microbialites forming in mesotrophic settings and megalodontid bivalve assemblages representing eutrophy [440]. It is difficult to apply modern nutrient gradients to Devonian reef builders [500]. There are no modern analogs for calcimicrobes like *Renalcis* and the dominant skeletal constructers, stromatoporoid sponges, were most likely filter feeders [454] that may have required more nutrients than modern 'photozoan' reef communities. Regardless, it is critical to separate calcimicrobes and micritic microbialites in terms of their environmental implications and proxy value [98] and to discriminate between benthic and cryptic microbialite settings [469,479]. Modern micritic microbialites may represent higher nutrients [439,501], but cryptic varieties, which are common in modern and Devonian reefs, may still occur in oligotrophic settings because they form in microenvironments where nutrients and alkalinity are governed by co-occurring organisms, rather than ambient seawater [130,479].

Additionally, all microbialites are not the products of cyanobacteria. Calcification in many microbialites is driven mainly by heterotrophy, and in particular sulfate-reducing bacteria [472,502], which also have been implicated in modern cryptic reefal microbialites [503]. Whereas benthic microbialites may have cyanobacteria and algae as their primary producers with heterotrophs driving calcification beneath them, cryptic microbialite heterotrophs probably consume eukaryote organic matter produced by other cryptic organisms like sponges [399] or ascidians [132]. Calcimicrobes are less well understood, but *Epiphyton* (which has ecological similarities to *Renalcis*) is thought to be a cyanobacterium, and was recently shown to have carried out oxygenic photosynthesis [455]. However, its common ability to thrive in fully cryptic habitats suggests that it was a mixotroph [455]. Hence, calcimicrobe communities may have adapted to a wide range of nutrient settings and bathymetry. In the DGBR, *Renalcis* (cryptic and benthic) and cryptic thrombolite shared trace element geochemistry (rare earth elements, REE) with associated early marine cements, which are consistent with well-oxygenated, shallow seawater [504]. Early marine cements are the best proxies for ambient seawater trace element distributions as they incorporate most of their signature from dissolved ions, rather than from trapped and bound detritus [505]. Comparison to modern open ocean seawater suggests that Leonard Shelf seawater had greater terrigenous input [504]. Average CF3 stromatoporoid–*Renalcis* frameworks formed in waters equivalent to normal salinity, inshore waters in the GBR lagoon (based on light REE depletion levels—a seawater indicator: mean $(Nd/Yb)_n$ = ~0.43 for DGBR early cement compared to mean $(Nd/Yb)_n$ = ~0.40–0.50 for well-mixed estuarine seawater in the modern GBR lagoon [506]). For comparison, Holocene cryptic microbialites from Heron Reef, southern GBR, which occurs >70 km offshore near the continental margin, have much greater light REE depletion typical of more open marine (oligotrophic) seawater (mean $(Nd/Yb)_n$ = 0.236) [507]. Inshore DGBR microbialites that occur with siliciclastic sediments, on the other hand, have trace element signatures dominated by terrigenous inputs as expected and show no light REE depletion [504]. Thus, *Renalcis*-containing Devonian reef margins may have required more nutrients than modern reefs, thus being less susceptible both to nutrification and drowning [440].

Even Devonian stromatoporoid–coral reefs from platform edges may not represent the oligotrophic settings that modern reefs prefer, as that interpretation is based largely on the tabulate corals having established photosymbionts, e.g., [72], a hypothesis that remains controversial. However, quantitative relationships between nutrient flux and terrestrial input may have differed in the Late Devonian, as terrestrial floras and weathering rates and processes were different than today [253,254]. Late Devonian continental runoff, although containing greater nutrients than in the Early and Middle Devonian [253] may have contained fewer nutrients relative to modern runoff entering the GBR lagoon. Limited palynological data suggest that the coastal lowlands bordering the DGBR were occupied by a well-developed terrestrial vegetation [90] and adequate nutrients supported a diverse plankton including abundant radiolarians [508], acritarchs, microalgae and cyanobacteria [233]. Givetian to Frasnian oils generated from Lennard Shelf source rocks (Gogo and Fairfield formations) were sourced from plankton and suggest episodes of water stratification with anoxic deposition due to eutrophication during transgressions with high primary productivity from eukaryotes (e.g., algae) [426]. Stratification is indicated by biomarkers for bacterivorous ciliates that fed at the oxic–anoxic interface, and that interpretation is supported by the normal marine fauna preserved in the anoxic to suboxic deposits that hosted the Gogo Formation Lagerstätte [419]. Across the Fitzroy Trough to the south, oils obtained from the Barbwire Terrace are dominated by bacterial sources in a more continuous suboxic environment that may reflect restriction from more open waters of the Fitzroy Trough behind the reef complexes developed along its edge [426] (Figure 12).

Somewhat paradoxically, although the prominent association between CF4 and CF5 micritic microbialites and terrigenous sediment is well documented and provides strong support for increased nutrients through the regressive (Nullara) part of the supercycle, the DGBR lacks evidence for the abundant shallow anoxic facies [509] that characterize the interval and Late Devonian extinction events elsewhere in the world. One of the major models for Late Devonian extinction events involves eutrophication and pervasive shallow anoxia within the photic zone resulting from the very high influx of terrestrial nutrients associated with the newly deeper rooting trees and greater soil thicknesses and weathering profiles through the Devonian [261,510]. The lack of shallow anoxia in the DGBR seems paradoxical being set within a relatively enclosed epicontinental embayment and argues against extreme eutrophication from the terrestrial nutrient flux through the Famennian. The global dominance of microbial communities in reefs after the Frasnian–Famennian extinction of most skeletal reef builders is well documented [511], but there was also a potentially global decrease in the amount of reefal calcimicrobes during the Famennian [236,406]. More importantly, early marine cement volumes decreased, and thus so did biocementstones [498].

Early marine cement declined in the DGBR from CF3 to CF4 and CF5, when terrestrial flux was empirically greatest. Alkalinity in the form of HCO_3^-, like terrigenous nutrients, results from continental weathering and is delivered in runoff, so they might be expected to correlate. They do not correlate well in the DGBR. It is even possible that increased CF3 marine cement resulted from lower nutrients, wherein there was a lower volume of active microbial organic matter to promote nucleation, leaving abundant alkalinity available for thicker cement. Could the much more frequent nucleation of micritic carbonate in CF4 and CF5 stromatolites and thrombolites effectively remove adequate alkalinity to suppress other cements? Alternatively, the reduced CF5 cements may reflect the decrease in larger framework cavities to host large cement volumes. The lack of abundant early marine cements in CF1 facies does not support the first hypothesis, but comparisons of carbonate volumes through time are needed to test the concept. Regardless, other factors may have been significant. Salinity is a major control on shallow marine alkalinity [512,513] and would be subject to temporal changes in climate (both precipitation/evaporation balance and circulation). The arid hypersaline deposits across the Fitzroy Trough extended into the Famennian, so temporal changes to broader basin circulation could have altered salinity gradients, even without changes to local precipitation on the adjoining continent. Unfortunately, conodont-based $\delta^{18}O$ isotope data are absent for the DGBR [514], but could help indicate relative salinity, and thus inform intervals of aridity versus terrestrial runoff [251]. Regardless of whether the abundant reefal microbialites and biocementstones of the DGBR were constrained more by nutrients or local alkalinity, the current OA trend poses a serious risk to the role of cryptic microbialites in unification of modern reefs. At the same time, modern reefs are probably far more susceptible to nutrient-invoked drowning (sensu [440]) than were Late Devonian reefs.

4.5. Evaluating the Analogs

The concept of uniformitarianism [515] is well known in Earth history. The present is the key to the past, and increasingly, the past is the key to the future. This is especially the case for climate studies, which depend more and more on numerical modeling to make climate change predictions. The conservation of coral reefs is intimately tied to understanding their past and present as their futures are similarly tied to changing climate. One of the unique contributions of geology to such studies is the introduction of real Earth system time scales where cyclic phenomena, such as Milankovitch orbital forcing and broader tectonic cycles can be recognized and differentiated from more stochastic events, such as bolide impacts, and unidirectional changes, such as increasing continent size [516], increasing solar luminosity [331,517] and changes associated with the evolution and diversification of life. The reef systems compared above show clear evidence of Milankovitch-scale sea level cycles but are overlain on different phases of the broader icehouse—greenhouse climate cycle and the calcite versus aragonite seas cycle (Figure 2) (Table 2). At the same time, the biology of the reef builders has changed unidirectionally with extinctions and radiations imposing very different carbonate factories on the two reef provinces (Table 1). As the evolution of oxygenic photosynthetic metabolisms in cyanobacteria unidirectionally changed the redox states of shallow Earth systems, the evolution and diversification of different biomineralizers may have altered the marine carbonate system such that even marine cement and calcimicrobe abundance have been significantly reduced since the Devonian.

Table 2. Comparison of environmental forcing attributes of the Great Barrier Reef Province (GBRP) and the exposed Devonian Great Barrier Reef (DGBR).

	GBRP	DGBR
plate tectonic setting	intracratonic rift movement through latitude post rifting deposition	intracratonic rift latitudinal stability syn-rift deposition
latitude	mostly tropical	tropical
climate	mostly icehouse mainly low	greenhouse very high
parasequence amplitude	>100 m	m—10s m
preferred mineralogy	aragonite	calcite
modeled calcite saturation	low	high
nutrients	oligotrophic	mesotrophic

Through Earth history, successive reef communities developed integrated systems, adapting to a steadily changing Earth, that is, when given time. It is best not to think of them piecemeal. Modern coral reefs are more than just healthy coral communities, more than just corals and CCA. They are the end products of a suite of closely integrated processes: construction, destruction, sediment transport, sediment deposition and unification. Reef communities dominate some processes and impact on each of the others, but every process has constraints imposed by external forcing that potentially affect sustainable reef growth. The ability of modern and ancient reef communities to consistently reform after major sea level changes with each parasequence is a testament to their resilience [70,197], but also to their integration and interdependencies [62,518]. Regardless, the communities must respond to environmental forcing [70]. Although reef communities engineer and construct the environment where they live, that environment must be maintained constantly through time or the entire system may break down, i.e., undergo a phase shift to a very different ecosystem [108,167,519]. To be successful, reef communities have required huge adaptability through time. A possible key to success for major intervals of reefs in the Phanerozoic is the combination of skeletal constructors, typically, fast-growing hypercalcifiers, with effective binders, such as CCA and microbialites, the key agents of unification [62]. This combination provides both capacity for aggradation and for rapid acquisition of rigidity. CF3 margin frameworks show that abundant marine cement can be added to the encrusters and microbialites. However, reef systems can collapse and be terminated or changed significantly when external forcing pushes constructors or key unifiers beyond their tolerances. The DGBR and a few other major Late Devonian reef systems (Figure 1b) were the last holdouts from a much 'reefier' Middle Devonian world, where many equally large and larger reef systems were distributed widely across the globe [85]. Even within the more depleted surviving DGBR the major changes in reef communities (CF1 to CF5) and resulting reef environments and geomorphology that occurred through time would have appeared 'catastrophic' if viewed from any one endmember to the next. However, even major changes, such as the replacement of most skeletal organisms by microbes that characterizes the Frasnian–Famennian extinction event, may have been more gradual than 'catastrophic' when viewed in higher resolution in the DGBR [402,408,416]. Each parasequence and lower-frequency sea level fluctuation causes at least a local 'catastrophic' change, making the recognition of gradual change more difficult (e.g., a temporal scale issue [70]). Regardless, Frasnian–Famennian changes followed other major changes in Devonian reefs, particularly affecting corals, near the end of the Givetian [85]. Long-lived, low-relief CF1–CF2 carbonate banks with rich, widespread coral and stromatoporoid biostromes shifted to back-stepping, and then increasingly higher-relief CF3 margins dominated by stromatoporoids, calcimicrobes and early marine cement. CF1 contained abundant hypercalcifying constructors, but lacked effective binders and abundant cement, thus failing to unify. The margins of most cycles never aggraded to sea level. Only abundant marine cement and rapid calcimicrobe calcification provided the unification that allowed CF3 margins to aggrade through the increasing accommodation of the Pillara transgression. The shift to prograding, low-relief CF4 and CF5 margins dominated by micritic microbialites and calcimicrobes had few hypercalcifiers, unless microbialites take that roll, but they prograded more than three km basinward through the lower accommodation Nullara regression. Each carbonate factory differed in important ways, but in each case, they engineered, constructed and maintained an environment that provided habitat for countless organisms, both on the reefs and in associated coastal areas behind them. Reef communities have changed even more through the Phanerozoic, but reef forming processes have built similar reefs when and where possible [35]. The way the carbonate factories of the GBRP and DGBR 'engineered' and constructed their reefs is similar, producing similar geomorphology at times (Figures 27 and 28), and with ample evidence of comparable storms and waves producing both reef flat and fore-reef rubble (Figure 32a–f), relatively high tidal ranges (Figure 32g,h), and typical coastal currents (Figure 29).

Even with such basic similarities, to study reefs as 'timeless' entities it is important to focus on the functions and processes that govern their construction (e.g., Figure 5). However, a common tendency to evaluate ancient reefs against the 'gold standard' of modern reefs can lead to an (over)emphasis on skeletal contributors. The 'pull of the recent' is very strong, despite the great complexity and diversity of modern reef systems and the very different biology and external forcing through time (climate, marine chemistry, etc.) (Table 2). Bias against microbialite reef framework is discussed above, but many ancient reefs in high-energy settings simply do not contain adequate skeletal organisms to make a skeletal framework, e.g., [465], yet they were in every other way analogous to modern reefs. This perceived dominance or primacy of skeletal frameworks also has led to a preoccupation with corals, almost to the exclusion of the critical binders, CCA [61], never mind cryptic microbialites. Unification must happen for reefs to form hard substrates at sea level; carbonate production by constructors, even in situ skeletons, currently is not enough (e.g., Figure 32a,c,e). However, many studies on modern reefs focus primarily on skeletal organisms, commonly just corals, to predict if current reefs are healthy and likely to flourish. At the same time, some of the best recent ecological studies aimed at reef evolution in the Devonian, e.g., [69,70,72,410] also have focused primarily on corals. In the case of the DGBR, late Givetian to early Frasnian CF1–CF2 corals were abundant and thrived through each parasequence highstand, yet they did not make a reef that could aggrade adequately to keep up with sea level (Figure 21). Rather, they were completely subordinate to stromatoporoids, which occupied the shallower, higher energy facies and were capable of aggrading at the margin adequately to protect a lagoon facies near Menyous Gap (Figure 22c). Corals were not major reef builders at the time (in the DGBR). Constratal stromatoporoids made significant skeletal frameworks (CF2 margins) but still could not keep up with the Pillara transgression. Aggrading reefs geomorphologically similar to modern reefs occurred only when CF3 communities consisting mainly of constratal massive tabular and delicate laminar and branching stromatoporoids were joined by potent binders (calcimicrobes) and abundant marine cement. The reefs were made, not by hypercalcifiers, but by a consortium of, in some cases delicate constructors, but with adequate binders and effective early cementation to ensure unification in ecological time. Thus, rigid substrates (frameworks) were constructed upon which additional reef growth could occur in high-energy settings. Corals had little if anything to do with it, whether or not they had associated photosymbionts. The more likely controls on reef evolution in the DGBR were marine chemistry to support, or not, early and voluminous cementation, and thus allow for the formation of calcimicrobe binders [98,475]. This degree of unification supported the steep upright CF3 and CF4 margins that grew at sea level as evidenced by the extreme winnowing of well-cemented *Stachyodes* fore-reef slope rudstone–biocementstone (Figure 25h). Despite the morphological and taphonomic similarities between branching *Acropora* and *Stachyodes*, there are no modern equivalents to such Late Devonian rudstones–cementstones. Despite the Givetian and Frasnian corals preferring deeper settings, they did not thrive during the Pillara transgression when cycles were getting progressively deeper. Rather, the new CF3 stromatoporoid–calcimicrobe carbonate factory dominated the margins and relegated the more limited corals to back-reef environments, such as at Mowanbini, where *Disphyllum* continued in a limited role to make biostromes behind the margin.

Another major difference is nutrient flux [440,469], as the Leonard shelf waters were likely to be more nutrient-rich than are offshore GBR reefs today. This may have better supported the filter-feeding stromatoporoids and microbial communities, but its effect on corals remains speculative. Regardless, the restriction of corals to shore attached lagoons does not suggest that nutrients were the immediate cause of their decline. Increasing nutrients through the regressive CF4 and CF5 times may have reduced both corals and stromatoporoids leaving calcimicrobes and other microbialites to dominate the reefs. Increasing nutrients and local alkalinity associated with lowering sea levels during Famennian cooling are more likely causes for the decline of stromatoporoids and corals than the changing temperatures themselves, which despite falling in the late Famennian, were still very high by today's or any icehouse climate standards. Additionally, local alkalinity may have increased on the Leonard Shelf owing to higher salinity in the Famennian. Unfortunately, the isolation of Leonard Shelf reefs within the intracratonic Canning Basin renders correlations to global processes more difficult. The degree of circulation with the Paleotethys Ocean is poorly constrained [233] and even the global shallow anoxia events of the Kellwaser and Hangenberg events are not apparent in the DGBR [509]. Thus, minor climate changes that drove precipitation patterns and upwelling in the basin could have altered nutrient flux, delivery of HCO_3^- from terrestrial weathering and salinity, all of which would have affected the carbonate factories. The uncertainties are exacerbated by the very thick stratigraphic succession and difficulties in correlation across the paleogeography. Additional detailed studies are required, especially to put both stable isotope and trace element geochemistry into the context of different carbonate factories and global trends and events, and this is true both for the DGBR and the GBRP (particularly the older buried parts to the north). There is much more to learn by comparing reefs such as the GBRP and DGBR.

5. Conclusions

Coral reefs are among the most significant modern habitats in terms of biodiversity and the ecological, economic and cultural services they provide. The reefs of the modern Great Barrier Reef Province (GBRP) are built by corals and crustose coralline algae in oligotrophic settings within a regime of high bioerosion, high-amplitude sea level change, relatively low temperatures and low marine carbonate saturation. Reefs and banks of the ancient Devonian Great Barrier Reef (DGBR) of western Australia were built by stromatoporoid sponges, microbialites, corals and marine cements within a regime of higher nutrients, low bioerosion, low-amplitude sea level change, higher temperatures and high marine carbonate saturation. Modern and fossil reefs share significant commonalities but also differences that can be related to specific external constraints. The study of one can usefully inform the other. Comparison of these two reef provinces allows the following conclusions to be drawn.

5.1. Reefs in General

- Biological reefs (today, coral reefs) are not just biological communities and ecosystems, they are the relief-bearing structures that are bio-engineered and constructed by those communities. Those structures provide many ecological services that extend far beyond the reefs themselves.
- Repeated reassembly of biological communities after successive disruptions, (e.g., between parasequences, even tracking sea level down and back up through glacial–interglacial cycles) in both reef provinces suggests that community integration and interdependency is a general characteristic of reef-building communities and provides further evidence for their specialized nature. Reef communities are more than random associations of organisms that can coexist in the same shallow carbonate environments.
- Despite bio-engineering their own geomorphological habitats, reef communities are subject to numerous external constraints (climate, sea level, wave energy, nutrients, marine alkalinity, to name a few). These constraints occur as both temporal and spatial unidirectional trends (e.g., latitudinal gradients, increasing continentality through time) and heterogeneous and stochastic distributions (e.g., intermediate disturbance to glacial cycles). In the face of these constraints, reef building depends on communities managing or balancing five key processes (Figure 5): (1) construction of calcium carbonate by a carbonate factory; (2) destruction of existing structures by physical or biological erosion; (3) transport of particulate sediment; (4) deposition of particulate sediment; and finally, (5) unification, where both in situ and transported carbonate skeletal material is bound and cemented into a hard reef substrate. It is that hard substrate that allows continued reef growth and underpins reef geomorphology and thus, survival of the broader reef system. Changes to the external constraints on, or biological roles in, any of the five key processes may cause major changes within the other processes and, thus, to the entire reef system.
- Very different reef communities can make very similar reef geomorphologies. However, they also can create very different structures and platforms where communities and/or external constraints differ. Understanding the differences can provide information on fundamental aspects of reef communities that might be applicable to reefs across time and space.

5.2. Modern Reef Workers and Conservationists

- The study of ancient reefs provides significant insights into how modern reef communities have come to be, not just at the assembly level, but the evolutionary level. For example, the dominance of many (but not all) Indo-Pacific reefs by fast growing acroporids has been attributed to the high amplitude of modern icehouse sea level oscillations [27]. Over longer time scales the rise of rasping bioeroders, like scarid fishes, may have favored the evolution of modern CCA at the expense of the earlier, less well-calcified solenoporoid algae [447], which occurred in the DGBR.

- Abundant carbonate production in an appropriate shallow setting is inadequate to build a reef. The abundant growth of CF1 corals and stromatoporoids constructed only low-relief carbonate banks. What was lacking was adequate unification to allow the banks to aggrade to sea level at their margins. Subsequent CF3 communities on the other hand, had far fewer and less massive skeletal constructors, but had effective binders in calcimicrobes and early marine cement to support unification. They constructed steep, high-energy reef margins similar to those of modern reefs from less volumetrically abundant, fragile stromatoporoids with virtually no corals at all. The example highlights that successful reef growth requires both construction, although not necessarily by so-called 'hypercalcifiers,' and unification from binders [62]. Importantly, these two processes may be constrained independently from each other by external forcing.
- While coral bleaching is a major threat to important members of the carbonate factory today, it is far from the only threat. Many skeletal organisms can manage calcification in lower pH waters, but the larval and juvenile forms of many taxa in reef carbonate factories are at risk, although under-studied [520]. These include CCA, which is more sensitive to carbonate saturation state than are corals [521,522]. The effects of OA on binders, both CCA and microbialites, could severely restrict modern reef unification, and thus development of the hard substrates required for recruitment and renewed coral growth after even small disturbance events. The scale and frequency of disturbance events are projected to increase, with increased wave energy and cyclone activity and higher sea levels allowing more wave energy across reef platforms. Critically, modern reefal microbialites that increase the strength of porous reef carbonates by filling pore space at many scales are non-obligate calcifiers; their occurrence and distribution is almost completely constrained by ambient water chemistry. Reduced marine pH associated with anthropogenic OA could inhibit their precipitation, thus removing a major component of modern reef unification.
- From a conservation perspective, it is critical to remember that it takes more than a coral community to raise a reef. Corals are without doubt the most important skeletal framework constructors on modern reefs, but all five of the key reef-building processes must balance for reefs to be maintained. The role of bioerosion and source and fate of reef sediment is increasingly acknowledged and studied (from conservation and carbonate budget–carbon cycle perspectives), but much of the reef conservation and monitoring effort is directed only to corals and commonly only coral cover on the slope. This partly reflects the relative ease of measuring coral cover remotely, but it is not the only measure of reef health and studies of other critical reef growth processes have lagged behind. Through modern reef coring we are only beginning to grasp the importance of cryptic microbial communities in framework unification and know very little about their metabolic processes, trophic structures or even rates of growth. Fortunately, reef cores through the last glacial cycle have provided a natural laboratory to understand the effects of changing pCO_2 on modern reefal microbialite growth [148], but the results are alarming in the face of OA.

5.3. Geologists and Paleobiologists

- Ancient reefs provide a means to explore temporal differences in marine chemistry, such as carbonate saturation state, as it provides major external constraints on carbonate factories, reef unification and carbonate platform development. Marine alkalinity is also intimately related to Earth's climate through time. While direct comparison of modern and Late Devonian unification processes and modeled marine alkalinity appear to be consistent (explaining the lack of abundant calcimicrobes and marine cement today), major inconsistencies in expectations are evident in other time intervals since the Mesozoic suggesting major gaps in our understanding of marine alkalinity through time.

- Comparison of the GBRP and DGBR provides strong support for the hypothesis that 'empty bucket' carbonate platform morphologies are favored by, if not limited to, icehouse conditions [460]. This association likely results from the interplay between the amplitude of accommodation creation versus the relative aggradation potential of the carbonate factory across the geographical profile (Figure 15). Where the amplitude is high, geographically variable carbonate aggradation rates can lead to deep lagoons and steeper barriers and margins, as in the GBRP. However, low parasequence amplitude provides a vertical 'equalizer' that does not allow the greater aggradation potential at the margin to reach fruition. Higher-aggrading communities are limited by sea level, spilling sediment laterally to help fill accommodation with lower-aggrading communities in the back-reef. This may have limited development of deeper lagoons behind aggrading CF3 margins in the DGBR. Where aggradation potential is equivalent across the geographic profile, the role of accommodation is decreased, but still, no deep lagoon should form. The CF1 banks of the DGBR had similar aggradation potential across the bank mostly even to the margin and thus, did not create deep lagoons even as accommodation increased through the early Frasnian.
- The DGBR is not a single reef geographically or through time any more than is the modern GBR. To understand it, it must be separated into different temporal units that reflect the changing carbonate factories and external constraints.
- Despite the strong 'pull of the recent', corals are not the answer to understanding the DGBR or other Late Devonian reefs. Stromatoporoids were the best functional analogs for modern reef corals, but their trophic structures and general biology are less well understood. Importantly, the key binders in the DGBR, calcimicrobes, are even less well understood. Recent work suggests that some were mixotrophs capable of photosynthesis where exposed, but other types of metabolism in cryptic settings [455]. We know little about them but cannot afford to treat all reefal microbialites as the products of cyanobacteria.
- Trace element geochemistry suggests that the DGBR occurred in a higher nutrient setting than the outer GBR today (i.e., not oligotrophic). However, the source of the high alkalinity that drove cementation and calcimicrobe growth in CF3–5, but not in CF1, is less easy to identify.
- Isolation of the exposed DGBR in an intracontinental setting and absence of abundant shallow anoxia [509] suggests that local marine chemistry could have differed in many important ways from that of the 'global ocean' where coeval reefs around the world occurred. However, broad stratigraphy shows that most or all Late Devonian reefs were affected by a major eustatic sea level cycle through the Frasnian and Famennian. Hence, the DGBR may be typical of reefs of its time in some ways, but not in others.

5.4. Recommendations

- Conservation of the GBR and other modern reefs requires transdisciplinary study that addresses threats that affect changes to any and all of the five key reef processes (Figure 5).

- Additional deep reef coring is needed in the northern GBR and in the offshore plateaus to better understand how the older reef history (Miocene to Pleistocene) was constrained, as that history led to the reefs we know today.
- Additional Holocene and Pleistocene reef coring in the GBR is needed to provide direct analogues for modern reefs that had changing local conditions, and to recover both paleoclimate data to inform modern climate models and projections and data of the specific responses of individual organisms, communities and whole reefs to past environmental–climatic changes. All will inform conservation strategies for modern reefs.
- Current understanding of the DGBR is limited by difficulties in correlating the shallow successions. Hence, increased shallow biostratigraphy (e.g., microvertebrates) and chemostratigraphy (especially stable isotopes) are needed for addition to the ever-improving sequence stratigraphy.
- Additional geochemical investigations are also needed in the Canning Basin to better understand the patterns and causes of changes to marine alkalinity and nutrients that affected carbonate factories, hydrocarbon systems and base metal emplacement. Trace element geochemistry including REEs, redox sensitive metals and halogens along with stable C and N isotopes derived from carbonates, shales and organic matter would be useful.
- Additional study of subsurface examples of the DGBR (offshore western Australia and Barbwire Terrace) and the exposures in the Bonaparte Gulf would also shed more light on this large reef province.

Funding: This review was funded by the Dorothy Hill Chair of Palaeontology and Stratigraphy, The University of Queensland.

Acknowledgments: I am grateful to Markes E. Johnson for inviting this review and encouraging me throughout as well as the very patient editors at *JMSE* and for the critical reviews of reviewers, which helped improve the paper. Although I am alone responsible for the content above, I must also acknowledge and thank the many people who have spent time with me in the field, both on the GBR and in the Kimberley, discussing science and helping me understand and formulate the ideas presented above. These people include, but are not limited to: John S. Jell, Patrick Sutherland, Markes E. Johnson, Luke D. Nothdurft, Jody Webster, Scott Brownlaw, Shen Jianwei, Annette George, Joachim Reitner, Trevor Graham, Nicole Leonard, Josh Reid, Tania Kenyon, Adam Miethke, Narottam Saha, James Sadler, Jennine McCutcheon, Marcos Salas-Saavedra, Belinda Dechnik, Atifeh Sansoleimani, Dan Harris and Kelsey Sanborn. I also thank the late Phil Playford for his initial help in getting me started in the DGBR. I am also grateful to Vikram Vakil for aiding with drafting of some figures. I am also very grateful to Joe Ross and the other Bunuba elders for their kind access to, knowledge about and aid working in their beautiful country in the Kimberley, and for the many years working in the southern Great Barrier Reef in the Sea Country of the First Nations Bailai, Gurang, Goreng Goreng, and Taribelang Bunda Peoples.

Conflicts of Interest: The author declares no conflicts of interest.

References

1. Knowlton, N.; Brainard, R.E.; Fisher, R.; Moews, M.; Plaisance, L.; Caley, M.J. Coral reef biodiversity. In *Life in the World's Oceans*; McIntyre, A.D., Ed.; John Wiley & Sons: Chichester, UK, 2010; pp. 65–77. [CrossRef]
2. Burke, L.; Reytar, K.; Spalding, M.; Perry, A. *Reefs at Risk Revisited*; World Resources Institute: Washington, DC, USA, 2011; pp. 1–130. ISBN 978-1-56973-762-0.
3. Harris, D.L.; Rovere, A.; Casella, E.; Power, H.; Canavesio, R.; Collin, A.; Pomeroy, A.; Webster, J.M.; Parravicini, V. Coral reef structural complexity provides important coastal protection from waves under rising sea levels. *Sci. Adv.* **2018**, *4*, eaao4350. [CrossRef]

4. Woodhead, A.J.; Hicks, C.C.; Norström, A.V.; Williams, G.J.; Graham, N.A.J. Coral reef ecosystem services in the Anthropocene. *Funct. Ecol.* **2019**, *33*, 1023–1034. [CrossRef]
5. Reguero, B.G.; Storlazzi, C.D.; Gibbs, A.E.; Shope, J.B.; Cole, A.D.; Kumming, K.A.; Beck, M.W. The value of US coral reefs for flood risk reduction. *Nat. Sustain.* **2021**, *4*, 688–698. [CrossRef]
6. Bindoff, N.L.; Cheung, W.W.; Kairo, J.G.; Arístegui, J.; Guinder, V.A.; Hallberg, R.; Williamson, P. Changing Ocean, Marine Ecosystems, and Dependent Communities. In *IPCC Special Report on the Ocean and Cryosphere in a Changing Climate*; Intergovernmental Panel on Climate Change: Cambridge University Press: Cambridge, UK; New York, NY, USA, 2019; pp. 477–587. [CrossRef]
7. De Valck, J.; Rolfe, J. Reviewing the use of proxies to value coastal and marine biodiversity protection: The Great Barrier Reef in Australia. *Mar. Policy* **2022**, *136*, 104890. [CrossRef]
8. Barker, B. *The Sea People: Late Holocene Maritime Specialisation in the Whitsunday Islands, Central Queensland*; Pandanus Books: Canberra, Australia, 2004; pp. 1–236.
9. Narchie, N.E.; Price, L.L. *Ethnobiology of Corals and Coral Reefs*; Springer: Cham, Switzerland, 2015; pp. 1–240. [CrossRef]
10. Marrie, H.L. Indigenous Coral Reef Tourism. In *Coral Reefs: Tourism, Conservation and Management*; Prideaux, B., Pabel, A., Eds.; Routledge: London, UK, 2018; pp. 227–246.
11. Bradley, J. with Yanyuwa families. Three coral reefs in Yanyuwa country, meaningful and powerful. In *Coral Reefs of Australia: Perspectives from Beyond the Water's Edge*; Hamylton, S.M., Hutchings, P., Hoegh-Guldberg, O., Eds.; CSIRO Publishing: Clayton South, VIC, Australia, 2023; pp. 53–57.
12. Rowland, M.J.; Lambrides, A.B.J.; McNiven, I.J.; Ulm, S. Great Barrier Reef Indigenous archaeology and occupation of associated reef and continental islands. *Aust. J. Environ. Manag.* **2024**, *32*, 22–45. [CrossRef]
13. Reyes-Nivia, C.; Diaz-Pulido, G.; Kline, D.; Hoegh-Guldberg, O.; Dove, S. Ocean acidification and warming scenarios increase microbioerosion of coral skeletons. *Glob. Change Biol.* **2013**, *19*, 1919–1929. [CrossRef]
14. Hoegh-Guldberg, O.; Pendleton, L.; Kaup, A. People and the changing nature of coral reefs. *Reg. Stud. Mar. Sci.* **2019**, *30*, 100699. [CrossRef]
15. Hughes, T.P.; Kerry, J.T.; Baird, A.H.; Connolly, S.R.; Dietzel, A.; Eakin, C.M.; Heron, S.F.; Hoey, A.S.; Hoogenboom, M.O.; Liu, G.; et al. Global warming transforms coral reef assemblages. *Nature* **2018**, *556*, 492–496. [CrossRef] [PubMed]
16. Walsh, K.J.E.; McBride, J.L.; Klotzbach, P.J.; Balachandran, S.; Camargho, S.J.; Holland, G.; Knutson, T.R.; Kossin, J.P.; Lee, T.-C.; Sobel, A.; et al. Tropical cyclones and climate change. *WIREs Clim. Chang.* **2016**, *7*, 65–89. [CrossRef]
17. Cheal, A.J.; MacNeil, M.A.; Emslie, M.J.; Sweatman, H. The threat to coral reefs from more intense cyclones under climate change. *Glob. Change Biol.* **2017**, *23*, 1511–1524. [CrossRef]
18. Eyre, B.D.; Cyronak, T.; Drupp, P.; De Carlo, E.H.; Sachs, J.P.; Andersson, A.J. Coral reefs will transition to net dissolving before end of century. *Science* **2018**, *359*, 908–911. [CrossRef] [PubMed]
19. Hill, T.S.; Hoogenboom, M.O. The indirect effects of ocean acidification on corals and coral communities. *Coral Reefs* **2022**, *41*, 1557–1583. [CrossRef]
20. Pandolfi, J.; Bradbury, R.H.; Sala, E.; Hughes, T.P.; Bjorndal, K.A.; Cooke, R.G.; McArdle, D.; McClenachan, L.; Newman, M.J.H.; Paredes, G.; et al. Global trajectories of the long-term decline of coral reef ecosystems. *Science* **2003**, *301*, 955–958. [CrossRef]
21. Riegl, B.M.; Glynn, P.W. Population dynamics of the reef crisis: Consequences of the growing human population. *Adv. Mar. Biol.* **2020**, *87*, 2–30. [CrossRef]
22. Castro-Sanguino, C.; Ortiz, J.C.; Thompson, A.; Wolff, N.H.; Ferrari, R.; Robson, B.; Magno-Canto, M.M.; Puotinen, M.; Fabricius, K.E.; Uthicke, S. Reef state and performance as indicators of cumulative impacts on coral reefs. *Ecol. Indic.* **2021**, *123*, 107335. [CrossRef]
23. Lewis, S.; McCloskey, G.; Bainbridge, Z.; Davis, A.; Bartley, R.; Turner, R. Sediment and nutrient flux from land. In *Oceanographic Processes of Coral Reefs, Physical and Biological Links in the Great Barrier Reef*, 2nd ed.; Wolanski, E., Kingsford, M., Eds.; CRC Press: Boca Raton, FL, USA, 2024; pp. 126–142.
24. Darwin, C.R. *The Structure and Distribution of Coral Reefs. Being the First Part of the Geology of the Voyage of the Beagle, Under the Command of Capt. Fitzroy, R.N. During the Years 1832 to 1836*; Smith Elder and Co.: London, UK, 1842; pp. 1–214.
25. Montaggioni, L.F.; Braithwaite, C.J.R. *Quaternary Coral Reef Systems: History, Development, Processes and Controlling Factors*; Elsevier: Amsterdam, The Netherlands, 2009; pp. 1–532.
26. Pandolfi, J.M.; Kiessling, W. Gaining insights from past reefs to inform understanding of coral reef response to global climate change. *Curr. Opin. Environ. Sustain.* **2014**, *7*, 52–58. [CrossRef]
27. Renema, W.; Pandolfi, J.M.; Kiessling, W.; Bosellini, F.R.; Klaus, J.S.; Korpanty, C.; Rosen, B.R.; Santodomingo, N.; Wallace, C.C.; Webster, J.M.; et al. Are coral reefs victims of their own past success? *Sci. Adv.* **2016**, *2*, e1500850. [CrossRef] [PubMed]
28. O'Dea, A.; Lepore, M.; Altieri, A.H.; Chan, M.; Morales-Saldaña, J.M.; Muñoz, N.-H.; Pandolfi, J.M.; Toscano, M.A.; Zhao, J.X.; Dillon, E.M. Defining variation in pre-human ecosystems can guide conservation: An example from a Caribbean coral reef. *Sci. Rep.* **2020**, *10*, 2922. [CrossRef]

29. Dimitrijević, D.; Santodomingo, N.; Kiessling, W. Reef refugia in the aftermath of past episodes of global warming. *Coral Reefs* **2024**, *43*, 1431–1442. [CrossRef]
30. Webb, G.E. Advances in understanding climate change on the Great Barrier Reef using coral-based proxies, Chapter 5. In *Oceanographic Processes of Coral Reefs: Physical and Biological Links in the Great Barrier Reef*, 2nd ed.; Wolanski, E., Kingsford, M.J., Eds.; CRC Press: Boca Raton, Fl, USA, 2024; pp. 62–85.
31. Fagerstrom, J.A. *The Evolution of Reef Communities*; Wiley: New York, NY, USA, 1987; pp. 1–600.
32. Fagerstrom, J.A. A structural model for reef communities. *Palaios* **1988**, *3*, 217–220. [CrossRef]
33. Fagerstrom, J.A. Reef-building guilds and a checklist for determining guild membership. *Coral Reefs* **1991**, *10*, 47–52. [CrossRef]
34. Machel, H.G.; Hunter, I.G. Facies models for Middle to Late Devonian shallow-marine carbonates, with comparisons to modern reefs: A guide for facies analysis. *Facies* **1994**, *30*, 155–176. [CrossRef]
35. Wood, R. *Reef Evolution*; Oxford University Press: Oxford, UK, 1999; pp. 1–414.
36. Stanley, G.D., Jr. *The History and Sedimentology of Ancient Reef Systems*; Kluwer Academic/Plenum Publishers: New York, NY, USA, 2001; pp. 1–458.
37. Kiessling, W.; Flügel, E.; Golonka, J. (Eds.) *Phanerozoic Reef Patterns*; SEPM: Tulsa, OK, USA, 2002; pp. 1–775.
38. Kiessling, W. Long-term relationships between ecological stability and biodiversity in Phanerozoic reefs. *Nature* **2005**, *433*, 410–413. [CrossRef]
39. Kiessling, W. Geologic and biologic controls on the evolution of reefs. *Ann. Rev. Ecol. Evol. Syst.* **2009**, *40*, 73–192. [CrossRef]
40. Berger, W.H. Increase of carbon dioxide in the atmosphere during deglaciation: The coral reef hypothesis. *Naturwissenschaften* **1982**, *69*, 87–88. [CrossRef]
41. Ryan, D.A.; Opdyke, B.N.; Jell, J.S. Holocene sediments of Wistari Reef: Towards a global quantification of coral reef related neritic sedimentation in the Holocene. *Palaeogeogr. Palaeoclimatol. Palaeoecol.* **2001**, *175*, 173–184. [CrossRef]
42. Felis, T.; Hinstrosa, G.; Köhler, P.; Webster, J. Role of the Deglacial Buildup of the Great Barrier Reef for the Global Carbon Cycle. *Geophys. Res. Lett.* **2022**, *49*, e2021GL096495. [CrossRef]
43. James, K.; Macreadie, P.I.; Burdett, H.L.; Davies, I.; Kamenos, N.A. It's time to broaden what we consider a 'blue carbon ecosystem. ' *Glob. Change Biol.* **2024**, *30*, e17261. [CrossRef]
44. Greenstein, B.J.; Pandolfi, J.M. Escaping the heat: Range shifts of reef coral taxa in coastal Western Australia. *Glob. Change Biol.* **2008**, *14*, 513–528. [CrossRef]
45. Camoin, G.F.; Webster, J.M. Coral reef response to Quaternary sea-level and environmental changes: State of the science. *Sedimentology* **2015**, *62*, 401–428. [CrossRef]
46. Webster, J.M.; Braga, J.C.; Humblet, M.; Potts, D.C.; Iryu, Y.; Yokoyama, Y.; Fujita, K.; Bourillot, R.; Esat, T.M.; Fallon, S.; et al. Response of the Great Barrier Reef to sea-level and environmental changes over the past 30,000 years. *Nat. Geosci.* **2018**, *11*, 426–432. [CrossRef]
47. Davies, P.J. Subsurface solution unconformities at Heron Island, Great Barrier Reef. In *Proceedings of the Second International Symposium on Coral Reefs*; Cameron, A.M., Campbell, B.M., Cribb, A.B., Endean, R., Jell, J.S., Jones, O.A., Talbot, F.H., Eds.; The Great Barrier Reef Committee: Brisbane, Australia, 1974; Volume 2, pp. 573–578.
48. Macintyre, I.G. Modern coral reefs of western Atlantic: New geologic perspectives. *Am. Assoc. Pet. Geol. Bull.* **1988**, *72*, 1360–1369.
49. International Consortium for Great Barrier Reef Drilling; Alexander, I.; Andres, M.S.; Braithwaite, C.J.R.; Braga, J.C.; Davies, P.J.; Elderfield, H.; Gilmour, M.A.; Kay, R.L.; Kroon, D.; et al. New constraints on the origin of the Australian Great Barrier Reef: Results from an international project of deep coring. *Geology* **2001**, *29*, 483–486. [CrossRef]
50. Birkeland, C.; Green, A.; Lawrence, A.; Coward, G.; Vaeoso, M.; Fenner, D. Different resiliencies in coral communities over ecological and geological time scales in American Samoa. *Mar. Ecol. Prog. Ser.* **2021**, *673*, 55–68. [CrossRef]
51. Webb, G.E.; Webster, J.M. Scientific drilling on the Great Barrier Reef: Unlocking the history of the reef. In *Coral Reefs of Australia: Perspectives from Beyond the Water's Edge*; Hamylton, S.M., Hutchings, P., Hoegh-Guldberg, O., Eds.; CSIRO Publishing: Clayton South, VIC, Australia, 2023; pp. 124–130.
52. Pingitore, N.E. Vadose and phreatic diagenesis: Processes, products and their recognition in corals. *J. Sediment. Petrol.* **1976**, *46*, 985–1006.
53. Hendy, E.J.; Gagan, M.K.; Lough, J.M.; McCulloch, M.; de Menocal, P. Impact of skeletal dissolution and secondary aragonite on trace element and isotopic climate proxies in *Porites* corals. *Paleoceanography* **2007**, *22*, PA4101. [CrossRef]
54. Webb, G.E.; Nothdurft, L.D.; Kamber, B.S.; Kloprogge, J.T.; Zhao, J.-X. Rare earth element geochemistry of scleractinian coral skeleton during meteoric diagenesis: A sequence through neomorphism of aragonite to calcite. *Sedimentology* **2009**, *56*, 1433–1463. [CrossRef]
55. Gischler, E.; Thomas, A.L.; Droxler, A.W.; Webster, J.M.; Yokoyama, Y.; Schöne, B.R. Microfacies and diagenesis of older Pleistocene (pre-last glacial maximum) reef deposits, Great Barrier Reef, Australia (IODP Expedition 325): A quantitative approach. *Sedimentology* **2013**, *60*, 1432–1466. [CrossRef]

56. Booker, S.; Jones, B.; Li, L. Diagenesis in Pleistocene (80 to 500 ka) corals from the Ironshore Formation: Implications for paleoclimate reconstruction. *Sed. Geol.* **2020**, *399*, 105615. [CrossRef]
57. Salles, T.; Ding, X.; Webster, J.M.; Vila-Concejo, A.; Brocard, G.; Pall, J. A unified framework for modelling sediment fate from source to sink and its interactions with reef systems over geological times. *Sci. Rep.* **2018**, *8*, 5252. [CrossRef]
58. Pall, J.; Chandra, R.; Azam, D.; Salles, T.; Webster, J.M.; Scalzo, R.; Cripps, S. Bayesreef: A Bayesian inference framework for modelling reef growth in response to environmental change and biological dynamics. *Environ. Model. Softw.* **2020**, *125*, 104610. [CrossRef]
59. Pandolfi, J.M. Limited membership in Pleistocene reef coral assemblages from the Huon Peninsula, Papa New Guinea: Constancy during global change. *Paleobiology* **1996**, *22*, 152–176. [CrossRef]
60. Pandolfi, J.M.; Connolly, S.R.; Marshall, D.J.; Cohen, A.L. Projecting coral reef futures under global warming and ocean acidification. *Science* **2011**, *333*, 418–422. [CrossRef] [PubMed]
61. Weiss, A.; Martindale, R.C. Crustose coralline algae increased framework and diversity on ancient coral reefs. *PLoS ONE* **2017**, *12*, e0181637. [CrossRef]
62. Raja, N.B.; Pandolfi, J.M.; Kiessling, W. Modularity explains large-scale reef booms in earth's history. *Facies* **2023**, *69*, 15. [CrossRef]
63. Fagerstrom, J.A.; Weidlich, O. Strengths and Weaknesses of the Reef Guild Concept and Quantitative Data: Application to the Upper Capitan-Massive Community (Permian), Guadalupe Mountains, New Mexico-Texas. *Facies* **1999**, *40*, 131–156. [CrossRef]
64. Cheal, A.J.; MacNeil, M.; Cripps, E.; Emslie, M.J.; Jonker, M.; Schaffelke, B.; Sweatman, H. Coral–macroalgal phase shifts or reef resilience: Links with diversity and functional roles of herbivorous fishes on the Great Barrier Reef. *Coral Reefs* **2010**, *29*, 1005–1015. [CrossRef]
65. Steneck, R.S. The ecology of coralline algal crusts: Convergent patterns and adaptive strategies. *Annu. Rev. Ecol. Syst.* **1986**, *17*, 273–303. [CrossRef]
66. Teichert, S.; Steinbauer, M.; Kiessling, W. A possible link between coral reef success, crustose coralline algae and the evolution of herbivory. *Sci. Rep.* **2020**, *10*, 17748. [CrossRef]
67. Teichert, S. Attached and free-living crustose coralline algae and their functional traits in the geological record and today. *Facies* **2024**, *70*, 8. [CrossRef]
68. Jackson, J.B.C. Pleistocene perspectives on coral reef community structure. *Am. Zool.* **1992**, *32*, 719–731. [CrossRef]
69. Zapalski, M.K.; Baird, A.H.; Bridge, T.; Jakubowicz, M.; Daniell, J. Unusual shallow water Devonian coral community from Queensland and its recent analogues from the inshore Great Barrier Reef. *Coral Reefs* **2021**, *40*, 417–431. [CrossRef]
70. Godbold, A.; Clark, N.; Cunningham, E.T.; Bottjer, D.J.; Pandolfi, J.M. Temporal dynamics of Devonian reef communities: Insights into natural phase shifts and long-term resilience in the face of environmental variability. *Palaeogeogr. Palaeoclimat. Palaeoecol.* **2024**, *648*, 112264. [CrossRef]
71. Bridge, T.C.; Baird, A.H.; Pandolfi, J.M.; McWilliam, M.J.; Zapalski, M.K. Functional consequences of Palaeozoic reef collapse. *Sci. Rep.* **2022**, *12*, 1386. [CrossRef]
72. Bellwood, D.R.; Brandl, S.J.; McWilliam, M.; Streit, R.P.; Yan, H.F.; Tebbett, S.B. Studying functions on coral reefs: Past perspectives, current conundrums, and future potential. *Coral Reefs* **2024**, *43*, 281–297. [CrossRef]
73. Arzey, A.K.; McGregor, H.V.; Clark, T.R.; Webster, J.M.; Lewis, S.E.; Mallela, J.; McKay, N.P.; Fahey, H.W.; Chakraborty, S.; Razak, T.B.; et al. Coral skeletal proxy records database for the Great Barrier Reef, Australia. *Earth Syst. Sci. Data* **2024**, *16*, 4869–4930. [CrossRef]
74. Brodie, J.; Kroon, F.; Schaffelke, B.; Wolanski, E.; Lewis, S.; Devlin, M.; Bohnet, I.; Bainbridge, Z.; Waterhouse, J.; Davis, A. Terrestrial pollutant runoff to the Great Barrier Reef: An update of issues, priorities and management responses. *Mar. Pollut. Bull.* **2012**, *65*, 81–100. [CrossRef] [PubMed]
75. De'ath, G.; Fabricius, K.E.; Sweatman, H.; Puotinen, M. The 27–year decline of coral cover on the Great Barrier Reef and its causes. *Proc. Nat Acad. Sci. USA* **2012**, *109*, 17995–17999. [CrossRef] [PubMed]
76. Smith, J.N.; Mongin, M.; Thompson, A.; Jonker, M.J.; De'ath, G.; Fabricius, K.E. Shifts in coralline algae, macroalgae, and coral juveniles in the Great Barrier Reef associated with present-day ocean acidification. *Glob. Change Biol.* **2020**, *26*, 2149–2160. [CrossRef]
77. Dixon, A.M.; Puotinen, M.; Ramsay, H.A.; Beger, M. Coral Reef Exposure to Damaging Tropical Cyclone Waves in a Warming Climate. *Earth's Futur.* **2022**, *10*, e2021EF002600. [CrossRef]
78. Duke, N.C.; Canning, A.; Mackenzie, A. More intense severe tropical cyclones in recent decades cause greater impacts on mangroves bordering Australia's Great Barrier Reef. In *Oceanographic Processes of Coral Reefs: Physical and Biological Links in the Great Barrier Reef*; Wolanski, E., Kingsford, M.J., Eds.; CRC Press: Boca Raton, FL, USA, 2024; pp. 226–253.
79. Emslie, M.J.; Ceccarelli, D.M.; Logan, M.; Blandford, M.I.; Bray, P.; Campili, A.; Jonker, M.J.; Parker, J.G.; Prenzlau, T.; Sinclair-Taylor, T.H. Changing dynamics of Great Barrier Reef hard coral cover in the Anthropocene. *Coral Reefs* **2024**, *43*, 747–762. [CrossRef]

80. Hopley, D.; Smithers, S.G.; Parnell, K. *The Geomorphology of the Great Barrier Reef: Development, Diversity and Change*; Cambridge University Press: Cambridge, UK, 2007; pp. 1–532.
81. Wolanski, E.; Kingsford, M.; Lambrechts, J.; Marmorino, G. The physical oceanography of the Great Barrier Reef: A review, Chapter 2. In *Oceanographic Processes of Coral Reefs: Physical and Biological Links in the Great Barrier Reef*, 2nd ed.; Wolanski, E., Kingsford, M.J., Eds.; CRC Press: Boca Raton, FL, USA, 2024; pp. 9–34.
82. Davies, P.J.; Symonds, P.A.; Feary, D.A.; Pigram, C.J. Horizontal plate motion: A key allocyclic factor in the evolution of the Great Barrier Reef. *Science* **1987**, *238*, 1697–1700. [CrossRef] [PubMed]
83. Feary, D.A.; Davies, P.J.; Pigram, C.J.; Symonds, P.A. Climatic evolution and control on carbonate deposition in northeast Australia. *Palaeogeogr. Palaeoclimat. Palaeoecol.* **1991**, *89*, 341–361. [CrossRef]
84. Copper, P. Silurian and Devonian reefs: 80 million years of global greenhouse between two ice ages. In *Phanerozoic Reef Patterns*; Kiessling, W., Flügel, E., Golonka, J., Eds.; SEPM Special Publication; SEPM Society for Sedimentary Geology: Tulsa, OK, USA, 2002; Volume 72, pp. 181–238.
85. Copper, P.; Scotese, C.R. Megareefs in Middle Devonian Supergreenhouse Climates. In *Extreme depositional environments: Mega end members in geologic time*; Chan, M.A., Archer, A.W., Eds.; Geological Society of America Special Paper 370; Geological Society of America: Boulder, CO, USA, 2003; pp. 209–230.
86. Webb, G.E. Latest Devonian and Early Carboniferous reefs: Depressed reef building after the middle Paleozoic collapse. In *Phanerozoic Reef Patterns*; Kiessling, W., Flügel, E., Golonka, J., Eds.; SEPM Special Publication; SEPM Society for Sedimentary Geology: Tulsa, OK, USA, 2002; Volume 72, pp. 239–269.
87. Playford, P.E.; Lowry, D.C. Devonian reef complexes of the Canning Basin, Western Australia. *Geol. Surv. West. Aust. Bull.* **1966**, *118*, 1–150.
88. Playford, P.E. Devonian 'Great Barrier Reef' of Canning Basin, Western Australia. *Am. Ass. Petrol. Geol. Bull.* **1980**, *62*, 814–840. [CrossRef]
89. Playford, P.E.; Hurley, N.F.; Kerans, C.; Middleton, M.F. Reefal platform development, Devonian of the Canning Basin, Western Australia. In *Controls on Carbonate Platform and Basin Development*; Crevello, P.D., Wilson, J.L., Sarg, J.F., Read, J.F., Eds.; SEPM Special Publication; SEPM Society for Sedimentary Geology: Tulsa, OK, USA, 1989; Volume 44, pp. 187–202.
90. Playford, P.E.; Hocking, R.M.; Cockbain, A.E. (Eds.) *Devonian Reef Complexes of the Canning Basin, Western Australia*; Bulletin 145; Geological Survey of Western Australia: Perth, WA, Australia, 2009; Volume 444, pp. 415–440.
91. Sepkoski, J.J. Patterns of Phanerozoic extinction: A perspective from global data bases. In *Global Events and Event Stratigraphy in the Phanerozoic*; Walliser, O.H., Ed.; Springer: Berlin/Heidelberg, Germany, 1996; pp. 35–51.
92. Johnson, M.E. *Islands in Deep Time, Ancient Landscapes Lost and Found*; Columbia University Press: New York, NY, USA, 2023; pp. 1–294.
93. Porter, S.M. Calcite and aragonite seas and the de novo acquisition of carbonate skeletons. *Geobiology* **2010**, *8*, 256–277. [CrossRef]
94. Sandberg, P.A. An oscillating trend in Phanerozoic non-skeletal carbonate mineralogy. *Nature* **1983**, *305*, 19–22. [CrossRef]
95. Vérard, C.; Hochard, C.; Baumgartner, P.O.; Stampfli, G.M. 3D palaeogeographic reconstructions of the Phanerozoic versus sea-level and Sr-ratio variations. *J. Palaeogeogr.* **2015**, *4*, 64–84. [CrossRef]
96. Scotese, C.R.; Song, H.; Mills, B.J.; van der Meer, D.G. Phanerozoic paleotemperatures: The earth's changing climate during the last 540 million years. *Earth-Sci. Rev.* **2021**, *215*, 103503. [CrossRef]
97. Kiessling, W. Secular variations in the Phanerozoic reef system. In *Phanerozoic Reef Patterns*; Kiessling, W., Flügel, E., Golonka, J., Eds.; SEPM: Tulsa, OK, USA, 2002; pp. 625–690.
98. Webb, G.E. Biologically induced carbonate Precipitation in reefs through time, Chapter 5. In *Sedimentology of Ancient Reef the History and Systems*; Stanley, G.D., Jr., Ed.; Academic/Plenum Publishers: New York, NY, USA, 2001; pp. 159–203.
99. Riding, R.; Liang, L. Geobiology of microbial carbonates: Metazoan and seawater saturation state influences on secular trends during the Phanerozoic. *Palaeogeog. Palaeoclimat. Palaeoecol.* **2005**, *219*, 101–115. [CrossRef]
100. Cohen, K.M.; Finney, S.C.; Gibbard, P.L.; Fan, J.-X. The ICS International Chronostratigraphic Chart. *Episodes* **2013**, *35*, 199–204. [CrossRef]
101. Heckel, P.H. Carbonate buildups in the geologic record: A review. In *Reefs in Time and Space*; Laporte, L.F., Ed.; SEPM Special Publication; SEPM Society for Sedimentary Geology: Tulsa, OK, USA, 1974; Volume 18, pp. 90–155.
102. Stanley, G.D., Jr. (Ed.) Introduction to reef ecosystems and their evolution, Chapter 1. In *The History and Sedimentology of Ancient Reef Systems*; Academic/Plenum Publishers: New York, NY, USA, 2001; pp. 1–39.
103. Riding, R. Structure and composition of organic reefs and carbonate mud mounds: Concepts and categories. *Earth-Sci. Rev.* **2002**, *58*, 163–231. [CrossRef]
104. Hubbard, D.K.; Gill, I.P.; Burke, R.B. The role of framework in modern reefs and its application to ancient systems. In *The History and Sedimentology of Ancient Reef Systems*; Stanley, G.D., Jr., Ed.; Academic/Plenum Publishers: New York, NY, USA, 2001; pp. 351–388.

105. Kleypas, J.A.; Buddemeier, R.W.; Gattuso, J.-P. The future of coral reefs in an age of global change. *Int. J. Earth Sci.* **2001**, *90*, 426–437. [CrossRef]
106. Hopley, D.; Smithers, S. Geomorphology of coral reefs with special reference to the Great Barrier Reef. In *The Great Barrier Reef: Biology, Environment and Management*; Hutchings, P., Kingsford, M., Hoegh-Guldberg, O., Eds.; CSIRO Publishing: Canberra, Australia, 2008; pp. 9–24.
107. Wolfe, K.; Anthony, K.; Babcock, R.C.; Bay, L.; Bourne, D.G.; Burrows, D.; Byrne, M.; Deaker, D.J.; Diaz-Pulido, G.; Frade, P.R.; et al. Priority species to support the functional integrity of coral reefs. In *Oceanography and Marine Biology, An Annual Review*; Hawkins, S.J., Allcock, A.L., Bates, A.E., Evans, A.J., Firth, L.B., McQuaid, C.D., Russell, B.D., Smith, I.P., Swearer, S.E., Todd, P.A., Eds.; CRC Press: Boca Raton, FL, USA, 2021; Volume 58, pp. 179–318. ISBN 978-0-367-36794-7.
108. Perry, C.T.; Spencer, T.; Kench, P.S. Carbonate budgets and reef production states: A geomorphic perspective on the ecological phase-shift concept. *Coral Reefs* **2008**, *27*, 853–866. [CrossRef]
109. De Carlo, T.M.; Cohen, A.L. Dissepiments, density bands and signatures of thermal stress in Porites skeletons. *Coral Reefs* **2017**, *36*, 749–761. [CrossRef]
110. Connell, J.H. Diversity in tropical rain forests and coral reefs. *Science* **1978**, *199*, 1302–1310. [CrossRef] [PubMed]
111. Yu, K.-F.; Zhao, J.-X.; Roff, G.; Lybolt, M.; Feng, Y.-X.; Clark, T.; Li, S. High-precision U-series ages of transported coral blocks on Heron Reef (southern Great Barrier Reef) and storm activity during the past century. *Palaeogeogr. Palaeoclim. Palaeoecol.* **2012**, *337–338*, 23–36. [CrossRef]
112. Clark, T.R.; Zhao, J.X.; Roff, G.; Feng, Y.-X.; Done, T.J.; Nothdurft, L.D.; Pandolfi, J.M. Discerning the timing and cause of historical mortality events in modern *Porites* from the Great Barrier Reef. *Geochim Cosmochim Acta* **2014**, *138*, 57–80. [CrossRef]
113. Chen, T.; Li, S.; Zhao, J.X.; Feng, Y.X. Uranium-thorium dating of coral mortality and community shift in a highly disturbed inshore reef (Weizhou Island, northern South China Sea). *Sci. Tot. Environ.* **2021**, *752*, 141866. [CrossRef] [PubMed]
114. Hendy, E.J.; Gagan, M.K.; Alibert, C.A.; McCulloch, M.T.; Lough, J.M.; Isdale, P.J. Abrupt decrease in tropical Pacific sea surface salinity at end of Little Ice Age. *Science* **2002**, *295*, 1511–1514. [CrossRef]
115. Lough, J.M.; Lewis, S.E.; Cantin, N.E. Freshwater impacts in the central Great Barrier Reef: 1648–2011. *Coral Reefs* **2015**, *34*, 739–751. [CrossRef]
116. Saha, N.; Webb, G.E.; Zhao, J.-X. Coral skeletal geochemistry as a monitor of inshore water quality. *Sci. Tot. Environ.* **2016**, *566–567*, 652–684. [CrossRef]
117. Felis, T. Extending the Instrumental Record of Ocean-Atmosphere Variability into the Last Interglacial Using Tropical Corals. *Oceanography* **2020**, *33*, 68–79. [CrossRef]
118. Thompson, D.M. Environmental records from coral skeletons: A decade of novel insights and innovation. *Wiley Interdiscip. Rev. Clim. Chang.* **2022**, *13*, e745. [CrossRef]
119. Shirai, K.; Sowa, K.; Watanabe, T.; Sano, Y.; Nakamura, T.; Clode, P. Visualization of sub-daily skeletal growth patterns in massive *Porites* corals grown in Sr-enriched seawater. *J. Struct. Biol.* **2012**, *180*, 47–56. [CrossRef] [PubMed]
120. Mohtar, A.M.; Hughen, K.A.; Goodkin, N.F.; Streanga, I.-M.; Ramos, R.D.; Samanta, D.; Cervino, J.; Switzera, A.D. Coral-based proxy calibrations constrain ENSO-driven sea surface temperature and salinity gradients in the Western Pacific Warm Pool. *Palaeogeogr. Palaeoclimat. Palaeoecol.* **2021**, *561*, 110037. [CrossRef]
121. Rodriguez-Ramirez, A.; Grove, C.A.; Zinke, J.; Pandolfi, J.M.; Zhao, J.X. Coral luminescence identifies the pacific decadal oscillation as a primary driver of river runoff variability impacting the southern Great Barrier Reef. *PLoS ONE* **2014**, *9*, e84305. [CrossRef]
122. Abram, N.J.; Hargreaves, J.A.; Wright, N.M.; Thirumalai, K.; Ummenhofer, C.C.; England, M.H. Palaeoclimate perspectives on the Indian Ocean Dipole. *Quat. Sci. Rev.* **2020**, *237*, 106302. [CrossRef]
123. Chappell, J.; Omura, A.; Esat, T.; McCulloch, M.; Pandolfi, J.; Ota, Y.; Pillans, B. Reconciliation of late Quaternary sea levels derived from coral terraces at Huon Peninsula with deep sea oxygen isotope records. *Earth Planet. Sci. Lett.* **1996**, *141*, 227–236. [CrossRef]
124. Roehl, P.O.; Choquette, P.W. *Carbonate Petroleum Reservoirs*; Springer: New York, NY, USA, 1985; pp. 1–622.
125. Macintyre, I.G.; Marshall, J.F. Submarine lithification in coral reefs: Some facts and misconceptions. In Proceedings of the 6th International Coral Reef Symposium, Townsville, Australia, 8–12 August 1988; Choat, J.H., Ed.; 1988; Volume 1, pp. 263–272.
126. Harrington, L.; Fabricius, K.; De'ath, G.; Negri, A. Recognition and selection of settlement substrata determine post-settlement survival in corals. *Ecology* **2004**, *85*, 3428–3437. [CrossRef]
127. Madin, J.S.; Dell, A.I.; Madin, E.M.P.; Nash, M.C. Spatial variation in mechanical properties of coral reef substrate and implications for coral colony integrity. *Coral Reefs* **2013**, *32*, 173–179. [CrossRef]
128. Ceccarelli, D.M.; McLeod, I.M.; Boström-Einarsson, L.; Bryan, S.E.; Chartrand, K.M.; Emslie, M.J.; Gibbs, M.T.; Rivero, M.G.; Hein, M.Y.; Heyward, A.; et al. Substrate stabilisation and small structures in coral restoration: State of knowledge, and considerations for management and implementation. *PLoS ONE* **2020**, *15*, e0240846. [CrossRef]

129. Webb, G.E.; Baker, J.C.; Jell, J.S. Inferred syngenetic textural evolution in Holocene cryptic reefal microbialites, Heron Reef, Great Barrier Reef, Australia. *Geology* **1998**, *26*, 355–358. [CrossRef]
130. Nothdurft, L.D.; Webb, G.E. Earliest diagenesis in scleractinian coral skeletons: Implications for palaeoclimate-sensitive geochemical archives. *Facies* **2009**, *55*, 161–201. [CrossRef]
131. Westphal, H.; Heindel, K.; Brandano, M.; Peckmann, J. Genesis of microbialites as contemporaneous framework components of deglacial coral reefs, Tahiti (IODP 310). *Facies* **2010**, *56*, 337–352. [CrossRef]
132. Braga, J.C.; Puga-Bernabeua, A.; Heindel, K.; Patterson, M.A.; Birgeld, D.; Peckmann, J.; Sanchez-Almazo, I.M.; Webster, J.M.; Yokoyama, Y.; Riding, R. Microbialites in Last Glacial Maximum and deglacial reefs of the Great Barrier Reef (IODP Expedition 325, NE Australia). *Palaeogeogr. Palaeoclimat. Palaeoecol.* **2019**, *514*, 1–17. [CrossRef]
133. Nash, M.C.; Troitzsch, U.; Opdyke, B.N.; Trafford, J.M.; Russell, B.D.; Kline, D.I. First discovery of dolomite and magnesite in living coralline algae and its geobiological Implications. *Biogeosciences* **2011**, *8*, 3331–3340. [CrossRef]
134. Cornwall, C.E.; Carlot, J.; Branson, O.; Courtney, T.A.; Harvey, B.P.; Perry, C.T.; Andersson, A.J.; Diaz-Pulido, G.; Johnson, M.D.; Kennedy, E.; et al. Crustose coralline algae can contribute more than corals to coral reef carbonate production. *Commun. Earth Environ.* **2023**, *4*, 105. [CrossRef]
135. Maxwell, W.G.H.; Jell, J.S.; McKellar, R.G. Differentiation of carbonate sediments on the Heron Island Reef. *J. Sedim. Petrol.* **1964**, *34*, 294–308. [CrossRef]
136. Scoffin, T.P.; Tudhope, A.W. Sedimentary environments of the Central Region of the Great Barrier Reef of Australia. *Coral Reefs* **1985**, *4*, 81–93. [CrossRef]
137. Dawson, J.L.; Smithers, S.G. Carbonate sediment production, transport, and supply to a coral cay at Raine Reef, northern Great Barrier Reef, Australia: A facies approach. *J. Sed. Res.* **2014**, *84*, 1120–1138. [CrossRef]
138. Kench, P.S. Effects of environmental and climatic changes on coral reef islands. *Ann. Rev. Mar. Sci.* **2025**, *17*, 301–324. [CrossRef]
139. Ladd, H.S. The problem of coral reefs. *Sci. Mon.* **1949**, *69*, 297–305.
140. Blanchon, P.; Jones, B.; Kalbfleisch, W. Anatomy of a fringing reef around Grand Cayman; storm rubble, not coral framework. *J. Sediment. Res.* **1997**, *67*, 1–16. [CrossRef]
141. Hubbard, D.K.; Burke, R.B.; Gill, I.P. Where's the reef: The role of framework in the Holocene. *Carbonates Evaporites* **1998**, *13*, 3–9. [CrossRef]
142. Braithwaite, C.J.R.; Montaggioni, L.F.; Camoin, G.F.; Dalmasso, H.; Dullo, W.-C.; Mangini, A. Origins and development of Holocene coral reefs: A revisited model based on reef boreholes in the Seychelles, Indian Ocean. *Int. J. Earth Sci.* **2000**, *89*, 431–445. [CrossRef]
143. Riegl, B. Inhibition of reef framework by frequent disturbance: Examples from the Arabian Gulf, South Africa, and the Cayman Islands. *Palaeogeog. Palaeoclimat. Palaeoecol.* **2001**, *175*, 79–101. [CrossRef]
144. Grotzinger, J.P. Introduction to Precambrian reefs. In *Reefs, Canada and Adjacent Areas*; Geldsetzer, H.H.J., James, N.P., Tebbutt, G.E., Eds.; Canadian Society of Petroleum Geologists: Calgary, AB, Canada, 1989; Volume 13, pp. 9–12.
145. Schlager, W. Benthic carbonate factories of the Phanerozoic. *Int. J. Earth Sci.* **2003**, *92*, 445–464. [CrossRef]
146. Hallock, P.; Schlager, W. Nutrient excess and the demise of coral reefs and carbonate platforms. *Palaios* **1986**, *1*, 389–398. [CrossRef]
147. Riding, R. Temporal variation in calcification in marine cyanobacteria. *J. Geol. Soc.* **1992**, *149*, 979–989. [CrossRef]
148. Riding, R.; Liang, L.; Braga, J.C. Millennial-scale ocean acidification and late Quaternary decline of cryptic bacterial crusts in tropical reefs. *Geobiology* **2014**, *12*, 387–405. [CrossRef]
149. Reijmer, J.J.G. Marine carbonate factories: Review and update. *Sedimentology* **2021**, *68*, 1729–1796. [CrossRef]
150. Rasser, M.W.; Riegl, B. Holocene coral reef rubble and its binding agents. *Coral Reefs* **2002**, *21*, 57–72. [CrossRef]
151. Blanchon, P.; Richards, S.; Bernal, J.P.; Cerdeira-Estrada, S.; Ibarra, M.S.; Corona-Martínez, L.; Martell-Dubois, R. Retrograde accretion of a Caribbean fringing reef controlled by hurricanes and sea-level rise. *Front. Earth Sci.* **2017**, *5*, 78. [CrossRef]
152. Wolfe, K.; Kenyon, T.M.; Mumby, P.J. The biology and ecology of coral rubble and implications for the future of coral reefs. *Coral Reefs* **2021**, *40*, 1769–1806. [CrossRef]
153. Kenyon, T.M.; Doropoulos, C.; Wolfe, K.; Webb, G.E.; Dove, S.; Harris, D.; Mumby, P.J. Coral rubble dynamics in the Anthropocene and implications for reef recovery. *Limnol. Oceanogr.* **2023**, *68*, 110–147. [CrossRef]
154. Kenyon, T.M.; Mumby, P.J.; Webb, G.E.; Dove, S.; Newman, S.P.; Doropoulos, C. Trajectories and agents of binding in stabilized and unstabilized coral rubble across environmental gradients. *Ecosphere* **2025**, *16*, e70195. [CrossRef]
155. Littler, M.M.; Doty, M.S. Ecological components structuring the seaward edges of tropical Pacific reefs: The distribution, communities, and productivity of *Porolithon*. *J. Ecol.* **1975**, *63*, 117–129. [CrossRef]
156. Adey, W.H. Algal ridges of the Caribbean Sea and West Indies. *Phycologia* **1978**, *17*, 361–367. [CrossRef]
157. Insalaco, E. The descriptive nomenclature and classification of growth fabrics in fossil scleractinian reefs. *Sediment. Geol.* **1998**, *118*, 159–186. [CrossRef]
158. Webb, G.E. Was Phanerozoic reef history controlled by the distribution of non-enzymatically secreted reef carbonates (microbial carbonates and biologically induced cement)? *Sedimentology* **1996**, *43*, 947–971. [CrossRef]

159. Dunham, R.J. Classification of carbonate rocks according to depositional texture. In *Classification of Carbonate Rocks*; Ham, W.E., Ed.; American Association of Petroleum Geologists: Tulsa, OK, USA, 1962; Volume 1, pp. 108–121.
160. Embry, A.F.; Klovan, J.E. A Late Devonian reef tract on northeastern Banks Island, Northwest Territories. *Bull. Can. Pet. Geol.* **1971**, *33*, 730–781.
161. Wulff, J.L. Sponge-mediated coral reef growth and rejuvenation. *Coral Reefs* **1984**, *3*, 157–163. [CrossRef]
162. Wahlman, G.P. Upper Carboniferous-lower Permian (Bashkirian-Kungurian) mounds and reefs. In *Phanerozoic Reef Patterns*; Kiessling, W., Flügel, E., Golonka, J., Eds.; SEPM Special Publication; SEPM Society for Sedimentary Geology: Tulsa, OK, USA, 2002; Volume 72, pp. 271–338.
163. De Bakker, D.M.; Perry, C.T.; Magaña-Gallegos, E.; Pérez-Cervantes, E.; Alvarez-Filip, L. Fine-grained sediment production by endolithic sponges on Caribbean coral reefs. *Limnol. Oceanogr.* **2024**, *69*, 2015–2028. [CrossRef]
164. McCutcheon, J.; Nothdurft, L.D.; Webb, G.E.; Paterson, D.; Southam, G. Beachrock formation via microbial dissolution and re-precipitation of carbonate minerals. *Mar. Geol.* **2016**, *382*, 122–135. [CrossRef]
165. Scoffin, T.P.; Stearn, C.W.; Boucher, D.; Frydl, P.; Hawkins, C.M.; Hunter, I.G. Calcium carbonate budget of a fringing reef on the west coast of Barbados: Part II, Erosion, sediments and internal structure. *Bull. Mar. Sci.* **1980**, *30*, 475–508.
166. Yarlett, R.T.; Perry, C.T.; Wilson, R.W. Quantifying production rates and size fractions of parrotfish-derived sediment: A key functional role on Maldivian coral reefs. *Ecol. Evol.* **2021**, *11*, 16250–16265. [CrossRef] [PubMed]
167. Hughes, T.P.; Rodrigues, M.J.; Bellwood, D.R.; Ceccarelli, D.; Hoegh-Guldberg, O.; McCook, L.; Moltschaniwskyj, N.; Pratchett, M.S.; Steneck, R.S.; Willis, B. Phase Shifts, Herbivory, and the Resilience of Coral Reefs to Climate Change. *Curr. Biol.* **2007**, *17*, 360–365. [CrossRef] [PubMed]
168. MacNeil, A.J.; Jones, B. Stromatoporoid growth forms and Devonian reef fabrics in the Upper Devonian Alexandra Reef System, Canada—Insight on the challenges of applying Devonian reef facies models. *Sedimentology* **2016**, *63*, 1425–1457. [CrossRef]
169. Riding, R.; Virgone, A. Hybrid Carbonates: In situ abiotic, microbial and skeletal co-precipitates. *Earth-Sci. Rev.* **2020**, *208*, 103300. [CrossRef]
170. Szilagyi, Z.; Webster, J.M.; Patterson, M.A.; Hips, K.; Riding, R.; Foley, M.; Humblet, M.; Yokoyama, Y.; Liang, L.; Gischler, E.; et al. Controls on the spatio-temporal distribution of microbialite crusts on the Great Barrier Reef over the past 30,000 years. *Mar. Geol.* **2020**, *429*, 106312. [CrossRef]
171. Hubbard, D.K. Reef biology and geology—Not just a matter of scale. In *Coral Reefs in the Anthropocene*; Birkeland, C., Ed.; Springer: Dordrecht, The Netherlands, 2015; pp. 43–66. [CrossRef]
172. Camoin, G.F.; Montaggioni, L.F. High energy coralgal-stromatolite frameworks from Holocene reefs (Tahiti, French Polynesia). *Sedimentology* **1994**, *41*, 655–676. [CrossRef]
173. Tsien, H.H. Ancient reefs and reef carbonates. In *Proceedings of the 4th International Coral Symposium*; Gomez, E.D., Birkeland, C.E., Eds.; University of Philippines: Manila, Philippines, 1981; Volume 1, pp. 601–609.
174. Wilson, J.L. *Carbonate Facies in Geologic History*; Springer: New York, NY, USA, 1975; pp. 1–471.
175. Jell, J.S.; Flood, P.G. *Guide to the Geology of Reefs of the Capricorn and Bunker Groups, Great Barrier Reef Province, with Special Reference to Heron Reef*; Department of Geology, University of Queensland: Brisbane, QLD, Australia, 1978; Volume 8, pp. 1–85.
176. Jell, J.S.; Webb, G.E. Geology of Heron Island and adjacent reefs, Great Barrier Reef, Australia. *Episodes* **2012**, *35*, 110–119. [CrossRef]
177. Playford, P.E. Platform-margin and marginal slope relationships in Devonian reef complexes of the Canning Basin. In *The Canning Basin, Proceedings of the Geological Society of Australia and Petroleum Exploration Society of Australia Symposium, Perth, WA, Australia, 27–29 June 1984*; Purcell, P.G., Ed.; Geological Society of Australia and Petroleum Exploration Society of Australia: Perth, WA, Australia, 1984; pp. 189–215.
178. Davies, P.J.; Marshall, J.F.; Hopley, D. Relationships between reef growth and sea level in the Great Barrier Reef. In Proceedings of the Fifth International Coral Reef Congress, Tahiti, France, 27 May–1 June 1985; Gabrie, C., Toffart, J.L., Salvat, B., Eds.; Antenne Museum-Ephe: Moorea, France, 1985; Volume 3, pp. 95–103.
179. Neumann, A.C.; Macintyre, I.G. Reef response to sea level rise: Keep up catch up or give up. In Proceedings of the Fifth International Coral Reef Congress, Tahiti, France, 27 May–1 June 1985; Gabrie, C., Toffart, J.L., Salvat, B., Eds.; Antenne Museum-Ephe: Moorea, France, 1985; Volume 3, pp. 105–110.
180. Blanchon, P. Reef demise and back-stepping during the last interglacial, northeast Yucatan. *Coral Reefs* **2010**, *29*, 481–498. [CrossRef]
181. Maxwell, W.G.H. *Atlas of the Great Barrier Reef*; Elsevier: Amsterdam, The Netherlands, 1968; pp. 1–258.
182. Braithwaite, C.J.; Dalmasso, H.; Gilmour, M.A.; Harkness, D.D.; Henderson, G.M.; Kay, R.L.F.; Kroon, D.; Montaggioni, L.F.; Wilson, P.A. The Great Barrier Reef: The chronological record from a new borehole. *J. Sediment. Res.* **2004**, *74*, 298–310. [CrossRef]
183. Lehrmann, D.J.; Goldhammer, R.K. Secular Variation in Parasequence and Facies Stacking Patterns of Platform Carbonates: A Guide to Application of Stacking-Patterns Analysis in Strata of Diverse Ages and Settings. In *Advances in Carbonate Sequence*

Stratigraphy: Application to Reservoirs, Outcrops and Models; Harris, P.M., Simo, J.A., Eds.; SEPM Special Publication; SEPM Society for Sedimentary Geology: Tulsa, OK, USA, 1999; Volume 63, pp. 187–225. [CrossRef]
184. Woodroffe, C.D.; Webster, J.M. Coral reefs and sea-level change. *Mar. Geol.* **2014**, *352*, 248–267. [CrossRef]
185. Toscano, M.A.; Lundberg, J. Submerged Late Pleistocene reefs on the tectonically stable S.E. Florida margin: High-precision geochronology, stratigraphy, resolution of Substage 5a sea-level elevation, and orbital forcing. *Quat. Sci. Rev.* **1999**, *18*, 753–767. [CrossRef]
186. Lidz, B.H. Pleistocene Corals of the Florida Keys: Architects of Imposing Reefs—Why? *J. Coast. Res.* **2006**, *224*, 750–759. [CrossRef]
187. Muhs, D.R.; Simmons, K.R.; Schumann, R.R.; Halley, R.B. Sea-level history of the past two interglacial periods: New evidence from U-series dating of reef corals from south Florida. *Quat. Sci. Rev.* **2011**, *30*, 570–590. [CrossRef]
188. Church, J.A.; Clark, P.U.; Cazenave, A.; Gregory, J.M.; Jevrejeva, S.; Levermann, A.; Merrifield, M.A.; Milne, G.A.; Nerem, R.S.; Nunn, P.D.; et al. Sea Level Change. In *Climate Change 2013: The Physical Science Basis. Contribution of Working Group I*; Stocker, T.F., Qin, D., Plattner, G.-K., Tognor, M., Allen, S.K., Boschung, J., Nauels, A., Zia, Y., Bex, V., Midgley, P.M., Eds.; Fifth Assessment Report of the Intergovernmental Panel on Climate Change; Cambridge University Press: Cambridge, UK, 2013; pp. 1137–1216.
189. Van Wagoner, J.C.; Posamentier, H.W.; Mitchum, R.M.; Vail, P.R.; Sarg, J.F.; Loutit, T.S.; Hardenbol, J. An overview of sequence stratigraphy and key definitions. In *Sea Level Changes—An Integrated Approach*; Wilgus, C.K., Hastings, B.S., Kendall, C.G.S.C., Posamentier, H.W., Ross, C.A., Van Wagoner, J.C., Eds.; SEPM Special Publication; SEPM Society for Sedimentary Geology: Tulsa, OK, USA, 1988; Volume 42, pp. 39–45.
190. Symonds, P.A.; Davies, P.J.; Parisi, A. Structure and stratigraphy of the Central Great Barrier Reef. *Bur. Min. Resour. J. Austral. Geol. Geophys.* **1983**, *8*, 277–291.
191. Larcombe, P.; Woolfe, K.J. Increased sediment supply to the Great Barrier Reef will not increase sediment accumulation at most coral reefs. *Coral Reefs* **1999**, *18*, 163–169. [CrossRef]
192. Smithers, S.; Larcombe, P. Late Holocene initiation and growth of a nearshore turbid-zone coral reef: Paluma Shoals, central Great Barrier Reef, Australia. *Coral Reefs* **2003**, *22*, 499–505. [CrossRef]
193. Perry, C.T.; Smithers, S.G.; Gulliver, P.; Browne, N.K. Evidence of very rapid reef accretion and reef growth under high turbidity and terrigenous sedimentation. *Geology* **2012**, *40*, 719–722. [CrossRef]
194. Sansoleimani, A.; Webb, G.E.; Harris, D.L.; Phinn, S.R.; Roelfsema, C.M. Antecedent topography and active tectonic controls on Holocene reef geomorphology in the Great Barrier Reef. *Geomorphology* **2022**, *413*, 108354. [CrossRef]
195. Deep Reef Explorer. Available online: www.deepreef.org/2010/07/06/3dgbr-bathy/ (accessed on 28 March 2025).
196. Marshall, J.F.; Davies, P.J. Last interglacial reef growth beneath modern reefs in the southern Great Barrier Reef. *Nature* **1984**, *307*, 44–46. [CrossRef]
197. Webster, J.M.; Davies, P.J. Coral variation in two deep drill cores: Significance for the Pleistocene development of the Great Barrier Reef. *Sed. Geol.* **2003**, *159*, 61–80. [CrossRef]
198. Richards, H.C.; Hill, D. Great Barrier Reef bores, 1926 and 1937. Descriptions, analyses and interpretations. *Rep. Great Barrier Reef Com.* **1942**, *5*, 1–111.
199. Derrington, S.S. *Completion Report: H.B.R. No. 1 Bore, Wreck Island, Queensland*; Petroleum Search Subsidy Acts Publication No. 4; Bureau of Mineral Resources, Geology and Geophysics, Commonwealth of Australia: Canberra, Australia, 1960; pp. 1–14.
200. McKenzie, J.A.; Davies, P.J.; Palmer-Julson, A.A.; Betzler, C.G.; Brachert, T.C.; Chen, M.-P.P.; Crumière, J.-P.; Dix, G.R.; Droxler, A.W.; Feary, D.A.; et al. *Proceedings of the Ocean Drilling Program, Scientific Results, Northeast Australian Margin*; Ocean Drilling Program: College Station, TX, USA, 1993; Volume 132–133, pp. 1–903.
201. Davies, P.J.; Symonds, P.A.; Feary, D.A.; Pigram, C.J. The evolution of the carbonate platforms of northeast Australia. In *Controls on Carbonate Platform and Basin Development*; Crevello, P.D., Wilson, J.L., Sarg, J.F., Read, J.F., Eds.; SEPM Special Publication; Society of Economic Paleontologists and Mineralogists: Tulsa, OK, USA, 1989; Volume 44, pp. 233–258.
202. Betzler, C. Ecological controls on geometries of carbonate platforms: Miocene/ Pliocene shallow-water microfaunas and carbonate biofacies from the Queensland Plateau (NE Australia). *Facies* **1997**, *37*, 147–166. [CrossRef]
203. Betzler, C.; Hübscher, C.; Lindhorst, S.; Lüdmann, T.; Hincke, C.; Beaman, R.J.; Webster, J.M. Seismic stratigraphic and sedimentary record of a partial carbonate platform drowning, Queensland Plateau, north-east Australia. *Mar. Geol.* **2024**, *470*, 107255. [CrossRef]
204. Ehrenberg, S.N.N.; McArthur, J.M.M.; Thirlwall, M.F.F. Growth, demise, and dolomitization of Miocene carbonate platforms on the Marion Plateau, offshore NE Australia. *J. Sediment. Res.* **2006**, *76*, 91–116. [CrossRef]
205. Isern, A.R.; McKenzie, J.A.; Feary, D.A. The role of sea-surface temperature as a control on carbonate platform development in the western Coral Sea. *Palaeogeogr. Palaeoclimat. Palaeoecol.* **1996**, *124*, 247–272. [CrossRef]
206. Petrick, B.; Reuning, L.; Auderset, A.; Pfeiffer, M.; Auer, G.; Schwark, L. High sea surface temperatures were a prerequisite for the development and expansion of the Great Barrier Reef. *Sci. Adv.* **2024**, *10*, eado2058. [CrossRef]
207. Petrick, B.; Reuning, L.; Pfeiffer, M.; Auer, G.; Schwark, L. Impact of the Late Miocene Cooling on the loss of coral reefs in the Central Indo-Pacific. *Clim. Past* **2025**, *21*, 405–417. [CrossRef]

208. Dubois, N.; Kindler, P.; Spezzaferri, S.; Coric, S. The initiation of the southern central Great Barrier Reef: New multiproxy data from Pleistocene distal sediments from the Marion Plateau (NE Australia). *Mar. Geol.* **2008**, *250*, 223–233. [CrossRef]
209. Paleomap Project, Christopher, R. Scotese. Available online: www.scotese.com (accessed on 5 August 2024).
210. Scotese, C.R. An atlas of Phanerozoic paleogeographic maps: The seas come in and the seas go out. *Annu. Rev. Earth Planet. Sci.* **2021**, *49*, 679–728. [CrossRef]
211. Tcherepanov, E.N.; Droxler, A.W.; Dickens, G.R.; Bentley, S.J.; Peterson, L.C.; Beaufort, L.; Daniell, J. Neogene evolution of the mixed carbonate- siliciclastic system in the Gulf of Papua. *J. Geophys. Res.* **2008**, *113*, F01S21. [CrossRef]
212. Tcherepanov, E.N.; Droxler, A.W.; Lapointe, P.; Mohn, K. Carbonate seismic stratigraphy of the Gulf of Papua mixed depositional system: Neogene stratigraphic signature and eustatic control. *Basin Res.* **2008**, *20*, 185–209. [CrossRef]
213. Pigram, C.J.; Davies, P.J.; Feary, D.A.; Symonds, P.A. Tectonic controls on carbonate platform evolution in southern Papua New Guinea: Passive margin to foreland basin. *Geology* **1989**, *17*, 199–202. [CrossRef]
214. DiCaprio, L.; Müller, R.D.; Gurnis, M. A dynamic process for drowning carbonate reefs on the northeastern Australian margin. *Geology* **2010**, *38*, 11–14. [CrossRef]
215. Jell, J.S. Eastern continental margin, sect. 8.3. In *Geology of Queensland*; Jell, P.A., Ed.; Geological Survey of Queensland: Brisbane, QLD, Australia, 2013; pp. 606–629.
216. Quigley, M.C.; Clark, D.; Sandiford, M.J.G.S. Tectonic geomorphology of Australia. *Geol. Soc. Lond. Spec. Publ.* **2010**, *346*, 243–265. [CrossRef]
217. Dechnik, B.; Webster, J.M.; Webb, G.E.; Nothdurft, L.; Dutton, A.; Braga, J.C.; Zhao, J.-X.; Duce, S.; Sadler, J. The evolution of the Great Barrier Reef during the Last Interglacial Period. *Glob. Planet. Chang.* **2017**, *149*, 53–71. [CrossRef]
218. Salas-Saavedra, M.; Dechnik, B.; Webb, G.E.; Webster, J.M.; Zhao, J.-x.; Nothdurft, L.D.; Clark, T.R.; Graham, T.; Duce, S. Holocene reef growth over irregular Pleistocene karst confirms major influence of hydrodynamic factors on Holocene reef development. *Quat. Sci. Rev.* **2018**, *180*, 157–176. [CrossRef]
219. Dechnik, B.; Webster, J.M.; Davies, P.J.; Braga, J.-C.; Reimer, P.J. Holocene "turn-on" and evolution of the Southern Great Barrier Reef: Revisiting reef cores from the Capricorn Bunker Group. *Mar. Geol.* **2015**, *363*, 174–190. [CrossRef]
220. Kleypas, J.A.; Hopley, D. Reef development across a broad continental shelf, southern Great Barrier Reef, Australia. In Proceedings of the 7th International Coral Reef Symposium, Guam, Micronesia, 22–27 June 1992; Richmond, R.H., Ed.; University of Guam Press: Gaum, Micronesia; Volume 2, pp. 1129–1141.
221. Rynn, J.M.W. *Commentary on Seismic Risk Estimates and Related Uncertainties for Northeastern Australia (Queensland and Northeastern New South Wales)*; Special Paper Report; University of Queensland: Brisbane, QLD, Australia, 1989; pp. 1–90.
222. Sansoleimani, A.; Webb, G.E.; Harris, D.L.; Phinn, S.R.; Roelfsema, C.M. The impact of neotectonics on the geomorphology of the northern Great Barrier Reef. *Coral Reefs* **2025**, *44*, 775–789. [CrossRef]
223. Murray-Wallace, C.; Belperio, A. The last interglacial shoreline in Australia—A review. *Quat. Sci. Rev.* **1991**, *10*, 441–461. [CrossRef]
224. Sandiford, M. The tilting continent: A new constraint on the dynamic topographic field from Australia. *Earth Planet. Sci. Lett.* **2007**, *261*, 152–163. [CrossRef]
225. Rovere, A.; Pico, T.; Richards, F.D.; O'Leary, M.J.; Mitrovica, J.X.; Goodwin, I.D.; Austermann, J.; Latychev, K. Influence of reef isostasy, dynamic topography, and glacial isostatic adjustment on sea-level records in Northeastern Australia. *Com. Earth Environ.* **2023**, *4*, 328. [CrossRef]
226. Lin, Y.; Whitehouse, P.L.; Hibbert, F.D.; Woodroffe, S.A.; Hinestrosa, G.; Webster, J.M. Relative sea level response to mixed carbonate-siliciclastic sediment loading along the Great Barrier Reef margin. *Earth Planet. Sci. Lett.* **2023**, *607*, 118066. [CrossRef]
227. Hopley, D. *The Geomorphology of the Great Barrier Reef: Quaternary Development of Coral Reefs*; John Wiley: New York, NY, USA, 1982; pp. 1–453.
228. Mory, A.J.; Beere, G.M. Geology of the Onshore Bonaparte and Ord Basins in Western Australia. Geological Survey of Western Australia: Bulletin; State Printing Division: Nashville, TN, USA, 1988; Volume 134, pp. 1–184.
229. Yule, C.T.G.; Daniell, J.; Edwards, D.S.; Rollet, N.; Roberts, E.M. Reconciling the onshore/offshore stratigraphy of the Canning Basin and implications for petroleum prospectivity. *Aust. J. Earth Sci.* **2023**, *70*, 691–715. [CrossRef]
230. Lehmann, P.R. The stratigraphy, palaeogeography and petroleum potential of the Lower to lower Upper Devonian sequence in the Canning Basin. In *The Canning Basin, W.A., Proceedings of the Geological Society of Australia and Petroleum Exploration Society of Australia Symposium, Perth, WA, Australia, 27–29 June 1984*; Purcell, P.G., Ed.; Geological Society of Australia and Petroleum Exploration Society of Australia: Perth, WA, Australia, 1984; pp. 253–275.
231. Begg, J. Structuring and controls on Devonian reef development on the north-west Barbwire and adjacent terraces, Canning Basin. *APPEA J.* **1987**, *27*, 137–151. [CrossRef]
232. Wallace, M.W. Origin of dolomitization on the Barbwire Terrace, Canning Basin, Western Australia. *Sedimentology* **1990**, *37*, 105–122. [CrossRef]

233. Spaak, G.; Edwards, D.S.; Allen, H.J.; Grotheer, H.; Summons, R.E.; Coolen, M.J.L.; Grice, K. Extent and persistence of photic zone euxinia in Middle–Late Devonian seas—Insights from the Canning Basin and implications for petroleum source rock formation. *Mar. Petrol. Geol.* **2018**, *93*, 33–56. [CrossRef]
234. Hardman, E.T. *Report on the Geology of the Kimberley District, Western Australia*; Western Australian Parliamentary Papers number 31 of 1884; Government Printer: Perth, WA, Australia, 1884.
235. Wade, A. *Petroleum Prospects Kimberley District of Western Australia and Northern Territory*; Commonwealth of Australia: Melbourne, Australia, 1924; pp. 1–63.
236. Kerans, C. *Petrology of Devonian and Carboniferous Carbonates of the Canning and Bonaparte Basins*; Report 12; Western Australia Mining and Petroleum Research Institute: Perth, WA, Australia, 1985; pp. 1–203. ISBN 0730900731.
237. Moors, H.T.; Gardner, W.E.; Davis, J. Geology of the Blina oilfield. In *The Canning Basin, W.A., Proceedings of the Geological Society of Australia and Petroleum Exploration Society of Australia Symposium, Perth, WA, Australia, 27–29 June 1984*; Purcell, P.G., Ed.; Geological Society of Australia and Petroleum Exploration Society of Australia: Perth, WA, Australia, 1984; pp. 277–283.
238. Wallace, M.W.; Middleton, H.A.; Johns, B.; Marshallsea, S. Hydrocarbons and Mississippi Valley-type sulfides in the Devonian Reef Complexes of the eastern Lennard Shelf, Canning Basin, Western Australia. In *The Sedimentary Basins of Western Australia, Volume 3*; Keep, M., Moss, S.J., Eds.; Petroleum Exploration Society of Australia: Perth, WA, Australia, 2002; pp. 1–21.
239. Brown, S.A.; Boserio, I.M.; Jackson, K.S.; Spence, K.W. The geological evolution of the Canning Basin; implications for petroleum exploration. In *The Canning Basin W.A.: Proceedings of the Geological Society of Australia/Petroleum Exploration Society of Australia Canning Basin Symposium, Perth, W.A.*; Purcell, P.G., Ed.; Geological Society of Australia/Petroleum Exploration Society of Australia: Perth, WA, Australia, 1984; pp. 85–96.
240. Drummond, B.J.; Etheridge, M.A.; Davies, P.J.; Middleton, M.F. Half-graben model for the structural evolution of the Fitzroy Trough, Canning Basin, and implications for resource exploration. *APEA J.* **1988**, *28*, 76–86. [CrossRef]
241. Mory, A.J.; Haines, P.A. A Paleozoic Perspective of Western Australia. In Proceedings of the West Australian Basins Symposium 2013, Perth, WA, Australia, 18–21 August 2013; Petroleum Exploration Society of Australia: Perth, WA, Australia, 2013; pp. 1–25.
242. Jell, J.S. Lower and Middle Devonian of Queensland, Australia. In *Devonian of the World*; McMillan, N.J., Embry, A.F., Glass, D.J., Eds.; Canadian Society of Petroleum Geologists: Calgary, AB, Canada, 1988; Volume 1, pp. 755–772.
243. Webby, B.D.; Zhen, Y.Y. Silurian and Devonian clathrodictyids and other stromatoporoids from the Broken River region, North Queensland. *Alcheringa* **1997**, *21*, 1–56. [CrossRef]
244. Jell, J.S.; Zhen, Y.Y. Middle Devonian rugose coral biostratigraphy of the Fanning River Group, north Queensland, Australia. *Cour. Forschungsinst. Senckenberg.* **1994**, *172*, 1–6.
245. Zhen, Y.-Y. Succession of coral associations during a Givetian transgressive-regressive cycle in Queensland. *Acta Palaeont. Pol.* **1996**, *41*, 59–88.
246. Zhen, Y.Y.; Wright, A.J.; Jell, J.S. Rugose coral diversifications and migrations in the Devonian of Australasia. *Hist. Biol.* **2001**, *15*, 61–76. [CrossRef]
247. Henderson, R.A.; Withnall, I.W. Broken River Province. In *Geology of Queensland*; Jell, P.A., Ed.; Geological Survey of Queensland: Brisbane, QLD, Australia, 2013; pp. 250–279.
248. Henderson, R.A.; Donchak, P.J.T. Hodgkinson Province. In *Geology of Queensland*; Jell, P.A., Ed.; Geological Survey of Queensland: Brisbane, QLD, Australia, 2013; pp. 229–249.
249. Becker, R.T.; Marshall, J.E.A.; Da Silva, A.-C.; Agterberg, F.P.; Gradstein, F.M.; Ogg, J.G. The Devonian Period. In *Geologic Time Scale 2020*, 1st ed.; Gradstein, F.M., Ogg, J.G., Schmitz, M.D., Ogg, G.M., Eds.; Elsevier: Amsterdam, The Netherlands, 2020; Volume 2, pp. 733–810. [CrossRef]
250. Brugger, J.; Hofmann, M.; Petri, S.; Feulner, G. On the sensitivity of the Devonian climate to continental configuration, vegetation cover, orbital configuration, CO_2 concentration, and insolation. *Paleoceanogr. Paleoclimat.* **2019**, *34*, 1375–1398. [CrossRef]
251. Chen, B.; Ma, X.; Mills, B.J.W.; Qie, W.; Joachimski, M.M.; Shen, S.; Wang, C.; Xu, H.; Wang, X. Devonian paleoclimate and its drivers: A reassessment based on a new conodont $\delta^{18}O$ record from South China. *Earth-Sci. Rev.* **2021**, *222*, 103814. [CrossRef]
252. Algeo, T.J.; Berner, R.; Maynard, J.; Scheckler, S. Late Devonian oceanic anoxic events and biotic crises: "Rooted" in the evolution of vascular plants. *GSA Today* **1995**, *5*, 64–66.
253. Algeo, T.J.; Scheckler, S. Land plant evolution and weathering rate changes in the Devonian. *J. Earth Sci.* **2010**, *21*, 75–78. [CrossRef]
254. Davies, N.S.; Gibling, M.R. Cambrian to Devonian evolution of alluvial systems: The sedimentological impact of the earliest land plants. *Earth-Sci. Rev.* **2010**, *98*, 171–200. [CrossRef]
255. Capel, E.; Cleal, C.J.; Xue, J.; Monnet, C.; Servais, T.; Cascales-Miñana, B. The Silurian-Devonian terrestrial revolution: Diversity patterns and sampling bias of the vascular plant macrofossil record. *Earth Sci. Rev.* **2022**, *231*, 104085. [CrossRef]
256. Long, J.A. *The Rise of Fishes*; University of New South Wales Press: Sydney, Australia, 1995; pp. 1–223.

257. Dahl, T.W.; Hammarlund, E.U.; Anbar, A.D.; Bond, D.P.G.; Gill, B.C.; Gordon, G.W.; Knoll, A.H.; Nielsen, A.T.; Schovsbo, N.H.; Canfield, D.E. Devonian rise in atmospheric oxygen correlated to the radiations of terrestrial plants and large predatory fish. *Proc. Natl. Acad. Sci. USA* **2010**, *107*, 17911–17915. [CrossRef]
258. Clack, J.A. Devonian climate change, breathing, and the origin of the tetrapod stem group. *Integr. Compar. Biol.* **2007**, *47*, 510–523. [CrossRef]
259. Becker, R.T.; Kaiser, S.I.; Aretz, M. Review of chrono-, litho- and biostratigraphy across the global Hangenberg crisis and Devonian–Carboniferous boundary. *Geol. Soc. Lond. Spec. Publ.* **2016**, *423*, 355–386. [CrossRef]
260. Kaiser, S.I.; Becker, R.T.; Aretz, M. The global Hangenberg Crisis (Devonian-Carboniferous transition): Review of a first order mass extinction. In *Devonian Climate, Sea Level and Evolutionary Events*; Becker, R.T., Konigshof, P., Brett, C.E., Eds.; Special Publications; Geological Society: London, UK, 2016; Volume 423, pp. 387–437. [CrossRef]
261. Carmichael, S.K.; Waters, J.A.; Konigshof, P.; Suttner, T.J.; Kido, E. Paleogeography and paleoenvironments of the Late Devonian Kellwasser Event: A review of its sedimentological and geochemical expression. *Glob. Planet. Chang.* **2019**, *183*, 102984. [CrossRef]
262. Kabanov, P.; Hauck, T.E.; Gouwy, S.A.; Grasby, S.E.; van der Boon, A. Oceanic anoxic events, photic-zone euxinia, and controversy of sea-level fluctuations during the Middle-Late Devonian. *Earth-Sci. Rev.* **2023**, *241*, 104415. [CrossRef]
263. Playford, P.E.; Cockbain, A.E. *Revised Stratigraphic and Facies Nomenclature in Devonian Reef Complexes of the Canning Basin*; Annual Report for 1975; Geological Survey of Western Australia: Perth, WA, Australia, 1976; pp. 52–56.
264. George, A.D.; Seyedmehdi, Z.; Chow, N. Late Devonian–Early Carboniferous tectonostratigraphic framework for northern Canning Basin carbonate platform evolution. In *Sedimentary Basins of Western Australia IV: Proceedings of the Petroleum Exploration Society of Australia Symposium*; Keep, M., Moss, S.J., Eds.; Petroleum Exploration Society of Australia: Perth, WA, Australia, 2013; pp. 1–16.
265. Playton, T.E.; Kerans, C. Late Devonian carbonate margins and foreslopes of the Lennard Shelf, Canning Casin, Western Australia, part a: Development during backstepping and the aggradation-to-progradation transition. *J. Sed. Res.* **2015**, *85*, 1334–1361. [CrossRef]
266. Playton, T.E.; Kerans, C. Upper Devonian carbonate margins and foreslopes of the Lennard Shelf, Canning Basin, Western Australia, Part B: Development during progradation and across the Frasnian–Famennian biotic crisis. *J. Sed. Res.* **2015**, *85*, 1362–1392. [CrossRef]
267. Shen, J.; Webb, G.E.; Jell, J.S. Platform margins, reef facies, and microbial carbonates; a comparison of Devonian reef complexes in the Canning Basin, Western Australia. *Earth-Sci. Rev.* **2008**, *88*, 33–59. [CrossRef]
268. Shen, J.-W.; Zhao, N.; Young, A.; Mao, Y.-J.; Wang, Y. Upper Devonian reefs and microbialite at Maoying, South China—Implications for paleoenvironmental changes. *Palaeogeogr. Palaeoclimat. Palaeoecol.* **2017**, *474*, 98–112. [CrossRef]
269. Hillbun, K.; Playton, T.E.; Katz, D.A.; Tohver, E.; Trinajstic, K.; Haines, P.W.; Hocking, R.M.; Roelofs, B.; Montgomery, P. Correlation and sequence stratigraphic interpretation of Upper Devonian carbonate slope facies using carbon isotope chemostratigraphy, Lennard Shelf, Canning Basin, Western Australia. In *New Advances in Devonian Carbonates: Outcrop Analogs, Reservoirs, and Chronostratigraphy*; Playton, T.E., Kerans, C., Weissenberger, J.A.W., Eds.; SEPM Special Publication; Society of Economic Paleontologists and Mineralogists: Tulsa, OK, USA, 2016; Volume 107, pp. 248–301. [CrossRef]
270. George, A.D.; Trinajstic, K.M.; Chow, N. Frasnian reef evolution and palaeogeography, SE Lennard Shelf, Canning Basin, Australia. In *Devonian Change: Case Studies in Palaeogeography and Palaeoecology*; Konighsof, P., Ed.; Special Publications; Geological Society: London, UK, 2009; Volume 314, pp. 73–107. [CrossRef]
271. Playton, T.E.; Hocking, R.M.; Tohver, E.; Hillbun, K.; Haines, P.W.; Trinajstic, K.M.; Roelofs, B.; Katz, D.A.; Kirschvink, J.L.; Grice, K.; et al. Integrated stratigraphic correlation of Upper Devonian platform-to-basin carbonate sequences, Lennard Shelf, Canning Basin, Western Australia: Advances in carbonate margin-to-slope sequence stratigraphy and stacking patters. In *New Advances in Devonian Carbonates: Outcrop Analogs, Reservoirs, and Chronostratigraphy*; Playton, T.E., Kerans, C., Weissenberger, J., Eds.; Special Publication; SEPM (Society for Sedimentary Geology): Tulsa, OK, USA, 2016; Volume 107. [CrossRef]
272. Klapper, G. Upper Devonian conodonts in the Canning Basin, Appendix 1. In *Devonian Reef Complexes of the Canning Basin, Western Australia*; Playford, P.E., Hocking, R.M., Cockbain, A.E., Eds.; Bulletin 145; Geological Survey of Western Australia: Perth, WA, Australia, 2009; pp. 405–413.
273. Becker, R.T.; House, M.R. Devonian ammonoid biostratigraphy of the Canning Basin, Appendix 2. In *Devonian Reef Complexes of the Canning Basin, Western Australia*; Playford, P.E., Hocking, R.M., Cockbain, A.E., Eds.; Bulletin 145; Geological Survey of Western Australia: Perth, WA, Australia, 2009; pp. 415–439.
274. Trinajstic, K.; George, A.D. Microvertebrate biostratigraphy of Upper Devonian (Frasnian) carbonate rocks in the Canning and Carnarvon basins of Western Australia. *Palaeontology* **2009**, *52*, 641–659. [CrossRef]
275. Read, J.F. Carbonate cycles, Pillara Formation (Devonian), Canning Basin, Western Australia. *Canad. Petrol. Geol. Bull.* **1973**, *21*, 38–51.
276. Brownlaw, R.L.S.; Hocking, R.M.; Jell, J.S. High frequency sea-level fluctuations in the Pillara Limestone, Guppy Hills, Lennard Shelf, northwestern Australia. *Hist. Biol.* **1996**, *11*, 187–212. [CrossRef]

277. Brownlaw, R.L.S.; Hearn, S.J.; Jell, J.S. Spectral analysis of the back-reef limestones of the 'Devonian Great Barrier Reef', Western Australia. *Proc. R. Soc. Qld.* **1998**, *107*, 99–107.
278. Ferguson, I.J.; Da Silva, A.-C.; Chow, N.; George, A.D. Interplay of eustatic, tectonic and autogenic controls on a late Devonian carbonate platform, northern Canning Basin, Australia. *Basin Res.* **2019**, *33*, 312–341. [CrossRef]
279. Holmes, A.E.; Christie-Blick, N. Origin of sedimentary cycles in mixed carbonate-siliciclastic systems: An example from the Canning Basin, Western Australia. In *Carbonate Sequence Stratigraphy*; Loucks, R.G., Sarg, J.F., Eds.; American Association of Petroleum Geologists: Tulsa, OK, USA, 1993; Volume 57, pp. 181–212.
280. George, A.D.; Chow, N.; Trinajstic, K.M. Syndepositional fault control on lower Frasnian platform evolution, Lennard Shelf, Canning Basin, Australia. *Geology* **2009**, *37*, 331–334. [CrossRef]
281. Southgate, P.N.; Kennard, J.M.; Jackson, M.J.; O'Brien, P.E.; Sexton, M.J. Reciprocal lowstand clastic and highstand carbonate sedimentation, subsurface Devonian reef complex, Canning Basin, Western Australia. In *Carbonate Sequence Stratigraphy*; Loucks, R.G., Sarg, J.F., Eds.; American Association of Petroleum Geologists: Tulsa, OK, USA, 1993; Volume 57, pp. 157–179. [CrossRef]
282. George, A.D.; Playford, P.E.; Powell, C.M.; Tornatora, P.M. Lithofacies and sequence development on an Upper Devonian mixed carbonate-siliciclastic fore-reef slope, Canning Basin, Western Australia. *Sedimentology* **1997**, *44*, 843–867. [CrossRef]
283. Ward, W.B. Tectonic control on backstepping sequences revealed by mapping of Frasnian backstepped platforms, Devonian reef Complexes, Napier Range, Canning Basin, Western Australia. In *Advances in Carbonate Sequence Stratigraphy: Application to Reservoirs, Outcrops and Models*; Harris, P.M., Simo, J.A., Eds.; SEPM Special Publication; Society of Economic Paleontologists and Mineralogists: Tulsa, OK, USA, 1999; Volume 63, pp. 47–74.
284. Chow, N.; George, A.D.; Trinajstic, K.M. Tectonic control on development of a Frasnian–Famennian (Late Devonian) palaeokarst surface, Canning Basin reef complexes, northwestern Australia. *Aust. J. Earth Sci.* **2004**, *51*, 911–917. [CrossRef]
285. Chow, N.; George, A.D.; Trinajstic, K.M.; Chen, J.-Q. Stratal architecture and platform evolution of an early Frasnian syn-tectonic carbonate platform, Canning Basin, Australia. *Sedimentology* **2013**, *60*, 1583–1620. [CrossRef]
286. Playford, P.E. Devonian reef complexes of the Canning Basin, Western Australia. In *Geological Society of Australia, 5th Australian Geological Convention, Perth, Field Excursion Guidebook*; Geological Society of Australia: Perth, WA, Australia, 1981; pp. 1–64.
287. George, A.D.; Playford, P.E.; Powell, C.M. Platform-margin collapse during Famennian reef evolution, Canning Basin, Western Australia. *Geology* **1995**, *23*, 691–694. [CrossRef]
288. Frost, E.L., III; Kerans, C. Platform-margin trajectory as a control on syndepositional fracture patterns, Canning Basin, Western Australia. *J. Sed. Res.* **2009**, *79*, 44–55. [CrossRef]
289. Frost, E.L., III; Kerans, C. Controls on syndepositional fracture patterns, Devonian reef complexes, Canning Basin, Western Australia. *J. Struct. Geol.* **2010**, *32*, 1231–1249. [CrossRef]
290. Johnson, M.E.; Webb, G.E. Outer rocky shores of the Mowanbini Archipelago, Devonian Reef Complex, Canning Basin, Western Australia. *J. Geol.* **2007**, *115*, 583–600. [CrossRef]
291. Playford, P.E. Palaeokarst, pseudokarst, and sequence stratigraphy in Devonian reef complexes of the Canning Basin, Western Australia. In *The Sedimentary Basins of Western Australia, Volume 3, Proceedings of the Petroleum Exploration Society of Australia Symposium: Perth, WA, Australia, 20–23 October 2002*; Keep, M., Moss, S.J., Eds.; Petroleum Exploration Society of Australia: Perth, WA, Australia, 2002; pp. 763–793.
292. Johnson, M.E.; Webb, G.E.; Baarli, B.G.; Walsh, D.R. Upper Devonian shoal-water delta integrated with cyclic backreef facies off the Mowanbini Archipelago (Canning Basin), Western Australia. *Facies* **2013**, *59*, 991–1009. [CrossRef]
293. Baarli, G.; Webb, G.E.; Johnson, M.E.; Cook, A.G.; Walsh, D.R. Shoal-water dynamics and coastal biozones in a sheltered island setting: Upper Devonian Pillara Limestone (Western Australia). *Lethaia* **2016**, *49*, 507–523. [CrossRef]
294. Waterhouse, J.; Gruber, R.; Logan, M.; Petus, C.; Howley, C.; Lewis, S.; Tracey, D.; James, C.; Mellors, J.; Tonin, H.; et al. Marine Monitoring Program: Annual Report for Inshore Water Quality Monitoring 2019–20. In *Report for the Great Barrier Reef Marine Park Authority*; Great Barrier Reef Marine Park Authority: Townsville, QLD, Australia, 2021; pp. 1–13. Available online: https://hdl.handle.net/11017/3826 (accessed on 4 May 2025).
295. Saha, N.; Rodriguez-Ramirez, A.; Nguyen, A.D.; Clark, T.R.; Zhao, J.; Webb, G.E. Seasonal to decadal scale influence of environmental drivers on Ba/Ca and Y/Ca in coral aragonite from the southern Great Barrier Reef. *Sci. Total Environ.* **2018**, *639*, 1099–1109. [CrossRef]
296. Wolff, N.H.; Wong, A.; Vitolo, R.; Stolberg, K.; Anthony, K.R.N.; Mumby, P.J. Temporal clustering of tropical cyclones on the Great Barrier Reef and its ecological importance. *Coral Reefs* **2016**, *35*, 613–623. [CrossRef]
297. Humblet, M.; Potts, D.C.; Webster, J.M.; Braga, J.C.; Iryu, Y.; Yokoyama, Y.; Bourillot, R.; Séard, C.; Droxler, A.; Fujita, K.; et al. Late glacial to deglacial variation of coralgal assemblages in the Great Barrier Reef, Australia. *Glob. Planet. Chang.* **2019**, *174*, 70–91. [CrossRef]
298. Sloss, C.R.; Murray-Wallace, C.V.; Jones, B.G. Holocene sea-level change on the southeast coast of Australia: A review. *Holocene* **2007**, *17*, 999–1014. [CrossRef]

299. Lewis, S.E.; Sloss, C.R.; Murray-Wallace, C.V.; Woodroffe, C.D.; Smithers, S.G. Postglacial sea-level changes around the Australian margin: A review. *Quat. Sci. Rev.* **2013**, *74*, 115–138. [CrossRef]
300. Leonard, N.D.; Lepore, M.L.; Zhao, J.x.; Rodriguez-Ramirez, A.; Butler, I.R.; Clark, T.R.; Roff, G.; McCook, L.; Nguyen, A.D.; Feng, Y.-x.; et al. Re-evaluating mid-Holocene reef "turn-off" on the inshore Southern Great Barrier Reef. *Quat. Sci. Rev.* **2020**, *244*, 106518. [CrossRef]
301. Leonard, N.D.; Welsh, K.J.; Clark, T.R.; Feng, Y.x.; Pandolfi, J.M. New evidence for "far-field" Holocene sea level oscillations and links to global climate records. *Earth Planet. Sci. Lett.* **2018**, *487*, 67–73. [CrossRef]
302. Leonard, N.D.; Zhao, J.-x.; Welsh, K.J.; Feng, Y.-x.; Smithers, S.G.; Pandolfi, J.M.; Clark, T.R. Holocene sea level instability in the southern Great Barrier Reef, Australia: High-precision U-Th dating of fossil microatolls. *Coral Reefs* **2016**, *35*, 625–639. [CrossRef]
303. Perry, C.T.; Smithers, S.G. Evidence for the episodic "turn on" and "turn off" of turbid-zone coral reefs during the late Holocene sea-level highstand. *Geology* **2010**, *38*, 119–122. [CrossRef]
304. Perry, C.; Smithers, S. Cycles of coral reef 'turn-on', rapid growth and 'turn-off' over the past 8500 years: A context for understanding modern ecological states and trajectories. *Glob. Change Biol.* **2011**, *17*, 76–86. [CrossRef]
305. Dechnik, B.; Webster, J.M.; Webb, G.E.; Nothdurft, L.; Zhao, J.X. Successive phases of Holocene reef flat development: Evidence from the mid-to outer Great Barrier reef. *Palaeogeogr. Palaeoclim. Palaeoecol.* **2017**, *466*, 221–230. [CrossRef]
306. Toth, L.T.; Aronson, R.B. The 4.2-ka event, ENSO, and coral-reef development. *Clim. Past* **2018**, *15*, 105–119. [CrossRef]
307. Felis, T.; McGregor, H.V.; Linsley, B.K.; Tudhope, A.W.; Gagan, M.K.; Suzuki, A.; Inoue, M.; Thomas, A.L.; Esat, T.M.; Thompson, W.G.; et al. Intensification of the meridional temperature gradient in the Great Barrier Reef following the Last Glacial Maximum. *Nat. Commun.* **2014**, *5*, 4102. [CrossRef]
308. Brenner, L.D.; Linsley, B.K.; Webster, J.M.; Potts, D.; Felis, T.; Gagan, M.K.; Inoue, M.; McGregor, H.; Suzuki, A.; Tudhope, A.; et al. Coral Record of Younger Dryas Chronozone Warmth on the Great Barrier Reef. *Paleoceanogr. Paleoclimat.* **2020**, *35*, e2020PA003962. [CrossRef]
309. Sadler, J.; Webb, G.E.; Leonard, N.D.; Nothdurft, L.D.; Clark, T.R. Reef core insights into mid-Holocene water temperatures of the southern Great Barrier Reef. *Paleoceanography* **2016**, *31*, 1395–1408. [CrossRef]
310. Gagan, M.K.; Ayliffe, L.K.; Hopley, D.; Cali, J.A.; Mortimer, G.E.; Chappell, J.; McCulloch, M.T.; Head, M.J. Temperature and surface-ocean water balance of the mid-Holocene tropical Western Pacific. *Science* **1998**, *279*, 1014–1018. [CrossRef]
311. McCulloch, M.; Mortimer, G.; Esat, T.; Li, X.H.; Pillans, B.; Chappell, J. High resolution windows into early Holocene climate: Sr/Ca coral records from the Huon Peninsula. *Earth Planet. Sci. Lett.* **1996**, *138*, 169–178. [CrossRef]
312. Abram, N.J.; McGregor, H.V.; Gagan, M.K.; Hantoro, W.S.; Suwargadi, B.W. Oscillations in the southern extent of the Indo-Pacific Warm Pool during the mid-Holocene. *Quat. Sci. Rev.* **2009**, *28*, 2794–2803. [CrossRef]
313. Roche, R.C.; Perry, C.T.; Smithers, S.G.; Leng, M.J.; Grove, C.A.; Sloane, H.J.; Unsworth, C.E. Mid-Holocene Sea surface conditions and riverine influence on the inshore Great Barrier Reef. *Holocene* **2014**, *24*, 885–897. [CrossRef]
314. Lough, J.M.; Llewellyn, L.E.; Lewis, S.E.; Turney, C.S.M.; Palmer, J.G.; Cook, C.G.; Hogg, A.G. Evidence for suppressed mid-Holocene northeastern Australian monsoon variability from coral luminescence. *Paleoceanography* **2014**, *29*, 581–594. [CrossRef]
315. Hua, Q.; Webb, G.E.; Zhao, J.-X.; Nothdurft, L.D.; Lybolt, M.; Price, G.J.; Opdyke, B.N. Large variations in the Holocene marine radiocarbon reservoir effect reflect ocean circulation and climatic changes. *Earth Planet. Sci. Lett.* **2015**, *422*, 33–44. [CrossRef]
316. Hua, Q.; Ulm, S.; Yu, K.; Clark, T.R.; Nothdurft, L.D.; Leonard, N.D.; Pandolfi, J.M.; Jacobsen, G.E.; Zhao, J.-X. Temporal variability in the Holocene marine radiocarbon reservoir effect for the Tropical and South Pacific. *Quat. Sci. Rev.* **2020**, *249*, 106613. [CrossRef]
317. McNeil, M.; Nothdurft, L.; Erler, D.; Hua, Q.; Webster, J.M. Variations in mid- to late Holocene nitrogen supply to northern Great Barrier Reef *Halimeda* macroalgal bioherms. *Paleoceanogr. Paleoclimat.* **2021**, *36*, e2020PA003871. [CrossRef]
318. Denniston, R.F.; Wyrwoll, K.-H.; Polyak, V.J.; Brown, J.R.; Asmerom, Y.; Wanamaker, A.D., Jr.; LaPointe, Z.; Ellerbroek, R.; Barthelmes, M.; Cleary, D. A stalagmite record of Holocene Indonesian–Australian summer monsoon variability from the Australian tropics. *Quat. Sci. Rev.* **2013**, *78*, 155–168. [CrossRef]
319. Salas-Saavedra, M.; Webb, G.E.; Sanborn, K.L.; Zhao, J.-x.; Webster, J.M.; Nothdurft, L.D.; Nguyen, A. Holocene microbialite geochemistry records > 6000 years of secular influence of terrigenous flux on water quality for the southern Great Barrier Reef. *Chem. Geol.* **2022**, *604*, 120871. [CrossRef]
320. Sanborn, K.L.; Webster, J.M.; Erler, D.; Webb, G.E.; Salas-Saavedra, M.; Yokoyama, Y. The impact of elevated nutrients on the Holocene evolution of the Great Barrier Reef. *Quat. Sci. Rev.* **2024**, *332*, 108636. [CrossRef]
321. Nott, J.; Forsyth, A. Punctuated global tropical cyclone activity over the past 5000 years. *Geophys. Res. Lett.* **2012**, *39*, L14703. [CrossRef]
322. Ryan, E.J.; Smithers, S.G.; Lewis, S.E.; Clark, T.R.; Zhao, J.-X. The influence of sea level and cyclones on Holocene reef flat development: Middle Island, central Great Barrier Reef. *Coral Reefs* **2016**, *35*, 805–818. [CrossRef]
323. Hardie, L.A. Secular variation in seawater chemistry: An explanation for the coupled secular variation in the mineralogies of marine limestones and potash evaporites over the past 600 m.y. *Geology* **1996**, *24*, 279–283. [CrossRef]
324. Balthasar, U.; Cusack, M. Aragonite-calcite seas—Quantifying the gray area. *Geology* **2015**, *43*, 99–102. [CrossRef]

325. Stanley, M.S.; Hardie, L.A. Secular oscillations in the carbonate mineralogy of reef-building and sediment-producing organisms driven by tectonically forced shifts in seawater chemistry. *Palaeogeogr. Palaeoclimat. Palaeoecol.* **1998**, *144*, 3–19. [CrossRef]
326. De Vleeschouwer, D.; Percival, L.M.E.; Wichern, N.M.A.; Batenburg, S.J. Pre-Cenozoic cyclostratigraphy and palaeoclimate responses to astronomical forcing. *Nat. Rev. Earth Environ.* **2024**, *5*, 59–74. [CrossRef]
327. Caputo, M.V. Late Devonian glaciation in South America. *Palaeogeogr. Palaeoclimat. Palaeoecol.* **1985**, *51*, 291–317. [CrossRef]
328. Caputo, M.V.; de Melo, J.H.G.; Streel, M.; Isbell, J.L. Late Devonian and Early Carboniferous glacial records of South America. In *Resolving the Late Paleozoic Ice Age in Time and Space*; Fielding, C.R., Frank, T.D., Isbell, J.L., Eds.; Geological Society of America: Ithaca, NY, USA, 2008; Volume 441, pp. 161–174.
329. McGhee, G.R. The search for sedimentary evidence of glaciation during the Frasnian/Famennian (Late Devonian) biodiversity crisis. *Sediment. Rec.* **2014**, *12*, 4–8. [CrossRef]
330. Veizer, J.; Prokoph, A. Temperatures and oxygen isotopic composition of Phanerozoic oceans. *Earth-Sci. Rev.* **2015**, *146*, 92–104. [CrossRef]
331. Shaviv, N.J.; Svensmark, H.; Veizer, J. The Phanerozoic climate. *Ann. N. Y. Acad. Sci.* **2023**, *1519*, 7–19. [CrossRef] [PubMed]
332. Li, Z.X.; Powell, C. An outline of the palaeogeographic evolution of the Australasian region since the beginning of the Neoproterozoic. *Earth-Sci. Rev.* **2001**, *53*, 237–277. [CrossRef]
333. Hocking, R.M.; Copp, I.A.; Playford, P.E.; Kempton, R.H. The Cadjebut Formation: A Givetian evaporitic precursor to Devonian reef complexes of the Lennard Shelf, Canning Basin, Western Australia. In *Geological Survey of Western Australia: Annual Review 1995–96*; Nowak, I.R., Ed.; Geological Survey of Western Australia: Perth, WA, Australia, 1996; Volume 1, pp. 48–55.
334. Marsaglia, K.M.; Klein, G.D. The paleogeography of Paleozoic and Mesozoic storm depositional systems. *J. Geol.* **1983**, *91*, 117–141. [CrossRef]
335. Lu, M.; Lu, Y.-H.; Ikejiri, T.; Sun, D.; Carroll, R.; Blair, E.H.; Algeo, T.J.; Sun, Y. Periodic oceanic euxinia and terrestrial fluxes linked to astronomical forcing during the Late Devonian Frasnian–Famennian mass extinction. *Earth Planet. Sci. Lett.* **2021**, *562*, 116839. [CrossRef]
336. Handford, C.R.; Loucks, R.G. Carbonate depositional sequences and systems tracts—Responses of carbonate platforms to relative sea-level changes. In *Carbonate Sequence Stratigraphy, Recent Developments and Applications*; Loucks, R.G., Sarg, J.F., Eds.; AAPG Memoir; American Association of Petroleum Geologists: Tulsa, OK, USA, 1993; Volume 57, pp. 3–41.
337. Flügel, E. *Microfacies of Carbonate Rocks*, 2nd ed.; Springer: Berlin/Heidelberg, Germany, 2010; pp. 1–924.
338. Pitcher, C.R.; Doherty, P.; Arnold, P.; Hooper, J.; Gribble, N.; Bartlett, C.; Browne, M.; Campbell, N.; Cannard, T.; Cappo, M.; et al. *Seabed Biodiversity on the Continental Shelf of the Great Barrier Reef World Heritage Area*; AIMS/CSIRO/QM/QDPI CRC Reef Research Task Final Report; CSIRO Marine and Atmospheric Research: Caberra, Australia, 2007; pp. 1–320.
339. GBRMPA. *Great Barrier Reef Outlook Report 2014*; Great Barrier Reef Marine Park Authority: Townsville, QLD, Australia, 2019; pp. 1–354.
340. Perry, C.T.; Hepburn, L.J. Syn-depositional alteration of coral reef framework through bioerosion, encrustation and cementation: Taphonomic signatures of reef accretion and reef depositional events. *Earth-Sci. Rev.* **2008**, *86*, 106–144. [CrossRef]
341. Orme, G.R.; Flood, P.G.; Ewart, A. An investigation of the sediments and physiography of Lady Musgrave Reef—A preliminary account. In Proceedings of the Second International Symposium on Coral Reefs, Brisbane, QLD, Australia, 22 June–2 July 1974; Volume 2, pp. 371–386.
342. Mather, P.; Bennett, I. (Eds.) *A Coral Reef Handbook*, 3rd ed.; Surrey Beatty & Sons Pty Ltd.: Chipping Norton, NSW, Australia, 1993; pp. 1–264.
343. Hamylton, S.; Carvalho, R.C.; Duce, S.; Roelfsema, C.M.; Vila-Concejo, A. Linking pattern to process in reef sediment dynamics at Lady Musgrave Island, southern Great Barrier Reef. *Sedimentology* **2016**, *63*, 1634–1650. [CrossRef]
344. Tanner, J.E.; Connell, J.H. Coral community data Heron Island, Great Barrier Reef 1962–2016. *Sci. Data* **2022**, *9*, 617. [CrossRef]
345. McNeil, M.; Firn, J.; Nothdurft, L.D.; Pearse, A.R.; Webster, J.M.; Pitcher, C.R. Inter-reef Halimeda algal habitats within the Great Barrier Reef support a distinct biotic community and high biodiversity. *Nat. Ecol. Evol.* **2021**, *5*, 647–655. [CrossRef]
346. Sanborn, K.L.; Webster, J.M.; Webb, G.E.; Braga, J.C.; Humblet, M.; Nothdurft, L.; Patterson, M.A.; Dechnik, B.; Warner, S.; Graham, T. A new model of Holocene reef initiation and growth in response to sea-level rise on the Southern Great Barrier Reef. *Sediment. Geol.* **2020**, *397*, 105556. [CrossRef]
347. De'ath, G.; Fabricius, K.E. Water quality as a regional driver of coral biodiversity and macroalgae on the Great Barrier Reef. *Ecol. Appl.* **2010**, *20*, 840–850. [CrossRef] [PubMed]
348. González-Rivero, M.; Bongaerts, P.; Beijbom, O.; Pizarro, O.; Friedman, A.; Rodriguez-Ramirez, A.; Upcroft, B.; Laffoley, D.; Kline, D.; Bailhache, C.; et al. The Catlin Seaview Survey—Kilometre-scale seascape assessment, and monitoring of coral reef ecosystems. *Aquat. Conserv. Mar. Freshw. Ecosyst.* **2014**, *24*, 184–198. [CrossRef]
349. Roelfsema, C.M.; Kovacs, E.; Stetner, D.; Phinn, S.R. Georeferenced benthic photoquadrats captured annually from 2002-2017, distributed over Heron Reef flat and slope areas. *PANGAEA* **2018**, 894801. [CrossRef]

350. Roelfsema, C.M.; Kovacs, E.; Markey, K.; Vercelloni, J.; Rodriguez-Ramirez, A.; Lopez-Marcano, S.; Gonzalez-Rivero, M.; Hoegh-Guldberg, O.; Phinn, S.R. Benthic and coral reef community field data for Heron Reef, Southern Great Barrier Reef, Australia, 2002–2018. *Sci. Data* **2021**, *8*, 84. [CrossRef]
351. Phinn, S.R.; Roelfsema, C.M.; Mumby, P.J. Multi-scale image segmentation for mapping coral reef geomorphic and benthic community zone. *Int. J. Remote Sens.* **2012**, *33*, 3768–3797. [CrossRef]
352. Hamylton, S.M. Mapping coral reef environments: A review of historical methods, recent advances and future opportunities. *Prog. Phys. Geogr.* **2017**, *41*, 803–833. [CrossRef]
353. Hamylton, S.M.; Duce, S.; Vila-Concejo, A.; Roelfsema, C.M.; Phinn, S.R.; Carvalho, R.C.; Shaw, E.C.; Joyce, K.E. Estimating regional coral reef calcium carbonate production from remotely sensed seafloor maps. *Remote Sens. Environ.* **2017**, *201*, 88–98. [CrossRef]
354. Barve, S.; Webster, J.M.; Chandra, R. Reef-Insight: A framework for reef habitat mapping with clustering methods using remote sensing. *Information* **2023**, *14*, 373. [CrossRef]
355. Carrasco Rivera, D.E.; Diederiks, F.F.; Hammerman, N.M.; Staples, T.; Kovacs, E.; Markey, K.; Roelfsema, C.M. Remote Sensing Reveals Multidecadal Trends in Coral Cover at Heron Reef, Australia. *Remote Sens.* **2025**, *17*, 1286. [CrossRef]
356. Braga, J.C.; Aguirre, J. Coralline algae indicate Pleistocene evolution from deep, open platform to outer barrier reef environments in the northern Great Barrier Reef margin. *Coral Reefs* **2004**, *23*, 547–558. [CrossRef]
357. McCaffrey, J.; Wallace, M.W.; Gallagher, S.J. A Cenozoic Great Barrier Reef on Australia's North West Shelf. *Glob. Planet. Chang.* **2020**, *184*, 1030148. [CrossRef]
358. McCaffrey, J.; Wallace, M.W.; Gallagher, S.J.; Averes, T.; Fabian, S.G.; Lindhorst, K.; Reuning, L.; Krastel, S. The Rowley Shoals atolls: Remnants of a Miocene great barrier reef on the north-west Australian margin. *Glob. Planet. Chang.* **2025**, *245*, 104688. [CrossRef]
359. Williams, C.; Paumard, V.; Webster, J.M.; Leonard, J.; Salles, T.; O'Leary, M.; Lang, S. Environmental controls on the resilience of Scott Reefs since the Miocene (North West Shelf, Australia): Insights from 3D seismic data. *Mar. Pet. Geol.* **2023**, *151*, 106188. [CrossRef]
360. Flood, P.G. Geological history of the reef. In *A Coral Reef Handbook*, 3rd ed.; Mather, P., Bennett, I., Eds.; Surrey Beatty & Sons Pty Ltd.: Chipping Norton, NSW, Australia, 1993; pp. 3–6.
361. Palmieri, V. Correlation and environmental trends of the subsurface Tertiary Capricorn Basin. *Rep. Geol. Surv. Qld.* **1974**, *86*, 1–14.
362. Maxwell, W.G.H. Sediments of the Great Barrier Reef Province. In *Biology and Geology of Coral Reefs, Vol. 3, Geology 1*; Jones, O.A., Endean, R., Eds.; Academic Press: New York, NY, USA, 1973; pp. 299–345.
363. Orme, G.R. Aspects of sedimentation in the coral reef environment. In *Biology and Geology of Coral Reefs, Vol. 4, Geology 2*; Jones, O.A., Endean, R., Eds.; Academic Press: New York, NY, USA, 1973; pp. 129–182.
364. Doo, S.S.; Hamylton, S.; Finfer, J.; Byrne, M. Spatial and temporal variation in reef-scale carbonate storage of large benthic foraminifera: A case study on One Tree Reef. *Coral Reefs* **2017**, *36*, 293–303. [CrossRef]
365. Veron, J.E.N. *Corals of the World*; Australian Institute of Marine Sciences: Townsville, QLD, Australia, 2000; Volume 1–3, pp. 1–1382.
366. DeVantier, L.M.; De'ath, G.; Turak, E.; Done, T.J.; Fabricius, K.E. Species richness and community structure of reef-building corals on the nearshore Great Barrier Reef. *Coral Reefs* **2006**, *25*, 329–340. [CrossRef]
367. Done, T.J. Patterns in the distribution of coral communities across the central Great Barrier Reef. *Coral Reefs* **1982**, *1*, 95–107. [CrossRef]
368. Bridge, T.C.L.; Done, T.J.; Friedman, A.; Beaman, R.J.; Williams, S.B.; Pizarro, O.; Webster, J.M. Variability in mesophotic coral reef communities along the Great Barrier Reef. *Mar. Ecol. Prog Ser.* **2011**, *428*, 63–75. [CrossRef]
369. Frade, P.R.; Bongaerts, P.; Englebert, N.; Rogers, A.; Gonzalez-Rivero, M.; Hoegh-Guldberg, O. Deep reefs of the Great Barrier Reef offer limited thermal refuge during mass coral bleaching. *Nat. Commun.* **2018**, *9*, 3447. [CrossRef]
370. Fabricius, K.; De'ath, G. Environmental factors associated with the spatial distribution of crustose coralline algae on the Great Barrier Reef. *Coral Reefs* **2001**, *19*, 303–309. [CrossRef]
371. Hurrey, L.P.; Pitcher, C.R.; Lovelock, C.E.; Schmidt, S. Macroalgal species richness and assemblage composition of the Great Barrier Reef seabed. *Mar. Ecol. Prog. Ser.* **2013**, *492*, 69–83. [CrossRef]
372. Dean, A.J.; Steneck, R.S.; Tager, D.; Pandolfi, J.M. Distribution, abundance and diversity of crustose coralline algae on the Great Barrier Reef. *Coral Reefs* **2015**, *34*, 581–594. [CrossRef]
373. Marshall, J.F.; Davies, P.J. *Halimeda* bioherms of the northern Great Barrier Reef. *Coral Reefs* **1988**, *6*, 139–148. [CrossRef]
374. Orme, G.R.; Salama, M.S. Form and seismic stratigraphy of *Halimeda* banks in part of the northern Great Barrier Reef Province. *Coral Reefs* **1988**, *6*, 131–137. [CrossRef]
375. Rees, S.A.; Opdyke, B.N.; Wilson, P.A.; Henstock, T.J. Significance of Halimeda bioherms to the global carbonate budget based on a geological sediment budget for the Northern Great Barrier Reef, Australia. *Coral Reefs* **2007**, *26*, 177–188. [CrossRef]
376. Vila-Concejo, A.; Hamylton, S.M.; Webster, J.M.; Duce, S.J.; Fellowes, T.E. Lagoon infilling by coral reef sand aprons as a proxy for carbonate sediment productivity. *Geology* **2022**, *50*, 1427–1431. [CrossRef]

377. Bellwood, D.R. Production and reworking of sediment by parrotfishes (family Scaridae) on the Great Barrier Reef, Australia. *Mar. Biol.* **1996**, *125*, 795–800. [CrossRef]
378. Lange, I.D.; Perry, C.T.; Alvarez-Filip, L. Carbonate budgets as indicators of functional reef "health": A critical review of data underpinning census-based methods and current knowledge gaps. *Ecol. Indic.* **2020**, *110*, 105857. [CrossRef]
379. Yamano, H.; Miyajima, T.; Koike, I. Importance of foraminifera for the formation and maintenance of a coral sand cay: Green Island, Australia. *Coral Reefs* **2000**, *19*, 51–58. [CrossRef]
380. Montaggioni, L.F.; Camoin, G.F. Stromatolites associated with coralgal communities in Holocene high energy reefs. *Geology* **1993**, *21*, 149–152. [CrossRef]
381. Webb, G.E.; Nothdurft, L.D.; Zhao, J.-X.; Opdyke, B.; Price, G. Significance of shallow core transects for reef models and sea-level curves, Heron Reef, Great Barrier Reef. *Sedimentology* **2016**, *63*, 1396–1424. [CrossRef]
382. Marino, G.; Rohling, E.J.; Rodríguez-Sanz, L.; Grant, K.M.; Heslop, D.; Roberts, A.P.; Stanford, J.D.; Yu, J. Bipolar seesaw control on last interglacial sea level. *Nature* **2015**, *522*, 197–201. [CrossRef] [PubMed]
383. Meziere, Z.; Popovic, I.; Prata, K.; Ryan, I.; Pandolfi, J.; Riginos, C. Exploring coral speciation: Multiple sympatric *Stylophora pistillata* taxa along a divergence continuum on the Great Barrier Reef. *Evol. Appl.* **2024**, *17*, e13644. [CrossRef] [PubMed]
384. Tribollet, A.; Golubic, S. Cross-shelf differences in the pattern and pace of bioerosion of experimental carbonate substrates exposed for 3 years on the northern Great Barrier Reef, Australia. *Coral Reefs* **2005**, *24*, 422–434. [CrossRef]
385. Schönberg, C.H.L. Small-scale distribution of Great Barrier Reef bioeroding sponges in shallow water. *Ophelia* **2001**, *55*, 39–54. [CrossRef]
386. Copp, I.A. *Subsurface Facies Analysis of Devonian Reef Complexes, Lennard Shelf, Canning Basin Western Australia*; Geological Survey of Western Australia Bulletin: Perth, Australia, 2000; Volume 58, pp. 1–127.
387. Wood, R. Palaeoecology of a Late Devonian back reef, Canning Basin, Western Australia. *Palaeontology* **2000**, *43*, 671–703. [CrossRef]
388. Stephens, N.P.; Sumner, D.Y. Famennian microbial reef facies, Napier and Oscar Ranges, Canning Basin, Western Australia. *Sedimentology* **2003**, *50*, 1283–1302. [CrossRef]
389. Cockbain, A.E. *Stromatoporoids from the Devonian Reef Complexes, Canning Basin, Western Australia*; Geological Survey of Western Australia Bulletin: Perth, Australia, 1984; Volume 129, pp. 1–108.
390. Hill, D.; Jell, J.S. *Devonian Corals from the Canning Basin, Western Australia*; Geological Survey of Western Australia Bulletin: Perth, WA, Australia, 1970; Volume 123, pp. 1–158.
391. Brownlaw, R.L.S.; Jell, J.S. Middle and Upper Devonian rugose corals from the Canning Basin, Western Australia. *Assoc. Australas. Paleontol. Mem.* **2008**, *35*, 1–126.
392. Wray, J.L. *Upper Devonian Calcareous Algae from the Canning Basin, Western Australia*; Professional Contributions of the Colorado School of Mines: Golden, CO, USA, 1967; Volume 3, p. 49.
393. Veevers, J.J. *Devonian Brachiopods from the Fitzroy Basin, Western Australia*; Bulletin Department of National Development/Bureau of Mineral Resources, Geology and Geophysics: Canberra, Australia, 1959; Volume 45, pp. 1–173.
394. Grey, K. *Devonian Atrypid Brachiopods from the Reef Complexes of the Canning Basin*; Bulletin Geological Survey of Western Australia: Perth, Australia, 1978; Volume 5, pp. 1–70.
395. Cook, A.G.; Blodgett, R.B.; Becker, R.T. Late Devonian gastropods from the Canning Basin, Western Australia. *Alcheringa* **2003**, *27*, 181–207. [CrossRef]
396. Playford, P.E.; Cockbain, A.E.; Druce, E.C.; Wray, J.L. Devonian stromatolites from the Canning Basin, Western Australia. In *Stromatolites*; Walter, M.R., Ed.; Elsevier: Amsterdam, The Netherlands, 1976; Developments in Sedimentology Volume 20, pp. 543–563.
397. George, A.D. Deep-water stromatolites, Canning Basin, northwestern Australia. *Palaios* **1999**, *14*, 493–505. [CrossRef]
398. Rigby, J.K. *Late Devonian Sponges of Western Australia*; Bulletin Geological Survey of Western Australia: Perth, WA, Australia, 1986; Volume 18, pp. 1–59.
399. Reitner, J.; Gautret, P.; Maurin, F.; Neuweiler, F. Automicrites in a modern marine microbialite. Formation model via organic matrices (Lizard Island, Great Barrier Reef, Australia). In *Biomineralization 93: Proceedings of the 7th International Symposium on Biomineralization*; Allemand, D., Cuif, J.-P., Eds.; Bulletin de l'Institut océanographique, Numéro spécial 14; Institut océanographique: Monaco, 1995; pp. 237–263.
400. Ma, X.; Gong, Y.; Chen, D.; Racki, G.; Chen, X.; Liao, W. The late Devonian Frasnian-Famennian event in South China—Patterns and causes of extinctions, sea level changes, and isotope variations. *Palaeogeogr. Palaeoclimatol. Palaeoecol.* **2016**, *448*, 224–244. [CrossRef]
401. Stephens, N.P.; Sumner, D.Y. Late Devonian carbon isotope stratigraphy and sea level fluctuations, Canning Basin, Western Australia. *Palaeogeogr. Palaeoclimat. Palaeoecol.* **2003**, *19*, 203–219. [CrossRef]
402. George, A.D.; Chow, N. The depositional record of the Frasnian/Famennian boundary interval in a fore-reef succession, Canning Basin, Western Australia. *Palaeogeogr. Palaeoclimat. Palaeoecol.* **2002**, *181*, 347–374. [CrossRef]

403. Benn, C.J. Facies changes and development of a carbonate platform, east Pillara Range. In *The Canning Basin, W.A., Proceedings of the Geological Society of Australia and Petroleum Exploration Society of Australia Symposium, Perth, WA, Australia, 27–29 June 1984*; Purcell, P.G., Ed.; Geological Society of Australia and Petroleum Exploration Society of Australia: Perth, WA, Australia, 1984; pp. 223–228.
404. Pratt, B.R. *Epiphyton* and *Renalcis*: Diagenetic microfossils from calcification of coccoid blue-green algae. *J. Sed. Petrol.* **1984**, *54*, 948–971. [CrossRef]
405. Stephens, N.P.; Sumner, D.Y. Renalcids as Fossilized Biofilm Clusters. *Palaios* **2002**, *17*, 225–236. [CrossRef]
406. Feng, Q.; Gong, Y.-M.; Riding, R. Mid-Late Devonian calcified marine algae and cyanobacteria, South China. *J. Paleont.* **2010**, *84*, 569–587. [CrossRef]
407. Hall, W.D.M. The stratigraphic and structural development of the Givetian–Frasnian reef complex, Limestone Billy Hills, Western Pillara Range, W.A. In *The Canning Basin, W.A., Proceedings of the Geological Society of Australia and Petroleum Exploration Society of Australia Symposium, Perth, WA, Australia, 27–29 June 1984*; Purcell, P.G., Ed.; Geological Society of Australia and Petroleum Exploration Society of Australia: Perth, WA, Australia, 1984; pp. 215–222.
408. Webb, G.E. Famennian mud-mounds in the proximal fore-reef slope, Canning Basin, Western Australia. *Sed. Geol.* **2001**, *144*, 295–315. [CrossRef]
409. Cooper, R.W.; Hall, W.D.W.; Styles, G.R. The Devonian Stratigraphy of the Central Pillara Range. In *The Canning Basin, W.A., Proceedings of the Geological Society of Australia and Petroleum Exploration Society of Australia Symposium, Perth, WA, Australia, 27–29 June 1984*; Purcell, P.G., Ed.; Geological Society of Australia and Petroleum Exploration Society of Australia: Perth, WA, Australia, 1984; pp. 229–234.
410. Zapalski, M.K.; Nowicki, J.; Jakubowicz, M.; Berkowski, B. Tabulate corals across the Frasnian/Famennian boundary: Architectural turnover and its possible relation to ancient photosymbiosis. *Palaeogeogr. Palaeoclimatol. Palaeoecol.* **2017**, *487*, 416–429. [CrossRef]
411. Wood, R.; Oppenheimer, C. Spur and groove morphology from a Late Devonian reef. *Sed. Geol.* **2000**, *133*, 185–193. [CrossRef]
412. Diaz, M.R.; Eberli, G.P.; Blackwelder, P.; Phillips, B.; Swart, P.K. Microbially mediated organomineralization in the formation of ooids. *Geology* **2017**, *45*, 771–774. [CrossRef]
413. Li, F.; Yan, J.; Burne, R.V.; Chen, Z.-Q.; Algeo, T.J.; Zhang, W.; Tian, L.; Gan, Y.; Liu, K.; Xie, S. Paleo-seawater REE compositions and microbial signatures preserved in laminae of Lower Triassic ooids. *Palaeogeogr. Palaeoclimatol. Palaeoecol.* **2017**, *486*, 96–107. [CrossRef]
414. Batchelor, M.T.; Burne, R.V.; Henry, B.I.; Li, F.; Paul, J. A biofilm and organomineralisation model for the growth and limiting size of ooids. *Sci. Rep.* **2018**, *8*, 559. [CrossRef]
415. Adams, E.W.; Hasler, C.-A. The intrinsic effect of shape on the retrogradation motif and timing of drowning of carbonate patch reef systems (Lower Frasnian, Bugle Gap, Canning Basin, Western Australia). *Sedimentology* **2010**, *57*, 956–984. [CrossRef]
416. Shen, J.-W.; Webb, G.E.; Qing, H. Microbial mounds prior to the Frasnian-Famennian mass extinctions, Hantang, Guilin, South China. *Sedimentology* **2010**, *57*, 1615–1639. [CrossRef]
417. Chow, N.; George, A.D. Tepee-shaped agglutinated microbialites: An example from a Famennian carbonate platform on the Lennard Shelf, northern Canning Basin, Western Australia. *Sedimentology* **2004**, *51*, 253–265. [CrossRef]
418. Long, J.A.; Trinajstic, K. The Late Devonian Gogo Formation Lägerstatte—Exceptional preservation and diversity in early vertebrates. *Ann. Rev. Earth Plane. Sci.* **2010**, *38*, 665–680. [CrossRef]
419. Trinajstic, K.; Briggs, D.E.G.; Long, J.A. The Gogo Formation Lagerstätte: A view of Australia's first great barrier reef. *J. Geol. Soc.* **2022**, *179*, jgs2021-105. [CrossRef]
420. Aitchison, J.C. Devonian (Frasnian) radiolarians from Gogo Formation, Canning Basin, Western Australia. *Palaeontogr. A* **1993**, *228*, 105–128.
421. Briggs, D.E.; Rolfe, W.I.; Butler, P.D.; Liston, J.J.; Ingham, J.K. Phyllocarid crustaceans from the Upper Devonian Gogo Formation, Western Australia. *J. Syst. Palaeont.* **2011**, *9*, 399–424. [CrossRef]
422. Tetlie, O.E.; Braddy, S.J.; Butler, P.D.; Briggs, D.E.G. A new eurypterid (Chelicerata: Eurypterida) from the Upper Devonian Gogo Formation of Western Australia, with a review of the Rhenopteridae. *Palaeontology* **2004**, *47*, 801–809. [CrossRef]
423. Long, J.A.; Trinajstic, K.M. A review of recent discoveries of exceptionally preserved fossil fishes from the Gogo sites (Late Devonian, Western Australia). *Earth Environ. Sci. Trans. R. Soc. Edinb.* **2018**, *108*, 111–117. [CrossRef]
424. Trinajstic, K.; Marshall, C.; Long, J.; Bifield, K. Exceptional preservation of nerve and muscle tissues in Late Devonian placoderm fish and their evolutionary implications. *Biol. Lett.* **2007**, *3*, 197–200. [CrossRef]
425. Long, J.A.; Trinajstic, K.; Young, G.C.; Senden, T. Live birth in the Devonian period. *Nature* **2008**, *453*, 650–652. [CrossRef]
426. Tulipani, S.; Grice, K.; Greenwood, P.F.; Haines, P.W.; Sauer, P.E.; Schimmelmann, A.; Summons, R.E.; Foster, C.B.; Böttcher, M.E.; Playton, T.; et al. Changes of palaeoenvironmental conditions recorded in Late Devonian reef systems from the Canning Basin, Western Australia: A biomarker and stable isotope approach. *Gondwana Res.* **2015**, *28*, 1500–1515. [CrossRef]
427. Viohl, G. Die Solnhofener Plattenkalke—Entstehung und Lebensräume. *Archaeopteryx* **1998**, *16*, 37–68.

428. Munnecke, A.; Westphal, H.; Kölbl-Ebert, M. Diagenesis of plattenkalk: Examples from the Solnhofen area (Upper Jurassic, southern Germany). *Sedimentology* **2008**, *55*, 1931–1946. [CrossRef]
429. Barthel, K.W.; Swinburne, N.H.M.; Morris, S.C. *Solnhofen, A Study in Mesozoic Palaeontology*; Cambridge University Press: Cambridge, UK, 1994; pp. 1–246.
430. Richards, F.D.; Kalnins, L.M.; Watts, A.B.; Cohen, B.E.; Beaman, R.J. The morphology of the Tasmantid Seamounts: Inter-actions between tectonic inheritance and magmatic evolution. *Geochem. Geophys. Geosyst.* **2018**, *19*, 3870–3891. [CrossRef]
431. Johnson, M.E. Ecology of Intertidal Rocky Shores Related to Examples of Coastal Geology across Phanerozoic Time. *J. Mar. Sci. Eng.* **2024**, *12*, 1399. [CrossRef]
432. Yokoyama, Y.; Webster, J.M.; Cotterill, C.; Braga, J.C.; Jovane, L.; Mills, H.; Morgan, S.; Suzuki, A. IODP Expedition 325 Scientists IODP Expedition 325: Great Barrier Reef reveals past sea-level, climate and environmental change since the last ice age. *Sci. Drill.* **2011**, *12*, 32–45. [CrossRef]
433. Cuif, J.-P. The Rugosa–Scleractinia gap re-examined through microstructural and biochemical evidence: A tribute to H.C. Wang. *Palaeoworld* **2014**, *23*, 1–14. [CrossRef]
434. Webb, G.E. Morphological variation and homoplasy: The challenge of Paleozoic coral systematics. *Paleontol. Soc. Pap.* **1996**, *1*, 135–157. [CrossRef]
435. Coates, A.G.; Jackson, J.B.C. Clonal growth, algal symbiosis, and reef formation by corals. *Paleobiology* **1987**, *13*, 363–378. [CrossRef]
436. Majchrzyk, A.; Jakubowicz, M.; Berkowski, B.; Król, J.J.; Zatón, M.; Zapalski, M.K. Modern-type reef in ancient time—Palaeoecology of a Middle Devonian coral community from Mad'ene el Mrakib (Anti-Atlas, Morocco). *Palaeogeogr. Palaeoclimat. Palaeoecol.* **2024**, *633*, 111876. [CrossRef]
437. Pearson, R.G. Recovery and recolonization of coral reefs. *Mar. Ecol. Prog. Ser.* **1981**, *4*, 105–122. [CrossRef]
438. Adey, W.H.; Burke, R.B. Holocene bioherms (algal ridges and bank-barrier reefs) of the eastern Caribbean. *Geol. Soc. Am. Bull.* **1976**, *87*, 95–109. [CrossRef]
439. Camoin, G.F.; Gautret, P.; Montaggioni, L.F.; Cabioch, G. Nature and environmental significance of microbialites in Quaternary reefs: The Tahiti paradox. *Sed. Geol.* **1999**, *126*, 271–304. [CrossRef]
440. MacNeil, A.J.; Jones, B. Nutrient-gradient controls on Devonian reefs: Insight from the ramp-situated Alexandra Reef System (Frasnian), northwest territories, Canada. In *Controls on Carbonate Platform and Reef Development*; Lukasik, J., Simo, J.A.T., Eds.; SEPM Special Publication; Society of Economic Paleontologists and Mineralogists: Tulsa, OK, USA, 2008; Volume 89, pp. 271–289.
441. Wood, R. The changing biology of reef-building. *Palaios* **1995**, *10*, 517–529. [CrossRef]
442. Wood, R. The ecological evolution of reefs. *Annu. Rev. Ecol. Syst.* **1998**, *29*, 179–206. [CrossRef]
443. Webb, G.E. Paleokarst, paleosol, and rocky-shore deposits at the Mississippian—Pennsylvanian unconformity, northwestern Arkansas. *Geol. Soc. Am. Bull.* **1994**, *106*, 634–648. [CrossRef]
444. Taylor, P.D.; Wilson, M.A. Palaeoecology and evolution of marine hard substrate communities. *Earth-Sci. Rev.* **2003**, *62*, 1–103. [CrossRef]
445. Kobluk, D.R.; James, N.P.; Pemberton, S.G. Initial diversification of macroboring ichnofossils and exploitation of the macroboring niche in the Lower Paleozoic. *Paleobiology* **1978**, *4*, 163–170. [CrossRef]
446. Fagerstrom, J.A. Paleozoic brachiopod symbioses: Testing the limits of modern analogues in paleoecology. *Geol. Soc. Am. Bull.* **1996**, *108*, 1393–1403. [CrossRef]
447. Steneck, R.S. Escalating herbivory and resulting adaptive trends in calcareous algal crusts. *Paleobiology* **1983**, *9*, 44–61. [CrossRef]
448. Hallock, P. Coral reefs, carbonate sediments, nutrients, and global change, Chapter 11. In *The History and Sedimentology of Ancient Reef Systems*; Stanley, G.D., Jr., Ed.; Academic/Plenum Publishers: New York, NY, USA, 2001; pp. 387–427.
449. Zapalski, M.K. Evidence of photosymbiosis in Palaeozoic tabulate corals. *Proc. R. Soc. Lond. B Biol. Sci.* **2014**, *281*, 20132663. [CrossRef] [PubMed]
450. Majchrzyk, A.; Jakubowicz, M.; Bongaerts, P.; Zapalski, M.K. Different times, similar mechanism? Convergent patterns in light-induced phenotypic plasticity in Devonian and modern corals. *Coral Reefs* **2023**, *42*, 893–903. [CrossRef]
451. Jung, J.; Zoppe, S.F.; Söte, T.; Moretti, S.; Duprey, N.N.; Foreman, A.D.; Wald, T.; Vonhof, H.; Haug, G.H.; Sigman, D.M.; et al. Coral photosymbiosis on Mid-Devonian reefs. *Nature* **2024**, *636*, 647–653. [CrossRef]
452. Król, J.J.; Berkowski, B.; Denayer, J.; Zapalski, M.K. Deducing photosymbiosis in extinct heliolitid corals. *Coral Reefs* **2024**, *43*, 91–105. [CrossRef]
453. Kershaw, S.; Munnecke, A.; Jarochowska, E. Understanding Palaeozoic stromatoporoid growth. *Earth-Sci. Rev.* **2018**, *187*, 53–76. [CrossRef]
454. Kershaw, S.; Jeon, J. Stromatoporoids and extinctions: A review. *Earth-Sci. Rev.* **2024**, *252*, 104721. [CrossRef]
455. Li, Y.-F.; Li, F.; Webb, G.E.; Chen, J. Intense intrusion of low-oxygen waters into mid-Cambrian surface ocean carbonate factories. *Chem. Geol.* **2024**, *669*, 122360. [CrossRef]

456. Littler, M.M.; Littler, D.S.; Hanisak, M.D. Deep-water rhodolith distribution, productivity, and growth history at sites of formation and subsequent degradation. *J. Exp. Mar. Biol. Ecol.* **1991**, *150*, 163–182. [CrossRef]
457. Wilson, M.E.J. Equatorial carbonates: An earth systems approach. *Sedimentology* **2012**, *59*, 1–31. [CrossRef]
458. Hoegh-Guldberg, O. Low coral cover in a high-CO_2 world. *J. Geophys. Res.* **2005**, *110*, C09S06. [CrossRef]
459. Yokoyama, Y.; Esat, T.M.; Thompson, W.G.; Thomas, A.L.; Webster, J.M.; Miyairi, Y.; Sawada, C.; Aze, T.; Matsuzaki, H.; Okuno, J.; et al. Rapid glaciation and a twostep sea-level plunge into the Last Glacial Maximum. *Nature* **2018**, *559*, 603–607. [CrossRef]
460. Purdy, E.G.; Gischler, E. The transient nature of the empty bucket model of reef sedimentation. *Sed. Geol.* **2005**, *175*, 35–47. [CrossRef]
461. George, A.D.; Chow, N. Palaeokarst development in a lower Frasnian (Devonian) platform succession, Canning Basin, northwestern Australia. *Aust. J. Earth Sci.* **1999**, *46*, 905–913. [CrossRef]
462. Backshall, D.G.; Barnett, J.; Davies, P.J.; Duncan, D.C.; Harvey, N.; Hopley, D.; Isdale, P.; Jennings, J.N.; Moss, R. Drowned dolines-the blue holes of the Pompey Reefs, Great Barrier Reef. *BMR J. Aust. Geol. Geophys.* **1979**, *4*, 99–109.
463. Kershaw, S.; Munnecke, A.; Jarochowska, E.; Young, G. Palaeozoic stromatoporoid diagenesis: A synthesis. *Facies* **2021**, *67*, 20. [CrossRef]
464. Wallace, M.W.; Kerans, C.; Playford, P.E.; McManus, A. Burial diagenesis in the Upper Devonian reef complexes of the Geikie Gorge Region, Canning Basin, Western Australia. *Am. Assoc. Petrol. Geol. Bull.* **1991**, *75*, 1018–1038. [CrossRef]
465. Webb, G.E. Quantitative analysis and paleoecology of earliest Mississippian microbial reefs, Gudman Formation, Queensland, Australia: Not just post-disaster phenomena. *J. Sed. Res.* **2005**, *75*, 877–896. [CrossRef]
466. Ladd, H.S. Recent reefs. *Am. Assoc. Petrol. Geol. Bull.* **1950**, *34*, 203–214. [CrossRef]
467. Schlager, W. The paradox of drowned reefs and carbonate platforms. *Geol. Soc. Am. Bull.* **1981**, *92*, 197–211. [CrossRef]
468. McNeil, M.A.; Webster, J.M.; Beaman, R.J.; Graham, T.L. New constraints on the spatial distribution and morphology of the *Halimeda* bioherms of the Great Barrier Reef, Australia. *Coral Reefs* **2016**, *35*, 1343–1355. [CrossRef]
469. Whalen, M.T.; Day, J.; Eberli, G.P.; Homewood, P.W. Microbial carbonates as indicators of environmental change and biotic crises in carbonate systems: Examples from the Late Devonian, Alberta basin, Canada. *Palaeogeogr. Palaeoclimat. Palaeoecol.* **2002**, *181*, 127–151. [CrossRef]
470. Riding, R. Calcified cyanobacteria. In *Calcareous Algae and Stromatolites*; Riding, R., Ed.; Springer: Berlin/Heidelberg, Germany, 1991; pp. 55–87.
471. Arp, G.; Reimer, A.; Reitner, J. Photosynthesis-induced biofilm calcification and calcium concentrations in Phanerozoic oceans. *Science* **2001**, *292*, 1701–1704. [CrossRef]
472. Dupraz, C.; Reid, R.P.; Braissant, O.; Decho, A.W.; Norman, R.S.; Visscher, P.T. Processes of carbonate precipitation in modern microbial mats. *Earth-Sci. Rev.* **2009**, *96*, 141–162. [CrossRef]
473. Pratt, B.R. Calcification of cyanobacterial filaments: *Girvanella* and the origin of Lower Paleozoic lime mud. *Geology* **2001**, *29*, 763–766. [CrossRef]
474. Saint Martin, J.-P. The *Girvanella*-like remains from Messinian marine deposits (Sardinia, Italy): Lagerstätten paradigm for microbial biota? *Annal. Paléontol.* **2010**, *96*, 33–50. [CrossRef]
475. Zhang, H.S.; Dai, M.-Y.; Qi, Y.-A.; Han, L.-L.; Yin, Z.-L.; Chen, S.-H.; Lin, L.-B. *Girvanella* fossils from the Phanerozoic: Distribution, evolution and controlling factors. *J. Palaeogeogr.* **2024**, *13*, 924–938. [CrossRef]
476. Kamenos, N.A.; Perna, G.; Gambi, M.C.; Micheli, F.; Kroeker, K.J. Coralline algae in a naturally acidified ecosystem persist by maintaining control of skeletal mineralogy and size. *Proc. R. Soc. B* **2016**, *283*, 20161159. [CrossRef]
477. Cornwall, C.E.; Comeau, S.; McCulloch, M.T. Coralline algae elevate pH at the site of calcification under ocean acidification. *Glob. Change Biol.* **2017**, *23*, 4245–4256. [CrossRef] [PubMed]
478. Kawahata, H.; Fujita, K.; Iguchi, A.; Inoue, M.; Iwasaki, S.; Kuroyanagi, A.; Maeda, A.; Manaka, T.; Moriya, K.; Takagi, H.; et al. Perspective on the response of marine calcifiers to global warming and ocean acidification—Behavior of corals and foraminifera in a high CO_2 world "hot house". *Prog. Earth Planet. Sci.* **2019**, *6*, 5. [CrossRef]
479. Webb, G.E.; Kamber, B.S. Trace element geochemistry as a tool for interpreting microbialites. In *Earliest Life on Earth: Habitats, Environments and Methods of Detection*; Golding, S., Glickson, M., Eds.; Springer: Dordrecht, The Netherlands, 2011; pp. 127–170. [CrossRef]
480. Ezaki, Y. Secular fluctuations in Palaeozoic and Mesozoic reef-forming organisms during greenhouse periods: Geobiological interrelations and consequences. *Paleont. Res.* **2009**, *13*, 23–38. [CrossRef]
481. Riding, R. An atmospheric stimulus for cyanobacterial-bioinduced calcification ca. 350 million years ago? *Palaios* **2009**, *24*, 685–696. [CrossRef]
482. Riding, R.; Liang, L.; Lee, J.-H.; Virgone, A. Influence of dissolved oxygen on secular patterns of marine microbial carbonate abundance during the past 490 Myr. *Palaeogeog. Palaeoclimat. Palaeoecol.* **2019**, *514*, 135–143. [CrossRef]
483. Middelburg, J.J.; Soetaert, K.; Hagens, M. Ocean alkalinity, buffering and biogeochemical processes. *Rev. Geophys.* **2020**, *58*, e2019RG000681. [CrossRef]

484. Farkaš, J.; Wallmann, K.; Mosley, L.; Staudigel, P.; Zheng, X.-Y.; Leyden, E.; Shao, Y.; Frýda, J.; Holmden, C.; Eisenhauer, A. Alkalinity and elemental cycles in present and past ocean: Insight from geochemical modeling and alkali and alkaline earth metal isotopes. In *Treatise on Geochemistry*, 3rd ed.; Elsevier Science: Amsterdam, The Netherlands, 2025; Volume 5, pp. 33–85. [CrossRef]
485. Riding, R.; Liang, L. Seawater chemistry control of marine limestone accumulation over the past 550 million years. *Rev. Española Micropaleont.* **2005**, *37*, 1–11.
486. Pratt, B.R. Stromatolite decline—A reconsideration. *Geology* **1982**, *10*, 512–515. [CrossRef]
487. Griffin, K.M. Microbial reefs on a carbonate ramp: A case study from western North America with global perspective. In *Cavalcade of Carbonates*; Cooper, J.D., Ed.; Volume and Guidebook for Annual Meeting: SEPM, Pacific Section, Field Trip 3; Society of Economic Paleontologists and mineralogists: Tulsa, OK, USA, 1989; pp. 101–110.
488. Hönisch, B.; Ridgwell, A.; Schmidt, D.N.; Thomas, E.; Gibbs, S.J.; Sluijs, A.; Zeebe, R.; Kump, L.; Martindale, R.C.; Greene, S.E.; et al. The geological record of ocean acidification. *Science* **2012**, *335*, 1058–1063. [CrossRef] [PubMed]
489. Kiessling, W.; Simpson, C. On the potential for ocean acidification to be a general cause of ancient reef crises. *Glob. Change Biol.* **2011**, *17*, 56–67. [CrossRef]
490. Rowland, S.M.; Gangloff, R.A. Structure and paleoecology of Lower Cambrian reefs. *Palaios* **1988**, *3*, 111–135. [CrossRef]
491. Pratt, B.R.; James, N.P. Cryptalgal-metazoan bioherms of early Ordovician age in the St George Group, western Newfoundland. *Sedimentology* **1982**, *29*, 543–569. [CrossRef]
492. Adams, A.E. Development of algal-foraminiferal-coral reefs in the Lower Carboniferous of Furness, northwest England. *Lethaia* **1984**, *17*, 233–249. [CrossRef]
493. Mountjoy, E.W.; Jull, R.K. Fore-reef carbonate mud bioherms and associated reef margin, Upper Devonian, Ancient Wall reef complex, Alberta. *Can. J. Earth Sci.* **1978**, *15*, 1304–1325. [CrossRef]
494. Mountjoy, E.W.; Riding, R. Foreslope stromatoporoid-renalcid bioherm with evidence of early cementation, Devonian Ancient Wall reef complex, Rocky Mountains. *Sedimentology* **1981**, *28*, 299–319. [CrossRef]
495. Tsien, H.H. Paleoecology of algal-bearing facies in the Devonian (Couvinian to Frasnian) reef complexes of Belgium. *Paleogeog. Palaeoclim. Palaeoecol.* **1979**, *27*, 103–127. [CrossRef]
496. Antoshkina, A.I. Organic buildups and reefs on the Palaeozoic carbonate platform margin, Pechora Urals, Russia. *Sed. Geol.* **1998**, *118*, 187–211. [CrossRef]
497. Antoshkina, A.I. Palaeoenvironmental implications of *Palaeomicrocodium* in Upper Devonian microbial mounds of the Chernyshev Swell, Timan-northern Ural Region. *Facies* **2006**, *52*, 611–625. [CrossRef]
498. Klimenko, S.S.; Anischenko, L.A.; Antoshkina, A.I. The Timan-Pechora sedimentary basin: Palaeozoic reef formations and petroleum systems, Chapter 13. In *Arctic Petroleum Geology*; Spencer, A.M., Embry, A.F., Gautier, D.L., Stoupakova, A.V., Sørensen, K., Eds.; Memoirs; Geological Society: London, UK, 2011; Volume 35, pp. 223–236. [CrossRef]
499. Ulmishek, G.F. Upper Devonian-Tournaisian facies and oil resources of the Russian craton's eastern margin. In *Devonian of the World*; McMillan, N.J., Embry, A.F., Glass, D.J., Eds.; Canadian Society of Petroleum Geologists: Calgary, AB, Canada, 1988; Volume 1, pp. 527–549.
500. Shen, J.-W.; Webb, G.E.; Yang, H.-Q. Reply to the Discussion by Alex MacNeil on "Microbial mounds prior to the Frasnian–Famennian mass extinctions, Hantang, Guilin, South China" by Shen et al., Sedimentology, 57, 1615–1639. *Sedimentology* **2011**, *58*, 2066–2071. [CrossRef]
501. Sprachta, S.; Camoin, G.; Golubic, S.; Le Campion, T. Microbialites in a modern lagoonal environment: Nature and distribution (Tikehau atoll, French Polynesia). *Palaeogeogr. Palaeoclimatol. Palaeoecol.* **2001**, *175*, 103–124. [CrossRef]
502. Visscher, P.T.; Reid, R.P.; Bebout, B.M. Microscale observations of sulfate reduction: Correlation of microbial activity with lithified micritic laminae in modern marine stromatolites. *Geology* **2000**, *28*, 919–922. [CrossRef]
503. Heindel, K.; Birgel, D.; Peckmann, J.; Kuhnert, H.; Westphal, H. Formation of deglacial microbialites in coral reefs off Tahiti (IODP 310) involving sulfate-reducing bacteria. *Palaios* **2010**, *25*, 618–635. [CrossRef]
504. Nothdurft, L.D.; Webb, G.E.; Kamber, B.S. Rare earth element geochemistry of Late Devonian reefal carbonates, Canning Basin, Western Australia: Confirmation of a seawater REE proxy in ancient limestones. *Geochim. Cosmochim. Acta* **2004**, *68*, 263–283. [CrossRef]
505. Della Porta, G.; Webb, G.E.; McDonald, I. REE patterns of microbial carbonate and cements from Sinemurian (Lower Jurassic) siliceous sponge mounds (Djebel Bou Dahar, High Atlas, Morocco). *Chem. Geol.* **2015**, *400*, 65–86. [CrossRef]
506. Saha, N.; Webb, G.E.; Zhao, J.-X.; Lewis, S.E.; Nguyen, A.D.; Feng, Y. Spatiotemporal variation of rare earth elements from river to reef continuum aids monitoring of terrigenous sources in the Great Barrier Reef. *Geochim. Cosmochim. Acta* **2021**, *299*, 85–112. [CrossRef]
507. Webb, G.E.; Kamber, B.S. Rare earth elements in Holocene reefal microbialites: A new shallow seawater proxy. *Geochim. Cosmochim. Acta* **2000**, *64*, 1557–1565. [CrossRef]

508. Won, M.-Z. Review of family Entactiniidae (Radiolaria), and taxonomy and morphology of Entactiniidae in the late Devonian (Frasnian) Gogo Formation, Australia. *Micropaleont* **1997**, *43*, 333–369. [CrossRef]
509. George, A.D.; Chow, N.; Trinajstic, K.M. Oxic facies and the Late Devonian mass extinction, Canning Basin, Australia. *Geology* **2014**, *42*, 327–330. [CrossRef]
510. Chen, J.; Lu, Y.-H.; Sun, Y.-G.; Hogancamp, N.; Lu, M. Oceanic euxinia and seafloor oxygenation linked to continental weathering and influxes during the Late Devonian Frasnian–Famennian bio-crisis and Annulata bio-event. *Chem. Geol.* **2024**, *662*, 122213. [CrossRef]
511. Riding, R. Phanerozoic reefal microbial carbonate abundance: Comparisons with metazoan diversity, mass extinction events, and seawater saturation state. *Rev. Española Micropaleont.* **2005**, *37*, 23–39.
512. Jiang, Z.-P.; Tyrrell, T.; Hydes, D.J.; Dai, M.; Hartman, S.E. Variability of alkalinity and the alkalinity-salinity relationship in the tropical and subtropical surface ocean. *Glob. Biogeochem. Cycles* **2014**, *28*, 729–742. [CrossRef]
513. Fry, C.H.; Tyrrell, T.; Haina, M.P.; Bates, N.R.; Achterberg, E.P. Analysis of global surface ocean alkalinity to determine controlling processes. *Mar. Chem.* **2015**, *174*, 46–57. [CrossRef]
514. Joachimski, M.M.; Breisig, S.; Buggisch, W.; Talent, J.A.; Mawson, R.; Gereke, M.; Morrow, J.R.; Day, J.; Weddige, K. Devonian climate and reef evolution: Insights from oxygen isotopes in apatite. *Earth Planet. Sci. Lett.* **2009**, *284*, 599–609. [CrossRef]
515. Playfair, J. *Illustrations of the Huttonian Theory of the Earth*; William Creech: Edinburgh, UK, 1802; p. 528.
516. Kocsis, A.T.; Scotese, C.R. Mapping paleocoastlines and continental flooding during the Phanerozoic. *Earth-Sci. Rev.* **2021**, *213*, 103463. [CrossRef]
517. Gough, D.O. Solar interior structure and luminosity variations. *Solar Phys.* **1981**, *74*, 21–34. [CrossRef]
518. Weiss, A.M.; Martindale, R.C. Paleobiological traits that determined scleractinian coral survival and proliferation during the late Paleocene and early Eocene hyperthermals. *Paleoceanogr. Paleoclimat.* **2019**, *34*, 252–274. [CrossRef]
519. Done, T.J. Phase shifts in coral reef communities and their ecological significance. *Hydrobiologia* **1992**, *247*, 121–132. [CrossRef]
520. Espinel-Velasco, N.; Hoffmann, L.; Agüera, A.; Byrne, M.; Dupont, S.; Uthicke, S.; Webster, N.S.; Lamare, M. Effects of ocean acidification on the settlement and metamorphosis of marine invertebrate and fish larvae: A review. *Mar. Ecol. Prog. Ser.* **2018**, *606*, 237–257. [CrossRef]
521. Kuffner, I.B.; Andersson, A.J.; Jokiel, P.L.; Rodgers, K.S.; Mackenzie, F.T. Decreased abundance of crustose coralline algae due to ocean acidification. *Nat. Geosci.* **2008**, *1*, 114–117. [CrossRef]
522. Jokiel, P.L.; Rodgers, K.S.; Kuffner, I.B.; Andersson, A.J.; Cox, E.F.; Mackenzie, F.T. Ocean acidification and calcifying reef organisms: A mesocosm investigation. *Coral Reefs* **2008**, *27*, 473–483. [CrossRef]

Journal of
Marine Science and Engineering

MDPI

Review

Rhodoliths as Global Contributors to a Carbonate Ecosystem Dominated by Coralline Red Algae with an Established Fossil Record

Markes E. Johnson

Department of Geosciences, Williams College, Williamstown, MA 01267, USA; markes.e.johnson@williams.edu;
Tel.: +1-413-597-2329

Abstract

Rhodoliths (from Greek etymology meaning red + stone) are spheroidal accretions composed of various types of crustose coralline red algae that dwell in relatively shallow waters where sunlight allows for photosynthesis. Unlike most other kinds of algae that are attached to the seabed by a holdfast, rhodoliths are free to roll about by circumrotary movements stimulated mainly by gentle wave action and bottom currents, as well as by disruptions by associated fauna. Frequent movement exposes every part of the algal surface to an equitable amount of sunlight, which generally results in an evenly concentric pattern of growth over time. Individual structures may attain a diameter of 10 to 20 cm, representing 100 years of growth or more. Initiation typically involves encrustation by founder cells on a rock pebble or shell fragment. In life, the functional outer surface is red or pink in complexion, whereas the structure's inner core amounts to dead weight. Chemically, rhodoliths are composed of high magnesium calcite [$(Ca,Mg)CO_3$], with examples known around many oceanic islands and virtually all continental shelves in the present world. The oldest fossil rhodoliths appeared during the early Cretaceous, 113 million years ago. Geologically, rhodoliths may occur in massive limestone beds composed of densely packed accumulations. Living rhodoliths commonly occur in waters as shallow as -2 to -10 m, as well as seaward in mesophotic waters up to -100 m under exceptional conditions of water clarity. Especially in shallower waters, rhodoliths are vulnerable to transfer by storm waves to supratidal settings, which result in bleaching under direct sunlight and death. Increasingly, marine biologists recognize that rhodolith beds represent a habitat that offers shelter to a community of other algae and diverse marine invertebrates.

Keywords: maërl beds; depth limitation; latitudinal range; taphonomy; fossilization potential; storm deposits

1. Introduction

Phycology is the study of the world's living marine algae, which entails detailed descriptions of species belonging to the major divisions of intertidal to shallow subtidal algae commonly referred to as the green, red, and brown algae. Like land plants, these algae are distant relatives that thrive by the same process of photosynthesis studied more generally by botanists. Paleobotany is concerned with the fossil history of plants through geologic time, and paleophycology is a subdiscipline devoted to the fossil record of marine algae. Early on, during the 1940s, it was thought that studies on fossil algae were slow to gain traction compared with other branches of paleontology due to discouragement by the leading phycologists of that day [1]. A lack of hard parts is the fundamental reason for the poor showing of marine algae as fossils, especially among the foliaceous green and brown algae. Most spectacularly represented by the kelps with their long, fleshy stipes, the preservation of brown algae as fossils might seem unlikely. However, evidence from 38-million-year-old Oligocene strata was reported as late as 2024 on the basis of kelp holdfasts preserved as fossils [2]. Green algae fare somewhat better because some incorporate calcium-carbonate materials for internal skeletal support. The codiacian green alga *Halimeda*, for example, has articulated skeletal segments that are heavily calcified. Their fossil representatives are known by paleontologists to first occur in upper Jurassic or lower Cretaceous strata [3]. In contrast, the crustose coralline red algae that accrete as rhodoliths are heavily fortified by an amalgam of high magnesium calcite [$(Ca,Mg)CO_3$] and exhibit a robust fossil record.

The rhodoliths (red + stone from Greek etymology) are distinctly spheroidal in construction and capable of circumrotary movement on the seabed under the influence mainly of bottom currents and waves that touch the bottom under normal, fair-weather conditions in subtidal habitats typically at depths between −2 and −10 m [4]. Under conditions of exceptional water clarity, rhodoliths also accumulate in mesophotic waters up to a depth of −100 m [5]. Charles Darwin was among the earliest naturalists to appreciate the rock-forming characteristics of rhodoliths, which he identified as *nullipora* (without pores), as distinguished from the small exoskeletal openings occupied by polyps in colonial corals. The Pleistocene rhodoliths described by Darwin from Santiago Island in the Cape Verde Archipelago were collected during his famous voyage on the HMS *Beagle* from 1831 to 1836 [6]. The term rhodolith was not effectively promoted until long afterward [7,8]. That Darwin clearly understood their algal origins is a certainty, because specimens personally collected by him and sent to the Sedgwick Museum at Cambridge University remain available for study [9]. Re-examination of the same localities on Santiago Island first studied by Darwin confirms that the Tertiary age for rhodolith layers reported by him corresponds more precisely to a Late Pleistocene age of about 700,000 years [10]. A formal taxonomy by phycologist W.J. Woelkerling became more widely available in 1988 that incorporates 110 living species of crustose coralline red algae arrayed among 15 genera [11].

In life, rhodoliths appear as thin drapes over relatively shallow flats often referred to as maërl beds in Europe [12]. Recognition by geologists that such marine flats are the source for transferal to thick limestone deposits was slow to develop [13], but the known occurrence of such deposits representing a range of geologic ages continues to expand with ongoing exploration and research. Major storms are linked to the build-up of substantial carbonate formations through before-and-after observations of coastal zones that accumulate rhodoliths in great numbers. Barely a few months after the passing of a Category 3 hurricane along the gulf shores of Mexico's Baja California peninsula in August 2006, the author witnessed a large accumulation of rhodoliths transferred to a supratidal zone where none existed previously [14]. Storm waves also provide the necessary energy to reduce rhodoliths by frictional agitation into coarse, sand-size particles pushed landward and trapped by embayments, shoved directly onto beaches, and even beyond, to sand dunes, by onshore winds [15]. The study of taphonomy includes the post-mortem disintegration of rhodoliths into byproducts when subjected to such physical agencies. The process may be rephrased: how do living rhodoliths become fossils after death [16,17]? The goals of this review are organized in four parts to: (1) summarize previous work on the global distribution of rhodolith flats or maërl beds, (2) represent the efforts by paleontologists to record the developmental history of crustose coralline red algae through geologic time, (3) offer a digest from the best examples of fossil rhodolith beds and other products from the published literature, and (4) promote further study of

 https://doi.org/10.3390/jmse14020169

rhodolith taphonomy in relationship to physical disturbances caused by storms. Although effectively global in terms of present-day marine realms and across geologic time, parts of Asia remain poorly represented where additional research is sure to yield future results. The ultimate objective of this review is to encourage fresh studies on rhodoliths, especially in the western Pacific Ocean and throughout the islands and borderlands of the Indian Ocean.

2. Background Ecology, Taphonomy, and Geological History

The global range of living rhodoliths found in dense aggregates is identified by numbered triangles (Figure 1) that extend south from subtropical settings around volcanic islands in the northwestern Atlantic Ocean (triangle 1) to the islands in the eastern Indian Ocean off Western Australia (triangle 7) and north to islands above the Arctic Circle off the coasts of Norway and Greenland (triangle 6). Water temperature is not a factor in this regard, although certain genera of crustose coralline red algae appear better adapted to colder than warmer seawater. The controlling factor is more likely the strength of sunlight that is strongest when penetrating water from directly above. Due to the curvature of the Earth, sunlight enters water at ever-greater angles, shifting poleward from latitudes around 30° north or south of the equator.

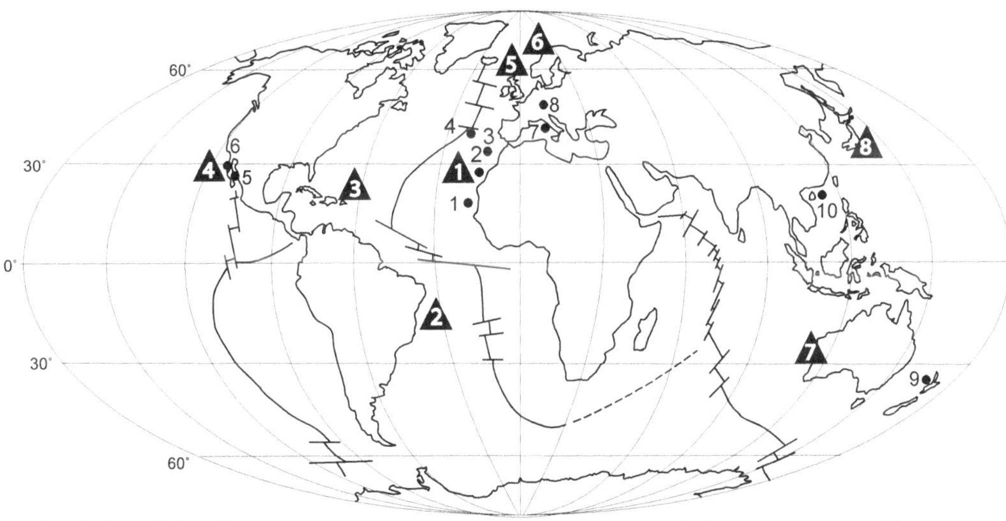

Figure 1. Global representation of today's continents and oceans on a Mollweide projection showing major ocean spreading zones. Numerals on black triangles denote the locations of shelf regions with contemporary rhodolith banks. The numbered black dots mark the location of selected sites rich in fossil rhodoliths, ranging in age from the late Cretaceous to the late Pleistocene.

The growth style of rhodoliths varies owing to levels of ambient energy at different water depths [4]. In waters shallower than −10 m, rhodoliths tend to be dense and generally lumpy, whereas in deeper waters below normal wave base, rhodoliths often exhibit more delicate, open branches attributed to a style of fruticose growth (Figure 2).

 https://doi.org/10.3390/jmse14020169

Figure 2. Two different growth styles commonly acquired by rhodoliths living at water depths that reflect different levels of ambient energy relative to normal wave base. Also fragile in composition, a third style known as foliose growth [12] is not illustrated. Examples come from the author's collection, represented here from a beach on the Caribbean Island of Aruba.

Rhodoliths undergo varied taphonomic transformations after removal by natural agencies from shallow, subtidal conditions at depths often between −2 and −10+ m [18]. The most important variations are summarized in a flow chart (Figure 3).

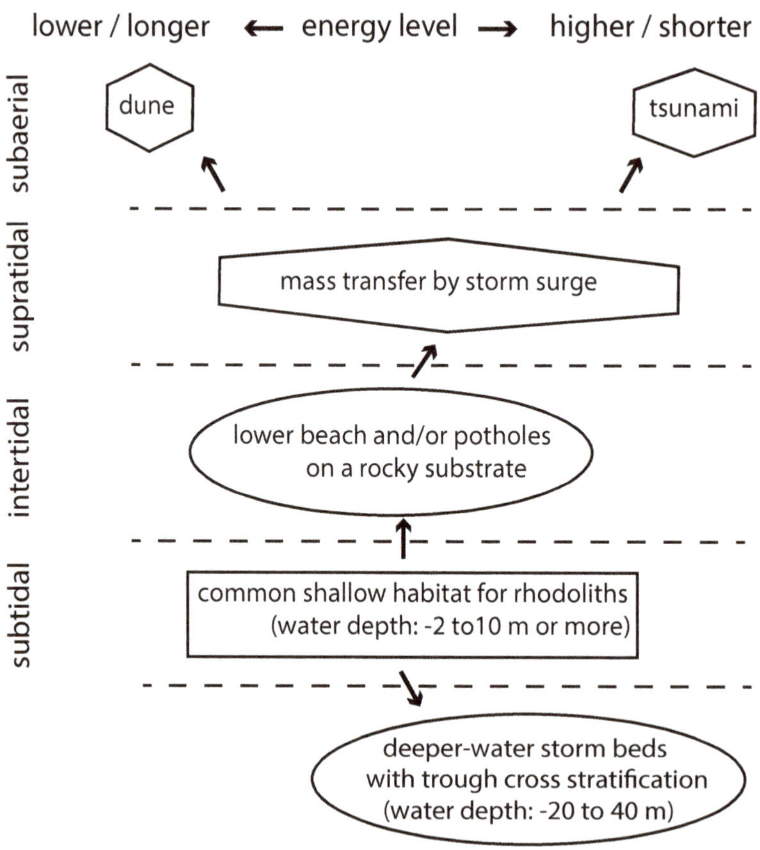

Figure 3. Schematic flow chart summarizing the post-mortem transition from the common habitat of crustose coralline red algae, where rhodoliths typically accrue under life conditions, to the various kinds of deposits where they are buried as geological formations.

https://doi.org/10.3390/jmse14020169

Sea squalls are sufficient in energy to sweep rhodoliths shoreward where they may be caught in rocky tide pools or stranded on a beach. Storm surge during a hurricane may lift the largest rhodoliths normally immune to lesser waves by mass transfer onto a supratidal flat. The same storm may carry rhodoliths in the opposite direction seaward to water depths between −20 to −40 m. Under intense turbulence, rhodoliths are abraded and pulverized as contributors to sediments characterized by trough cross stratification (Figure 2). On land, stiff onshore winds, often seasonal in nature, may blow across beach sand rich in rhodolith debris and remove the finer sand fraction by deflation to adjacent dunes [15]. Tsunami events triggered by coastal landslides or underwater earthquakes are among the most violent events to disturb coastal settings. By nature, the deposits carried far inland by tsunami events are highly chaotic, bearing a mixture of poorly sorted materials with an enormous range in size, including rhodoliths [19].

Maio Island is situated close to Santiago Island in the Cape Verde Archipelago (Figure 4), where Darwin first encountered fossil rhodoliths [6]. It is one of the smaller islands in the archipelago and among the most insightful places to view the extreme taphonomic range shown by recent and fossil rhodolith deposits [20–22]. Specific examples of rhodolith transfer and mechanical reduction derive from field observations on Maio Island (Figure 4). A vast sheet of recent rhodoliths in a supratidal deposit is draped over an area of 250,000 m^2, barely a meter above sea level at Praia Real at the north end of Maio Island [20]. The largest rhodolith was found to have a radius of 20 cm, but the average density with smaller rhodoliths was determined to be 450 rhodoliths/m^2. A most unusual specimen was found nucleated around a ceramic fragment, which indicates that the deposit is historic and likely connected in time to the sinking of a supply ship. An image of the surface drape shows a cluster of worn rhodoliths mixed with large grains eroded from what were the distal tips of rhodolith branches (Figure 5a). The ruins of several lime kilns also occur at Praia Real, revealing an aspect of the local economy for the manufacture of mortar as a by-product of slaked lime from rhodoliths [20].

The northeast trade winds that constantly buffet Maio's windward shores typically register between 5 and 6 on the Beaufort Wind Scale, defined as a moderate to fresh breeze. The resulting wind energy is adequate to lift the finer fraction of beach sand for transport inland to accumulate in dune fields. A large Pleistocene dune at Lomba Greija near Pilão Cão village covers some 330,300 m^2, with well-defined stoss slopes that dip 8° to 10° NE and leeward slip faces that dip 28° to 32° SW (Figure 5b). Point counts on thin-section samples reveal that basalt and other non-carbonate materials account for 5% of grains, but bioclasts derived from rhodoliths account for 74%, as differentiated from tiny shell fragments and other carbonate debris [21]. The fossil dune is one of the largest such structures known to be dominated by rhodolith grains that average 0.5 mm in diameter.

Powerful sea storms not only cause over-wash deposits on land, as at Praia Real, but contribute to the offshore mobilization of sand. A classic example of trough cross-stratification preserved in Miocene sandstone is exposed in a gulley at Ribeira de Calhetinha in Vila do Maio on the island's south shore [22]. Low-angle layers at the center of the image incorporate coarse rhodolith debris (Figure 5c). Overall, the combination of extensive supratidal deposits, sand dunes, and storm beds provides evidence for the existence of living rhodolith banks off the north and east shores of Maio Island today and in the geologic past.

https://doi.org/10.3390/jmse14020169

Beyond the geologic history of rhodoliths elsewhere in the Cape Verde Archipelago and others in the Macaronesian Realm (Figures 1, 4 and 5), the earliest confirmed appearance of crustose coralline red algae as rhodoliths traces back to strata dating from the early Cretaceous, about 113 million years ago [23]. Evidence any deeper in geologic time is obscure, although a claim is made for the occurrence of the earliest coralline red algae during the Ordovician Period more than 450 million years ago [24,25].

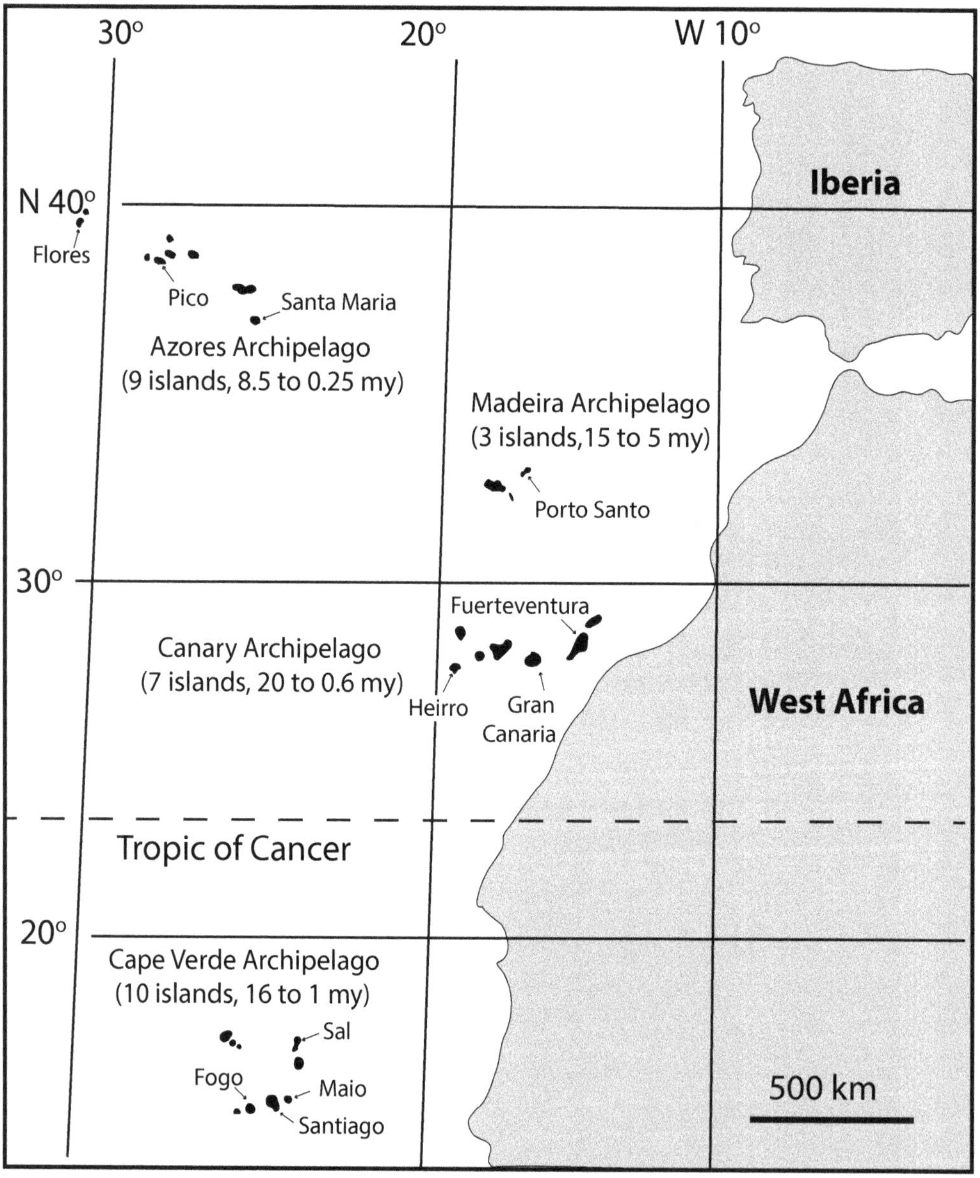

Figure 4. Map of the Macaronesian Realm in the northeast Atlantic Ocean off the west coast of Africa.

https://doi.org/10.3390/jmse14020169

Figure 5. Examples of extreme transfer and mechanical reduction of rhodoliths found on Maio Island in the Cape Verde Archipelago: (**a**) Recent rhodoliths redeposited in a huge supratidal setting around Praia Real at the north end of the island (pen for scale); (**b**) Massive Pleistocene dune formation near Pilão Cão constructed by rhodolith sand in the eastern part of the island (person for scale); and (**c**) Miocene strata with characteristic trough cross-stratification including rhodolith debris at Ribeira de Calhetinha at the south end of the island (scale bar: 3 cm).

3. Review of Contemporary Rhodolith Distributions

A thorough review regarding the global distribution of rhodoliths in contemporary marine settings was authored in 2021 by Rebelo et al. [26]. It remains the first such survey to consider the overall diversity of rhodoliths accreted by crustose coralline red algae as well as their biogeography. Analysis is based on a total of 106 species arrayed in 21 genera and 10 families. All twelve ocean realms defined by Spalding [27] register the presence of rhodoliths. Although the Temperate South African Realm was over-looked in the survey, a report on rhodoliths from the Amathole Offshore Marine Protected Area documents rhodoliths recovered from depths of −30 to −65 m [28]. A key finding from the Rebelo survey is that among the 106 species, 57% are geographically restricted to a single biogeographic province [26]. This suggests that the present-day spread of species is limited

https://doi.org/10.3390/jmse14020169

by biogeographic barriers. The highest diversity is found in the Tropical Southwest Atlantic (TSWA) and the Tropical Eastern Pacific (TEP). A high of 30 species appears in the TSWA as opposed to 23 in the TEP. For example, the Isthmus of Panama is an effective barrier to interchange between the two oceans, but an open passage remained in place until about 3.5 million years ago [29]. In terms of latitudinal range, most of the species among the 13 genera canvassed are limited to tropical settings, whereas the range of species belonging to the genera *Lithophyllum* and *Lithothamnion* extends to higher, more temperate latitudes.

3.1. Contemporary Rhodoliths of the Macaronesian Realm

The Macaronesian realm (Figure 1, triangle 1) includes the Cabo Verde, Canary, Madeira, and Azores archipelagos, all of which originated as volcanic islands. Except for Madeira, which remains quiescent, recent volcanic eruptions occurred in all other island groups, including Hierro Island in 2011, Fogo Island in 1995, and Flores Island in 1951 (Figure 3). Except for rare fossil examples on Porto Santo and Sal (see Figure 4) [30,31], modern coral reefs are notably absent around North Atlantic islands. Rhodolith banks commonly occupied eco-space around many such islands as they aged and stabilized. Located in the eastern part of the Canaries Archipelago, Fuerteventura Island provides an ideal example where different stages in the coastal development of subaerial rhodolith deposits are found along a continuous stretch of shoreline [18,32]. In particular, the north shore of Fuerteventura is that part of the island most directly impacted by the northeast trade winds, just the same as Maio Island in the Cape Verde Archipelago. A distance of 10 km from east to west separates the settlements of Majanicho and El Colito, where a complete gradation in rhodolith taphonomy appears on contemporary beaches (Figure 6). *Lithothamnion* cf. *corallioides* is the most prominent species throughout [32].

The size of rhodoliths transferred by waves from their life habitat to post-mortem subaerial deposits on Fuerteventura is notably smaller than that found on Maio in the Cape Verde archipelago. On the flats behind Praia Real at the north end of Maio Island [20], rhodoliths are the size of oranges (Figure 5a), whereas on Majanicho at the north end of Fuerteventura Island, rhodoliths are "popcorn"-sized (Figure 6a,b). Given that the size of the spheroid accreted by a crustose coralline red alga is proportional to the time span of development, it follows that the smaller rhodoliths from Fuerteventura were harvested, or removed from their normal habitat, at an earlier stage in their life span than those from Maio Island. In both settings, waves driven by the same persistent trade winds are insufficient to effectively disturb the rhodolith beds that normally live at water depths of −10 m or more. Living rhodoliths occur at a depth of −12 m off the north shores of Fuerteventura Island [32]. There are no comparable reports on the living rhodolith beds around Maio Island, but their average size clearly indicates a longer survival span between sea storms that may vary from rare hurricanes to more frequent gales. Essentially, the deeper the habitat for a living rhodolith bed, the lower the likelihood of a disturbance by storm waves that feel bottom.

Elsewhere in the Azores, investigations on living rhodoliths are reported from Pico Island (Figure 4), involving both a shallow-water habitat and much deeper offshore accumulations [33]. In a lagoonal setting at a depth less than −4 m, the coralline red algae *Phymatolithon calcareum* encrusts small basalt pebbles and accretes to a diameter of 2.5 cm. Smaller rhodoliths are reported from dredge samples recovered off Pico Island from depths between −64 and −73 m, where an initial encrustation of *P. calcareum* is succeeded by *Lithophyllum incrustans*. Rhodoliths from a mid-shelf setting were not detected but are assumed to have been present for transport to a deeper part of the shelf by storm waves.

https://doi.org/10.3390/jmse14020169

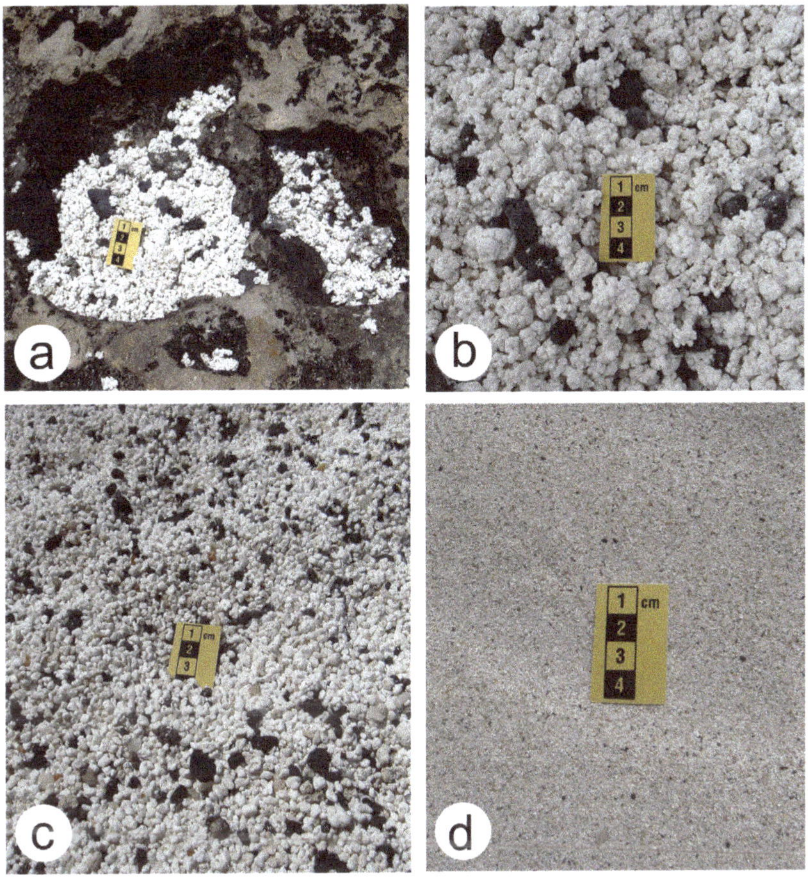

Figure 6. Subaerial variations in rhodolith taphonomy on north-facing beaches of Fuerteventura Island in the Canary Archipelago (same scale card): (**a**) Small rhodoliths captured in a tide pool eroded in a basalt shelf at Majanicho; (**b**) Same "popcorn"-sized rhodoliths deposited in a berm on the landward side of the same platform; (**c**) Crushed rhodoliths forming coarse beach sand west of Majanicho; (**d**) Dune sand dominated by rhodolith debris at El Colito.

3.2. Contemporary Rhodoliths off Brazil in the South Atlantic

The highest biodiversity of living coralline red algae as rhodoliths occurs on the Abrolhos Bank in the South Atlantic Ocean off Brazil [34]. The region (Figure 1, triangle 2) features the most extensive zone documented for the occurrence of rhodoliths, covering much of the Abrolhos Bank over a vast area of 20,900 km^2 at depths up to −80 m in exceptionally clear water within the mesophotic zone (Figure 7).

Images recovered from remotely operated vehicles (ROVs) exploring the Abrolhos Bank show the vast extent of densely crowded rhodoliths formed chiefly by *Lithothamnion crispatus*, attaining an average diameter of 6 cm, but also including other species from the genera *Lithophyllum*, *Hydrolithon*, *Mesophyllum*, *Sporolithon*, and *Neogoniolithon* [34]. The biodiversity of other marine algae and marine invertebrates is very high because rhodoliths perform the function of an umbrella organism. At such great depths where storms have no effect on bottom conditions, tilefish are known to plow the bank and heap rhodoliths into mounds. Alternatively, the phenomenon of internal waves due to disruption between water layers of different density may potentially affect deep-water rhodolith banks [35].

https://doi.org/10.3390/jmse14020169

Closer to the shorelines of Espirito Santo and Bahia states [36,37], rhodoliths occupy much shallower waters, where various brown algae characteristically attach to them. The foliose growth of brown algae, which has the appearance of plumes, catches onshore winds during storms. The phenomenon is known in Brazil as the *arribadas* or "the arrived." It means that the rhodoliths dwelling in well-lighted waters where brown algae co-exist arrive on a seasonal basis when washed ashore (Figure 8).

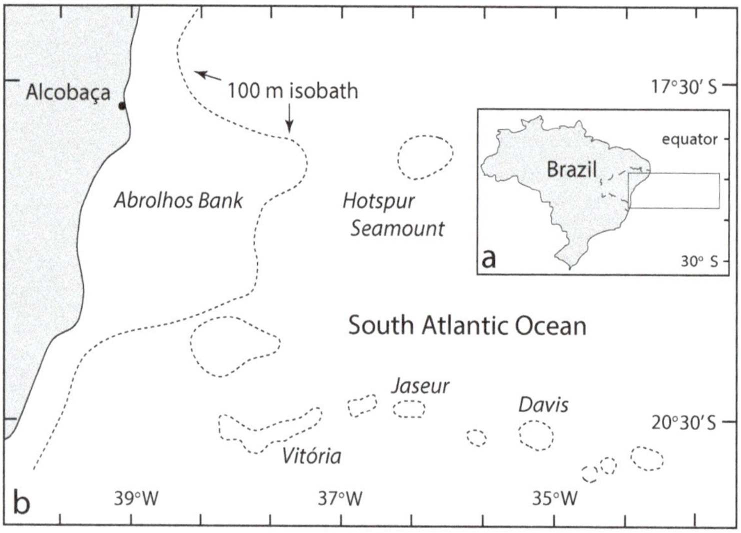

Figure 7. Maps for Brazil and seamounts off the South Atlantic coast: (**a**) Brazil with dashed line outlining the State of Bahia and rectangular box outlining location of seamounts, (**b**) Detailed map showing the extent of the Abrolhos Bank off the State of Bahia and several of the seamounts that extend over five degrees of longitude across the South Atlantic.

Figure 8. Example of a rhodolith carried onto a Brazilian beach in Bahia State by the hydrodynamics of attached brown algae during an *arribadas* event (scale bar = 1 cm).

https://doi.org/10.3390/jmse14020169

Seamounts form a discontinuous ridge eastward from the Abrolhos Bank (Figure 7b). The flat tops of several, including Vitória, Jaseur, and Davis, have been mapped remotely by side-scan sonar, explored by ROVs, and even sampled for rhodoliths at depths around −60 m [38]. The size, species composition, and density of rhodoliths are similar to those found on the Abrolhos Bank. Farther to the northeast from the mainland, the Brazilian islands belonging to the tiny Fernando de Noronha Archipelago also yield rhodolith beds at several localities ranging in water depths between −10 m and −100 m [39]. Again, tilefish are reported as active disrupters of rhodolith cover, responsible for turning over individual spheroids. The Rocas Atoll, located near the Fernando de Noronha islands, is the only such feature in the Atlantic Ocean, amounting to 378 km^2 as defined by its surrounding 1000 m isobath. A combination of rhodoliths and scattered coral reefs occurs in the mesophotic zone to a depth of −80 m [5]. Around the steep margins of the atoll, the coalescence of rhodoliths is abetted by the inter-connected growth of sponges. In the upper 30 m of waters across the atoll, rhodoliths are free-living and subject to movement. Altogether, the extraordinary size and richness of the Abrolhos Bank, related seamounts, as well as the Fernando de Noronha islands and Rocas Atoll, contribute to making marine scientists from Brazil among the leading experts in the study of rhodoliths and the rhodolith ecosystem [5,34–39].

3.3. Contemporary Rhodoliths in the Caribbean Realm

The Caribbean region (Figure 1, triangle 3) is better known for the extensive fringing and barrier reefs developed especially along the windward margins of many islands [40]. Rhodoliths accreted by the coralline red alga *Cruoriella armorica* are reported from shallow-water, back reef settings in Puerto Rico [41]. In this case, the rhodoliths are nucleated around coral debris, often with a cylindrical shape up to 7 cm in length and with concentric laminations as much as 3 cm in thickness. Elongated fragments of various reef corals, identified as derived from *Acropora palmata*, *A. cervicornis*, and *Porites furcate*, form the core at the center of these rhodoliths. Maximum density was found to exceed 500 rhodoliths per square meter. In September 1998, Hurricane Georges passed directly over one of the study sites as a Category 2 event with wind speeds up to 177 km/h. Visiting after the storm, it was found that the site suffered a nearly complete removal of rhodoliths from lagoon waters less than −1.3 m in depth [41]. The same level of detailed research on rhodoliths in reef settings accomplished on Puerto Rico remains to be replicated elsewhere in the Caribbean region. The author points to personal experience from Aruba off the coast of Venezuela, where rhodoliths were found to occur in a leeward setting as a major contributor to beach sand (see Figure 2).

3.4. Contemporary Rhodoliths in Mexico's Baja California Peninsula

Mexico's Gulf of California is a marginal sea covering an area amounting to 210,000 km^2 that fits between the Mexican mainland and the 1100 km long Baja California peninsula adjacent to the Pacific Ocean. Due to a subtropical setting and regional hydrodynamics that entail seasonal upwelling, the gulf hosts one of the most productive ecosystems in the world, with rich biotas including diverse marine invertebrates, fish stocks, and marine mammals [42]. It is a young sea, dating back in geologic time to less than 12 million years ago, with initial flooding into a rift zone related to seafloor spreading along the East Pacific Rise. The gulf region has a recorded history of hurricanes that enter the gulf from the south during El Niño years with a periodic frequency of every six to eight years. During such events, the orbital energy generated by storm waves touches bottom at a greater depth beyond fair-weather wave base and has a destructive impact on regional

 https://doi.org/10.3390/jmse14020169

rhodolith banks. Beaches and dunes adjacent to rhodolith banks have a propensity to be enriched by rhodolith sand following major storms [15].

Extensive survey work on modern rhodolith banks was initiated throughout the lower Gulf of California during the 1990s and expanded thereafter during the next decade by phycologists from the Moss Landing Marine Laboratories in southern California [43–46]. Extensive rhodolith banks typically inhabit the seabed at water depths from −2 to 10+ m and often occur side-by-side with fossil rhodolith deposits exposed along the peninsular coast of Baja California (Figure 9).

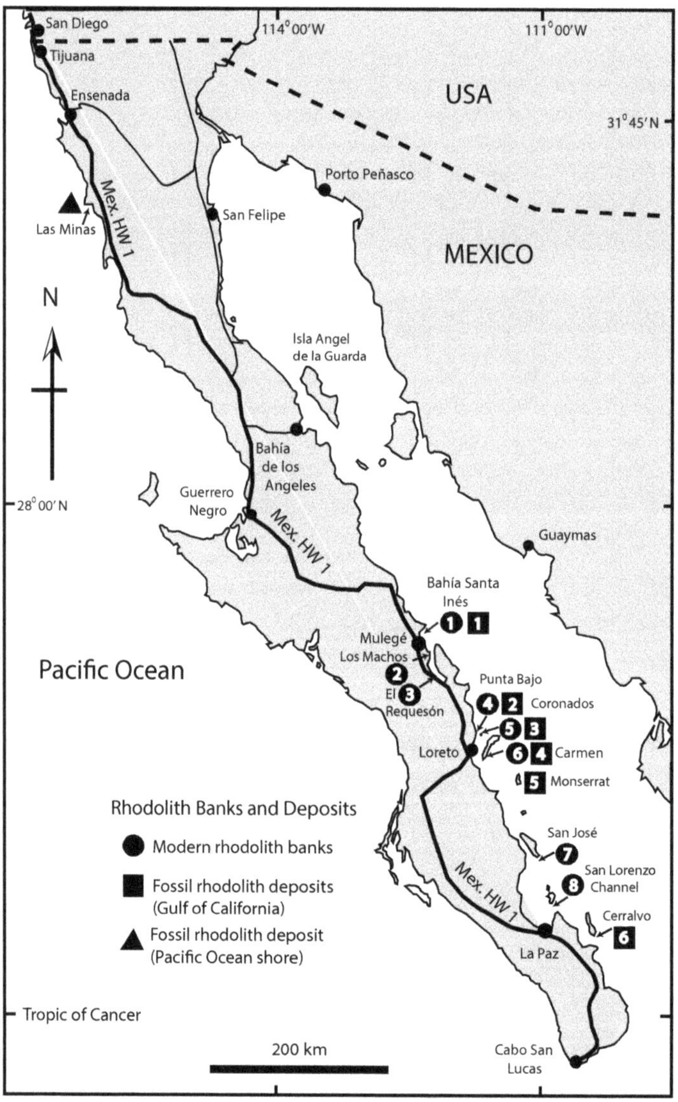

Figure 9. Map over the Gulf of California between Mexico's mainland and the Baja California peninsula, showing multiple localities where contemporary rhodolith banks occur (numbered circles) as well as localities with fossil rhodolith deposits (numbered squares) from the Pliocene and Pleistocene. Localities in each category are numbered from north to south. A triangle marks the occurrence of Cretaceous rhodoliths at Las Minas.

https://doi.org/10.3390/jmse14020169

Early studies focused on a locality called El Requeson (Figure 8, circle 3), where a large rhodolith bank sits off Isla Requeson (Figure 9) [44]. Requeson is a small, narrow island less than 1000 m in length, connected to the peninsular mainland by a 350 m long tombolo submerged during high tides (Figure 10).

Figure 10. View over Isla Requeson and the tombolo, exposed only during low tide, that connects the island with the peninsular mainland in Bahía Concepción.

Concepción Peninsula forms the outer arm of the greater bay. The Spanish name *requeson* refers to "cheese curds" that mimic the same texture as the broken distal tips of rhodolith branches. Anecdotal history relates that rhodoliths and rhodolith sand are added to the tombolo after major storms.

Based on video surveillance in conjunction with deployment of current meters, movement was monitored on the crescent-shaped rhodolith bank situated off the windward shore of Isla El Requesón [44]. Rhodoliths from this wave-swept bed occur in dense concentrations from 4 m to 12 m below the surface, with 89% of the forms attributed to *Lithophyllum margaritae*. Recorded at a depth of 5 m, small rhodoliths with a diameter of 2 cm start to roll under the influence of wind-driven waves when the water current reaches speeds of 25–30 cm/second. At higher velocities of 30–35 cm/second, rhodoliths up to 3 cm in diameter start to roll away [44]. The Moss Landing team of phycologists continued to expand studies to other areas, such as Cabo Las Mochas (Figure 9, circle 2), still within Bahía Concepción north of El Requesón [45]. They also produced a major review over the living rhodolith banks at Punta Baja, Isla Coronados, and the San Lorenzo Channel (Figure 9, circles 4, 5, 8) [46].

Other teams have contributed notable observations on major rhodolith banks around Punta Chivato in Bahía Santa Ines (Figure 9, circle 1) [47] and off the southern tip of Isla del Carmen within the refuge of the Loreto Marine Park (Figure 9, circle 6) [48]. A detailed study of carbonates dominated 40% by rhodoliths off the south end of Isla San José (Figure 8, circle 7) features one of the largest banks in the Gulf of California, covering 45 km^2 [49].

3.5. Contemporary Rhodoliths off the Scottish Western Isles

High-latitude temperate settings with maërl beds tend to attract less attention and dedicated research than the more extensive rhodolith banks in subtropical settings [34,46]. Even so, some of the largest maërl beds in Europe occur around the Western Isles of Scotland

 https://doi.org/10.3390/jmse14020169

and other parts of the British Isles (Figure 1, triangle 5) [50]. Key factors affecting the disposition of the Scottish maërl beds appear to be tide-generated currents that are typically moderate to strong. These beds consist of living rhodoliths combined with post-mortem debris scattered across shallow, subtidal flats often rippled by wave action. Though growth is slow [51], the interlocking structure of rhodoliths dominated by *Lithothamnion glacial* and *L. corallioides* provides a wide range of niches for infaunal and epifaunal invertebrates. A detailed study shows how the role of Scottish maërl beds functions as a nursery for juvenile queen scallops *Aequipecten opercularis* and other invertebrates [52]. Dunes in what is called the machair ecosystem occur on Scottish coasts inland from maërl beds, which transfer considerable carbonate input to a terrestrial system [53].

3.6. Contemporarhy Rhodoliths off the Coast of Arctic Norway and Greenland

Research on contemporary rhodoliths from high-latitude regions in Norway, Svalbard, and Greenland (Figure 1, triangle 6) is as poorly represented in the phycological literature as the maërl beds of Scotland in the British Isles. Fixed, reef-like buildups formed by crustose coralline red algae are described from the south end of Rebbenesøy in the Troms district at a latitude of North 70° [54]. Different phases of algal development occur over 500 m across the Storvoll shelf, with free rhodoliths dominated by *Lithothamnion* at depths between −12 and −15 m. A rhodolith pavement and maërl beds occur closer to shore. The process of rhodolith turnover is suggested on the basis of the predator–prey relationship between the brittle star *Ophiopholis aculeata* and the wolffish *Anarhichas lupus,* living as members in the ecosystem. Farther north, rhodolith beds are documented in Norway's Svalbard Archipelago at latitudes close to North 80°, where the coralline red algal flora is dominated by *Lithothamnion glaciale* and *Phymatolithon tenue,* living at depths between −27 and −47 m [55].

Opportunities for rhodolith studies are open to fresh exploration along the shores of sub-Arctic Norway. This assessment is based on the author's personal observations along the shores of the National Norwegian Geological Monument on Leka Island within the greater Trollfjell Geopark (Figure 11a,b). Abundant rhodoliths with the same branching structure typical of *Lithothamnion* occur on the northeast shores of Leka Island (Figure 11c) and are readily collected during low tide.

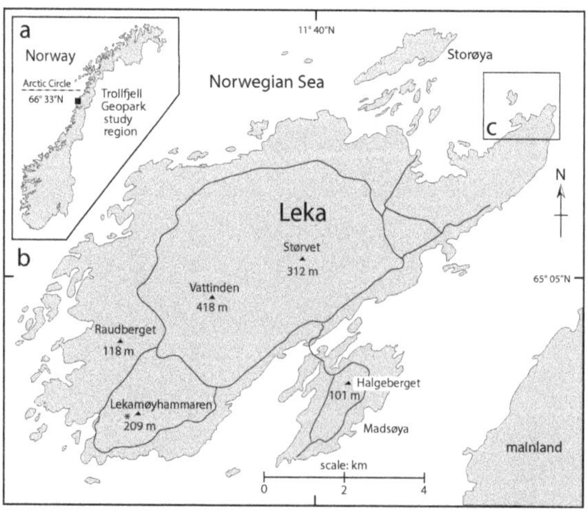

Figure 11. The west coast of Norway and Leka Island: (**a**) Norway, marking the Arctic Circle and Trollfjell Geopark (black box); (**b**) Leka Island showing roads and highland peaks; (**c**) Area with rhodoliths stranded in the intertidal zone.

 https://doi.org/10.3390/jmse14020169

While the interior of Leka Island exhibits significant topography (Figure 11b), the northeastern end of the island is low and close to sea level (Figure 11a). It is noted that individual rhodoliths may be found along that shore with the remnants of the holdfasts from foliose brown algae (Figure 12).

Figure 12. Coastal landscape on Leka Island and typical rhodolith: (**a**) View to the northeast on Leka Island showing low terrain at the side of a sheltered embayment (see Figure 11c), (**b**) Example of a branching-type rhodolith characteristic of *Lithothamnion* with the intact holdfast from a foliose brown alga.

The crustose coralline red algae are long-lived, but brown algae grow during a single season and die off annually. This temperate example of biological co-existence is much the same in tropical settings as found in the seasonal *arribadas* of coastal Brazil (see Figure 8). There are clear implications to suggest that a cessation in rhodolith mobility occurs during stabilization during the seasonal overgrowth of brown algae.

Other work on rhodolith banks at multiple localities, many above the Arctic Circle on both the east and west coasts of Greenland, provides details on marine communities dominated by *Clathromorphu compactum* and *Lithothamnion glacial*, up to 3 cm in diameter, living at depths of −18 to −25 m with percent coverage on the seabed as much as 40% [56]. Associated faunas typically include as many as 40 species of marine invertebrates, dominated by mollusks, polychaetes, and echinoderms.

https://doi.org/10.3390/jmse14020169

3.7. Contemporary Rhodoliths off the Coast of Western Australia

Published information on rhodoliths from Australia remains limited in scope but offers some detailed reporting from the Indian Ocean shores of Western Australia (Figure 1, triangle 7). Areas with substantial rhodolith development commonly are found at depths from −30 to −100 m along the western shores of that province [57]. The greater Shark Bay region, which is famous for its living stromatolites [58], belongs to a World Heritage UNESCO (United Nations Educational Scientific and Cultural Organization) site with full conservation protection. Dirk Hartog Island (Figure 13a,b) is part of a barrier system that shelters Shark Bay from the open Indian Ocean. The island's east shore and adjacent coastal waters were the location for a preliminary study on contemporary rhodoliths that looked at species composition and density, as well as the contribution of rhodolith detritus to beach and dune sand [59]. Sunday Island Bay at the island's south end (Figure 13b) hosts a dense concentration of rhodolith-forming species, including *Neogoniolithon brassica-florida, Hydrolithon reinboldii*, and *Lithophyllum* (Figure 13c).

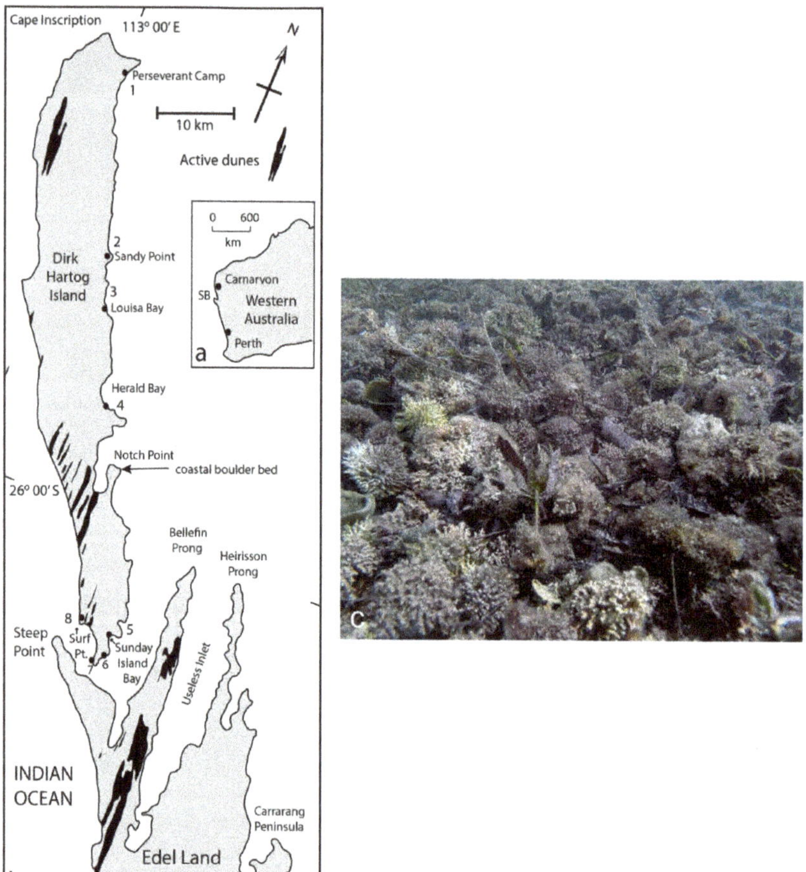

Figure 13. Maps over Western Australia and Dirk Hartog Island on the outer edge of Western Australia's Shark Bay and subtidal image: (**a**) Western Australia (note: SB marks the location of Shark Bay); (**b**) Dirk Hartog Island study site for contemporary rhodoliths and detrital contributions to beach and dune sand; (**c**) Dense accumulation of rhodoliths formed by *Neogoniolithon brassica-florida* occurs at a water depth of −1.5 m in Sunday Island Bay (locality 5 in (**b**)).

https://doi.org/10.3390/jmse14020169

The Dirk Hartog project [59] canvassed ten beach and adjacent dune localities mostly on the east shore to assess the scale of rhodolith-derived sediments transferred to those settings. In addition to carbonate grains derived from coralline red algae, categories included fine shell debris, quartz sand, and dark minerals. The highest input from coralline red algae occurred at Sandy Point (Figure 13b, locality 2) with beach sediments enriched by 14% and the adjacent dune at 9%. Quartz sand was the primary constituent at most localities, ranging between 32 and 86% by point counts [59]. The result indicates a more spotty, irregular placement of rhodolith concentrations along the island's east shore.

3.8. *Contemporary Rhodoliths from the Ryukyu Islands of Japan*

A chain of islands belonging to the Ryukyu Archipelago is part of Japan, stretching nearly to Taiwan along an arc following a zone of oceanic plate-to-plate subduction from the latitude of N 30° to N 24°. With Okinawa being the largest at 1200 km^2 in size (Figure 1, triangle 8), many are known for a combination of marine ecosystems that include seagrass beds, inner coral flats, outer reefs, and coralline red algae occupying the outer reef slope [60]. Okinawa and its many small satellite islands have been the subject of detailed work on rhodoliths that occupy water depths typically between −50 and −135 m [61]. Based on underwater photos, rhodoliths cover 45% to 78% of the seabed, mainly with spherical forms having an average diameter of 8 cm. The most common accretionary species of coralline red algae include *Lithothamnion australe* and *L. pulchrum*, but preliminary work also confirms the presence of at least five species belonging to the genus *Sporolithon* [61].

4. Major Fossil Rhodolith Deposits

Labeled numerically in Figure 1, nine localities provide a sampling of fossil rhodolith deposits from around the world, chosen to reflect substantial thicknesses of whole rhodoliths, as well as the accumulated debris of broken and disintegrated rhodoliths.

4.1. *Fossil Deposits from the Macaronnesian Realm in the Northeast Atlantic*

The distinctive rhodolith layer in the harbor area at Praia on Santiago Island (Figure 1, locality 1) was originally described by Darwin as 20 feet (~6 m) in thickness [6]. His observation is accurate, although the actual thickness varies laterally as later measured in a series of stratigraphic profiles from the port toward the island's southeast end [10]. The variation is about a meter due to uneven erosion of the underlying basalt shelf on which the rhodoliths accumulated in late Pleistocene time. The outcrops studied by Darwin remain largely undisturbed. A typical exposure from the middle of the rhodolith layer on Ilhéu de Santa Maria in Praia harbor shows spheroids from 4 to 6 cm in diameter, some of which are partially eroded to exhibit nucleation around basalt cores (Figure 14).

The Santiago rhodolith layer is sufficiently thick that rhodoliths at the bottom of the deposit had no opportunity to receive the sunlight required for photosynthesis. Hence, the layer is interpreted as a storm deposit that washed rhodoliths closer to land. The basalt layers that followed Darwin's rhodolith layer arrived about 700,000 years ago, flowing to the sea from nearby volcanic cones [10].

The Canaries Archipelago is centered 1300 km northeast of the Cape Verde Archipelago (Figure 3). Among the fossil rhodolith beds from the Canary Islands, the best documented occurs at Los Rehoyas in urban Las Palmas on Gran Canaria Island (Figure 1, locality 2; Figure 4) [62]. Three separate rhodolith layers are preserved as conglomerate deposits that vary in thickness from 0.25 to 0.5 m and consist of densely packed spheroids from 2.0 to 3.0 cm in diameter. The taxonomy of coralline red algae was found to consist of species in multiple genera, including *Lithoporella*, *Hydrolithon*, *Sporolithon*, *Boreolithathamnion*, and possibly *Lithophyllum*. Sample size was sufficient to determine that no single species

https://doi.org/10.3390/jmse14020169

dominates. Trace fossils from associated layers indicate that deposition occurred within the middle shoreface and shoreface [55]. Scores of rhodoliths were broken open from each layer to reveal none were nucleated around basalt cores. It means that the rhodoliths originated well offshore in waters on the order of −50 m in depth. As in the contemporary "popcorn" rhodoliths from Fuerteventura Island [32], the rhodoliths at Los Rehoyas on Gran Canaria are interpreted as storm deposits. The age of the rhodolith layers is latest Miocene or early Pliocene based on the dating of related basalt flows.

Figure 14. Darwin's rhodolith layer from Ilhéu de Santa Maria in Praia harbor on Santiago, Cape Verde Archipelago. Some are eroded to show nucleation around a basalt core (white arrows).

Still within the Macaronesian Realm, the Madeira Archipelago is centered 500 km due north of the Canaries Archipelago (Figure 4), where an islet off the coast of Porto Santo is the site of an extraordinary rhodolith deposit dated to the middle Miocene, from about 15 million years ago, on the basis of microfossils (Figure 1, locality 3) [63]. As at Los Rehoyas in the Canary Islands, multiple rhodolith layers are repeated stratigraphically at *Cabeço das Laranjas* (Hill of Oranges, so called compared to size and color of the fruit) on Ilhéu de Cima off Porto Santo. The basal rhodolith layer is 2.6 m thick (Figure 15) and occupies what was a former rocky shore, with multiple sea stacks that poke through the rhodolith deposit, forming a stabilizing catchment zone. Approximately 90,000 rhodoliths are exposed on the upper surface of the basal layer, amounting to 450 per square meter. The average diameter of spheroids in the thick basal layer was found to be 10 cm, with many nucleated around basalt cores up to 3 cm in diameter. Thin-section analyses identified three genera of coralline red algae: *Sporolithon*, *Lithothamnion*, and *Neogoniolithon*. The overall depositional setting indicates that a major sea storm was the agency that transferred so many large rhodoliths onto the shore from a deeper, offshore habitat [63].

https://doi.org/10.3390/jmse14020169

Figure 15. View in cross section showing the maximum thickness of the basal rhodolith layer sitting on a basalt shelf at *Cabeço das Laranjas* (Hill of Oranges) on Ilhéu de Cima off Porto Santo in the Madeira Archipelago of Portugal.

Centered 1500 km northwest from Madeira, the Azores Archipelago includes several islands, only one of which, Santa Maria Island, hosts extensive fossil deposits (Figure 3). Rhodoliths from basal Pliocene strata dated to 4.25 million years ago are buried in multiple surge channels with 0.8 m or more of relief among pillow basalts traced along 150 m of exposure in the high cliffs of Malbusca on the south coast [64]. The rhodoliths are small, generally less than 3 cm in diameter and include species from the genera *Lithophyllum*, *Spongites*, *Hydrolithon*, and *Phymatolithon*. The side walls of the surge channels are partially encrusted by barnacles, and in some cases, rhodoliths also are encrusted by barnacles. Many rhodoliths show evidence of abrasion and breakage interpreted as storm-related. Cross-bedding within some channels is a strong indicator that the rhodoliths were introduced from deeper, offshore waters into shallower and higher-energy settings in surge channels. A transgressive trend in Pliocene strata overlying the rhodoliths records a gradual rise in sea level that eventually flooded the island until renewed volcanism and tectonic uplift restored subaerial conditions during later Pliocene and Pleistocene times [64].

4.2. Fossil Deposits from Mexico's Baja California Peninsula

Limestone-forming layers dominated by thick rhodolith deposits are well represented at multiple localities along the shores and islands in Mexico's Gulf of California, ranging in age from the Pliocene to the late Pleistocene [65]. Located near the town of Loreto in Baja California Sur, Isla Coronados is especially rich in geological features that include fossil rhodoliths (Figure 8, square 3). This small island, 7.5 km^2 in area, is crowned by a volcano last active about 600,000 years ago, which continues to influence the dynamics of surrounding marine shores. A modern sand beach on the east shore as well as the adjacent dune field are composed of rhodolith sand [15]. On the south shore, a late Pleistocene

 https://doi.org/10.3390/jmse14020169

lagoon contains a fossil coral reef that grew in place about 121,000 years ago. Covering 3.5 ha, the former lagoon is well exposed in the eroded walls of Cañada Coronados to reveal a fixed outer barrier formed by volcanic basement rocks with only a narrow connection to the sea [66]. From a sedimentological standpoint, part of the lagoon demonstrates the repeated effects of storms relative to the recycling of rhodolith materials. A thick succession of rhodolith sand occurs in discrete layers that dip 20° uniformly off the fixed barrier toward the middle of the lagoon, stopping short of the reef complex (Figure 16a). The immediate source of the rhodolith debris derives from the opposite side of the barrier, where whole rhodoliths mixed with basalt cobbles laid down a cover on the island shelf (Figure 16b). Periodic storms of a similar intensity to hurricanes that today enter the Gulf of California from the south are hypothesized to have caused waves pushing rhodoliths against the barrier and sending their crushed debris upwards and over the top into the sheltered lagoon behind. Calculations that follow the same formula to calculate the number of contemporary rhodoliths feeding the island's contemporary dune field [15] estimate that the fossil lagoon deposit holds a volume equal to the disintegration of a little less than 50 billion whole rhodoliths [66].

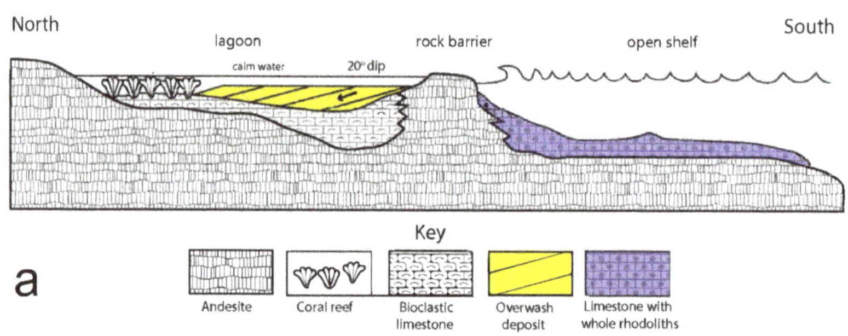

Figure 16. Aspects of a Pleistocene lagoon on Isla Corondos in Mexico's Gulf of California: (**a**) Cross section through the lagoon showing different environments including an outer shelf with whole rhodoliths forming a limestone layer, a rock barrier, and extensive deposits of crushed rhodoliths that washed over the rock barrier during storms; (**b**) View showing the limestone accumulation outside the lagoon (rock hammer for scale).

https://doi.org/10.3390/jmse14020169

Nearby, the much larger Isla del Carmen is situated across from Loreto and features an almost complete stratigraphic succession through the Pliocene and overlying Pleistocene, exposed at Arroyo Blanco on the island's east coast (Figure 8, square 4). Traced inland from the shore, the sequence is in the form of a shallow ramp dipping seaward that amounts to a total accumulated thickness of 61 m [67]. The margins of the 3.3 km^2 basin are well-defined by border faults and are estimated to be filled 64% by coarse- to sand-sized sediments derived from crushed rhodolith debris. For example, a massive layer easily accessible at the shore measures 9 m in thickness. The unit is fully dominated by rhodolith debris, with few whole rhodoliths preserved intact. Other layers interspersed throughout the sequence are formed by fossil shell beds deposited during calm interludes, but the thick rhodolith limestone layers are interpreted as storm-related [67].

Several other fossil rhodolith deposits are summarized elsewhere in detail; for example, the 3 m thick capstone of Pliocene rhodolith limestone now exposed on the uplifted plateau of Monserrat Island (Figure 8, square 5) [65]. The much-diminished capstone has a present area of 35.6 ha but is interpreted to have formerly covered most of the island's 18.5 km^2 surface at a shallow water depth prior to tectonic uplift. The number of rhodoliths that contributed to the capstone is incalculable, but enormous beyond that estimated for the much smaller basin on Isla Coronados [66].

Yet another example of a massive rhodolith limestone occurs at *Paredones Blancos* on Isla Cerralvo, located in the far southern opening of the Gulf of California (Figure 8, square 5) [68]. The locality name translates from Spanish as "white walls", an appropriate description for a 10 m thick band of white rocks that extends for three-quarters of a kilometer along the rocky west coast of Isla Cerralvo (Figure 17). Key index fossils from a lineage of echinoids recovered from the base of the limestone date the unit to a mid-Pliocene age. Some few whole rhodoliths are present, but otherwise, crushed rhodolith debris forms a coarse to very coarse carbonate sandstone. Based on point counts from multiple thin sections sampled through the main unit, the rhodolith content peaks at 88.7% against 11.3% for siliciclastic sediments. Local topography suggests that the coastal entrance to a large valley was inundated with rhodolith debris from an offshore bank during a rise in sea level [68].

Figure 17. View showing the massive white walls at *Paredones Blancos* on the western rocky coast of Isla Cerralvo in Mexico's lower Gulf of California. Enormous boulders of rhodolith limestone eroded from the cliff face line the shore. Cardon cacti, which appear as match sticks at the mouth of the nearby canyon, are 8 m in height (marked by adjacent white lines).

 https://doi.org/10.3390/jmse14020169

Older geological formations appear along the Pacific shores of Mexico's Baja California peninsula, where outcrops belonging to the Upper Cretaceous Rosario formation occur at Las Minas south of Ensenada (Figure 1, locality 9; 8, see triangle marker). A paleoisland formed by a volcanic ridge with a topographic relief of 65 m is flanked by limestone layers containing distinctive fossil components that dip off in opposite directions [69]. These are interpreted as the direct result of a persistent setting against sea swells in one direction and a more sheltered setting in the other direction. Fossil oysters and delicate, rock-encrusting bryozoans are the prominent components in a relatively calm-water environment. The windward setting is dominated by abundant rhodoliths, typically 1.5 to 2.5 cm in diameter, nucleated around volcanic rock cores up to 0.8 cm in diameter. The spherical shapes with algal rinds at a consistent thickness are interpreted as characteristic of a turbulent-water environment. The only coralline red alga identified from thin-section analysis belongs to the genus *Sporolithon* [69].

4.3. Fossil Deposits from the Western Mediterranean and Southern Europe

Rhodolith-bearing limestone beds deposited during the Miocene and Pliocene epochs are well represented throughout southern Europe and various islands in the western Mediterranean Sea (Figure 1, locality 7). The development and spread of rhodoliths in the genus *Sporolithon* throughout the Mediterranean realm have been thoroughly reviewed [70]. In particular, Minorca, as part of Spain's Balearic Archipelago, is the location of a study on a major carbonate ramp with shore-proximal facies formed by foraminifera and mollusks and more distal facies dominated by rhodoliths [71,72]. The adjacent "foramol" and "rhodalgal" facies preserve an exposed Miocene shelf covering roughly the southern third of the island abutted against Miocene land represented by older Triassic and Jurassic rocks. The scenario represents one of the largest, best preserved paleogeographic entities in the western Mediterranean. Subsequent studies elsewhere in the region focused on other Pliocene rhodolith deposits in southeastern Spain [73,74] and the Pliocene of the Monte Cetona region in northern Italy [75], as well as the older Miocene rhodolith deposits at Vitulano in the Southern Apennines of Italy [76]. Rhodolith materials from the Miocene Pietra da Cantoni limestone in the Piedmont area of northwestern Italy are described as a "lost carbonate factory" displaced by transport to a deeper-water setting from its source area [77]. Significant rhodolith deposits from the Eocene of the Colli Berici platform of northeastern Italy also are described as influenced by storm events [78].

Elsewhere during the Miocene, the Paratethys seaway existed as a more expansive Mediterranean precursor that penetrated deeper into southern and eastern Europe, forming a broader connection to the Atlantic Ocean [79,80]. For example, carbonate banks up to 40 m thick with abundant rhodoliths were deposited around the Leitha Mountains of present-day Styria in southwestern Austria (Figure 1, locality 8). The so-called Leitha Limestone (or *Leithakalk*) enjoyed a particularly long history as a favored building stone dating back to Roman times, when the cities of Carnuntum and Flavia Solva were constructed [81]. Today, the use of rhodolith limestone continues unabated in Vienna, where the Leitha Limestone is used as facing stone for metro entrances (Figure 18).

4.4. Fossil Deposits from New Zealand

Limestone units rich in rhodoliths occur in several deposits of Oligocene age exposed on New Zealand's North Island (Figure 1, locality 9). Basal layers in these formations are as much as 5 m in thickness and are recognized as stratigraphically useful marker beds equated with rising sea level [82].

https://doi.org/10.3390/jmse14020169

Figure 18. Entrance to the Volkstheater metro station on Vienna's Ringstrasse, faced with the rhodolith-rich Leitha Limestone. Individual rhodoliths are up to 8 cm in diameter.

5. Discussion

Consideration of the global significance of rhodolith beds as contributors to the enhancement of biodiversity in the recent world as well as their economic importance based on geological deposits is worthy of further discussion.

5.1. Role of Contemporary Rhodolith Flats as Biodiversity Multipliers

Although rhodolith flats, or maërl beds, constitute relatively thin drapes on the seabed during life, they provide refuge for an extensive community of other algae and marine invertebrates that dwell dispersed among the spheroids, as well as a surprising number of species that are hosted within individual spheroids either nested between branching forms or as borers in more solid forms. Previously cited is the relationship between brittle star populations belonging to the species *Ophiopholis aculeata*, preyed on by the wolffish *Anarhichas lupus*, in the maërl beds off Rebbenesøy in the Troms district of Norway [48]. More extensive biological relationships are reported in the deep-water rhodolith beds on Brazil's great Abrolhos Bank, including an enormous number of species divided between other algae (308 species), invertebrates (469 species), fish (252 species), and foraminifera (24 species) [83]. Considering that less than 5% of Brazil's ocean shelf is protected for conservation, there is serious concern over the potential loss of biodiversity due to deep-sea mining of rhodoliths for the possible application of lime fertilizer in land-based agriculture. Similar data from Mexico's Gulf of California are available, showing that mollusk species are the most conspicuous co-inhabitants of the gulf's rhodolith banks, likely exceeding 200 species [84]. Related work emphasizes the importance of cryptofaunal organisms living within individual rhodoliths, mainly crustaceans and polychaetes [15,85]. The same concern felt in Brazil exists that conservation practice is currently inadequate to protect against the damage of rhodolith beds due to trawling by fishing fleets in Mexico's Gulf of California [86,87].

5.2. Role of Rhodolith Deposits to the World Economy

Use of rhodolith limestone for construction in southern Europe is well documented; for example, the medieval Papal Palace in Avignon, southern France [80]. Surviving lime kilns scattered throughout many of the volcanic islands in the northeastern Atlantic Ocean testify

https://doi.org/10.3390/jmse14020169

to a history of economic use during the seventeenth and eighteenth centuries for supply of slaked lime in that region [20]. One of the largest such sites exploited on a commercial scale for slaked lime anywhere in Macaronesia was established on Isla de Lobos off the northeast coast of Fuerteventura in the Canary Islands. Further depletion of European maërl beds for use as agricultural fertilizer is now checked due to regulations by the European Union that list the dominant rhodolith species of Atlantic maërl as a non-renewable resource [86]. Moreover, licensing for maërl extraction has ceased altogether in the United Kingdom. Due to the high porosity of rhodolith limestone, it acts as a superior reservoir rock capable of retaining large volumes of oil and gas. For example, the Asmari Formation in Iran is a rhodolith-rich rock dating from the late Oligocene to early Miocene and it is claimed to be the reservoir for 90% of Iranian oil in production [80]. Likewise, the submarine strata of the Miocene Zujian Formation in the South China Sea may hold a reserve of as much as a billion barrels of oil [80].

Prospects against the commercial over-exploitation of both living and fossil rhodolith beds all around the world depend on the ongoing research efforts by marine biologists and geologists to fully register the extent of those features everywhere. Great strides have been made in this regard by specialists working in Europe, throughout the many islands of Macaronesia in the northeast Atlantic Ocean, and other critical areas such as Mexico's Gulf of California (Figure 1). Despite progress, the number of marine phycologists and paleobotanists actively doing research on living and fossil rhodoliths is extremely small compared to the number of researchers dedicated to studies on living and fossil coral reefs. Large parts of the world remain poorly represented with regard to rhodolith research, especially in the western Pacific and Indian Oceans. Only a single published report on contemporary and fossil rhodoliths from India is widely available [88]. In regions previously well studied by phycologists, new work is conducted on rhodoliths and the diversity of rhodolith faunal associations off the shores of Malta in the Central Mediterranean [89]. While the fragility of coral reefs due to climate change and overfishing is widely understood by the general public throughout the world, the comparable dangers related to the degradation of rhodolith banks or maërl beds deserve greater publicity.

6. Conclusions

Most often found as spheroidal constructions, rhodoliths embody a peculiar structure for the accretion of crustose, coralline red algae, lacking any kind of holdfast for attachment to the seabed in sunlit waters as shallow as -2 m but as deep as -100 m under conditions of extreme clarity in the water column. Like any other algae, those that grow as rhodoliths require sunlight for photosynthesis. The shape promotes circumrotary movement that provides every part of the curved surface an equal share of sunlight over time due to ongoing turnover stimulated by waves, bottom currents, or disruption by marine animals sharing the same habitat. The following points emphasize the main elements of rhodolith ecology as related to living aggregates (also known as maërl beds) as well as fossil deposits.

- Rhodoliths occur today in high concentrations on seabeds that cover parts of all continental shelves around the world, as well as the marine shelves of many oceanic islands. The latitudinal range of living rhodoliths extends from tropical settings in equatorial zones to polar zones as extreme as the Svalbard Archipelago at latitudes between N 76° and N 80°, as well as the shores of Greenland. In the Southern Hemisphere, contemporary rhodoliths are reported at a latitude of S 33.5° off the coast of South Africa.
- Rhodoliths possess an excellent fossil record that began in geological time during the Cretaceous, at least 113 million years ago. Maximum rhodolith development

 https://doi.org/10.3390/jmse14020169

appears to have coincided with the Middle Miocene Climatic Optimum, approximately 15 million years ago, when extensive deposits accumulated in southern Europe and around islands in today's Western Mediterranean Sea.

- Today and during the geologic past, rhodoliths are and were vulnerable to major sea storms capable of disturbing the seabed at a considerable depth, whereby great numbers might be pushed onto land in supratidal settings or farther offshore into deeper-water settings. Agitation of rhodoliths against one another during violent storms leads to fragmentation and the development of massive deposits of rhodolith debris. These are useful indicators of storm events. Moreover, rhodoliths in an extreme state of pulverization may contribute to beach and coastal dune deposits.
- Contemporary rhodolith beds are recognized by marine ecologists as biodiversity multipliers capable of hosting a vast number of associated invertebrates dwelling both among and embedded within individual spheroids. The role is similar to that of reef-forming corals that function as umbrella species to support a more diverse marine community. The concern among marine ecologists is that commercial exploitation for use in agriculture as lime fertilizer will lead to the destruction of an important marine ecosystem, especially in Brazil.
- The number of phycologists engaged in studies on crustose coralline algae that accrete as rhodoliths is small compared to the community of marine biologists who work on coral reefs. The same is true for paleobotanists studying fossil rhodolith deposits. Much remains to be learned, particularly in continental shelf regions of the Western Pacific Ocean and the Indian Ocean, as well as related oceanic islands.

It is hoped this review may encourage future research and cooperation between marine ecologists and paleoecologists, especially in those parts of Asia where such studies are certain to provide fresh data on rhodoliths and their natural history.

Funding: This research received no external funding.

Data Availability Statement: No new data were created or analyzed in this study.

Acknowledgments: M.E.J. is grateful to B. Gudveig Baarli for assistance in editing the line drawings for this paper. Images provided in Figures 13c and 18 were made available by colleagues Adela Harvey (LaTrobe University, Australia) and Carlos M. da Silva (Lisbon University, Portugal), respectively. Two anonymous readers contributed peer reviews with useful comments that led to the improvement of this manuscript.

Conflicts of Interest: The author declares no conflicts of interest.

References

1. Elias, M.K. Fossil symbiotic algae in comparison with other fossil and living algae. *Am. Midl. Nat.* **1946**, *36*, 282–290. [CrossRef]
2. Kiel, S.; Goedert, J.L.; Huynh, T.L.; Krings, M.; Parkinson, D.; Romero, R.; Looy, C.V. Early Oligocene kelp holdfasts and stepwise evolution of the kelp ecosystem in the North Pacific. *Proc. Natl. Acad. Sci. USA* **2024**, *121*, e22317054121. [CrossRef] [PubMed]
3. Wray, J.L. *Calcareous algae*; Developments in Palaeontology and Stratigraphy; Elsevier Scientific Publication Co.: Amsterdam, The Netherlands, 1977; 185p.
4. Foster, M.S.; Riosmena-Rodríguez, R.; Steller, D.L.; Woelkerling, W. Living rhodolith beds in the Gulf of California and their implications for paleoenvironmenmtal interpretations. In *Pliocene Carbonates and Related Facies Flanking the Gulf of California, Baja California*; Johnson, M.E., Ledesma-Vázquez, J., Eds.; Geological Society of America Special Paper; Geological Society of America: Washington, DA, USA, 1997; Volume 318, pp. 127–139.
5. Amado-Filho, G.M.; Moura, R.L.; Bastos, A.C.; Francini-Filho, R.B.; Pereira-Filho, G.I.H.; Bahia, R.G.; Moraes, F.C.; Motta, F.S. Mesophotic ecosystems of the unique South Atlantic atoll are composed by rhodolith beds and scattered consolidated reefs. *Mar. Biodivers.* **2016**, *46*, 933–936. [CrossRef]
6. Darwin, C. *Geological Observations on the Volcanic Islands Visited During the Voyage of the H.M.S. Beagle. Together with Some Brief Notices on the Geology of Australia and the Cape of Good Hope. Being the Second Part of the Geology of the Voyage of the Beagle, Under the Command of Capt. FitzRoy, R.N.N. During the Years 1832 to 1836*; Smith, Elder & Co.: London, UK, 1844; 175p.

https://doi.org/10.3390/jmse14020169

7. Bosellini, A.; Ginsburg, R.N. Form and internal structure of recent algal nodules (rhodolites) from Bermuda. *J. Geol.* **1971**, *79*, 669–682. [CrossRef]
8. Bosence, D.W.I. Description and classification of rhodoliths (Rhodoids, Rhodolites). In *Classification of Coated Grains*; Peryt, T.M., Ed.; Springer: Berlin/Heidelberg, Germany, 1983; pp. 217–224.
9. Herbert, S. *Charles Darwin Geologist*; Cornell University Press: Ithica, NY, USA, 2005; 485p.
10. Johnson, M.E.; Baarli, B.G.; Cachão, M.; da Silva, C.M.; Ledesma-Vásquez, J.; Ramalho, R.S.; Mayoral, E.; Santos, A. Rhodoliths, uniformitarianism, and Darwin: Pleistocene and Recent carbonate deposits in the Cape Verde and Canary archipelagos. *Palaeogeogr. Palaeoclim. Palaeoecol.* **2012**, *329–330*, 83–100. [CrossRef]
11. Woelkerling, W.J. *The Coralline Red Algae: An Analysis of the Genera and Subfamilies of Nongeniculate Corallinacea*; British Museum of Natural History Oxford University Press: London, UK, 1988; 268p.
12. Riosmena-Rodríguez, R.; Nelson, W.; Aguirre, J. (Eds.) *Rhodolith/Maërl Beds: A Global Perspective*; Coastal Research Library 15; Springer International Publishing: Cham, Switzerland, 2017; 368p.
13. Adey, P.I.; Macintyre, I.G. Crustose coralline algae: A re-evaluation in the geological sciences. *Bull. Geol. Soc. Am.* **1973**, *84*, 883–904. [CrossRef]
14. Johnson, M.E. *Baja California's Coastal Landscapes Revealed*; University Arizona Press: Tucson, AZ, USA, 2021; 239p.
15. Sewell, A.A.; Johnson, M.E.; Backus, D.H.; Ledesma-Vázquez, J. Rhodolith detritus impounded by a coastal dune on Isla Coronados, Gulf California. *Cienc. Mar.* **2007**, *33*, 483–494. [CrossRef]
16. Bassi, D.; Iryu, Y.; Nebelsick, J.H. To be or not to be a fossil rhodolith? Analytical methods for studying fossil rhodolith deposits. *J. Coast. Resh.* **2012**, *28*, 288–295. [CrossRef]
17. Bassi, D.; Nebelsick, J.H.; Checconi, A.; Hohenegger, J.; Iryu, Y. Present-day and fossil rhodolith pavements compared: Their potential for analyzing shallow-water carbonate deposits. *Sediment. Geol.* **2009**, *214*, 74–84. [CrossRef]
18. Johnson, M.E.; Ledesma-Vázquez, J.; Ramalho, R.S.; da Silva, C.M.; Rebelo, A.C.; Santos, A.; Baarli, B.G.; Mayoral, E.; Cachão, M. Taphonomic range and sedimentary dynamics of modern and fossil rholdolith beds (North Atlantic Ocean). In *Rhodolith/Maërl "Beds: A Global Perspective*; Riosmena-Rodríguez, R., Nelson, W., Aguirre, J., Eds.; Coastal Research Library; Springer International Publishing: Cham, Switzerland, 2017; Volume 15, pp. 221–261.
19. Paris, R.; Ramalho, R.S.; Madeira, J.; Ávila, S.P.; May, S.M.; Rixhon, G.; Engel, M.; Brückner, H.; Herzog, M.; Schukraft, G.; et al. Megatsunami conglomerates and flank collapses of ocean island volcanoes. *Mar. Geol.* **2018**, *395*, 168–187. [CrossRef]
20. Johnson, M.E.; Baarli, B.G.; da Silva, C.M.; Cachão, M.; Ramalho, R.S.; Santos, A.; Mayoral, E.J. Recent rhodolith deposits stranded on the windward shores of Maio (Cape Verde Islands): Historical Resource for the Local Economy. *J. Coast. Res.* **2016**, *32*, 735–743. [CrossRef]
21. Johnson, M.E.; Baarli, B.G.; da Silva, C.M.; Cachão, M.; Ramalho, R.S.; Ledesma-Vázques, J.; Mayoral, E.J.; Santos, A. Coastal dunes with high content of rhodolith (coralline red algae) bioclasts: Pleistocene formations on Maio and São Nicolau in the Cape Verde archipelago. *Aeolian Res.* **2013**, *8*, 1–9. [CrossRef]
22. Mayoral, E.; Ledesma-Vázquez, J.; Baarli, B.G.; Santos, A.; Ramalho, R.; Cachão, M.; da Silva, C.M.; Johnson, M.E. Ichnology in oceanic islands: Case studies from the Cape Verde Archipelago. *Palaeogeogr. Palaeoclim. Palaeoecol.* **2013**, *381–382*, 47–66. [CrossRef]
23. Aquirre, J.; Braga, J.C. Rhodolith beds in a shifting world: A palaeontological perspective. *Mar. Freshw. Ecosyst.* **2024**, *34*, e70015. [CrossRef]
24. Riding, R.; Braga, J.C. Halysis Hø, 1932—An Ordovician corralline red alga? *J. Paleontol.* **2005**, *79*, 835–841. [CrossRef]
25. Teichert, S. Attached and free-living crustose coralline algae and their functional traits in the geological record and today. *Facies* **2024**, *70*, 8. [CrossRef]
26. Rebelo, A.C.; Johnson, M.E.; Rasser, M.W.; Silva, L.; Melo, C.S.; Ávila, S.P. Global biodiversity and biogeography of rhodolith-forming species. *Front. Biogeogr.* **2021**, *13*, e50646. [CrossRef]
27. Spalding, M.D.; Fox, H.E.; Allen, G.R.; Davidson, N.; Ferdaña, Z.A.; Finlayson, M.; Halpern, B.S.; Jorge, M.A.; Lombana, A.; Lourie, S.A.; et al. Marine ecoregions of the world: A bioregionalization of coastal and shelf areas. *BioScience* **2007**, *57*, 573–583. [CrossRef]
28. Adams, L.A.; Maneveldt, G.W.; Green, A.; Karenyi, N.; Parker, D.; Samaai, T.; Kerwath, S. Rhodolith bed discovered off the South African Coast. *Diversity* **2020**, *12*, 125. [CrossRef]
29. Coates, A.G.; Stallard, R.F. How old is the Isthmus of Panama? *Bull. Mar. Sci.* **2013**, *89*, 801–813. [CrossRef]
30. Baarli, B.G.; Cachão, M.; Da Silva, C.M.; Johnson, M.E.; Mayoral, E.J.; Santos, A. A Middle Miocene carbonate embankment on an active volcanic slope: Ilhéu de Baixo, Madeira Archipelago, Eastern Atlantic. *Geol. J.* **2014**, *49*, 90–106. [CrossRef]
31. Johnson, M.E.; Uchman, A.; Ramalho, R.S.; Martín-González, E.; Martins, G.M.; Hipólito, A.; Marques, S.; Ávila, G.C.; Madeira, P.; Doukani, M.A.; et al. Basalt sea stack and related facies from the Middle Pleistocene on Sal Island (Cabo Verde Archipelago, NE Atlantic Ocean). *Facies* **2025**, *71*, 20. [CrossRef]

 https://doi.org/10.3390/jmse14020169

32. Rebelo, A.C.; Martín-González, E.; Melo, C.S.; Johnson, M.E.; González-Rodríguez, A.; Galindo, I.; Quartau, R.; Baptista, L.; Ávila, S.P.; Rasser, M.W. Rhodolith beds and their onshore transport in Fuerteventura Island (Canary Archipelago Spain). *Front. Mar. Sci.* **2022**, *9*, 917883. [CrossRef]
33. Rebelo, A.C.; Johnson, M.E.; Quartau, R.; Rasser, M.W.; Melo, C.S.; Neto, A.L.; Temera, R.; Madeira, P.; Ávila, S.P. Modern coralline algal nodules from the insular shelf of Pico in the Azores (northeast Atlantic Ocean). *Estuauraine Coast. Shelf Sci.* **2018**, *210*, 7–17. [CrossRef]
34. Amando-Filho, G.M.; Moura, R.L.; Bastos, A.C.; Salgado, L.T.; Sumida, P.Y.; Guth, A.Z.; Francini-Filho, R.B.; Pereira-Filho, G.H.; Abrantes, D.P.; Brasileiro, P.S.; et al. Rhodolith beds are major $CaCO_3$ bio-factories in the tropical South West Atlantic. *PLoS ONE* **2012**, *7*, e35171. [CrossRef]
35. Pomar, L.; Morsilli, M.; Hallock, P.; Bádenas, B. Internal waves, an under-explored source of turbulence events in the sedimentary record. *Earth-Sci. Rev.* **2012**, *111*, 56–81. [CrossRef]
36. Henriques, M.C.; Villas-Boas, A.; Rodriguez, R.R.; Figueiredo, M.A.O. New records of rhodolith-forming species (Corallinales, Rhodophyta) from deep water in Espírito Santo State, Brazil. *Helgol. Mar. Res.* **2012**, *66*, 219–231. [CrossRef]
37. Bahia, R.G.; Abranttes, D.P.; Brasileiro, P.S.; Pereora-Filho, G.H.; Amado-Filho, G.M. Rhodolith bed structure along a depth gradient on the northern coast of Bahia State, Brazil. *Braz. J. Oceanog.* **2010**, *58*, 323–337. [CrossRef]
38. Pereira-Filho, G.H.; Amado-Filho, G.M.; de Moura, R.L.; Bastos, A.C.; Guimaraes, S.M.P.B.; Salgado, L.T.; Fracinia-Filho, R.B.; Bahia, R.G.; Abrantes, D.P.; Guth, A.Z.; et al. Extensive rhodolith beds over the summits of southwestern Atlantic Ocean seamounts. *J. Coast. Res.* **2012**, *28*, 261–269. [CrossRef]
39. Amado-Filho, G.M.; Pereira-Filho, G.H.; Bahia, R.G.; Abrantes, D.P.; Veras, P.C.; Matheus, Z. Occurrence and distribution of rhodolith beds on the Fernando de Noronha Archipelago of Brazil. *Aquat. Bot.* **2022**, *101*, 41–45. [CrossRef]
40. Roberts, H.H.; Wilson, P.A.; Lugo-Fernández, A. Biologic and geologic responses to physical processes: Examples from modern reef systems of the Caribbean-Atlantic region. *Cont. Shelf Res.* **1992**, *12*, 809–834. [CrossRef]
41. Ballantine, D.L.; Bowden-Kerby, A.; Aponte, N.E. *Cruoriella* rhodoliths from shallow-water back reef environments in La Parguera Puerto Rico (Caribbean Sea). *Coral Reefs* **2000**, *19*, 75–81. [CrossRef]
42. Ledesma-Vázquez, J.; Johnson, M.E.; Gonzalez-Yajimovich, O.; Santamaria-del-Angel, E. Gulf of California geography, geological origins, oceanography and sedimentation patterns. In *Atlas of Coastal Ecosystems in the Western Gulf of California*; Johnson, M.E., Ledesma-Vázquez, J., Eds.; University Arizona Press: Tucson, AZ, USA, 2009; pp. 1–10.
43. Steller, D.L.; Foster, M.S. Environmental factors influencing distribution and morphology of rhodoliths in Bahía Concepción, BCS, México. *J. Exp. Mar. Biol. Ecol.* **1995**, *194*, 201–212. [CrossRef]
44. Marrack, E.C. The relationship between water motion and living rhodolith beds in the southwestern Gulf of California, Mexico. *Palaios* **1999**, *14*, 159–171. [CrossRef]
45. McConnico, L.A.; Foster, M.S.; Steller, D.L.; Riosmena-Rodríguez, R. Population biology of a long-lived rhodolith: The consequences of becoming old and large. *Mar. Ecol. Prog. Ser.* **2014**, *504*, 109–118. [CrossRef]
46. Steller, D.L.; Riosmena-Rodríguez, R.; Foster, M.S. Living rhodolith bed ecosytems in the Gulf of California. In *Atlas of Coastal Ecosystems in the Western Gulf of California*; Johnson, M.E., Ledesma-Vázquez, J., Eds.; University Arizona Press: Tucson, AZ, USA, 2009; pp. 71–82.
47. Halfar, J.; Eisele, M.; Riegl, B.; Hetzinger, S.; Godinez-Orta, L. Modern rhodolith-dominated carbonates at Punta Chivato, Mexico. *Geodiversitas* **2012**, *34*, 99–113. [CrossRef]
48. Riosmena-Rodríguez, R.; López-Calderón, J.M.; Mariano-Meléndez, E.; Sánchez-Rodrígues, A.; Fernández-Garcia, C. Size and distribution of rhodolith beds in the Loreto Marine Park: Their role in coastal processes. *J. Coast. Res.* **2012**, *28*, 255–260. [CrossRef]
49. Hetzinger, S.; Halfar, J.; Riegl, B.; Godinez-Orta, L. Sedimentology and acoustic mapping of modern rhodolith facies on a non-tropical carbonate shelf (Gulf of California, Mexico). *J. Sediment. Res.* **2006**, *76*, 670–682. [CrossRef]
50. Birkett, D.A.; Maggs, C.A.; Dring, M.J. Maerl: An overview of dynamic and sensitivity characteristics for conservation management of marine SCAs. *Scott. Assoc. Mar. Sci.* **1998**, *5*, 116.
51. Kamenos, N.A.; Law, A. Temperature controls on coralline algal skeletal growth. *J. Phycol.* **2009**, *46*, 331–335. [CrossRef]
52. Kamenos, N.A.; Moore, P.G.; Hall-Spencer, J.M. Nursery-area function of maerl grounds for juvenile queen scallops *Aequipecten opercularis* and other invertebrates. *Mar. Ecol. Prog. Ser.* **2004**, *274*, 183–189. [CrossRef]
53. Backus, D.H. Contemporary issues and advancements in coastal eolianite research. *J. Mar. Sci. Eng.* **2025**, *13*, 321. [CrossRef]
54. Freiwwald, A. Sedimentological and biological aspects in the formation of branched rhodoliths in northern Norway. *Beir. Paläont.* **1995**, *20*, 7–19.
55. Teichert, S.; Woelkerling, W.; Rüggeberg, A.; Wisshak, M.; Piepenburg, D.; Meyerhöfer, M.; Form, A.; Freiwald, A. Arctic rhodolith beds and their environmental controls (Spitsbergen, Norway). *Facies* **2014**, *60*, 15–37. [CrossRef]
56. Jørgensbye, H.I.; Halfar, J. Overview of coralline red algal crusts and rhodolith beds (Corallinales, Rhodophyta) and their possible ecological importance in Greenland. *Polar Biol.* **2017**, *40*, 517–531. [CrossRef]

57. Harvey, A.S.; Harvey, R.M.; Merton, E. The distribution significance and vulnerability of Australian rhodolith bds: A review. *Mar. Freshw. Res.* **2017**, *68*, 411–428. [CrossRef]
58. Landing, E.; Johnson, M.E. Stromatolites and their "kin"as living microbialites in contemporary settings linked to a long fossil record. *J. Mar. Sci. Eng.* **2024**, *12*, 2127. [CrossRef]
59. Harvey, A.; Johnson, M.E.; Harvey, R. Heterozoan carbonate enriched beach sand and coastal dunes: Dirk Hartog Island, Shark Bay, Western Australia. *Facies* **2018**, *64*, 23. [CrossRef]
60. Iryu, Y.; Nakamori, T.; Matsuda, S.; Abe, O. Distribution of marine organisms and its geological significance in the modern reef complex of the Ryukyu Islands. *Sediment. Geol.* **1995**, *99*, 243–258. [CrossRef]
61. Matuda, S.; Iryu, Y. Rhodloliths from deep fore-reef to shelf areas around Okinawa-jima, Ryukyu Islands, Japan. *Mar. Geol.* **2011**, *282*, 215–230. [CrossRef]
62. Rebelo, A.C.; Uchman, A.; Johnson, M.E.; Melo, C.S.; Vegas, J.; Galindo, I.; Mayoral, E.J.; Santos, A.; González-Rodríguez, A.; Afonso-Carrillo, J.; et al. Rhodoliths and trace fossils record stabilization of a fan-delta system: An example from the Mio-Pliocene deposits of Gran Canaria (Canary Islands, Spain). *J. Paleogeog.* **2025**, *14*, 100266. [CrossRef]
63. Johnson, M.E.; da Silva, C.M.; Santos, A.; Baarli, B.G.; Cachão, M.; Mayoral, E.J.; Rebelo, A.C.; Ledesma-Vázquez, J. Rhodolith transport and immobilization on a volcanically active rocky shore: Middle Miocene at Cabeço das Laranjas on Ilhéu de Cima (Madeira Archipelago Portugal. *Palaeogeogr. Palaeoclim. Palaeoecol.* **2011**, *300*, 113–127. [CrossRef]
64. Rebelo, A.C.; Rasser, M.W.; Kroh, A.; Johnson, M.E.; Ramalho, R.S.; Melo, C.; Uchman, A.; Berning, B.; Silva, L.; Zanon, V.; et al. Rocking around a volcanic island shelf: Pliocene rhodolith beds from Malbusca, Santa Maria Island (Azores, NE Atlantic). *Facies* **2016**, *62*, 22. [CrossRef]
65. Johnson, M.E.; Backus, D.H.; Riosmena-Rodríguez, R. Contribution of rhodoliths to the generation of Pliocene-Pleistocene limestone in the Gulf of California. In *Atlas of Coastal Ecosystems in the Western Gulf of California*; Johnson, M.E., Ledesma-Vázquez, J., Eds.; University Arizona Press: Tucson, AZ, USA, 2009; pp. 83–94.
66. Ledesma-Vázquez, J.; Johnson, M.E.; Backus, D.H.; Mirabal-Davila, C. Coastal evolution from transgressive barrier deposit to marine terrace on Isla Coronados, Baja California Sur, Mexico. *Cienc. Mar.* **2007**, *33*, 335–351. [CrossRef]
67. Eros, J.M.; Johnson, M.E.; Backus, D.H. Rocky shores and development of the Pliocene-Pleistocene Arroyo Blanco Basin on Isla Carmen in the gulf of California Mexico. *Can. J. Earth Sci.* **2006**, *43*, 1149–1164. [CrossRef]
68. Emhoff, K.F.; Johnson, M.E.; Backus, D.H.; Ledesma-Vázquez, J. Pliocene stratigraphy at Paredones Blancos: Significance of a massive crushed-rhodolith deposit on Isla Cerralvo Baja California Sur (Mexico). *J. Coast. Res.* **2012**, *28*, 234–243. [CrossRef]
69. Johnson, M.E.; Hayes, M.L. Dichotomous facies on a Late Cretaceous rocky island as related to wind and wave patterns (Baja California, Mexico. *Palaios* **1993**, *8*, 385–395. [CrossRef]
70. Braga, J.C.; Bassi, D. Neogene history of *Sporolithon* Heydrich (Corallnales Rhodphyta) in the Mediterranean region. *Palaeogeogr. Palaeoclim. Palaeoecol.* **2007**, *243*, 189–203. [CrossRef]
71. Pomar, L. Ecological control of sedimentary accommodation: Evolution from a carbonate ramp to rimmed shelf Upper Miocene Balearic Islands. *Palaeogeogr. Palaeoclim. Palaeoecol.* **2001**, *175*, 249–272. [CrossRef]
72. Brandano, M.; Vannucci, G.; Pomar, L.; Obrador, A. Rhodolith assemblages from the lower Tortonian carbonate ramp of Menorca (Spain): Environmental and paleoclimatic implications. *Palaeogeogr. Palaeoclim. Palaeoecol.* **2005**, *226*, 307–323. [CrossRef]
73. Aguirre, J.; Braga, J.; Martín, J. Algal nodules in the Upper Pliocene deposits at the coast of Cadiz (S Spain). In *Studies on Fossil Benthic Algae*; Barattolo, F., De Castro, P., Parente, M., Eds.; Bollettino della Società Paleontologica Italiana: Modena, Italy, 1993; Volume 1, pp. 1–7. ISBN 9788870002171.
74. Aguirre, J.; Braga, J.C.; Martín, J.M.; Betzler, C. Palaeoenvironmental and stratigraphic significance of Pliocene rhodolith beds and coralline algal bioconstructions from the Caroneras Basin (SE Spain). *Geodiversitas* **2012**, *34*, 115–136. [CrossRef]
75. Checconi, A.; Bassi, D.; Passeri, L.; Rettori, R. Corallline red algal assemblage from the Middle Pliocene shallow-water temperate carbonates of the Monte Cetona (Northrn Apennines Italy). *Facies* **2007**, *53*, 57–66. [CrossRef]
76. Checconi, A.; Bassi, D.; Monaco, P.; Carannante, G. Re-deposited rhodoliths in the Middle Miocene hemipelagic deposits of Vitulano (Southern Apennines, Italy): Coralline assemblage characterization and related trace fossils. *Sediment. Geol.* **2020**, *225*, 50–66. [CrossRef]
77. Coletti, G.; Basso, D.; Frixa, A.; Corselli, C. Transported rhodoliths witness the lost carbonate factory: A case history from the Miocene Pietra da Cantoni limestone (NW Italy). *Riv. Ital. Paleontol. Stratigr.* **2015**, *121*, 345–368.
78. Bassi, D. Larger foraminiferal and coralline algal facies in an Upper Eocene storm-influenced, shallow-water carbonate platform (Colli Berici, north-eastern Italy). *Palaeogeogr. Palaeoclim. Palaeoecol.* **2005**, *226*, 17–35. [CrossRef]
79. Harzhauser, M.; Landau, B.; Mandic, O.; Neubauer, T.A. The Central Paratethys Sea–Part of the tropical eastern Atlantic rather than gate into the Indian Ocean. *Glob. Planet. Change* **2024**, *243*, 104595. [CrossRef]
80. Moshammer, B.; Uhlir, C.; Rohatsch, A.; Unterwurzacher, M. Adnet 'Marble', Untersberg 'Marble' and Leitha Limestone—Best examples expressing Austria's physical cultural heritage. In *Engineering Geology for Society and Territory*; Lollino, G., Manconi, A., Guzzetti, F., Culshaw, M., Bobrowsky, P., Luino, F., Eds.; Springer: Berlin/Heidelberg, Germany, 2014; Volume 5, pp. 253–257.

81. Coletti, G.; Basso, D.; Frixa, A. Economic importance of coralline carbonates. In *Rhodolith/Maërl Beds: A Global Perspective*; Riosmena-Rodríguez, R., Nelson, W., Aguirre, J., Eds.; Coastal Research Library; Springer International Publishing: Cham, Switzerland, 2017; pp. 87–101.
82. Nalin, R.; Nelson, C.S.; Basso, D.; Massari, F. Rhodolith-bearing limestones as transgressive marker beds; fossil and modern examples from North Island New Zealand. *Sedimentology* **2008**, *55*, 249–274. [CrossRef]
83. Schubert, N.; Magris, R.A.; Berchez, F.; Bernardino, A.F.; Ferreira, C.E.; Francini-Filho, R.B.; Gaspar, T.L.; Pereira-Filho, G.H.; Rossi, S.; Silva, J.; et al. Rhodolith beds in Brazil—A Natural Heritage in need of Conservation. *Divers. Distrib.* **2025**, *31*, e13960. [CrossRef]
84. Cintra-Buenrostro, C.E.; Foster, M.S.; Meldahl, K.H. Response of nearshore marine assemblages to global change: A comparison of molluscan assemblages in Pleistocene and modern rhodolith beds in the southwestern Gulf of California, México. *Palaeogeogr. Palaeoclim. Palaeoecol.* **2002**, *183*, 299–320. [CrossRef]
85. Steller, D.L.; Riosmena-Rodríguez, R.; Foster, M.S.; Roberts, C.A. Rhodolith bed diversity in the Gulf of California: The importance of rhodolith structure and consequences of disturbance. *Aquat. Conserv. Mar. Freshw. Ecosyst.* **2003**, *13*, S5–S20. [CrossRef]
86. de Los Ángles Carvajal, M.; Robles, A.; Ezcurra, E. Ecological conservation in the Gulf of California. In *The Gulf of California Biodiversity and Conservation*; Brusca, R.C., Ed.; University Arizona Press: Tucson, AZ, USA, 2010; pp. 219–250.
87. Barbera, C.; Bordehore, C.; Borg, J.A.; Glémarec, M.; Grall, J.; Hall-Spencer, J.M.; De La Huz, C.H.; Lanfranco, E.; Lastra, M.; Moore, P.G.; et al. Conservation and management of northeast Atlantic and Mediterranean maerl beds. *Aquat. Conserv. Mar. Freshw. Ecosyst.* **2003**, *13*, S65–S76. [CrossRef]
88. Sarkar, S. Ecology of coralline red algae and their fossil evidences from India. *Thalassas* **2016**, *33*, 15–28. [CrossRef]
89. Tabone, L.; Knittweis, L.; Aguilar, R.; Alvarez, H.; Borg, J.A.; Garcia, S. Habitat characterization, anthropogenic impacts and conservation of rhodolith beds off southeastern Malta. *Aquat. Conserv. Mar. Freshw. Ecosyst.* **2024**, *34*, e4148. [CrossRef]

 https://doi.org/10.3390/jmse14020169

Journal of
Marine Science and Engineering

MDPI

Review

Trace Fossils on Oceanic Volcanic Islands of Macaronesia: Current State of Knowledge

Alfred Uchman

Faculty of Geography and Geology, Institute of Geological Sciences, Jagiellonian University, Gronostajowa 3a, PL 30-387 Kraków, Poland; alfred.uchman@uj.edu.pl

Abstract

Ichnological research on trace fossils from the volcanic islands of Macaronesia (North Atlantic) is reviewed in light of significant advances over the past two decades. These studies contribute to the interpretation of paleoenvironments and enhance our understanding of the biota preserved in Miocene–Holocene shallow marine and non-marine deposits across the Azores, Cape Verde, Canary, Madeira, and Salvagens archipelagos. Trace fossils provide evidence of organisms not always known from body fossils, or whose potential tracemakers are absent from the extant island fauna. They include sedimentary burrows, borings in hard substrates, and traces of plant–insect interactions. Some ichnotaxa are widespread and common (e.g., *Bichordites monastiriensis, Dactyloidites ottoi, Macaronichnus segregatis, Ophiomorpha nodosa, Thalassinoides* isp.), whereas others are rare. Several new ichnotaxa have also been described from the islands, including *Alaichnus kabuverdiensis* (cumulative trace of bivalve siphons), *Centrichnus dentatus* (attachment trace of verrucid barnacles), *Diopatrichnus santamariaensis* (polychaete tubes armored with shell debris), *Ericichnus bromleyi* and *E. asgaardi* (bioerosion grooves of regular echinoids), and *Rebuffoichnus guanche* (coleopteran pupation chambers). Despite these advances, ichnological research in Macaronesia remains uneven, with many topics still underexplored and significant gaps in the geographic and inventory record.

Keywords: Ichnology; ichnotaxonomy; ichnofossils; borings; burrows; Cenozoic

1. Introduction

Since Darwin's time, islands isolated from continents by deep seas have been central to understanding the history of life, particularly speciation and the dispersal of biota. Among them, oceanic volcanic islands hold special significance because they arise directly from submarine volcanic activity rather than from continental drift or submergence. Unlike continental fragments, which may already carry biota at the moment of separation, oceanic volcanic islands are initially sterile and must be colonized entirely by incoming organisms [1,2].

Sedimentary rocks are rare or poorly exposed on the emerged surfaces of volcanic islands. For a long time, this led to relatively little attention from geologists and paleontologists, who traditionally focused on sedimentary successions. However, during the past three decades, remarkable progress has been achieved, particularly in the Atlantic archipelagos of Macaronesia, which include the Azores, Madeira, the Salvagens, the Canary Islands, and Cape Verde (Figure 1). Studies have shown that marine deposits on these islands are often fossiliferous. Numerous works have documented body fossils from diverse taxonomic groups in Miocene, Pliocene, and Quaternary deposits, forming a solid basis for palaeoecological and palaeogeographical interpretations [3–8].

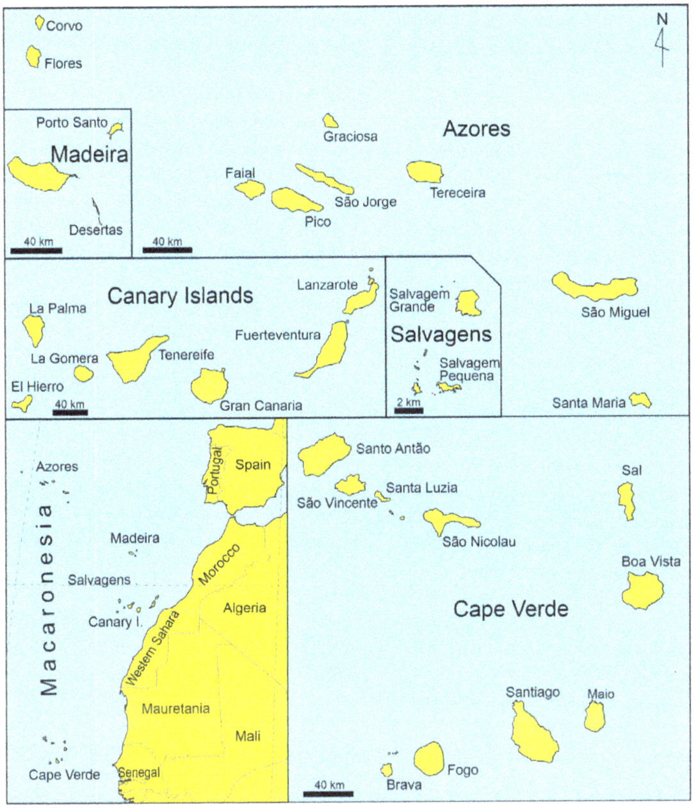

Figure 1. Simplified maps of Macaronesia.

Complementing the record of body fossils are trace fossils—the preserved evidence of biological activity that modifies the substrate and typically exhibits recurrent morphological patterns. The organisms that produce them are often absent as body fossils, meaning trace fossils provide essential, and sometimes unique, insights into past environments and biotic interactions. They therefore represent a powerful tool for paleoenvironmental reconstruction [9–11]. Over the last two decades, the ichnological record from Macaronesia has expanded considerably, including the description of new ichnotaxa. This paper reviews recent advances in the study of trace fossils from these islands and highlights their importance for understanding colonization processes and the ecological history of oceanic volcanic islands.

2. Synopsis of Selected Trace Fossils

The trace fossils preserved in the sedimentary successions of Macaronesia show broadly comparable assemblages across the archipelagos. This chapter outlines the most common and distinctive ichnotaxa. As elsewhere, the ichnological record can be grouped into two main categories: burrows, produced in soft and unconsolidated substrates, and borings, produced in hard substrates such as rocks or skeletal material.

For clarity, the discussion is organized into three parts: marine burrows, marine borings, and trace fossils from continental deposits. The latter include a more diverse range of structures, such as borings, burrows, root traces, and plant-damage traces attributable to insect activity.

2.1. Burrows in Marine Deposits

Bichordites monastiriensis Plaziat & Mahmoudi, 1988 [12] (Figure 2A) is a horizontal, straight to gently winding, complex burrow characterized by meniscate backfill surrounding a central core, typically with a heart-shaped cross section. The upper surface of the core bears a narrow median furrow, whereas the lower surface displays a longitudinal crest. The core, interpreted as the drainage structure of a burrowing irregular echinoid—most likely of the *Echinocardium* group [12–15]—is preferentially cemented and thus usually much better preserved than the surrounding meniscate fill.

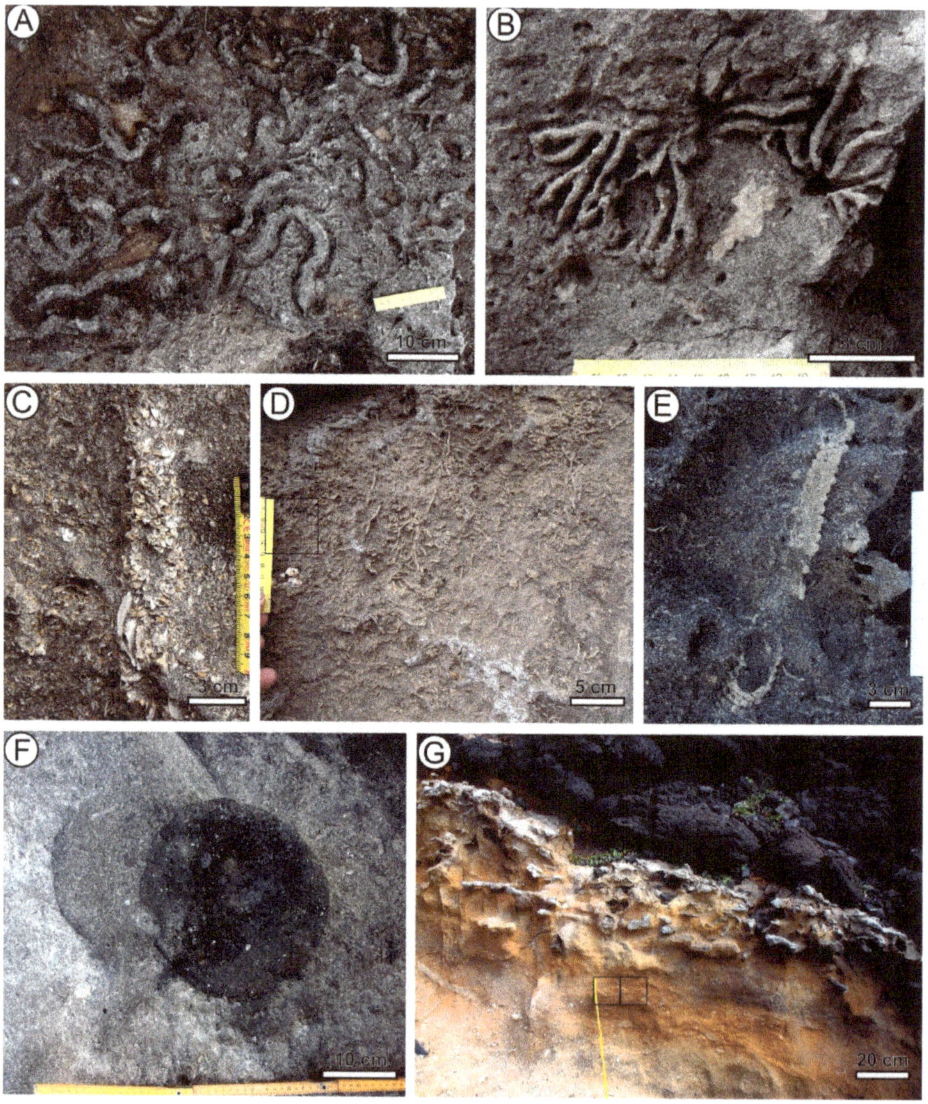

Figure 2. Trace fossils (fossil burrows) in Pliocene calcarenites–sandstones of Santa Maria, Azores. (**A**) *Bichordites monastiriensis.* (**B**) *Dacyloidites ottoi.* (**C**) *Diopatrichnus santamariaensis.* (**D**) *Macaronichnus segregatis.* (**E**) *Ophiomorpha nodosa.* (**F**) *Piscichnus waitemata.* (**G**) *Thalassinoides isp.* (**A**–**F**) Malbusca; (**G**) Castillo.

Dactyloidites ottoi (Geinitz, 1849) in [16] (Figure 2B) forms nearly complete rosette structures composed of outwardly radiating, horizontal to subhorizontal cylindrical elements, occasionally showing spreiten displacement. It is interpreted as a feeding structure [13], most likely produced by polychaete worms [17–19].

Diopatrichnus santamariaensis Uchman, Quintino & Rodrigues, 2017, in [20] (Figure 2C) is a vertical to inclined tube, armored predominantly with crushed or uncrushed bivalve shell valves. The valves are arranged concave-upward and oriented mostly perpendicular to the burrow axis. It is attributed to the activity of polychaetes, comparable to the burrowing behavior of extant *Owenia* [20].

Macaronichnus segregatis Clifton & Thompson, 1978 in [21] (Figure 2D) is a variably oriented, usually straight to winding cylindrical burrow characterized by a light-grained core surrounded by a mantle of darker grains. It is produced by opheliid polychaetes that segregate sand grains, ingesting the lighter fraction—presumably richer in microbial content—while rejecting darker grains [21–24]. Two ichnosubspecies are recognized: the smaller *M. s. segregatis* and the larger *M. s. degiberti* [23].

Ophiomorpha nodosa Lundgren, 1891 in [25] (Figure 2E) is a tubular, rarely branched burrow with walls bearing a distinctly knobby external texture. In some cases, abundant interconnecting branches form a boxwork network. It is interpreted as a dwelling and feeding, or facultative farming, structure produced by decapod crustaceans [9,26–30].

Palaeophycus isp. consists of predominantly horizontal, tubular burrows with smooth walls. It is interpreted as a dwelling and feeding structure, produced by various deposit-feeding or carnivorous invertebrates, most likely polychaetes [31,32].

Piscichnus waitemata Gregory, 1991 in [33] (Figure 2F) is a bowl-shaped depression interpreted as a feeding trace, produced primarily by rays and, in some cases, by mammals hunting invertebrates within the sediment [33].

Thalassinoides isp. (Figure 2G) comprises predominantly horizontal, cylindrical, branched burrows that may form complex networks. It is interpreted as a semi-permanent dwelling and feeding structure, produced mainly by obligatory or facultative deposit-feeding decapod crustaceans [9,34,35].

2.2. Marine Borings

Caulostrepsis ispp. (Figure 3A) are U-shaped galleries, commonly interconnected by a vane and forming pouch-shaped structures. They are borings with a single entrance or embedment structure, wider in the distal portion than at the apertural part and typically display elliptical to dumbbell-shaped cross sections. These structures are produced mainly by spionid polychaetes of the genus *Polydora*, and less frequently by some eunicid polychaetes [36,37].

Centrichnus dentatus Uchman, Wisshak, Madeira, Melo, Sacchetti, Ávila & Ávila, 2025 in [38] (Figure 3B) is a small, symmetric or asymmetric depression on shells, circular or elliptical in outline, and bounded by a groove and/or a series of pits. It is interpreted as an attachment trace produced by verrucid barnacles [38].

Circolites kotoucensis Mikuláš, 1992 in [39] (Figure 3C) are bowl-shaped structures formed in rocky substrates, including basalts. They are interpreted as traces produced by regular echinoids, and in modern Macaronesia by *Paracentrotus lividus* [39,40].

Entobia ispp. (Figure 3D) are branched systems of galleries composed of small chambers interconnected by tunnels and opening to the surface. They are produced by clionaid sponges in calcareous hard substrates, most commonly in mollusk shells. Several ichnospecies of *Entobia* are distinguished based on morphological details [41–43].

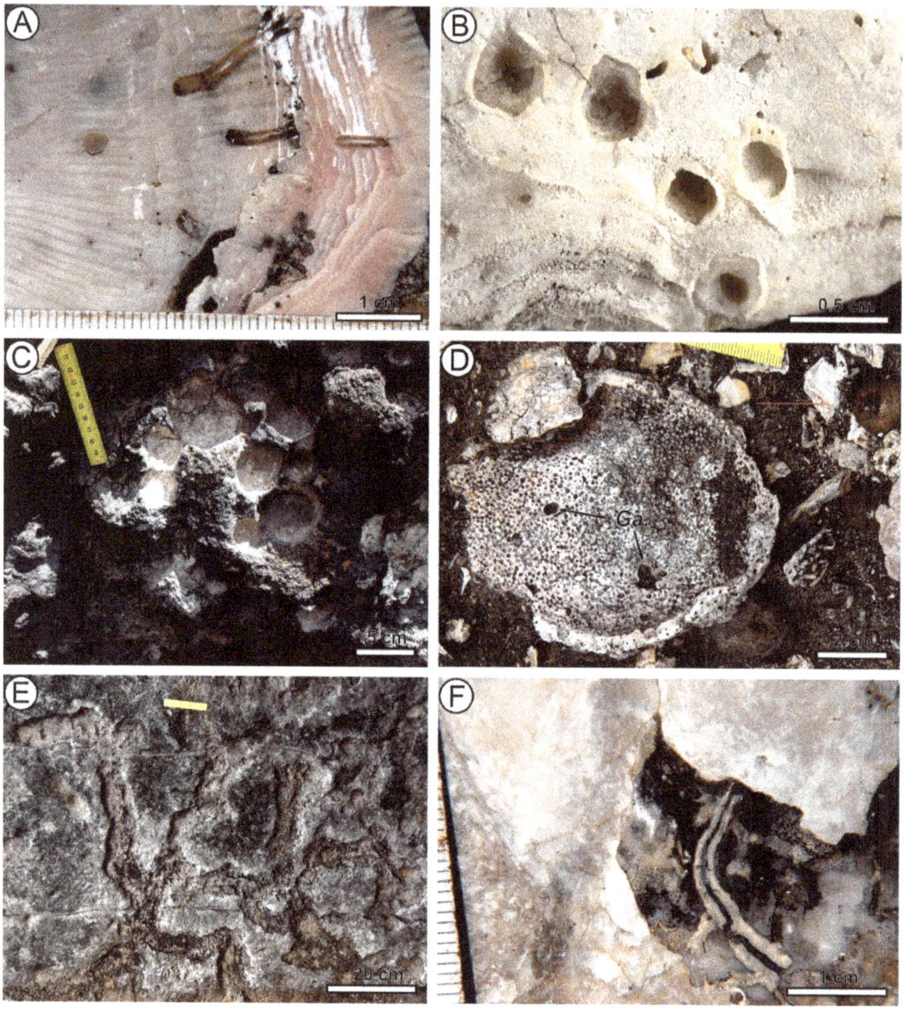

Figure 3. Trace fossils (borings) from Santa Maria, Azores. (**A**) *Caulostrepsis* isp. in *Spondylus* shell valve, Pliocene, Baía de Nossa Senhora. (**B**) *Centrichnus dentatus* in an oyster shell valve, Pliocene, Pdara-que-pica. (**C**) *Circolites kotoucensis* in basalt, Pleistocene, Prainha. (**D**) *Entobia* isp. and *Gastrochaenolits* isp. (*Ga*) in a bivalve shell, Pliocene, Malbusca. (**E**) *Ericichnus bromleyi* in calcarenite, Plaistocene, Ichnofossil Cave. (**F**) *Maendropolydora* isp., Pliocene, Baía de Nossa Senhora.

Ericichnus bromleyi Santos et al., 2015 in [44] (Figure 3E) is an irregularly sinuous groove on the rock surface, which may converge or diverge, and form irregular networks. It is produced by regular echinoids [44].

Gastrochaenolites ispp. (Figure 3D) are flask-shaped cavities in hard substrates, consisting of a chamber and a neck. They are interpreted as borings produced by bivalves [45,46].

Maeandropolydora ispp. (Figure 3F) are long, sinuous, and typically contorted cylindrical galleries in hard calcareous substrates, opening to the surface. The limbs of the galleries may be interconnected. They are interpreted as borings produced by suspension-feeding spionid polychaetes [36,47].

2.3. Trace Fossils in Continental Deposits

Cuniculonomus isp. is a linear, undulating to sinuous leaf mine, of uniform width or with a slight terminal widening [48,49]. It is interpreted as the feeding trace of flies (*Agromyzidae*) or moths (*Gracillariidae*) [50].

Palmiraichnus castellanosi (Roselli, 1987) in [51] is an ovoid to subcylindrical structure with a smooth, shiny chamber. It is straight or slightly curved, concave or truncated on one side, and rounded on the other, where a spiral closure separates the main chamber from the ante-chamber. The ante-chamber may be filled with structureless sediment. This trace fossil occurs in palaeosols and is interpreted as the nesting burrow of solitary bees of the family Andrenidae [52,53].

Phagophytichnus ekowskii Van Amerom, 1966 in [54] consists of semicircular excisions along the margins of plant leaves that do not reach the midrib, forming an incomplete oval to elliptical outline. It is interpreted as a feeding trace produced by butterfly caterpillars or locust larvae [49].

Rebuffoichnus guanche Genise et al., 2013 in [53] (Figure 4A) comprises subovoid to subcylindrical structures composed of a chamber and a wall. The outer wall surface is rough and lumpy, whereas the inner surface is smooth or nearly smooth. The chamber is ellipsoid in outline and circular in cross section. The wall may exhibit a rounded opening [55].

Figure 4. Some continental trace fossils and biogenic structures. (**A**) specimens of *Rebuffoichnus guanche*. Top view in the upper row and side view in the lower row. Pleistocene, El Altillo, Gran Canaria, Canary Islands. (**B**) Root structures in Quaternary aeolianites, road on side of the road from Deserto de Viana to Ervatão, Boa Vista island, Cape Verde.

Xylonichnus trypetus Genise, 1995 in [55] consists of a boxwork of anastomosing longitudinal, tangential, and radial tunnels, rectangular to circular in cross section, developed at various depths within the xylem. The tunnels may be filled with frass. This trace is interpreted as the work of buprestid beetles or termites [49].

3. Ichnological Investigations in Macaronesia

3.1. Burrows in Marine Deposits

Exposed marine deposits in the Macaronesian islands are almost exclusively restricted to shallow-marine Miocene–Holocene facies, influenced primarily by wave and storm activity. These deposits, typically lacking quartz grains, consist mainly of rocks ranging from calcarenites to lithic sandstones, with variable contributions from volcanic-derived components. Conglomerates also occur, mostly in fan deltas and fossil beach deposits. Characteristic depositional architectures include shoreface-to-backshore sheet-like covers or patches (at outcrop scale) preserved in cliffs and marine terraces, clinoforms prograding into the open ocean, and fan-delta cones. In several cases, sedimentary lithosomes are encapsulated between lava flows. Trace fossils are common and locally abundant in Miocene and Pleistocene calcarenite–sandstone exposures. The most widespread ichnotaxa include *Macaronichnus segregatis*, *Bichordites monastiriensis*, *Thalassinoides* isp., *Ophiomorpha nodosa*, and *Dactyloidites ottoi*, with other forms occurring more sporadically.

The first comprehensive ichnological study of marine sediments on volcanic islands was conducted by Mayoral et al. [56], who examined Miocene and Pleistocene deposits of the Cape Verde Archipelago. Some ichnotaxa had already been reported in related papers from the same period [57–60]. For example, *Sinusichnus* isp. was documented from São Nicolau Island [60]. Mayoral et al. [56] recognized two principal ichnoassociations: (1) a Miocene *Thalassinoides* ichnoassociation (restricted to Maio Island) and (2) a Pleistocene *Macaronichnus–Dactyloidites* ichnoassociation (Santiago, Maio, São Nicolau). Within the latter, mostly horizontal trace fossils such as *Dactyloidites ottoi*, *Dactyloidites* isp., *Bichordites monastiriensis*, *Bichordites* isp., *Cardioichnus* isp., *Phycodes* isp., *Teichichnus rectus*, aff. *Taenidium* isp., and *Palaeophycus tubularis* were assigned to the proximal Cruziana ichnofacies. In contrast, *Macaronichnus* and vertical burrows including *Skolithos*, *Conichnus conicus*, *Ophiomorpha nodosa*, and *Ophiomorpha* isp. were referred to the Skolithos ichnofacies. Importantly, based on the geometric relationship between submarine and subaerial lava flows, as well as the stratigraphic distribution of ichnofossil-bearing deposits, Mayoral et al. [56] calibrated the transition from the proximal Cruziana ichnosubfacies to the Skolithos ichnofacies at water depths of approximately 12–18 m.

The Miocene *Thalassinoides* ichnoassociation comprises *Thalassinoides suevicus*, *Thalassinoides* isp., *Conichnus conicus*, *Dactyloidites ottoi*, and *Macaronichnus segregatis*, and belongs to the proximal Cruziana ichnofacies. In addition, a Psilonichnus ichnofacies, represented by a monospecific assemblage of *Psilonichnus upsilon* (produced by crabs), was recognized in a Pleistocene section of Maio Island, indicating an upper foreshore–backshore setting. At the Nossa Senhora da Luz section (Santiago), protected-bay sediments contain almost exclusively *Thalassinoides* [61].

Mayoral et al. [62] described thin pockets of Upper Pleistocene biocalcirudites encapsulated between lava flows on Santiago Island. These contained *Thalassinoides*, *Rhizocorallium commune irregulare*, *Rhizocorallium jenense versum* (a new ichnosubspecies), *Alaichnus kabuverdiensis* (a new ichnogenus and ichnospecies, interpreted as a cumulative trace of bivalve siphons), and cf. *Dactyloidites*. Taphonomic analysis allowed interpretation of erosion–sedimentation events, colonization sequences, and estimates of erosion depth during deposition.

In the Canary Islands, ichnological investigations are less numerous. Sánchez-Pinto et al. [63] reported *Dactyloidites ottoi* from Pliocene deposits of Fuerteventura, also mentioned by Martín González [64] and Martín González & Castillo [65]. Schneider et al. [66] described Skolithos ichnofacies traces in Miocene–Pliocene fan-delta deposits of the Cuevas los Guinchos section, Las Palmas Detritic Formation, Gran Canaria. Mayoral et al. [67] recognized *Rosselia socialis*, *Macaronichnus segregatis*, *Bichordites monastiriensis*, and *Ophiomor-*

pha nodosa in the same formation, assigning them to the Skolithos and proximal Cruziana ichnofacies. Uchman et al. [68] reinterpreted these assemblages in terms of sedimentation dynamics within a transgressive–regressive cycle of fan-delta progradation, rather than as indicators of bathymetric change. At the Las Rehoyas section of the same formation, *Macaronichnus segregatis, Ophiomorpha nodosa, Rosselia socialis, Skolithos linearis*, and indeterminate *Thalassinoides* have been described, assigned to the Skolithos ichnofacies, and interpreted as evidence of seafloor stabilization within sand-dominated intervals of a conglomeratic fan-delta system [69]. *Ophiomorpha nodosa* was described from Mio-Pliocene deposits in a few localities of Gran Canaria and Fuertenventura [70–72].

By contrast, extensive ichnological studies have been carried out on Santa Maria Island (Azores), well known for its fossiliferous formations. Meireles et al. [73] reported *Diplocraterion, Crossopodia, Rhizocorallium, Ophiomorpha*, and *Thalassinoides* from storm-influenced Miocene–Pliocene sandy deposits of the Castello section, though the presence of the first three ichnotaxa is doubtful based on later observations. In the Ichnofossil Cave, Santos et al. [44] documented *Arenicolites* isp., *Bichordites* isp., *Macaronichnus segregatis, Dactyloidites ottoi*, and *Ophiomorpha nodosa* from Miocene–Pliocene deposits, see also [74]. In the Pedra-que-pica section, associated with an exceptional coquina, assemblages include *Macaronichnus segregatis*, ?*Ophiomorpha* isp. (misidentified as *Thalassinoides* isp. by Kirby et al. [75]), *Asterosoma* isp., and *Bichordites* isp. [76,77]. These traces helped refine the palaeoenvironmental interpretation and stratigraphic position of the coquina.

Uchman et al. [78,79] described unusual vertically oriented *Macaronichnus segregatis* from Miocene–Pliocene calcarenites and sandstones of Santa Maria. These deposits, formed on an erosionally truncated island surface and prograding clinoforms on its southern coast, are interpreted as reflecting vertical migration of trace makers following the mixing zone of fresh and marine waters. This ichnotaxon cooccurs with *Piscichnus* isp., *Diopatrichnus* isp., ?*Palaeophycus* isp., ?*Parmaichnus* isp., *Thalassinoides* isp., *Ophiomorpha nodosa*, ?*Psilonichnus* isp., *Asterosoma* isp., and *Dactyloidites ottoi*, which belong to the Psilonichnus, Skolithos, and Cruziana ichnofacies, confirming shallow-water deposition [80]. A similar assemblage, including ?*Scolicia* isp. and ?*Curvolithus* isp., is linked to a major hurricane bed (Figure 5) in the Malbusca section [81]. *Diopatrichnus santamariaensis* [20], a new ichnospecies from Santa Maria, is a vertical tube armored with shell fragments, interpreted as the trace of owenid polychaetes. *Piscichnus waitemata*, produced mainly by hunting rays, is abundant in the sediment cover of the truncated island but less common in clinoform deposits on the island's flanks, probably due to lower temperatures and reduced preservation potential [82]. Subtidal clinoforms encapsulated between lava flows on the eastern margin of the island yielded *Thalassinoides, Scolicia* isp., *Palaeophycus* isp., *Diopatrichnus santamariaensis, Macaronichnus segregatis, Ophiomorpha* isp., and *Piscichnus waitemata*, which refine interpretations of depositional environments [83]. The presence of crustacean burrows (*Ophiomorpha* isp., *Ophiomorpha nodosa, Thalassinoides* isp., ?*Psilonichnus* isp., ?*Parmaichnus* isp.) has further clarified the diversity of burrowing decapods in the Pliocene–Pleistocene of Santa Maria [84].

In contrast, information about fossil burrows from Madeira and the Salvagens is scarce. From middle Miocene calcarenites of the Madeira Archipelago, *Thalassinoides* was reported from Ilhéu de Cima [40], and *Bichordites* isp. and *Dactyloidites* isp. from Ilhéu de Baixo [85].

Figure 5. Major hurricane bed (Pliocene calcareous sandy deposits) in between volcanic and volcanogenic rocks at Malbusca, Santa Maria, Azores.

3.2. Marine Bioerosion Structures

Marine bioerosion structures in Macaronesia occur mostly on rocky substrates—typically calcareous, more rarely volcanic—and on skeletal elements, especially mollusk shells.

In the Cape Verde Archipelago, Santos et al. [86] reported *Gastrochaenolites* isp. in Pliocene–Pleistocene limestones, corals, and basalt boulders at Tarrafal outcrop and Ponta das Bicudas (Santiago Island). Mayoral et al. [56] described *Circolites kotoucensis* (echinoid boring, also found in basalt), bivalve borings (*Gastrochaenolites torpedo*, *G. cluniformis*, *Gastrochaenolites* isp.), and sponge borings (*Entobia* isp.) developed on the Miocene–Pleistocene unconformity of Maio Island and on basaltic paleocliffs at Ponta das Bicudas, Santiago, see also [62]. These assemblages belong to the *Entobia* ichnofacies. Baarli et al. [61] also documented *Gastrochaenolites* isp. and *C. kotoucensis* on Santiago, noting that *Gastrochaenolites* penetrated through hydrozoan *Millepora alcicornis* into the underlying basalt. *C. kotoucensis* has also been recorded from São Nicolau [44,59], Sal, and Santo Antão (in basalts), and Brava (in phonolites) [74,87].

In the Canary Islands, Martín-González et al. [88] reported *Entobia*, *Gastrochaenolites*, and probable vermetid borings in Pliocene mollusk shells from Fuerteventura. Verde et al. [89] introduced *Santichnus mayorali* as a new ichnogenus and ichnospecies representing the attachment trace of vermetid gastropods, identified in Miocene–Pliocene mollusk shells from Fuerteventura and La Graciosa (near Lanzarote). *Circolites kotoucensis*, *Gastrochaenolites torpedo*, *Gastrochaenolites* isp., and *Maeandropolydora sulcans*, *Caulostrepsis taeniola*, *Entobia ovunamesla*, and *Entobia* isp. were described from Mio-Pliocene deposits of Gran Canaria and Fuerteventura [70], and *Entobia* from the Pleistocene of Fuerteventura [90].

In the Madeira Archipelago, Santos et al. [91] documented *Gastrochaenolites torpedo*, *G. vivus*, *G. lapidicus*, and *Entobia* isp. in Miocene algal crusts and corals on a basaltic rocky shore at Ilhéu de Cima near Porto Santo; notably, *G. lapidicus* also penetrated into basalt. Subsequent work at the same locality revealed *G. ornatus*, which likewise entered basalt [86]. Additional *Gastrochaenolites* isp. were found in Miocene corals and rhodoliths [92]. Santos et al. [93] also described *Imbutichnus costatum*, a bioclaustration structure resulting from overgrowth of a Miocene hermatypic coral on a pyrgomatid barnacle. However, bioclaustration features are no longer regarded as trace fossils [94,95].

In the Azores, records are concentrated on Santa Maria Island, with only rare finds from other islands. Boring traces by regular echinoids—now recognized as *Circolites kotoucensis*—were first reported from Holocene basalts of São Miguel [96,97]. Ávila et al. [98] described Pleistocene *Circolites kotoucensis* (Figure 2C) in basalts and *Gastrochaenolites* produced by *Leiosolenus ariestatus* in algal limestones at Prainha (Santa Maria). In the Ichnofossil Cave (Santa Maria), Santos et al. [44] distinguished two new ichnospecies of echinoid grazing grooves, *Ericichnus bromleyi* (Figure 3E) and *E. asgaardi*, as well as *C. kotoucensis*. Similar grooves were reported from tuffites in Sal (Cape Verde). Ávila et al. [74,77]) noted *Gastrochaenolites* isp., *Entobia* isp., and *Caulostrepsis* isp. in bivalve shells from the Pliocene coquina of the Pedra-que-pica geosite and the Ponta do Cedro section [83]. *Gastrochaenolites* and *Entobia* also occur in a paleocliff below a large storm bed at Malbusca [81] and in shells armoring *Diopatrichnus santamariaensis* (Figure 2C) [20]. *Gastrochaenolites* cf. *torpedo* (produced by *Leiosolenus ariestatus*) was reported in bioclasts, rhodoliths, and algal encrustations at Ponta do Cedro [83]. This boring, along with *Entobia*, is also present in rhodoliths of Malbusca [81] and in Pleistocene (MIS 5e) algal buildups [99].

A new ichnospecies, *Centrichnus dentatus* (Figure 3B), was described from the Pedra-que-pica coquina [38]. Produced by the barnacle *Verruca spengleri* in oyster shells, it co-occurs with *Talpina* isp., *Entobia* isp., *Maeandropolydora* isp., and *Gastrochaenolites* isp. The assemblage was assigned to the *Gnathichnus* ichnofacies, reflecting early colonization of hard substrates, often within months.

Dávid et al. [100] recorded 22 ichnotaxa of bioerosion structures in Lower Pliocene deposits of the Baía de Nossa Senhora section (Santa Maria). The assemblage is dominated by well-preserved *Entobia*, but also includes *Gastrochaenolites*, *Caulostrepsis*, *Maeandropolydora*, *Trypanites* (sipunculid worms), *Talpina* (phoronid worms), and *Iramena* (ctenostome bryozoans). These borings, belonging to the *Entobia* ichnofacies, closely resemble assemblages from the Paratethys, Mediterranean, and eastern Atlantic regions—consistent with the general Neogene shallow-water faunal affinities of the Azores.

Modern bioerosion has also been studied. Wisshak et al. [101] conducted in situ experiments on the submarine slope of Faial Island, spanning intertidal to bathyal depths. They identified 43 ichnotaxa, including microborings produced by cyanobacteria, chlorophytes, fungi, and other chemotrophs; attachment scars of diatoms, foraminifera, bryozoans, cirripeds, see also [38], gastropods, bivalves, and brachiopods; grazing traces from gastropod radulae and echinoids; and several macroborings (*Entobia* isp., *Caulostrepsis taeniola*, *Maeandropolydora decipiens*, *Talpina ramosa*).

3.3. Trace Fossils in Continental Environments

Insect nests have been reported from the Canary Islands for over a century, primarily in Pleistocene and Holocene sandy dunes, sandy silts associated with palaeosols, and calcretes of Lanzarote and Fuerteventura [53,102–112]. These structures were initially attributed to *Celliforma* and interpreted as nests of solitary bees. Genise [113], however, suggested alternative interpretations as coleopteran pupal chambers or lepidopteran cocoons.

Further studies by Meco et al. [114,115] described mass occurrences of these traces in the Pliocene, Pleistocene, and Holocene of Fuerteventura, interpreting them as evidence of episodic locust invasions. Later analyses of material from the island recognized bee nests attributed to *Palmiraichnus* or to an undescribed ichnogenus of the Celliformidae, as well as coleopteran pupal cells (*Rebuffoichnus*), both produced under semiarid paleoclimatic conditions during wetter intervals [116]. *Rebuffoichnus* isp. was subsequently documented in association with megarhizoliths in Pleistocene aeolianites of Gran Canaria [117]. Genise et al. [53] introduced the new ichnospecies *Rebuffoichnus guanche*, interpreted as pupation chambers of coleopterans (Curculionidae or Scarabaeidae). Moreover, *R. casamiquelai* Roselli, 1987 [51] is recognized in Fuerteventura and Lanzarote [118]. *Palmiraichnus castellanosi* has also been reported from the Pleistocene and Holocene of Fuerteventura,

Lanzarote, and Gran Canaria [52]. Some *Rebuffoichnus guanche* specimens show holes or pits in their walls, which belong to the ichnogenus *Tombownichnus* Mikúlaš & Genise, 2003 [119], with two ichnospecies *T. plenus*, *T. parabolicus* produced by some parasitic insects [118].

The distribution of insect nests across the archipelago is uneven. Genise et al. [53] noted abundant occurrences in the eastern islands (Fuerteventura and Lanzarote) but only sporadic records in the western islands (Tenerife, La Palma, La Gomera, El Hierro), where suitable deposits are scarce. Comparable fossils from the more remote Macaronesian archipelagos remain undocumented.

A more recent ichnological development in Macaronesia concerns traces of plant–insect interactions. Pokorný & Borges [49] reported *Phagophytichnus ekowskii* (marginal feeding traces on leaves), *Cuniculonomus* isp. (leaf-mining traces; see also [120]), and *Xylonichnus trypetus* (wood borings) from Quaternary plant collections in the Azores (São Miguel, Terceira, Faial). These findings have since been critically reviewed by Gois-Marques et al. [121]. These traces belong to the Palaeoscolytus ichnofacies.

Borings produced by insects in bones have been examined experimentally in Gran Canaria [122]. However, no evidence of insect borings in fossil bones has yet been reported from Macaronesia.

Plant root structures occupy a conceptual "grey zone" between trace fossils, plant body fossils, and diagenetic features (Figure 4B). Root-induced deformations can represent genuine bioturbation, but differentiating them from abiotic structures is often challenging [94,95]. This issue has been addressed in studies of Pleistocene root structures in Gran Canaria and Tenerife [117,123–125]. Similar interpretative difficulties have been noted for rhizobial nodules from the Pleistocene of Madeira [126].

4. Concluding Remarks

The review demonstrates that ichnological research in Macaronesia has made substantial progress over the past two decades, establishing the archipelago as the best-ichnologically studied among volcanic islands. Many islands have been investigated for the first time, often through initiatives led by the University of the Azores. Consequently, the known list of ichnotaxa has expanded considerably, including numerous new ichnogenera and ichnospecies, and understanding of the distribution of trace fossils and their palaeoenvironmental significance has improved markedly. These studies also show that a variety of burrowing and boring animals can colonize newly emerged volcanic islands, even when their extant representatives are not present locally.

Nevertheless, important gaps remain. Microborings in the fossil record of Macaronesia have not yet been studied. Continental ichnology is largely restricted to insect nests in the Canary Islands [52,53] and plant–insect interaction traces [49]. Neoichnological research—such as studies of bioerosion traces [101] and ray-fish burrows [82]—has begun, but much more remains to be done. Fossil burrows are entirely unknown from the Madeira and Salvagens archipelagos. It is also unclear whether the widespread occurrence of the ray-fish hunting burrow *Piscichnus waitemata* in Santa Maria extends to other islands. In general, geographic trends in trace fossil distribution are poorly understood, and several islands remain completely unexplored ichnologically.

These gaps highlight the need for continued ichnological studies, integrated with other palaeobiological and sedimentological research, to better understand the colonization history and palaeoecology of biota on Macaronesia's volcanic islands.

Funding: The fieldwork of the author on Santa Maria Island, Azores (2014–2019, 2021, 2013), was supported by the International Workshop Palaeontology in Atlantic Islands. Fieldwork in the Canary Islands (2023) was supported by the PAI2023-CAN International Workshop, sponsored by the Regional Directorate for Science and Technology, Azores Government, through the project Base de Dados da PaleoBiodiversidade da Macaronésia (M1.1.A/INFRAEST CIENT/A/001/2021), within the funding program PRO-SCIENTIA: Eixo 1—Valorizar, Ação 1.1—Capacitar as entidades do SCTA e valorizar as suas atividades, and by the Museums and Centers Autonomous Organism of the Cabildo of Tenerife. Fieldwork in Cape Verde (2024) was supported by FEDER funds through the Operational Program for Competitiveness Factors—COMPETE, and by national funds through FCT (UIDB/50027/2020; POCI-01-0145-FEDER-006821; UIDB/00153/2020; LA/P/0048/2020), as well as through the Regional Government of the Azores (M1.1.a/005/Funcionamento-C-/2016, CIBIO-A; M1.1.A/INFRAEST CIENT/A/001/2021; M3.3.B/ORG.R.C./005/2021). This work also received support from FEDER funds (85%) and the Regional Government of the Azores (15%) through project M1.1.A/INFRAEST CIENT/A/001/2021—Base de Dados da PaleoBiodiversidade da Macaronésia. Additional support was provided by a grant from the Faculty of Geography and Geology under the Strategic Programme Excellence Initiative at Jagiellonian University.

Data Availability Statement: No new data were created or analyzed in this study. Data sharing is not applicable to this article.

Conflicts of Interest: The author declares no conflicts of interest.

References

1. Ramalho, R.S.; Quartau, R.; Trenhaile, A.S.; Mitchell, N.C.; Woodroffe, C.D.; Ávila, S.P. Coastal evolution on volcanic oceanic islands: A complex interplay between volcanism, erosion, sedimentation, sea level change and biogenic production. *Earth-Sci. Rev.* **2013**, *127*, 140–170. [CrossRef]
2. Johnson, M.E.; Baarli, B.G.; Cachão, M.; Mayoral, E.; Ramalho, R.S.; Santos, A.; da Silva, C.M. On the rise and fall of oceanic islands: Towards a global theory following the pioneering studies of Charles Darwin and James Dwight Dana. *Earth-Sci. Rev.* **2018**, *180*, 17–36. [CrossRef]
3. Ávila, S.P.; Madeira, P.; Zazo, C.; Kroh, A.; Kirby, M.; da Silva, C.M.; Cachão, M.; Martins, A.d.F. Palaeoecology of the Pleistocene (MIS 5.5) outcrops of Santa Maria Island (Azores) in a complex oceanic tectonic setting. *Palaeogeogr. Palaeoclimatol. Palaeoecol.* **2009**, *274*, 18–31. [CrossRef]
4. Ávila, S.P.; Ramalho, R.; Vullo, R. Systematics, palaeoecology and palaeobiogeography of the Neogene fossil sharks from the Azores (Northeast Atlantic). *Ann. Paleontol.* **2012**, *98*, 167–189. [CrossRef]
5. Ávila, S.P.; Azevedo, J.M.; Madeira, P.; Cordeiro, R.; Melo, C.S.; Baptista, L.; Torres, P.; Johnson, M.E.; Vullo, R. Pliocene and late Pleistocene actinopterygian fishes from Santa Maria Island, Azores (NE Atlantic Ocean): Palaeoecological and palaeobiogeographical implications. *Geol. Mag.* **2020**, *157*, 1526–1542. [CrossRef]
6. Freitas, R.; Romeiras, M.; Silva, L.; Cordeiro, R.; Madeira, P.; González, J.A.; Wirtz, P.; Falcón, J.M.; Brito, A.; Floeter, S.R.; et al. Restructuring of the 'Macaronesia'biogeographic unit: A marine multi-taxon biogeographical approach. *Sci. Rep.* **2019**, *9*, 15792. [CrossRef]
7. Florencio, M.; Patiño, J.; Nogué, S.; Traveset, A.; Borges, P.A.; Schaefer, H.; Amorim, I.R.; Arnedo, M.; Ávila, S.P.; Cardoso, P.; et al. Macaronesia as a fruitful arena for ecology, evolution, and conservation biology. *Front. Ecol. Evol.* **2021**, *9*, 718169. [CrossRef]
8. Ávila, S.P.; Paris, R.; Ramalho, R.S.; Melo, C.S.; Martín-González, E.; Rolán, E.; Madeira, P.; Ávila, G.C.; Porteiro, J.M.; Medeiros, A.M.; et al. Megatsunami deposits and range expansion of cold-temperate marine species towards the tropics in glacial times. *Front. Biogeogr.* **2025**, *18*, e138319. [CrossRef]
9. Bromley, R.G. *Trace Fossils: Biology, Taphonomy and Applications*; Chapman and Hall: London, UK, 1996; 361p.
10. Buatois, L.A.; Mángano, M.G. *Ichnology: Organism–Substrate Interactions in Space and Time*; Cambridge University Press: Cambridge, UK, 2011.
11. Knaust, D.; Bromley, R.G. (Eds.) *Trace Fossils as Indicators of Sedimentary Environments*; Developments in Sedimentology; Elsevier: Amsterdam, The Netherlands, 2012; Volume 64.
12. Plaziat, J.-C.; Mahmoudi, M. Trace fossils attributed to burrowing echinoids: A revision including new ichnogenus and ichnospecies. *Geobios* **1988**, *21*, 209–233. [CrossRef]
13. Uchman, A. Taxonomy and palaeoecology of flysch trace fossils: The Marnoso-arenacea Formation and associated facies (Miocene, Northern Apennines, Italy). *Beringeria* **1995**, *15*, 3–115.
14. Uchman, A.; Krenmayr, H.G. Trace fossils from Lower Miocene (Ottnangian) molasse deposits of Upper Austria. *Paläontol. Z.* **1995**, *69*, 503–524. [CrossRef]

15. D'Alessandro, A.; Uchman, A. *Bichordites* and *Bichordites–Rosselia* ichnoassemblages from the Lower Pleistocene Tursi Sandstone (southern Italy). In *Sediment–Organism Interactions: A Multifaceted Ichnology*; Bromley, R.G., Buatois, L.A., Mángano, M.G., Genise, J.F., Melchor, R.N., Eds.; SEPM (Society for Sedimentary Geology) Special Publication 88; SEPM Society for Sedimentary Geology: Tulsa, OK, USA, 2007; pp. 213–221.
16. Geinitz, H.B. *Das Quadersandsteingebirge oder Kreidegebirge in Deutschland*; Graz & Gerlach: Freiberg, Germany, 1849–1850; 292p.
17. Fürsich, F.T.; Bromley, R.G. Behavioural interpretation of a rosetted spreite trace fossil: *Dactyloidites ottoi* (Geinitz). *Lethaia* **1985**, *18*, 199–207. [CrossRef]
18. de Gibert, J.M.; Martinell, J.; Domènech, R. The rosetted feeding trace fossil *Dactyloidites ottoi* (Geinitz) from the Miocene of Catalonia. *Geobios* **1995**, *28*, 769–776. [CrossRef]
19. Curran, H.A.; Glumac, B. Dactyloidites ottoi (Geinitz, 1849) in Bahamian Pleistocene carbonates: A shallowest-marine indicator. In *The Ichnology of Shallow-Marine and Transitional Environments*; Cónsole-Gonella, C., de Valais, S., Díaz-Martínez, I., Citton, P., Verde, M., McIlroy, D., Eds.; Special Publications; Geological Society: London, UK, 2023. [CrossRef]
20. Uchman, A.; Quintino, V.; Rodrigues, A.M.; Johnson, M.E.; Melo, C.; Cordeiro, R.; Ramalho, R.S.; Ávila, S.P. The trace fossil *Diopatrichnus santamariensis* isp. nov. —A shell-armored tube from Pliocene sediments of Santa Maria Island, Azores (NE Atlantic Ocean). *Geobios* **2017**, *50*, 459–469. [CrossRef]
21. Clifton, H.E.; Thompson, J.K. *Macaronichnus segregatis*: A feeding structure of shallow marine polychaetes. *J. Sediment. Petrol.* **1978**, *48*, 1293–1302. [CrossRef]
22. Nara, M.; Seike, K. *Macaronichnus segregatis*-like traces found in the modern foreshore sediments of the Kujukurihama Coast, Japan. *J. Geol. Soc. Jpn.* **2004**, *110*, 545–551. [CrossRef]
23. Nara, M.; Seike, K. Palaeoecology of *Macaronichnus segregatis degiberti*: Reconstructing the infaunal lives of the travisiid polychaetes. *Palaeogeogr. Palaeoclimat. Palaeoecol.* **2019**, *516*, 284–294. [CrossRef]
24. Seike, K.; Yanagishima, S.; Nara, M.; Sasaki, T. Large *Macaronichnus* in modern shoreface sediments: Identification of the producer, the mode of formation, and palaeoenvironmental implications. *Palaeogeogr. Palaeoclimatol. Palaeoecol.* **2011**, *311*, 224–229. [CrossRef]
25. Lundgren, B. Studier öfver fossilförande lösa block. *Geol. Föreningen I Stockh. Förhandlinger* **1891**, *13*, 111–121. [CrossRef]
26. Knaust, D. *Atlas of Trace Fossils in Well Core: Appearance, Taxonomy and Interpretation*; Springer: Cham, Switzerland, 2017; 271p.
27. Frey, R.W.; Howard, J.D.; Pryor, W.A. *Ophiomorpha*: Its morphologic, taxonomic, and environmental significance. *Palaeogeogr. Palaeoclimatol. Palaeoecol.* **1978**, *23*, 199–229. [CrossRef]
28. Bromley, R.G. *Trace Fossils. Biology and Taphonomy*; Special Topics in Palaeontology Series; Unwin Hyman: London, UK; Boston, MA, USA; Sydney, Australia; Wellington, New Zealand, 1990; 280p.
29. Schlirf, M. Upper Jurassic trace fossils from the Boulonnais (northern France). *Geol. Palaeontol.* **2000**, *34*, 145–213.
30. Chrząstek, A. Palaeoenvironmental interpretation of the Late Cretaceous Idzików Conglomerate Member (SW Poland, Sudetes, Idzików Quarry) based on analysis of trace fossils. *Ann. Soc. Geol. Pol.* **2020**, *90*, 149–194. [CrossRef]
31. Pemberton, S.G.; Frey, R.W. Trace fossil nomenclature and the *Planolites–Palaeophycus* dilemma. *J. Paleontol.* **1982**, *56*, 843–881.
32. Keighley, D.G.; Pickerill, R.K. The ichnotaxa *Palaeophycus* and *Planolites*: Historical perspectives and recommendations. *Ichnos* **1995**, *3*, 301–309. [CrossRef]
33. Gregory, M.R. New trace fossils from the Miocene of Northland, New Zealand: *Roschachichnus amoeba* and *Piscichnus waitemata*. *Ichnos* **1991**, *1*, 195–205. [CrossRef]
34. Frey, R.W.; Curran, A.H.; Pemberton, G.S. Trace-making activities of crabs and their environmental significance: The ichnogenus *Psilonichnus*. *J. Paleontol.* **1984**, *58*, 333–350.
35. Giannetti, A.; Monaco, P.; Caracuel, J.E.; Soria, J.; Yébenes, A. Functional morphology and ethology of decapod crustaceans gathered by *Thalassinoides* branched burrows in Mesozoic shallow-water environments. *Mem. Soc. Ital. Sci. Nat. Mus. Civ. Stor. Nat. Milano* **2007**, *35*, 48–52.
36. Bromley, R.G.; D'Alessandro, A.S. Bioerosion in the Pleistocene of southern Italy: Ichnogenera *Caulostrepsis* and *Maeandropolydora*. *Riv. Ital. Paleontol. Stratigr.* **2020**, *89*, 283–309.
37. Gaaloul, N.; Uchman, A.; Ben Ali, S.; Janiszewska, K.; Stolarski, J.; Kołodziej, B.; Riahi, S. In vivo and post-mortem bioerosion traces in solitary corals from the upper Pliocene deposits of Tunisia. *Acta Palaeontol. Pol.* **2023**, *68*, 659–681. [CrossRef]
38. Uchman, A.; Wisshak, M.; Madeira, P.; Melo, C.S.; Sacchetti, C.; Ávila, G.C.; Ávila, S.P. A new attachment trace of a verrucid barnacle on Pliocene bivalve shells, Santa Maria Island, Azores (Central Atlantic). *Acta Palaeontol. Pol.* **2025**, *70*, 143–157. [CrossRef]
39. Mikuláš, R. Early Cretaceous borings from Štramberk (Czechoslovakia). *Čas. Mineral. Geol.* **1992**, *37*, 297–312.
40. Johnson, M.E.; da Silva, C.M.; Santos, A.; Baarli, B.G.; Cachão, M.; Mayoral, E.J.; Rebelo, A.C.; Ledesma-Vázquez, J. Rhodolith transport and immobilization on a volcanically active rocky shore: Middle Miocene at Cabeço das Laranjas on Ilhéu de Cima (Madeira Archipelago, Portugal). *Palaeogeogr. Palaeoclimatol. Palaeoecol.* **2011**, *300*, 113–127. [CrossRef]
41. Bromley, R.G. Borings as trace fossils and *Entobia cretacea* Portlock, as an example. *Geol. J.* **1970**, *3*, 49–90.

42. Bromley, R.G.; D'Alessandro, A. Ichnological study of shallow marine endolithic sponges from the Italian coast. *Riv. Ital. Paleontol. Strat.* **1989**, *95*, 279–340.
43. Garilli, V.; Dávid, Á.; Dominici, S. Natural casts of Entobia from the late Caenozoic of Sicily. *Riv. Ital. Paleontol. Strat.* **2022**, *128*, 211–228. [CrossRef]
44. Santos, A.; Mayoral, E.; Dumont, C.P.; da Silva, C.M.; Ávila, S.P.; Baarli, B.G.; Cachão, M.; Johnson, M.E.; Ramalho, R.S. Role of environmental change in rock-boring echinoid trace fossils. *Palaeogeogr. Palaeoclimatol. Palaeoecol.* **2015**, *432*, 1–14. [CrossRef]
45. Kelly, S.R.A.; Bromley, R.G. Ichnological nomenclature of clavate borings. *Palaeontology* **1984**, *27*, 793–807.
46. Donovan, S.K. Club-shaped borings: *Gastrochaenolites*. *Geol. Today* **2025**, *41*, 61–64. [CrossRef]
47. de Gibert, J.M.; Domènech, R.; Martinell, J. Bioerosion in shell beds from the Pliocene Roussillon Basin, France: Implications for the (macro) bioerosion ichnofacies model. *Acta Palaeont. Pol.* **2007**, *52*, 783–798.
48. Straus, A. Gallen, Minen und andere Fraßspuren im Pliokän von Willershauen am Harz. *Verh. Bot. Ver. Prov. Brandenbg.* **1977**, *113*, 43–80.
49. Pokorný, R.; Borges, P.A. Uma descoberta interessante de folhas fósseis com vestígios de interações planta-animal nas colecções do Museu Vulcanoespeleológico "Os Montanheiros". *Pingo Lava* **2022**, *44*, 76–78.
50. Robledo, J.M.; Sarzetti, L.C.; Anzótegui, L.M. New records and ichnospecies of linear leaf mines from the Late Miocene–Pliocene from Argentina and the establishment of leaf-mining ichnotaxobases. *Riv. Ital. Paleontol. Stratigr.* **2016**, *122*, 55–70. [CrossRef]
51. Roselli, F.L. Paleoicnología: Nidos de insectos fósiles de la cubertura Mesozoica del Uruguay. *Publ. Mus. Municip. Nueva Palmira* **1987**, *1*, 1–56.
52. La Roche, F.; Genise, J.F.; Castillo, C.; Quesada, M.L.; García-Gotera, C.M.; De la Nuez, J. Fossil bee cells from the Canary Islands: Ichnotaxonomy, palaeobiology and palaeoenvironments of *Palmiraichnus castellanosi*. *Palaeogeogr. Palaeoclimatol. Palaeoecol.* **2014**, *409*, 249–264. [CrossRef]
53. Genise, J.F.; Alonso-Zarza, A.M.; Verde, M.; Meléndez, A. Insect trace fossils in aeolian deposits and calcretes from the Canary Islands: Their ichnotaxonomy, producers, and palaeoenvironmental significance. *Palaeogeogr. Palaeoclimatol. Palaeoecol.* **2013**, *377*, 110–124. [CrossRef]
54. Van Amerom, H.W.J. *Phagophytichnus ekowskii* nov. ichnogen. & nov. ichnosp., eine Missbildung infolge von Insektenfrass, aus dem Spanischen Stephanien (Provinz Leon). *Leidse Geol. Meded.* **1966**, *38*, 181–184.
55. Genise, J.F. Upper Cretaceous trace fossils in permineralized plant remains from Patagonia, Argentina. *Ichnos* **1995**, *3*, 287–299. [CrossRef]
56. Mayoral, E.; Ledesma-Vázquez, J.; Baarli, B.G.; Santos, A.; Ramalho, R.; Cachão, M.; da Silva, C.M.; Johnson, M.E. Ichnology in oceanic islands: Case studies from the Cape Verde Archipelago. *Palaeogeogr. Palaeoclimatol. Palaeoecol.* **2013**, *381*, 47–66. [CrossRef]
57. Johnson, M.E.; Baarli, B.G.; Cachão, M.; da Silva, C.M.; Ledesma-Vázquez, J.; Mayoral, E.J.; Ramalho, R.S.; Santos, A. Rhodoliths, uniformitarianism, and Darwin: Pleistocene and Recent carbonate deposits in the Cape Verde and Canary archipelagos. *Palaeogeogr. Palaeoclimatol. Palaeoecol.* **2012**, *329*, 83–100. [CrossRef]
58. Johnson, M.E.; Baarli, B.G.; da Silva, C.M.; Cachão, M.; Ramalho, R.S.; Ledesma-Vázquez, J.; Mayoral, E.J.; Santos, A. Coastal dunes with high content of rhodolith (coralline red algae) bioclasts: Pleistocene formations on Maio and São Nicolau in the Cape Verde Archipelago. *Aeolian Res.* **2013**, *8*, 1–9. [CrossRef]
59. Johnson, M.E.; Ramalho, R.S.; Baarli, B.G.; Cachão, M.; da Silva, C.M.; Mayoral, E.J.; Santos, A. Miocene–Pliocene rocky shores on São Nicolau (Cape Verde Islands): Contrasting windward and leeward biofacies on a volcanically active oceanic island. *Palaeogeogr. Palaeoclimatol. Palaeoecol.* **2014**, *395*, 131–143. [CrossRef]
60. Baarli, B.G.; Santos, A.G.; Mayoral, E.J.; Ledesma-Vázquez, J.; Johnson, M.E.; da Silva, C.M.; Cachão, M. What Darwin did not see: Pleistocene fossil assemblages on a high-energy coast at Ponta das Bicudas, Santiago, Cape Verde Islands. *Geol. Mag.* **2013**, *150*, 183–189. [CrossRef]
61. Melo, M.; Madeira, J.; Ramalho, R.S.; Rebelo, A.C.; Rasser, M.; González, E.; Uchman, A.; Madeira, P.; Rolán, E.; Silva, L.; et al. Last Interglacial fossiliferous sequences from Santiago Island (Cabo Verde Archipelago): The palaeoecology of the Nossa Senhora da Luz section, a rare example of a protected bay in volcanic oceanic islands. In Proceedings of the EGU General Assembly 2020, EGU2020-10423, Online, 4–8 May 2020. [CrossRef]
62. Mayoral, E.; Santos, A.; Vintaned, J.G.; Ledesma-Vázquez, J.; Baarli, B.G.; Cachão, M.; da Silva, C.M.; Johnson, M.E. Upper Pleistocene trace fossils from Ponta das Bicudas, Santiago, Cape Verde Islands: Systematics, taphonomy and palaeoenvironmental evolution. *Palaeogeogr. Palaeoclimatol. Palaeoecol.* **2018**, *498*, 83–98. [CrossRef]
63. Sanchez-Pinto, L.; García-Talavera, F.; López Rondón, J.; Martín Oval, M. Sobre la presencia del icnofósil *Dactyloidites ottoi* (Geinitz, 1849) en sedimentos neógenos de la costa occidental de Fuerteventura (Islas Canarias). In *Homenaje al Profesor Dr. Wolfredo Wildpret de la Torre*; Beltrán Tejera, E., Afonso-Carrillo, J., García Gallo, A., Rodríguez Delgado, O., Eds.; Instituto de Estudios Canarios: La Laguna, Spain, 2009; Monografía LXXVIII; pp. 229–244.
64. Martín González, E. Paleontología de Canarias: Los yacimientos marinos fósiles. *Macaronesia* **2009**, *11*, 70–87.

65. Martín-González, E.; Castillo, C. Los yacimientos marinos del Neógeno de Lanzarote y Fuerteventura (Islas Canarias). Correlaciones con el sur de la Península Ibérica y otros archipiélagos de la Macaronesia. In *III Congreso Ibérico de Paleontología*; Universidad De Zaragoza: Lisboa, Portugal, 2010; Volume 1, 5p.
66. Schneider, J.L.; Torrado, F.J.P.; Torrente, D.G.; Wassmer, P.; Santana, M.D.C.C.; Carracedo, J.C. Sedimentary signatures of the entrance of coarse-grained volcaniclastic flows into the sea: The example of the breccia units of the Las Palmas Detritic Formation (Mio–Pliocene, Gran Canaria, Eastern Atlantic, Spain). *J. Volcanol. Geotherm. Res.* **2004**, *138*, 295–323. [CrossRef]
67. Mayoral, E.; Santos, A.; Galindo, I.; Martín-González, E.; Mangas, J. Contenido icnológico del Miembro medio de la Formación Detrítica Las Palmas (Mio–Plioceno) en el yacimiento de Cuevas del Guincho, Gran Canaria. In *Libro de Resúmenes, XXXV Jornadas de la SEP*; Martínez-Navarro, B., Ed.; Universidad de Málaga: Málaga, Spain, 2019; pp. 183–185.
68. Uchman, A.; Martín-González, E.; Madeira, J.; Madeira, P.; Hipólito, A.; Ávila, S.P. Ichnofabrics controlled by transgressive-regressive regimes in marine Neogene deposits from Gran Canaria, Canary Islands, Spain. In *XVIII International Ichnofabric Workshop, New Trends in Ichnofabrics Analysis*; Dorador, J., Rodríguez-Tovar, F.J., Eds.; Universidad de Granada: Granada, Spain, 2025; pp. 93–94.
69. Rebelo, C.; Uchman, A.; Johnson, M.E.; Melo, C.S.; Vegas, J.; Galindo, I.; Mayoral, E.J.; Santos, A.; González-Rodríguez, A.; Afonso-Carrillo, J.; et al. Rhodoliths and trace fossils record stabilization of a fan-delta system: An example from the Mio–Pliocene deposits of Gran Canaria (Canary Islands, Spain). *J. Paleogeogr.* **2025**, *14*, 100266. [CrossRef]
70. Betancort Lozano, J.F. Fósiles marinos del Neógeno de Canarias (Colección de la ULPGC): Dos neotipos, catálogo y nuevas aportaciones (sistemática, paleoecología y paleoclimatología). Doctoral Dissertation, Universidat del Las Palmas de Gran Canaria, Las Palmas de Gran Canaria, Spain, 2012; 377p.
71. Tuya, F.; Betancort, J.F.; Haroun, R.; Espino, F.; Lomoschitz, A.; Meco, J. Seagrass paleo-biogeography: Fossil records reveal the presence of Halodule cf. in the Canary Islands (eastern Atlantic). *Aquatic Bot.* **2017**, *143*, 1–7. [CrossRef]
72. Lozano, J.F.B. El patrimonio paleontológico de Gran Canaria: Ventanas abiertas a la historia del Atlántico Norte. In *Actas XV Semana Científica Telesforo Bravo*; Lozano, J.F.B., Ed.; Instituto de Estudios Hispánicos de Canarias: Puerto de la Cruz, Spain, 2020; pp. 41–59.
73. Meireles, R.P.; Quartau, R.; Ramalho, R.S.; Rebelo, A.C.; Madeira, J.; Zanon, V.; Ávila, S.P. Depositional processes on oceanic island shelves: Evidence from storm-generated Neogene deposits from the mid-North Atlantic. *Sedimentology* **2013**, *60*, 1769–1785. [CrossRef]
74. Ávila, S.P.; Ramalho, R.S.; da Silva, C.M.; Uchman, A.; Berning, B.; Johnson, M.E.; Melo, C.S.; Madeira, P.; Rebelo, A.C.; Câmara, R.; et al. *Os Fósseis de Santa Maria (Açores). 3. O Trilho Marítimo da Rota dos Fósseis*; OVGA—Observatório Vulcanológico e Geotérmico dos Açores: Lagoa, Portugal, 2022; 120p, ISBN 978-989-8164-27-8.
75. Kirby, M.X.; Jones, D.S.; Ávila, S.P. Neogene shallow-marine paleoenvironments and preliminary strontium isotope chronostratigraphy of Santa Maria Island, Azores. In *Proceedings of the 1st Atlantic Islands Neogene*; Ávila, S.P., de Frias Martins, A.M., Eds.; Açoreana Suplemento; University of Florida: Gainesville, FL, USA, 2007; (Suppl. 5), pp. 112–125.
76. Ávila, S.P.; Ramalho, R.S.; Habermann, J.M.; Quartau, R.; Kroh, A.; Berning, B.; Johnson, M.E.; Kirby, M.X.; Zanon, V.; Titschack, J.; et al. Palaeoecology, taphonomy, and preservation of a lower Pliocene shell bed (coquina) from a volcanic oceanic island (Santa Maria Island, Azores, NE Atlantic Ocean). *Palaeogeogr. Palaeoclimatol. Palaeoecol.* **2015**, *430*, 57–73. [CrossRef]
77. Ávila, S.P.; Ramalho, R.S.; da Silva, C.M.; Johnson, M.E.; Uchman, A.; Berning, B.; Quartau, R.; Madeira, P.; Melo, C.S.; Rebelo, A.C.; et al. *Os fósseis de Santa Maria (Açores). 2. Pedra-Que-Pica: Uma História Com 5 Milhões de Anos*; OVGA—Observatório Vulcanológico e Geotérmico dos Açores: Ponta Delgada, Portugal, 2022; 160p.
78. Uchman, A.; Johnson, M.E.; Rebelo, A.C.; Melo, C.; Cordeiro, R.; Ramalho, R.S.; Ávila, S.P. Vertically-oriented trace fossil *Macaronichnus segregatis* from Neogene of Santa Maria Island (Azores, NE Atlantic). In Proceedings of the 13th International Ichnofabric Workshop, Ichnofabric Studies Linking Past, Present, and Future, Kochi, Japan, 14–21 May 2015; Nara, M., Ed.; Abstract Book; p. 27.
79. Uchman, A.; Johnson, M.E.; Rebelo, A.C.; Melo, C.; Cordeiro, R.; Ramalho, R.S.; Ávila, S.P. Vertically-oriented trace fossil *Macaronichnus segregatis* from Neogene of Santa Maria Island (Azores, NE Atlantic) records a specific palaeohydrological regime on a small oceanic island. *Geobios* **2016**, *49*, 229–241. [CrossRef]
80. Rebelo, A.C.; Rasser, M.W.; Kroh, A.; Johnson, M.E.; Melo, C.; Ramalho, R.S.; Uchman, A.; Zanon, V.; Silva, L.; Neto, A.I.; et al. Rocking around a volcanic island shelf: Pliocene rhodolith beds from Malbusca, Santa Maria Island (Azores, NE Atlantic). *Facies* **2016**, *62*, 1–22. [CrossRef]
81. Johnson, M.E.; Uchman, A.; Costa, P.J.M.; Ramalho, R.S.; Ávila, S.P. Intense hurricane transports sand onshore: Example from the Pliocene Malbusca section on Santa Maria Island (Azores, Portugal). *Mar. Geol.* **2017**, *385*, 244–249. [CrossRef]
82. Uchman, A.; Torres, T.; Johnson, M.E.; Berning, B.; Ramalho, R.S.; Rebelo, A.C.; Melo, C.S.; Baptista, L.; Madeira, P.; Cordeiro, R.; et al. Feeding traces of recent ray fish and abundant occurrences of the trace fossil *Piscichnus waitemata* from the Pliocene of Santa Maria Island, Azores (Northeast Atlantic). *Palaios* **2018**, *33*, 361–375. [CrossRef]

83. Uchman, A.; Johnson, M.E.; Ramalho, R.S.; Quartau, R.; Berning, B.; Hipólito, A.; Melo, C.; Rebelo, A.C.; Cordeiro, R.; Ávila, S.P. Neogene marine sediments and biota encapsulated between lava flows on Santa Maria Island (Azores, NE Atlantic): An interplay between sedimentary, erosional, and volcanic processes. *Sedimentology* **2020**, *67*, 3595–3618. [CrossRef]
84. Hyžný, M.; Melo, C.S.; Ramalho, R.S.; Cordeiro, R.; Madeira, P.; Baptista, L.; Rebelo, A.C.; Gómez, C.; Torres, P.; Uchman, A.; et al. Pliocene and late Pleistocene (MIS 5e) decapod crustaceans from Santa Maria Island (Azores Archipelago: NE Atlantic): Systematics, palaeoecology and palaeobiogeography. *J. Quat. Sci.* **2021**, *36*, 91–109. [CrossRef]
85. Baarli, B.G.; Cachão, M.; da Silva, C.M.; Johnson, M.E.; Mayoral, E.J.; Santos, A. A Middle Miocene carbonate embankment on an active volcanic slope: Ilhéu de Baixo, Madeira Archipelago, Eastern Atlantic. *Geol. J.* **2014**, *49*, 90–106. [CrossRef]
86. Santos, A.; Mayoral, E.; Johnson, M.E.; Baarli, B.G.; Cachão, M.; da Silva, C.M.; Ledesma-Vázquez, J. Extreme habitat adaptation by boring bivalves on volcanically active paleoshores from North Atlantic Macaronesia. *Facies* **2012**, *58*, 325–338. [CrossRef]
87. Ramalho, R.; Helffrich, G.; Schmidt, D.N.; Vance, D. Tracers of uplift and subsidence in the Cape Verde Archipelago. *J. Geol. Soc. Lond.* **2010**, *167*, 519–538. [CrossRef]
88. Martín-González, E.; Castillo, C.; Castillo Ruiz, C.; Gutiérrez González, M.; Aguirre, J. Estudio paleoambiental de los depósitos litorales someros del Plioceno inferior de Fuerteventura (Islas Canarias). *Rev. Esp. Paleontol.* **2001**, *16*, 47–57. [CrossRef]
89. Verde, M.; Castillo, C.; Martín-González, E.; Cruzado-Caballero, P.; Mayoral, E.; Santos, A. A new Miocene–Pliocene ichnotaxon for vermetid anchoring bioerosion structures. *Front. Earth Sci.* **2022**, *10*, 906493. [CrossRef]
90. González Delgado, J.Á.; Zazo, C.; Goy, J.L.; Civis, J.; Dabrio, C.J.; Cabero, A.; Bardaji, T.; Silva, P.G. Paleontología de las terrazas marinas del valle de Jorós (Fuerteventura, España). *Geo-Temas* **2008**, *10*, 1265–1268.
91. Santos, A.; Mayoral, E.J.; da Silva, C.M.; Cachão, M.; Johnson, M.E.; Baarli, B.G. Miocene intertidal zonation on a volcanically active shoreline: Porto Santo in the Madeira Archipelago, Portugal. *Lethaia* **2011**, *44*, 26–32. [CrossRef]
92. Santos, A.; Mayoral, E.; Johnson, M.E.; Baarli, B.G.; da Silva, C.M.; Cachão, M.; Ledesma-Vázquez, J. Basalt mounds and adjacent depressions attract contrasting biofacies on a volcanically active Middle Miocene coastline (Porto Santo, Madeira Archipelago, Portugal). *Facies* **2012**, *58*, 573–585. [CrossRef]
93. Santos, A.; Mayoral, E.; Baarli, B.G.; da Silva, C.M.; Cachão, M.; Johnson, M.E. Symbiotic association of a pyrgomatid barnacle with a coral from a volcanic Middle Miocene shoreline (Porto Santo, Madeira Archipelago, Portugal). *Palaeontology* **2012**, *55*, 173–182. [CrossRef]
94. Bertling, M.; Braddy, S.; Bromley, R.G.; Demathieu, G.D.; Genise, J.F.; Mikuláš, R.; Nielsen, J.-K.; Nielsen, K.S.S.; Rindsberg, A.K.; Schlirf, M.; et al. Names for trace fossils: A uniform approach. *Lethaia* **2006**, *39*, 265–286. [CrossRef]
95. Bertling, M.; Buatois, L.A.; Knaust, D.; Laing, B.; Mángano, M.G.; Meyer, N.; Mikuláš, R.; Minter, N.J.; Neumann, C.; Rindsberg, A.K.; et al. Names for trace fossils 2.0: Theory and practice in ichnotaxonomy. *Lethaia* **2022**, *55*, 1–19. [CrossRef]
96. John, G. Über bohrende Seeigel. In *Inaugural Dissertation zur Erlangung des Doctorgrades der Philosophischen Facultät der Universität Leipzig*; Adolph Mehnert: Leipzig, Germany, 1888; 46p.
97. Otter, G.W. Rojazidoring echinoids. *Biol. Rev.* **1932**, *7*, 89–107. [CrossRef]
98. Ávila, S.P.; Rebelo, A.; Medeiros, A.; Melo, C.; Gomes, C.; Bagaço, L.; Madeira, P.; Borges, P.A.; Monteiro, P.; Cordeiro, R.; et al. *Os Fósseis de Santa Maria (Açores). 1. A Jazida da Prainha*; Marine PalaeoBiogeography Working Group: Ponta Delgada, Portugal, 2010; 105p.
99. Rebelo, A.C.; Rasser, M.W.; Ramalho, R.S.; Johnson, M.E.; Melo, C.S.; Uchman, A.; Quartau, R.; Berning, B.; Neto, A.I.; Mendes, A.R.; et al. Pleistocene coralline algal buildups on a mid-ocean rocky shore—insights into the MIS 5e record of the Azores. *Palaeogeogr. Palaeoclimatol. Palaeoecol.* **2021**, *579*, 110598. [CrossRef]
100. Dávid, Á.; Uchman, A.; Ramalho, R.S.; Madeira, J.; Melo, C.S.; Madeira, P.; Rebelo, A.C.; Berning, B.; Johnson, M.E.; Ávila, S.P. Diverse bioerosion structures in lower Pliocene deposits from a volcanic oceanic island: Baía de Nossa Senhora section, Santa Maria Island, Azores (central North Atlantic). *Palaeogeogr. Palaeoclimatol. Palaeoecol.* **2021**, *569*, 110284. [CrossRef]
101. Wisshak, M.; Tribollet, A.; Golubic, S.; Jakobsen, J.; Freiwald, A. Temperate bioerosion: Ichnodiversity and biodiversity from intertidal to bathyal depths (Azores). *Geobiology* **2011**, *9*, 492–520. [CrossRef]
102. Hernández-Pacheco, E. Exploración geológica de Lanzarote y de las Isletas Canarias. *Bol. R. Soc. Esp. Hist. Nat.* **1907**, *7*, 339–348.
103. Hernández-Pacheco, E. Estudio geológico de Lanzarote y de las Islas Canarias. *Mem. Real Soc. Esp. Hist. Nat.* **1909**, *6*, 107–342.
104. Aranda Millán, F. Sobre Moluscos de Lanzarote (Canarias). *Bol. Real Soc. Esp. Hist. Nat.* **1909**, *9*, 112–114.
105. Ellis, W.N.; Ellis-Adam, A.C. Fossil brood cells of solitary bees on Fuerteventura and Lanzarote, Canary Islands (Hymenoptera: Apoidea). *Entomol. Ber.* **1993**, *53*, 161–173.
106. Bravo, T. *Geografía General de las Islas Canarias 1*; Editorial Goya: Santa Cruz de Tenerife, Spain, 1954; 410p.
107. Machado, A. Introduction to a faunal study of the Canary Islands laurisilva, with special reference to ground beetles. In *Biogeography and Ecology in the Canary Islands*; Kunkel, G., Ed.; W. Junk: The Hague, The Netherlands, 1976; pp. 347–411.
108. Báez, M.; Bacallado, J.J. Los fósiles de Canarias. In *Fauna (marina y terrestre) del Archipiélago Canario*; Bacallado, J.J., Ed.; Edirca: Las Palmas de Gran Canaria, Spain, 1984; pp. 343–347.

109. Rothe, P. Kanarische Inseln. Lanzarote, Fuerteventura, Gran Canaria, Tenerife, Gomera, La Palma, Hierro. *Samml. Geol. Führer* **1986**, *81*, 1–226.
110. Edwards, N.; Meco, J. Morphology and palaeoenvironment of brood cells of Quaternary ground-nesting solitary bees (Hymenoptera, Apidae) from Fuerteventura, Canary Islands, Spain. *Proc. Geol. Assoc.* **2000**, *111*, 173–183. [CrossRef]
111. Alonso-Zarza, A.M.; Silva, P.G. Quaternary laminar calcretes with bee nests: Evidences of small-scale climatic fluctuations, Eastern Canary Islands, Spain. *Palaeogeogr. Palaeoclimatol. Palaeoecol.* **2002**, *178*, 119–135. [CrossRef]
112. Von Suchodoletz, H.; Kühn, P.; Hambach, U.; Dietz, M.; Zöller, L.; Faust, D. Loesslike and palaeosol sediments from Lanzarote (Canary Islands / Spain)—indicators of palaeoenvironmental change during the Late Quaternary. *Palaeogeogr. Palaeoclimatol. Palaeoecol.* **2009**, *278*, 71–87. [CrossRef]
113. Genise, J.F. The ichnofamily Celliformidae for *Celliforma* and allied ichnogenera. *Ichnos* **2000**, *7*, 267–284. [CrossRef]
114. Meco, J.; Petit-Maire, N.; Ballester, J.; Betancort, J.F.; Ramos, A.J.G. The Acridian plagues, a new Holocene and Pleistocene palaeoclimatic indicator. *Glob. Planet. Change* **2010**, *72*, 318–320. [CrossRef]
115. Meco, J.; Muhs, D.R.; Fontugne, M.; Ramos, A.J.G.; Lomoschitz, A.; Patterson, D. Late Pliocene and Quaternary Eurasian locust infestations in the Canary Archipelago. *Lethaia* **2011**, *44*, 440–454. [CrossRef]
116. Genise, J.F.; Edwards, N. Ichnotaxonomy, origin, and paleoenvironment of Quaternary insect cells from Fuerteventura, Canary Islands, Spain. *J. Kans. Entomol. Soc.* **2003**, *76*, 320–327.
117. Alonso-Zarza, A.M.; Genise, J.F.; Cabrera, M.C.; Mangas-Viñuela, J.; Martín-Pérez, A.; Valdeolmillos-Rodríguez, A.; Dorado-Valiño, M. Megarhizoliths in Pleistocene aeolian deposits from Gran Canaria (Spain): Ichnological and palaeoenvironmental significance. *Palaeogeogr. Palaeoclimatol. Palaeoecol.* **2008**, *265*, 39–51. [CrossRef]
118. Genise, J.F. *Ichnoentomology: Insect Traces in Soils and Paleosols*; Topics in Geobiology; Springer: Berlin/Heidelberg, Germany, 2017; Volume 37, 695p.
119. Mikúlaš, R.; Genise, J.F. Traces within traces: Holes, pits and galleries in walls and filling of insect trace fossils in paleosols. *Geol. Acta* **2003**, *1*, 339–348. [CrossRef]
120. Pokorný, R.; Borges, P.A. Plant–insect interactions in the Quaternary fossil record of the Azores Archipelago (Portugal). *J. Quat. Sci.* **2023**, *38*, 597–607. [CrossRef]
121. Gois-Marques, C.A.; Madeira, J.; de Sequeira, M.M. Comment on: Plant–insect interactions in the Quaternary fossil record of the Azores Archipelago (Portugal). Pokorný and Borges (2023). *J. Quat. Sci.* **2024**, *39*, 340–344. [CrossRef]
122. Henríquez-Valido, P.; Brito-Mayor, A. Taphonomy on the beach: Experimental approach to bone modifications made by insects on an island (Gran Canaria, Canary Island, Spain). *Archaeol. Anthropol. Sci.* **2024**, *16*, 192. [CrossRef]
123. Alonso-Zarza, A.M.; Genise, J.; Valdeolmillos Rodríguez, A.; Cabrera Santana, M.D.C.; Mangas, J.; Martín-Pérez, A.; Dorado-Valiño, M. Los megarrizolitos de las eolianitas de Tufia (Gran Canaria): Procesos de formación, ichnología y paleoambiente. *Geotemas* **2008**, *10*, 1377–1380.
124. Coello Bravo, J.J. Escrito en la arena: Tubos y diques sedimentarios en paleodepósitos costeros de El Médano (Granadilla de Abona, Tenerife). *Macaronesia* **2012**, *14*, 124–135.
125. Alonso-Zarza, A.M.; Casillas Ruiz, R. El Médano megarhizoliths field, Tenerife: Origin and paleoenvironmental significance. *Geogaceta* **2019**, *65*, 27–30.
126. Pokorný, R. Rhizobial root nodules in aeolian sandstones on Madeira (Piedade Beds, Pleistocene) and their significance for palaeoenvironmental and "originator" hypotheses. *Rev. Palaeobot. Palynol.* **2022**, *306*, 104740. [CrossRef]

Journal of
*Marine Science
and Engineering*

MDPI

Review

Distributional Range Shifts Caused by Glacial–Interglacial Cycles: A Review on Timing, Main Processes, and Patterns of Late Pleistocene Marine Dispersal by Invertebrates in the NE Atlantic

Sérgio P. Ávila [1,2,3,4,5]

[1] MPB—Marine Palaeontology and Biogeography Lab, University of the Azores, Rua da Mãe de Deus, 9501-801 Ponta Delgada, Açores, Portugal; avila@uac.pt
[2] CIBIO—Açores, Centro de Investigação em Biodiversidade e Recursos Genéticos, InBIO Laboratório Associado, Pólo dos Açores, 9501-801 Ponta Delgada, Azores, Portugal
[3] BIOPOLIS/CIBIO, Centro de Investigação em Biodiversidade e Recursos Genéticos, Rua do Crasto, 765, 4485-684 Vairão, Vila do Conde, Portugal
[4] Departamento de Biologia, Faculdade de Ciências e Tecnologia, Universidade dos Açores, 9501-801 Ponta Delgada, Açores, Portugal
[5] UNESCO Chair—Land Within Sea: Biodiversity and Sustainability in Atlantic Islands, Universidade dos Açores, R. Mãe de Deus 13A, 9500-321 Ponta Delgada, Açores, Portugal

Abstract

The fossil record of marine molluscs has been used as evidence to test marine island biogeography theories and to complement the evolutionary patterns of biodiversity and endemicity found in oceanic islands and archipelagos. Although long-distance dispersal patterns have been the subject of several studies, important questions may still be posed. For instance, are these processes random, or do they show distinct patterns? And are they restricted to "windows of opportunities", or do they occur continuously? In the NE Atlantic, the dispersal of tropical species towards higher latitudes associated with the last interglacial period is a well-known phenomenon. However, the most probable dispersal route remains a matter of debate. To test these ideas, we used the Atlantic and Mediterranean last interglacial fossil records, and compared the present geographic distribution of shallow-water marine molluscs with that registered during the last interglacial episode, aiming to detect changes. Our results show that 27 species became extinct during the course of the last glacial episode, and that 55 marine mollusc species (43 gastropods and 12 bivalves) are ecostratigraphic indicators for the MIS 5e fossiliferous deposits in the Macaronesian archipelagos, the Atlantic coasts of Morocco, and the Mediterranean. Finally, we provide arguments for the timing of dispersal, which occurred during a restricted "window of opportunity" associated with the inception of the last interglacial, and for the most probable route of dispersal of the tropical species, most of them denoting a Cabo Verdean origin.

Keywords: marine island biogeography; marine invertebrates; long-distance dispersal; MIS 5e; "windows of opportunity"; ecostratigraphic indicators

1. Introduction

Isolated oceanic islands and archipelagos are key places to study the biological evolution of shallow-water marine organisms [1–4]. Patterns of biogeography, biodiversity, and endemicity are commonly the result of such evolution [5–12]. One of the most important problems solved by the theory of marine island biogeography during recent decades was how marine species colonised faraway islands through long-distance dispersal. The issue of dispersal was pursued by many authors, but Rudolf Scheltema (1926–2019), working from the Woods Hole Oceanographic Institute in Massachusetts, was particularly successful in demonstrating the underlying patterns. For almost three decades, he explored the dispersal of marine organisms in detail [13–15]. As a result, he was able to correlate both the duration of the larval phase of marine gastropods with the dispersal ability of the species and with their geographic distributions, as well as the mode of larval development with the longevity of the species. In particular, Scheltema demonstrated that (1) marine mollusc gastropod species with planktotrophic larvae (i.e., larvae that feed during their more or less prolonged stay in the water column) have higher dispersal abilities and larger geographic ranges than non-planktotrophic species (including lecithotrophic larvae, i.e., larvae that do not feed during their short stay in the water column, and species with direct, intra-capsular development) [16–22]; and that (2) marine gastropod planktotrophic species generally possess longer lifespans (in a geological sense) compared to non-planktotrophic species [16,23–25].

Inspired by this framework, the author postulated [26] that relationships must be found between the mode of larval development, the dispersal process, the ecological zonation (i.e., the species' usual bathymetric range), and the geographic range of the species. The resulting hypothesis suggests that, for congeneric non-planktotrophic gastropod species, those living in the intertidal zone should have wider geographic ranges and have speciated during a shorter time span than those living at greater depths [27]. These assumptions were later tested against both the fossil record [28,29] and molecular/phylogenetic data [30], and the results supported the theory.

Questions related to the long-distance dispersal of species are pertinent, and may still be posed. For example, are these processes random, or do they adhere to discernible patterns? Do they occur continuously, or are they more common during rare "windows of opportunities"? To test these ideas, the Atlantic and Mediterranean last interglacial fossil records are reviewed in order to detect range alterations on the geographic distribution of shallow-water marine molluscs, and to infer the most probable dispersal directions. Accordingly, this paper aims to address such issues, which are key to understanding the mechanisms responsible for the expansion and contraction of the geographic ranges of marine species during glacial/interglacial cycles, and to explain the patterns that emerge from migratory pathways across marine sweepstake routes that open during "windows of opportunity", crossing former marine barriers in times of natural, global climate changes.

2. Materials and Methods

This study reviews and adds to previous work regarding the Marine Isotopic Stage 5e (MIS 5e, i.e., the warmest period of the last interglacial) deposits of the Macaronesia, the NW Atlantic coasts of Africa (from Senegal North to the Straits of Gibraltar), and the Mediterranean. Macaronesia is a geographic region that encompasses the Azores, Madeira, Selvagens, Canaries, and Cabo Verde archipelagos (Figures 1 and 2). For the Macaronesian archipelagos, we reviewed the most relevant works by Meco [31,32], García-Talavera et al. [33,34], García-Talavera [35–37], Meco et al. [38–41], Callapez and Soares [42], García-Talavera and Sánchez-Pinto [43], Ávila et al. [44–48], Ávila [49], Cabero [50], Cabero et al. [51], Martín-González et al. [52,53], and Melo et al. [54,55]. For the African Atlantic coastal MIS 5e deposits, we critically reviewed the works of Lecointre [56–58], Ortlieb [59], Brebion [60], Brebion et al. [61], Weisrock et al. [62], Plaziat et al. [63], and Chakroun et al. [64]. Finally, the review benefitted from the compilation by [65] that summarises the mollusc record of the MIS 5e deposits of the Mediterranean, adding new MIS 5e mollusc records from recent important works [66,67].

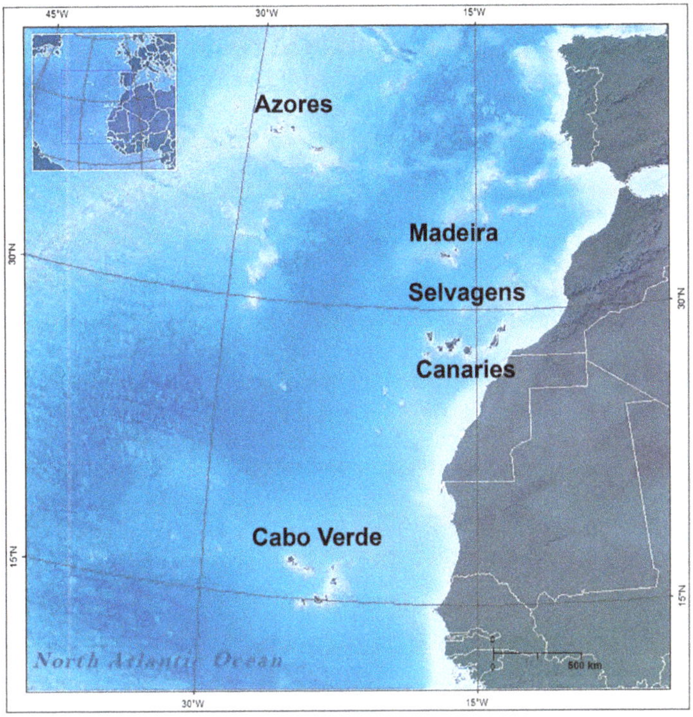

Figure 1. Location of the Macaronesian archipelagos within the North Atlantic (insert) and geographic placement of the Macaronesian archipelagos.

Figure 2. General views of MIS 5e outcrops in the Macaronesian archipelagos. (**A**): Vinha Velha (Santa Maria Island, Azores archipelago). (**B**): Zimbralinho (Porto Santo Island, Madeira archipelago). (**C**): Matas Blancas (Fuerteventura Island, Canaries archipelago), a famous outcrop renowned by the abundance of large specimens of the mollusc gastropod *Thetystrombus latus* (Gmelin, 1791). (**D**): Praia Lacacão (Boavista Island, Cabo Verde archipelago). All photos by the author.

Over the last ten years, more than 1500 primary sources (mainly research articles and monographs) and "grey literature" (about 30 Ph.D. and M.Sc. theses) have been consulted in the process of assembling databases on recent, and on MIS 5e shallow-water Atlantic (from the Arctic to the Antarctic) and Mediterranean mollusc species (gastropods and bivalves). For this work, two MIS 5e databases were formatted, both curated by the author, that include information on the geographic distribution and functional traits for a total of 1406 MIS 5e mollusc taxa: 887 gastropods and 519 bivalves. Information on the geographical distribution and functional traits of the present-day marine mollusc species was retrieved from two additional databases, also curated by the author: the first deals with the modern shallow-water (from the supralittoral down to 50 m depth) Atlantic and Mediterranean gastropods, with 5720 species [68], whereas the second database contains information on 1815 recent shallow-water (<100 m depth) Atlantic and Mediterranean bivalve species [69].

In order to detect possible species' geographic range changes among the marine molluscs from the Macaronesian archipelagos, the Mediterranean, and along the Atlantic African coast, their present-day range was compared against their MIS 5e geographical distribution. This comparison facilitated the compilation of a checklist of MIS 5e molluscs that became extinct (Table 1), as well as of MIS 5e mollusc species/taxa that were extirpated, i.e., that locally disappeared from the Macaronesian archipelagos, NW Atlantic African coasts, and Mediterranean shores as a result of the last glacial episode (Supplementary Table S1). The cross-checking of mollusc species' past and present-day geographic distributions also helped to determine the most probable sites of origin of the MIS 5e colonisers for each species, assuming that the most probable source of these colonisers was from the nearest archipelago or continental shores.

Table 1. List of MIS 5e species reported from the Macaronesian archipelagos and from the Mediterranean that became extinct (†) during the course of the last glacial episode. GAS: Gastropoda; BIV: Bivalvia; MED: Mediterranean; SEL: Selvagens archipelago; CAN: Canaries archipelago.

Class	MIS 5e Extinct Species	Site	Geological Timespan					
			Eocene	**Oligocene**	**Miocene**	**Pliocene**	**Pleistocene**	**Recent**
GAS	*Acteon bovettensis* G. Seguenza, 1880 †	MED				1	1	
GAS	*Alvania calliope* Chirli & U. Linse, 2011 †	MED					1	
GAS	*Alvania curta* (Dujardin, 1837) †	MED					1	
GAS	*Alvania mariae* (A. d'Orbigny, 1852) †	MED					1	
GAS	*Alvania unica* Amati & Quaggiotto, 2019 †	MED					1	
GAS	*Cancilla alligata* (Defrance, 1825) †	MED					1	
GAS	*Capulus laevis* (Bronn, 1831) †	MED					1	
GAS	*Cerithium diblasii* †	MED					1	
GAS	*Clathromangelia clathrata* (Serres, 1829) †	MED				1	1	
GAS	*Clathromangelia quadrillum* (Dujardin, 1837) †	MED					1	
GAS	*Episcomitra fusiformis* (Brocchi, 1814) †	MED					1	
GAS	*Dizoniopsis bilineata* (M. Hörnes, 1848) †	MED				1	1	
GAS	*Fusinus rudis* (R. A. Philippi, 1844) †	MED				1	1	
GAS	*Helminthia triplicata* (Brocchi, 1814) †	MED					1	
GAS	*Jujubinus bullula* (P. Fischer, 1877) †	MED					1	
GAS	*Metula deshayesi* (Michelotti, 1847) †	MED			1		1	
GAS	*Ocinebrina scalaris* (Brocchi, 1814) †	MED				1	1	
GAS	*Petaloconchus intortus* (Lamarck, 1818) †	MED	1		1	1	1	
BIV	*Europicardium multicostatum* (Brocchi, 1814) †	MED					1	
BIV	*Glycymeris inflata* (Brocchi, 1814) †	MED				1	1	
BIV	*Macoma obliqua* (J. Sowerby, 1817) †	MED					1	
BIV	*Megaxinus transversus* (Bronn, 1831) †	MED				1	1	
BIV	*Plicatula mytilina* (R. A. Philippi, 1836) †	MED				1	1	
BIV	*Pycnodonte squarrosa* de Serres 1843 †	MED			1		1	
GAS	*Acanthina dontelei* Garcia-Talavera & Sanchez-Pinto, 2002 †	SEL					1	
GAS	*Acanthina dontelei* Garcia-Talavera & Sanchez-Pinto, 2002 †	CAN					1	
GAS	*Spinucella plessisi* (Lecointre, 1952) †	CAN						
BIV	*Lucina columbella* Lamarck, 1818 †	CAN					1	

Regarding the NW Atlantic African coasts, it is necessary to exercise care, as doubts arise on correctly assigning some of the fossil mollusc species to the corresponding interglacials. Until recently, the major problem was the absence of fossil remains representing the large gastropod *Thetystrombus latus* (Gmelin, 1791) from the Atlantic coast of Morocco that would clarify this issue [59]. The solid sedimentary descriptions and stratigraphic analysis conducted by [70,71] provided such data for the Pleistocene (MIS 5e) coastal deposits of Rabat-Témara on the NW Atlantic coast of Morocco that, together with a precautionary use of older studies (mostly from the SW of Morocco, from Tan Tan beach to Tarfaya), allowed the compilation of a relevant set of MIS 5e Moroccan mollusc species to be included in our MIS 5e databases.

The data shown in Supplementary Materials Table S1 was additionally screened for the selection of MIS 5e ecostratigraphic indicators. As defined by [54], MIS 5e ecostratigraphic indicators are species that, cumulatively, fulfil the following conditions: (1) they are reported for the MIS 5e of the area in study (in our case, the Macaronesian archipelagos, the Atlantic coasts of Morocco and/or the Mediterranean); (2) these species do not occur today in the areas under study; and (3) these colonisers originate from a marine tropical region, i.e., an area located between 23.5° N and 23.5° S [72].

Finally, information on the functional traits of the ecostratigraphic indicator species was compiled, including data on the modes of larval development (planktotrophic/non-planktotrophic) and on the type of substrate (algae, gravel, rock, sand).

3. Results

From the 1406 MIS 5e species/taxa of molluscs (887 gastropods and 519 bivalves) reported from the entire Atlantic and Mediterranean, 27 species became extinct during the course of the last glacial episode (cf. Table 1): 18 gastropods and six bivalves in the Mediterranean; the gastropod *Acanthina dontelei* Garcia-Talavera & Sanchez-Pinto, 2002 in the Selvagens archipelago and in the Canaries; and one bivalve and one other gastropod species in the Canaries archipelago.

Besides the species that became extinct, another 137 MIS 5e species/taxa (38 Bivalvia, 99 Gastropoda) were extirpated from some of the areas where they occurred during the last interglacial (Supplementary Materials Table S1). Table 2 summarises the number of such species/taxa by area (for detailed information, please see Supplementary Materials Table S1). The Canary Islands, with 45 species/taxa, was the most affected archipelago, followed by the Mediterranean (35), Azores (27), Morocco (23), and Madeira (22). At the Selvagens archipelago, only two MIS 5e gastropods were extirpated and no longer live in the region: *Isara nigra* (Gmelin, 1791) and *Patella ordinaria* Mabille, 1888. The gastropod *Gemophos viverratus* (Kiener, 1834) was the only species that was extirpated from four different regions: the Azores and Madeira archipelagos, the Mediterranean, and the Atlantic coasts of Morocco. Three gastropod species—*Conus ermineus* Born, 1778, *Conus ventricosus* Bronn, 1831, and *Linatella caudata* (Gmelin, 1791)—were each extirpated from three regions: the first one from the Azores, Canaries, and Mediterranean; the second one from the Azores, Canaries, and Morocco; and the last one from the Azores, the Mediterranean, and Morocco (Supplementary Materials Table S1).

Table 3 displays the species/taxa of molluscs selected as MIS 5e ecostratigraphic indicators for the Azores, Madeira, Canaries, Cabo Verde, the Mediterranean, and Morocco, which are region-specific, based on information compiled in Supplementary Table S1, whereas Figures 3 and 4 illustrate the dispersal directions of the MIS 5e mollusc ecostratigraphic indicators. Some species formerly included in the Gignoux "warm-water Senegalese fauna" [73], e.g., *Monoplex trigonus* (Gmelin, 1791) and *Gemophos viverratus*, were not considered ecostratigraphic indicators because their present geographic distribution is not restricted to tropical areas (both species occur today in the Canaries archipelago).

Table 2. Total number of MIS 5e species/taxa that were extirpated, per region, during the course of the last glacial episode. For further details, please see Supplementary Materials Table S1.

	AZO	MAD	SEL	CAN	CAB	MOR	MED
Gastropoda	22	13	2	32	6	22	25
Bivalvia	5	9	0	13	6	1	10
Total	27	22	2	45	12	23	35

Table 3. The 55 species/taxa of molluscs selected as MIS 5e ecostratigraphic indicators for the Macaronesian archipelagos, Morocco Atlantic coasts, and the Mediterranean. The inferences made on the most probable origin of the colonising species results from crosschecking information regarding the present and MIS 5e species' geographic distribution. (*) corresponds to unlikely but still probable sites of origin of the colonisers. BIV: Bivalvia; GAS: Gastropoda; AZO: Azores; MAD: Madeira; SEL: Selvagens; CAN: Canaries; CAB: Cabo Verde; MED: Mediterranean; BER: Bermuda; MOR: Morocco; WES: Western Sahara; MAU: Mauritania; SEN: Senegal; IVO: Ivory Coast; GAB: Gabon; POR: Portugal; CAR: Caribbean.

Region	Class	Species/Taxa	Most Probable Source/Origin		
			Range Expansion Towards Higher Latitudes	Range Expansion Towards Lower Latitudes	Longitudinal Range Expansion
AZO	GAS	*Claremontiella nodulosa*	CAB/SEN/BER		
AZO	GAS	*Conus ambiguus*	CAB/MAU		
AZO	GAS	*Conus ermineus*	CAB/WES/CAR *		
AZO	GAS	*Conus miruchae*	CAB		
AZO	GAS	*Conus roeckeli*	CAB		
AZO	GAS	*Conus venulatus*	CAB		
AZO	GAS	*Zonaria picta*	CAB/SEN		
MAD	BIV	*Cardium* sp.	MAU		MOR *
MAD	GAS	*Acliceratia carinata*	MAU		
MAD	GAS	*Conus* sp.	CAB		
MAD	GAS	*Thais nodosa*	CAB		
MAD	GAS	*Zebina vitrea*	CAR		
CAN	BIV	*Acar olivercoseli*	CAB/SEN		
CAN	BIV	*Brachidontes puniceus*	CAB/MAU		
CAN	BIV	*Cardium* sp.	MAU		MOR *
CAN	BIV	*Chlamys* sp.	CAR		
CAN	BIV	*Codakia* sp.	CAR	BER	
CAN	BIV	*Ctena eburnea*	CAB/SEN		
CAN	GAS	*Cassis* sp.	CAB/BER *		
CAN	GAS	*Cheilea equestris*	CAB/BER *		
CAN	GAS	*Claremontiella nodulosa*	CAB/SEN/BER *		
CAN	GAS	*Conus ermineus*	CAB		WES
CAN	GAS	*Conus guinaicus*	SEN		
CAN	GAS	*Cumia intertexta*	SEN		
CAN	GAS	*Harpa doris*	CAB/SEN		
CAN	GAS	*Heliacus cylindricus*	CAB/BER *		
CAN	GAS	*Hexaplex rosarium*	CAB/SEN		
CAN	GAS	*Nerita senegalensis*	CAB/SEN		
CAN	GAS	*Prunum olivaeforme*	MAU		
CAN	GAS	*Thais nodosa*	CAB		

Table 3. *Cont.*

Region	Class	Species/Taxa	Most Probable Source/Origin Range Expansion Towards Higher Latitudes	Range Expansion Towards Lower Latitudes	Longitudinal Range Expansion
CAN	GAS	*Thetystrombus latus*	CAB/SEN		
CAN	GAS	*Thylacodes arenarius*	CAB		
CAN	GAS	*Vaughtia gruveli*	CAB		WES
CAN	GAS	*Zonaria zonaria*	SEN		
CAB	BIV	*Saccostrea cucculata*	IVO		SEN
CAB	GAS	*Fasciolaria* sp.			CAR
CAB	GAS	*Inermicosta inermicosta*			SEN
MED	BIV	*Acar olivercoseli*	CAB/SEN		
MED	BIV	*Anadara geissei*	CAB/MAU		
MED	BIV	*Arcopsis afra*	CAB/MOR		
MED	BIV	*Brachidontes puniceus*	CAB/MAU		
MED	BIV	*Dendostrea cristata*	CAR		
MED	BIV	*Parvilucina crenella*	CAR		
MED	GAS	*Acteon maltzani*	CAB/SEN		
MED	GAS	*Acteon senegalensis*	MAU		
MED	GAS	*Conus ermineus*	CAB/WES/CAR *		
MED	GAS	*Conus guinaicus*	SEN		
MED	GAS	*Epitonium trochoides*	MAU		
MED	GAS	*Hexaplex rosarium*	CAB/SEN		
MED	GAS	*Imbricariopsis carbonacea*	CAB/SEN		
MED	GAS	*Simnia senegalensis*	CAB/SEN		
MED	GAS	*Thetystrombus latus*	CAB/SEN		
MED	GAS	*Turbonilla kerstinae*	CAB/MAU		
MED	GAS	*Turbonilla secernenda*	SEN		
MED	GAS	*Xenophora senegalensis*	CAB/SEN		
MED	GAS	*Zonaria petitiana*	MAU		
MED	GAS	*Zonaria angolensis*	GAB		
MOR	BIV	*Crassatina contraria*	MAU		
MOR	GAS	*Anachis aurantia*	SEN		
MOR	GAS	*Cymbium gracile*	WES		
MOR	GAS	*Hexaplex angularis*	SEN		

As expected, when all MIS 5e mollusc species/taxa that locally disappeared from the Macaronesian archipelagos, the Atlantic shores of Morocco, and/or from the Mediterranean coasts during the last glacial episode are included in the analysis, it becomes clear that most of the dispersal movements take place from tropical regions towards northern latitudes (Supplementary Materials Figures S1 and S2). Moreover, as shown in Figure 5, although the percentage of tropical gastropods that reached the Macaronesian archipelagos is similar, ranging from 45.8% in the Azores to 54.5% in the Canary Islands, it is much higher in the Mediterranean (77.8%) and much lower in the Atlantic Morocco region (30.4%). No tropical bivalves reached the Azores during the MIS 5e, only 20% reached Madeira, and about 60% reached Canaries; this contrasts with the MIS 5e bivalves that dispersed to the Atlantic coasts of Morocco and to the Mediterranean, all being tropical in provenience (Figure 5). As previously noted by [54], many longitudinal dispersal events were also detected, as well as range expansion towards lower latitudes, although fewer in number (Supplementary Materials Figures S1 and S2).

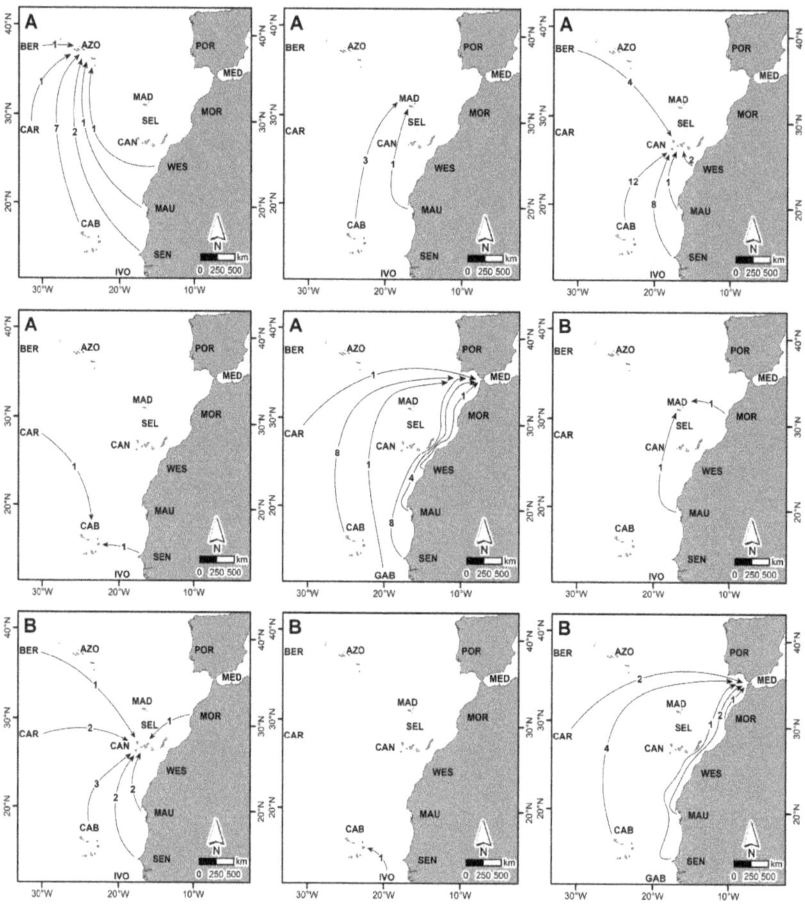

Figure 3. The arrows indicate the number and most probable origin of the MIS 5e mollusc ecostratigraphic indicators for the Azores, Madeira, Canaries, Cabo Verde, and the Mediterranean, according to data in Table 3. (**A**): Gastropoda; (**B**): Bivalvia. Abbreviations as in Table 3.

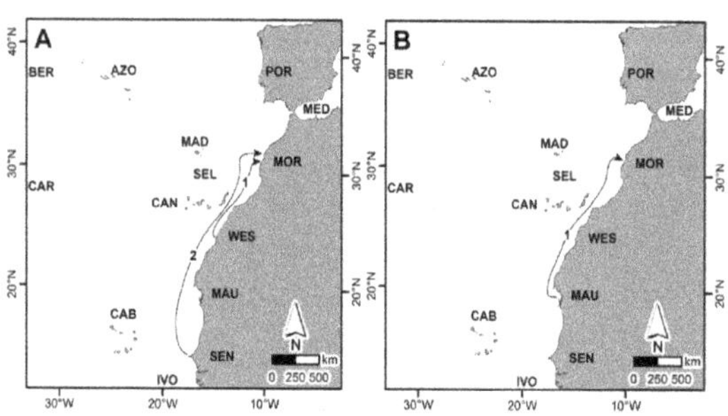

Figure 4. The arrows indicate the number and most probable origin of the MIS 5e mollusc ecostratigraphic indicators for Morocco, according to data in Table 3. (**A**): Gastropoda; (**B**): Bivalvia. Abbreviations as in Table 3.

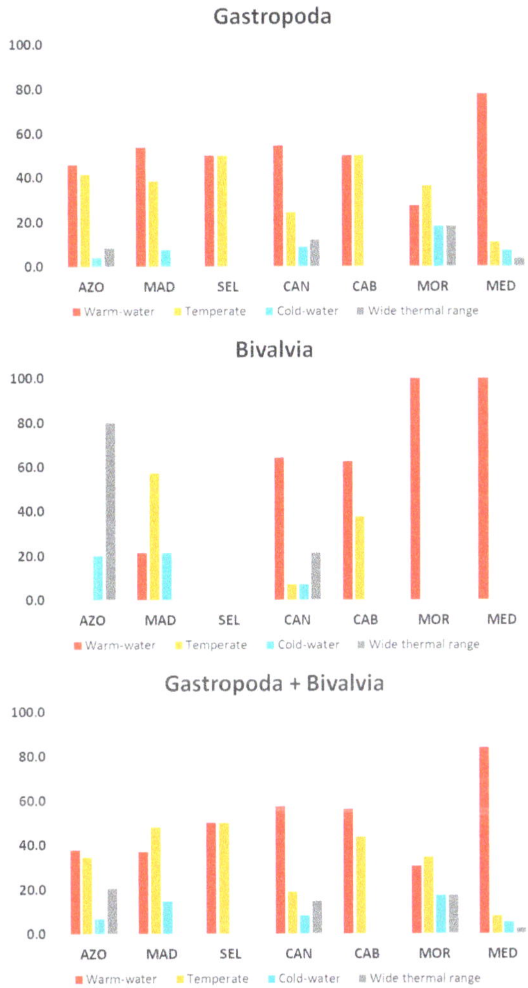

Figure 5. Relative comparison of the warm water (tropical), temperate, cold water, and wide thermal range number of MIS 5e species (in percentage) for the Macaronesian archipelagos, the Atlantic coast of Morocco, and the Mediterranean. Abbreviations as in Table 3.

4. Discussion

4.1. Modes of Larval Development and Long-Distance Dispersal of Shallow-Water Marine Species: The Main Processes

Long-distance dispersal is one of the best-studied biogeographic processes by which marine species are able to break marine barriers to dispersal such as wide stretches of deep sea, strong upwelling regions, or large rivers and long estuaries [74–77]. To cross these barriers and succeed in reaching isolated islands/archipelagos, marine species use a variety of strategies including self-motility, foresy, rafting, and larval drift. Self-motility is well represented by the example of fishes [78,79] or marine mammals, such as the pinniped members of the Phocidae family [80]. Foresy is the term applied to hitchhiking on migratory shorebirds, attached to their feathers, bills, and legs [81–83] or within their digestive tract [84,85]. Floating rafts that enhance the dispersal of marine invertebrates is provided by algal masses [86–94], floating wood [95], or other natural objects such as floating rocks attached to kelps [96], ice [97], or floating pelagic crustacean barnacles [98], the carapaces of turtles [99], and even on pumice [100–102]. Perhaps the most common dispersal process occurs in the form of larvae [103,104], which are passively transported by trade and storm winds and by sea-surface currents.

The author's database [68] includes information on the geographic distribution and 38 functional traits (longevity, shell composition, locomotion, life habit, diet, depth, maximum length, type of substrate) for 5720 recent Atlantic and Mediterranean shallow-water marine gastropod molluscs (shelled heterobranch gastropods are also included in the database). Out of these 5720 species, there are 161 gastropods (2.8%) that are amphi-Atlantic in geographic distribution. Of these 161, it was not possible to find information on the modes of larval development of 74 species; for the remaining species, the majority (52) are planktotrophic, 33 are non-planktotrophic, and two are brooding species [*Capulus ungaricus* (Linnaeus, 1758), and *Littorina saxatilis* (Olivi, 1792)]. This pattern is congruent with Scheltema rules [13–15]. Non-planktotrophic species are usually considered poor dispersers across sea barriers. Consequently, biologists have attributed the disjunct geographic distributions of non-planktotrophic species of marine gastropods to represent rare successful cases of long-distance dispersal on rafts of suitable, natural items by associated eggs, juveniles, or even small adult species [91]. Disjunct distributions are not restricted to molluscs, as examples are known from a variety of marine groups such as cnidarian hydrozoans [105], annelid polychaetes [106], echinoderms [107], and crustaceans [108,109].

Generally, for any given interval of time, it is expected that the number of planktotrophic gastropod species reaching isolated islands or archipelagos exceeds that of non-planktotrophic gastropod species. Moreover, as the geological lifespan of planktotrophic (p) gastropod species is usually longer (about 6–10 millions of years (Ma) on average; [110,111]) compared with that of non-planktotrophic (np) gastropod species (about 2–5 Ma on average; [110,111]), it is also expected that, for archipelagos/isolated islands older than 5 Ma, the number of planktotrophic species will be higher than that of non-planktotrophic gastropod species. However, our analysis, based on the database in [68], does not support these predictions, as for ten Atlantic archipelagos/islands studied (Table 4), only Bermuda clearly has a higher number of planktotrophic gastropods (150) in relation to the number of non-planktotrophic species (76). In the Azores, Madeira, Fernando de Noronha Island, and Trindade and Martim Vaz, the number of p-gastropods is similar to that of np-gastropods, and in the remaining archipelagos/islands, the proportion of np-gastropods is higher; therefore, other evolutionary factors must also be in action.

Table 4. Total number of shallow-water (from the supralittoral down to 50 m depth) species of gastropods (# spp.), and percentage of species (% spp.) with unknown mode of larval development reported for each archipelago/island in the Atlantic Ocean. p: number of planktotrophic gastropod species; np: number of non-planktotrophic gastropod species. Data extracted from [68]. Age of the oldest island in each archipelago in millions of years (Ma), according to [10].

Archipelago/Island	# spp.	p	np	% spp. with Unknown Mode of Larval Development	Age (Ma)
Azores	224	79	81	28.6	6
Madeira	332	98	101	40.1	18.8
Selvagens	156	45	73	24.4	29.5
Canaries	697	180	214	43.5	25
Cabo Verde	524	130	144	47.7	15.8
Bermuda	423	150	76	46.6	47
São Tomé and Príncipe	369	88	80	54.5	9
Saint Helena Island	37	7	10	54.1	15
Fernando de Noronha Island	152	62	60	19.7	12.3
Trindade and Martim Vaz	115	50	47	15.7	3.7

Several genera of marine gastropods portray relevant in situ adaptive radiations in the Atlantic islands (cf. Table 5), the most impressive being *Euthria* J. E. Gray, 1850 (92.9% of the 14 species are endemic); *Mirpurina* Ortea, Moro & Espinosa, 2019 (91.7% of the 12 species are endemic); *Manzonia* Brusina, 1870 (67.9% of the 28 species are endemic); *Schwartziella* G. Nevill 1881 (62.2% of the 37 species are endemic); *Plesiocystiscus* G. A. Coovert & H. K. Coovert, 1995 (46.7% of the 15 species are endemic); *Chrysallida* P. P. Carpenter, 1856 (45.5% of the 22 species are endemic); *Fissurella* Bruguière, 1789 (34.5% of the 29 species are endemic); *Marginella* Lamarck, 1799 (33.3% of the 30 species are endemic); *Mitromorpha* P. P. Carpenter, 1865 (33.3% of the 27 species are endemic); *Crisilla* Monterosato, 1917 (27.8% of the 36 species are endemic); *Mitrella* Risso, 1826 (25.0% of the 36 species are endemic); *Conus* Linnaeus, 1758 (21.1% of the 247 species reported from the Atlantic and Mediterranean are endemic to the Atlantic archipelagos and islands); *Gibberula* Swainson, 1840 (18.1% of the 116 species are endemic); and *Odostomia* J. Fleming, 1813 (14.3% of the 63 species are endemic). The Canaries is one of the oldest (25 Ma) and least isolated archipelagos in the Atlantic (presently just 98 km from the African continent). Together with Madeira and Selvagens, it forms the core of the Webbnesia biogeographic ecoregion [112], with many shared endemic and non-endemic species among these three archipelagos. Taken as a whole, these factors (age of the islands, insular littoral area, in situ speciation, and isolation) might explain the higher number of non-planktotrophic species in both Canaries and Selvagens. For further details, please see [10,11]. However, these arguments must be interpreted with some caution, as the number of gastropod species with unknown modes of larval development is still quite high, reaching 54.5% and 54.1% in São Tomé and Príncipe archipelago and Saint Helena Island, respectively (Table 4).

Table 5. Genera of Mediterranean and Atlantic gastropod molluscs with higher numbers of endemic species in the Atlantic archipelagos and islands. STP: São Tomé and Príncipe archipelago; STH: Saint Helena Island; ASC: Ascension Island; TRD: Trindade and Martim Vaz archipelago; ASP: São Pedro and São Paulo archipelago; FNO: Fernando de Noronha archipelago; FAL: Falkland Islands. For other abbreviations, see the legend of Table 3.

Genus	Total Number of Species	Archipelagos/Islands													% of Endemic Insular Species
		AZO	MAD	SEL	CAN	CAB	BER	STP	STH	ASC	TRD	ASP	FNO	FAL	
Alvania Risso, 1826	136	7		1	5	8	2	3							19.1
Ammonicera Vayssière, 1893	16							2							12.5
Anachis H. Adams & A. Adams, 1853	21					2		1							14.3
Barleeia W. Clark, 1853	12					3		1							33.3
Bittium Leach, 1847	15	1									5				40.0
Botryphallus Ponder, 1990	3	1				1									66.7
Caecum J. Fleming, 1813	83	2			2	3	6		1						16.9
Chrysallida P. P. Carpenter, 1856	22					10									45.5
Conus Linnaeus, 1758	247				1	49			1		1				21.1
Coralliophila H. Adams & A. Adams, 1853	23				1			1	1						13.0
Crisilla Monterosato, 1917	36				4	6									27.8
Diodora J. E. Gray, 1821	40				1	2		1							10.0
Eatonina Thiele, 1912	11					2									18.2
Euthria J. E. Gray, 1850	14					13									92.9

Table 5. *Cont.*

Genus	Total Number of Species	Archipelagos/Islands													% of Endemic Insular Species
		AZO	MAD	SEL	CAN	CAB	BER	STP	STH	ASC	TRD	ASP	FNO	FAL	
Fissurella Bruguière, 1789	29					6						1	1	2	34.5
Gibberula Swainson, 1840	116	1			10	3		4		3					18.1
Gibbula Risso, 1826	20	1				4									25.0
Hastula H. Adams & A. Adams, 1853	22				1	2		1							18.2
Jujubinus Monterosato, 1884	23	1			3	1									21.7
Manzonia Brusina, 1870	28	2	1	1	7	7		1							67.9
Marginella Lamarck, 1799	30							10							33.3
Metaxia Monterosato, 1884	15					2			1						20.0
Mirpurina Ortea, Moro & Espinosa, 2019	12					11									91.7
Mitrella Risso, 1826	36					4		5							25.0
Mitromorpha P. P. Carpenter, 1865	27	1			4			4							33.3
Muricopsis Bucquoy & Dautzenberg, 1882	19							4							21.1
Odostomia J. Fleming, 1813	63	2	1			3		1	2						14.3
Onoba H. Adams & A. Adams, 1852	35	1	1		1									1	11.4
Parviturbo Pilsbry & T. L. McGinty, 1945	14	1				2									21.4
Patella Linnaeus, 1758	13		1		1	1									23.1
Phorcus Risso, 1826	10			1		1									20.0
Plesiocystiscus G. A. Coovert & H. K. Coovert, 1995	15					1		3	3						46.7
Pradoxa F. Fernandes & Rolán, 1993	4							4							100.0
Pseudoscilla O. Boettger, 1902	5					1		1							40.0
Putzeysia Sulliotti, 1889	3		1		2										100.0
Pyrgocythara Woodring, 1928	10					1									10.0
Rissoa Desmarest, 1814	37	1			3	1									13.5
Rissoella J.E. Gray, 1847	31	3			1	1									16.1
Schwartziella G. Nevill 1881	37					19			3	1					62.2
Volvarina Hinds, 1844	266	1	1		12	6	1	1						1	8.6
TOTAL		**25**	**5**	**3**	**57**	**174**	**8**	**48**	**12**	**4**	**6**	**1**	**1**	**3**	

4.2. Relevance of the Fossil Record for Establishing the Times of Dispersal of Marine Species in the Macaronesian Archipelagos and in the Mediterranean

Both planktotrophic and non-planktotrophic marine species (these associated with rafts) are best able to cope with long-distance dispersal processes, thus reaching isolated islands and archipelagos. The fossil record helps to elucidate the time of arrival and past migratory pathways. In Macaronesia, fossiliferous sediments from the MIS 5e, with an attributed age of 129 to 116 ka [113], are well represented in all archipelagos: the Azores [10,28,35,42,44,45,47,48,114–122], Madeira [123], Selvagens [43], Canaries [31–33,36, 38–40,50–53,124,125], and Cabo Verde [37,126].

The Mediterranean is the region with the highest number of MIS 5e fossil mollusc species/taxa reported (378) and is certainly one of the best-studied regions in the world for this time interval due to a very high publication effort relative to the marine biodiversity of the last interglacial period. This only highlights the significance of the MIS 5e fossil gastropods from the Azores, where over half of the present-day species were also recovered from the last interglacial deposits (55.6%) and, to a lesser degree, the Canaries and Atlantic Moroccan gastropod fossil record (26.0% and 26.8%, respectively; cf. Table 6). In contrast, the number of MIS 5e fossil bivalves of the Mediterranean (181 species/taxa; 48.4%) is impressive when compared with the number of shallow-water species of bivalves presently reported from the area (Table 6). These numbers also attest to the importance of continuous, several-decades-long studies, thus helping to lessen the Pleistocene taxonomic gap [48].

Table 6. Total number of species of Gastropoda and Bivalvia reported from the MIS 5e deposits of the Macaronesian archipelagos, the Atlantic coasts of Morocco, and the Mediterranean, and comparison with the present fauna in the same sites (in %). Abbreviations as in Table 3.

Gastropoda	**AZO**	**MAD**	**SEL**	**CAN**	**CAB**	**MOR**	**MED**
MIS 5e	125	53	18	182	53	44	378
Present-day	225	333	156	699	528	164	1069
MIS 5e/Present-day (%)	55.6	15.9	11.5	26.0	10.0	26.8	35.4
Bivalvia	**AZO**	**MAD**	**SEL**	**CAN**	**CAB**	**MOR**	**MED**
MIS 5e	28	29	2	79	30	22	181
Present-day	150	135	39	233	130	287	374
MIS 5e/Present-day (%)	18.7	21.5	5.1	33.9	23.1	7.7	48.4

The marine Mollusca (especially the Gastropoda and the Bivalvia) is the best represented marine group in all Macaronesian archipelagos, followed by less common invertebrates (e.g., cnidarians, bryozoans, crustaceans, annelids, and echinoids), Chromista foraminifers, and vertebrate marine groups (e.g., bony fishes and cetaceans), as well as a few calcareous algal species. Quantitative data obtained from the fossil record of marine gastropods and bivalves derived from standardised 1 kg samples has been intensively used in the Macaronesia region for palaeoecological studies [46,47,127]. Additionally, the comparison between the insular faunas from the MIS 5e and those from the present day for each archipelago demonstrates the dispersal pathway patterns of many marine mollusc gastropods and bivalves from the continental shores of both the western and eastern Atlantic (i.e., from American, European, and African shores) to the Macaronesian archipelagos, as well as among the Macaronesian archipelagos. Moreover, these migrations took place during both interglacial [31,36,41,46,49,51,54,55,65,124,128] and glacial periods [127].

The MIS 5e is considered to have been the second-warmest interglacial period during the Quaternary, only behind the MIS 11c (424–397 ka; [129–131]. During the MIS 5e, the mean sea level was 6 to 9–10 m (MIS 5e) higher than today [132–134], with mean sea surface temperatures up to 3 °C higher than modern values [72,124,135,136]. Along the shores of the Macaronesian islands, the American and African coastlines, and in many coastal sites along the Mediterranean, marine terraces occur, ranging in altitude from 1 to over 10 m. Many of these erosional features include sediments rich in fossils from the last interglacial period (cf. Figure 2), and some of the molluscs found in these MIS 5e deposits in the Mediterranean and Atlantic Morocco and the Canaries, Madeira, and the Azores are warm-water species that do not occur today at these locations. This outcome suggests different climatic conditions during the last interglacial in comparison to those registered today.

Prior to the discovery by Milanković [137] of the orbital cycles that force the Quaternary climate, and based on the fossils collected and studied by the Italian geologist Domenico Lovisato (1842–1916), Arturo Issel described the "Tyrrhenian" stage of the Quaternary [138]. Originally called "Tirreno" [139], this stage refers to coastal marine fossiliferous sediments that were deposited during the last interglacial highstand at Cala Mosca (Gulf of Cagliari, Italy), now above present sea level. The fauna is typified by a warm-water malacofauna best represented by the iconic and large gastropod "*Strombus bubonius*" Lamarck, 1822, now accepted as *Thetystrombus latus*. Numerous authors reported on this warm-water fauna that characterises the MIS 5e deposits in various outcrops around the Mediterranean coasts, with the most relevant studies for this region being the Balearic Islands [140–146], Cyprus [147], Egypt [148], France [149], Greece [150,151], Israel [152], Italy [73,139,153–170], Lebanon [171], Southeast Spain [172–174], Tunisia [175–183], and Spain [184].

The MIS 5e interglacial period follows the glacial MIS 6, a period spanning from 191 to 130 ka [185]. The alternating interglacial and glacial periods, with warmer mean sea-surface temperature (SST) and humid climates for the former, and colder SST and arid climatic conditions for the latter, severely impacted the coastal marine biotas worldwide. One of the most important consequences was the effect that the changes in sea level had in the insular littoral area available for the settlement of shallow-water marine organisms. In insular habitats, and for over 80% of the islands [12], the littoral area reaches maximum values during interglacials and minimum values during glacials [10,11]. During the glacial Termination II (the small period of time between the end of the glacial MIS 6 and the inception of the MIS 5e), the sea level rose from about −130 m to +6 to +9 m in relation to the present sea level, and the end of this period is perceived by marine biogeographers as a "window of opportunity" for the range expansion of tropical species to higher latitudes [45,47,49]. In the NE Atlantic, some of these tropical species were able to enter into the Mediterranean via the Strait of Gibraltar, colonising its coastal habitats and establishing viable populations. The tropical species that migrated northwards have been informally called "Senegalese fauna". Several authors have used this designation (e.g. [38,39,50,124,186–188] or a similar one (e.g., the well-known "couches a *Strombes*" used by Gignou [73,153]), with most of the authors referring to species that today thrive in a tropical area that ranges from Mauritania/Senegal south to Angola. In comparison, only a few authors [47,51] explicitly included species that today inhabit the Cabo Verde archipelago as members of the "Senegalese fauna". Overall, these tropical Cabo Verdean/Senegalese species were subsequently extirpated from the northern Macaronesian archipelagos, the Mediterranean, and the Atlantic coasts of Morocco during the course of the last glacial episode.

Cabero [50] compiled a list of 23 "Senegalese" species (i.e., mollusc species associated with MIS 5e deposits in the Canary Islands and/or in the Mediterranean Sea, but that do not occur nowadays in either area), whereas Ávila and colleagues [47] identified 22 "southward ranging" species (sensu [39]), i.e., species associated with MIS 5e deposits in the Azores but that do not occur today in this archipelago. Later, Melo and colleagues [54] published a list of 23 ecostratigraphic indicators (10 bivalves, 13 gastropods) for the Macaronesian archipelagos. This list was criticised by Meco et al. [41], who rightly questioned the inclusion of some species. The opportunity has been taken to correct and update the list of MIS 5e ecostratigraphic indicators accepted as valid for the Macaronesian archipelagos, the Mediterranean, and Atlantic Morocco. The following species of bivalves were excluded from the list in [54] because of a present geographic distribution outside the tropical area (therefore not fulfilling the condition of being restricted to the tropics): *Barbatia barbata* (Linnaeus, 1758), *Cardita rufescens* Lamarck, 1819, *Ostrea stentina* Payeaudeau, 1826, *Panopea* sp., *Pitar tumens* (Gmelin, 1791), *Senilia senilis* (Linnaeus, 1758), and *Spondylus gaederopus* Linnaeus, 1758. In light of the current state of our knowledge, the author considers 12 to be the valid number of MIS 5e bivalve ecostratigraphic indicators for the Macaronesian archipelagos, the Atlantic shores of Morocco, and the Mediterranean coasts (see Table 7). Seven of these bivalve species (58.3%) are exclusively associated with hard grounds (rocks, shells, corals, coralline algae forming rhodoliths, gravel, pebbles or coarse sand), three bivalves (25.0%) are exclusively associated with soft, mobile bottoms (e.g., fine sand, sandy mud, or muddy sand), and two species (16.7%) are associated with mixed bottoms (cf. Table 7).

Table 7. List of the selected MIS 5e bivalve ecostratigraphic indicators for the Macaronesian archipelagos, Atlantic Morocco, and the Mediterranean, and respective type of substrate. Abbreviations as in Table 3.

Bivalve Species/Taxa	MIS 5e Geographic Range Expansion to:	SUBSTRATE				
		Hard Grounds	Gravel, Pebbles	Coarse Sand	Fine Sand	Sandy Mud-Muddy Sand
Acar olivercoseli	CAN/MED	1				
Anadara geissei	MED		1			
Arcopsis afra	MED	1				
Brachidontes puniceus	CAN/MED	1				
Cardium sp.	MAD/CAN				1	1
Chlamys sp.	CAN	1	1		1	
Codakia sp.	CAN				1	1
Crassatina contraria	MOR					
Ctena eburnea	CAN		1	1		
Dendostrea cristata	MED	1				
Parvilucina crenella	MED				1	1
Saccostrea cuccullata	CAB	1	1			

In contrast, of the 13 gastropod ecostratigraphic indicators for the MIS 5e of the Macaronesian archipelagos, only *Cypraecassis testiculus* (Linnaeus, 1758) and *Turritella bicingulata* Lamarck, 1822 are herein excluded from the list from [54], again because of a present geographic distribution outside the tropical area. Thus, the 43 gastropod ecostratigraphic indicators now accepted as valid for the MIS 5e include 22 species/taxa that are exclusive to the Macaronesian archipelagos, 3 species exclusive to the Atlantic coasts of Morocco, 13 species exclusive to the Mediterranean, and 5 species in common for both the Macaronesian archipelagos and the Mediterranean (Table 8). It was not possible to determine the type of substrate associated with 17 out of the 43 gastropod ecostratigraphic indicators. Of the remaining 26 species/taxa of gastropods, 13 (50.0%) are exclusively associated with hard bottoms, eight (30.8%) are exclusively associated with soft, mobile bottoms, and five species (19.2%) are associated with mixed bottoms (cf. Table 8).

The author's databases [68,69] also allow for an interesting hypothesis to be put forward: the tropical gastropod species *Alanbuella corrugata* (Perry, 1811) (reported from the MIS 5e fossil record of the Canaries, Cabo Verde, Atlantic Morocco, and the Mediterranean), *Monoplex trigonus* (reported from the MIS 5e fossil record of the Canaries, Cabo Verde, and the Mediterranean), and *Gemophos viverratus* (reported from the MIS 5e fossil record of the Azores, Madeira, Canaries, Cabo Verde, Atlantic Morocco, and the Mediterranean) are tropical gastropod species in their present geographical distribution, except for the fact that they also occur today in the Canary Islands under a subtropical climate. It might have been the case that these three species, along with many others, expanded their geographical distribution during the last interglacial, but, in contrast with all other MIS 5e species, their Canary populations were able to cope with the global climate deterioration that followed the last interglacial and were not extirpated during the last glacial episode, with their existence in the archipelago becoming permanent.

Table 8. List of the selected MIS 5e gastropod ecostratigraphic indicators for the Macaronesian archipelagos, Atlantic Morocco, and the Mediterranean, and respective type of substrate. Abbreviations as in Table 3.

Gastropod Species/Taxa	**MIS 5e Geographic Range Expansion to:**	**SUBSTRATE**				
		Hard Ground	**Gravel, Pebbles**	**Coarse Sand**	**Fine Sand**	**Sandy Mud-Muddy Sand**
Acliceratia carinata	MAD					
Acteon maltzani	MED					
Acteon senegalensis	MED					
Anachis aurantia	MOR	1				
Aporrhais senegalensis	MED					
Cassis sp.	CAN	1		1	1	
Cheilea equestres	CAN	1				
Claremontiella nodulosa	AZO/CAN	1				
Conus sp.	MAD	1				
Conus ambiguus	AZO	1	1			
Conus ermineus	AZO/CAN/MED	1			1	
Conus guinaicus	CAN/MED					
Conus miruchae	AZO					
Conus roeckeli	AZO					
Conus venulatus	AZO					
Cumia intertexta	CAN					
Cymbium gracile	MOR				1	
Epitonium trochoides	MED				1	
Fasciolaria sp.	CAB					
Gemophos viverratus	AZO/MAD/MED	1				
Harpa doris	CAN				1	
Heliacus cylindricus	CAN			1	1	1
Hexaplex angularis	MOR	1				
Hexaplex rosarium	CAN/MED	1				
Imbricariopsis carbonacea	MED					
Inermicosta inermicosta	CAB	1				
Monoplex trigonus	MED	1			1	
Nerita senegalensis	CAN	1				
Prunum olivaeforme	CAN					
Simnia senegalensis	MED	1				
Terebra corrugata	MED				1	
Thais nodosa	MAD/CAN	1				
Thetystrombus latus	CAN/MED	1			1	
Thylacodes arenarius	CAN					
Turbonilla kerstinae	MED					
Turbonilla secernenda	MED					
Vaughtia gruveli	CAN					
Xenophora senegalensis	MED				1	
Zebina vitrea	MAD					
Zonaria angolensis	MED				1	
Zonaria petitiana	MED				1	
Zonaria picta	AZO				1	
Zonaria zonaria	CAN	1				

As for the timing of the range expansion bringing these tropical mollusc species to higher latitudes, the author agrees with the mechanism proposed to break the previous marine barrier—the Canaries Current—between Cabo Verde and the northern Macaronesian archipelagos, and with the time for dispersal proposed by previous authors [11,45,47,49,54,55,116,189]. All of these authors agree that the detected northward range expansions of tropical fauna with a Cabo Verdean and west African origin occurred during a short "window of opportunity" just before the end of Termination II (the second Last Glacial termination, i.e., the period of time, c. 10 ka (140–130 ka), between the end of the MIS 6 glacial episode and the inception of the MIS 5e interglacial episode), or early in the last interglacial. Muhs et al. [189] clearly elucidated the process by suggesting that an orbitally forced mechanism triggered higher insolation in the early part of MIS 5e, and that this caused the northward shift of the intertropical convergence zone (ITCZ), complemented by weakened trade winds and reduced upwelling along the African coasts. According to [47], the average values for the summer insolation are inferred to have been about 11% higher than those of today. Ávila et al. [47] also stated that the oceanographic conditions prevalent during the earliest phase of the MIS 5e "moved the summer position of the Azores High to E of its present location, where it possibly persisted during the peak of the last interglacial, between the Azores and Iberia", and that this shifted the biogeographic boundaries northward, a situation fully explored by [65] and its consequences explained in detail by these authors. To end this issue on the time of dispersal, it is necessary to explain the mechanism and the time for the southward migration of temperate and cold-water mollusc species during the MIS 5e that was captured by our data (see Supplementary Materials Figure S1C for a resume), first detected by [189] and later by [54]. We herein confirm their results and agree with the hypothesis put forward by the former authors that attributed this phenomenon to reduced insolation at the end of last interglacial, which caused a cascade of events that allowed for the southward dispersal of temperate and cold-water species: the southward displacement of the ITCZ, strengthening of the Canaries Current (as suggested by [65]), strengthening of the trade winds, and re-establishment of upwelling along the NW African coasts.

4.3. Main Routes of Dispersal in the East Atlantic During the MIS 5e

In the East Atlantic, two main routes are possible for the dispersal of a restricted number of tropical species that underwent latitudinal range expansions to higher latitudes in the Northern Hemisphere. This phenomenon occurred during the final phase of glacial Termination II/the early stages of the MIS 5e, as hypothesised by [11,41,45,47,49,65,189]. The first possible route of dispersal is an archipelagic route (either in a continuous stepping-stone fashion, or via a "jump-dispersal mode", bypassing some of the archipelagos), with species originating in the Cabo Verde archipelago and migrating northward along the Macaronesian archipelagos, from Cabo Verde to the Canaries, Selvagens, and then to Madeira, with some species then entering the Mediterranean area, whereas other species were able to reach and colonise the Azores (see [31,35,37,39,42,43,45–47,50,52–55,123]). The second route of dispersal is along the NW continental shores of Africa, originating in the considered tropical areas (23.5° N and 23.5° S, i.e., from the south end of Western Sahara to Angola), and with adults creeping and/or larvae dispersing northward in times of a hypothesised reduction in the operation of the Mauritanian upwelling system and associated trade winds, finally entering and colonising the Mediterranean geographic area. Today, the Mauritania upwelling system, best expressed off Cap Blanc (Râs Nouâdhibou), constitutes a powerful marine biogeographic barrier for the dispersal of species and for the gene flow between populations of the same species inhabiting in the Canaries and Cabo Verde archipelagos, with an SST ranging from 14–15 °C in June to 21–22 °C in September–November [190]. The almost permanent regime of the Mauritania upwelling system results from the complex interactions between the coastal, shelf, and slope topography and the trade winds and currents that occur in the area [190]. The NNW-to-NNE trade winds drive the coastal upwelling off Mauritania, which takes place in the confluence zone where the northward Cabo Verde Current and Poleward Undercurrent meet the southward Canary Current [191,192].

These currents are deflected offshore and originate the SW directed Cabo Verde Frontal Zone, which displays a remarkable, giant Mauritanian chlorophyll filament [193]. As shown in Table 3 and in Supplementary Table S1, the longitudinal migration of species originating in the African shores towards similar Macaronesian archipelago latitudes also occur, an occurrence first detected by [54].

In order to choose between the two possible routes—the "Macaronesian archipelagic" or the "coastal African" route—detailed studies are necessary regarding the MIS 5e deposits of a large stretch of the African coast, from Senegal in the south, to the Atlantic Morocco Straits of Gibraltar in the north. The late Pleistocene fossiliferous deposits in the NW of Africa were studied by [56–63], just to name a few. A critical review of these works resulted in the compilation of a relevant set of 67 MIS 5e Moroccan mollusc species (23 bivalves and 44 gastropods) to be included in our MIS 5e databases. The comparison of the geographic distribution of the MIS 5e molluscs with their present-day occurrence yielded one species of bivalve and 22 species of gastropods that were extirpated from the MIS 5e Morocco Atlantic malacofauna during the last glacial episode. Although some gastropod species might have colonised the Atlantic coasts of Morocco from the Canary Islands and Madeira, or from Senegal and the Western Sahara, the majority seem to have most likely originated from the Mediterranean or from Iberian (Portugal) populations (cf. Supplementary Materials Figure S2). Thus, the geographic position of the Atlantic coasts of Morocco appears to render it a cornerstone position for the migration of marine species during the MIS 5e, with dispersal patterns much different to those registered at the Macaronesian archipelagos and the Mediterranean (cf. Supplementary Materials Figure S1C).

The functional traits of gastropods and bivalves may clarify the most important MIS 5e dispersal route for marine species. As shown in Tables 7 and 8, most of the ecostratigraphic indicator bivalves and gastropods are associated with hard grounds (58.3% and 50%, respectively), with fewer bivalves and gastropods associated with soft, mobile bottoms (25% and 30.8%, respectively). In making this determination, one must take into account the following facts: (1) soft bottoms predominate in the long stretch of African coastline from Senegal to Atlantic Morocco; (2) hard grounds are extremely rare along the NW African coastline; (3) soft, mobile bottoms are available even in islands classified as "young" sensu Ávila et al. ([11], their Table 2), e.g., Terceira Island in the Azores, with just 0.4 Ma for its oldest subaerial age [194]; (4) hard substrata are common and sometimes even predominant in the islands that compose the Macaronesian archipelagos, in particular in those islands classified as "young" or "immature" sensu Ávila et al. [11]; (5) the existence of weakened trade winds and a reduced Mauritania upwelling system along the African coasts during the MIS 5e, as suggested by [189], implies that the SST should be lower in the continental coasts than those registered at similar latitudes in the southern Macaronesian archipelagos, in particular at the Canaries; (6) as suggested by [65], the transition zone between Tropical and Subtropical climatic zones was positioned at a higher latitude during the MIS 5e, resulting in the Canaries and the Cabo Verde archipelagos being located within the same tropical climatic zone; (7) the iconic ecostratigraphic indicator *Thetystrombus latus* is a thermophilic marine gastropod that is associated with sandy–muddy bottoms, favours waters of normal salinity and SSTs of around 25 °C [195], but with a temperature tolerance ranging from 15 °C to 31 °C [188]; (8) so far, no MIS 5e fossils of *T. latus* have been found in the NW African coasts between the Strait of Gibraltar and Senegal; and (9) MIS 5e fossils of *T. latus* are reported from Cabo Verde, the Canaries, and the Mediterranean.

When these previous data are coupled with the majority of the ecostratigraphic-indicator marine molluscs being associated with hard bottoms (cf. Tables 7 and 8), the "Macaronesian archipelagic" route is most favourable in contrast to the less probable "coastal African" route. Although not with such strong and detailed arguments as those provided here, this "Macaronesian archipelagic" route was previously suggested by [45,47,49,54,55,65,128].

5. Conclusions

The last interglacial was one of the warmest interglacials during the last million years. Detailed knowledge of the MIS 5e fossil fauna from key sites such as the Macaronesian islands, the NW African coast, and the Mediterranean is crucial for a clear understanding of the various biogeographic processes and patterns that take place during glacial–interglacial transitions. This review suggests that the dispersal of tropical species to higher latitudes in the northern hemisphere, more specifically in the NE Atlantic, as registered in the last interglacial fossil record, is associated with a "window of opportunity", i.e., a short period of time that comprises the ending of the glacial Termination II and/or the early stages of the MIS 5e. During this event, oceanographic and climatic conditions were different from those registered today, with a northern displaced ITCZ, a summer position of a weakened Azores High easterly displaced to a region between the Azores and Iberia, a weaker Canaries Current, reduced trade winds, and a reduced Mauritanian upwelling system. In addition to these conditions [196] also suggest that a persistent negative NAO (North Atlantic Oscillation) occurred at the end of glacial Termination II. Together, these conditions allowed tropical shallow-water marine molluscs to cross over the prevalent marine barrier formed by the Canaries Current and the Mauritanian upwelling system, thus allowing for the northward long-distance dispersal and geographic range expansion of Cabo Verdean marine species and, to a lesser degree, also of west African tropical species (mainly from Senegal, Mauritania, and Western Sahara). A thorough analysis was made of large datasets (5720 and 1815 recent gastropods and bivalves, respectively; and 887 and 519 MIS 5e gastropods and bivalves, respectively). The comparison of their geographic distributions during the MIS 5e and the present interglacial allowed for the detection of 27 species that became extinct due to the last glacial episode and the selection of 55 MIS 5e ecostratigraphic indicators specific for the Macaronesian archipelagos, the Atlantic coast of Morocco, and the Mediterranean. Finally, the fossil record data coupled with the information on the functional traits of the ecostratigraphic indicator species suggest a preferential archipelagic route along the Macaronesian archipelagos for the species that underwent long-distance dispersal processes, with some of them reaching latitudes as high as the Azores archipelago, whereas others, such as the large gastropod *Thetystrombus latus*, were able to reach and colonise the Mediterranean coasts.

Supplementary Materials: The following supporting information can be downloaded at https://www.mdpi.com/article/10.3390/jmse13112024/s1: Figure S1: Total number (including ecostratigraphic and non-ecostratigraphic indicator species/taxa) and most probable origin of the MIS 5e mollusc species/taxa that underwent geographical range changes for the Azores, Madeira, Canaries, Cabo Verde, and the Mediterranean, according to Table S1. A: Gastropoda; B: Bivalvia. C: Sketch summarising the total number of molluscs (gastropods+bivalves) that underwent geographical range changes within the selected regions. Abbreviations as in Table S1; Figure S2: Total number (including ecostratigraphic and non-ecostratigraphic indicator species/taxa), and most probable origin of the MIS 5e mollusc species/taxa that underwent geographical range changes for Morocco Atlantic coasts, according to Table S1. A: Gastropoda; B: Bivalvia. Abbreviations as in Table S1; Table S1: MIS 5e mollusc species/taxa that locally disappeared from the Macaronesian archipelagos and/or from the Mediterranean coasts during the last glacial episode, with reference to their probable source/origin. The inferred most probable origin results from crosschecking information regarding the present and MIS 5e species' geographic distribution. (*) corresponds to unlikely but still probable sites of origin of the colonisers. BIV: Bivalvia; GAS: Gastropoda; AZO: Azores; MAD: Madeira; SEL: Selvagens; CAN: Canaries; CAB: Cabo Verde; MED: Mediterranean; BER: Bermuda; MOR: Morocco; WES:

Western Sahara; MAU: Mauritania; SEN: Senegal; IVO: Ivory Coast; POR: Portugal; CAR: Caribbean; VIR: Virginean biogeographic province (Atlantic shores of the USA, from Cape Cod (42° N) to Cape Hatteras, North Carolina (35°35′ N)).

Funding: SPA acknowledges his current FCT/2023.07418 CEEECIND research contract with BIOPOLIS (https://doi.org/10.54499/2023.07418.CEECIND/CP2845/CT0001). This work was supported by National Funds through FCT-Fundação para a Ciência e a Tecnologia in the scope of the project UID/50027-Rede de Investigação em Biodiversidade e Biologia Evolutiva. This work also benefitted from FEDER funds, through the Operational Program for Competitiveness Factors—COMPETE; National Funds through FCT (POCI-01–0145-FEDER-006821, UIDB/00153/2020, LA/P/0048/2020); and through the Regional Government of the Azores (M1.1.a/005/Funcionamento-C-/2016, CIBIO-A; M1.1.A/INFRAEST CIENT/A/001/2021; M3.3.B/ORG.R.C./005/2021, M3.3.B/ORG.R.C./008/2022/EDIÇÃO 1, M3.3.G/EXPEDIÇÕES CIENTÍFICAS/005/2022 and M3.3.G/EXPEDIÇÕES CIENTÍFICA S/004/2022). Finally, this work was also supported by FEDER funds (85%) and by funds from the Regional Government of the Azores (15%) through the Azores 2020 Operational Program under the VRPROTO project: Virtual Reality PROTOtype: the geological history of "Pedra- que-pica": ACORES-01-0145-FEDER-000078, and by M1.1.A/INFRAEST CIENT/A/001/2021—Base de Dados da PaleoBiodiversidade da Macaronésia.

Data Availability Statement: The original contributions presented in this study are included in the article/Supplementary Materials. Further inquiries can be directed to the corresponding author.

Conflicts of Interest: The author declares no conflicts of interest. The funders had no role in the design of the study; in the collection, analyses, or interpretation of data; in the writing of the manuscript; or in the decision to publish the results.

References

1. Cronin, T.M. Geographical Isolation in Marine Species: Evolution and Speciation in Ostracoda, I. *Dev. Palaeontol. Stratigr.* **1988**, *11*, 871–889.
2. Vermeij, G. Island life: A view from the sea. In *Frontiers of Biogeography: New Directions in the Geography of Nature*; Lomolino, M.W., Heaney, L.R., Eds.; Sinauer: Sunderland, MA, USA, 2004; pp. 239–254.
3. Baptista, L.; Meimberg, H.; Ávila, S.P.; Santos, A.M.; Curto, M. Dispersal ability, habitat characteristics, and sea-surface circulation shape population structure of *Cingula trifasciata* (Gastropoda: Rissoidae) in the remote Azores Archipelago. *BMC Ecol. Evol.* **2021**, *21*, 128. [CrossRef]
4. Sinigaglia, L.; Baptista, L.; Alves, C.; Feldmann, F.; Sacchetti, C.; Rupprecht, C.; Vijayan, T.; Martín-González, E.; Ávila, S.P.; Santos, A.M.; et al. Oceanic islands act as drivers for the genetic diversity of marine species: *Cardita calyculata* (Linnaeus, 1758) in the NE Atlantic as a case-study. *BMC Ecol. Evol.* **2024**, *24*, 138.
5. Hachich, N.F.; Bonsall, M.B.; Arraut, E.M.; Barneche, D.R.; Lewinsohn, T.M.; Floeter, S.R. Island biogeography: Patterns of marine shallow-water organisms in the Atlantic Ocean. *J. Biogeogr.* **2015**, *45*, 1871–1882. [CrossRef]
6. Hachich, N.F.; Bonsall, M.B.; Arraut, E.M.; Barneche, D.R.; Lewinsohn, T.M.; Floeter, S.R. Marine island biogeography. Response to comment on 'Island biogeography: Patterns of marine shallow-water organisms'. *J. Biogeogr.* **2016**, *43*, 2517–2519. [CrossRef]
7. Hachich, N.F.; Ferrari, D.S.; Quimbayo, J.P.; Pinheiro, H.T.; Floeter, S.R. Island biogeography of marine shallow-water organisms. *Encycl. World's Biomes* **2020**, *1*, 61–75. [CrossRef]
8. Pinheiro, H.T.; Bernardi, G.; Simon, T.; Joyeux, J.-C.; Macieira, R.M.; Gasparini, J.L.; Rocha, C.; Rocha, L.A. Island biogeography of marine organisms. *Nature* **2017**, *82*, 82–86. [CrossRef] [PubMed]
9. Ávila, S.P.; Cordeiro, R.; Haroun, R.; Wirtz, P. Comment on 'Island biogeography: Patterns of marine shallow-water organisms' by Hachich et al., Journal of Biogeography (2015). *J. Biogeog.* **2016**, *43*, 2515–2516. [CrossRef]
10. Ávila, S.P.; Cordeiro, R.; Madeira, P.; Silva, L.; Medeiros, A.; Rebelo, A.C.; Melo, C.; Neto, A.I.; Haroun, R.; Monteiro, A.; et al. Global change impacts on large-scale biogeographic patterns of marine organisms on Atlantic oceanic islands. *Mar. Pollut. Bull.* **2018**, *126*, 101–112. [CrossRef]
11. Ávila, S.P.; Melo, C.; Berning, B.; Sá, N.; Quartau, R.; Rijsdijk, K.F.; Ramalho, R.S.; Cordeiro, R.; De Sá, N.C.; Pimentel, A.; et al. Towards a "Sea-Level Sensitive" dynamic model: Impact of island ontogeny and glacio-eustasy on global patterns of marine island biogeography. *Biol. Rev.* **2019**, *94*, 1116–1142. [CrossRef]
12. Ávila, S.P.; Santos, A.M.; Melo, C.S.; Porteiro, J.M.; Medeiros, A.M.; Baptista, L.; Pimentel, A.; Madeira, P.; Rebelo, A.C.; Hipólito, A.; et al. Extending the Sea-Level Sensitive dynamic model of marine island biogeography to include fusion-fission islands. *Front. Biogeog.* **2025**, *18*, e141200. [CrossRef]

13. Scheltema, R.S. Larval dispersal as a means of genetic exchange between geographically separated populations of shallow-water benthic marine gastropods. *Biol. Bull. Mar. Biol. Lab. Woods Hole Massachussets* **1971**, *140*, 284–322. [CrossRef]
14. Scheltema, R.S. The dispersal of larvae of shoal-water benthic invertebrate species over long distances by ocean currents. In *Fourth Marine Biology Symposium*; Crisp, D.J., Ed.; Cambridge University Press: Cambridge, UK, 1971; pp. 7–28.
15. Scheltema, R.S. The relevance of passive dispersal for the biogeography of Caribbean mollusks. *Am. Malacol. Bull.* **1995**, *11*, 99–115.
16. Scheltema, R.S. Dispersal of marine invertebrate organisms: Paleobiogeographic and biostratigraphic implications. In *Concepts and Methods of Biostratigraphy*; Kauffmann, R.G., Hazel, J.E., Eds.; Dowden, Hutchinson & Ross: Stroudsburg, PA, USA, 1977; pp. 73–108.
17. Scheltema, R.S. Long distance dispersal by planktonic larvae of shoal-water benthic invertebrates among central Pacific Islands. *Bull. Mar. Sci.* **1986**, *39*, 241–256.
18. Scheltema, R.S. On dispersal and planktonic larvae of benthic invertebrates: An eclectic overview and summary of problems. *Bull. Marine Sci.* **1986**, *39*, 290–322.
19. Scheltema, R.S. Planktonic and non-planktonic development among prosobranch gastropods and its relationship to the geographic range of species. In *Reproduction, Genetics and Distribution of Marine Organisms*; Ryland, J.S., Tyler, P.A., Eds.; International Symposium Series; Olsen & Olsen: Hamburg, Germany, 1989; pp. 183–188.
20. Jablonski, D.; Lutz, R.A. Larval ecology of marine benthic invertebrates: Paleobiological implications. *Biol. Rev.* **1983**, *58*, 21–89. [CrossRef]
21. Scheltema, R.S.; Williams, I.P. Long-distance dispersal of planktonic larvae and the biogeography and evolution of some Polynesian and western Pacific mollusks. *Bull. Marine Sci.* **1983**, *33*, 545–565.
22. Scheltema, R.S.; Williams, I.P.; Lobel, P.S. Retention around and long-distance dispersal between oceanic islands by planktonic larvae of benthic gastropod mollusca. *Am. Malacol. Bull.* **1996**, *12*, 67–75.
23. Scheltema, R.S. On the relation between dispersal of pelagic larvae and the evolution of marine prosobranch gastropods. In *Marine Organisms: Genetics, Ecology, and Evolution*; Battaglia, B., Beardmore, J.A., Eds.; NATO Conference Series; Series IV: Marine Sciences; Plenum Press: New York, NY, USA, 1978; pp. 303–322.
24. Scheltema, R.S. Dispersal of pelagic larvae and the zoogeography of Tertiary marine benthic gastropods. In *Historical Biogeography, Plate Tectonics and the Changing Environment*; Gray, J., Boucot, A., Eds.; Oregon State University Press: Corvallis, OR, USA, 1979; pp. 391–397.
25. Jablonski, D.; Hunt, G. Larval ecology, geographic range, and species survivorship in Cretaceous mollusks: Organismic versus species-level explanations. *Am. Nat.* **2006**, *168*, 556–564. [CrossRef]
26. Ávila, S.P. Oceanic islands, rafting, geographical range and bathymetry: A neglected relationship? *Occas. Publ. Ir. Biogeogr. Soc.* **2006**, *9*, 22–39.
27. Ávila, S.P. Unravelling the patterns and processes of evolution of marine life in oceanic islands: A global framework. In *Climate Change Perspectives from the Atlantic: Past, Present and Future*; Fernández-Palacios, J.M., de Nascimento, L., Hernández, J., Clemente, S., González, A., Díaz-González, J.P., Eds.; Universidad de La Laguna, Tenerife: La Laguna, Spain, 2013; pp. 95–125.
28. Ávila, S.P.; Rebelo, A.C.; Medeiros, A.; Melo, C.; Gomes, C.; Bagaço, L.; Madeira, P.; Borges, P.A.; Monteiro, P.; Cordeiro, R.; et al. *Os fósseis de Santa Maria (Açores): 1. A jazida da Prainha*; OVGA—Observatório Vulcanológico e Geotérmico dos Açores: Lagoa, Portugal, 2010; 103p, ISBN 978-989-8164-09-04.
29. Meireles, R.P.; Quartau, R.; Ramalho, R.; Madeira, J.; Rebelo, A.C.; Zanon, V.; Ávila, S.P. Depositional processes on oceanic island shelves—evidence from storm-generated Neogene deposits from the mid-North Atlantic. *Sedimentology* **2013**, *60*, 1769–1785. [CrossRef]
30. Baptista, L.; Santos, A.M.; Cabezas, M.P.; Cordeiro, R.; Melo, C.; Ávila, S.P. Intertidal or subtidal/circalittoral species: Who appeared first? A phylogenetic approach to the evolution of non-planktotrophic species in Atlantic Archipelagos. *Mar. Biol.* **2019**, *166*, 88. [CrossRef]
31. Meco, J. Los *Strombus* Neógenos y Cuaternarios del Atlántico Euroafricano (Taxonomía, Biostratigrafía y Paleoecología). In *Paleontología de Canarias. Tomo I. Excmo*; Cabildo insular de Gran Canaria: Las Palmas de Gran Canaria, Spain, 1977; 142p.
32. Meco, J. Neogastrópodos fósiles de las Canarias orientales. *Anu. Estud. Atl.* **1981**, *27*, 601–615.
33. García-Talavera, F.; Kardas, S.J.; Richards, H.G. Quaternary marine mollusks from Tenerife, Canary Islands. *Nautilus* **1978**, *92*, 97–102.
34. García Talavera, F.G.; Paredes, R.; Martín, M. *Catálogo-Inventario: Yacimientos paleontológicos Provincia de Santa Cruz de Tenerife*; Instituto de Estudios Canarios: La Laguna, Spain, 1989; 76p.
35. García-Talavera, F. Fauna tropical en el Neotirreniense de Santa Maria (I. Azores). *Lav. Soc. Ital. Di Malacol.* **1990**, *23*, 439–443.
36. García-Talavera, F. Sobre la presencia de Strombus latus Gmel. (Mollusca, Megagasteropoda) en el Cuaternario marino de Tenerife. In *Homenaje al Prof. Dr. Telesforo Bravo., Secretariado de Publicaciones*; Universidad de la Laguna: La Laguna, Spain, 1990; pp. 375–382.

37. García-Talavera, F. Fauna malacológica del Cuaternario marino de Cabo Verde. *Rev. Académica Canar. De Las Ciências* **1999**, *11*, 9–25.
38. Meco, J.; Petite-Maire, N.; Fontugne, M.; Shimmield, G.; Ramos, A.J. The Quaternary deposits in Lanzarote and Fuerteventura (eastern Canary Islands, Spain): An overview. In *Climates of the Past*; Meco, J., Petite-Maire, N., Eds.; IUGS-UNESCO-Universidad de Las Palmas de Gran Canaria: Las Palmas de Gran Canaria, Spain, 1997; pp. 123–136.
39. Meco, J.; Gillou, H.; Carracedo, J.C.; Lomoschitz, A.; Ramos, A.J.G.; Rodríguez-Yánez, J.J. The maximum warmings of the Pleistocene world climate recorded in the Canary Islands. *Palaeogeog. Palaeoclim. Palaeoecol.* **2002**, *185*, 197–210. [CrossRef]
40. Meco, J.; Ballester, J.; Betancort, J.F.; Cilleros, A.; Scaillet, S.; Guillou, H.; Carracedo, J.C.; Lomoschitz, A.; Petit-Maire, N.; Ramos, A.J.G.; et al. *Historia geologica del clima en Canarias*; Las Palmas de Gran Canaria: Las Palmas de Gran Canaria, Spain, 2008; 296p.
41. Meco, J.; Lomoschitz, A.; Betancort, J.F.; Sendino, C. Comment on "Range expansion of tropical shallow-water marine molluscs in the NE Atlantic during the last interglacial (MIS 5e): Causes, consequences and utility of ecostratigraphic indicators for the Macaronesian archipelagos", by C.S. Melo, E. Martín-González, C.M. da Silva, I. Galindo, A. González-Rodríguez, L. Baptista, A.C. Rebelo, P. Madeira, A. Voelker, M. Johnson, S.A. Arruda and S.P. Ávila, Quaternary Science Reviews 278 (2022), 107377. *Quat. Sci. Rev.* **2022**, *288*, 107534. [CrossRef]
42. Callapez, P.; Soares, A.F. Late Quaternary warm marine mollusks from Santa Maria (Azores) paleoecologic and paleobiogeographic considerations. *Ciências Terra* **2000**, *14*, 313–322.
43. García-Talavera, F.; Sánchez-Pinto, L. Moluscos marinos fósiles de Selvagem Pequenha e Ilheu de Fora (Islas Salvajes). Descripción de una nueva especie de neogasterópodo. *Vieraea* **2001**, *13*, 9–21.
44. Ávila, S.P.; Amen, R.; Azevedo, J.M.N.; Cachão, M.; García-Talavera, F. Checklist of the Pleistocene marine molluscs of Prainha and Lagoinhas (Santa Maria Island, Azores). *Açoreana.* **2002**, *9*, 343–370.
45. Ávila, S.P.; da Silva, C.M.; Schiebel, R.; Cecca, F.; Backeljau, T.; Martins, A.M.F. How did they get here? Palaeobiogeography of the Pleistocene marine molluscs of the Azores. *Bull. Geol. Soc. Fr.* **2009**, *180*, 201–213. [CrossRef]
46. Ávila, S.P.; Madeira, P.; Zazo, C.; Kroh, A.; Kirby, M.; da Silva, C.M.; Cachão, M.; Martins, A.M.F. Palaeoecology of the Pleistocene (MIS 5.5) outcrops of Santa Maria Island (Azores) in a complex oceanic tectonic setting. *Palaeogeog. Palaeoclim. Palaeoecol.* **2009**, *274*, 18–31. [CrossRef]
47. Ávila, S.P.; Melo, C.; Silva, L.; Ramalho, R.; Quartau, R.; Hipólito, A.; Cordeiro, R.; Rebelo, A.C.; Madeira, P.; Rovere, A.; et al. A review of the MIS 5e highstand deposits from Santa Maria Island (Azores, NE Atlantic): Palaeobiodiversity, palaeoecology and palaeobiogeography. *Quat. Sci. Rev.* **2015**, *114*, 126–148. [CrossRef]
48. Ávila, S.P.; Landau, B.; Ávila, G.C.; Hipólito, A.; Uchman, A.; Johnson, M.E.; Madeira, P. A call to adress the taxonomic gap in the Pleistocene of Santa Maria Island (Azores Archipelago). *Acta Geologica Polonica.* **2025**, *75*, e40. [CrossRef]
49. Ávila, S.P. Processos e Padrões de Dispersão e Colonização nos Rissoidae (Mollusca: Gastropoda) dos Açores. Ph.D. Thesis, Universidade dos Açores, Ponta Delgada, Portugal, 2005; p. 329.
50. Cabero, A. Registro Costero de los Cambios Eustáticos y Climáticos Durante los Interglaciares Recientes Cuaternarios: Sur y Sureste Peninsular, Islas Baleares, Canarias y Cabo Verde. Ph.D. Thesis, University of Salamanca, Salamanca, Spain, 2009; p. 451.
51. Cabero, A.; González-Delgado, J.Á.; Zazo, C.; Goy, J.L.; Dabrio, C.J.; Lario, J.; Bardají, T.; Hillaire-Marcel, C.; Ghaleb, B. Major migration of the Senegalese warm fauna in the Mediterranean during MIS 5. In Proceedings of the International Conference on Geoevents, Geological Heritage, and the Role of the IGCP, Caravaca de la Cruz, Spain, 15–18 September 2010; Lamolda, M.A., Díaz, E., Moreno, G.J., Maurrasse, F.J.-M.R., Meléndez, G., Paul, C.R.C., Tovar, F.J.R., Eds.; pp. 116–118.
52. Martín-González, E.; González-Rodríguez, A.; Vera-Peláez, J.L.; Lozano-Francisco, M.d.C.; Castillo, C. Asociaciones de moluscos de los depósitos litorales del Pleistoceno Superior de Tenerife (Islas Canarias, España). *Vieraea* **2016**, *44*, 87–106.
53. Martín-González, E.; Galindo, I.; Mangas, J.; Romero-Ruiz, C.; Sánchez, N.; González- Rodríguez, A.; Coello, J.J.; Márquez, A.; de Vera, A.; Vegas, J.; et al. Revisión de los depósitos costeros del estadio isotópico marino 5e (MIS5e) de Fuerteventura (islas Canarias). *Vieraea* **2019**, *4*, 667–688. [CrossRef]
54. Melo, C.S.; Martín-González, E.; da Silva, C.M.; Galindo, I.; González-Rodríguez, A.; Baptista, L.; Rebelo, A.C.; Madeira, P.; Voelker, A.H.L.; Johnson, M.E.; et al. Range expansion of tropical shallow-water marine molluscs in the NE Atlantic during the Last Interglacial (MIS 5e): Causes, consequences and utility of ecostratigraphic indicators for the Macaronesian archipelagos. *Quat. Sci. Rev.* **2022**, *278*, 107377. [CrossRef]
55. Melo, C.S.; Martín-González, E.; da Silva, C.M.; Galindo, I.; González-Rodríguez, A.; Baptista, L.; Rebelo, A.C.; Madeira, P.; Voelker, A.H.L.; Johnson, M.E.; et al. Reply to the comment by Meco et al. on "Range expansion of tropical shallow-water marine molluscs in the NE Atlantic during the last interglacial (MIS 5e): Causes, consequences and utility of ecostratigraphic indicators for the Macaronesian archipelagos". *Quat. Sci. Rev.* **2022**, *28*, 107535.
56. Lecointre, G. Recherches sur le Néogène et le Quaternaire marin de la côte atlantique du Maroc. *Notes Mémoires Serv. Géologique Maroc.* **1952**, *99*, 173–198.

57. Lecointre, G. Recherches sur le Néogène et le Quaternaire marin de la côte atlantique du Maroc. Tome III, Les acquisitions nouvelles durant la période de 1952 à 1962 (stratigraphie et paléontologie). *Notes Mémoires Serv. Géologique Maroc.* **1963**, *174*, 1–75.
58. Lecointre, G. Le Quaternaire marin de l'Afrique du nord-ouest. *Quaternaria* **1965**, *7*, 9–28.
59. Ortlieb, L. Recherches sur les formations Plio-Quaternaires du littoral Ouest-Saharien (28°30′–20°40′ lat. N). *Travaux Documents de l'O.R.S.T.O.M.* **1975**, *48*, 1–267.
60. Brebion, P. Etude biostratigraphique et paléoécologique du Quaternaire marocain. *Ann. Paléontologie Invertébrés* **1979**, *65*, 1–42.
61. Brebion, P.; Hoang, C.T.; Weisrock, A. L'intérêt des coupes d'Agadir-Port pour l'étude du Pléistocène supérieur marin du Maroc. *Bull. Muséum Natl. D'histoire Nat. Paris* **1984**, *4*, 129–151.
62. Weisrock, A.; Occhietti, S.; Hoang, C.; Lauriat-Rage, A.; Brebion, P.; Pichet, P. Les séquences littorales pléistocènes de l'Atlas atlantique entre Cap Rhir et Agadir, Maroc. *Quaternaire* **1999**, *10*, 227–244. [CrossRef]
63. Plaziat, J.-C.; Aberkan, M.; Ahmamou, M.; Choukri, A. The quaternary deposits of Morocco. In *The geology of Morocco*; Michar, A., Ed.; Springer: Berlin/Heidelberg, Germany, 2008; pp. 359–379.
64. Chakroun, A.; Chahid, D.; Boudad, L.; Campmas, E.; Lenoble, A.; Nespoulet, R.; El Hajraoui, M.-A. The Pleistocene of Rabat (Morocco): Mollusks, coastal environments and human behavior. *Afr. Archaeol. Rev.* **2017**, *34*, 493–510. [CrossRef]
65. Melo, C.S.; da Silva, C.M.; Scarponi, D.; González, E.; Rolán, E.; Rojas, A.; Martinez, S.; Silva, L.; Johnson, M.; Rebelo, A.C.; et al. Palaeobiogeography of NE Atlantic archipelagos during the Last Interglacial (MIS 5e): A molluscan approach to the conundrum of Macaronesia as a marine biogeographic unit. *Quat. Sci. Rev.* **2023**, *319*, 108313. [CrossRef]
66. La Perna, R.; De Santis, V.; Caldara, M. Two West African molluscan species (Gastropoda, Acteonidae) from MIS 5.5 in the Taranto area (Southern Italy). *Riv. Ital. Paleontol. Stratigr.* **2025**, *131*, 515–527. [CrossRef] [PubMed]
67. Psarras, C.; Thiavaiou, D.; Quillévéré, F.; Conée, J.-J.; Moissette, P.; Philippon, M.; Fietzke, J.; Bosch, D.; Condomines, M.; Bruguier, O.; et al. Integrating Aegean Last Interglacial faunas into the Mediterranean palaeobiogeographic framework: New evidence from Karpathos (Greece). *J. Quat. Sci. in press*.
68. Ávila, S.P. Atlantic and Mediterranean Gastropoda: Geographical Distribution and Functional Traits Database. Version 1.1. 2025. Available online: https://www.researchgate.net/publication/385907293_Atlantic_and_Mediterranean_Gastropoda_geographical_distribution_and_functional_traits_database_Version_11 (accessed on 17 August 2025).
69. Ávila, S.P. Atlantic and Mediterranean Bivalvia: Geographical Distribution and Functional Traits Database. Version 1.0. 2024. Available online: https://www.researchgate.net/publication/387085775_Atlantic_and_Mediterranean_Bivalvia_geographical_distribution_and_functional_traits_database_Version_10 (accessed on 23 August 2025).
70. Chahid, D.; Boudad, L.; Lenoble, A.; El Hmaidi, A.; Chakroun, A.; Jacobs, Z. Nouvelles données morphostratigraphiques et géochronologiques sur le cordon littoral externe (SIM 5-c) de Rabat–Témara, Maroc. *Géomorphologie* **2016**, *22*, 253–264.
71. Chakroun, A.; Zaghbib-Turki, D. Facies and fauna proxies used to reconstruct the MIS 5 and MIS 7 coastal environments in eastern Tunisia. *Geol. Q.* **2017**, *61*, 186–204. [CrossRef]
72. Hoffman, J.S.; Clark, P.U.; Parnell, A.C.; He, F. Regional and global sea-surface temperatures during the last interglaciation. *Science* **2017**, *355*, 276–279. [CrossRef] [PubMed]
73. Gignoux, M. Les formations marines pliocènes et quaternaires de l'Italie du sud et de la Sicile. *Ann. L'université Lyon* **1913**, *36*, 1–693.
74. Carlqvist, S. The biota of long-distance dispersal. I. Principles of dispersal and evolution. *Quart. Rev. Biol.* **1966**, *41*, 247–270. [CrossRef]
75. van den Hoek, C. The possible significance of long-range dispersal for the biogeography of seaweeds. *Helgoländer Meeresunters.* **1987**, *41*, 261–272. [CrossRef]
76. Jokiel, P.L. Long-distance dispersal by rafting: Reemergence of an old hypothesis. *Endeav. New Ser.* **1990**, *14*, 66–73. [CrossRef]
77. Nikula, R.; Spencer, H.G.; Waters, J.M. Passive rafting is a powerful driver of transoceanic gene flow. *Biol. Lett.* **2013**, *9*, 20120821. [CrossRef]
78. Afonso, P.; Porteiro, F.M.; Fontes, J.; Tempera, F.; Morato, T.; Cardigos, F.; Santos, R.S. 2013. New and rare coastal fishes in the Azores islands: Occasional events or tropicalization process? *J. Fish Biol.* **2013**, *83*, 272–294. [CrossRef] [PubMed]
79. González, J.A.; Espino, F.; González-Lorenzo, J.G. Changes in biogeographic patterns of coastal fishes: Indicators of tropicalization in the Canary Islands over the last 40 years. *Mar. Environ. Res.* **2025**, *205*, 107002. [CrossRef]
80. Barcelos, L.M.D.; Barreiros, J.P. Pinniped (Carnivora, Phocidae) occurrences in the Azores Archipelago (NE Atlantic). *Biodivers. Data J.* **2022**, *10*, e96342. [CrossRef]
81. Rees, W.J. The aerial dispersal of Mollusca. *J. Molluscan Stud.* **1965**, *36*, 269–282. [CrossRef]
82. Wesselingh, F.P.; Cadée, G.C.; Renema, W. Flying high: On the airborne dispersal of aquatic organisms as illustrated by the distribution histories of the gastropod genera Tryonia and Planorbarius. *Geol. Mijnb.* **1999**, *7*, 165–174. [CrossRef]
83. Gittenberger, E.; Groenenberg, D.; Kokshoorn, B.; Preece, R. Molecular trails from hitch-hiking snails. *Nature* **2006**, *43*, 409. [CrossRef]

84. Sousa, W.P. Size-dependent predation on the saltmarsh snail *Cerithidea californica* Haldeman. *J. Exper. Mar. Biol. and Ecol.* **1993**, *166*, 19–37. [CrossRef]
85. Burger, J.; Gochfeld, M.; Ridgely, R. Migratory behavior of Franklin's gulls (*Larus pipixcan*) in Peru. *Energy Power Eng.* **2010**, *2*, 143–147. [CrossRef]
86. Edgar, G.J. Dispersal of fauna and floral propagules associated with drifting Macrocystis pyrifera plants. *Mar. Biol.* **1987**, *95*, 599–610.
87. Helmuth, B.; Veit, R.R.; Holberton, R. Long-distance dispersal of a subantarctic brooding bivalve (*Gaimardia trapesina*) by kelp-rafting. *Mar. Biol.* **1994**, *120*, 421–426. [CrossRef]
88. Ingólfsson, A. Floating clumps of seaweed around Iceland: Natural microcosms and a means of dispersal for shore fauna. *Mar. Biol.* **1995**, *122*, 13–21. [CrossRef]
89. Ingólfsson, A. Dynamics of macrofaunal communities of floating seaweed clumps off western Iceland: A study patches on the surface of the sea. *J. Exp. Mar. Biol. Ecol.* **1998**, *231*, 119–137. [CrossRef]
90. Thiel, M. Rafting of benthic macrofauna: Important factors determining the temporal succession of the assemblage on detached macroalgae. *Hydrobiologia* **2003**, *503*, 49–57. [CrossRef]
91. Thiel, M.; Gutow, L. The ecology of rafting in the marine environment. II. The rafting organisms and community. *Oceanog. Mar. Biol. Ann. Rev.* **2005**, *43*, 279–418.
92. Alsos, I.G.; Eidesen, P.B.; Ehrich, D.; Skrede, I.; Westergaard, K.; Jacobsen, G.H.; Landvik, J.Y.; Taberlet, P.; Brochmann, C. Frequent long-distance plant colonization in the changing Arctic. *Science* **2007**, *316*, 1606–1609. [CrossRef]
93. Fraser, C.I.; Nikula, R.; Waters, J.M. Oceanic rafting by a coastal community. *Proc. R. Soc. B* **2011**, *278*, 649–655. [CrossRef]
94. Gutow, L.; Beermann, J.; Buschbaum, C.; Rivadeneira, M.M.; Thiel, M. Castaways can't be choosers—Homogenization of rafting assemblages on floating seaweeds. *J. Sea Res.* **2014**, *95*, 161–171. [CrossRef]
95. Doong, D.J.; Chuang, H.C.; Shieh, C.L.; Hu, J.H. Quantity, distribution, and impacts of coastal driftwood triggered by a typhoon. *Mar. Pollution Bull.* **2011**, *62*, 1446–1454. [CrossRef] [PubMed]
96. Garden, C.J.; Craw, D.; Waters, J.M.; Smith, A. Rafting rocks reveal marine biological dispersal: A case study using clasts from beach-cast macroalgal holdfasts. *Estuar. Coast. Shelf Sci.* **2011**, *95*, 388–394. [CrossRef]
97. Johansen, S.; Hytteborn, H. A contribution to the discussion of biota dispersal with drift ice and driftwood in the North Atlantic. *J. Biogeogr.* **2001**, *28*, 105–115. [CrossRef]
98. Baptista, L.; dos Santos, A.M.; Melo, C.S.; Rebelo, A.C.; Madeira, P.; Cordeiro, R.; Botelho, A.Z.; Hipólito, A.; Pombo, J.; Voelker, A.H.L.; et al. Untangling the origin of the newcomer *Phorcus sauciatus* (Mollusca: Gastropoda) in a remote northeast Atlantic archipelago. *Mar. Biol.* **2021**, *168*, 9. [CrossRef]
99. Frazier, J.; Margaritoulis, D.; Muldoon, K.; Potter, C.W.; Rosewater, J.; Ruckeschel, C.; Sales, S. Epizoan communities on marine turtles. I. Bivalves and gastropod Mollusca. *Mar. Ecol. Prog. Ser. Z. N. I.* **1985**, *6*, 127–140. [CrossRef]
100. Jokiel, P.L. Long-distance dispersal of reef corals by rafting. *Coral Reefs.* **1984**, *3*, 69–76.
101. Bryan, S.E.; Cook, A.; Evans, J.P.; Colls, P.W.; Wells, M.G.; Lawrence, M.G.; Jell, J.S.; Greig, A.; Leslie, R. Pumice rafting and faunal dispersion during 2001–2002 in the southwest Pacific: Record of a dacitic submarine explosive eruption from Tonga. *Earth Planet Sci. Lett.* **2004**, *227*, 135–154. [CrossRef]
102. Bryan, S.E.; Cook, A.G.; Evans, J.P.; Hebden, K.; Hurrey, L.; Colls, P.; Jell, J.S.; Weatherley, D.; Firn, J. Rapid, long-distance dispersal by pumice rafting. *PLoS ONE* **2012**, *7*, e40583. [CrossRef]
103. Highsmith, R.C. Floating and larval rafting as potential dispersal mechanisms in brooding invertebrates. *Mar. Ecol. Prog. Ser.* **1985**, *25*, 169–179.
104. Parker, T.; Tunnicliffe, V. Dispersal strategies of the biota on an oceanic seamount: Implications for ecology and biogeography. *Biol. Bull.* **1994**, *187*, 336–345. [CrossRef]
105. Cornelius, P.F.S. The Azores hydroid fauna and its origin, with discussion of rafting and medusa suppression. *Arquipélago* **1992**, *10*, 75–99.
106. Knight-Jones, P.; Knight-Jones, E.W. Systematics, ecology and distribution of Southern Hemisphere spirorbids. In *First International Polychaete Conference*; Hutchings, P.A., Ed.; Linnean Society: Sydney, Australia, 1984; pp. 196–210.
107. Madeira, P.; Kroh, A.; Cordeiro, R.; Martins, A.M.F.; Ávila, S.P. The echinoderm fauna of the Azores (NE Atlantic Ocean). *Zootaxa* **2019**, *4639*, 1–231. [CrossRef]
108. Svavarsson, J. *Limnoria borealis* (Isopoda, Flabellifera) and its commensal, *Caecijaera borealis* (Isopoda, Asellota), found in Icelandic waters. *Sarsia* **1982**, *67*, 223–226. [CrossRef]
109. Peck, S.B. Diversity and zoogeography of the non-oceanic Crustacea of the Galapagos-Islands, Ecuador (Excluding Terrestrial Isopoda). *Can. J. Zool.-Rev. Can. Zool.* **1994**, *72*, 54–69.
110. Gili, C.; Martinell, J. Relationship between species longevity and larval ecology in nassariid gastropods. *Lethaia* **1994**, *27*, 291–299. [CrossRef]

111. Friend, D.S.; Anderson, B.M.; Allmon, W.D. Geographic contingency, not species sorting, dominates macroevolutionary dynamics in an extinct clade of neogastropods (Volutospina; Volutidae). *Paleobio* **2021**, *47*, 236–250. [CrossRef]
112. Freitas, R.; Romeiras, M.; Silva, L.; Cordeiro, R.; Madeira, P.; González, J.A.; Wirtz, P.; Falcón, J.M.; Brito, A.; Floeter, S.R.; et al. Restructuring of the "Macaronesia" biogeographic unit: A marine multi-taxon biogeographical approach. *Sci. Rep.* **2019**, *9*, 15792. [CrossRef] [PubMed]
113. Shackleton, S.; Baggenstos, D.; Menking, J.A.; Dyonisius, M.N.; Bereiter, B.; Bauska, T.K.; Rhodes, R.H.; Brook, E.J.; Petrenko, V.V.; McConnell, J.R.; et al. Global ocean heat content in the Last Interglacial. *Nat. Geosci.* **2020**, *13*, 77–81. [CrossRef]
114. Zbyszewsky, G.; Ferreira, O.d.V. La faune marine des basses plages quaternaires de Praia et Prainha dans l'ile de Santa Maria (Açores). *Comun. Dos Serviços Geológicos Port.* **1961**, *45*, 467–478.
115. Ávila, S.P.; Cordeiro, R.; Rodrigues, A.R.; Rebelo, A.C.; Melo, C.; Madeira, P.; Pyenson, N.D. Fossil mysticeti from the Pleistocene of Santa Maria island, Azores (NE Atlantic ocean), and the prevalence of fossil cetaceans on oceanic islands. *Palaeontol. Electron.* **2015**, *18.2.27A*, 1–12.
116. Ávila, S.P.; Azevedo, J.M.N.; Madeira, P.; Cordeiro, R.; Melo, C.S.; Baptista, L.; Torres, P.; Johnson, M.E.; Vullo, R. Pliocene and Late-Pleistocene actinopterygian fishes from Santa Maria Island (Azores: NE Atlantic Ocean): Systematics, palaeoecology and palaeobiogeography. *Geol. Mag.* **2020**, *157*, 1526–1542. [CrossRef]
117. Estevens, M.; Ávila, S.P. Fossil whales from the Azores. *Açoreana* **2007**, (Suppl. 5), 140–161.
118. Madeira, P.; Kroh, A.; Martins, A.M.F.; Ávila, S.P. The marine fossils from Santa Maria Island (Azores, Portugal): An historical overview. *Açoreana Supl.* **2007**, *5*, 5–73.
119. Madeira, P.; Kroh, A.; Cordeiro, R.; Meireles, R.; Ávila, S.P. The fossil echinoids of Santa Maria Island, Azores (northern Atlantic Ocean). *Acta Geol. Pol.* **2011**, *61*, 243–264.
120. Hyžný, M.; Melo, C.S.; Ramalho, R.S.; Cordeiro, R.; Madeira, P.; Baptista, L.; Rebelo, A.C.; Gómez, C.; Torres, P.; Uchman, A.; et al. Pliocene and Late Pleistocene (MIS 5e) decapod crustacean crabs from Santa Maria Island (Azores Archipelago: NE Atlantic): Systematics, palaeoecology and palaeobiogeography. *J. Quat. Sci.* **2021**, *36*, 91–109. [CrossRef]
121. Rebelo, A.C.; Rasser, M.W.; Ramalho, R.S.; Johnson, M.E.; Melo, C.S.; Uchman, A.; Quartau, R.; Berning, B.; Neto, A.I.; Mendes, A.R.; et al. Pleistocene coralline algal build-ups on a mid-ocean rocky shore—insights into the MIS 5e record of the Azores. *Palaeobiogeog. Palaeoclim Palaeoecol.* **2021**, *579*, 110598.
122. Ávila, S.P.; Madeira, P.; Mendes, N.; Rebelo, A.; Medeiros, A.; Gomes, C.; García-Talavera, F.; da Silva, C.M.; Cachão, M.; Hillaire-Marcel, C.; et al. Mass extinctions in the Azores during the last glaciation: Fact or myth? *J. Biogeogr.* **2008**, *35*, 1123–1129. [CrossRef]
123. Gerber, J.; Hemmen, J.; Gröh, K. Eine pleistozäne marine Molluskenfauna von Porto Santo (Madeira-Archipel). *Mitteilungen Der Dtsch. Malakozool. Ges. Heft.* **1989**, *44–45*, 10–30.
124. Montesinos, M.; Ramos, A.J.G.; Lomoschitz, A.; Coca, J.; Redondo, A.; Betancort, J.F.; Meco, J. Extralimital Senegalese species during marine isotope stages 5.5 and 11 in the Canary Islands (29° N): Sea Surface temperature estimates. *Palaeogeogr. Palaeoclimatol. Palaeoecol.* **2014**, *410*, 153–163. [CrossRef]
125. González-Rodríguez, A.; Melo, C.S.; Galindo, I.; Mangas, J.; Sánchez, N.; Coello, J.; Lozano-Francisco, M.C.; Johnson, M.E.; Romero, C.; Vegas, J.; et al. Historia geológica y reconstrucción paleobiológica de los depósitos paleontológicos de la playa de El Confital (Gran Canaria, Islas Canarias). *Cuad. Del Mus. Geomin.* **2018**, *27*, 491–499.
126. Zazo, C.; Goy, J.L.; Hillaire-Marcel, C.; Dabrio, C.J.; González-Delgado, J.A.; Cabero, A.; Bardají, T.; Ghaleb, B.; Soler, V. Sea level changes during the last and present interglacials in Sal Island (Cape Verde archipelago). *Glob. Planet. Change* **2010**, *72*, 302–317. [CrossRef]
127. Ávila, S.P.; Paris, R.; Ramalho, R.S.; Melo, C.S.; Martín-González, E.; Rolán, E.; Madeira, P.; Ávila, G.C.; Porteiro, J.M.; Medeiros, A.M.; et al. Mega-tsunami deposits and range expansion of cold-temperate marine species towards the tropics in glacial times. *Front. Biogeog.* **2025**, *18*, e138319. [CrossRef]
128. Cabero, A.; González-Delgado, J.A.; Zazo, C.; Goy, J.L.; Dabrio, C.J.; Lario, J.; Bardají, T.; Hillaire-Marcel, C.; Ghaleb, B. Distribution of warm Senegalese species during MIS 5 in the Atlantic-Mediterranean linkage area: Coastal settings. In *Decoding the Last Interglacial in Western Mediterranean*; INQUA Project 0911-CMP Commission: Sardinia, Italy, 2010; pp. 15–16.
129. Rohling, E.J.; Braun, K.; Grant, K.; Kucera, M.; Roberts, A.P.; Siddall, M.; Trommer, G. Comparison between Holocene and Marine Isotope Stage-11 sea-level histories. *Earth Planet. Sci. Lett.* **2010**, *291*, 97–105. [CrossRef]
130. Govin, A.; Capron, E.; Tzedakis, P.C.; Verheyden, S.; Ghabelb, B.; Hillaire-Marcel, C.; St-Onge, G.; Stoner, J.S.; Bassinot, F.; Bazin, L.; et al. Sequence of events from the onset to the demise of the Last Interglacial: Evaluating strengths and limitations of chronologies used in climatic archives. *Quat. Sci. Rev.* **2015**, *129*, 1–36. [CrossRef]
131. Past Interglacials Working Group of PAGES. Interglacials of the last 800,000 years. *Rev. Geophys.* **2016**, *54*, 162–219.
132. Raymo, M.; Mitrovica, J.X. Collapse of Polar Ice Sheets During the Stage 11 Interglacial. *Nature* **2012**, *483*, 453–456. [CrossRef]
133. Dutton, A.; Carlson, A.E.; Long, A.J.; Milne, G.A.; Clark, P.U.; DeConto, R.; Horton, B.P.; Rahmstorf, S.; Raymo, M.E. Sea-level rise due to polar ice-sheet mass loss during past warm periods. *Science* **2015**, *349*, aaa4019. [CrossRef]

134. Hearty, P.J.; Tormey, B.R. Sea-level change and superstorms: Geologic evidence from the last interglacial (MIS 5e) in the Bahamas and Bermuda offers ominous prospects for a warming Earth. *Mar. Geol.* **2017**, *390*, 347–365. [CrossRef]
135. Clark, P.U.; Huybers, P. Interglacial and future sea level. *Geophys. Res. Lett.* **2009**, *462*, 856–857.
136. Maréchal, C.; Boutier, A.; Mélières, M.-A.; Clauzel, T.; Betancort, J.F.; Lomoschitz, A.; Meco, J.; Fourel, F.; Barral, A.; Amiot, R.; et al. Last interglacial Sea Surface warming during the Sea-level highstand in the Canary Islands: Implications for the Canary current and the upwelling off African coast. *Quat. Sci. Rev.* **2020**, *234*, 106246. [CrossRef]
137. Milanković, M. Kanon der Erdbestrahlung und seine Anwendung auf das Eiszeitenproblem. *R. Serbian Acad. Spec. Publ.* **1941**, *133*, 1–633.
138. Cita Sironi, M.B.; Capotondi, L.; Asioli, A. The Tyrrhenian stage in the Mediterranean: Definition, usage and recognition in the deep-sea record. A proposal. *Accad. Naz. Lincei Rend. Cl. Sci. Fis. Mat. Nat.* **2005**, *9*, 297–310.
139. Issel, A. Lembi fossiliferi quaternari recenti osservati nella Sardegna meridionale dal prof. D. Lovisato. *Rend. Della Accad. Naz. Dei Lincei* **1914**, *23* (Suppl. 5), 759–770.
140. Cuerda Barceló, J. Fauna marina del Tirreniense de la bahía de Palma (Mallorca). *Bolletí De La Soc. D'història Nat. De Les Balear.* **1957**, *3*, 3–75.
141. Cuerda Barceló, J. Moluscos Marinos y Salobres del Pleistoceno balear. Caja de Baleares 'Sa Nostra' Mallorca: Mallorca, Spain, 1987; 421p.
142. Cuerda Barceló, J. *Los Tiempos Cuaternarios en Baleares*, 2nd ed.; Conselleria de Cultura, Educacio i Esports del Govern Balear: Palma, Spain, 1989; 305p.
143. Hillaire-Marcel, C.; Gariepy, C.; Ghaleb, B.; Goy, J.L.; Zazo, C.; Cuerda Barceló, J. U-series measurements in Tyrrhenian deposits from Mallorca. Further evidence for two last-interglacial high sea levels in the Balearic islands. *Quat. Sci. Rev.* **1996**, *15*, 53–62. [CrossRef]
144. Gàsser, Z. Jaciments paleontològics marins del Miocè i Quaternari d'es Ram (Formentera, Illes Pitiüses). *Boll. Soc. Hist. Nat. Balears.* **2002**, *45*, 87–92.
145. Vicens, D. El registre paleontològic dels dipòsits litorals Quaternaris a l'illa de Mallorca (Illes Balears, Mediterrània Occidental). Ph.D. Thesis, University of the Balearic Islands, Palma, Spain, 2015; 1011p.
146. del Valle, L.; Pons, G.X.; Fornos, J.J. Upper Pleistocene Marine Levels of the *Es Copinar–Es Estufadors* (Formentera, Balearic Islands, West Mediterranean). *Quaternary* **2025**, *8*, 38. [CrossRef]
147. Galili, E.; Şevketoğlu, M.; Salamon, A.; Zviely, D.; Mienis, H.K.; Rosen, B.; Moshkovitz, S. Late Quaternary beach deposits and archaeological relicts on the coasts of Cyprus, and the possible implications of sea-level changes and tectonics on the early populations. *Geol. Soc. Lond. Spec. Publ.* **2016**, *411*, 179–218. [CrossRef]
148. Hegab, O.A.; Hesman El Asmar, M. Last interglacial stratigraphy in the burg El-Arab region of the northwestern coast of Egypt. *Quat. Int.* **1995**, *29*, 23–30. [CrossRef]
149. Barriere, J. Le rivage Tyrrhénien de L'étang de Bages et de Sigean (Aude). *Bull. l'Assoc. Française Pour L'etude Quat.* **1966**, *4*, 251–283.
150. Anapliotis, K.A. Les couches à Strombes à l'île Armathia (région de Cassos). *Proc. Acad. Athens* **1963**, *38*, 137–142.
151. Angelier, J.; Gigout, M.; Hogrel, M. A propos du gisement Tyrrhénien d'Arvi (Crete): Cadre stratigraphique, faune, esquisse paléoécologique. *Ann. Géologiques Des Pays Helléniques* **1977**, *28*, 471–488.
152. Issar, A.; Picard, L. Sur le Tyrrhenien des cotes d'Israel et du Liban. *Bull. L'association Française Pour L'étude Du Quat.* **1969**, *6*, 35–41. [CrossRef]
153. Gignoux, M. Les couches à *Strombus bubonius* (LMK) dans la Méditerranée occidentale. *Comptes-Rendus De L'académie Des Sci. Paris.* **1911**, *15*, 1–339.
154. Blanc, A.C. Una spiaggia pleistocenica a *Strombus bubonius* presso Palidoro (Roma). *Rend. Dell'accademia Naz. Dei Lincei* **1936**, *23*, 200–204.
155. Trevisan, L.; di Napoli, E. Tirreniano, Siciliano e Calabriano nella Sicilia Sud-Occidentale. Note di Stratigrafia, Paleontologia e Morfologia. *G. Di Sci. Nat. Ed Econ.* **1938**, *39*, 1–39.
156. Fabiani, R. Tracce di Tirreniano (strati a *Strombus bubonius* Lk) entro la citta di Palermo. *Boll. Della Soc. Di Sci. Nat. Ed Econ. Palermo* **1941**, *19*, 1–7.
157. Segre, A.G. Molluschi del Tirreniano di Porto Torres e di Golfo Aranci (Sardegna). *Boll. Del Serv. Geol. D'italia* **1952**, *73*, 269–290.
158. Segre, A.G. Il Tirreniano del Golfo di Terranova Pausania (Olbia) e la sua fauna malacologica. *Boll. Del Serv. Geol. D'italia* **1954**, *76*, 43–84.
159. Comaschi-Caria, A. Nuovi lembi del quaternario in Sardegna. In *Rendiconti Seminari della Facoltà di Scienze dell'Università di Cagliari*; University of Cagliari: Cagliari, Italy, 1954; Volume 24, pp. 1–4.
160. Malatesta, A. Risultati del rilevamento del Foglio 192 (Alghero-Isola di Sardegna): Fossili delle spiagge tirreni,ane. *Bull. Serv. Geol. Ital.* **1954**, *76*, 9–17.

161. Bonifay, E.; Mars, P. Le Tyrrhenien dans le cadre de la chronologie Quaternaire Mediterraneenne. *Bull. Société Géologique Fr.* **1959**, *7*, 62–78. [CrossRef]
162. Ruggieri, G.; Buccheri, G.; Rendina, M. Segnalazione di Tirreniano fossilifero a Trapani. *Riv. Mineraria Sicil.* **1968**, *112–114*, 216–219.
163. Bonfiglio, L. Il Tirreniano di Bovetto e Ravagnese presso Reggio Calabria. *Quaternaria* **1972**, *6*, 137–148.
164. Ozer, A.; Ulzega, A. Livret-guide de l'excursion-table ronde sur le Tyrrhenian de Sardaigne. In *Annales de Géographie*; Armand Colin: Cagliari, France, 1980; 88p.
165. Vazzana, A. Le conchiglie del periodo Tirreniano nei dintorni di Reggio Calabria. *La Conchiglia* **1988**, *20*, 25–26.
166. Abate, B.; Ferruzza, G.; Incandela, A.; Renda, P. Ritrovamento di depositi a *Strombus bubonius* Lmk (Tirreniano) nell'isola di Favignana. *Riv. Mineraria Sicil.* **1992**, *162*, 37–46.
167. Buccheri, G.; Renda, P.; Morreale, C.; Sorrentino, G. Il Tirreniano dell'Isola di Lampedusa (Arcipelago Pelagiano, Agrigento, Italia). *Boll. Società Geol. Ital.* **1999**, *118*, 361–373.
168. Santagati, P.; Perri, E.; Bernasconi, M.P.; Borrelli, M.; Guerrieri, S.; Critelli, S. MIS 5e sea surface temperature estimation; a multi-proxy approach using a marine macrofossil assemblage (Mar Piccolo, Gulf of Taranto, Southern Italy). *J. Palaeogeogr.* **2024**, *13*, 327–350. [CrossRef]
169. Nalin, R.; Bracchi, V.A.; Basso, D.; Massari, F. *Persististrombus latus* (Gmelin) in the upper Pleistocene deposits of the marine terraces of the Crotone peninsula (southern Italy). *Italian J. Geosci.* **2012**, *131*, 95–101.
170. Amorosi, A.; Antonioli, F.; Bertini, A.; Marabini, S.; Mastronuzzi, G.; Montagna, P.; Negri, A.; Rossi, V.; Scarponi, D.; Taviani, M.; et al. The Middle-Late Quaternary Fronte Section (Taranto, Italy): An exceptionally preserved marine record of the Last Interglacial. *Glob. Planet. Change* **2014**, *119*, 23–38. [CrossRef]
171. Fleisch, H.J.; Comati, P.R.; Elouard, P. Gisement a *Strombus bubonius* Lmk. (Tyrrhenien) a Naame (Liban). *Quaternaria* **1973**, *15*, 217–237.
172. Lario, C.J.; Zazo, C.; Goy, J.L.; Hoyos, M.; Hillaire-Marcel, C. Episodios marinos del último Interglacial (Estadio isotópico 5) del litoral de Málaga (SE peninsular). In *Elementos de los paisajes de la provincia de Málaga, González, J.M., Ferre Bueno, E., Eds.*; (coord. Senciales), Servicio de publicaciones, Universidad de Málaga: Málaga, Spain, 1999; pp. 63–75.
173. Vera-Peláez, J.L.; Lozano-Francisco, M.C.; Ramos Fernández, J.R.; Cortés Sánchez, M. Moluscos del Tirreniense (Pleistoceno Superior) de la Playa la Araña-Cala del Moral (Málaga). *Rev. Española Paleontol.* **2004**, *19*, 251–259. [CrossRef]
174. Bardaji, T.; Goy, J.L.; Zazo, C.; Hillaire-Marcel, C.; Dabrio, C.J.; Cabero, A.; Ghaleb, B.; Silva, P.G.; Lario, J. Sea level and climate changes during OIS 5e in the Western Mediterranean. *Geomorphology* **2009**, *104*, 22–37. [CrossRef]
175. Castany, G. Le niveau à strombes de Tunisie. Sa place dans la chronologie préhistorique du Quaternaire. *Compte-Rendu Somm. Des Séances Société Géologique Fr.* **1954**, 55–56.
176. Castany, G. Données nouvelles sur la stratigraphie du Quaternaire de Jerba. *Bull. de la Soc. des Sci, Nat. de Tunisie.* **1955**, *8*, 135–144.
177. Castany, G. Données nouvelles sur le Quaternaire marin de Monastir (*Tunisie orientale*). *Comptes-Rendus De L'académie Des Sci. Paris* **1956**, *242*, 533–536.
178. Herm, D.; Karray, R.; Pascoff, R.; Sanlaville, P. Sur deux depots a Strombus bubonius du golfe de Tunis. *Bull. Soc. Géologique Fr.* **1975**, *17* (Suppl. 1), 21–22.
179. Ozer, A.; Paskoff, R.; Sanlaville, P.; Ulzega, A. Essai de corrélation du Pléistocène supérieur de la Sardaigne et de la Tunisie. *Comptes Rendus L'academie Des Sci. Paris Série D* **1980**, *291*, 801–804.
180. Chakroun, A. Étude sédimentologique et paléontologique des affleurements du Quaternaire le long de la côte Nord-orientale de la Tunisie. Ph.D. Thesis, Université de Tunis, Tunis, Tunisie, 2006; 400p.
181. Chakroun, A.; Zaghbib Turki, D.; Moncef Turki, M. The Upper Pleistocene deposits in Rafraf (North eastern Tunisia): New data on the *Persististrombus latus* level. *Arabian Journal of Geosciences* **2016**, *9*, 1–13.
182. Chakroun, A.; Zaghbib-Turki, D.; Miskovsky, J.-C.; Davaud, E. Two Tyrrhenian transgressive cycles in coastal deposits of the Cap Bon Peninsula. *Quaternaire* **2009**, *20*, 215–226. [CrossRef]
183. Chakroun, A.; Zaghbib-Turki, D.; Turki, M.-M. Synthèse sur le Pléistocène de la zone côtière de Tunisie. *Mémoire Du Serv. Géologique De L'algérie* **2021**, *22*, 109–128.
184. de Porta, J.; Martinell, J. *El Tyrrheniense Catalan, Sintesis y Nuevas Aportaciones*; Department de Paleontologia Universitat de Barcelona: Barcelona, Spain, 1981; 27p.
185. Lisiecki, L.E.; Raymo, M.E. A Pliocene-Pleistocene stack of 57 globally distributed benthic δ^{18}O records. *Paleoceanography* **2005**, *20*, PA1003. [CrossRef]
186. Meco, J. Los niveles con "*Strombus*" de Jandia (Fuerteventura, Islas Canarias). *Anu. Estud. Atl.* **1975**, *21*, 643–660.
187. Meco, J.; Ballester, J.; Betancourt, J.F.; Cilleros, A.; Scaillet, S.; Guillou, H.; Carracedo, J.C.; Lomoschitz, A.; Petit-Maire, N.; Ramos, A.J.G.; et al. *Paleoclimatologia del Neógeno en las Islas Canarias: Geliense, Pleistoceno y Holoceno*; Ministerio de Medio Ambiente, Universidad de las Palmas de Gran Canaria: Las Palmas, Spain, 2006; p. 203.
188. Taviani, M. Unpersisting *Persististrombus*: A Mediterranean story. *Vieraea* **2014**, *42*, 9–18. [CrossRef]

189. Muhs, D.R.; Meco, J.; Simmons, K.R. Uranium-series ages of corals, sea level history, and palaeozoogeography, Canary Islands, Spain: An exploratory study for two Quaternary interglacial periods. *Palaeogeogr. Palaeoclimatol. Palaeoecol.* **2014**, *394*, 99–118. [CrossRef]
190. Versteegh, G.J.M.; Zonneveld, K.A.F.; Hefter, J.; Romero, O.E.; Fischer, G.; Mollenhauer, G. Performance of temperature and productivity proxies based on long-chain alkane-1, mid-chain diols at test: A 5-year sediment trap record from the Mauritanian upwelling. *Biogeosciences* **2022**, *19*, 1587–1610. [CrossRef]
191. Romero, O.E.; Baumann, K.-H.; Zonneveld, K.A.F.; Donner, B.; Hefter, J.; Hamady, B.; Pospelova, V.; Fischer, G. Flux variability of phyto- and zooplankton communities in the Mauritanian coastal upwelling between 2003 and 2008. *Biogeosciences* **2020**, *17*, 187–214. [CrossRef]
192. Lovecchio, E.; Gruber, N.; Münnich, M. Mesoscale contribution to the long-range offshore transport of organic carbon from the Canary Upwelling System to the open North Atlantic. *Biogeosciences* **2018**, *15*, 5061–5091. [CrossRef]
193. Gabric, A.J.; Garcia, L.; Camp, L.V.; Nykjaer, L.; Eifler, W.; Schrimpf, W. Offshore export of shelf production in the Cape Blanc (Mauritania) giant filament as derived from coastal zone color scanner imagery. *J. Geophys. Res.* **1993**, *98*, 4697–4712. [CrossRef]
194. Hildenbrand, A.; Weis, D.; Madureira, P.; Marques, F.O. Recent plate reorganization at the Azores triple junction: Evidence from combined geochemical and geochronological data on Faial, S. Jorge and Terceira volcanic islands. *Lithos* **2014**, *210–211*, 27–39. [CrossRef]
195. Cornu, S.; Pätzold, J.; Bard, E.; Meco, J.; Cuerda-Barcelo, J. Paleotemperature of the last interglacial period based on $\delta^{18}O$ of *Strombus bubonius* from the western Mediterranean Sea. *Palaeogeogr. Palaeoclimatol. Palaeoecol.* **1993**, *103*, 1–20. [CrossRef]
196. Hevia-Cruz, F.; Sheldon, N.D.; Hildenbrand, A.; Hren, M.T.; Marques, F.O.; Carlut, J.; Chabauz, F. Regional variations of the Azores High across glacial-interglacial timescales. *Paleoceanogr. Palaeoclimatol.* **2024**, *39*, e2023PA004810. [CrossRef]

Journal of
Marine Science
and Engineering

MDPI

Review

Tsunamites Versus Tempestites: A Comprehensive Review from the Precambrian to Recent Times

Mohamed Amine Doukani [1,2,3,*], José Madeira [3,4], Linda Satour [1] and Sérgio P. Ávila [2,3,5,6]

1 Laboratoire de Paléontologie Stratigraphique et Paléoenvironnement, Faculté des Sciences de la Terre et de l'Univers, University of Oran 2 Mohamed Ben Ahmed, BP.1524 El M'Naouer, Oran 31000, Algeria; satourlindas@gmail.com
2 CIBIO, Centro de Investigação em Biodiversidade e Recursos Genéticos, InBIO Laboratório Associado, BIOPOLIS Program in Genomics, Biodiversity and Land Planning, Pólo dos Açores, 9500-801 Ponta Delgada, Portugal; sergio.pa.marques@uac.pt
3 MPB-Marine Palaeontology and Biogeography Laboratory, University of the Azores, Rua da Mãe de Deus, 9501-801 Ponta Delgada, Portugal; jmadeira@ciencias.ulisboa.pt
4 Instituto Dom Luiz, Faculdade de Ciências, Universidade de Lisboa, Campo Grande, 1749-016 Lisboa, Portugal
5 Departamento de Biologia, Universidade dos Açores, 9501-801 Ponta Delgada, Portugal
6 UNESCO Chair—Land Within Sea: Biodiversity & Sustainability in Atlantic Islands, University of the Azores, Rua da Mãe de Deus, 9500-321 Ponta Delgada, Portugal
* Correspondence: doukanimohamedamine@gmail.com or doukani.mohamed@univ-oran2.dz

Abstract

Insight regarding the overall geological history of tsunamis and their impacts requires information gained from preserved deposits. Although recent decades have seen a rise in tsunami deposit studies overall, most reviews focus on specific time intervals, such as the Paleozoic, the K–Pg boundary, the Quaternary, or historical and recent events, while others concentrated on particular depositional settings, including lacustrine, offshore, or onshore environments. This review paper provides a comprehensive synthesis of tsunami deposits spanning the geological record from the Precambrian to recent times based on a global compilation of onshore, offshore, and lacustrine examples. Selections from the available evidence is traced from the oldest known tsunamites in the Archaean through major extinction boundaries such as the K–Pg, to the well-preserved Holocene and historical deposits. The findings indicate that while the fundamental sedimentological signatures of tsunamis have remained broadly consistent over geological time, their recognition in ancient strata remains challenging due to difficulty in differentiating between storm deposits (tempestites) and other high-energy facies. A central aspect of this review is the critical assessment of diagnostic criteria proposed to differentiate tsunamites from tempestites. By using a multidisciplinary approach, integrating sedimentological, paleontological, geochemical, and geomorphological evidence in palaeotsunami research, this review provides a detailed framework to improve the confidence in identifying tsunami deposits. This, in turn, enhances palaeotsunami reconstructions, which are valuable for advancing hazard assessment along vulnerable coastlines.

Keywords: tsunami deposits; geological record; storm deposits; tempestites; earthquakes; sedimentology; taphonomy

1. Introduction

As a scientific discipline, palaeotsunami research remains relatively young. It is still in the process of refining its methodological frameworks and establishing robust criteria for the identification of tsunami deposits. As the field evolves, there has been an increasing emphasis on integrating multidisciplinary approaches, including sedimentology, palaeontology, geomorphology, geochemistry, geophysics and recent/historical records—to enhance the accuracy of tsunami reconstructions. Seminal works by Atwater [1] and Dawson et al. [2] were pivotal in initiating a fundamental paradigm shift in the understanding of tsunami generation and associated sedimentation processes. This shift was further accelerated by the catastrophic tsunamis of the Indian Ocean (2004), South Pacific (2009), Chile (2010), and Japan (2011) [3,4], which profoundly transformed scientific perspectives on tsunami hazards and underscored the critical need to reassess existing models (e.g., [5–7]). In the aftermath of these events, researchers focused on sedimentological evidence, coastal geomorphology, and the integration of geological and historical data in tsunami hazard assessment [8–10].

Over the last three decades, researchers have shown a surge of interest in studying tsunami related deposits with case studies spanning worldwide regions [4–25]. Consequently, this rapid build-up of different datasets and perspectives motivated updated comprehensive reviews aimed at synthesizing and summarizing the state-of-knowledge [19,26–29].

Despite the increasing volume of research on tsunami deposits from the Quaternary, historical, and recent times, evidence from older geological periods, particularly from the Precambrian (>541 Ma) to the Neogene (23.03–2.588 Ma), remain scarce and poorly documented [30–32]. According to Dawson et al. [26], this gap can be attributed to two primary factors: (i) the low preservation potential of tsunami deposits, as the coastal and shallow marine environments most impacted by such events are frequently subjected to continuous sediment reworking; and (ii) the inherent difficulty in distinguishing sedimentary structures generated by tsunamis from those produced by other high-energy marine processes, such as storms. The latter factor presents a major challenge for researchers in the field. That is why recently, many authors have increasingly focused on the comparative analysis of tsunamites and tempestites across various depositional settings, including onshore [17,20,21,24,25,33–38], offshore [11,15,19,39–41], and lacustrine systems [42–45]. These efforts aimed to refine diagnostic criteria and improve the reliability of distinguishing between these two types of high-energy sedimentary deposits. According to Sztanó et al. [43] and Doukani et al. [25], an additional challenge in palaeotsunami research is to accurately determine the triggering mechanism of a tsunami based only on the sedimentological and taphonomic characteristics of its deposits. However, recent and historical tsunami deposits are often directly linked to well-documented triggering events, such as large-magnitude earthquakes, submarine landslides or volcanic eruptions. However, establishing these cause-and-effect relationships becomes much more complex when it comes to older geological records, for which direct observational data are lacking. In these cases, interpreting the origin of tsunami deposits requires careful integration of sedimentological, stratigraphic and tectonic (seismic, volcanic, etc.) information. An additional, though less frequent, triggering mechanism for tsunami events is extraterrestrial impacts, such as the strike of a meteorite. Such events can generate large-scale tsunamis, and this mechanism is often associated with tsunamites preserved near the Cretaceous-Paleogene (K–Pg) boundary [46–52] (Figure 1).

The aim of this paper is to present a comprehensive review of the current state of knowledge regarding tsunami deposits preserved in onshore, offshore, and lacustrine environments, spanning from the earliest known geological records to recent historical events. Special attention is given to the methodological advances and challenges associated with the identification and interpretation of tsunami deposits, particularly in distinguishing them from other high-energy event deposits, such as those generated by storms (tempestites). By synthesizing recent research findings and addressing key diagnostic criteria, this review seeks to highlight existing gaps, ongoing debates, and emerging perspectives in palaeot-

https://doi.org/10.3390/jmse14010049

sunami studies, with the goal of contributing to a more robust framework for recognizing and interpreting tsunami-generated sedimentary records across diverse geological settings.

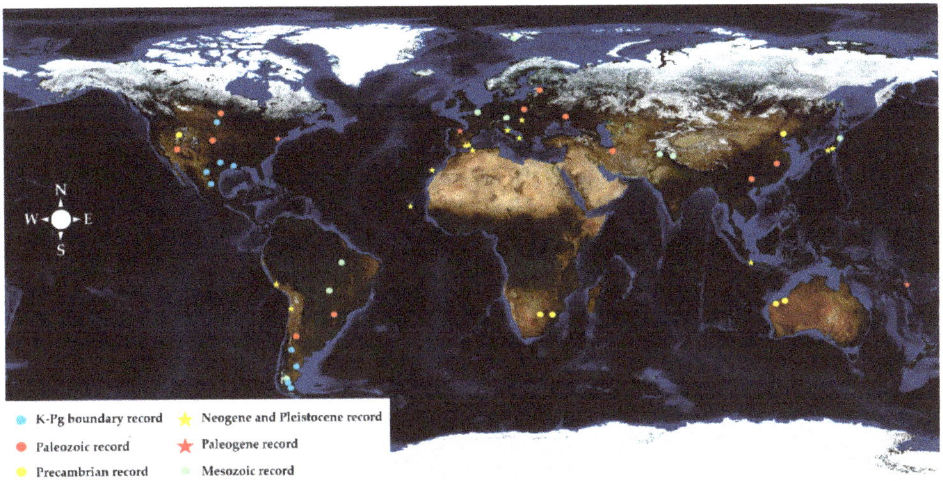

Figure 1. World map showing global distribution of tsunami records from the Precambrian to the Pleistocene compiled in this review.

2. Methodology

This review adheres to the geological time framework established by the International Commission on Stratigraphy, spanning from the Precambrian through the Holocene [53]. The synthesis is structured chronologically, beginning with the Precambrian and progressing through the Paleozoic and Mesozoic. A dedicated subsection addresses the Cretaceous-Paleogene boundary, given its global significance and the abundance of impact-related tsunami studies. From the Cenozoic onward, a finer subdivision is applied, treating the Paleogene and Neogene together and devoting special emphasis to the Pleistocene and Holocene to recent time, where the volume and quality of evidence are greatest.

The present review is based on a comprehensive and critical literature search, namely published articles, book chapters, and monographs, and also "grey literature" (e.g., unpublished PhD and MSc thesis), selected where they explicitly referred to tsunami deposits (tsunamite, tsunamiite) or where they employed broader terminology related to high-energy or catastrophic events, including homogenites, megaturbidites, megatsunamis, mass flows, seismites, and related concepts. Such a broad set of descriptions was adopted to ensure full coverage of the diverse terminologies used across different sub-disciplines of sedimentology, stratigraphy, and geochemistry.

Bibliographic data were compiled through systematic searches of major scientific databases, including Web of Science, Scopus, and GeoRef, supplemented by publishers' platforms such as Elsevier, Springer, and MDPI. Additional references were retrieved through Google Scholar and ResearchGate, especially for grey literature and regionally focused works. Search queries were applied across titles, abstracts, keywords, and, where possible, full texts. Following the removal of duplicates and works of limited relevance, a final corpus of literature was assembled. Although not exhaustive, this review is considered representative of the present state of research, recognizing that variable terminology and the incomplete preservation of ancient deposits inevitably limit coverage.

The collected works were analyzed with three main objectives: (i) to document the sedimentological and paleontological evidence used to infer tsunami deposition (e.g., erosional

 https://doi.org/10.3390/jmse14010049

bases, grading patterns, sedimentary structures, mixture of organisms, taphonomy); (ii) to evaluate the interpretive certainty, which was classified qualitatively: cases interpreted exclusively as tsunami deposits are regarded as highly reliable; those where tsunami origin is proposed alongside alternative mechanisms (such as storms or debris flows) are treated as moderately reliable; weakly supported cases are those where attribution to tsunamis remains tentative, studies explicitly rejecting a tsunami interpretation are also considered, as they highlight the methodological challenges in differentiating tsunamites from other high-energy deposits; (iii) to examine the proposed triggers across different ages, including seismic events, submarine landslides, volcanic activity, extraterrestrial impacts, and climate-driven processes.

Whenever possible, localities were consolidated into broader paleogeographic domains to reduce heterogeneity and highlight regional trends. For example, closely spaced sites within the same basin or coastal setting are treated together, even when multiple case studies are available. However, we do not combine events with different ages into a single example. Instead, we present together studies from the same region (e.g., Cabo Verde) or from nearby sites within the same basin or coastal environment, especially when they fall within the same age interval. This allows us to reduce heterogeneity and highlight regional trends without merging distinct events. Similarly, temporal clusters were identified in intervals where tsunami evidence is most abundant, such as at major extinction horizons, the Permian–Triassic boundary (P–T), the Cretaceous–Paleogene boundary (K–Pg), or within well-dated Quaternary sequences.

Ultimately, this review aims not only to compile occurrences of tsunami deposits through time but also to assess their diagnostic criteria, evaluate the certainty of their interpretation, and compare them against alternative high-energy depositional processes. By integrating cases across multiple geological periods and depositional settings, the study provides a global framework to understand how tsunami deposits are recognized, interpreted, and differentiated from other catastrophic sedimentary records.

3. Tsunami Deposits in the Geological Record

Extreme events such as tsunamis have frequently inundated the world's coastlines [29] and can have a major impact on marine environments [54,55]. According to Schnyder et al. [56], the geological record should reflect a comparable or even higher frequency than that observed in modern and historical times. However, evidence of such deposits is rarely preserved or identified in older strata [26,57]. Despite the challenges of identifying tsunamites in the geological record, due to factors such as post-depositional alteration, limited preservation potential, and the similarity of tsunami deposits to those produced by other high-energy processes, several authors have interpreted some sedimentary sequences as the result of tsunami events across different geological periods [58] (see Figure 2). These interpretations are typically based on a combination of sedimentological and palaeontological criteria, integrated with tectonic and volcanic context, as well as evidence of abrupt marine inundation extending into the onshore environment during run-up, or offshore deposits associated with backwash flows (Table 1).

https://doi.org/10.3390/jmse14010049

Table 1. Listing of tsunami records from the Precambrian to recent times, including age, triggering mechanism, location, and environmental settings. Most of the records are cited and discussed in the main text.

Age	Location	Criteria	Trigger	Environment	Reference
Archaean (3.48 Ga)	Australia	Large imbricated clasts, hummocky bedding, Bouma-type graded sequences.	Unknown	Below storm-wave base	[59]
Archaean (3.2–3.5 Ga)	South Africa	Impact spherules, shock-deformed minerals.	Meteorite impact	Shallow-marine	[60]
Late Archaean (2.6 Ga)	Australia	Anorbital wave ripples.	Meteorite impact	Deep marine shelf	[61]
Palaeoproterozoic (2.5–1.6 Ga)	India	Tidal successions with seismites and tsunami beds; includes deformed cross-bedding, convolute lamination, slump folds, and faults.	Seismic activity	Tidal to shallow-marine basin	[62]
Mesoproterozoic (1.8–1.6 Ga)	China	Rip-up clasts, poorly sorted gravels, fining-upward units, erosional bases.	Seismic activity	Coastal to shallow-marine	[63]
Mesoproterozoic (ca. 1.45 Ga)	USA	Molar-tooth structures, granular lime mud with terrigenous silt, swept ooids and rounded, coarse-grained feldspathic quartz sand, and erosive current with gutter casts.	Seismic activity	Below storm-wave base	[64]
Cambrian (541–485 Ma)	Argentina	Coarse-grained strata, allochthonous input, conglomerate interbedded within burrowed mudstones, and deep scour.	Faulting or rifting beyond the shelf margin	Continental shelves and epeiric sea	[65]
Cambrian (541–485 Ma)	USA	Deformation cracks, intraclast angularity, polygonal plan-view geometries, and intense scouring	Seismic activity	Shallow subtidal intracratonic shelf	[66]
Ordovician (485.4–443.8 Ma)	Poland	Single siliciclastic turbidite lobe,	Unknown	Outer ramp carbonates	[67]
Ordovician (485.4–443.8 Ma)	Canada	Widespread carbonate conglomerate, argillaceous marker beds, and lack of quartz sand/silt (not storm-related).	Seismic activity	Shallow, tropical epeiric (carbonate) shelf within an intracratonic basin	[68]
Ordovician (485.4–443.8 Ma)	Argentina	Graded beds overlying sharply scoured surfaces, anomalous intraclastic rudstone and laminated dolostone, lack of evidence for storms or eustatic sea-level changes.	Seismic activity	Shallow subtidal	[69]

https://doi.org/10.3390/jmse14010049

Table 1. *Cont.*

Age	Location	Criteria	Trigger	Environment	Reference
Silurian (443.8–419.62 Ma)	Ukraine	Sand to boulders size clasts and grainstone to mudstone laminae.	Unknown	Tidal flat/shallow-marine carbonates	[70]
Devonian (419.62–358.86 Ma)	USA	Carbonate megabreccia with shocked quartz, ejecta spherules, and multiple graded tsunami units.	Meteorite impact	Outer carbonate platform to slope	[71]
Devonian (419.62–358.86 Ma)	Poland	Stromatoporoid accumulations (allobiostrome and parabiostrome) with distinctive morphometric and taphonomic features indicating deposition by a high-energy extreme event.	Unknown	Shallow-water carbonate platform	[72]
Devonian (419.62–358.86 Ma)	China	Isochronous event deposits (rudstone, calcirudite, turbidite, homogenite) across platform-margin slope and inter-platform trough facies	Bolide impact	Platform-margin slope to inter-platform	[73]
Permian (298.90–252.17 Ma)	Brazil	Extensive soft-sediment deformation (seismites), thick debritic conglomeratic breccia bed with irregular, scoured base and fining-upward structure, chaotically oriented, imbricated clasts (10–400 cm) of chert, siltstone, and sandstone derived from underlying strata, shock-metamorphosed zircon grains.	Bolide impact	Shallow epicontinental marine to lacustrine conditions	[74]
Permian (298.90–252.17 Ma)	Brazil	Shell-rich rudstone and conglomerate beds, bioclastic concentrations with taphonomic signatures of sudden reworking.	Seismic activity	Lacustrine system	[45]
P-Tr boundary (~252 Ma)	India	Fining-upward bioclastic grainstone beds interbedded with argillites, hummocky cross-stratification, grading, and coarse clasts.	Seismic activity	Outer shelf to upper slope marine setting	[75]

https://doi.org/10.3390/jmse14010049

Table 1. *Cont.*

Age	Location	Criteria	Trigger	Environment	Reference
Triassic (251.9–201.3 Ma)	Germany	Laterally extensive crinoid-bearing bioclastic bed, Heterogeneous facies association with grain-size sorting and lateral transitions, Chaotic orientation of terebratulids and crinoids, mixed faunas from various ramp zones.	Seismic activity	Outer ramp to basin setting	[76]
Jurassic–Cretaceous boundary (~143 Ma)	France	Basal erosional surface, soft-sediment deformation structures, lateral facies variation, and erosional conglomerate overlain by wood fragments and clay.	Unknown	Coastal to nearshore	[56]
Cretaceous (143.1–66.0 Ma)	Pacific margin	Extraordinary amber concentrations, flame-like structures, terrestrial material (resin) over long distances from continental sources, and Absence of subaerial exposure before deposition.	Massive submarine landslide	Pelagic deep-sea	[77]
K–Pg boundary (~66 Ma)	Argentina	Presence of erosional and reworked horizons and mixed continental and marine fossils.	Chicxulub impact	Shallow marine to marginal marine	[78]
K–Pg boundary (~66 Ma)	Cuba	Ripple cross-laminations, changes in detrital provenance, and maximum grain-size variations.	Chicxulub impact	Shallow marine to coastal shelf	[47]
K–Pg boundary (~66 Ma)	Chile	Single iridium anomaly, Palynofacies disturbance: abrupt transition from fresh cuticles to degraded plant matter and spike in spore abundance.	Chicxulub impact	Fluvio-deltaic setting	[51]
Eocene (56–33.9 Ma)	USA	Sandy carbonaceous clay with carbon glass and rock fragments, sandy-matrix breccia with terrestrial (paleosol rip-ups, wood) and marine clasts (fossiliferous chert)	Chesapeake Bay impact	Shallow marine/coastal plain setting	[79]

https://doi.org/10.3390/jmse14010049

Table 1. *Cont.*

Age	Location	Criteria	Trigger	Environment	Reference
Eocene (56–33.9 Ma)	Spain	Typical deposits represent large, disintegrative submarine landslides, but with reduced occurrence during the thermal maximum	Initial Eocene Thermal Maximum (IETM)	Deep-sea continental slope and abyssal settings	[80]
Late Oligocene–Early Miocene (27.82–20.44 Ma)	Southwest Pacific	Massive spilite-rich rudite, graded coarse and fine rudite–arenite, intraformational rudite, brown/grey siltstone, minor calcarenite, airfall tuff, and Bouma-type turbidites, Dominance of massive and graded rudite–arenite beds from high-density turbidity currents.	Volcanic activity and slope failure	Deep-sea fan system	[81]
Miocene (23.03–5.333 Ma)	Chile	Boulder-bearing breccia and poorly sorted sandstone, inverse to normal grading, mixed subaerial, beach, and marine sources.	Seafloor fault	Shoreface	[82]
Miocene (23.03–5.333 Ma)	Hungary	Cobble- to boulder-grade gravel, subangular clasts derived from local Cretaceous sandstone, landward imbrication, mixed clayey-sandy matrix with articulated molluscs and ostracods	Seismic activity	Lacustrine system	[43]
Miocene (23.03–5.333 Ma)	Spain	Folded layers overlain by convex-upward, stratified megahummocks, thick shell-debris bed, erosional surfaces, inflow/backflow structures, and bioclastic redeposition	Seismic activity	Outer-ramp carbonate setting	[11]
Late Miocene–Early Pliocene (11.63–3.60 Ma)	Indonesia	Normal-graded sandstone with disturbed structures (siltstone rip-ups, clay clasts, flame structures), bimodal to multimodal grain-size distribution, mixed marine microfossils (inner to middle neritic).	Seismic activity	Shallow-marine	[83]

https://doi.org/10.3390/jmse14010049

Table 1. *Cont.*

Age	Location	Criteria	Trigger	Environment	Reference
Pliocene (5.333–2.588 Ma)	Italy	Composite shell bed, chaotic taxonomic mixing, draping geometry (thicker in lows), articulated valves, limited breakage, reworked deep-infaunal bivalves, and rapid burial	Unknown	Shallow-marine carbonate	[15]
Pleistocene (2.588–0.0117 Ma)	Maio Island (Cabo Verde)	Multiple stacked tsunami conglomerates and sandstones, erosive bases, rip-up clasts, high runups (>60 m a.s.l.), coarse clasts reworked from coastal settings.	Volcanic flank collapses	Coastal setting	[20]
Pleistocene (2.588–0.0117 Ma)	Santiago Island (Cabo Verde)	Erosive base, rip-up clasts of paleosol, high runup (>250 m a.s.l.), marine bioclasts out of life position, poorly to very poorly sorted conglomerate with boulders.	Volcanic flank collapse	Coastal setting	[16]
Pleistocene (2.588–0.0117 Ma)	Algeria	Irregular erosive base, lateral facies variation, wide grain-size range (clay to boulders), normal and inverse grading, angular boulders, fragile and hard-rock rip-up clasts, injection and deformation structures, imbricated and chaotic shell orientations, well-preserved mixed fauna from supralittoral to circalittoral zones, good sorting and sharp-edged fragments.	Seismic activity	Coastal setting	[25]
Pleistocene (2.588–0.0117 Ma)	Spain	Erosive unconformity, random accumulations of *Glycymeris* and *Acanthocardia* shells; well-preserved, disarticulated bivalves; and chaotic fabric.	Seismic activity	Coastal setting	[37]
Holocene (11.7–0.0 ka)	Ionian Sea	Thick (up to 10–20 m) graded mud turbidites and megaturbidites.	Collapse of the Santorini caldera.	Deep-sea to abyssal plain	[84]

https://doi.org/10.3390/jmse14010049

Table 1. *Cont.*

Age	Location	Criteria	Trigger	Environment	Reference
Holocene (11.7–0.0 ka)	Spain	Erosion of beaches and shoreline retreat, breaching and overwash of sandy barriers, deposition of bioclastic layers, and geomorphological reshaping of estuarine mouths.	Seismic activity	Estuarine and coastal setting	[85]
Holocene (11.7–0.0 ka)	Japan	Multiple sandy and gravelly sub-layers with scoured bases, inverse and normal grading, hummocky cross-stratification (HCS), alternations of sand and mud drapes.	Seismic activity	Shallow marine bay/nearshore	[4]
Holocene (11.7–0.0 ka)	Japan	Marine geochemical and paleontological signatures, erosive basal contacts, fining upward trends, presence of marine fossils, and geophysical (GPR) evidence of lateral continuity.	Seismic activity	Coastal setting	[86]
Holocene (11.7–0.0 ka)	Portugal	Coarse-grained layers with erosive bases, abrupt contrast with background fine sediments, distinct internal structures; element composition anomalies.	1755 CE Lisbon	Offshore shelf	[41]

https://doi.org/10.3390/jmse14010049

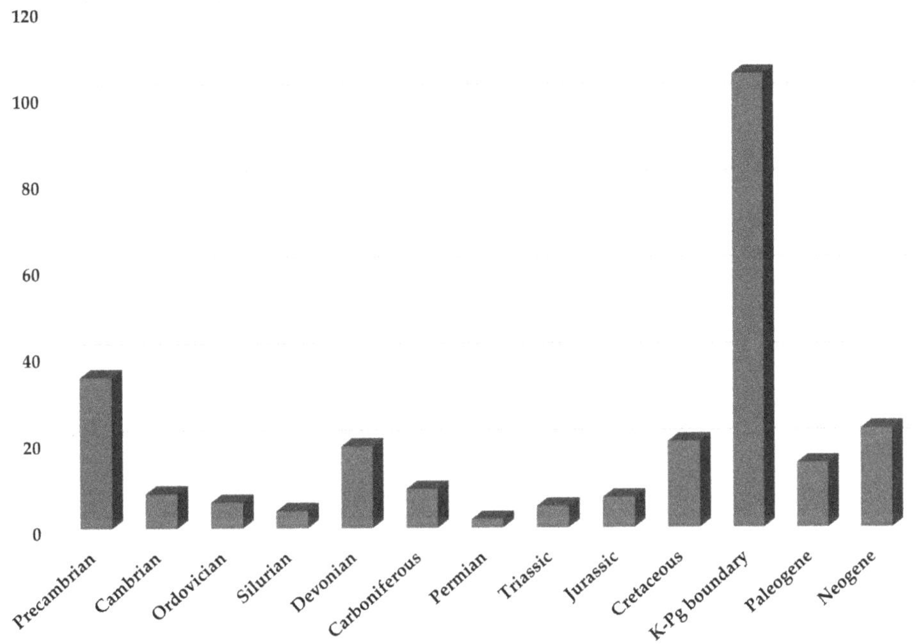

Figure 2. Number of published studies on tsunami deposits by geological age, from the Precambrian to Neogene. Data were compiled from major scientific databases and supplemented by publisher platforms and Google Scholar. Only studies explicitly addressing tsunami deposits or related high-energy sedimentary processes were included.

3.1. Precambrian

The oldest tsunami deposits ever reported on Earth dates to Early Archaean (ca. 3.48 Ga) and is located in the Pilbara Craton, Western Australia [59]. The authors link this catastrophic event to local faulting associated with volcanic activity, although an asteroid impact or volcanic eruption is also possible. Field features, include large imbricated clasts, hummocky bedding, and turbidity-type graded layers, indicate bidirectional currents. The lower sedimentary deposits were produced by the incoming tsunami waves, while the upper deposits testify the backwash process. However, previously the Hamersley Group of Western Australia provided the geological record of what was considered until recently the oldest known tsunami (~2.6 Ga), attributed to the Late Archaean–Palaeoproterozoic [61]. Here, sedimentary ripples are preserved in deep-marine shelf deposits containing impact-melt spherules—small glassy droplets produced by asteroid impacts during the Late Archaean. Numerical simulations supporting this interpretation indicate that a bolide with a diameter of 4.9 km, a density of 2.6 g/cm^3, and an impact velocity of 20 km/s could generate a crater approximately 60.4 km in diameter. Modelling of tsunami waves produced by the subsequent crater collapse, using shallow-water wave theory, suggests that impacts occurring between 2000 and 8000 km away would have been capable of generating waves strong enough to form the observed ripple structures in the Hamersley Basin. Byerly et al. [60] documented another Archaean example from the Barberton Greenstone Belt in South Africa. Both events are associated with extraterrestrial impacts, which occurred frequently during this period of early Earth history [87]. These impact-related tsunamites contain diagnostic features such as impact spherules, minerals showing shock-

https://doi.org/10.3390/jmse14010049

induced deformation (e.g., shocked quartz), and distinctive cosmochemical signatures (microtektites) [60,88,89].

During the Mesoproterozoic, within the Changcheng Group (1.8 to 1.6 Ga) of the North China Craton, Wang et al. [63] identified tsunami deposits representing one of the oldest recognized examples of tsunamites in the Northern Hemisphere. This event was attributed to a large earthquake in an active extensional setting during the breakup of the Columbia (Nuna) supercontinent. Although these deposits are approximately 1.5 billion years old, they display sedimentological characteristics essentially identical to those documented from the Quaternary and even from historical to recent tsunami deposits in onshore settings. Key features that distinguish them from normal shallow-marine siliciclastic facies include abundant rip-up clasts, poor sorting with a wide range of grain sizes from clay to gravel, fining-upward sequences indicative of waning flow, reworked and redeposited material eroded from underlying strata suggesting a complex provenance from multiple source units, and erosional scours at the base of beds produced by strong incoming waves and backwash flows. Similar features have been reported from Pleistocene and Holocene outcrops, e.g., [17,20,25] (Figure 3A,B), collectively providing compelling evidence for a high-energy tsunami origin, demonstrating that the fundamental depositional signatures of tsunami events have remained consistent over geological time.

Figure 3. Characteristics of tsunamigenic sediments deposited inland from Santiago island and Northwestern Algeria. (**A**) Rip-up clasts from the underlying soil in a tsunami deposit at Chão do Porto, Santiago Island (Cabo Verde). (**B**) Abundant mud and paleosol clasts ripped up from the underlying substrate (Port de Menard, Northwestern Algeria). (**C**) Tsunami deposit on Santiago Island over tuffites; ~80-ton basalt block with mixed boulders and pebbles in biogenic sand matrix. Walking stick (1.4 m) for scale. (**D**) Highly heterometric tsunami deposit on Santiago Island filling an erosive surface in tuffites; normal grading with larger clasts at base. Person (1.75 m) for scale. (**E**) Accumulation of variously sized materials in the coastal area of the Kef Boughetar site (northwestern Algeria). (**F**) Irregular erosional surface between the tsunami deposits and the underlying Pliocene basement, with the presence of parallel lamination within the tsunami deposits at Sid El Adjel beach (Northwestern Algeria).

https://doi.org/10.3390/jmse14010049

In Western North America, tsunami deposits from a previously offshore marine setting were recognized in the Mesoproterozoic Helena Formation of the Belt Supergroup (~1.45 Ga) [64]. The deposits associated with this event are characterized by the widespread occurrence of molar-tooth structures, which are composed of offshore swept ooids and rounded, coarse-grained, feldspath and quartz sand. These beds display an erosive base that testify to the action of strong, high-energy currents, producing gutter casts and other scour features. Such sedimentary structures indicate deposition under the influence of powerful tsunami-generated flows, attributed to major submarine fault surface ruptures during episodes of tectonic subsidence.

3.2. Palaeozoic

The Palaeozoic (ca. 541–252 Ma) is known as a period of major tectonic and climatic changes, as well as the proliferation of more complex biological forms. Intense tectonic activity created more outcrops and exposures that remain accessible today, which likely contributed to the greater number of published studies on tsunami deposits (~40 works), compared to those documented from the Precambrian (Figure 2). Despite the difficulty in identifying and dating depositional signatures of catastrophic events, evidence suggests that such records (tsunamites and seismites) extend deep into geological time (e.g., Early Paleozoic). For example, Guo et al. [90] documented deformation structures induced by earthquakes of magnitude 7.0–7.6 in the Early Cambrian of western Zhejiang, China, which are classified as seismites. Such strong tectonic activity highlights the potential for tsunami generation during the Early Paleozoic. During the Cambrian, many shallow-marine deposits from both South and North America were traditionally attributed to storm activity [65,66]. However, new interpretation suggests that some of these record tsunami events instead. In the Middle Cambrian La Laja Formation (Argentine Precordillera), coarse-grained strata on an epeiric shelf include lenticular beds of intraclastic conglomerate interbedded within burrowed mudstones. These beds lack stratigraphic patterns, allochthonous input, or evidence of sea-level fluctuation, and instead record brief, anomalous episodes of deep scour and strong oscillatory flow that are best explained by tsunami waves rather than storms [65]. Similarly, the Upper Cambrian Deadwood Formation (Montana, USA) provides one of the earliest confident identifications of tsunamites in the Paleozoic. Here, conglomerates composed of angular to subangular sandstone intraclasts with a high degree of scouring, were long interpreted as storm deposits [91]. However, Pratt [66] proposed recognition criteria that distinguish these deposits from storm-generated counterparts, including deformation cracks, intraclast angularity, polygonal plan-view geometries, and intense scouring, thereby linking them to tsunami activity. These case studies demonstrate that tsunamites can be reliably differentiated from tempestites in ancient shallow-marine successions. Taken together, they challenge the conventional tempestite paradigm and emphasize the importance of recognizing tsunamites as a distinct class of high-energy deposits in the geological record.

Landslides, slumps, and density-based avalanches can generate sediment gravity flows (turbidity current) [92]. According to Shanmugam [29] earthquakes, tsunamis, and cyclones are additional powerful triggers of turbidity currents. A comparable case was reported by Põldsaar et al. [67] from the Middle Ordovician shallow-marine deposits of the Baltoscandian palaeobasin (Northern Europe), where sedimentological evidence indicates a single turbidite event displaying a Bouma sequence. This deposit was interpreted as the product of a rare tsunami, which eroded and transported sediments from land into deeper parts of the basin. Such evidence demonstrates that tsunami-triggered turbidites can also occur in epicontinental settings, not only along active continental margins. Ruban [93] synthesized Ordovician tsunami deposits (485.4–443.8 Ma) from various paleocontinent do-

https://doi.org/10.3390/jmse14010049

mains, including Gondwana, Laurentia, Baltica, North China, and Siberia, and concluded that our knowledge of ancient (pre-Quaternary) tsunamis is extremely limited, likely representing less than 0.00001% of all events. Pratt and Sproat [68] documented a widespread carbonate conglomerate in the latest Ordovician Stonewall Formation (Williston Basin, central Laurentia), composed of dolomudstone intraclasts, and interpreted as the result of successive tsunami waves and backwash. This interpretation may also account for other anomalous marker beds in shallow-marine epeiric seas, particularly in settings where storm activity or sea-level fluctuations are unlikely [69].

In Sweden, Parnell [94] concluded that the unusually high concentration of meteorites over a short time interval indicates a flux roughly 100 times greater than present-day rates. These events may have triggered a tsunami during the Middle Ordovician. Concurrently, sedimentary megabreccias (i.e., large rock fragments dispersed in a fine matrix) were deposited in multiple locations. These deposits likely reflect downslope movement of sediments along continental margins on a global scale. The megabreccias may have been triggered by seismic activity and slope destabilization caused by the meteorite bombardment, suggesting that the anomalous occurrence of such deposits in the geological record could serve as a proxy for other episodes of enhanced meteorite delivery to Earth.

Contrary to other systems in the Paleozoic, Silurian tsunami deposits (ca. 443–419 Ma) are still poorly documented. One of the rare examples is reported by Jarochowska and Munnecke [70], who interpreted the Makarivka Member of the Ustya Formation (Ukraine) as a tsunamite, with the run-up phase represented by conglomerate facies, while the heterolith unit composed of grainstone and mudstone laminae reflects alternating high-density landward currents, stagnant intervals that allow mud and land-derived debris to settle, and backwash flows. This study provides rare insights into the interplay between high-energy redeposition and carbon isotope stratigraphy in the geological record, where a tsunami event mixes sediments from different depths and stratigraphic levels, producing clasts with variable $\delta^{13}C$ values.

In contrast, there are a remarkably large number of deposits related to tsunami events documented for the Devonian period across different domains [71–73,95,96]. This abundance may be explained by the fact that the Devonian coincides with major orogenic events, such as the Acadian orogeny (Laurentia-Avalonia collision) and the ongoing Caledonian-Variscan tectonics, which created seismically active continental margins—ideal settings for large earthquakes capable of generating tsunamis. This intense seismic activity has been linked by Du et al. [73] to the formation of seismite structures, including seismic-cracks, sandstone dykes, syn-depositional faults, microfolds (micro-corrugated lamination), fluidized veins, load casts, flame structures, pillow structures, and brecciation, resulting from the collision between the North China Plate and the Qaidam microplate during the Devonian. Extraterrestrial bolide impacts are another trigger of Devonian tsunami deposits as reported by Warme and Kuehner [71] from southern Nevada, where shallow-water carbonate-platform deposits are represented by a carbonate breccia—an unusual sedimentary unit containing shocked quartz grains, unique ejecta spherules, and an iridium anomaly interpreted as impact-generated tsunami event. Additional evidence suggests that the event occurred over a few hours to days, with the impact producing a seismic shock that delaminated the upper 50–100 m of the platform, generating large carbonate blocks. Taphonomic analysis of skeletal concentrations is often used to identify depositional processes in onshore settings [24–26,34,97], or from offshore environments [11,15,19]. From the central Poland, within the Devonian Kowala Formation, Łuczyński [72] applied the same methods to examine the sedimentary history of two types of stromatoporoid accumulations—an allobiostrome and a parabiostrome. By combining sedimentological observations with taphonomic analyses, he concluded that the large-scale onshore redeposi-

https://doi.org/10.3390/jmse14010049

tion of stromatoporoid skeletons could only have occurred through erosion at considerable depths, most likely caused by a tsunami event.

Despite the long duration of the Carboniferous (~60 Ma) and its complex tectonic setting, this system represents a paradoxical interval in tsunami research [93], with only rare scientific papers reporting evidence of deposits attributed to tsunami events, most of them discussed only marginally, e.g., [98–102]. This rarity likely reflects a research bias rather than a genuine absence of tsunamites or their poor preservation. Similarly, what may be the only high-energy extreme event recorded in deposits from the Late Permian from the intracratonic Paraná Basin (Brazil), is documented by Tohver et al. [74] which were interpreted as being related to an impactogenic earthquake.

3.3. Mesozoic

The Permian–Triassic boundary (P–Tr), represents one of the most challenging limits in the geological records, characterized by the largest biological crisis in Earth's history [103,104], during which approximately 90% of marine species and about 70% of terrestrial species were eliminated [105]. Based on limited occurrences of shocked quartz, Retallack et al. [106] proposed that an extraterrestrial impact was the cause of this event. However, according to Reichow et al. [107] and Korte et al. [108], the eruption of the Siberian Traps flood basalts is currently regarded as the most probable trigger for the extinction event, as the environmental disturbances it generated are widely considered decisive. Within this debate and close to the paleontologically defined P–Tr boundary at Guryul Ravine in Kashmir (India), Brookfield et al. [75] reported sedimentary structures interpreted as seismites, including convoluted bedding and fluid-escape structures, followed by three lenticular, fining-upward bioclastic grainstone beds interbedded with argillites and interpreted as tsunamites. According to the authors, the presence of hummocky cross-stratification and grading within these beds, combined with the associated physical processes and faunal assemblages, indicates deposition by waning irregular waves at water depths exceeding 100 m. The recurrence of such deposits suggests multiple successive tsunami events in a large open-ocean setting. However, the absence of geochemical evidence for an extraterrestrial impact, supports a terrestrial origin, likely related to tectonic or volcanic activity. The same types of sedimentary structures have also been observed in offshore tsunami deposits associated with large earthquakes from Spain [19].

At the P–Tr boundary (~251 Ma), seismites and tsunamites have been reported in multiple sections globally [109–114], suggesting that voluminous volcanic activity, earthquakes, bolide impacts, and related tsunamis may have been significant contributors to the environmental stress during the largest mass extinction in Earth history. A recent review of the extant literature on Triassic tsunamis [114], highlighted that most reported examples cluster around two major intervals: the Permian–Triassic transition and the end-Triassic crisis, close to the Triassic–Jurassic boundary. The certainty of these interpretations remains limited, as many of the proposed tsunami deposits could alternatively be attributed to storm activity. This ambiguity arises from two main factors. First, distinguishing between storm- and tsunami-generated facies in the rock record is challenging due to the inherent variability of such sedimentary features. Second, the paucity of detailed sedimentological studies available for the Triassic further complicates matters. Differentiation between tsunami and storm deposits thus relies primarily on the development and application of robust diagnostic criteria rather than the number of examples available from a specific period. Therefore, although the potential role of tsunamis during these major biotic crises is intriguing, the existing evidence should be considered provisional until further investigations, involving more precise sedimentological, stratigraphic, and geochemical analyses, like those reported near the K–Pg boundary, are conducted. The difficulty in

https://doi.org/10.3390/jmse14010049

unequivocally identifying the triggering mechanism also means that most interpretations of tsunamis-related deposits in the Mesozoic remain uncertain. A representative example is provided by Schnyder et al. [56], who attributed sedimentary units deposited around the Jurassic–Cretaceous boundary from northern France to a tsunami event. Their interpretation was based on comparison with modern tsunami deposits and the recognition of diagnostic features, including a basal erosional surface, soft-sediment deformation, the presence of wood fragments, and clay layers containing a mixed assemblage of continental and marine fauna. This last criterion, faunal mixing across different marine environments or with continental fauna, has often been considered a benchmark indicator of tsunami deposits [15,19,24–26,34,37]. Nevertheless, despite the presence of these diagnostic features, the absence of direct evidence for the triggering source led the authors to classify these sediments as possible rather than definitively tsunamites.

Since the Cretaceous, and in contrast to the other systems of the Mesozoic, research on palaeotsunami deposits has expanded considerably. More robust interpretations are now supported by multiple lines of evidence, including several diagnostic criteria. These include anomalous concentrations of amber in deep-sea settings far from its continental source and the occurrence of deformation features identical to soft-sediment structures in surrounding deposits [77], rapid burial of mixed faunas without time for reworking, taphonomic contrasts such as articulated shells versus fragmented shells, the intermixing of freshwater and marine faunas within the same layer [115], and the exceptional preservation of molluscan fossils from different shallow-marine settings [116]. Additional diagnostic features include the large lateral extent and facies uniformity of the deposits, their multi-scale facies heterogeneity, and their predictable petrophysical properties [117]. The latter case-study highlights how tsunami deposits can be identified not only through sedimentological and taphonomic features, but also by their large-scale architecture and internal facies organization, with implications that extend beyond sedimentology (e.g., for reservoir characterization).

The triggering mechanisms of these deposits are linked to a range of geological processes, including major earthquakes [118,119], submarine landslides [77], and even bolide impacts [120]. This diversification of recognized triggers reflects not only the improved preservation potential and accessibility of Cretaceous tsunamites, but also the increasing application of modern sedimentological and palaeontological approaches, which enable a more confident distinction between tsunami deposits and other high-energy event beds, such as storm deposits.

3.4. Near the K–Pg Boundary

The search for the ultimate cause behind the extinction of the dinosaurs and nearly 72% of all other species stands as one of the most widely debated questions in Earth's history [121]. About sixty-six million years ago, a bolide impacted at Chicxulub on the northwestern corner of the Yucatán Peninsula, Mexico [78,122–125]. This impact is now largely accepted as the primary cause of the mass extinction that marks the Cretaceous–Paleogene (K–Pg) boundary ([125] and references therein). It generated a complex cascade of hydrodynamic and mass-transport processes around the Gulf of Mexico and beyond, including tsunamis, impact-seiche waves, margin collapse, and sediment-gravity flows [50].

The K–Pg boundary provides a particularly compelling case for studying tsunamis generated by catastrophic extraterrestrial impacts. Consequently, tsunami deposits from this interval have been the focus of extensive research worldwide and studied for more than 30 years [97]. Well-documented examples are found across the Gulf of Mexico and Caribbean regions, including the Brazos River in Texas, northeastern Mexico, Cuba, and Haiti [49,126–129].

 https://doi.org/10.3390/jmse14010049

In Western Cuba, the Moncada Formation preserves a tsunamite characterized by shocked quartz, impact glass, an iridium anomaly, and complex ripple cross-lamination interpreted as evidence of oscillatory tsunami palaeocurrents [47]. Similarly, in the La Popa Basin of Northeastern Mexico, an 8-m-thick chaotic unit contains Chicxulub ejecta overlain by graded sandstones that reflect multiple tsunami surges and hyperconcentrated flows generated by slope failure [49]. High-altitude and continental deposits at Tanis (ND, USA) have provided complementary evidence, when a remarkable seiche-induced inland surge deposit containing fish fossils with impact spherules embedded in their gills, capturing the near-instantaneous arrival of seismic waves and ejecta fallout [50]. In Patagonia, an iridium anomaly, platinum group enrichment, and palynofacies changes at the boundary horizon further attest to the impact signal [51].

Identifying tsunami-related deposits in offshore settings is challenging due to limited access to shelf archives [22]. Preservation potential strongly depends on the depositional environment, sandy tsunami deposits in areas shallower than ~65 m are particularly prone to storm reworking [39], providing a framework to assess which K-Pg deposits are likely to be preserved. However, another deep-water record from the southeastern Gulf of Mexico (DSDP Sites 536 and 540) reveals a five-unit succession at the boundary, interpreted as deposits of a giant wave and associated backflow in a basin-floor to lower-slope setting [130]. These cores are characterized by graded sand, rip-up clasts, and reworked bioclasts, which extend the tsunami signal into deeper parts of the basin and demonstrate rapid, high-energy emplacement immediately following ejecta deposition. In Argentina, Scasso et al. [48] documented a coarse-grained sandstone bed within the Jagüel Formation showing erosive bases, grading, and hummocky cross-stratification, features interpreted as products of tsunami flows in a distal neritic setting.

Following the impact-extinction hypothesis, tsunami deposits became central to testing and documenting catastrophic sedimentary responses to the Chicxulub event, and subsequent resistance to the impact theory stimulated detailed sedimentological, geochemical, and paleontological investigations [121]. Hydrodynamic models further confirm the extraordinary reach of the Chicxulub tsunami, with velocities capable of eroding sediments at distances exceeding 10,000 km from the crater [131]. Collectively, these records demonstrate that K–Pg tsunamites are globally distributed and diverse in sedimentological character, ranging from chaotic breccias and graded marine sandstones to inland surge deposits. Their identification relies on the integration of sedimentology, paleontology, geochemistry, and geophysical modelling, and these collectively provide compelling evidence that the Chicxulub impact tsunami was not only a local catastrophe but also a globally significant sedimentary phenomenon that played a major role in the End-Cretaceous mass extinction.

3.5. Paleogene and Neogene

Compared with K–Pg, tsunami deposits from the Paleogene are fewer and patchier, reflecting both preservation and recognition challenges as well as research bias toward younger successions. Despite this, several major tectonic phases occurred worldwide during this System [132–136], many of which strongly influenced sedimentation, basin evolution, and even potential tsunami generation. Documented Paleogene tsunamites are concentrated in the Eocene, when two large impacts (Chesapeake Bay, USA) occurred within ~0.5 Myr of each other [79,137,138]. In the Atlantic Coastal Plain of North Carolina (USA), a sandy-matrix breccia with mixed terrestrial (paleosol rip-up clasts, petrified wood) and marine (fossiliferous chert, meter-scale clasts) components, interpreted as high-energy tsunami surges and a capping quartz sand with gravel, interpreted as a subsequent tsunami wave or backwash [79]. This extreme event related to Late Eocene Chesapeake

https://doi.org/10.3390/jmse14010049

Bay impact illustrates the preservation of coupled impact ejecta and tsunami deposits in a shallow-marine to marginal-terrestrial setting. From the New Jersey continental margin at ODP Sites 903 and 904, the same impact generated both strong ground motion and tsunami waves [137,138]. These impact-related tsunamites highlight the importance of extraterrestrial forcing, whereas other Paleogene catastrophic events underscore tectonic and climatic trigger.

Climatic perturbation also played an indirect role in triggering tsunami. Rapid warming events, such as the Initial Eocene Thermal Maximum (IETM), which represents the warmest period on Earth during the Cenozoic [139], influenced sedimentary processes by affecting slope stability and promoting landslide and turbidity current [140,141]. When slope failure occurs, the sudden pressure drop or sediment remobilization can destabilize hydrates within or beneath the failed mass, leading to rapid dissociation and the release of methane [142,143]. This process not only contributes to weakening slope sediments, thereby enhancing the likelihood and magnitude of failure-generated destructive tsunamis that can produce a turbidity current activity on non-glacially influenced margins [80,144,145]. A similar case is documented from the Eocene of Zumaia in Northeastern Spain, demonstrating that such findings are highly relevant for assessing future landslide-triggered tsunami hazards and the vulnerability of seafloor infrastructures, highlighting the value of recent geological records for recurrence estimates [80,145–148].

An earthquake may generate tsunami waves that subsequently induce turbidity currents, may trigger turbidity currents directly without the involvement of a tsunami [29], or in the case of large submarine landslide, turbidity current and slope failures may themselves act as the primary mechanism for tsunami generation. This overlap often leads to confusion between triggering mechanisms and depositional processes, as illustrated by the Late Oligocene–Early Miocene from the Lower Sarava Formation in the Southwestern Pacific. Deposited within a deep-sea fan at depths > 4.25 km and fed by volcanic material from the Vitiaz arc, its thick rudite–arenite beds have been interpreted as tsunami-related deposits [81]. This interpretation is supported by their unusual thickness, coarse grading, and rapid emplacement, features consistent with high-energy flows most likely triggered by strong seismicity, volcanic activity, or slope failure within the interarc basin.

The research on Neogene (23–2.58 Ma) tsunami deposits has progressed considerably. These deposits have been identified in a wide range of settings, from onshore to offshore environments [11,15,19,82,83,149–157], as well as within lacustrine systems [43]. Multi-proxy approaches are commonly used, with sedimentary evidence documenting both the run-up and the backwash phases.

From the Miocene Makran accretionary complex (Iran), a chaotic unit is represented by reworked blocks of various lithologies and sizes dispersed within a muddy matrix, previously interpreted as a tectonic mélange diapirically emplaced from depth [158,159]. According to Burg et al. [160], the presence of ophiolitic blocks and reworked fragments of underlying turbidites, together with soft-sediment deformation structures within this unit, supports a sedimentary rather than tectonic origin, making it comparable to large debris flows observed along modern continental margins or unstable volcanic edifices. Another debated case is reported from northern Chile, where Miocene shoreface sandstones of the Caleta Herradura half-graben were firstly interpreted as deposits formed by tsunami surges [82]. The succession consists of two erosively based units, likely emplaced by successive waves, and displaying diagnostic features attributed to tsunami backwash processes. These include exceptionally coarse grain size compared to surrounding deposits, sharp erosional bases, mixed sediment sources, abrupt lateral facies changes, and grading from chaotic to massive textures. However, this interpretation was later re-evaluated by Bahlburg et al. [161], who argued that the deposits are better explained as debris flow

https://doi.org/10.3390/jmse14010049

units representing physically continuous distal subaqueous equivalents of graben flank alluvial fan breccias, rather than tsunami deposits. For Son et al. [157], discriminating specific triggering events from sedimentary gravity-flow deposits is particularly challenging, especially in ancient records where diagnostic features are often subtle, overprinted, or poorly preserved. Such sedimentary features may be produced by both tsunamis and the other type of events mentioned above; therefore, distinguishing between the triggering mechanism and the depositional processes is crucial to avoid misinterpretation [29].

Nevertheless, despite such interpretative challenges, particularly in offshore settings, the Miocene deposits of Spain [11] and the Pliocene successions of Italy [15] provide well-preserved examples that, together with model cases, serve as methodological benchmarks. These studies highlight the value of multi-proxy approaches in improving the reliability of interpretations and in distinguishing tsunami deposits from other high-energy events such as storms and mass-flow processes. The Hackcheon Chogok Megaturbidite from the Miocene Pohang-Youngouk basin (Republic of Korea) represents one of the best-documented examples where sedimentological, structural, and provenance evidence point to a tsunami-triggered megaturbidite [157], helping to distinguish it from other, similar looking gravity flow deposits in the geological record. Recently, Di Celma et al. [156] reported that in the Lower Miocene Chilcatay Formation of southern Peru, offshore tsunami deposits are characterized by several key features: an irregular basal erosion surface with multiple asymmetric and symmetric scours, unusually coarse grain-size compared to the surrounding background sediments, absence of internal bioturbation, the presence of multiple graded beds, and a mixture of both well-rounded and angular clasts derived from nearshore and subaerial coastal areas. Additionally, the abundance of vertebrate skeletal remains (sedimentological concentration *sensu* Kidwell et al. [162]) embedded within these deposits suggests that such catastrophic tsunami events could be responsible for mass mortality of fishes, turtles, and dolphins, as observed in modern tsunami events [82]. The preservation potential of tsunamites in offshore settings is enhanced by their emplacement below the storm wave base, where they remain largely unaffected by typical wave reworking. For Puga-Bernabéu and Aguirre [19], shell beds generated by tsunami backwash are characterized by a distinctive taphonomic signature, including the admixture of organisms from different ramp zones, reduced fragmentation, sharper shell edges, oblique to perpendicular skeletal orientations, chaotic stacking of concave shells, and a lower incidence of encrustations and borings.

Within the Late Miocene–Early Pliocene Bentang Formation at Tegal Buleud (Indonesia), where active subduction between the Australian and Eurasian plates frequently generated strong earthquakes, the first and oldest geological evidence of a tsunami has recently been identified in the region [83]. This discovery not only extends the tsunami record of Java and Indonesia into the Neogene but also confirms some benchmarks usually considered the most critical criteria marking tsunamites across the Phanerozoic, such as clay clasts, siltstone rip-up clasts, normally graded sandstones, bimodal to multimodal grain-size distributions, and mixed marine fossils from different environmental settings (e.g., inner shelf to middle neritic). Another criterion, reported by Le Roux et al. [163] within deposits linked to a large tsunami event affecting the southern Chilean coastline during the Pliocene, includes inverse grading, planar laminations, ripple or trough cross-lamination, and the injection of sand-rich material into the underlying cohesive substrate. Similar characteristics have been documented in tsunami-related deposits from foreshore and backshore settings in the Mediterranean [25] and the Atlantic Ocean [16,17,20] (Figure 4).

https://doi.org/10.3390/jmse14010049

Figure 4. Representative photographs showing injection of tsunami sediment into the underlying cohesive substrates. (**A–C**) Examples from Maio Island (Cabo Verde Archipelago). (**A**) Injection of tsunami sediment into a basaltic lava flow (Ponta Pedrenau). (**B**) Sediments injected into conglomerate facies (Ponta dos Flamengos). (**C**) Downward injections of tsunami sediment into weathered lava (Ponta dos Flamengos). (**D**) Tsunami materials injected into terrestrial sediments from Gran Canaria (Canary Islands). (**E**,**F**) Downward injections of tsunami sediment from northwestern Algeria.

3.6. Pleistocene

In contrast to other Phanerozoic times, the Pleistocene (2.588–0.0117 Ma) represents the most crucial interval for understanding palaeotsunami deposits. This epoch provides the most continuous, accessible, and well-preserved sequences, which can be directly compared with historical and recently documented events across diverse environmental settings. Interest in palaeotsunami deposits from this interval has grown not only because of their significance for reconstructing palaeocoastline evolution, palaeogeomorphology, and reservoir characterization [26,117], but also because studies of Quaternary tsunamis highlight the history of volcanism and tectonic activity, particularly earthquakes and plate-boundary processes, essential for hazard assessment. Many Pleistocene tsunami-related deposits (tsunamites), originally triggered by earthquakes or landslides, were later overlain by additional tsunami deposits of the same age [17,20,164,165] or by younger Holocene and even historical events [25,37,38]. This stratigraphic superposition indicates that seismic activity and tsunami generation persisted from the Pleistocene into the Holocene and historical periods. However, as several previously cited examples demonstrate, tsunami events in this interval were not exclusively linked to tectonic activity; flank collapses and other non-seismic triggers in tectonically less active regions also played a significant role. Together, these records underscore the long-term recurrence of tsunamigenic processes across a range of geodynamic settings.

The Pleistocene Epoch was characterized by repeated glacio-eustatic sea-level fluctuations, tectonic activity that generated marine terraces, and large-scale slope instabilities [25,166–174]. These conditions were highly favorable for tsunami generation. Several well-documented Pleistocene examples highlight the diversity of depositional settings associated with catastrophic events, often linked to seismic activity and volcanic collapses. Such occurrences have been reported worldwide, including the

https://doi.org/10.3390/jmse14010049

Pacific Basin—particularly Japan [175] and New Zealand [176,177], as well as in the Atlantic regions [16,17,20,178] and the Mediterranean [25,34,37,84].

Oceanic volcanic islands are prime settings for investigating tsunami deposits, as volcanism, flank instability, erosion, and vertical crustal movements (subsidence or uplift) have driven rapid morphological changes during the Quaternary [17,168,179–181]. Compared to regions located along active subduction zones, such as the margins of the Pacific and Indian Oceans, tsunami hazards in the NE Atlantic archipelagos, including Madeira, the Canary Islands, and Cabo Verde, may be significantly underestimated [20]. While subduction zones are recognized as the primary sources of large, destructive tsunamis, oceanic volcanic islands can also generate catastrophic events through mechanisms such as massive flank collapses, volcanic explosions, and associated landslides. These mechanisms can generate a mega-tsunami [17]. The term mega-tsunami refers to a wave amplitude exceeding 50 m [182]. In comparison, the 1958 Lituya Bay tsunami [183] is often cited as the only historical mega-tsunami, with a maximum run-up of 524 m, however, its effects were spatially limited and rapidly diminished away from the landslide source [17]. The limited number of historical tsunami records in the Central Atlantic may create a false sense of security; however, geological evidence demonstrates that these islands have experienced multiple large-scale events during the Pleistocene [16,24,164,165,184].

One of the best examples interpreted as evidence of a mega-tsunami generated by catastrophic volcanic flank collapse is represented by chaotic marine conglomerates and large boulders found at a wide range of elevations on the flanks of the Cabo Verde islands [16,20,164,165,178] and Canary Islands [184,185]. These deposits, often perched tens to hundreds of meters above present sea level, display sedimentological features consistent with high-energy emplacement. Their interpretation as tsunami deposits was previously debated [17]. Previous studies proposed that these accumulations could instead represent uplifted littoral deposits [186–189]. However, more recent investigations integrating detailed sedimentological, geomorphological, and chronological data [17–20] have provided robust evidence supporting their interpretation as tsunami deposits. These advances highlight the importance of combining multiple lines of evidence to accurately distinguish tsunami deposits from uplifted coastal facies in island settings.

Dating tsunami deposits using methods such as U-Th, radiocarbon and/or amino acid racemization in shells, cosmogenic exposure techniques, or stratigraphically constraining the deposits between the underlying and overlying units is critical, as it not only constrains the timing of extreme events but also helps to evaluate whether they had a regional impact. By establishing the timing of events at multiple sites, it is possible to correlate deposits across the region and assess the extent and impact of individual tsunamis. This is particularly relevant for the Last Interglacial (MIS 5e), when tsunami deposits have been reported across the Western Mediterranean. In Southwestern Spain, MIS 5e deposits at La Mata and Torrevieja reflect coastal uplift that occurred between MIS 7 and MIS 5e [37]. These successions include accumulations of randomly arranged and well-preserved *Glycymeris* and *Acanthocardia* mollusc bivalve shells, as well as the transport of enormous boulders onto stepped abrasion platforms. These features are interpreted as tsunami surge deposits generated by an offshore earthquake. Comparable successions interpreted as the result of an offshore earthquake-induced tsunami are also documented from Northwestern Algeria [25], where the *Glycymeris*-rich Unit of the Hachacha Plateau is characterized by mixed molluscan assemblages, including the so-called Senegalese fauna, excellent shell preservation, and diagnostic sedimentological features such as erosive bases, lateral facies variations, wide grain-size ranges (from clay to boulders), rip-up clasts, and soft-sediment deformation structures. Both Spanish and Algerian examples emphasize the importance of precise dating for identifying synchronous tsunami events, and their shared faunal content

https://doi.org/10.3390/jmse14010049

(notably the dominance of *Glycymeris*), combined with geomorphological, sedimentological, and taphonomic characteristics, suggest a regional-scale impact during MIS 5e, directly linked to seismic activity along the Africa–Europe plate boundary.

From Central Japan, anomalous deposits are interpreted as earthquake-induced tsunami deposits from the Pleistocene [175]. These show a rare example where successions collectively suggest a three-stage depositional model, with the preservation of a lower unit that shows the presence of contorted structured beds (convolute structures) interpreted as earthquake event deposits (seismites), which may testify to the triggering mechanism of this catastrophic event or help to relate these tsunamites directly to seismic activity. The middle unit suggests tsunami current-dominated processes, whereas the upper unit represents a suspension-settling event characterized by the presence of cross-laminated sand with mud drapes. The same succession has been reported within tsunami deposits related to seismic activity documented from northwestern Algeria [25], where seismite structures were found near the erosion surface. In this case, the authors described soft-sediment deformation structures (SSDS) rather than convolute bedding, and a mud layer instead of the mud drapes. The features reported above, including the presence of unit presenting seismite structures, overlain by tsunami deposits, which are themselves characterized by rapid lateral facies variation, changes in thickness, and in some cases the presence of large boulders and shell beds, can therefore be used as indicative of an earthquake-induced tsunami sequence in similar tectonically active settings.

4. Holocene to Recent Times Records

The Holocene (the last ~11,700 years) is characterized by relatively stable coastlines, following the last major post-glacial sea-level rise. This stability has created ideal conditions for the preservation and dating of palaeotsunami deposits [190–192]. The good preservation and dating of these deposits allow for direct correlation with historically documented events. Their significance extends far beyond academic interest, as they are fundamental for extending the often short and incomplete record of tsunamigenic events in any given region [18,22,33,37]. Moreover, the study of tsunami deposits from the Holocene to recent times enables the scientific community to better characterize the full range of diagnostic features of these deposits by applying the principle of actualism [193]. This approach also strengthens the assessment of geological hazards, particularly those triggered by seismic activity [37].

From the Holocene, the identification and interpretation of tsunami deposits are based on an integrated combination of multi-proxy approaches, including sedimentology, paleontology/micropaleontology (e.g., foraminifera, diatoms, mollusks, corals, and other fauna), geochemistry, and geomorphology with precise dating techniques comprising radiocarbon (^{14}C), optically stimulated luminescence (OSL), and amino acid racemization (AAR) [12,34,35,37,40,194–199].

Several studies have emphasized the value of estuaries as natural archives for tsunami research, since they can preserve a wide range of geological evidence [85,199,200]. The estuaries are excellent coastal systems for recording tsunami deposits because they preserve a variety of geological indicators, including beach erosion and shoreline retreat, breaching and overtopping of sandy spits, and the deposition of bioclastic layers above the normal action of sea waves and storms [85]. The multiple fining layers or units reflect multiple successive waves during the extreme events (tsunamis) [15,25,82]. These features have also been reported from the 2004 Indian Ocean tsunami [201], where the catastrophic event generated multiple patterns in vertical textural trends represented, from bottom to top, by upward fining, upward coarsening, and uniform units. In the same context, from eastern Japan, a Holocene depositional model demonstrates that, while individual structures such

 https://doi.org/10.3390/jmse14010049

as hummocky cross-stratification and graded bedding resemble storm deposits, the vertical stacking forms a distinctive signature of tsunami wave trains [4]. This interpretation allowed the authors to propose a genetic vertical facies model organized into four units (Tna–Tnd), reflecting changes in wave size and energy, and providing a key criterion to distinguish tsunamites from tempestites in the same setting. However, this interpretation has been challenged. For Shanmugam [29], a single depositional event, such as a tsunami cannot reasonably account for the formation of both traction carpets and hummocky cross-stratification (HCS) based on hydrodynamic consideration. This is because these features form via genetically unrelated processes, dispersive pressure and oscillatory flow, respectively. This controversy remains a challenge not only in distinguishing storm from tsunami deposits, but also in determining whether these succession units are related to a single or to multiple tsunami events that occurred close together or separated in time.

Many terms such as turbidite, megaturbidite, debrite, gravitite, and homogenite have been used with overlapping meanings and are often collectively referred to as tsunamites [29]. A classic example is found in the eastern Mediterranean, near the Calabrian and Hellenic arcs. There, a thick marl layer composed of clay, quartz, nannofossils, and foraminifera has been described. This thick unit is characterized by a sharp basal contact, a normally graded sandy base, and overall homogeneity-showing no visible sedimentary structures in core sections or X-radiographs. Since its discovery, it has been interpreted as a homogenite [202]. According to Cita et al. [84], this deposit is a key piece of evidence for the tsunami generated by the Late Bronze age collapse of the Santorini caldera (~3500 years ago). The event triggered large-scale slope failures that, in turn, generated sediment gravity flows. These flows deposited turbidites and megaturbidites in the abyssal plains of the Ionian Sea, while the distinctive homogenite was laid down across the subdued topography of the Mediterranean Ridge, an area often referred to as Cobblestone Topography. This interpretation suggests that tsunamis, turbidites, mass flows, and slope failures are closely interconnected processes. A comparable relationship has been reported from the northern flank of Tenerife, where a massive volcanic collapse mobilized a volume of ~200 km^3 offshore, producing three debris lobes and seven turbidites, and was associated with a Pleistocene mega-tsunami [17,203].

The study of both historical and modern tsunami deposits, particularly those from the Lisbon earthquake tsunami (1755), Indian Ocean tsunami (2004), and Tohoku-Oki tsunami (2011), has served as a crucial calibration tool for interpreting older events [18,204–213]. These modern analogues allowed refining diagnostic features, such as landward fining, inverse-to-normal grading, the presence of intraclasts (rip-up clasts), and the transport of marine sediments (sand, shells, microfossils) far inland [214,215]. Geochemical signatures, including elevated concentrations of salt-derived ions (e.g., Cl^-, Na^+, S) in soil and groundwater, have been established as a reliable indicator of seawater inundation, even when the sedimentary evidence is subtle [18,216–219]. Furthermore, taphonomic analyses of molluscan and foraminiferal assemblages in modern tsunamites have confirmed faunal mixing across different ecological zones as a key indicator [195,220–222].

Well-documented Holocene tsunami records have been established in numerous tectonically active regions worldwide. In the Pacific Rim, extensive research has been conducted along the coasts of Japan [35,86,223,224], Chile [225–227], the Cascadia subduction zone of North America [228–230], and New Zealand [176,231]. In the Atlantic Ocean, the 1755 Lisbon tsunami serves as a benchmark event, with its deposits identified in Spain, Portugal, and Morocco [211,212,218,219,232–239]. The Mediterranean Sea, a basin with a long history of seismic and volcanic activity [240], also preserves a rich Holocene tsunami record, with studies in Turkey [241], Greece [242–244], Italy [245,246], the Levant [247,248], Malta [249], Algeria [14,38], and Spain [250] revealing repeated large-scale events.

 https://doi.org/10.3390/jmse14010049

Despite their higher preservation potential, interpreting Holocene tsunami deposits still presents challenges. These include distinguishing the deposits of closely spaced events [229], differentiating tsunami signals from intense storm deposits (tempestites) in areas where both occur [27,33,251], and understanding the complex effects of local geomorphology and anthropogenic activity on sediment deposition and preservation [252,253]. Nevertheless, the Holocene record, including tsunamis from the last few decades, remains an indispensable reference for interpreting older events. It provides a long-term perspective on tsunami recurrence, magnitude, and source mechanisms, thereby enabling more accurate association between triggers and depositional outcomes, and supporting robust hazard and risk assessments [254–256].

5. Tsunamites vs. Tempestites: Key Challenges

5.1. Sedimentological Criteria

Tsunamites and tempestites are sedimentary deposits formed by extreme wave events, such as tsunamis and major storms, respectively. These events are powerful geological agents capable of transporting and depositing vast amounts of sediment both inland and across shallow marine environments [4,17,25,29,57,181,257]. However, distinguishing between tsunami and storm deposits remains one of the most persistent and challenging problems in sedimentary geology [25,28,34]. Both events occur in similar environmental settings (e.g., coastal plains and shallow marine areas) and can share several sedimentary characteristics, such as erosive bases, coarse-grained layers, and internal stratification. Misidentification can therefore lead to significant errors in interpreting the geological record, for example, by attributing a seismic history to a region based on deposits that were actually formed by storms, or vice versa. In fact, several sedimentological and palaeontological characteristics have recently been used to discriminate between these two extreme events across the Phanerozoic, with examples provided from different oceans [12,20,24–27,37,241].

From the Storegga tsunami deposits in Norway, Dawson et al. [2] were the first to describe the presence of a sharp erosional contact at the base of tsunamigenic layers [198]. Since their pioneering work and following their contribution, this type of erosion has been observed later at the base of deposits related to tsunami events from Japan [258], Papua New Guinea [259], Thailand [260–262], and Indonesia [263]. However, Morton et al. [27] reported that this feature can also occur in storm deposits. This overlap highlights one of the major challenges in distinguishing between these two types of extreme events [85], and underscores the need for a multi-proxy approach when interpreting the deposits left by this type of events. The erosional surfaces produced by the 2004 Indian Ocean Tsunami in Tamil Nadu, India, were described as an "erosional unconformity" by Bahlburg and Weiss [264]. This terminology is stratigraphically misleading, as a true unconformity classically represents a significant gap in the depositional record, often involving millions of years [265,266]. Since the 2004 event lacks such a substantial temporal hiatus, its erosional features should not be conflated with stratigraphic unconformities [29]. Instead, such short-lived, high-energy features are more accurately classified as "event surfaces" to prevent confusion in sequence stratigraphic interpretation.

Beyond sharp erosional contacts, other erosive features have also been reported in tsunami-related successions. From northwestern Algeria, within deposits interpreted as the result of a catastrophic tsunami event, Doukani et al. [25] described an irregular erosional surface separating the event deposits from the underlying bedrock (Figure 3F). Here the authors linked this irregular erosion to the lithology and heterogeneity of the eroded basements. According to Bryant and Young [267] and Aalto et al. [268], the bedrock sculpting is related to a tsunami activity. However, Bourgeois [87] (p. 18) has challenged

 https://doi.org/10.3390/jmse14010049

this interpretation and emphasizing that "*There is no fundamental basis for the argument that tsunamis are more powerful sculptors than storm waves*".

Tsunami deposits can show a wide grain-size spectrum in sediments, ranging from clay to boulders [20,25,198] (Figure 3C–E). Such conditions are characteristic of catastrophic depositional events, during which extreme hydrodynamic energy rapidly entrains, transports, and reworks sediments from diverse environments in a chaotic manner [269]. According to Bourgeois [87], the grain-size distribution within tsunami deposits is controlled primarily by the availability of sediment sources rather than by the hydraulic competence or flow direction. In many cases, the presence of imbricated boulders is considered another benchmark criteria of tsunami deposits [14,16,17,20,25,270–274], reflecting rapid and high-energy depositional processes. This contrasts with deposits linked to storm events, where sediments are formed under more gradual and oscillatory flow conditions.

Both fragile soil fragments and hard-rock rip-up angular clasts have been reported from the oldest known tsunami event on Earth [59] to more recent records across diverse environmental settings, ranging from coastal lowlands to deep-marine environments [17,20,25,27,79,82,126,130,185,259,275–278]. These features are typically concentrated in the lower portions of tsunami deposits and along erosional surfaces, particularly where fine-grained sediments are present. In contrast, they are generally absent or rarely documented in storm deposits [27]. Sediment injections, resulting from the extreme dynamic pressure of tsunami surges, provide another key criterion for identifying tsunamites, as they are commonly emplaced into both soft and consolidated underlying strata and reported from several tsunami deposits (Figure 4) [17,20,25].

The identification of tsunami deposits based only on internal sedimentary structures remains a significant challenge. Several tsunamites display features such as cross-stratification, parallel and oblique lamination, hummocky cross-stratification (HCS), and mud drapes [4,25,27,62,145,259,277,279–281]. However, many of these structures, particularly parallel and cross-lamination, as well as HCS, are also widely documented in storm-related deposits [281,282], reducing their diagnostic reliability. For instance, parallel lamination has been identified in deposits from the 2004 Indian Ocean tsunami in Sri Lanka and Tamil Nadu [29,201]. Similarly, hummocky cross-stratification has been reported in tsunami deposits from the Precambrian of Australia [61], the Cretaceous–Paleogene boundary in Argentina [48], the Neogene of Spain [13], the Pleistocene of Algeria [25], and the Holocene of Japan [4]. Low-angle cross-stratification has also been described not only in fossil and recent tsunami deposits [15,283] but also in estuarine successions and tidal facies [284]. Unlike tsunami deposits, storm deposits frequently display a wider variety of sedimentary structures beyond simple planar stratification [27]. These include foreset bedding [258,285–288], backset lamination [289], and climbing ripple structures [282–291].

In addition to cross-stratification and lamination, both normal and reverse grading have been reported from tsunami deposits at multiple localities worldwide [15,17,20,25,292]. According to Massari et al. [15], the occurrence of these grading patterns reflects sedimentation from prolonged, high-energy flow conditions, which can produce both upward fining and upward coarsening depending on the turbulence and sediment supply. Spiske and Jaffe [293] further emphasized that inverse grading is commonly associated with deposits from tropical cyclones, where oscillatory storm surges and strong backwash can generate upward coarsening trends. By contrast, normal grading has frequently been reported within tsunami deposits attributed to deep environment beneath the storm wave base [83,84], reflecting the waning energy conditions typical of decelerating tsunami flows.

Within this ongoing debate, it is evident that no single sedimentary structure can serve as a definitive indicator of tsunami deposits. Instead, recent studies emphasize that a more reliable identification requires the integration of multiple diagnostic features. These include

https://doi.org/10.3390/jmse14010049

the association of graded bedding with mud drapes, the presence of rip-up clasts, layers enriched in heavy minerals, sediment injections into both hard and soft substrates, and evidence of both landward and seaward-directed transport, often considered alongside geomorphological context and macro to microfossil assemblages [17,20,24,25,27,37,43,216]. Such multi-proxy approaches substantially improve the reliability of distinguishing tsunamites from tempestites and highlight the importance of integrating sedimentological, palaeontological, geochemical, and geomorphological lines of evidence in paleo-tsunami research (Table 2).

Table 2. Tsunami versus storm deposit features based on sedimentological, palaeontological and taphonomic, geomorphological, and geochemical criteria.

Criteria	Tsunami Deposits (Tsunamites)	Storm Deposits (Tempestites)	Key References
Trigger and energy source	Generated by long-period seismic sea waves (earthquakes, landslides, volcanic activity), extremely high energy with multiple run-up and backwash phases.	Generated by short-period oscillatory waves and storm surges (cyclones, hurricanes, etc.), energy concentrated nearshore.	[57,265]
Erosional features	Sharp, irregular erosional bases, deep scours and event surfaces, erosion may cut into bedrock or soil, extensive inland erosion.	Shallow scours and erosional truncations near beach/shoreface, usually limited to berms and dunes.	[2,25,27,264]
Grain-size	Very wide spectrum (clay to boulders), often poor sorting, rapid energy fluctuations, possible imbricated boulders.	Moderate to good sorting; coarser sands and gravels typical.	[16,20,25,270]
Sedimentary structures	Combination of HCS, parallel and cross-lamination, normal and reverse grading, mud drapes; stacked sublayers reflecting multiple waves.	Dominated by HCS, cross-bedding, ripple lamination, foresets, climbing ripples; grading mainly normal or inverse depending on wave energy.	[27,190,294]
Rip-up clasts and sediment injections	Common, angular rip-up clasts and injections of sediment into underlying strata from extreme pressure.	Rare, weak injection or absent rip-up clasts.	[17,20,25]
Thickness and lateral extent	Thick, laterally continuous sheets; often thinning landward but extensive over kilometers; multiple subunits possible.	Thinner, lenticular layers restricted to nearshore bars and back-beach environments.	[27,207,261]
Facies variability	Strong lateral variation depending on topography and inundation pathways; may overlie non-marine units.	Typically confined to marine or shoreface settings; limited facies variability.	[20,25,261]
Macrofossil assemblages	Mixed allochthonous assemblages from multiple habitats (marine, continental); high richness and evenness, The abundance of large, heavy shells in molluscan assemblages	Autochthonous or parautochthonous assemblages from local habitats; lower richness.	[12,19,24,25,35,37]
Microfossils (foraminifera, diatoms, etc.)	Mixed assemblages of deep, shallow, and terrestrial taxa; offshore species transported inland.	Restricted to local shallow-marine assemblages; absence of deep-water species.	[295–297]
Taphonomy	Angular fragmentation, shell breakage, chaotic orientations (often >30–60°) with dominance of vertical and oblique orientations; evidence of strong backwash; limited bioerosion, low encrustation.	Rounded fragmentation, ordered shell orientation, concave-up stacking, higher bioerosion and encrustation, concave-up, subhorizontal stacking (<30°).	[15,19,25,37]
Geomorphological	Deep scour channels, stripped soil layers, uprooted vegetation, boulder ridges, landward sediment lobes extending far inland (hundreds of meters–km).	Washover fans, beach ridges, dune breaches, limited to low-lying coastal zones.	[28,57,263]

https://doi.org/10.3390/jmse14010049

Table 2. *Cont.*

Criteria	Tsunami Deposits (Tsunamites)	Storm Deposits (Tempestites)	Key References
Inland extent of erosion and deposition	Can extend several kilometers inland and tens hundreds of meters above sea level.	Typically confined to nearshore and back-beach areas (<few meters inland).	[259,298,299]
Geochemical signature	Enrichment in Sr, Ca, Mg; marine-derived saline and carbonate influx; distinct XRF peaks; possible marine environmental DNA (eDNA).	Less marine element enrichment; composition similar to local sediments; absence of eDNA marine signals.	[209,217,300]

This interpretation is further strengthened by incorporating additional sedimentological and taphonomic criteria, including a wide grain-size spectrum from clay to boulders, pronounced lateral variations in facies and thickness, and the presence of reworked biological remains. Taken together, these features provide compelling support for a high-energy tsunami origin rather than a storm-related process (Figure 5).

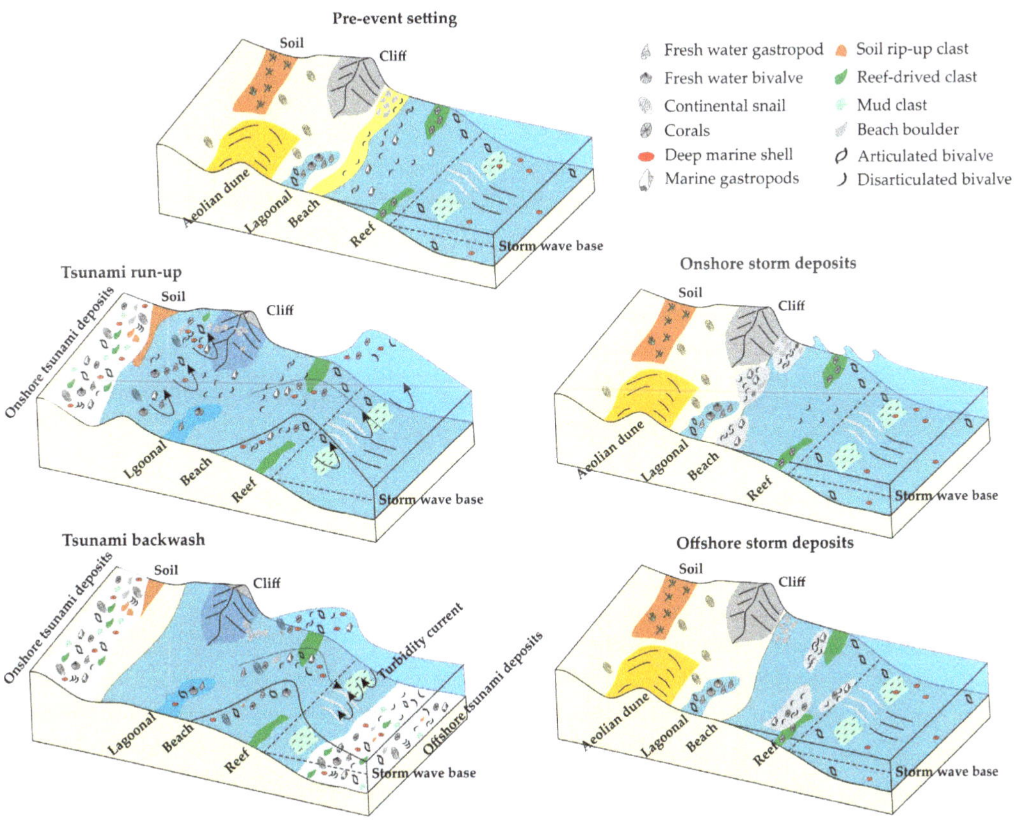

Figure 5. Conceptual model illustrating the differences between tsunamites and tempestites in both offshore and onshore settings, based on sedimentological, paleontological, and geomorphological criteria.

5.2. Paleontological and Taphonomic Criteria

Both tsunamis and storms events represent highly energetic physical processes capable of producing sedimentological shell beds (*sensu* Kidwell et al. [162]), although the resulting deposits display distinct sedimentological and taphonomic characteristics [25]. More recently, shell-bed concentrations have been widely used to infer their origins in both

https://doi.org/10.3390/jmse14010049

marine and lacustrine settings [12,15,19,24,25,37,43,45,76,257,292,301]. Studies on storm-generated shell beds (tempestites) have examined how taphonomic attributes such as fragmentation, orientation, and encrustation vary across different environments [259,286,294,302–305], while research on tsunami shell beds (tsunamites), has highlighted key taphonomic criteria related to extreme high-energy events (tsunami) [12,15,24–26,34,35,37,301].

From Oman, Donato et al. [12] present one the best examples applying bivalve-shell taphonomy to identification of tsunami deposits. They concluded that a key criterion characterizing such deposits is the mixture of allochthonous organisms derived from different environments (e.g., pelagic, shallow-water, and deep water, and even continental sources), coupled with a high degree of angular shell fragmentation. Similar signatures have been recognized in tsunamites from offshore settings associated with backwash flows. In this context, Puga-Bernabéu and Aguirre [19] compared the taphonomic features of two shell beds—one attributed to a tsunami and the other to a storm—and emphasized the importance of such comparisons for process discrimination. Comparable evidence has also been reported from coastal settings influenced by both run-up and backwash, including southwestern Spain [37] and northwestern Algeria [25].

Marine shell deposits linked to tsunamis are generally characterized by dense, thick, and laterally extensive beds [19]. The shells are often well to excellently preserved, frequently oriented obliquely or perpendicularly to bedding, and may be dominated by a single species [25,37]. Standardized 1 kg samples of fossiliferous sediments from tsunamigenic facies also yield consistently higher taxonomic richness and evenness than equivalent samples from tempestites [24], and these authors suggested that whenever fossils were present in tsunami deposits, this could be a decisive criterion to distinguish between tsunami and storm deposits. In addition, the low frequency of boring and encrustation combined at times with the presence of articulated bivalves in non-living position or disarticulated shells stacked like plates in dishwater (Figure 6), suggests that exceptionally powerful tsunami waves penetrated below the taphonomically active zone, eroding sediment masses that had previously buried the shells [13,15,20,25].

By contrast, tempestite shell beds usually consist of organisms derived from adjacent habitats, reflecting localized reworking by storm waves [19] (Figure 5). Their assemblages tend to exhibit lower taxonomic richness, higher rates of encrustation and bioerosion, and more ordered shell orientations. According to some authors [15,19,25,33,37] shells in tempestites are typically arranged subhorizontally, at angles of less than 30° to stratification, and concave-up stacking is common. Tsunamites, on the other hand, frequently display a wide range of depositional orientations, including chaotic fabrics, normal and reverse grading, as well as oblique to vertical alignments (>30–60°) (Figure 6). The abundance of large, heavy shells in molluscan assemblages has also been recognized as an additional criterion reflecting the extreme energy of tsunami waves in shallow coral reef lagoons [35]. Similarly, the chaotic and variable arrangement of concave shells in tsunamites, both from offshore [19] and onshore records [25,37], contrasts with the more systematic stacking observed in tempestites, further underscoring the diagnostic value of taphonomic features in distinguishing between these two types of extreme-wave deposits.

In addition to taphonomic criteria, comparison of microfossil taxonomic assemblages (e.g., ostracods, foraminifera, diatoms, and pollen) has also been successfully used to differentiate between storm and tsunami deposits [34,294–297,304,305]. Foraminiferal tests are frequently preserved in tsunami deposits and provide reliable evidence of sediment remobilization, often indicating the transport of seafloor material from depths greater than 100 m and from locations several kilometres offshore [221]. Similarly, diatom assemblages in tsunami deposits are often characterized by a chaotic mixture of freshwater and brackish–

 https://doi.org/10.3390/jmse14010049

marine species, reflecting the entrainment of material from different environmental settings during transport [198] (Figure 5).

Figure 6. Examples of the depositional structures of shells within tsunami deposits from Cabo Verde and Algeria that reflect their taphonomy. (**A**) Articulated bivalves in non-living position, imbricated within tsunami deposits at Nossa Senhora da Luz, Santiago Island (Cabo Verde). (**B**) Another example of closed bivalves in non-living position within tsunami deposits from Ponta Pedrenau, Maio Island (Cabo Verde). (**C**) Coquina related to a tsunami event from Port de Menard (northwestern Algeria), showing a chaotic shell orientation with a predominance of angles > 30° and a dominance of *Glycymeris* spp. (**D**) Disarticulated *Glycymeris* shells stacked like plates in dishwater.

5.3. Geomorphological Criteria

In addition to sedimentological and taphonomic evidence, geomorphological indicators play a crucial role in identifying and distinguishing tsunami from storm deposits. Features such as boulders ridges, and landward sediment lobes provide insights into the dynamics of extreme-wave events and their capacity to reshape coastal landscapes [17,18,25,57,181,257]. These landforms provide crucial insights into the hydrodynamic energy, directionality, and inundation distance of extreme-wave events [16,20,28]. When preserved in the stratigraphic record, such geomorphological signatures serve as important proxies for identifying and reconstructing past tsunami events in both modern and ancient contexts. For Doukani et al. [25], the variations in facies and thickness in tsunamites are largely controlled by the pre-existing topography and the lithological characteristics of the affected areas (Figure 7). According to Nishimura and Miyaji [306] topographic undulations often control the thickness of sediment layers produced by a tsunami event. Such features have been documented both from the Pleistocene record [25] and recently, during the 2004 event along the west coast of Thailand [261], tsunami deposits tending to be thicker in channelized areas, whereas they are comparatively thinner over locally convex topographic highs. For Morton et al. [27] the lateral variability is not considered a key diagnostic criterion, as both tsunami and storm deposits may exhibit lateral variation in geometry. This variability is often strongly influenced by local costal and nearshore topography, including features such as river channels, canyons, embayments, and irrefular shorelines. Such topographic controls can modify flow paths, sediment transport, and deposition patterns, producing lateral differences in thickness, grain size, and bedding architecture that are not uniquely indicative of either tsunamis or storms. Therefore, while topography affects the distribution of deposits, lateral variability alone cannot reliably distinguish between these high-energy events.

 https://doi.org/10.3390/jmse14010049

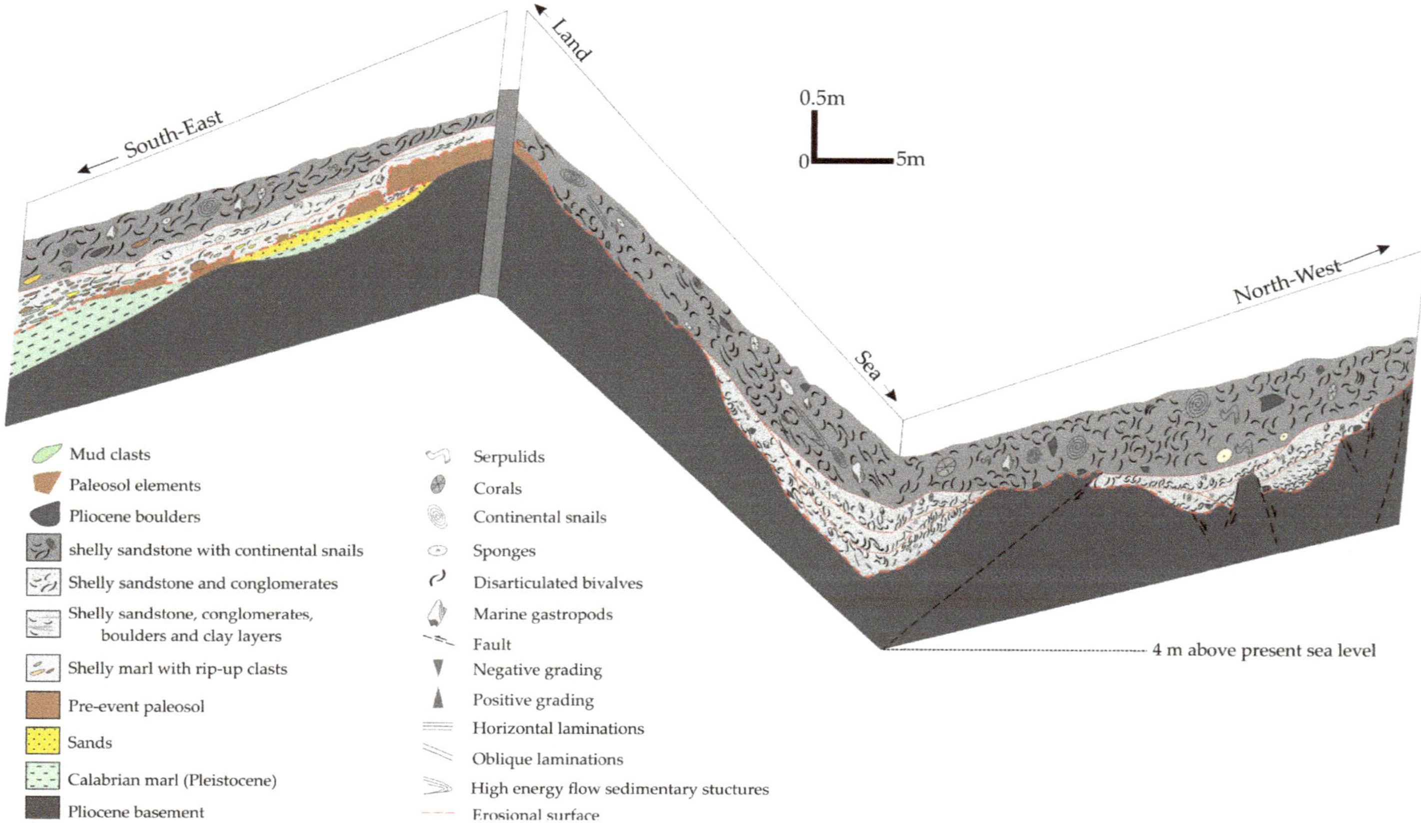

Figure 7. Simplified and modified example from [25] showing the lateral variation in thickness and facies, with details of the internal structures within tsunami deposits from Northwestern Algeria.

https://doi.org/10.3390/jmse14010049

Erosion zones produced by tsunami generally penetrate much farther inland and to much higher elevations above sea level than those generated by storms (Figure 5), owing to their greater wave amplitude, long period, and enhanced competence for sediment entrainment immediately beyond the shoreline [260,298,307]. By contrast, storm overwash deposits usually remain confined to the back-beach setting, often forming as extensions of the berm or beginning at an erosional scarp [299,308]. Broader erosion fields may develop under extreme storm conditions when wind stress substantially enhances flow [309]. Tsunamis also leave behind distinctive landform signatures, including deep scour channels, stripped soil layers, uprooted vegetation mats, and laterally continuous sand sheets or boulder ridges that may extend kilometers inland [25,28,34,43,263,277]. In similar depositional settings, the absence of marine sediments above and below tsunami layers is considered another key criterion for identifying tsunamites [25,37].

The transport of megaclasts onto coastal platforms or cliff tops is another hallmark rarely matched by storm activity [38,57,178,195]. Storms, in contrast, tend to generate localized geomorphic features such as overwash fans, washover terraces, and breaches through dune ridges, generally restricted to low-lying coastal barriers [299]. Taken together, blanketing of steep slopes by the sediments, often into elevations well above the reach of storms, the inland extent of erosion, the scale of sediment reworking, and the capacity to mobilize a metric megaclasts are key geomorphological criteria that help differentiate tsunami deposits from those generated by storms.

5.4. Geochemical Criteria

The distinctive geochemical signature has also been used to identify and interpret deposits related to tsunami events, e.g., [217,300]. The analysis of elemental geochemistry, often via X-ray fluorescence (XRF), further help the discrimination between tsunami and storm events. Tsunami deposits frequently exhibit enrichments in Strontium (Sr), Calcium (Ca), and Magnesium (Mg) related to the influx of shell fragments, carbonate sediments, and saline water [18,209,216,217] (Figure 8). Recent works analyzed the environmental DNA signatures for distinguishing between tsunami and storm deposits, e.g., [310]. Environmental DNA (eDNA) is a novel biogeochemical proxy for identifying tsunami deposits, with case studies from the Tohoku region (Japan) [300]. Here, the authors demonstrated that marine eDNA preserved in lacustrine sediments provided unequivocal evidence of seawater incursion during the 2011 Tohoku-oki tsunami, despite the inland setting of the deposits. Importantly, eDNA assemblages were also used to assess older events, including the 869 CE Jōgan tsunami and a prehistoric tsunami dated at 2400–2900 calibrated yr BP from the same region. In both cases, the eDNA composition of the event layers differed markedly from that of background sediments. Moreover, eDNA detected marine taxa in a thin muddy layer above the Jōgan deposit, which was visually indistinguishable from peat, suggesting that eDNA can reveal cryptic tsunami layers not recognized through sedimentological observation alone. This study highlights the potential of eDNA to complement classical geochemical indicators and to elucidate the origin of event deposits that are otherwise difficult to identify.

As mentioned above, it is often difficult to determine whether tsunami succession units are the result of a single event or of multiple events that occurred either close together or separated in time. Moreira et al. [218] demonstrated that high-resolution XRF core scanning, combined with detailed grain-size and image analyses, can help disentangle the complexity of tsunami deposition. Their study of the 1755 Lisbon tsunami revealed up to four distinct depositional pulses, interpreted as successive inundation and backwash phases, thereby illustrating how geochemical signatures and textural variability can be used to refine event stratigraphy and reconstruct depositional phases.

https://doi.org/10.3390/jmse14010049

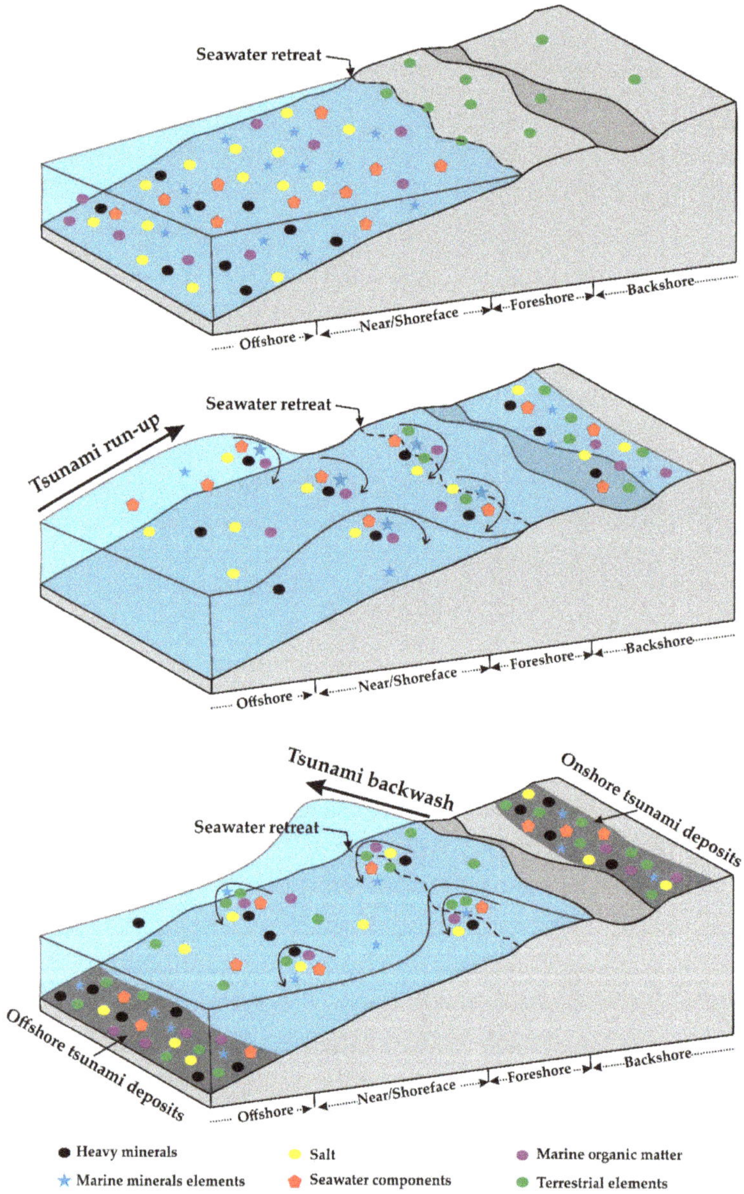

Figure 8. Schematic diagram simplified from [216,300], illustrating the processes responsible for the elevated concentrations of salts and marine geochemical signals in onshore settings, as well as the occurrence of terrestrial chemical signals within offshore tsunami deposits, and the transport and deposition of environmental DNA (eDNA) by tsunami processes.

6. Conclusions

This review covers the global and temporal distribution of tsunami deposits from the Precambrian to recent times, underscoring their value as persistent recorders of catastrophic events throughout Earth's history. The geological record, from the earliest tsunamites triggered by bolide impacts in the Precambrian to those associated with major mass extinctions

https://doi.org/10.3390/jmse14010049

and the well-calibrated events of the Holocene, provides an unparalleled long-term perspective. It reveals that tsunamis are a recurring geological phenomenon, generated by a variety of triggers including earthquakes, submarine landslides, volcanic flank collapses, volcanic eruptions, extraterrestrial impacts, and climate-driven processes. The preservation of these event-deposits across the geological times highlights the importance of tsunamites not only for understanding past dynamics of the Earth system, but also for assessing future coastal hazards.

Whereas the Pleistocene and Holocene records are relatively robust, older successions from the Precambrian to the Cenozoic remain underexplored, presenting a significant knowledge gap. Future studies should focus on applying advanced geochemical, taphonomic, sedimentological, geomorphological and, whenever possible, quantitative palaeontological approaches to these ancient deposits, while continuing to use modern analogues for calibration.

A principal finding of this synthesis is the critical importance and necessity of a multi-proxy methodology for the accurate identification of tsunami deposits and their discrimination from tempestites. As detailed across the sedimentological, taphonomic, geomorphological, geochemical, and palaeontological criteria, no single feature is unequivocally diagnostic. The most confident interpretations arise from the convergence of multiple lines of evidence: (i) the presence of erosive bases and rip-up clasts coupled with a chaotic grain-size distribution; (ii) taphonomic signatures such as mixed ecological assemblages and variable shell orientations; (iii) geomorphological indicators like far-inland penetration and megaclast emplacement; (iv) geochemical tracers of marine incursion, including ions, elemental ratios, and now, revolutionary environmental DNA; (v) and quantitative palaeontological methods. This integrated approach is paramount for avoiding misinterpretation in the geological record and for building reliable paleotsunami databases.

Author Contributions: M.A.D.: Writing—original draft, conceptualization, methodology, software, validation, formal analysis, investigation, data curation, preparation, writing—review and editing, visualization. J.M.: Formal analysis, investigation, visualization, validation, writing—review and editing. L.S.: Project administration, supervision, funding acquisition, validation, writing—review and editing. S.P.Á.: Supervision, funding acquisition, conceptualization, methodology, formal analysis, validation, writing—review and editing. All authors have read and agreed to the published version of the manuscript.

Funding: This work is carried out thanks to the support of the DGRSDT (General Direction of the Scientific Research and Technologic Development, Ministry of Higher Education and Scientific Research, Algeria). SPA acknowledges his former contract with project M1.1. A/INFRAEST CIENT/A/001/2021—Base de Dados da PaleoBiodiversidade da Macaronésia, funded by the Regional Government of the Azores. SPA acknowledges his current FCT/2023.07418 CEEECIND research contract with BIOPOLIS (https://doi.org/10.54499/2023.07418.CEECIND/CP2845/CT0001). This work also benefitted from FEDER funds, through the Operational Program for Competitiveness Factors—COMPETE, and from National Funds, through FCT (UIDB/50027/2020, POCI-01–0145-FEDER-006821, UIDB/00153/2020, LA/P/0048/2020), as well as through the Regional Government of the Azores (M1.1.a/005/Funcionamento-C-/2016, CIBIO-A; M1.1.A/INFRAEST CIENT/A/001/2021). JM acknowledges Portuguese Fundação para a Ciência e Tecnologia, FCT, I.P./MCTES through national funds (PIDDAC): UID/50019/2023, LA/P/0068/2020 e UID/50019/2025; https://doi.org/10.54499/LA/P/0068/2020 and https://doi.org/10.54499/UID/PRR/50019/2025.

Data Availability Statement: No new data were created or analyzed in this study.

Acknowledgments: The authors thank Markes E. Johnson (Williams College) for the invitation to participate in the journal's Special Issue and for improving the English. We also sincerely thank Ouali Mehadji Abdelkader (University of Oran 2) for his valuable feedback during the preparation

https://doi.org/10.3390/jmse14010049

of this paper, especially regarding the Paleozoic section, given his expertise in this time interval. This study was conducted as part of the doctoral training program of the 3rd Cycle "Geology of Marine and Continental Environments: Integrated Stratigraphy, Chronology and Dynamics of Paleoenvironments".

Conflicts of Interest: The authors declare that they have no known competing financial interests or personal relationships that could have appeared to influence the work reported in this article.

References

1. Atwater, B.F. Evidence for great Holocene earthquakes along the outer coast of Washington State. *Science* **1987**, *236*, 942–944. [CrossRef] [PubMed]
2. Dawson, A.G.; Long, D.; Smith, D.E. The Storegga slides: Evidence from eastern Scotland for a possible tsunami. *Mar. Geol.* **1988**, *82*, 271–276. [CrossRef]
3. Costa, P.J.M.; Andrade, C. Tsunami deposits: Present knowledge and future challenges. *Sedimentology* **2020**, *67*, 1189–1206. [CrossRef]
4. Fujiwara, O.; Kamataki, T. Identification of tsunami deposits considering the tsunami waveform: An example of subaqueous tsunami deposits in Holocene shallow bay on southern Boso Peninsula, Central Japan. *Sediment. Geol.* **2007**, *200*, 295–313. [CrossRef]
5. Grezio, A.; Babeyko, A.; Baptista, M.A.; Behrens, J.; Costa, A.; Davies, G.; Geist, E.L.; Glimsdal, S.; González, F.I.; Griffin, J.; et al. Probabilistic tsunami hazard analysis: Multiple sources and global applications. *Rev. Geophys.* **2017**, *55*, 1158–1198. [CrossRef]
6. Geist, E.L.; Lynett, P.J. Source processes for the probabilistic assessment of tsunami hazards. *Oceanography* **2014**, *27*, 86–93. [CrossRef]
7. Mori, N.; Satake, K.; Cox, D.; Goda, K.; Catalan, P.A.; Ho, T.C.; Imamura, F.; Tomiczek, T.; Lynett, P.; Miyashita, T.; et al. Giant tsunami monitoring, early warning and hazard assessment. *Nat. Rev. Earth Environ.* **2022**, *3*, 557–572. [CrossRef]
8. Karkani, A.; Evelpidou, N.; Tzouxanioti, M.; Petropoulos, A.; Gogou, M.; Mloukie, E. Tsunamis in the Greek region: An overview of geological and geomorphological evidence. *Geosciences* **2021**, *12*, 4. [CrossRef]
9. Shiki, T.; Tachibana, T.; Fujiwara, O.; Goto, K.; Nanayama, F.; Yamazaki, T. Characteristic features of tsunamiites. In *Tsunamiites*, 2nd ed.; Shiki, T., Tsuji, Y., Yamazaki, T., Nanayama, F., Eds.; Elsevier: Amsterdam, The Netherlands, 2020; pp. 383–404. [CrossRef]
10. Reid, J.A.; Mooney, W.D. Tsunami occurrence 1900–2020: A global review, with examples from Indonesia. *Pure Appl. Geophys.* **2023**, *180*, 1549–1571. [CrossRef]
11. Puga-Bernabéu, Á.; Martín, J.M.; Braga, J.C. Tsunami-related deposits in temperate carbonate ramps, Sorbas Basin, southern Spain. *Sediment. Geol.* **2007**, *199*, 107–127. [CrossRef]
12. Donato, S.V.; Reinhardt, E.G.; Boyce, J.I.; Rothaus, R.; Vosmer, T. Identifying tsunami deposits using bivalve shell taphonomy. *Geology* **2008**, *36*, 199–202. [CrossRef]
13. Donato, S.V.; Reinhardt, E.G.; Boyce, J.I.; Pilarczyk, J.E.; Jupp, B.P. Particle-size distribution of inferred tsunami deposits in Sur Lagoon, Sultanate of Oman. *Mar. Geol.* **2009**, *257*, 54–64. [CrossRef]
14. Maouche, S.; Morhange, C.; Meghraoui, M. Large boulder accumulation on the Algerian coast evidence tsunami events in the western Mediterranean. *Mar. Geol.* **2009**, *262*, 96–104. [CrossRef]
15. Massari, F.; D'Alessandro, A.; Davaud, E. A coquinoid tsunamite from the Pliocene of Salento (SE Italy). *Sediment. Geol.* **2009**, *221*, 7–18. [CrossRef]
16. Paris, R.; Giachetti, T.; Chevalier, J.; Guillou, H.; Frank, N. Tsunami deposits in Santiago Island (Cape Verde archipelago) as possible evidence of a massive flank failure of Fogos volcano. *Sediment. Geol.* **2011**, *239*, 129–145. [CrossRef]
17. Paris, R.; Ramalho, R.S.; Madeira, J.; Ávila, S.P.; May, S.M.; Rixhon, G.; Engel, M.; Brückner, H.; Herzog, M.; Schukraft, G.; et al. Mega-tsunami conglomerates and flank collapses of ocean island volcanoes. *Mar. Geol.* **2018**, *395*, 168–187. [CrossRef]
18. Chagué-Goff, C.; Goff, J.; Nichol, S.L.; Dudley, W.; Zawadzki, A.; Bennett, J.W.; Mooney, S.D.; Fierro, D.; Heijnis, H.; Dominey-Howes, D.; et al. Multi-proxy evidence for trans-Pacific tsunamis in the Hawai'ian Islands. *Mar. Geol.* **2012**, *299*, 77–89. [CrossRef]
19. Puga-Bernabéu, A.; Aguirre, J. Contrasting storm-versus tsunami-related shell beds in shallow-water ramps. *Palaeogeogr. Palaeoclimat. Palaeoecol.* **2017**, *471*, 1–14. [CrossRef]
20. Madeira, J.; Ramalho, R.S.; Hoffmann, D.L.; Mata, J.; Moreira, M. A geological record of multiple Pleistocene tsunami inundations in an oceanic island: The case of Maio, Cape Verde. *Sedimentology* **2020**, *67*, 1529–1552. [CrossRef]
21. Costa, P.J.M.; Dawson, S.; Ramalho, R.S.; Engel, M.; Dourado, F.; Bosnic, I.; Andrade, C. A review on onshore tsunami deposits along the Atlantic coasts. *Earth Sci. Rev.* **2021**, *212*, 103441. [CrossRef]

https://doi.org/10.3390/jmse14010049

22. De Martini, P.M.; Bruins, H.J.; Feist, L.; Goodman-Tchernov, B.N.; Hadler, H.; Lario, J.; Mastronuzzi, G.; Obrocki, L.; Pantosti, D.; Paris, R.; et al. The Mediterranean Sea and the Gulf of Cadiz as a natural laboratory for paleotsunami research: Recent advancements. *Earth Sci. Rev.* **2021**, *216*, 103578. [CrossRef]
23. Coca, O.; Ramírez-Herrera, M.T. Tsunami deposits in tropical regions: A review. *Quat. Int.* **2024**, *698*, 11–25. [CrossRef]
24. Ávila, S.P.; Paris, R.; Ramalho, R.S.; Melo, C.S.; Martín-González, E.; Rolán, E.; Madeira, P.; Ávila, G.C.; Porteiro, J.M.; Medeiros, A.M.; et al. Mega-tsunami deposits and range expansion of cold-temperate marine species towards the tropics in glacial times. *Front. Biogeogr.* **2025**, *18*, e138319. [CrossRef]
25. Doukani, M.A.; Satour, L.; Ozer, C.K.; Belkebir, L.; Voelker, A.H.L.; Tedjeddine, H.; Landau, B.; Uchman, A.; Bessedik, M.; Johnson, M.E.; et al. A Glycymeris-rich Unit as evidence of late Pleistocene tsunami event from NW Algeria: A biostratigraphy, taphonomy, and sedimentological approach. *Quat. Sci. Rev.* **2025**, *368*, 109552. [CrossRef]
26. Dawson, A.G.; Stewart, I. Tsunami deposits in the geological record. *Sediment. Geol.* **2007**, *200*, 166–183. [CrossRef]
27. Morton, R.A.; Gelfenbaum, G.; Jaffe, B.E. Physical criteria for distinguishing sandy tsunami and storm deposits using modern examples. *Sediment. Geol.* **2007**, *200*, 184–207. [CrossRef]
28. Engel, M.; Brückner, H. The identification of palaeo-tsunami deposits—A major challenge in coastal sedimentary research. *Coastline Rep.* **2011**, *17*, 65–80.
29. Shanmugam, G. Process-sedimentological challenges in distinguishing paleo-tsunami deposits. *Nat. Hazards* **2012**, *63*, 5–30. [CrossRef]
30. Putra, P.S.; Nugroho, S.H.; Setyonegoro, W. Bibliometric analysis of paleotsunami research: Current trends and overview. *Curr. Sci.* **2023**, *125*, 1304–1309. [CrossRef]
31. Goodman-Tchernov, B.; Katz, T.; Shaked, Y.; Qupty, N.; Kanari, M.; Niemi, T.; Agnon, A. Offshore evidence for an undocumented tsunami event in the 'low risk' Gulf of Aqaba–Eilat, Northern Red Sea. *PLoS ONE* **2016**, *11*, e0145802. [CrossRef]
32. Chiu, W.T.; Ho, Y.S. Bibliometric analysis of tsunami research. *Scientometrics* **2007**, *73*, 3–17. [CrossRef]
33. Engel, M.; Oetjen, J.; May, S.M.; Brückner, H. Tsunami deposits of the Caribbean towards an improved coastal hazard assessment. *Earth Sci. Rev.* **2016**, *163*, 260–296. [CrossRef]
34. Khadraoui, A.; Kamoun, M.; Hamad, A.B.; Zaïbi, C.; Bonnin, J.; Viehberg, F.; Bahrouni, N.; Sghari, A.; Abida, H.; Kamoun, F. New insights from microfauna associations characterizing palaeoenvironments, sea level fluctuations and a tsunami event along Sfax Northern coast (Gulf of Gabes, Tunisia) during the Late Pleistocene-Holocene. *J. Afr. Earth Sci.* **2018**, *147*, 411–429. [CrossRef]
35. Kitamura, A.; Ito, M.; Ikuta, R.; Ikeda, M. Using molluscan assemblages from paleotsunami deposits to evaluate the influence of topography on the magnitude of late Holocene mega-tsunamis on Ishigaki Island, Japan. *Prog. Earth Planet Sci.* **2018**, *5*, 41. [CrossRef]
36. Dawson, S.; Costa, P.J.M.; Dawson, A.; Engel, M. Onshore archives of tsunami deposits. In *Geological Records of Tsunamis and Other Extreme Waves*; Elsevier Inc.: Amsterdam, The Netherlands, 2020; pp. 95–111. [CrossRef]
37. Torres, T.; Ortiz, J.E.; Mediavilla, R.; Sanchez-Palencia, Y.; Santisteban, J.I.; Vega-Panizo, R. Assessment of prospective geological hazards in Torrevieja-La Mata coast (western Mediterranean) based on Pleistocene and Holocene events. *Nat. Hazards* **2022**, *111*, 2721–2748. [CrossRef]
38. Benfedda, A.; Bouhadad, Y.; Maouche, S.; Bezzeghoud, M. Tsunami hazard along the Northwestern Algerian coast: Evidence of unusual deposits, boulders accumulation and tidal waves modeling. *J. Seismol.* **2025**. [CrossRef]
39. Weiss, R.; Bahlburg, H. A note on the Preservation of Offshore Tsunami Deposits. *J. Sediment. Res.* **2006**, *76*, 1267–1273. [CrossRef]
40. Kümmerer, V.; Drago, T.; Veiga-Pires, C.; Silva, P.F.; Magalhães, V.; Mena, A.; Lopes, A.; Rodrigues, A.I.; Schmidt, S.; Terrinha, P.; et al. Exploring offshore sediment evidence of the 1755 CE tsunami (Faro, Portugal): Implications for the study of outer shelf tsunami deposits. *Minerals* **2020**, *10*, 731. [CrossRef]
41. Feist, L.; Costa, P.J.M.; Bellanova, P.; Bosnic, I.; Santisteban, J.I.; Andrade, C.; Brückner, H.; Duarte, J.F.; Kuhlmann, J.; Schwarzbauer, J.; et al. Holocene offshore tsunami archive—Tsunami deposits on the Algarve shelf (Portugal). *Sediment. Geol.* **2023**, *448*, 106369. [CrossRef]
42. Freundt, A.; Kutterolf, S.; Wehrmann, H.; Schmincke, H.U.; Strauch, W. Eruption of the dacite to andesite zoned Mateare Tephra, and associated tsunamis in Lake Managua, Nicaragua. *J. Volcanol. Geotherm. Res.* **2006**, *149*, 103–123. [CrossRef]
43. Sztanó, O.; Budai, S.; Magyar, I.; Csillag, G.; Nadrai, J.; Fodor, L. Facies and implications of a coarse-grained lacustrine onshore paleo-tsunamiite: An integrated study of an upper Miocene bouldery cobble gravel. *Glob. Planet Change* **2020**, *195*, 103321. [CrossRef]
44. Kremer, K.; Anselmetti, F.S.; Evers, F.M.; Goff, J.; Nigg, V. Freshwater (paleo)tsunamis—A review. *Earth Sci. Rev.* **2021**, *212*, 103447. [CrossRef]
45. Rodrigues, M.G.; Giovanini Varejão, F.; Fürsich, F.T.; Christofoletti, B.; Matos, S.A.; Warren, L.V.; Inglez, L.; Cerri, R.I.; Assine, M.L.; Simões, M.G. Not all shell beds are made equal: Recognizing singular event-concentrations in megalakes. *Sedimentology* **2024**, *71*, 4–47. [CrossRef]

 https://doi.org/10.3390/jmse14010049

46. Gilbert, A.; Martins, P.P., Jr. A possible tsunami deposit at the Cretaceous-Tertiary boundary in Pernambuco, northeastern Brazil. *Sediment. Geol.* **1996**, *104*, 189–201. [CrossRef]
47. Tada, R.; Nakano, Y.; Iturralde-Vinent, M.A.; Yamamoto, S.; Kamata, T.; Tajika, E.; Toyoda, K.; Kiyokawa, S.; García Delgado, D.; Oji, T.; et al. Complex tsunami waves suggested by the Cretaceous-Tertiary boundary deposit at the Moncada section, western Cuba. *Geol. Soc. Am. Spec. Pap.* **2002**, *356*, 109–123. [CrossRef]
48. Scasso, R.; Viramonte, J.; Keller, G.; Stover, L. A tsunami deposit at the Cretaceous/Paleogene boundary in the Neuquén Basin of Argentina. *Cretac. Res.* **2005**, *26*, 283–297. [CrossRef]
49. Schulte, P.; Smit, J.; Deutsch, A.; Salges, T.; Friese, A.; Beichel, K. Tsunami backwash deposits with Chicxulub impact ejecta and dinosaur remains from the Cretaceous–Palaeogene boundary in the La Popa Basin, Mexico. *Sedimentology* **2012**, *59*, 737–765. [CrossRef]
50. DePalma, R.A.; Smit, J.; Burnham, D.A.; Kuiper, K.; Manning, P.L.; Oleinik, A.; Larson, P.; Maurrasse, F.J.; Vellekoop, J.; Richards, M.A.; et al. A seismically induced onshore surge deposit at the KPg boundary, North Dakota. *Proc. Natl. Acad. Sci. USA* **2019**, *116*, 8190–8199. [CrossRef]
51. Manríquez, L.M.E.; Krahl, G.; Carvalho, M.A.; Lavina, E.L.C.; Santiago, G.; Bom, M.H.H.; Fauth, G.; Leppe, M. The K/Pg event at high southern latitudes: New evidence from continental deposits in the Magallanes/Austral Basin, Patagonia, South America. *Cretac. Res.* **2024**, *129*, 105931. [CrossRef]
52. LeVeque, R.J.; DePalma, R.A.; Garrison Laney, C.; Maurya, S.; Smit, J.; Richards, M.A. Possible mechanisms for tsunami-like surge deposits due to the Chicxulub impact at the K-Pg boundary at the Tanis site, North Dakota. *J. Geophys. Res. Solid Earth* **2024**, *129*, e2023JB027643. [CrossRef]
53. Cohen, K.M.; Harper, D.A.T.; Gibbard, P.L.; Car, N. The ICS international chronostratigraphic chart this decade. *Episodes* **2025**, *48*, 105–115. [CrossRef] [PubMed]
54. Einsele, G.; Chough, S.K.; Shiki, T. Depositional events and their records—An introduction. *Sediment. Geol.* **1996**, *104*, 1–9. [CrossRef]
55. Bahlburg, H.; Spiske, M. Sedimentology of tsunami inflow and backflow deposits: Key differences revealed in a modern example: Tsunami deposits at Isla Mocha. *Sedimentology* **2012**, *59*, 1063–1086. [CrossRef]
56. Schnyder, J.; Baudin, F.; Deconinck, J.F. A possible tsunami deposit around the Jurassic—Cretaceous boundary in the Boulonnais area (northern France). *Sediment. Geol.* **2005**, *177*, 209–227. [CrossRef]
57. Scheffers, A.; Kelletat, D. Sedimentologic and geomorphologic tsunami imprints worldwide—A review. *Earth Sci. Rev.* **2003**, *63*, 83–92. [CrossRef]
58. Dunbar, P.; McCullough, H. Global tsunami deposits database. *Nat. Hazards* **2012**, *63*, 267–278. [CrossRef]
59. Runge, E.A.; Duda, J.P.; Van Kranendonk, M.J.; Reitner, J. Earth's oldest tsunami deposit? Early Archaean high-energy sediments in the ca. 3.48 Ga Dresser Formation (Pilbara, Western Australia). *Depos. Rec.* **2022**, *8*, 590–602. [CrossRef]
60. Byerly, G.R.; Lowe, D.R.; Wooden, J.L.; Xie, X.G. An Archean impact layer from the Pilbara and Kaapvaal cratons. *Science* **2002**, *297*, 1325–1327. [CrossRef]
61. Hassler, S.W.; Robey, H.F.; Simonson, B.M. Bedforms produced by impact-generated tsunami, ~2.6 Ga Hamersley Basin, Western Australia. *Sediment. Geol.* **2000**, *135*, 283–294. [CrossRef]
62. Bhattacharya, H.N.; Bandyopadhyay, S. Seismites in a Proterozoic tidal succession, Singhbhum, Bihar, India. *Sediment. Geol.* **1998**, *119*, 239–252. [CrossRef]
63. Wang, Y.N.; Zhang, M.S.; Li, X.B.; Yang, M. The sedimentary record of a Mesoproterozoic tsunami from the North China Craton. *Geol. J.* **2015**, *50*, 56–70. [CrossRef]
64. Pratt, B.R. Oceanography, bathymetry and syndepositional tectonics of a Precambrian intracratonic basin: Integrating sediments, storms, earthquakes and tsunamis in the Belt Supergroup (Helena Formation, ca. 1.45 Ga), western North America. *Sediment. Geol.* **2001**, *141–142*, 371–394. [CrossRef]
65. Pratt, B.R.; Bordonaro, O.L. Tsunamis in a stormy sea: Middle Cambrian inner shelf limestones of western Argentina. *J. Sedi. Res.* **2007**, *77*, 256–262. [CrossRef]
66. Pratt, B.R. Storms versus tsunamis: Dynamic interplay of sedimentary, diagenetic, and tectonic processes in the Cambrian of Montana. *Geology* **2002**, *30*, 423–426. [CrossRef]
67. Põldsaar, K.; Ainsaar, L.; Nemliher, R.; Tinn, O.; Stinkulis, G. A siliciclastic shallow-marine turbidite on the carbonate shelf of the Ordovician Baltoscandian palaeobasin. *Est. J. Earth Sci.* **2019**, *68*, 1–14. [CrossRef]
68. Pratt, B.R.; Sproat, C.D. A tsunami deposit in the Stonewall Formation (Upper Ordovician), northeastern margin of the Williston Basin, Canada, and its tectonic and stratigraphic implications. *Sediment. Geol.* **2023**, *457*, 106518. [CrossRef]
69. Pratt, B.R.; Raviolo, M.M.; Bordonaro, O.L. Carbonate platform dominated by peloidal sands: Lower Ordovician La Silla Formation of the eastern Precordillera, San Juan, Argentina. *Sedimentology* **2012**, *59*, 843–866. [CrossRef]
70. Jarochowska, E.; Munnecke, A. Silurian carbonate high-energy deposits of potential tsunami origin: Distinguishing lateral redeposition and time averaging using carbon isotope chemostratigraphy. *Sediment. Geol.* **2015**, *315*, 14–28. [CrossRef]

 https://doi.org/10.3390/jmse14010049

71. Warme, J.E.; Kuehner, H.C. Anatomy of an anomaly: The Devonian catastrophic Alamo impact breccia of southern Nevada. *Int. Geol. Rev.* **1998**, *40*, 189–216. [CrossRef]
72. Łuczyński, P. Tsunamites versus tempestites: Various types of redeposited stromatoporoid beds in the Devonian of the Holy Cross Mountains (Poland), a case study from the Ołowianka Quarry. *PLoS ONE* **2022**, *17*, e0268349. [CrossRef]
73. Du, Y.S.; Gong, Y.M.; Zeng, X.W.; Huang, H.W.; Yang, J.H.; Zhang, Z.; Huang, Z.Q. Devonian Frasnian–Famennian transitional event deposits of Guangxi, South China and their possible tsunami origin. *Sci. China D Earth Sci.* **2008**, *51*, 1570–1580. [CrossRef]
74. Tohver, E.; Schmieder, M.; Lana, C.; Mendes, P.S.T.; Jourdan, F.; Warren, L.; Riccomini, C. End-Permian impactogenic earthquake and tsunami deposits in the intracratonic Paraná Basin of Brazil. *Bull. Geol. Soc. Am.* **2018**, *130*, 1099–1120. [CrossRef]
75. Brookfield, M.E.; Algeo, T.J.; Hannigan, R.; Williams, J.; Bhat, G.M. Shaken and Stirred: Seismites and Tsunamites at the Permian-Triassic Boundary, Guryul Ravine, Kashmir, India. *Palaios* **2013**, *28*, 568–582. [CrossRef]
76. Diedrich, C. Tsunami killed and backwashed accumulated crinoids in Middle Triassic (Anisian) intracratonic Germanic Basin carbonates of central Europe. *Carbo. Evap.* **2017**, *32*, 435–458. [CrossRef]
77. Kubota, A.; Takeda, Y.; Yi, K.; Sano, S.I.; Iba, Y. Amber in the Cretaceous deep sea deposits reveals large-scale tsunamis. *Sci. Rep.* **2025**, *15*, 14298. [CrossRef]
78. Clyde, W.C.; Krause, J.M.; De Benedetti, F.; Ramezani, J.; Cúneo, N.R.; Gandolfo, M.A.; Haber, P.; Whelan, C.; Smith, T. New South American record of the Cretaceous–Paleogene boundary interval (La Colonia Formation, Patagonia, Argentina). *Cretac. Res.* **2021**, *126*, 104889. [CrossRef]
79. Ganis, G.R.; Willoughby, R.H.; Cicimurri, D.J.; Whittecar, G.R.; Hageman, S.J. Evidence for distal bolide impact and tsunami deposits in the upper Atlantic Coastal Plain of Moore County (North Carolina, USA) generated by the Eocene Chesapeake Bay bolide impact. *Southeast. Geol.* **2025**, *55*, 47–67.
80. Clare, M.A.; Talling, P.J.; Hunt, J.E. Implications of reduced turbidity current and landslide activity for the Initial Eocene Thermal Maximum—Evidence from two distal, deep-water sites. *Earth Planet Sci. Lett.* **2015**, *420*, 102–115. [CrossRef]
81. Neef, G.; Plimer, I.R.; Bottrill, R.S. Submarine-fan deposited sandstone and rudite in a mid-Cenozoic interarc basin in Maewo, Vanuatu (New Hebrides). *Sedimentology* **1985**, *32*, 519–542. [CrossRef]
82. Cantalamessa, G.; Di Celma, C. Sedimentary features of tsunami backwash deposits in a shallow marine Miocene setting, Mejillones Peninsula, northern Chile. *Sediment. Geol.* **2005**, *178*, 259–273. [CrossRef]
83. Rizal, Y.; Aswan, A.; Zaim, J.; Puspaningrum, M.R.; Santoso, W.D.; Rochim, N. Late Miocene to Pliocene tsunami deposits in Tegal Buleud, South Sukabumi, West Java, Indonesia. *Mod. Appl. Sci.* **2019**, *13*, 80–92. [CrossRef]
84. Cita, M.B.; Camerlenghi, A.; Rimoldi, B. Deep-sea tsunami deposits in the eastern Mediterranean: New evidence and depositional models. *Sediment. Geol.* **1996**, *104*, 155–173. [CrossRef]
85. Ruiz, F.; Abad, M.; Rodríguez-Vidal, J.; Cáceres, L.M.; Carretero, M.I.; Pozo, M.; Gómez-Toscano, F.; Izquierdo, T.; Font, E. Morphosedimentary evidence of Holocene tsunamis in the southwestern Spanish estuaries: A summary. In Proceedings of the VIII Reunión de Cuaternario Ibérico, La Rinconada, Sevilla, Spain, 3–6 October 2013.
86. Velasco-Reyes, E.R.; Goto, K.; Sugawara, D.; Nishimura, Y.; Shinohara, T.; Chiba, T. Paleotsunami history of Hachinohe, northern Japan: A multiproxy analysis and numerical modeling approach. *Prog. Earth Planet Sci.* **2022**, *9*, 77. [CrossRef]
87. Bourgeois, J. Geological effects and records of tsunamis. In *The Sea, Volume 15: Tsunamis*; Robinson, A.R., Bernard, E.N., Eds.; Harvard University Press: Boston, MA, USA, 2009; pp. 53–91.
88. Hassler, S.W.; Simonson, B.M. The sedimentary record of extraterrestrial impacts in deep-shelf environments: Evidence from the Early Precambrian. *J. Geol.* **2001**, *109*, 1–19. [CrossRef]
89. Glikson, A.Y.; Allen, C.; Vickers, J. Multiple 3.47-Ga-old asteroid impact fallout units, Pilbara Craton, Western Australia. *Earth Planet Sci. Lett.* **2004**, *221*, 383–396. [CrossRef]
90. Guo, F.S.; Yan, Z.B.; Chen, L.Q. Early Cambrian seismic events and their tectonic significance in western Zhejiang. *Adv. Mater. Res.* **2012**, *524–527*, 42–48. [CrossRef]
91. Hein, F.J.; Nowlan, G.S. Regional sedimentology, conodont biostratigraphy and correlation of Middle Cambrian–Lower Ordovician (?) strata of the "Finnegan" and Deadwood formations, Alberta subsurface, Western Canada Sedimentary Basin. *Bull. Can. Pet. Geol.* **1998**, *46*, 166–188.
92. Mutti, E.; Tinterri, R.; Remacha, E.; Mavilla, N.; Angella, S.; Fava, L. *An Introduction to the Analysis of Ancient Turbidite Basins from an Outcrop Perspective*; The American Association of Petroleum Geologists: Tulsa, OK, USA, 1999; 61p.
93. Ruban, D.A. An enigma of Carboniferous tsunamis: Synthesizing the almost invisible knowledge. *Nat. Hazards* **2025**, *121*, 16431. [CrossRef]
94. Parnell, J. Global mass wasting at continental margins during Ordovician high meteorite influx. *Nat. Geosci.* **2009**, *2*, 57–61. [CrossRef]
95. Bełka, Z.; Narkiewicz, M. Devonian. In *The Geology of Central Europe. Vol. 1: Precambrian and Paleozoic*; McCann, T., Ed.; The Geological Society: London, UK, 2008; pp. 383–411.

 https://doi.org/10.3390/jmse14010049

96. Vierek, A. The palaeogeographical background of Late Devonian storm events in the western part of the Holy Cross Mountains (Poland). *Geologos* **2013**, *19*, 257–272. [CrossRef]
97. Paris, R.; Goto, K.; Goff, J.; Yanagisawa, H. Advances in the study of mega-tsunamis in the geological record. *Earth Sci. Rev.* **2020**, *210*, 103381. [CrossRef]
98. Wagner, R.H.; Diez, J.B. Verdeña (Spain): Life and death of a Carboniferous forest community. *Comptes Rendus Palevol* **2007**, *6*, 495–504. [CrossRef]
99. Pfefferkorn, H.W.; Alleman, W.; Iannuzzi, R. A greenhouse interval between icehouse times: Climate change, long-distance plant dispersal, and plate motion in the Mississippian (late Visean–earliest Serpukhovian) of Gondwana. *Gondwana Res.* **2014**, *25*, 1338–1347. [CrossRef]
100. Gierlowski-Kordesch, E.H.; Cassle, C.F. The 'spirorbis' problem revisited: Sedimentology and biology of microconchids in marine–nonmarine transitions. *Earth Sci. Rev.* **2015**, *148*, 209–227. [CrossRef]
101. Gierlowski-Kordesch, E.H.; Falcon-Lang, H.J.; Cassle, C.F. Reply to comment on the paper of Gierlowski-Kordesch and Cassle "the 'spirorbis' problem revisited: Sedimentology and biology of microconchids in marine–nonmarine transitions". *Earth-Sci. Rev.* **2015**, *148*, 209–227. [CrossRef]
102. Nasiri, N.; Bayet-Goll, A.; Mahboubi, A.; Moussavi-Harami, R. Paleoenvironmental control on trace fossils across a Mississippian carbonate ramp succession, Mobarak Formation, east of central and eastern Alborz, Iran. *J. Afr. Earth Sci.* **2020**, *165*, 103800. [CrossRef]
103. Wignall, P.B.; Twitchett, R.J. Extent, duration and nature of the Permian–Triassic superanoxic event. In *Catastrophic Events and Extinctions; Impacts and Beyond*; Koeberl, C., MacLeod, K., Eds.; Geological Society of America Special Paper 356: Boulder, CO, USA, 2002; pp. 395–413.
104. Shen, S.Z.; Crowley, J.L.; Wang, Y.; Bowring, S.A.; Erwin, D.H.; Sadler, P.M.; Cao, C.Q.; Rothman, D.H.; Henderson, C.M.; Ramezani, J.; et al. Calibrating the end-Permian mass extinction. *Science* **2011**, *334*, 1367–1372. [CrossRef]
105. Erwin, D.H. *Extinction: How Life on Earth Nearly Died 250 Million Years Ago*; Princeton University Press: Princeton, NJ, USA, 2006; 296p.
106. Retallack, G.J.; Seyedolali, A.; Krull, E.S.; Holser, W.T.; Ambers, C.P.; Kyte, F.T. Search for evidence of impact at the Permian–Triassic boundary in Antarctica and Australia. *Geology* **1998**, *26*, 979–982. [CrossRef]
107. Reichow, M.K.; Pringle, M.S.; Al'Mukhamedov, A.I.; Allen, M.B.; Andreichev, V.L.; Buslov, M.M.; Davies, C.E.; Fedoseev, G.S.; Fitton, J.G.; Inger, S.; et al. The timing and extent of the eruption of the Siberian Traps large igneous province: Implications for the end-Permian environmental crisis. *Earth Planet Sci. Lett.* **2009**, *277*, 9–20. [CrossRef]
108. Korte, C.; Pande, P.; Kalia, P.; Kozur, H.W.; Joachimski, M.M.; Oberhänsli, H. Massive volcanism at the Permian–Triassic boundary and its impact on the isotope composition of the ocean and atmosphere. *J. Asian Earth Sci.* **2010**, *37*, 293–311. [CrossRef]
109. Tohver, E.; Cawood, P.A.; Riccomini, C.; Lana, C.; Trindade, R.I.F. Shaking a methane fizz: Seismicity from the Araguainha impact event and the Permian-Triassic global carbon isotope record. *Palaeogeogr. Palaeoclimatol. Palaeoecol.* **2013**, *387*, 66–75. [CrossRef]
110. Ghosh, N.; Basu, A.R.; Bhargava, O.N.; Shukla, O.K.; Ghatak, A.; Garzione, C.N.; Ahluwalia, A.D. Catastrophic environmental transition at the Permian-Triassic Neo-Tethyan margin of Gondwanaland: Geochemical, isotopic and sedimentological evidence in the Spiti Valley, India. *Gondwana Res.* **2016**, *34*, 324–345. [CrossRef]
111. Brookfield, M.E.; Wolbach, W.S.; Stebbins, A.G.; Gilmour, I.; Roegge, D.R. Organic carbon content and carbon isotope variations across the Permo-Triassic boundary in the Gartnerkofel-1 borehole, Carnic Alps, Austria. *Acta Geochim.* **2018**, *37*, 422–432. [CrossRef]
112. Zuchuat, V.; Sleveland, A.R.N.; Twitchett, R.J.; Svensen, H.H.; Turner, H.; Augland, L.E.; Jones, M.T.; Hammer, O.; Hauksson, B.T.; Haflidason, H.; et al. A new high-resolution stratigraphic and palaeoenvironmental record spanning the End-Permian Mass Extinction and its aftermath in central Spitsbergen, Svalbard. *Palaeogeogr. Palaeoclimatol. Palaeoecol.* **2020**, *554*, 109732. [CrossRef]
113. Mir, J.A.; Bhat, I.M.; Murtaza, K.O.; Qader, W.; Dar, R.A. Geological Heritage of the Kashmir Valley, North-Western Himalaya, India. *Geoheritage* **2023**, *15*, 26. [CrossRef]
114. Ruban, D.A. Tsunamis Struck Coasts of Triassic Oceans and Seas: Brief Summary of the Literary Evidence. *Water* **2023**, *15*, 1590. [CrossRef]
115. Varela, A.N.; Richiano, S.; Poiré, D.G. Tsunami vs Storm Origin for Shell Bed Deposits in a Lagoon Environment: An Example from the Upper Cretaceous of Southern Patagonia, Argentina. *Lat. Am. J. Sedimentol. Basin Anal.* **2011**, *18*, 63–85.
116. Le Roux, J.P.; Vargas, G. Structure and depositional processes of a gravelly tsunami deposit in a shallow marine setting: Lower Cretaceous Miyako Group, Japan—Discussion. *Sid. Geol.* **2007**, *201*, 485–487. [CrossRef]
117. Veloso, F.M.L.; Liesa, C.L.; Soria, A.R.; Meléndez, N.; Frykman, P. Facies and petrophysical modelling of a thick Lower Cretaceous tsunami deposit in E Spain: Up-scaling from sample to outcrop scales. *Sediment. Geol.* **2016**, *343*, 38–55. [CrossRef]
118. Rossetti, D.D.F.; Góes, A.M.; Truckenbrodt, W.; Anaisse, J. Tsunami-induced large-scale scour-and-fill structures in Late Albian to Cenomanian deposits of the Grajau Basin, northern Brazil. *Sedimentology* **2000**, *47*, 309–323. [CrossRef]
119. Fujino, S.; Maeda, H. Environmental changes and shallow marine fossil bivalve assemblages of the Lower Cretaceous Miyako Group, NE Japan. *J. Asian Earth Sci.* **2013**, *64*, 168–179. [CrossRef]

120. King, D.T.; Petruny, L.W.; Neathery, T.L. Paleobiologic effects of the Late Cretaceous Wetumpka marine impact, a 7.6-km-diameter impact structure, Gulf Coastal Plain, USA; In *Biological Processes Associated with Impact Events*; Cockell, C., Koeberl, C., Gilmour, I., Eds.; Springer: Berlin/Heidelberg, Germany, 2006; pp. 121–142. [CrossRef]
121. Smit, J. The KPg boundary Chicxulub impact-extinction hypothesis: The winding road towards a solid theory. In *From the Guajira Desert to the Apennines, and from Mediterranean Microplates to the Mexican Killer Asteroid: Honoring the Career of Walter Alvarez*; Koeberl, C., Claeys, P., Montanari, A., Eds.; Geological Society of America: Boulder, CO, USA, 2022; Volume 557, pp. 391–414. [CrossRef]
122. Alvarez, W.T. *Rex and the Crater of Doom*; Princeton University Press: Princeton, NJ, USA, 1997; 185p.
123. Gradstein, F.M.; Ogg, J.G.; Schmitz, M.D.; Ogg, G.M. *The Geologic Time Scale 2012*; Elsevier: Oxford, UK, 2012; ISBN 978-0-444-59425-9.
124. Renne, P.R.; Deino, A.L.; Hilgen, F.J.; Kuiper, K.F.; Mark, D.F.; Mitchell, W.S.; Morgan, L.E.; Mundil, R.; Smit, J. Time scales of critical events around the Cretaceous-Paleogene boundary. *Science* **2013**, *339*, 684–687. [CrossRef]
125. Kinsland, G.L.; Egedahl, K.; Strong, M.A.; Ivy, R. Chicxulub impact tsunami megaripples in the subsurface of Louisiana: Imaged in petroleum industry seismic data. *Earth Planet Sci. Lett.* **2021**, *570*, 117063. [CrossRef]
126. Bourgeois, J.; Hansen, T.A.; Wiberg, P.L.; Kauffman, E.G. A tsunami deposit at the Cretaceous–Tertiary boundary in Texas. *Science* **1988**, *241*, 567–570. [CrossRef] [PubMed]
127. Smit, J.; Montanari, A.; Swinburne, N.H.M.; Alvarez, W.; Hildebrand, A.R.; Margolis, S.V.; Claeys, P.; Lowrie, W.; Asaro, F. Tektite-bearing, deep-water clastic unit at the Cretaceous–Tertiary boundary in northeastern Mexico. *Geology* **1992**, *20*, 99–103. [CrossRef] [PubMed]
128. Stinnesbeck, W.; Barbarin, J.M.; Keller, G.; Lopez-Oliva, J.G.; Pivnik, D.A.; Lyons, J.B.; Officer, C.B.; Adatte, T.; Graup, G.; Rocchia, R.; et al. Deposition of channel deposits near the Cretaceous–Tertiary boundary in northeastern Mexico: Catastrophic or "normal" sedimentary deposits. *Geology* **1993**, *21*, 797–800. [CrossRef]
129. Schulte, P.; Alegret, L.; Arenillas, I.; Arz, J.A.; Barton, P.J.; Bown, P.R.; Bralower, T.J.; Christeson, G.L.; Claeys, P.; Cockell, C.S.; et al. The Chicxulub asteroid impact and mass extinction at the Cretaceous-Paleogene boundary. *Science* **2010**, *327*, 1214–1218. [CrossRef]
130. Alvarez, W.; Smit, J.; Lowrie, W.; Asaro, F.; Margolis, S.V.; Claeys, P.; Kastner, M.; Hildebrand, A.R. Proximal impact deposits at the Cretaceous–Tertiary boundary in the Gulf of Mexico: A restudy of DSDP Leg 77 Sites 536 and 540. *Geology* **1992**, *20*, 697–700. [CrossRef]
131. Range, M.M.; Arbic, B.K. The Chicxulub impact produced a powerful global tsunami. *AGU Adv.* **2022**, *3*, e2021AV000627. [CrossRef]
132. Dickinson, W.R.; Snyder, W.S. Plate tectonics of the Laramide Orogeny. *Geol. Soc. Am. Mem.* **1978**, *151*, 355–366. [CrossRef]
133. Vergés, J.; Millán, H.; Roca, E.; Muñoz, J.A.; Marzo, M.; Cirés, J.; Cloetingh, S. Eastern Pyrenees and related foreland basins: Pre-, syn- and post-collisional crustal-scale cross-sections. *Mar. Pet. Geol.* **1995**, *12*, 903–915. [CrossRef]
134. Najman, Y. The detrital record of orogenesis: A review of approaches and techniques used in the Himalayan sedimentary basins. *Earth Sci. Rev.* **2006**, *74*, 1–72. [CrossRef]
135. Pindell, J.; Kennan, L. Tectonic evolution of the Gulf of Mexico, Caribbean and northern South America in the mantle reference frame: An update. *Geol. Soc. Lond. Spec. Publ.* **2009**, *328*, 1–55. [CrossRef]
136. Tugend, J.; Manatschal, G.; Kusznir, N.J. Spatial and temporal evolution of hyperextended rift systems: Implication for the nature, kinematics, and timing of the Iberian–European plate boundary. *Geology* **2015**, *43*, 15–18. [CrossRef]
137. McHugh, C.M.; Snyder, S.W.; Miller, K.G. Upper Eocene ejecta of the New Jersey continental margin reveal dynamics of Chesapeake Bay impact. *Earth Planet Sci. Lett.* **1998**, *160*, 353–367. [CrossRef]
138. Poag, C.W. The Chesapeake Bay bolide impact: A convulsive event in Atlantic Coastal Plain evolution. *Sediment. Geol.* **1997**, *108*, 45–90. [CrossRef]
139. Schmitz, B.; Pujalte, V.; Nunez-Betelu, K. Climate and sea-level perturbations during the initial Eocene thermal maximum: Evidence from siliciclastic units in the Basque Basin (Ermua, Zumaia and Trabakua Pass), northern Spain. *Palaeo. Palaeoclim. Palaeoecol.* **2001**, *165*, 299–320. [CrossRef]
140. Kennett, J.P.; Cannariato, K.G.; Hendy, I.L.; Behl, R.J. *Methane Hydrates in Quaternary Climate Change: The Clathrate Gun Hypothesis*; American Geophysical Union: Washington, DC, USA, 2003; Volume 54, pp. 1–216.
141. Bock, J.; Martinerie, P.; Witrant, E.; Chappellaz, J. Atmospheric impacts and ice core imprints of a methane pulse from clathrates. *Earth Planet Sci. Lett.* **2012**, *349*, 98–108. [CrossRef]
142. Hornbach, M.J.; Lavier, L.L.; Ruppel, C.D. Triggering mechanism and tsunamogenic potential of the Cape Fear Slide complex, U.S. Atlantic margin. *Geochem. Geophys. Geosyst.* **2007**, *8*, Q12008. [CrossRef]
143. Maslin, M.; Owen, M.; Betts, R.; Day, S.; Jones, T.D.; Ridgwell, A. Gas hydrates: Past and future geohazard? *Philos. Trans. R. Soc. A Math. Phys. Eng. Sci.* **2010**, *368*, 2369–2393. [CrossRef]

 https://doi.org/10.3390/jmse14010049

144. Tappin, D.R.; Watts, P.; McMurtry, G.M.; Lafoy, Y.; Matsumoto, T. The Sissano, Papua New Guinea tsunami of July 1998—Offshore evidence on the source mechanism. *Mar. Geol.* **2001**, *175*, 1–23. [CrossRef]
145. Bondevik, S.; Løvholt, F.; Harbitz, C.; Mangerud, J.; Dawson, A.; Svendsen, J.I. The Storegga Slide tsunami—Comparing field observations with numerical simulations. *Mar. Pet. Geol.* **2005**, *22*, 195–208. [CrossRef]
146. ten Brink, U.S.; Barkan, R.; Andrews, B.D.; Chaytor, J.D. Size distributions and failure initiation of submarine and subaerial landslides. *Earth Planet Sci. Lett.* **2009**, *287*, 31–42. [CrossRef]
147. ten Brink, U.S.; Lee, H.J.; Geist, E.L.; Twichell, D. Assessment of tsunami hazard to the U.S. East Coast using relationships between submarine landslides and earthquakes. *Mar. Geol.* **2009**, *264*, 65–73. [CrossRef]
148. ten Brink, U.S.; Geist, E.L.; Andrews, B.D. Size distribution of submarine landslides and its implication to tsunami hazard in Puerto Rico. *Geophys. Res. Lett.* **2006**, *33*, L11307. [CrossRef]
149. Shiki, T.; Yamazaki, T. Tsunami-induced conglomerates in Miocene upper bathyal deposits, Chita Peninsula, central Japan. *Sediment. Geol.* **1996**, *104*, 175–188. [CrossRef]
150. Le Roux, J.P.; Vargas, G. Hydraulic behavior of tsunami backflows: Insights from their modern and ancient deposits. *Environ. Geol.* **2005**, *49*, 65–75. [CrossRef]
151. Fortuin, A.R.; Dabrio, C.J. Evidence for Late Messinian seismites, Nijar Basin, south-east Spain. *Sedimentology* **2008**, *55*, 1595–1622. [CrossRef]
152. Smit, J.; Laffra, C.; Meulenaars, K.; Montanari, A. Probable late Messinian tsunamiites near Monte Dei Corvi, Italy, and the Nijar Basin, Spain: Expected architecture of offshore tsunami deposits. *Nat. Hazards* **2012**, *63*, 241–266. [CrossRef]
153. Tachibana, T. Lonestones as indicators of tsunami deposits in deep-sea sedimentary rocks of the Miocene Morozaki Group, central Japan. *Sediment. Geol.* **2013**, *289*, 62–73. [CrossRef]
154. Tachibana, T.; Tsuji, Y. Geological and hydrodynamical examination of bathyal tsunamigenic origin of Miocene conglomerates in Chita Peninsula, central Japan. *Pure Appl. Geophys.* **2011**, *168*, 997–1014. [CrossRef]
155. Spiske, M.; Bahlburg, H.; Weiss, R. Pliocene mass failure deposits mistaken as submarine tsunami backwash sediments—An example from Hornitos, northern Chile. *Sediment. Geol.* **2014**, *305*, 69–82. [CrossRef]
156. Di Celma, C.; Bosio, G.; Pierantoni, P.P.; Gariboldi, K.; Malinverno, E.; Gioncada, A.; Varas-Malca, R.M.; Urbina, M.; Collareta, A.; Lambert, O.; et al. Allostratigraphy and paleontology of the lower Miocene Chilcatay Formation in the Zamaca area, East Pisco Basin, southern Peru. *J. Maps* **2019**, *15*, 393–405. [CrossRef]
157. Son, J.; Cho, S.; Hwang, I.G. The Hakcheon-Chogok Megaturbidite in the Miocene Pohang-Youngduk Basin, SE Korea: High-gradient slope failure probably triggered by a giant tsunami wave. *Geosci. J.* **2024**, *28*, 301–318. [CrossRef]
158. McCall, G.J.H.; Kidd, R.G.W. The Makran, southeastern Iran: The anatomy of a convergent margin active from Cretaceous to present. In *Trench-Forearc Geology: Sedimentation and Tectonics of Modern and Ancient Plate Margins*; Leggett, J.K., Ed.; Geological Society, London, Special Publication; Geological Society of London: London, UK, 1982; Volume 10, pp. 387–397. [CrossRef]
159. McCall, G.J.H. Mélanges of the Makran, southeastern Iran. In *Ophiolitic and Related Mélanges*; InMcCall, G.J.H., Ed.; Benchmark Papers in Geology; Hutchinson Ross Publ. Co.: Stroudsburg, PA, USA, 1983; Volume 66, pp. 292–299.
160. Burg, J.P.; Bernoulli, D.; Smit, J.; Dolati, A.; Bahroudi, A. A giant catastrophic mud-and-debris flow in the Miocene Makran. *Terra Nova* **2008**, *20*, 188–193. [CrossRef]
161. Bahlburg, H.; Spiske, M.; Weiss, R. Comment on "Sedimentary features of tsunami backwash deposits in a shallow marine Miocene setting, Mejillones Peninsula, northern Chile" by G. Cantalamessa and C. Di Celma [Sediment. Geol. 178 (2005) 259–273]. *Sediment. Geol.* **2010**, *228*, 77–80. [CrossRef]
162. Kidwell, S.M.; Fürsich, F.T.; Aigner, T. Conceptual framework of the analysis and classification of fossil concentrations. *Palaios* **1986**, *1*, 228–238. [CrossRef]
163. Le Roux, J.P.; Nielsen, S.N.; Kemnitz, H.; Henriquez, A. A Pliocene mega-tsunami deposits and associated features in the Ranquil Formation, Southern Chile. *Sediment. Geol.* **2008**, *203*, 164–180. [CrossRef]
164. Madeira, J.; Mata, J.; Moreira, M.; Hoffman, D.L. Deposits of a major Pleistocene tsunami in the Island of Maio (Cape Verde). *Geophys. Res. Abstr.* **2011**, *13*, EGU2011–1995.
165. Madeira, J.; Mata, J.; Moreira, M.; Ramalho, R.S. Tsunami deposits in the Island of Maio (Cape Verde): Paleocurrent markers and basal erosion features. In *Abstract Volume, Proceedings of the 5th International Tsunami Field Symposium, Lisbon, Portugal, 3–7 September 2017*; Costa, P.J.M., Andrade, C., Freitas, M.C., Eds.; Universidade de Lisboa: Lisboa, Portugal, 2017; pp. 76–77.
166. Matougui, R.; Haddoum, H. Mise en evidence de mouvements gravitaires sur la pente continentale de la marge du Dahra occidental (Tenes–Cap Ivi). *Bull. Serv. Geol. Natl.* **2008**, *19*, 267–285.
167. Ramalho, R.; Helffrich, G.; Schmidt, D.N.; Vance, D. Tracers of uplift and subsidence in the Cape Verde Archipelago. *J. Geol. Soc. Lond.* **2010**, *167*, 519–538. [CrossRef]
168. Ramalho, R.S.; Quartau, R.; Trenhaile, A.S.; Mitchell, N.C.; Woodroffe, C.D.; Ávila, S.P. Coastal evolution on volcanic oceanic islands: A complex interplay between volcanism, erosion, sedimentation, sea-level change, and biogenic production. *Earth Sci. Rev.* **2013**, *127*, 140–170. [CrossRef]

 https://doi.org/10.3390/jmse14010049

169. Andreucci, S.; Panzeri, L.; Martini, P.; Maspero, F.; Martini, M.; Pascucci, V. Evolution and architecture of a west Mediterranean upper Pleistocene to Holocene coastal apron-fan system. *Sedimentology* **2014**, *61*, 333–361. [CrossRef]
170. Benjamin, J.; Rovere, A.; Fontana, A.; Furlani, S.; Vacchi, M.; Inglis, R.H.; Galili, E.; Antonioli, F.; Sivan, D.; Miko, S.; et al. Late Quaternary sea-level changes and early human societies in the central and eastern Mediterranean Basin: An interdisciplinary review. *Quat. Int.* **2017**, *449*, 29–57. [CrossRef]
171. Cerrone, C.; Vacchi, M.; Fontana, A.; Rovere, A. Last Interglacial sea-level proxies in the western Mediterranean. *Earth Syst. Sci. Data* **2021**, *13*, 4485–4527. [CrossRef]
172. Melo, C.S.; Da Silva, C.M.; Scarponi, D.; González, E.; Rolán, E.; Rojas, A.; Martinez, S.; Silva, L.; Johnson, M.E.; Rebelo, A.C.; et al. Palaeobiogeography of NE Atlantic archipelagos during the Last Interglacial (MIS 5e): A molluscan approach to the conundrum of Macaronesia as a marine biogeographic unit. *Quat. Sci. Rev.* **2023**, *319*, 108313. [CrossRef]
173. del Valle, L.; Timar-Gabor, A.; Fornós, J.J. Chronology of Pleistocene sedimentary cycles in the western Mediterranean. *Quat. Sci. Rev.* **2024**, *330*, 108451. [CrossRef]
174. Johnson, M.E.; Uchman, A.; Ramalho, R.S.; Martín-González, E.; Martins, G.M.; Hipólito, A.; Marques, S.; Ávila, G.C.; Madeira, P.; Doukani, M.A.; et al. Basalt sea stack and related facies from the Middle Pleistocene on Sal Island (Cabo Verde Archipelago, NE Atlantic Ocean). *Facies* **2025**, *71*, 20. [CrossRef]
175. Takashimizu, Y.; Masuda, F. Depositional facies and sedimentary successions of earthquake-induced tsunami deposits of Upper Pleistocene incised valley fills, central Japan. *Sediment. Geol.* **2000**, *135*, 231–239. [CrossRef]
176. Goff, J.R.; Chagué-Goff, C.; Nichol, S.L. Palaeotsunami deposits: A New Zealand perspective. *Sediment. Geol.* **2001**, *143*, 1–6. [CrossRef]
177. Goff, J.R. *The New Zealand Palaeotsunami Database*; NIWA Technical Report (NIWA TR 131); National Institute of Water & Atmospheric Research: Wellington, New Zealand, 2008.
178. Ramalho, R.S.; Winclker, G.; Madeira, J.; Helffrich, G.R.; Hipólito, A.R.; Quartau, R.; Adena, K.; Schaefer, J.M. Hazard potential of volcanic flank collapses raised by new megatsunami evidence. *Sci. Adv.* **2015**, *1*, e1500456. [CrossRef] [PubMed]
179. Paris, R.; Pérez-Torrado, F.J.; Cabrera, M.C.; Schneider, J.L.; Wassmer, P.; Carracedo, J.C. Tsunami-induced conglomerates and debris flow deposits on the western coast of Gran Canaria (Canary Islands). *Acta Vulcanol.* **2004**, *16*, 133–136.
180. Paris, R.; Pérez-Torrado, F.J.; Carracedo, J.C. Massive flank failures and tsunamis in the Canary Islands: Past, present, future. *Z. Geomorphol.* **2005**, *140*, 37–54.
181. Johnson, M.E.; Uchman, A.; Costa, P.J.M.; Ramalho, R.S.; Ávila, S.P. Intense hurricane transport sand onshore: Example from the Pliocene Malbusca section on Santa Maria Island (Azores, Portugal). *Mar. Geol.* **2017**, *385*, 244–249. [CrossRef]
182. Goff, J.; Terry, P.; Chagué-Goff, C.; Goto, K. What is a mega-tsunami? *Mar. Geol.* **2014**, *358*, 12–17. [CrossRef]
183. Miller, D.J. Giant waves in Lituya Bay, Alaska. In *Geological Survey Professional Paper 354-C*; U.S. Government Printing Office: Washington, DC, USA, 1960.
184. Ferrer, M.; Gonzalez de Vallejo, L.; Seisdedos, J.; Coello, J.J.; García, J.C.; Hernandez, L.E.; Casillas, R.; Martín, C.; Rodríguez, J.A.; Madeira, J.; et al. Güímar and La Orotava mega-landslides (Tenerife) and tsunamis deposits in Canary Islands. In *Landslide Science and Practice, Volume 5: Complex Environment*; Margottini, C., Canuti, P., Sassa, K., Eds.; Springer: Berlin/Heidelberg, Germany, 2013; pp. 27–33.
185. Perez-Torrado, F.J.; Paris, R.; Cabrera, M.C.; Schneider, J.L.; Wasmer, P.; Carracedo, J.C.; Rodriguez Santana, A.; Santana, F. Tsunami deposits related to flank collapse in oceanic volcanoes: The Agaete Valley evidence, Gran Canaria, Canary Islands. *Mar. Geol.* **2006**, *227*, 137–147. [CrossRef]
186. Moore, J.G.; Moore, G.W. Deposit from a giant wave on the island of Lanai, Hawaii. *Science* **1984**, *226*, 1312–1315. [CrossRef]
187. Brückner, H.; Radtke, U. Fossile Strände und Korallenbänke auf Oahu, Hawaii. In *Neue Ergebnisse der Küstenforschung. Vorträge der AMK Jahrestagung Wilhelmshaven 18. und 19. Mai 1989*; Kelletat, D., Ed.; Essener Geographische Arbeiten: Essen, Germany, 1989; Volume 17, pp. 291–308.
188. Meco, J. Islas Canarias. In *Mapa del Cuaternario de España a Escala 1:100000*; Pérez-González, A., Cabra, G.P., Martín Serrano, A., Eds.; Instituto Tecnológico y Geominero de España: Madrid, Spain, 1989.
189. Grigg, R.W.; Jones, A.T. Uplift caused by lithospheric flexure in the Hawaiian archipelago, as revealed by elevated coral deposits. *Mar. Geol.* **1997**, *141*, 11–25. [CrossRef]
190. Fujiwara, O.; Masuda, F.; Sakai, T.; Irizuki, T.; Fuse, K. Tsunami deposits in Holocene bay mud in southern Kanto region, Pacific coast of central Japan. *Sediment. Geol.* **2000**, *135*, 219–230. [CrossRef]
191. Kortekaas, S.; Dawson, A.G. Distinguishing tsunami and storm deposits: An example from Martinhal, SW Portugal. *Sediment. Geol.* **2007**, *200*, 208–221. [CrossRef]
192. King, D.N.; Clark, K.; Chagué, C.; Li, X.; Lane, E.; McFadgen, B.G.; Hippolite, J.; Meihana, P.; Wilson, B.; Dobson, J.; et al. Tsunami or Storm Deposit? A Late Holocene Sedimentary Record from Swamp Bay, Rangitoto ki te Tonga/D'Urville Island, Aotearoa–New Zealand. *N. Z. J. Geol. Geophys.* **2023**, *66*, 629–645. [CrossRef]

 https://doi.org/10.3390/jmse14010049

193. Chagué-Goff, C.; Schneider, J.L.; Goff, J.R.; Dominey-Howes, D.; Strotz, L. Expanding the proxy toolkit to help identify past events—Lessons from the 2004 Indian Ocean Tsunami and the 2009 South Pacific Tsunami. *Earth Sci. Rev.* **2011**, *107*, 107–122. [CrossRef]
194. Boomer, I.; Waddington, C.; Stevenson, T.; Hamilton, D. Holocene coastal change and geoarchaeology at Howick, Northumberland, UK. *Holocene* **2007**, *17*, 89–104. [CrossRef]
195. Mamo, B.; Strotz, L.; Dominey-Howes, D. Tsunami sediments and their foraminiferal assemblages. *Earth Sci. Rev.* **2009**, *96*, 263–278. [CrossRef]
196. Pozo, M.; Ruiz, F.; Carretero, M.I.; Rodríguez Vidal, J.; Cáceres, L.M.; Abad, M.; González-Regalado, M.L. Mineralogical assemblages, geochemistry and fossil associations of Pleistocene–Holocene complex siliciclastic deposits from the Southwestern Doñana National Park (SW Spain): A palaeoenvironmental approach. *Sediment. Geol.* **2010**, *225*, 1–18. [CrossRef]
197. Dura, T.; Hemphill-Haley, E.; Sawai, Y.; Horton, B.P.; Ely, L.L.; Engelhart, S.E.; Shennan, I.; Garrett, E.; Kelsey, H.M.; Briggs, R.W.; et al. The application of diatoms to reconstruct the history of subduction zone earthquakes and tsunamis. *Earth Sci. Rev.* **2016**, *152*, 181–197. [CrossRef]
198. Costa, P.J.M.; Dawson, S. Tsunami Sedimentology. In *Complexity in Tsunamis, Volcanoes, and Their Hazards*; Tilling, R.I., Ed.; Springer: New York, NY, USA, 2022; pp. 135–151.
199. Kench, P.S.; Nichol, S.L.; Smithers, S.G.; McLean, R.F.; Brander, R.W. Tsunami as agents of geomorphic change in mid-ocean reef islands. *Geomorphology* **2008**, *95*, 361–383. [CrossRef]
200. Morales, J.A.; Borrego, J.; San Miguel, E.G.; López-González, N.; Carro, B. Sedimentary record of recent tsunamis in the Huelva Estuary (Southwestern Spain). *Quat. Sci. Rev.* **2008**, *27*, 734–746. [CrossRef]
201. Morton, R.A.; Goff, J.R.; Nichol, S.L. Hydrodynamic implications of textural trends in sand deposits of the 2004 tsunami in Sri Lanka. *Sediment. Geol.* **2008**, *207*, 56–64. [CrossRef]
202. Kastens, K.A.; Cita, M.B. Tsunami induced sediment transport in the Abyssal Mediterranean Sea. *Geol. Soc. Am. Bull.* **1981**, *89*, 591–604. [CrossRef]
203. Hunt, J.E.; Wynn, R.B.; Masson, D.G.; Talling, J.; Teagle, D.A.H. Sedimentological and geochemical evidence for multi-stage failure of volcanic island landslides: A case study from Icod landslide on north Tenerife, Canary Islands. *Geochem. Geophys. Geosyst.* **2011**, *12*, Q12007. [CrossRef]
204. Andrade, C. Tsunami generated forms in the Algarve Barrier Islands (South Portugal). *Sci. Tsunami Haz.* **1992**, *10*, 21–34.
205. Andrade, C.; Freitas, M.C.; Oliveira, M.A.; Costa, P.J.M. On the sedimentiological and historical evidences of seismic-triggered tsunamis on the Algarve Coast of Portugal. In *Plate Boundaries and Natural Hazards*; Duarte, J.C., Schellar, W.P., Eds.; American Geophysical Union: Washington, DC, USA; John Wiley & Sons: Hoboken, NJ, USA, 2016; pp. 219–238.
206. Paris, R.; Wassmer, P.; Sartohadi, J.; Lavigne, F.; Barthomeuf, B.; Desgages, E.; Grancher, D.; Baumert, P.; Vautier, F.; Brunstein, D.; et al. Coastal sedimentation associated with the December 26, 2004 tsunami in Lhok Nga, west Banda Aceh (Sumatra, Indonesia). *Mar. Geol.* **2007**, *238*, 93–106. [CrossRef]
207. Goto, K.; Chagué-Goff, C.; Fujino, S.; Goff, J.; Jaffe, B.; Nishimura, Y.; Richmond, B.; Sugawara, D.; Szczuciński, W.; Tappin, D.R.; et al. New insights of tsunami hazard from the 2011 Tohoku-oki event. *Mar. Geol.* **2011**, *290*, 46–50. [CrossRef]
208. Chagué-Goff, C.; Andrew, A.; Szczuciński, W.; Goff, J.; Nishimura, Y.; Richmond, B.; Sugawara, D.; Wallace, L. Environmental impact assessment of the 2011 Tohoku-oki tsunami on the Sendai Plain. *Sediment. Geol.* **2012**, *282*, 175–187. [CrossRef]
209. Szczuciński, W. The post-depositional changes of the onshore 2004 tsunami deposits on the Andaman Sea coast of Thailand. *Nat. Hazards* **2012**, *60*, 115–133. [CrossRef]
210. Richmond, B.; Szczuciński, W.; Chagué-Goff, C.; Goff, J.; Jaffe, B.; Fujino, S.; Nishimura, Y.; Hong, Y.; Gusman, A.; Dominey-Howes, D.; et al. Erosion, deposition and landscape change on the Sendai coastal plain, Japan, resulting from the March 11, 2011 Tohoku-oki tsunami. *Sediment. Geol.* **2012**, *282*, 27–39. [CrossRef]
211. Costa, P.J.M.; Andrade, C.; Dawson, A.G.; Mahaney, W.C.; Freitas, M.C.; Paris, R.; Taborda, R. Microtextural characteristics of quartz grains transported and deposited by tsunamis and storms. *Sediment. Geol.* **2012**, *275*, 55–69. [CrossRef]
212. Costa, P.J.M.; Costas, S.; González-Villanueva, R.; Oliveira, M.A.; Roelvink, D.; Andrade, C.; Freitas, M.C.; Cunha, P.P.; Martins, A.; Buylaert, J.P.; et al. How did the AD 1755 tsunami impact on sand barriers across the southern coast of Portugal? *Geomorphology* **2016**, *268*, 296–311. [CrossRef]
213. Abe, T.; Goto, K.; Sugawara, D. Relationship between the maximum extent of tsunami sand and the inundation limit of the 2011 Tohoku-oki tsunami on the Sendai Plain, Japan. *Sediment. Geol.* **2022**, *435*, 106138. [CrossRef]
214. Matsumoto, D.; Shimamoto, T.; Hirose, T.; Gunatilake, J.; Wickramasooriya, A.; DeLile, J.; Young, S.M.; Rathnayake, C.; Ranasooriya, J.; Murayama, M. Thickness and grain-size distribution of the 2004 Indian Ocean tsunami deposits in Periya Kalapuwa Lagoon, eastern Sri Lanka. *Sediment. Geol.* **2010**, *230*, 95–104. [CrossRef]
215. Naruse, H.; Abe, T. Inverse-to-normal grading of sand layer deposited by the 2011 Tohoku-oki tsunami at Sendai Plain, Japan. *Sediment. Geol.* **2017**, *353*, 1–12.

 https://doi.org/10.3390/jmse14010049

216. Chagué-Goff, C.; Szczuciński, W.; Shinozaki, T. Applications of geochemistry in tsunami research: A review. *Earth Sci. Rev.* **2017**, *165*, 203–244. [CrossRef]
217. Chagué-Goff, C. Chemical signatures of palaeotsunamis: A forgotten proxy? *Mar. Geol.* **2010**, *271*, 67–71. [CrossRef]
218. Moreira, S.; Costa, P.J.M.; Andrade, C.; Ponte Lira, C.; Freitas, M.C.; Oliveira, M.A.; Reichart, G.J. High resolution geochemical and grain-size analysis of the AD 1755 tsunami deposit: Insights on the inland extent and inundation phases. *Mar. Geol.* **2017**, *390*, 94–105. [CrossRef]
219. Vigliotti, L.; Andrade, C.; Freitas, M.C.; Capotondi, L.; Gallerani, A.; Bellucci, L.G. Paleomagnetic, rock magnetic and geochemical study of the 1755 tsunami deposit at Boca do Rio (Algarve, Portugal). *Palaeogeogr. Palaeocl.* **2019**, *514*, 550–566. [CrossRef]
220. Hawkes, A.D.; Bird, M.; Cowie, S.; Grundy-Warr, C.; Horton, B.P.; Tan Shau Hwai, A.; Law, L.; Macgregor, C.; Nott, J.; Eong Ong, J.; et al. Sediments deposited by the 2004 Indian Ocean tsunami along the Malaysia-Thailand Peninsula. *Mar. Geol.* **2007**, *242*, 169–190. [CrossRef]
221. Uchida, J.; Fujiwara, O.; Hasegawa, S.; Kamata, T. Sources and depositional processes of tsunami deposits: Analysis using foraminiferal tests and hydrodynamic verification. *Isl. Arc.* **2010**, *19*, 427–442. [CrossRef]
222. Pilarczyk, J.E.; Dura, T.; Horton, B.P.; Engelhart, S.E.; Kemp, A.C.; Sawai, Y. Microfossils from coastal environments as indicators of paleo-earthquakes, tsunamis and storms. *Palaeogeogr. Palaeoclimatol. Palaeoecol.* **2014**, *413*, 144–157. [CrossRef]
223. Sawai, Y.; Namegaya, Y.; Okamura, Y.; Satake, K.; Shishikura, M. Challenges of anticipating the 2011 Tohoku earthquake and tsunami using coastal geology. *Geophys. Res. Lett.* **2012**, *39*, L21309. [CrossRef]
224. Fujiwara, O.; Kamataki, T.; Fuse, K. Genesis of mixed molluscan assemblages in the tsunami deposits distributed in Holocene drowned valleys on the southern Kanto region, east Japan. *Quat. Res. (Daiyonki-Kenkyu)* **2003**, *42*, 389–412. (In Japanese with English Abstract) [CrossRef]
225. Cisternas, M.; Atwater, B.F.; Torrejón, F.; Sawai, Y.; Machuca, G.; Lagos, M.; Eipert, A.; Youlton, C.; Salgado, I.; Kamataki, T.; et al. Predecessors of the giant 1960 Chile earthquake. *Nature* **2005**, *437*, 404–407. [CrossRef] [PubMed]
226. Garrett, E.; Shennan, I.; Watcham, E.P.; Hawkes, A.D.; Sawai, Y. Reconstructing paleoseismic deformation, 2: 1000 years of great earthquakes at Chucalén, south central Chile. *Quat. Sci. Rev.* **2015**, *113*, 112–122. [CrossRef]
227. Dura, T.; Cisternas, M.; Horton, B.P.; Ely, L.L.; Nelson, A.R.; Wesson, R.L.; Pilarczyk, J.E. Coastal evidence for Holocene subduction-zone earthquakes and tsunamis in central Chile. *Quat. Sci. Rev.* **2015**, *113*, 93–111. [CrossRef]
228. Peters, R.; Bourgeois, J.; Hutchinson, I.; Nelson, A.R.; Engelhart, S.E. Distribution and sedimentary characteristics of tsunami deposits along the Cascadia margin of western North America. *Sediment. Geol.* **2007**, *200*, 372–386. [CrossRef]
229. Witter, R.C.; Peterson, C.D.; Bourgeois, J.; Engelhart, S.E.; Reynolds, J.M. Simulating tsunami inundation at Bandon, Coos County, Oregon, using hypothetical Cascadia and Alaska earthquake scenarios. *Oregon Dep. Geol. Miner. Ind. Spec. Pap.* **2012**, *43*, 57.
230. Kelsey, H.M.; Nelson, A.R.; Hemphill-Haley, E.; Witter, R.C. Tsunami history of an Oregon coastal lake reveals a 4600 yr record of great earthquakes on the Cascadia subduction zone. *Geol. Soc. Am. Bull.* **2005**, *117*, 1009–1032. [CrossRef]
231. Cochran, U.; Berryman, K.; Zachariasen, J.; Mildenhall, D.; Hayward, B.; Southall, K.; Hollis, C.; Barker, P.; Wallace, L.; Alloway, B.; et al. Paleoecological insights into subduction zone earthquake occurrence, eastern North Island, New Zealand. *Geosci. World* **2006**, *118*, 9–10. [CrossRef]
232. Andrade, C.; Freitas, M.C.; Miranda, J.M.; Baptista, M.A.; Cachão, M.; Silva, P.; Munhá, J. Recognizing possible tsunami sediments in the ultradissipative environment of the Tagus estuary (Portugal). In Proceedings of the Coastal Sediments'03—The 5th International Symposium on Coastal Engineering and Science of Coastal Sediment Processes, Clearwater Beach, FL, USA, 14–18 June 2003; 14p.
233. Whelan, F.; Kelletat, D. Boulder deposits on the southern Spanish Atlantic Coast: Possible evidence for the 1775 AD Lisbon Tsunami. *Sci. Tsunami Haz.* **2005**, *23*, 25–38.
234. Ramos-Pereira, A.; Trindade, J.M.N.; Borges, B. Indicadores geomorfológicos de tsunami no Parque Natural do SW Alentejano e Costa Vicentina (Malhão). In Proceedings of the VI Congresso da Associacao Portuguesa de Geomorfólogos, Braga, Portugal, 16–18 October 2007; pp. 51–56.
235. Medina, F.; Mhammdi, N.; Chiguer, A.; Akil, M.; Jaaidi, E.B. The Rabat and Larache boulder fields; new examples of high-energy deposits related to extreme waves in north-western Morocco. *Nat. Hazards* **2010**, *59*, 725–747. [CrossRef]
236. Hindson, R.A.; Andrade, C.; Dawson, A.G. Sedimentary processes associated with the tsunami generated by the 1755 Lisbon earthquake on the Algarve coast, Portugal. *Phys. Chem. Earth* **1996**, *21*, 57–63. [CrossRef]
237. Font, E.; Cuven, S.; Gutiérrez, F.; Sabatier, P.; Pinos, M.; Dias, J.A. Identification of tsunami-induced deposits using magnetic fabric and rock magnetic analyses: The 1755 Lisbon tsunami in Algarve, Portugal. *Earth Planet Sci. Lett.* **2013**, *377–378*, 218–230.
238. El Talibi, H.; El Moussaoui, S.; Zaghloul, M.N.; Aboumaria, K.; Wassmer, P.; Mercier, J.L. New sedimentary and geomorphic evidence of tsunami flooding related to an older events along the Tangier-Asilah coastal plain, Morocco. *Geoenviron. Disasters* **2016**, *3*, 14. [CrossRef]
239. Khalfaoui, O.; Joudar, I.; Chahid, N.E.; El Khalidi, K.; Minoubi, A.; Bouchkara, M.; Zourarah, B. The Safi boulders (Morocco): Evidence of past extreme wave events. *Mar. Geol.* **2024**, *470*, 107265. [CrossRef]

 https://doi.org/10.3390/jmse14010049

240. Álvarez-Gómez, J.A.; Aniel-Quiroga, Í.; González, M.; Olabarrieta, M.; Carreño, E. Scenarios for earthquake-generated tsunamis on a complex tectonic area of diffuse deformation and low velocity: The Alboran Sea, Western Mediterranean. *Mar. Geol.* **2011**, *284*, 55–73. [CrossRef]
241. Şahoğlu, V.; Sterba, J.H.; Katz, T.; Çayır, Ü.; Gündoğan, Ü.; Tyuleneva, N.; Tuğcu, İ.; Bichler, M.; Erkanal, H.; Goodman-Tchernov, B.N. Volcanic ash, victims, and tsunami debris from the Late Bronze Age Thera eruption discovered at Çeşme-Bağlararası (Turkey). *Proc. Natl. Acad. Sci. USA* **2022**, *119*, e2114213118. [CrossRef]
242. Vött, A.; Brückner, H.; May, S.M.; Sakellariou, D.; Nelle, O.; Lang, F.; Kapsimalis, S.; Jahns, R.; Herd, R.; Handl, M.; et al. The Lake Voulkaria (Akarnania, NW Greece) palaeoenvironmental archive—A sediment trap for multiple tsunami impact since the mid-Holocene. *Z. Geomorphol.* **2009**, *53*, 1. [CrossRef]
243. Scheffers, A.; Kelletat, D.; Vött, A.; May, S.M.; Scheffers, S. Late Holocene tsunami traces on the western and southern coastlines of the Peloponnesus (Greece). *Earth Planet Sci. Lett.* **2008**, *269*, 271–279. [CrossRef]
244. Paris, R.; Smedile, A.; Falvard, S.; Devidal, J.L.; Suchorski, K. A Minoan and a Neolithic tsunami recorded in coastal sediments of Ios Island, Aegean Sea, Greece. *Mar. Geol.* **2022**, *452*, 106908. [CrossRef]
245. De Martini, P.M.; Tinti, S.; Armigliato, A.; Pignone, S.; Gutiérrez, F. A unique 4000 yr long geological record of multiple tsunami inundations in the Augusta Bay (eastern Sicily, Italy). *Mar. Geol.* **2010**, *276*, 42–57. [CrossRef]
246. Smedile, A.; De Martini, P.M.; Pantosti, D. Combining inland and offshore paleotsunamis evidence: The Augusta Bay (eastern Sicily, Italy) case study. *Nat. Hazards Earth Syst. Sci.* **2012**, *12*, 2557–2567. [CrossRef]
247. Goodman-Tchernov, B.N.; Cisternas, M.; Kessel, R.; Friedrich, W.L.; Abbasi, R. Tsunami waves generated by the Santorini eruption reached Eastern Mediterranean shores. *Geology* **2009**, *37*, 943–946. [CrossRef]
248. Shah-Hosseini, M.; Saleem, A.; Mahmoud, A.A.; Morhange, C. Coastal boulder deposits attesting to large wave impacts on the Mediterranean coast of Egypt. *Nat. Hazards* **2016**, *83*, 849–865. [CrossRef]
249. Bellanova, P.; Feist, L.; Mathes-Schmidt, M.; Micallef, A.; Mottershead, D.N.; Reicherter, K.; Terry, J.P. Tsunami deposits preserved in coastal karst pockets on Malta (Central Mediterranean Sea). *Sedimentology* **2025**, *72*, 2060–2086. [CrossRef]
250. Roig-Munar, F.X.; Gelabert, B.; Rodríguez-Perea, A.; Martín-Prieto, J.A.; Vilaplana, J.M. Storm or tsunamis: Boulder deposits on the rocky coasts of the Balearic Islands (Spain). *Mar. Geol.* **2023**, *463*, 107112. [CrossRef]
251. Switzer, A.D.; Jones, B.G. Large-scale washover sedimentation in a freshwater lagoon from the southeast Australian coast: Tsunami or exceptionally large storm? *Holocene* **2008**, *18*, 787–803. [CrossRef]
252. Judd, K.; Chagué-Goff, C.; Goff, J.; Gadd, P.; Zawadzki, A.; Fierro, D. Multi-proxy evidence for small historical tsunamis leaving little or no sedimentary record. *Mar. Geol.* **2017**, *385*, 204–215. [CrossRef]
253. Spiske, M.; Tang, H.; Bahlburg, H. Post-depositional alteration of onshore tsunami deposits—Implications for the reconstruction of past events. *Earth Sci. Rev.* **2020**, *202*, 103068. [CrossRef]
254. Atwater, B.F.; Musumi-Rokkaku, S.; Satake, K.; Tsuji, Y.; Ueda, K.; Yamaguchi, D.K. *The Orphan Tsunami of 1700—Japanese Clues to a Parent Earthquake in North America*; University of Washington Press: Washington, DC, USA, 2005; Volume 1707, 144p.
255. Goff, J.; Chagué-Goff, C.; Nichol, S.; McFadgen, B. The elusive AD 1826 tsunami, South Westland, New Zealand. *Nat. Hazards Earth Syst. Sci.* **2020**, *20*, 1579–1591.
256. Clark, K.; Howarth, J.; Litchfield, N.; Cochran, U.; Turnbull, J.; Dowling, L.; Howell, A.; Berryman, K.; Wolfe, F. Geological evidence for past large earthquakes and tsunamis along the Hikurangi subduction margin, New Zealand. *Mar. Geol.* **2019**, *412*, 139–172. [CrossRef]
257. Ávila, S.P.; Ramalho, R.S.; Habermann, J.M.; Quartau, R.; Kroh, A.; Berning, B.; Jonson, M.; Kirby, M.X.; Zanon, V.; Titshack, J.; et al. Palaeoecology, taphonomy, and preservation of a lower Pliocene shell bed (coquina) from a volcanic oceanic island (Santa Maria Island, Azores). *Palaeogeogr. Palaeoclimatol. Palaeoecol.* **2015**, *430*, 57–73. [CrossRef]
258. Nanayama, F.; Shigeno, K.; Satake, K.; Shimokawa, K.; Koitabashi, S.; Miyasaka, S.; Ishii, M. Sedimentary differences between the 1993 Hokkaido-nansei-oki tsunami and the 1959 Miyakojima typhoon at Taisei, southwestern Hokkaido, northern Japan. *Sediment. Geol.* **2000**, *135*, 255–264. [CrossRef]
259. Gelfenbaum, G.; Jaffe, B. Erosion and sedimentation from the 17 July, 1998 Papua New Guinea tsunami. *Pure Appl. Geophys.* **2003**, *160*, 1969–1999. [CrossRef]
260. Szczuciński, W.; Chaimanee, N.; Niedzielski, P.; Rachlewicz, G.; Saisuttichai, D.; Tepsuwan, T.; Lorenc, S.; Siepak, J. Environmental and geological impacts of the 26 December 2004 tsunami in the coastal zone of Thailand—Overview of short- and long-term effects. *Pol. J. Environ. Stud.* **2006**, *15*, 793–810.
261. Hori, K.; Kuzumoto, R.; Hirouchi, D.; Umitsu, M.; Janjirawuttikul, N.; Patanakanog, B. Horizontal and vertical variation of 2004 Indian tsunami deposits: An example of two transects along the western coast of Thailand. *Mar. Geol.* **2007**, *239*, 163–172. [CrossRef]
262. Fujino, S.; Naruse, H.; Matsumoto, D.; Jarupongsakul, T.; Sphawajruksakul, A.; Sakakura, N. Stratigraphic evidence for pre-2004 tsunamis in southwestern Thailand. *Mar. Geol.* **2009**, *262*, 25–28. [CrossRef]

263. Paris, R.; Wassmer, P.; Sartohadi, J.; Lavigne, F.; Barthomeuf, B.; Desgages, E.; Grancher, D.; Baumert, P.; Vautier, F.; Brunstein, D.; et al. Tsunamis as geomorphic crises: Lessons from the December 26, 2004 tsunami in Lhok Nga, West Banda Aceh (Sumatra, Indonesia). *Geomorphology* **2009**, *104*, 59–72. [CrossRef]
264. Bahlburg, H.; Weiss, R. Sedimentology of the December 26, 2004, Sumatra tsunami deposits in eastern India (Tamil Nadu) and Kenya. *Int. J. Earth Sci.* **2007**, *96*, 1195–1209. [CrossRef]
265. Shanmugam, G. Origin, recognition and importance of erosional unconformities in sedimentary basins. In *New Perspectives in Basin Analysis*; Kleinspehn, K.L., Paola, C., Eds.; Springer: New York, NY, USA, 1988; pp. 83–108.
266. Catuneanu, O. *Principles of Sequence Stratigraphy*, 1st ed.; Elsevier B.V.: Amsterdam, The Netherlands, 2006; pp. 1–375.
267. Bryant, E.A.; Young, R.W. Bedrock-sculpting by tsunami, south coast New South Wales. *J. Geol.* **1996**, *104*, 565–582. [CrossRef]
268. Aalto, K.R.; Aalto, R.; Garrison-Laney, C.E.; Abramson, H.F. Tsunami(?) sculpting of the Pebble Beach wave-cut platform, Crescent City area, California. *J. Geol.* **1999**, *107*, 607–622. [CrossRef]
269. Scheffers, S.R.; Haviser, J.; Browne, T.; Scheffers, A. Tsunamis, hurricanes, the demise of coral reefs and shifts in prehistoric human populations in the Caribbean. *Quat. Int.* **2009**, *195*, 69–87. [CrossRef]
270. Goto, K.; Chavanich, S.A.; Imamura, F.; Kunthasap, P.; Matsui, T.; Minoura, K.; Sugawara, D.; Ynagisawa, H. Distribution, origin and transport process of boulders deposited by the 2004 Indian Ocean tsunami at Pakarang Cape, Thailand. *Sediment. Geol.* **2007**, *202*, 821–837. [CrossRef]
271. Scicchitano, G.; Monaco, C.; Tortorici, L. Large boulder deposits by tsunami waves along the Ionian coast of south-eastern Sicily (Italy). *Mar. Geol.* **2007**, *238*, 75–91. [CrossRef]
272. Barbano, M.S.; Pirrotta, C.; Gerardi, F. Large boulders along the south-eastern Ionian coast of Sicily: Storm or tsunami deposits? *Mar. Geol.* **2010**, *275*, 140–154. [CrossRef]
273. Sugawara, D.; Goto, K.; Jaffe, B.E. Numerical models of tsunami sediment transport—Current understanding and future directions. *Mar. Geol.* **2014**, *352*, 295–320. [CrossRef]
274. Brill, D.; May, S.M.; Mhammdi, N.; King, G.; Brückner, H. OSL surface exposure dating of wave-emplaced coastal boulders—Research concept and first results from the Rabat coast, Morocco. In Proceedings of the 19th EGU General Assembly, EGU2017, Viena, Austria, 23–28 April 2017.
275. Takayama, H.; Tada, R.; Matsui, T.; Iturralde-Vinent, M.; Oji, T.; Tajika, E.; Kiyokawa, S.; Garcia, D.; Okada, H.; Hasegawa, T.; et al. Origin of the Peñalver Formation in northwestern Cuba and its relation to K/T boundary impact event. *Sediment. Geol.* **2000**, *135*, 295–320. [CrossRef]
276. Le Roux, J.P.; Gómez, C.; Fenner, J.; Middleton, H. Sedimentological processes in a scarp-controlled rocky shoreline to upper continental slope environment, as revealed by unusual sedimentary features in the Neogene Coquimbo Formation, north-central Chile. *Sediment. Geol.* **2004**, *165*, 67–92. [CrossRef]
277. Goff, J.; McFadgen, B.G.; Chagué-Goff, C. Sedimentary differences between the 2002 Easter storm and the 15th-century Okoropunga tsunami, southeastern North Island, New Zealand. *Mar. Geol.* **2004**, *204*, 235–250. [CrossRef]
278. Matsumoto, D.; Naruse, H.; Fujino, S.; Urphawajruksakul, A.; Jarupongsakul, T.; Sakakura, N.; Murayama, M. Truncated flame structures within a deposit of the Indian Ocean tsunami: Evidence of synsedimentary deformation. *Sedimentology* **2008**, *55*, 1559–1570. [CrossRef]
279. Bondevik, S.; Svendsen, J.I.; Mangerud, J. Tsunami sedimentary facies deposited by the Storegga tsunami in shallow marine basins and coastal lakes, western Norway. *Sedimentology* **1997**, *44*, 1115–1131. [CrossRef]
280. Choowong, M.; Murakoshi, N.; Hisada, K.I.; Charusiri, P.; Charoentitirat, T.; Chutakositkanon, V.; Jankaew, K.; Kanjanapayont, P.; Phantuwongraj, S. 2004 Indian Ocean tsunami inflow and outflow at Phuket, Thailand. *Mar. Geol.* **2008**, *248*, 179–192. [CrossRef]
281. Harms, J.C.; Southard, J.B.; Spearing, D.R.; Walker, R.G. *Depositional Environments as Interpreted from Primary Sedimentary Structures and Stratification Sequences*; SEPM Short Course No. 2; SEPM: Dallas, TX, USA, 1975.
282. Benyoucef, M.; Bendella, M.; Brunetti, M.; Ferré, B.; Kocie, T.; Bouchemla, I.; Slami, F.; Ghenim, A.F. Upper Pliocene bivalve shell concentrations from the Lower Chelif Basin (NW Algeria): Systematics, sedimentologic and taphonomic framework. *Annal. Paleontol.* **2021**, *107*, 102509. [CrossRef]
283. Srinivasalu, S.; Rajeshwara Rao, N.; Thangadurai, N.; Jonathan, M.P.; Roy, P.D.; Ram Mohan, V.; Saravanan, P. Characteristics of 2004 tsunami deposits of the northern Tamil Nadu coast, southeastern India. *Bol. Soc. Geol. Mex.* **2009**, *61*, 111–118. [CrossRef]
284. Shanmugam, G. 50 years of the turbidite paradigm (1950s–1990s): Deep-water processes and facies models—A critical perspective. *Mar. Pet. Geol.* **2000**, *17*, 285–342. [CrossRef]
285. Hayes, M.O. Hurricanes as geological agents—Case studies of Hurricanes Carla, 1961, and Cindy, 1963. *Bur. Econ. Geol. Rep. Invest.* **1967**, *61*, 1–56.
286. Schwartz, R.K. *Nature and Genesis of Some Storm Washover Deposits*; Technical Memorandum No. 61; U.S. Army Engineer Research and Development Center: Vicksburg, MS, USA, 1975; pp. 1–109.
287. Morton, R.A. Large-scale rhomboid bed forms and sedimentary structures associated with hurricane washover. *Sedimentology* **1978**, *25*, 183–204. [CrossRef]

288. Leatherman, S.P.; Williams, A.T. Vertical sedimentation units in a barrier island washover fan. *Earth Surf. Process. Landf.* **1983**, *8*, 141–150. [CrossRef]
289. Barwis, J.H.; Hayes, M.O. Antidunes on modern and ancient washover fans. *J. Sediment. Pet.* **1985**, *55*, 907–916.
290. Bendella, M.; Benyoucef, M.; Mukilas, R.; Bouchemla, I.; Ferré, B. Shallow to marginal marine ichnoassemblages from the Upper Pliocene Slama Formation (Lower Chelif Basin, NW Algeria). *Geol. Carpathica* **2021**, *72*, 529–548. [CrossRef]
291. Nemra, A.; Tellal, J.; Knaust, D.; Bendella, M.; Belkhedim, S.; Ouali Mehadji, A. Middle Miocene trace fossils from the Tenes area (NW Algeria) and their palaeoenvironmental implications. *Palaeobio. Palaeoenv.* **2023**, *104*, 327–362. [CrossRef]
292. Naimi, M.N.; Cherif, A.; Mahboubi, C.Y.; Bachagha Bensaci, M.; Mazouzi, A.; Benkhedda, A.; Laouini, H. Palaeoecological, taphonomic and biosedimentary analysis of Mid-Holocene lacustrine deposits from Ouargla area (Lower Sahara, SE Algeria). *Neues Jahrb. Geol. Palaontol. Abh.* **2023**, *307*, 65–80. [CrossRef]
293. Spiske, M.; Jaffe, B.E. Sedimentology and hydrodynamic implications of a coarse-grained hurricane sequence in a carbonate reef setting. *Geology* **2009**, *37*, 839–842. [CrossRef]
294. Smith, D.E.; Shi, S.; Cullingford, R.A.; Dawson, A.G.; Dawson, S.; Firth, C.R.; Foster, I.D.L.; Fretwell, P.T.; Haggart, B.A.; Holloway, L.K.; et al. The Holocene Storegga Slide tsunami in the United Kingdom. *Quat. Sci. Rev.* **2004**, *23*, 2291–2321. [CrossRef]
295. Hindson, R.A.; Andrade, C. Sedimentation and hydrodynamic processes associated with the tsunami generated by the 1755 Lisbon earthquake. *Quat. Int.* **1999**, *56*, 27–38. [CrossRef]
296. Pilarczyk, J.E.; Sawai, Y.; Matsumoto, D.; Namegaya, Y.; Nishida, N.; Ikehara, K.; Fujiwara, O.; Gouramanis, C.; Dura, T.; Horton, B.P. Constraining sediment provenance for tsunami deposits using distributions of grain size and foraminifera from the Kujukuri coastline and shelf, Japan. *Sedimentology* **2020**, *67*, 1373–1392. [CrossRef]
297. Liu, B.C.; Dai, M.Y.; Qi, Y.A.; Sun, X.F.; Zheng, W.; Wang, M.; Xing, Z.F.; Li, D.; Fan, Y.C. Alternation of stromatolites and vertical burrows controlled by storm processes and sea-level changes in the middle Cambrian carbonates of central China. *J. Earth Syst. Sci.* **2022**, *131*, 65. [CrossRef]
298. Shi, S.Z.; Dawson, A.G.; Smith, D.E. Coastal sedimentation associated with the December 12th, 1992 tsunami in Flores, Indonesia. *Pure Appl. Geophys.* **1995**, *144*, 525–536. [CrossRef]
299. Morton, R.A.; Sallenger, A.H. Morphological impacts of extreme storms on sandy beaches and barriers. *J. Coast. Res.* **2003**, *19*, 560–573.
300. Shinozaki, T.; Iguchi, A.; Nishijima, M.; Goto, K.; Fujino, S. Identification of deposits from modern and ancient large tsunamis by means of environmental DNA. *Sci. Rep.* **2025**, *15*, 242. [CrossRef]
301. Reinhardt, E.G.; Pilarczyk, J.; Brown, A.L. Probable tsunami origin for a shell and sand sheet from marine ponds on Anegada, British Virgin Islands. *Nat. Hazards* **2012**, *63*, 101–117. [CrossRef]
302. Chakroun, A.; Zaghbib-Turki, D. Facies and fauna proxies used to reconstruct the MIS 5 and MIS 7 coastal environments in eastern Tunisia. *Geol. Quat.* **2017**, *61*, 131–142. [CrossRef]
303. Khallili, R.; Satour, L.; Mennad, S.; Tadjeddine, H. Bioérosion et encroûtement sur les huîtres Pliocène du bassin de Bas Chélif (Algérie nord-occidentale). *Rev. Paléobiol.* **2022**, *41*, 241–265.
304. Cundy, A.; Sprague, D.; Hopkinson, L.; Maroukian, H.; Gaki-Papanastassiou, K.; Papanastassiou, D.; Frogley, M.R. Geochemical and stratigraphic indicators of late Holocene coastal evolution in the Gythio area, southern Peloponnese, Greece. *Mar. Geol.* **2006**, *230*, 161–177. [CrossRef]
305. Belghouthi, F.; Zouari, H. Storm-dominated shallow-marine carbonates of the lower Eocene succession of the northwestern Tunisian domain. *J. Afr. Earth Sci.* **2019**, *160*, 103608. [CrossRef]
306. Nishimura, Y.; Miyaji, N. Tsunami deposits from the 1993 Southwest Hokkaido earthquake and the 1640 Hokkaido Komagatake eruption, northern Japan. *Pure Appl. Geophys.* **1995**, *144*, 719–733. [CrossRef]
307. Jaffe, B.; Gelfenbaum, G.; Rubin, D.; Peters, R.; Anima, R.; Swensson, M.; Olcese, D.; Bernales, L.; Gomez, J.; Riega, P. Tsunami deposits: Identification and interpretation of tsunami deposits from the June 23, 2001 Perú tsunami. In Proceedings of the International Conference on Coastal Sediments 2003, Corpus Christi, TX, USA, 18–23 May 2003; 13p.
308. Morton, R.A.; Paine, J.G. *Beach and Vegetation-Line Changes at Galveston Island, Texas: Erosion, Deposition, and Recovery from Hurricane Alicia*; The University of Texas at Austin: Austin, TX, USA, 1985; p. 85.
309. Morton, R.A. Subaerial storm deposits formed on barrier flats by wind-driven currents. *Sediment. Geol.* **1979**, *24*, 105–122. [CrossRef]
310. Yap, W.; Switzer, A.D.; Gouramanis, C.; Marzinelli, E.; Wijaya, W.; Yan, Y.T.; Dominey-Howes, D.; Labbate, M.; Srinivasalu, S.; Jankaew, K.; et al. Environmental DNA signatures distinguish between tsunami and storm deposition in overwash sand. *Commun. Earth Environ.* **2021**, *2*, 129. [CrossRef]

Journal of
Marine Science and Engineering

MDPI

Review

Volcanic Islands as Reservoirs of Geoheritage: Current and Potential Initiatives of Geoconservation

Esther Martín-González [1,*], Juana Vegas [2], Inés Galindo [2], Carmen Romero [3] and Nieves Sánchez [2]

1 Museo de Ciencias Naturales de Tenerife, Organismo Autónomo de Museos y Centros, Fuente Morales, 1, 38003 Santa Cruz de Tenerife, Spain
2 Instituto Geológico y Minero de España (IGME-CSIC), Unidad Territorial de Canarias, Alonso Alvarado, 43, 2A, 35003 Las Palmas de Gran Canaria, Spain; j.vegas@igme.es (J.V.); i.galindo@igme.es (I.G.); n.sanchez@igme.es (N.S.)
3 Departamento de Geografia, Campus de Guajara, Universidad de La Laguna, 38071 La Laguna, Spain; mcromero@ull.edu.es
* Correspondence: mmartin@museosdetenerife.org; Tel.: +34-922532742

Abstract

Volcanic islands host exceptional geological features that illustrate complex endogenic processes and interactions with climatic and marine forces, while also being particularly vulnerable to the impacts of climate change. Despite their scientific, educational, touristic, and aesthetic values, such islands remain underrepresented within the UNESCO Global Geoparks (UGGp). This study reviews current volcanic island geoparks and evaluates territories with potential for future designation, based on documented geoheritage, geosite inventories, and geoconservation frameworks. Geoparks are categorized according to their dominant narratives—ranging from recent Quaternary volcanism to broader tectonic, sedimentary, and metamorphic histories. Through an analysis of their distribution, management strategies, and integration into territorial planning, this work highlights the challenges that insular territories face, including vulnerability to global environmental change, limited legal protection, and structural inequalities in access to international resources recognition. It concludes that volcanic island geoparks represent strategic platforms for implementing sustainable development models, especially in ecologically and socially fragile contexts. Enhancing their global representation will require targeted efforts in ecologically and socially fragile contexts. Enhancing their global representation will require targeted efforts in capacity building, funding access, and regional cooperation—particularly across the Global South.

Keywords: insular geoheritage; protected volcanic landscapes; geoheritage education; geosite inventories; UNESCO Global Geoparks

1. Introduction

There is a notable increase in interest regarding geoheritage research and geoconservation, as well as in the enhancement and outreach of geoheritage knowledge based on a combination of field studies and territorial policy-management perspectives [1]. The geological heritage of a territory with international significance serves as the foundation for any proposal seeking the establishment of a UNESCO Global Geopark (UGGp). The primary pillars of such proposals are geoconservation, geoeducation, and sustainable development [2]. Volcanic landscapes are recognized as some of the most distinctive of the many protected areas within the World Heritage Program, the Global Geopark, and the

Man and Biosphere programs of UNESCO. These landscapes are valued for their biological, geological, and cultural significance to communities and societies worldwide. Preliminary reviews show that a significant number of World Heritage Sites, Biosphere Reserves, and Global Geoparks contain important volcanic features [3,4].

Many volcanic islands, particularly those formed over oceanic crust, are large structures that rise from the ocean floor to elevations well above sea level, showcasing exceptional geological features [5]. Their genesis is closely linked to plate tectonic processes, such as hotspot activity and seafloor spreading, which regulate magma supply over geological time [6]. The evolution of such islands is influenced by a combination of internal and external factors, during which growth and erosional processes interact to shape each stage of development [7]. While not all volcanic islands share these precise tectonic contexts, many of the principles derived from oceanic island research offer valuable insights for understanding insular volcanic geoheritage in general.

In the case of volcanic islands, geoheritage, or the aspects of a landscape that hold some combination of scientific, educational, cultural, or aesthetic values, is strongly influenced by eruption dynamics, different magma types, and the resulting morphology of cooled structures. In addition, local climate variations driven by latitude, altitude, and island orientation contribute to differential weathering and sedimentary dynamics [8,9]. Due to their limited area, geographic isolation, and anthropogenic pressures, small islands are particularly vulnerable to territorial anthropogenic impacts and global environmental changes. These conditions demand tailored, locally implemented conservation strategies [10,11]. Importantly, volcanic geoheritage can also serve as a socioeconomic advantage, particularly in island territories affected by recent eruptions, offering opportunities for sustainable development and community revitalization [12–14]. These differences affect volcanic landforms and result in features responding to a range of sedimentary processes and erosion over different time scales and areas [9].

The main objective of this study is to review a wide range of volcanic island settings in the context of their geoheritage relevance for inclusion in the UNESCO Global Geoparks, which places a premium on local socioeconomic development. In addition, other examples of volcanic islands that also attracted notable geological interest are reviewed, which merit potential for geopark designation and the accrual of benefits related to geoheritage [15]. While this study focuses on UNESCO Global Geoparks (UGGp), it is important to recognize that such designations are not the only valid frameworks for geoconservation and territorial development. Many countries have developed national or regional geopark programs that reflect their own priorities, governance models, and community needs—often without any immediate intention of joining the UNESCO network [16,17]. These initiatives can be equally effective in promoting sustainable development, education, and geoheritage conservation at a local scale. Consequently, the UGGp label should not be seen as a universal standard, but rather as one of several viable models within the broader geoconservation landscape.

2. Materials and Methods

This study is based on a structured literature review and comparative analysis. The bibliographic search focused on the period from 2000 to 2025 and used keywords such as geopark, geoheritage, geoconservation, geotourism, and volcanic islands. Searches were conducted in Scopus, Web of Science, and Google Scholar, as well as in official websites and documents from the UNESCO Global Geoparks Network [18,19] and national geological surveys.

The selection criteria for sources included peer-reviewed journal articles, conference proceedings, UNESCO reports, and institutional documents with explicit references to volcanic island contexts. In total, over 150 documents were initially retrieved, of which

73 were selected for detailed analysis based on relevance, data quality, and geographic coverage. The analysis focused on (a) identifying the geological attributes highlighted in official geopark narratives; (b) evaluating existing geosite inventories; and (c) assessing each geopark's management strategy and integration into territorial planning (Figure 1).

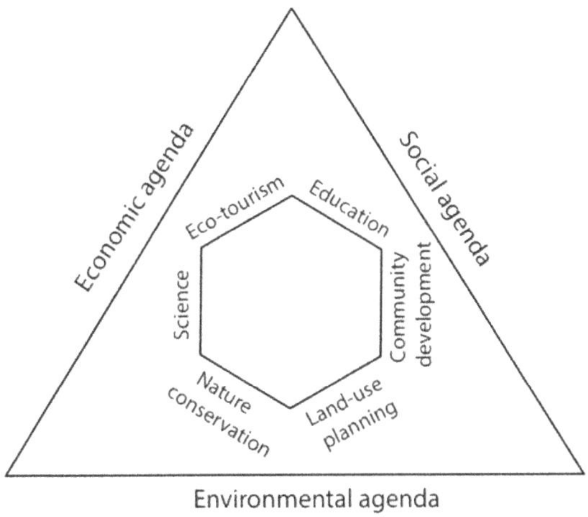

Figure 1. Interrelationship of tripart agendas that characterize successful UNESCO Global Geoparks (UGGp).

For volcanic islands with potential for geopark designation, we included only those territories with at least one peer-reviewed publication or official inventory referencing their geodiversity and geoheritage value. These were then systematized based on tectonic setting, base age, dominant volcanic features, and institutional context.

3. Results

3.1. Inventory of UGGp in Volcanic Island Territories

Volcanic islands are substantial geological structures formed by volcanic activity in several tectonic frameworks that often form distinctive landscapes with exceptional geoheritage value [20]. Although approximately 20,000 volcanic islands are known globally, there are only a limited number that host geoheritage sites recognized under the UNESCO Global Geoparks (UGGp). According to the most recent update, only 25 of the 229 UGGp are located on volcanic islands. Their distribution is geographically uneven. Only one is found in the Mediterranean Sea, five occur in the Atlantic Ocean, and the remainder are located in the Indo-Pacific region (Table 1). This distribution highlights the strategic importance of volcanic island geoparks within the framework of global geoconservation.

Table 1. List of officially recognized UNESCO Global Geoparks related to volcanic islands.

Geopark/Country	Year	Area (km^2)	Base Age	Key Geological Features
Toya-Usu, Japan	2008	1181	Quaternary	Pyroclastic flow forming caldera depression
Unzen Area, Japan	2009	459	Pliocene	Lava domes and pyroclastic deposits
Aso Area, Japan	2013	2	Quaternary	Lava flows and tephra layers
Oki Islands, Japan	2013	674	Permian-Triassic	Volcanic rocks and mantle xenoliths; aquatic fossils

Table 1. *Cont.*

Geopark/Country	Year	Area (km^2)	Base Age	Key Geological Features
Mt. Apoi, Japan	2015	364	Miocene	Peridotites from earth's mantle
Itoigawa, Japan	2015	750	Cambrian	East-West tectonic line of separation
San'lin Kaigan, Japan	2015	246	Miocene	Granite outcrops and other volcanics
Muroto, Japan	2015	248	Paleogene	Accretionary complex at plate boundary
Izu Peninsula, Japan	2016	2027	Miocene	Pillow lava; pumice; scoria cone, hot spring
Hakusan Tedorigawa, Japan	2023	755	Permian	Continental collision; dinosaur tracks
Batur, Indonesia	2012	370	Quaternary	Lava flows, fumaroles, lake calderas
Ciletuh-Palabuhanratu, Indonesia	2015	1260	Oligocene	Tectonic processes, hot springs; geysers
Rinjani-Lombock, Indonesia	2018	2800	Oligocene	Intrusive igneous rocks and volcanoes
Toba Caldera, Indonesia	2020	1165	Quaternary	Landscape from a plinnian eruption
Ijen Lake, Indonesia	2023	4723	Quaternary	The world's most acidic volcanic lake with sulfuric gases that ignite blue flames
Kebjumen, Indonesia	2025	1139	Early Cretaceous	Tectonic records dating back 119 million years; caves, and underground rivers
Meratus, Indonesia	2025	3645	Early Jurassic	Indonesia's oldest and most complete ophiolite sequence
Leiquiong, Hainan, China	2015	3050	Quaternary	Basaltic and phreatomagmatic craters
Katla, Iceland	2015	9542	Quaternary	Fissures and hyaloclastite ridges; ice-capped volcanoes and tuff deposits
Jeju Island, Republic of Korea	2010	1847	Quaternary	Basaltic—trachytic lavas; tuff rings and cones
Reykjanes, Iceland	2014	825	Quaternary	Fissure swarms and geothermal fields
Lesvos Island, Greece	2015	1636	Miocene	Volcanic hot springs and fossil sites
El Hierro Island, Spain	2014	595	Quaternary	Gravitational landslides; pyroclastic cones
Lanzarote and Chinijo Islands, Spain	2015	250	Miocene	Prehistoric and historic volcanism; fossils; aeolian deposits, submarine processes
Azores islands, Portugal	2013		Miocene	Caldera lakes, fumaroles, and hot springs; steam-cooked meals

The global distribution of volcanic island UNESCO Global Geoparks is shown in Figure 2. Whereas these island UGGPs share a common igneous or volcanic origin, their approach to geoheritage representation differs significantly. Despite its limited area, volcanic island UGGp typically exhibits a high geodiversity, which is reflected in the variety and uniqueness of its designated geosites. Some of them focus narrowly on features that emphasize their volcanic character with minimal reference to other aspects of geological evolution. In contrast, other islands UGGp adopt a more comprehensive strategy, incorporating geosites that represent a broad spectrum of geological processes and evolutionary stages. These examples typically base their selection on the diversity of geological formations and the relative ages of the exposed materials, thereby providing a more holistic view of an island's volcanic and post-volcanic history. This variability in geosite representation reflects differing strategies in how island UGGp valorize their geoheritage, ranging from purely volcanic features to integrated narratives of island formation, erosion, and sedimentation, and the impacts caused by volcanism.

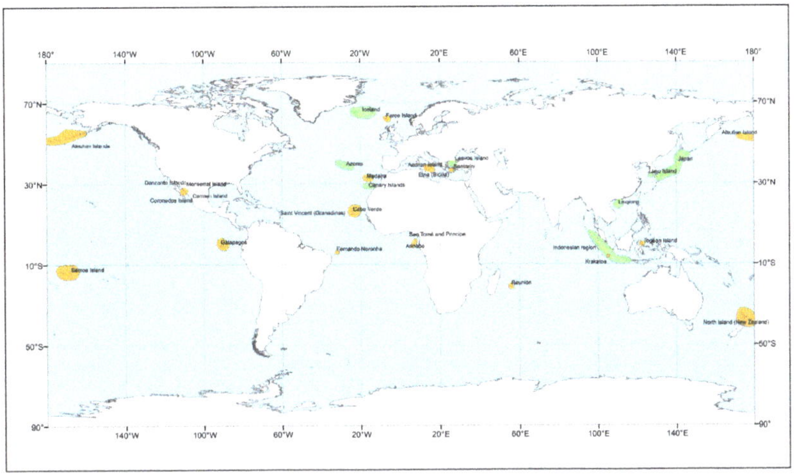

Figure 2. Map showing the global distribution of UNESCO Global Geoparks islands (patches in green) as well as the location of potential new geopark islands (patches in orange), some of which occur within territories already conserved by other means, such as national parks. Reference system: EPSG:32600-WGS 84. Map projection: UTM grid system (northern hemisphere).

3.2. UGGp Dominated by Quaternary Volcanism and Recent Activity

Several island geoparks are characterized by active or recently active volcanic processes, where Quaternary volcanism defines both their landscape and geosite selection. In these territories, geoheritage is closely linked to eruptive structures, lava fields, calderas, geothermal features, and, in some cases, ongoing volcanic activity. A clear example is the Reykjanes UGGp in Iceland, which is notable for its exceptional geothermal activity and fissure occurring presently, associated with the Mid-Atlantic Ridge [21]. Similarly, the Katla UGGp includes outstanding examples such as the Eldgjá fissure eruption, one of the largest historical flood-basalt events (Figure 3a), and the hyaloclastite ridges of Kattarhryggir, formed by subglacial eruptions [22]. Also linked to the North Atlantic mid-ocean ridge system, the Azores UGGp (Portugal) includes all nine islands of the archipelago. It features 121 geosites, six of international relevance, such as the Capelinhos volcano (Figure 3b), an iconic example of mid-twentieth-century volcanism listed in the IUGS "First 100 Geological Heritage Sites" [23], but this geopark's management is particularly complex due to its multi-island structure [10,24].

In the North Atlantic, the El Hierro UGGp became the first designated geopark in the Canary Islands (Spain). Its geological evolution was shaped by massive gravitational flank collapses, notably the "El Golfo" landslide (Figure 3c), recognized as Global Geosite VC010 [25]. The island gained renewed geological attention following the 2011 submarine eruption off its southern coast, which formed the underwater volcano Tagoro, still visible only in bathymetric records [26]. Among its 61 geosites, there are Holocene lava tubes preserving fossil bones of *Gallotia simonyi* and other extinct avifauna [27]. In Asia, the Toya-Usu UGGp (Japan) stands out as a natural laboratory for geohazard management, particularly after its twentieth-century eruptions [28], while the Unzen UGGp exemplifies community resurgence following the 1990–1995 eruption [29]. Centered around one of the world's largest calderas, the Aso UGGp continues to be monitored by the Japan Meteorological Agency due to ongoing volcanic activity [30]. The Toba Caldera UGGp (Indonesia) features the largest caldera lake on Earth, formed by a super-eruption approximately 74,000 years ago that may have had global climatic and anthropological consequences [31,32]. More recently, around 1257 CE, Mount Rinjani erupted in what is now the Rinjani-Lombok UGGp, having major regional climatic impacts [33]. Other Indonesian geoparks, such as Batur (Figure 3d) and Ijen, also emphasize Quaternary volcanism that showcases calderas and volcanic lakes with extremely acidic waters [34,35].

Figure 3. UNESCO Global Geoparks featuring Quaternary or recent volcanic activity; (**a**) Southern view of Katla Volcano, Katla UGGp in Iceland (courtesy of Jóhannes M. Jóhanneson, project manager); (**b**) Capelinhos Volcano, Azores UGGp (Esther Martín photo); (**c**) El Golfo gravitational landside, El Hierro UGGp (Carmen Romero photo); (**d**) Mount Batur Caldera, Batur UGGp (courtesy of Oka Agastya, geopark geoscientist).

In the Pacific Ocean, the Jeju Island UGGp (Republic of Korea) combines diverse volcanic morphologies that include Hawaiian, Strombolian, and phreatomagmatic styles with marine and aeolian erosional forms [36]. Its international significance is underscored by its triple UNESCO designation: Biosphere Reserve (2002), World Heritage Site (2007), and Global Geopark (2010) [37]. Similarly, the Leiqiong UNESCO Global Geopark (China), located on the Leizhou Peninsula in Northern Hainan Island, presents a remarkable assemblage of Quaternary basaltic volcanic landforms shaped by multiple eruptive phases [37]. The geopark features extensive lava plateaus, shield volcanoes, maars, and well-preserved volcanic cores, illustrating both effusive and explosive activity. This geodiversity is further enriched by the park's coastal settings and tectonic background linked to the South China Sea marginal basin evolution, granting Leiqiong a high value in terms of volcanic, tectonic, and marine geoheritage.

3.3. Geoparks Representing Broader Geological Narratives

In contrast to UGGp, which focused exclusively on recent volcanism, other island geoparks adopt a more comprehensive geological perspective, typically integrating longer geological time frames, fossil sites, tectonic and metamorphic features, or submerged volcanic structures. A key example is the Lanzarote and Chinijo Islands UGGp (Canary Islands, Spain), which was designated in 2015. It includes 82 geosites grouped into eight geological frameworks, several of which have international significance. More than 20 geosites are registered with the Global Geosite "Recent Volcanism of Timanfaya", whereas 15 geosites explore aspects of submarine volcanic heritage [38,39]. Although limited to the Upper Miocene, the fossil sites record includes both marine and terrestrial localities (Figure 4b), having well-documented conservation concerns [40].

Figure 4. Fossil geosites associated with geoparks in Greece and the Canary Islands of Spain. (**a**) The tallest standing petrified tree from the Lesvos Petrified Forest in the Lesvos UNESCO Global Geopark (courtesy of the Natural History Museum of the Lesvos Petrified Forest) in Greece; (**b**) Pliocene fossil site of Valle Chico in the Lanzarote and Chinijo Islands UNESCO Global Geopark in the Canary Islands (Esther Martín photo).

In Greece, the Lesvos UGGp is internationally renowned for its petrified forest, which contains an exceptional paleoflora record from the Miocene, complemented by volcanic, tectonic, and coastal geosites. Its early designation in 2004 was preceded in 1994 by the establishment of the Natural History Museum of the Lesvos Petrified Forest (Figure 4a). The museum became a model institution for geoscience and educational outreach [16,41].

In Japan, the Muroto UGGp documents neotectonic uplift related to the Nanjai Trough megathrust system [42], while the Izu Peninsula UGGp remains an active seismic zone shaped by arc-arc collision and frequent earthquake swarms [43]. In Indonesia, the Ciletuh-Palabuhanratu UGGp features Jurassic-Cretaceous oceanic crust together with fossil-bearing sediments alongside gravity-induced landforms such as the Ciletuh Amphitheatre [44].

3.4. Alternative Designations That Encompass the Formal Protected Status of Volcanic Island Geosites

Application of concepts and methodologies in geoconservation in the Azores archipelago also led to the Palaeopark on Santa Maria Island [45]. This approach stresses the scientific and touristic value of fossil sites, including two new geosites classified as the Ponta Castelo tempestite deposit and the Pedra-que-Pica coquina framed by volcanic deposits. These are classified as "Regional Natural Monuments" by the Regional Government of the Azores on account of their international relevance. The tempestite deposit at Ponta do Castelo is overlain by a coastal lava delta, which enables the inference of the site's precise water depth immediately before the time of deposition [46]. The Pedra-que-Pica site is the most extensive multi-specific fossil coquina yet reported from the shelf of any of the ~20,000 known volcanic islands worldwide [47].

3.5. Volcanic Islands with Geoheritage Potential as Future UGGp

Only a limited number of volcanic islands are officially registered as UNESCO Global Geoparks, but several other regions around the world exhibit a geological heritage of international relevance. Many of these islands hold strong potential for future inclusion in the UNESCO program (Table 2; Figure 2). These territories display significant geodiversity—including volcanic, sedimentary, tectonic, and geomorphological features—often combined with ecological and cultural values. In many cases, they have already been the focus of scientific assessments, geosite inventories, or preliminary conservation initiatives, positioning them as credible candidates for geopark designation.

Table 2. Volcanic island territories with potential for inclusion in UNESCO Global Geoparks.

Island	Base Age	Key Geological Features
Krakatoa (Indonesia)	Quaternary	Catastrophic historical volcanism
Togean Islands (Indonesia)	Neogene-Quaternary	Volcanoes, lava domes, karst, and ophiolites
Samoa (Samoa)	Quaternary	Active geodynamic processes (tsunamis, earthquakes, volcanoes)
Gwanmae Island (Republic of Korea)	Quaternary	Volcanic landscape
North Island (New Zealand)	Quaternary	Stratovolcanoes of Tongariro Area, Taupo geothermal fields
Fogo, Santo Antão and Santiago (Cabo Verde)	Neogene-Quaternary	Active volcanism, tsunami deposits
São Tomé and Príncipe (São Tomé and Príncipe)	Neogene-Quaternary	Alkaline volcanism
Annobón (Equatorial Guinea)	Neogene	Submerged geomorphology, coral platforms
Madeira (Portugal)	Neogene	Volcanism, coastal geomorphology, fossils deposit
Etna (Sicily) and Aeolian Islands (Italy)	Quaternary	Active and historic volcanoes, cultural heritage
Santorini Island (Greece)	Pliocene-Quaternary	Active caldera subsidence
Faroe Islands (Denmark)	Eocene	Ancient volcanic landscapes, unique geoforms
Aleutian Islands Arc (United State)	Neogene-Quaternary	More than fifty active volcanoes
Fernando Noronha (Brazil)	Neogene-Quaternary	Noronha hotspot
Réunion Island (France)	Quaternary	Pitons, cirques and ramparts
Saint Vicent (Saint Vincent and the Grenadines)	Pliocene	La Soufrière volcano, lahars
Galápagos (Ecuador)	Quaternary	Insular volcanism, calderas, active geosites
Coronados Island, Baja California (Mexico)	Quaternary	Volcanic cone and lagoon with fossil corals
Carmen Island, Baja California (Mexico)	Neogene	Tide-water delta in a volcanic landscape
Danzante Island, Baja California (Mexico)	Neogene	Andesite flows and coastal rock pediments
Monserrat Island, Baja California (Mexico)	Neogene	Tilted andesite covered by Pliocene limestone

The selection of potential geopark territories presented here was guided by the following combination of factors: documented scientific literature, presence of inventoried geosites, institutional or community-driven interest in geoconservation, and evidence of emerging regional initiatives. Notably, some geologically outstanding regions such as Hawaii or other Pacific islands were not included—not due to a lack of geodiversity or heritage value, but because they either lack formal geopark proposals, have no publicly accessible geosite inventories, or are already managed under alternative conservation frameworks (e.g., national parks or heritage areas). This study prioritizes territories with available documentation and realistic pathways toward integration into the UNESCO Global Geoparks Network.

In Southeast Asia, the Krakatoa complex stands out as a globally significant geoheritage site due to the catastrophic 1883 eruption and its continued volcanic activity. Recent evaluations using the Indonesian Standard Scientific Value Assessment have highlighted its exceptional scientific, historical, and educational value [48]. Also in Indonesia, the Togean Islands exhibit outstanding geodiversity, including the Uma-Una solitary volcano, lava domes, karst features, and ophiolitic complexes. Although their geological significance is evident, further institutional support is needed to move the area toward formal geopark candidacy [49].

In the Pacific, Samoa presents a combination of volcanic and tectonic hazards, such as tsunamis, earthquakes, and landslides, that have increasingly been interpreted as opportunities for geoeducation and disaster risk awareness [50].

The stratovolcanoes of the Tongariro area, together with the Taupō geothermal fields in New Zealand's central North Island, represent a geopatrimonial ensemble of exceptional scientific and educational value. These features are part of the Taupō Volcanic Zone, one of the most active rhyolitic volcanic systems on Earth, shaped by complex subduction-related processes. The region hosts iconic stratovolcanoes such as Mount Ngauruhoe and Mount Tongariro, alongside geothermal fields with geysers, hot springs, mud pools, and silica terraces. This dynamic landscape offers outstanding opportunities to study volcanism, geothermal activity, and landscape evolution, while also holding cultural significance for the Māori people. The combination of geological diversity, active processes, and cultural narratives makes this area a key candidate for geoconservation and geotourism within global geopatrimonial frameworks [51]. Its outstanding accessibility and educational potential provide a strong foundation for the promotion of geotourism and justify its candidacy for inclusion in the UNESCO Global Geopark network.

The islands of São Tomé and Príncipe, situated on the Cameroon Volcanic Line, also have undergone extensive geoheritage assessment. Key sites—such as phonolitic domes (Figure 5a), volcanic plugs, and fossil-rich marine platforms have been evaluated for their scientific and educational value, forming the basis for proposed geotourism and geoeducation strategies [52,53]. Relatedly, Annobon Island (Equatorial Guinea) has been the focus of a geoitinerary project that interprets the submerged geomorphology of its coral reef system as a living record of Darwinian evolution, linking biodiversity to volcanic island dynamics [53]. The Cabo Verde archipelago, particularly the islands of Fogo, Santo Antão, and Santiago, includes exceptional volcanic landscapes featuring well-preserved calderas, historic eruptions (e.g., Pico do Fogo), and sedimentary sequences that document long-term island evolution. Partial inventories of geosites have already been developed, and recent proposals emphasize both geotourism potential and community interest in geoheritage valorization [12]. Also on Santiago Island, extraordinary blocks sit in high relief as megatsunami deposits (Figure 5c).

The Fernando de Noronha archipelago, located in the South Atlantic Ocean, holds significant geoheritage value as a unique example of intraplate volcanism likely associated with a mantle hotspot. Its geodiversity includes phonolitic domes, basanitic lava flows, pyroclastic deposits, and well-preserved subvolcanic structures, which together allow the reconstruction of a polyphasic eruptive history spanning from 12 to 1.7 million years ago. This geological richness, combined with outstanding scenic, educational, and scientific relevance, underpins its designation as a Natural World Heritage Site and its current status as an aspiring UNESCO Global Geopark [54].

Figure 5. Potential island geoparks with adequate background information and scenery occur in the Atlantic Ocean. (**a**) Cão Grande, the most emblematic geosite of São Tomé Island (image courtesy of Maria Helena Henriques); (**b**) Lahar deposits from the 2021 eruption of La Soufrière volcano, Saint Vincent (Nieves Sánchez photo); (**c**) Megatsunami-deposited blocks on Santiago Island, Cabo Verde (Esther Martín photo).

In the Caribbean region, lahar deposits from the 2021 eruption of La Soufriére volcano on Saint Vincent remain well exposed (Figure 5b) and constitute a potential geopark resource. Réunion Island (France) is recognized as a UNESCO World Heritage Site for its "pitons, cirques and ramparts" and features the Piton de la Fournaise, one of the most active volcanoes on Earth. Despite its global visibility and institutional infrastructure, the island is not yet part of the UGGp. Recent studies have advocated for its inclusion, emphasizing the combination of active volcanism, deep erosion structures, and public education initiatives linked to the Réunion National Park [55].

The Galapagos Islands, belonging to Ecuador, are globally renowned for their biodiversity and recognized as both a World Heritage Site and a Biosphere Reserve. Increasingly, they are appreciated for their underexploited geological value. Recent proposals suggest incorporating volcanic geoheritage and landscape evolution into geotourism and educational programs. The potential for developing structured geological itineraries—particularly on islands such as San Cristobal in order to complement existing biological narratives and support sustainable local economies is under consideration [56].

Elsewhere in the eastern Pacific Ocean region, Mexico's Baja California peninsula includes multiple destinations with touristic value under the protection of national parks. The Loreto Bay National Park includes five islands in the southern part of the Gulf of California that achieved status as a Natural Protected Area in 1996. The islands of Coronados, Carmen, Danzante, Monserrat, and Catalina are well constrained within a setting of 206 square kilometers. Coronados is a volcanic island with a well-preserved edifice accessible by trail and a carbonate white-sand beach popular with tourists. It also features an uplifted lagoon containing a fossil coral reef. Isla del Carmen is the largest in the group with an area of 150 square kilometers dominated by Oligocene to Miocene andesite flows and a coastal exposure of an enormous tide-water delta dating from the Pliocene. Danzante is another andesite island popular with visitors for its coastal landscape that includes fragile rock pediments, otherwise known as mushroom rocks. More inaccessible is Isla Monserrat, which features a major unconformity between tilted andesite layers overlain by Pliocene limestone enriched by fossil rhodoliths. These four islands have come under considerable attention as potential candidates for the coexistence of a multi-island geopark under the protection of the existing national park [57].

In Europe, Madeira (Portugal) represents one of the most advanced volcanic island candidates in terms of heritage legislation and management. The region includes the Laurisilva World Heritage Site, a biosphere reserve, and a well-documented inventory of geosites. In addition, regional strategies for the conservation of geological heritage have already been adopted, although integration into a Global Geopark remains ongoing [58,59].

Sicily and the Aeolian Islands (Italy) are home to some of the most iconic active volcanoes—Mount Etna, Stromboli, and Vulcano—central to the historical development of volcanology. Despite their World Heritage designation, these territories have not yet been incorporated into the UGGp. Their rich geodiversity, combined with strong cultural integration, makes them highly eligible candidates [60,61]. Similarly, the island of Santorini in the Aegean Sea was shaped by the catastrophic Minoan eruption of 1613 BC [62] that destroyed much of the original landmass and created its present form, which lies within the tectonically active Hellenic Volcanic Arc. Characterized by subduction and crustal extension, this setting turns Santorini into a natural laboratory for studying Earth's geological evolution. The formation of its emblematic caldera and the analysis of its volcanic deposits illustrate how volcanic processes can profoundly influence both natural landscapes and human history, revealing a rich and complex geological and cultural narrative [63].

The Faroe Islands (Denmark), although shaped by ancient and now inactive volcanic activity, offer well-preserved volcanic landscapes. Such remote and geologically distinctive systems, despite being understudied, could gain recognition for their geomorphological uniqueness and growing geotourism appeal—even in the absence of formal conservation frameworks [64]. Altogether, these island territories underscore the breadth of underrepresented volcanic island geoheritage across the globe (Figure 2). Their inclusion in the geopark framework would not only enhance regional geoconservation but also contribute to the diversification and geographical balance of the UNESCO Global Geoparks.

The Aleutian Arc represents an outstanding example of volcano-tectonic evolution along an active oceanic subduction margin. Accordingly [65], the arc's dynamics reflect complex interactions between oblique subduction, volcanic front migration, and compositional variations in magmatism, controlled by changes in slab geometry and sediment input. These processes have generated remarkable geological diversity, reinforcing the Aleutians' potential for geoscientific valorization and their possible designation as a geopark.

4. Discussion

Volcanic island geoparks represent a geographically constrained yet highly significant subset within the UNESCO Geoparks Network (UGGp). Their value lies not only in the spectacular nature and diversity of volcanic landscapes but also in the opportunity to study active geological processes in insular contexts, where endogenic dynamics manifestation with an intensity and visibility rarely found in continental environments [66,67].

4.1. The Geodiversity Perspective

From the geodiversity perspective, these territories stand out for the coexistence of multiple volcanic features—cones, calderas, lava tubes, pyroclastic fields—as well as landforms shaped by the interaction between volcanism and marine or climatic processes, such as marine terraces, reworked pyroclastic deposits, and erosional forms in tuff and ash. This geological richness supports not only scientific research but also biodiversity conservation, cultural heritage, and local economic activities such as geotourism [68,69]. Many of these islands represent key chapters in the history of volcanology worldwide.

4.2. The Conservation Perspective

The systematic assessment of geological features in volcanic island landscapes is Essential for promoting their conservation and sustainable use, especially in territories under strong anthropogenic pressure [12]. This is particularly relevant in small islands, where geological heritage tends to be highly concentrated, unique, and exposed to rapid transformation processes. The management of natural heritage—specifically geological heritage—in insular territories is both sensitive and complex. These islands function as coastal systems and are therefore especially vulnerable to global changes such as climate variability and sea level rise, as well as to anthropogenic disturbances, including land-use change, unregulated tourism, and urban expansion [70]. Despite their limited surface area, many of these islands host geosites of considerable scientific, educational, and aesthetic value—some already situated within formally protected areas.

However, a preliminary analysis of their geoconservation status reveals persistent threats, including illegal extraction of rocks and fossils, infrastructure development, mass tourism, and extractive industries. These pressures have, in some cases, led to the degradation or irreversible loss of geoheritage elements. Although legal protection mechanisms exist in certain contexts, they are often weakly enforced or disconnected from spatial planning frameworks.

4.3. The Management Perspective

In response to these challenges, new management models are being proposed. One such approach involves the application of geosystems as a conceptual framework for geoconservation, particularly within geoparks. The Izu Peninsula Geopark in Japan offers a compelling case study. As described by [71], this model emphasizes the holistic evaluation of geodiversity, considers processes across broad spatiotemporal scales, and promotes the integration of abiotic, biotic, and cultural components. Such frameworks are especially relevant for island geoparks located near urban centers, where territorial conflicts and environmental pressures are most acute.

In terms of governance, insular geoparks face structural challenges: limited accessibility, dependency on external resources, seasonal fluctuations in tourism, and heightened exposure to geohazards. Nonetheless, some governance models—such as those implemented in the Azores UGGp, Lanzarote and Chinijo Islands UGGp, or Jeju UGGp—have successfully integrated local communities, scientists, and policymakers in territorial planning efforts aligned with conservation, education, and sustainable development objectives [72]. These integrated models, however, remain the exception rather than the norm.

Regarding representation within the UGGp, there exists a clear imbalance. Volcanic island UGGp are underrepresented in Africa, Oceania, and America, despite the evident geological potential of such regions (UNESCO as of 2023). This underrepresentation is not due to a lack of geodiversity and geoheritage, but rather to unequal access to institutional support, funding, and scientific expertise —factors that hinder many potential candidates from advancing toward formal designation.

5. Conclusions

In summary, safeguarding the geological heritage of volcanic island territories requires not only an accurate assessment and inventory but also adaptive management frameworks capable of integrating conservation goals with the complex socioecological realities of island systems. Volcanic island geoparks are not only natural laboratories of high scientific value; they are also strategic platforms for testing sustainable territorial models in ecologically and socially fragile environments under Climate Change scenarios. Their effective study and management demand integrative approaches that bridge geoscience with local governance. Expanding their presence within the UNESCO Global Geopark program will require overcoming structural barriers and fostering stronger international cooperation—particularly in the Global South.

This review is exploratory in nature and necessarily limited by the availability and accessibility of documentation on volcanic island geoparks, particularly those not yet formally integrated into the UNESCO Global Geoparks Network. The analysis relies primarily on published literature and institutional sources, which may overlook informal or emerging initiatives in less-documented regions. Furthermore, while the identification of potential territories is based on observable geodiversity and preliminary conservation actions, it does not account for local political, social, or economic constraints that may hinder actual geopark development.

Future research should aim to integrate field-based assessments, local stakeholder interviews, and geosite inventory validation to refine the evaluation of candidate territories. Comparative studies between designated and non-designated volcanic island geoparks could also help identify key success factors in geopark implementation. Lastly, examining how different governance models—national, regional, or international—affect geoconservation outcomes on islands would offer valuable insight for policy and planning.

Author Contributions: E.M.-G. initiated the project as a contribution to the Special Issue in the Journal of Marine Sciences and Engineering devoted to "Feature Review Papers in Geological Oceanography". Original draft preparation, E.M.-G. Additional writing and review; C.R., J.V., I.G. and N.S. All authors have read and agreed to the published version of the manuscript.

Funding: This research received no external funding.

Acknowledgments: The authors thank Markes E. Johnson (Williams College) for the invitation to participate in the journal's Special Issue and for his editorial assistance. We sincerely thank the anonymous reviewers for their thorough evaluation and constructive comments, which greatly improved the quality and clarity of this manuscript. Their insightful suggestions and critical observations have been invaluable in refining our arguments and strengthening the overall contribution of this work.

Conflicts of Interest: The authors declare no conflict of interest.

References

1. Pescatore, E.; Bentivenga, M.; Giano, S.I. Geoheritage and geoconservation: Some remarks and considerations. *Sustainability* **2023**, *15*, 5823. [CrossRef]
2. Website for the International Geoscience and Geoparks Programme. 2025. Available online: www.unesco.org/en/iggp/geoparks/about (accessed on 21 May 2025).
3. Németh, K.; Casadevall, T.; Moufti, M.R.; Martí, J. Volcanic geoheritage. *Geoheritage* **2017**, *9*, 251–254. [CrossRef]
4. Casadevall, T.; Tormey, D.; Roberts, J. *World Heritage Volcanoes: Classification, Gap Analysis, and Recommendations for Future Listings*; IUCN: Gland, Switzerland, 2019.
5. Pimentel, A.; Ramalho, R.S.; Becerril, L.; Larrea, P.; Brown, R.J. Editorial: Ocean island volcanoes: Genesis, Evolution and Impact. *Front. Earth Sci.* **2020**, *8*, 82. [CrossRef]
6. Menard, H.; Ladd, H. Oceanic islands, seamounts, guyots and atolls. In *The Sea*, 3rd ed.; Hill, M.N., Ed.; Wiley Interscience: New York, NY, USA, 1963; pp. 365–385.

7. Ramalho, R.S.; Quartau, R.; Trenhaile, A.S.; Mitchell, N.C.; Woodroffe, C.D.; Ávila, S.P. Coastal evolution on volcanic oceanic islands: A complex interplay between volcanism, erosion, sedimentation, sea-level change and biogenic production. *Earth Sci. Rev.* **2013**, *127*, 140–170. [CrossRef]
8. Perfit, M.R.; Davidson, J.P.; Sigurdsson, H. Plate tectonics and volcanism. In *Encyclopedia of Volcanoes*; Academic Press: Cambridge, MA, USA, 2000; pp. 89–113.
9. Dóniz-Páez, J.; Beltrán-Yanes, E.; Becerra-Ramírez, R.; Pérez, N.; Hernández, P.; Hernández, W. Diversity of volcanic geoheritage in the Canary Islands, Spain. *Geosciences* **2020**, *10*, 390. [CrossRef]
10. Lima, E.A.; Machado, M.; Guerreiro, M.; Nunes, J.C.; Costa, M.P. Geological heritage management in small islands: The example of the Azores UNESCO Global Geopark (Portugal). *Geoheritage* **2018**, *10*, 659–671. [CrossRef]
11. Pérez-Hernández, E.; Ferrer, N.; Hernández-Calvento, L. Lost and preserved coastal landforms after urban growth. The case of Las Palmas de Gran Canaria city (Canary Islands, Spain). *J. Coast. Conserv.* **2020**, *24*, 26. [CrossRef]
12. Alfama, V.; Henriques, M.H.; Barros, A. The challenging nature of volcanic heritage: The Fogo Island (Cabo Verde, W Africa). *Geoheritage* **2024**, *16*, 34. [CrossRef]
13. Ferrer, N.; Vegas, J.; Galindo, I.; Lozano, G.A. Geoheritage valuation to prevent environmental degradation of a new volcanic landscape in the Canary Islands. *Land Degrad. Dev.* **2023**, *34*, 2494–2507. [CrossRef]
14. Ramos, W.H.; Pérez, N.; Dóniz-Páez, J.; García-Hernández, R. Evaluation of sites of geotouristic interest on active volcanic island La Palma, Spain for potential volcanic tourism. *Geoheritage* **2024**, *16*, 102. [CrossRef]
15. García-Casco, A.; García-Delgado, D. Islands as natural laboratories for geological heritage. *Episodes* **2023**, *46*, 289–299.
16. Zouros, N. Lesvos petrified forest Geopark, Greece: Geoconservation, geotourism, and local development. *Geoparks* **2010**, *27*, 19–28.
17. Henriques, M.H.; Pena dos Reis, R.; Brilha, J.; Mota, T.S. Geoconservation as an emerging geoscience. *Geoheritage* **2011**, *3*, 117–128. [CrossRef]
18. Website for the United Nations Educational Scientific and Cultural Organization Program Related to Its Global Geopark Program in China. Available online: http://www.globalgeopark.org.cn/ (accessed on 15 May 2025).
19. Website for the UNESCO Program Related to Geoparks in the Asian Pacific Region. Available online: https://asiapacificgeoparks.org/ (accessed on 15 May 2025).
20. Vegas, J.; Galindo, I.; Romero, C.; Sánchez, N.; Díaz, G.A.; Martín-González, E.; Mateo, E. Geoconservation and geotourism in the Lanzarote and Chinijo Islands UNESCO Global Geopark. In *Lanzarote and Chinijo Islands Geopark: From Earth to Space, Geoheritage, Geoparks and Geotourism*; Mateo, E., Martínez-Frías, J., Vegas, J., Eds.; Springer: Cham, Switzerland, 2019. [CrossRef]
21. Jousset, P.; Mortensen, A.K.; Fridleifsson, G.O.; Ágústsoon, K.; Gudmundsson, M.T. Reykjanes, Iceland: Structure and dynamics of mid-oceanic ridge geo/hydrothermal systems. *J. Volcanol. Geotherm. Res.* **2020**, *391*, 106692. [CrossRef]
22. Jóhannesson, J.M.; Sigmundsdóttir, B.; Sigursveinsson, S. Volcanic features within Katla UNESCO Global Geopark. *Geoconserv. Res.* **2023**, *6*, 53–69.
23. Lima, E.A.; Meneses, S. Azores UNESCO Global Geopark: Where volcanoes tell us their stories. *Geoconserv. Res.* **2023**, *6*, 114–127.
24. Nunes, J.C. The Azores Archipelago: Islands of geodiversity. In *Volcanic Tourist Destinations*; Erfurt-Cooper, P., Ed.; Springer: New York, NY, USA, 2014; pp. 57–67.
25. Casillas, R.; Candelario, Y.P.; Fernández, C.F. El Hierro UNESCO Global Geopark: Geological heritage, geoconservation and geotourism. *Geoconservation Res.* **2023**, *6*, 128–138.
26. Vegas, J.; Galindo, I.; Vázquez, J.T.; León, R.; Sánchez, N.; Martín-González, E.; Romero, C. Tagoro, the Youngest Submarine Volcano in the Spanish Geoheritage Inventory: Scientific Value, Geoconservation and Opportunities for Geotourism. In *El Hierro Island*; Active Volcanoes of the World; González, P.J., Ed.; Springer: Cham, Switzerland, 2023; pp. 257–272. [CrossRef]
27. Martín-González, E.; Coello, J.J.; Castillo, C.; González-Rodríguez, A.; Galindo, I.; Vegas, J. Revisión del patrimonio paleontológico del Geoparque Mundial de la Unesco de El Hierro. *Geogaceta* **2019**, *65*, 11–17.
28. Jones, T.E. Evolving approaches to volcanic tourism crisis management: An investigation of long-term recovery models at Toya-Usu Geopark. *J. Hosp. Tour. Manag.* **2016**, *28*, 31–40. [CrossRef]
29. Gomez, C.; Wassmer, P. Evolution of the Unzen Volcano and the Shimabara Peninsula (Japan) during the last 60 years: The role of the 1990–1995 eruption in modifying the landscape. *Géomorphologie* **2015**, *21*, 205–216. [CrossRef]
30. Sato, E.; Fukui, K.; Shimbori, T. Aso volcano eruption on October 8, 2016, observed by weather radars. *Earth Planets Space* **2018**, *70*, 105. [CrossRef]
31. Petraglia, M.D.; Korisettar, R.; Pal, J.N. The Toba volcanic super-eruption of 74,000 years ago: Climate change, environments, and evolving humans. *Quat. Int.* **2012**, *258*, 1–4. [CrossRef]
32. Muzambiq, S.; Sibarani, R.; Nasution, Z.P.; Gustanto, G. Geoheritage and cultural heritage overview of the Toba caldera geosite, North Sumatra, Indonesia. *Smart Tour.* **2024**, *5*, 2517. [CrossRef]
33. Rachmat, H. *The Geomorphology of the Rinjani Volcanic Complex, Lombok Island, Indonesia: A Result of a Major Historic Eruption*; IAG: París, Francia, 2013.

34. Reubi, O.; Nicholls, I.A. Structure and dynamics of a silisic magmatic system associated with caldera-forming eruptions at Batur volcanic field, Indonesia. *J. Petrol.* **2005**, *46*, 1367–1391. [CrossRef]
35. Delmelle, P.; Bernard, A. Dowstream composition changes of acidic volcanic waters discharged into the Banyupahit stream, Ijen Caldera, Indonesia. *J. Volcanol. Geotherm. Res.* **2000**, *97*, 55–75. [CrossRef]
36. Jeon, Y.; Ki, J.; Southcott, D. Koren geoheritage: The volcanic landforms of the Jeju Island UNESCO Global Geopark. *Episodes* **2024**, *7*, 295–310. [CrossRef] [PubMed]
37. Wei, F.; Wei, W.; Yu, H. The Cenozoic volcanic fields in northern Hainan Island and Leizhou Peninsula, South China: Eruption history, magma source and dynamic background. *Geol. Soc. Lond. Spec.* **2020**, *510*, 179–196. [CrossRef]
38. Galindo, I.; Vegas, J.; Romero, C.; Llorente, M.; Martín-González, E.; Rubio, J.C.; Díaz, G.A.; Mangas, J.; Mateo, E.; Sánchez, N. Geoheritage inventory of the Lanzarote and Chinijo Islands UNESCO Global Geopark. In *Lanzarote and Chinijo Islands Geopark: From Earth to Space, Geoheritage, Geoparks and Geotourism*; Mateo, E., Martínez-Frías, J., Vegas, J., Eds.; Springer: Cham, Switzerland, 2019. [CrossRef]
39. Galindo, I.; Romero, C.; Llorente, M.; Rubio, J.C.; Díaz, G.A.; Sánchez, N.; Martín-González, E.; Mangas, J.; Vegas, J. Geoheritage in the shallow submarine slopes of fan oceanic volcanic edifice: A new option for diving geotourism. In *Lanzarote and Chinijo Islands Geopark: From Earth to Space, Geoheritage, Geoparks and Geotourism*; Mateo, E., Martínez-Frías, J., Vegas, J., Eds.; Springer: Cham, Switzerland, 2019. [CrossRef]
40. Martín-González, E.; Galindo, I.; Romero, C.; Sánchez, N.; Vegas, J. Evaluation of geoconservation in geosites of palaeontological interest from Lanzarote and Chinijo Islands UNESCO Global Geopark. In *Lanzarote and Chinijo Islands Geopark: From Earth to Space, Geoheritage, Geoparks and Geotourism*; Mateo, E., Martínez-Frías, J., Vegas, J., Eds.; Springer: Cham, Switzerland, 2019. [CrossRef]
41. Wang, J.; Zouros, N. Educational activities in Fangshan UNESCO Global Geopark and Lesvos Island UNESCO Global Geopark. *Geoheritage* **2021**, *13*, 51. [CrossRef]
42. Nakamura, Y.; Yuhora, K. Muroto Geopark: Understanding the moving Earth. In *Natural Heritage of Japan: Geoheritage, Geoparks and Geotourism*; Chakraborty, A., Mokudai, K., Cooper, M., Watanabe, M., Chakraborty, S., Eds.; Springer: Cham, Switzerland, 2018. [CrossRef]
43. Nanjo, K.Z.; Yukutake, Y.; Kumazawa, T. Changes in Seismicity in a Volcanically Active Region of the Izu Peninsula, Japan. 2024. Available online: https://ssrn.com/abstract=5097110 (accessed on 30 April 2025).
44. Irzon, R.; Samodra, H.; Permanadewi, S. Prospecting CCS project in Indonesia: A case study in Meratus Mountains, South Borneo. *J. Geol. Miner. Resour.* **2024**, *25*, 31–40. [CrossRef]
45. Ávila, S.P.; Cachão, M.; Ramalho, R.S.; Botelho, A.Z.; Madeira, P.; Rebelo, A.C.; Cordeiro, R.; Melo, C.; Hipólito, A.; Ventura, M.A.; et al. The palaeontological heritage of Santa Maria Island (Azores: NE Atlantic): A re-evaluation of geosites in geopark Azores and their use in geotourism. *Geoheritage* **2016**, *8*, 155–171. [CrossRef]
46. Meireles, R.P.; Quartau, R.; Ramalho, R.S.; Rebelo, A.C.; Madeira, J.; Zanon, V.; Ávila, S.P. Depositional processes on oceanic island shelves—Evidence from storm-generated Neogene deposits from the mid-North Atlantic. *Sedimentology* **2013**, *60*, 1769–1785. [CrossRef]
47. Rebelo, A.C.; Rasser, M.W.; Kroh, A.; Johnson, M.E.; Ramalho, R.S.; Melo, C.; Uchman, A.; Berning, B.; Silva, L.; Zanon, V.; et al. Rocking around a volcanic island shelf: Pliocene rhodolith beds from Malbusca, Santa Maria Island (Azores, NE Atlantic). *Facies* **2016**, *62*, 22. [CrossRef]
48. Harbowo, D.G. An assessment of the scientific value of Krakatoa, Indonesia from a geoheritage perspective. *J. Appl. Geosci. Eng.* **2023**, *2*, 11–25. [CrossRef]
49. Sendjaja, P.; Suparka, E.; Botjing, M.U. Geodiversity of the Togean Islands National Park, Central Sulawesi Province for Geopark Assessment. *IOP Conf. Ser. Earth Environ. Sci.* **2020**, *589*, 012023. [CrossRef]
50. Fepuleai, A.; Nemeth, K. Volcanic geoheritage of landslides and rockfalls on a tropical ocean island (Western Samoa, SW Pacific). *Geoheritage* **2018**, *11*, 577–596. [CrossRef]
51. Migon, P.; Pijet-Migon, E. Geoconservation and tourism at geothermal sites—Lessons learnt from the Taupo volcanic zone, New Zealand. *Proc. Geol. Assoc.* **2016**, *127*, 413–421. [CrossRef]
52. Henriques, M.H.; Neto, K. Geoheritage at the Equator: Selected geosites of Sao Tomé Island (Cameron Line, Central Africa). *Sustainability* **2015**, *7*, 648–667. [CrossRef]
53. Henriques, M.H.; Neto, K.A. Geo-itinerary to foster sustainable tourism in West African Islands: Storytelling the evolution of the ancient Cameroon Volcanic Line coral reefs. *Sustainability* **2023**, *15*, 16863. [CrossRef]
54. Moreira, J.C.; Vale, T.F.; Burns, R.C. Fernando de Noronha Archipelago (Brazil): A Coastal Geopark Proposal to Foster the Local Economy, Tourism and Sustainability. *Water* **2021**, *13*, 1586. [CrossRef]
55. Michon, L.; Di Muro, A.; Villeneuve, N.; Saomt-Marc, C.; Fadda, P.; Manta, F. Explosive activity of the summit of Piton de la Fournaise volcano (La Réunion Island): A historical and geological review. *J. Volcanol. Geotherm. Res.* **2013**, *264*, 117–133. [CrossRef]

56. Carrión-Mero, P.; Sánchez-Zambrano, E.; Mata-Perelló, J.; Jaya-Montalvo, M.; Herrera-Franco, G.; Berrezueta Espinel, R.; Baque, M.; Morante-Carballo, F. Geosites assessment in a volcanic hotsot environment and its impart on geoturism, Santa Cruz-Galapagos Islands, Ecuador. *Int. J. Geoherit. Parks* **2024**, *12*, 147–167. [CrossRef]
57. Johnson, M.E.; Ledesma-Vázquez, J. *Gulf of California Coastal Ecology: Insights from the Present and Patterns from the Past*; Sunbelt Publications: San Diego, CA, USA, 2016; 134p.
58. da Silveira, A.B.; Prada, S.; Ramalho, R.; Madeira, J.; Fonseca, P.; Canha, E.; Brilha, E. Inventariaço do património geológico da Ilha da Madeira. Secretaria Regional do Ambiente e Recursos Naturais—Relatório Final 2012, 414p. Available online: https://geodiversidade.madeira.gov.pt (accessed on 23 June 2025).
59. Sotiriou, P.; Nunes, P. The geoheritage of Madeira: Implications for Natural Heritage and geotourism. *Geoheritage* **2024**, *16*, 74. [CrossRef]
60. Francaviglia, V. Characterization of Mediterranean obsidian sources by classical petrochemical methods. *Preist. Alp.* **1984**, *20*, 311–332.
61. Keller, J. Lipari's fiery past: Dating the medieval pumice eruption of Monte Pelato. In Proceedings of the International Conference 'The Fire Between Air and Water', UNESCO-Regione, Siciliana, Lipari, Italy, 29 September–2 October 2002.
62. Friedrich, W.L.; Heinemeier, J. The Minoan eruption of Santorini radiocarbon dated to 1613 ± 13 BC—Geological and stratigraphic considerations. In *Time's Up! Dating the Minoan Eruption of Santorini*; Warburton, D.A., Ed.; Monographs of the Danish Institute: Athens, Greece, 2009; Volume 10, pp. 57–63.
63. Preine, J.; Karstens, J.; Hübscher, C.; Druitt, T.; Kutterolf, S.; Nomikou, P.; Manga, M.; Gertisser, R.; Pank, K.; Beethe, S.; et al. Hazardous explosive eruptions of a recharging multi-cyclic island arc caldera. *Nat. Geosci.* **2024**, *17*, 323–331. [CrossRef]
64. Raška, P.; Pokomý, R.; Krmíček, L.; Koboušková, S.; Mortensen, L. Basaltic dyke with specific volcanogenic structures and its geomorphic evolution: Unique geoheritage of the Faroe Islands (North Atlantic Ocean). *Geoheritage* **2019**, *11*, 417–426. [CrossRef]
65. Jicha, B.R.; Scholl, D.W.; Singer, B.; Yogodzinski, G.; Kay, S.M. Revised age of Aleutian Island Arc formation implies high rate of magma production. *Geology* **2006**, *34*, 661–664. [CrossRef]
66. Wood, C.A. Volcanoes of the World. In *Encyclopedia of Volcanoes*; Sigurdsson, H., Ed.; Academic Press: Cambridge, MA, USA, 2009; pp. 3–17.
67. Németh, K. Volcanic Geoheritage in the Light of Volcano Geology. In *El Hierro Island Global Geopark*; Dóniz-Paez, J., Pérez, N.M., Eds.; Springer: Berlin/Heidelberg, Germany, 2023; pp. 1–32. [CrossRef]
68. Gray, M. *Geodiversity: Valuing and Conserving Abiotic Nature*, 2nd ed.; Wiley-Blackwell: Hoboken, NJ, USA, 2013.
69. Brilha, J. Inventory and quantitative assessment of geosites and geodiversity sites: A review. *Geoheritage* **2016**, *8*, 119–134. [CrossRef]
70. Saffache, P.; Angelelli, P. Environmental challenges in the small islands of the Caribbean: The example of Martinique. *Caribb. Geogr.* **2010**, *15*, 83–95.
71. Chakraborty, A.; Mogi, K.; Takemura, K. Geosystem and geopark: A holistic approach to natural hazard studies in the Izu Peninsula Geopark, Japan. *J. Geol. Soc. India* **2015**, *86*, 525–533.
72. Reynard, E.; Brilha, J. (Eds.) *Geoheritage: Assessment, Protection, and Management*; Elsevier: Amsterdam, The Netherland, 2018. [CrossRef]

MDPI AG
Grosspeteranlage 5
4052 Basel
Switzerland
Tel.: +41 61 683 77 34

Journal of Marine Science and Engineering Editorial Office
E-mail: jmse@mdpi.com
www.mdpi.com/journal/jmse

www.ingramcontent.com/pod-product-compliance
Lightning Source LLC
Chambersburg PA
CBHW061549220926
51850CB00005B/291

* 9 7 8 3 7 2 5 8 7 9 2 5 0 *